AF598339

MATERIALS SCIENCE RESEARCH
Volume 17

EMERGENT PROCESS METHODS FOR HIGH-TECHNOLOGY CERAMICS

MATERIALS SCIENCE RESEARCH

Recent volumes in the series:

Volume 8 **CERAMIC ENGINEERING AND SCIENCE: Emerging Priorities**
Edited by V. D. Fréchette, L. D. Pye, and J. S. Reed

Volume 9 **MASS TRANSPORT PHENOMENA IN CERAMICS**
Edited by A. R. Cooper and A. H. Heuer

Volume 10 **SINTERING AND CATALYSIS**
Edited by G. C. Kuczynski

Volume 11 **PROCESSING OF CRYSTALLINE CERAMICS**
Edited by Hayne Palmour III, R. F. Davis, T. M. Hare

Volume 12 **BORATE GLASSES: Structure, Properties, Applications**
Edited by L. D. Pye, V. D. Fréchette and N. J. Kreidl

Volume 13 **SINTERING PROCESSES**
Edited by G. C. Kuczynski

Volume 14 **SURFACES AND INTERFACES IN CERAMIC AND CERAMIC–METAL SYSTEMS**
Edited by Joseph Pask and Anthony Evans

Volume 15 **ADVANCES IN MATERIALS CHARACTERIZATION**
Edited by David R. Rossington, Robert A. Condrate, and Robert L. Snyder

Volume 16 **SINTERING AND HETEROGENEOUS CATALYSIS**
Edited by G. C. Kuczynski, Albert E. Miller, and Gordon A. Sargent

Volume 17 **EMERGENT PROCESS METHODS FOR HIGH-TECHNOLOGY CERAMICS**
Edited by Robert F. Davis, Hayne Palmour III, and Richard L. Porter

Volume 18 **DEFORMATION OF CERAMICS II**
Edited by Richard E. Tressler and Richard C. Bradt

MATERIALS SCIENCE RESEARCH • Volume 17

EMERGENT PROCESS METHODS FOR HIGH-TECHNOLOGY CERAMICS

Edited by

Robert F. Davis

Hayne Palmour III

and

Richard L. Porter

North Carolina State University
School of Engineering
Raleigh, North Carolina

PLENUM PRESS • NEW YORK AND LONDON

Library of Congress Cataloging in Publication Data

Conference on Emergent Process Methods for High-Technology Ceramics (1982: North Carolina State University)
Emergent process methods for high-technology ceramics.

(Materials science research; v. 17)
"Proceedings of the Conference on Emergent Process Methods for High-Technology Ceramics, held November 8-10, 1982, at North Carolina State University, Raleigh, North Carolina"—T.p. verso.
Includes bibliographical references and index.
1. Ceramics—Congresses. I. Davis, Robert F. (Robert Foster), 1942- . II. Palmour, Hayne. III. Porter Richard L., 1946- . IV. Title. V. Series.
TP785.C693 1982 666 84-6941
ISBN 0-306-41677-8

Proceedings of the conference on Emergent Process Methods for High-Technology Ceramics, held November 8-10, 1982, at North Carolina State University, Raleigh, North Carolina

A Division of Plenum Publishing Corporation
233 Spring Street, New York, N.Y. 10013

Printed in the United States of America

PREFACE

This volume constitutes the Proceedings of the November 8-10, 1982 Conference on EMERGENT PROCESS METHODS FOR HIGH TECHNOLOGY CERAMICS, held at North Carolina State University in Raleigh. It was the nineteenth in a series of "University Conferences on Ceramic Science" initiated in 1964 by four institutions of which North Carolina State University is a charter member, along with the University of California at Berkeley, Notre Dame University, and the New York State College of Ceramics at Alfred University. More recently, ceramic-oriented faculty in departments at the Pennsylvania State University and Case-Western Reserve University have joined the four initial institutions as permanent members of the consortium. These research-oriented conferences, each uniquely concerned with a timely ceramic theme, have been well attended by audiences which typically were both international and interdisciplinary in character; their published Proceedings have been well received and are frequently cited.

This three day conference addressed the fundamental scientific background as well as the technological state-of-the-art of several novel methods which are beginning to influence present and future directions for non-traditional ceramic processing, thus affecting many of the advanced ceramic materials needed for a wide variety of research and industrial applications.

The number, the importance and the application of new ceramic processing techniques have expanded considerably during the last ten years. The reasons for this expansion have been the urgent need for (a) the advent or improvement in properties of a wide variety of modern dielectric, magnetic, electronic, nuclear and structural ceramics; (b) the increasing availability of high purity ceramic materials in extremely fine and highly reactive powder or dense poly-crystalline or single crystal form; (c) the processing of both quasi-traditional and exotic ceramic materials in unusual sizes and geom-etries which are capable of withstanding in-service environmental extremes; (d) the creation of nonequilibrium crystalline materials possessing improved properties; and (e) the dramatic alteration of the internal structure of the surface in an attempt to achieve increases in density or strength or abrasion resistance.

The 58 papers included in these Proceedings have been authored by practicing scientists and engineers from various materials-related disciplines whose expertise is contributing or may contribute to revolutionary advances in the processing of high technology ceramics. Thematically, the book is divided into ten sections. The first two sections are principally concerned with (1) The Science of Colloidal Processing and (2) Novel Powder-Forming and Powder-Processing Methods. The next three sections focus attention on recent developments in nonparticulate forming of ceramics via (3) Polymer Processing, (4) Chemical Vapor Deposition and (5) Ion Beam Deposition. Contemporary with the development of the techniques have been the attempts to alter the properties of the bulk material via (6) Laser and Ion Beam Modification of Surfaces. The pressure variable may also be effectively employed to directly enhance the densification of ceramics via techniques such as (7) Hot Isostatic Processing, (8) Dynamic Compaction, (9) Shock Conditioning and Subsequent Densification and (10) Very High Pressure Processing.

The List of Contributors formally acknowledges the considerable cooperation and assistance rendered the Co-Chairmen by (1) a distinguished Advisory Committee, (2) the Conference Staff and (3) the several Session Chairmen, as well as (4) the creative efforts of one hundred thirty two distinguished contributing authors representing many of the world's ceramic research centers. We extend our personal thanks to all of them collectively and individually, for their cooperative attitudes, timely responses and many helpful suggestions which have characterized all our relationships with them.

On behalf of the participants and the ceramic community, we gratefully acknowledge enabling and indispensable financial support provided this Conference from (1) federal sources and by the Army Research Office, by the Department of Energy and the Office of Naval Research; (2) industrial sources by the GTE Laboratories, the Martin-Marietta Laboratories, the Phillip M. McKenna Foundation and the TRW Foundation; and (3) university sources by the School of Engineering, North Carolina State University.

We also acknowledge support of, and participation in, many of the Conference activities by administrators and members of the faculty and staff at North Carolina State University. Special thanks are due to Dr. L. K. Monteith, Dean of Engineering and Dr. H. Conrad, Head, Department of Materials Engineering who welcomed attendees to the Conference. We also wish to express our appreciation to Dr. J. F. Wenckus, Ceres Corporation who presented a timely, interesting and humorous dinner address entitled "Cubic Zirconia: A Girl's Next Best Friend."

In an undertaking of this magnitude, the prospects of a favorable outcome really rest upon the skills, enthusiasms, and efforts of a modest group of persons who work with dedication, but largely behind the scenes. We wish to acknowledge our very special personal thanks to Dr. Bruce Winston and Mavis Stillman for coordinating our Conference, to Betty J. Randall, Marion S. Rand and Yvonne D. Maness for their secretarial assistance before, during and after the Conference; and to D. B. Stansel and Jane E. Hodge for assistance in adapting the excellent physical facilities of the McKimmon Center to the specific needs of this Conference. We also acknowledge our endebtedness to Mary N. Yionoulis for publicity; to A. Pressley Brower for serving as coordinator of projectionists; and to the several graduate and undergraduate students enrolled in the Ceramic Engineering option, Department of Materials Engineering at NCSU who served as aides, projectionists and pages at the conference. We pay special tribute to Lynn Kaufman for her special skills and experience in typing and/or revising these edited Proceedings. Finally, it is appropriate to acknowledge with real affection the patience, tolerance, and tangible and moral support we have been accorded by our colleagues, our families and our friends through those extended periods of time we have had to commit to planning, organizing and editing.

Raleigh, N.C.
December, 1983

Robert F. Davis
Hayne Palmour III
Richard L. Porter

CONTENTS

PART I: THE SCIENCE OF COLLOIDAL PROCESSING

Interfacial Electrochemistry of Disperse Systems 1
J. Lyklema

How Colloid Stability Affects the Behavior of Suspensions . 25
J. T. G. Overbeek

Formation and Stability of Colloidal Dispersions of Fine Particles in Water. 45
A. J. Rubin

Flocculation and Filtration of Colloidal Particles 59
J. Gregory

The Science of the Interactions of Colloidal Particles and Ceramics Processing 71
A. Bleier

Preparation of Shaped Glasses Through the Sol-Gel Method . 83
S. Sakka and K. Kamiya

Inorganic Oxide Gels and Gel-Monoliths: Their Crystallization Behavior. 95
S. P. Mukherjee

Boron Nitride Fiber Synthesis from Boric Oxide Precursors. 111
A. E. Lindemanis

PART II: NOVEL POWDER-FORMING AND POWDER-PROCESSING METHODS

Some Common Aspects of the Formation of Nonoxide Powders by the Vapor Reaction Method. 123
A. Kato, J. Hojo and T. Watari

Synthesis of Powders and Thin Films by Laser Induced Gas Phase Reactions 137
J. S. Haggerty

Preparation of Zirconia-Alumina Fine Powders by Hydrothermal Oxidation of Zr-Al Alloys 155
S. Somiya, M. Yoshimura and S. Kikugawa

Combustion Synthesis of Transition Metal Nitrides. 167
J. B. Holt and D. D. Kingman

The Influence of Powder Synthesis Techniques on Processes Occurring During Compact Formation and its Sintering 177
M. Paulus

Dispersion and Packing of Narrow Size Distribution Ceramic Powders. 193
R. L. Pober, E. A. Barringer, M. V. Parish, N. Levoy, and H. K. Bowen

Plasma Sintering of Ceramics 207
D. L. Johnson, V. A. Kramb, and D. C. Lynch

Plasma Melting of Selected Compositions in the Al_2O_3-ZrO_2-SiO_2 System. 213
J. V. Portugal and L. D. Pye

Liquid Phase Sintering of Ceramics 225
W. A. Kaysser and G. Petzow

Precision Digital Dilatometry: A Microcomputer-Based Approach to Sintering Studies 233
A. D. Batchelor, M. J. Paisley, T. M. Hare, and H. Palmour III

PART III: CERAMICS DERIVED BY POLYMER PROCESSING

The Conversion of Methylchloropolysilanes and Polydisilylazanes to Silicon Carbide and Silicon Carbide/Silicon Nitride Ceramics, Respectively. . 253
R. H. Baney, J. H. Gaul, Jr. and T. K. Hilty

Silicon-Nitrogen Polymers and Ceramics Dervied from Reactions of Dichlorosilane, H_2SiCl_2. 263
D. Seyferth, G. H. Wiseman and C. Prud'homme

Formation of Ceramic Composites and Coatings Utilizing Polymer Pyrolysis 271
W. S. Coblenz, G. H. Wiseman, P. B. Davis, and R. W. Rice

Gas Analysis During the Pyrolysis of Carbosilane 287
J. J. Poupeau, D. Abbe and J. Jamet

PART IV: CHEMICAL VAPOR DEPOSITION

Chemical Vapor Deposition of Ceramic Materials 299
J. M. Blocher, Jr., M. F. Browning and D. M. Barrett

The Application of Thermodynamic Calculations to Chemical Vapor Deposition Processes 317
A. I. Kingon and R. F. Davis

CVD of Si_3N_4 and its Composites. 329
T. Hirai

Preparation of Amorphous Si_3N_4-BN Composites by Chemical Vapor Deposition. 347
T. Hirai, T. Goto and T. Sakai

A Morphological Study of Silicon Borides Prepared by CVD . 359
R. R. Dirkx and K. E. Spear

A Morphological Study of Silicon Carbide Prepared by Chemical Vapor Deposition 371
P. Tsui and K. E. Spear

Low-Temperature Preparation of Pyrolytic Carbon. 381
R. W. Kidd, D. A. Seifert and M. F. Browning

Laser Chemical Vapor Deposition (LCVD) 397
S. D. Allen

PART V: ION BEAM DEPOSITION

Ion Beam Techniques for the Deposition of Ceramic Thin Films 415
J. M. E. Harper

Ionized-Cluster Beam Deposition and Epitaxy. 425
T. Takagi

Ion Beam Deposition of Ceramic-Like Coatings 447
C. Weissmantel, K. Bewilogua, K. Breuer, J. Erler, B. Rau, G. Reisse, and D. Roth

PART VI: LASER AND ION BEAM MODIFICATION OF SURFACES

Laser Surface Melting of Metals and Alloys 461
D. B. Snow

Laser Processing of Ceramics 473
J. R. Spann, R. W. Rice, W. S. Coblenz, and W. J. McDonough

Microstructural Analysis of Rapidly Solidified Alumina . . 505
J. P. Pollinger and G. L. Messing

Structure of Ceramic Surfaces Modified by Ion Beam Techniques. 519
C. J. McHargue, H. Naramoto, C. W. White, J. M. Williams, B. R. Appleton, P. S. Sklad, and P. Angelini

Microstructure and Mechanical Properties of Ion-Implanted Ceramics 533
C. S. Yust and C. J. McHargue

Microhardness of N-Implanted Yttria Stabilized ZrO_2. . . . 549
J. K. Cochran, K. O. Legg and G. R. Baldau

PART VII: HOT ISOSTATIC PRESSING

Hot Isostatic Pressing of Ceramic Materials. 559
R. R. Wills, M. C. Brockway and L. G. McCoy

Dense Ceramic Parts Hot Pressed to Shape by HIP. 571
H. T. Larker

Fabrication of Si_3N_4 Ceramics with Additives of Metal Nitrides by High Pressure Hot-Pressing and HIPing 583
M. Shimada, N. Uchida and M. Koizumi

Diffusion Bonding of Al_2O_3 and Si_3N_4 Ceramics by HIPing. . 591
M. Shimada, K. Tanihata, T. Kaba and M. Koizumi

Relationship Between Densification and High Temperature Mechanical Properties of HIPed Silicon Nitride. . 597
R. R. Wills, M. C. Brockway and G. K. Bansal

Microstructural Changes During Hot Isostatic Pressing of Sintered Lead Zirconate Titanate 609
K. G. Ewsuk and G. L. Messing

PART VIII: DYNAMIC COMPACTION

Dynamic Compaction of Powders. 621
R. Prummer

Dynamic Compaction of Ceramic Powders. 639
J. H. Adair, R. R. Wills and V. D. Linse

Explosive Consolidation of Aluminum Nitride Ceramic Powder: A Case History 657
W. H. Gourdin, S. L. Weinland, C. J. Echer and S. L. Huffsmith

Computer Simulation of Dynamic Compaction. 673
M. L. Wilkins and C. F. Cline

Investigation of a Method to Consolidate Hard Materials in a Tough Matrix 695
J. D. Mote and J. J. Fitzpatrick

PART IX: SHOCK SYNTHESIS: SHOCK CONDITIONING AND SUBSEQUENT DENSIFICATION

Modern Uses of Explosive Pressure--From Rock Blasting to Synthetic Diamond 711
O. R. Bergmann

Shock-Induced Modification of Inorganic Powders. 719
R. A. Graham, B. Morosin, E. L. Venturini, E. K. Beauchamp, and W. F. Hammetter

Densification Kinetics of Shock-Activated Nitrides 735
E. K. Beauchamp, R. E. Loehman, R. A. Graham, B. Morosin, and E. L. Venturini

Rate Controlled Sintering of Explosively Shock-Conditioned Alumina Powders 749
K. Y. Kim, A. D. Batchelor, K. L. More and H. Palmour III

PART X: VERY HIGH PRESSURE PROCESSING

High Pressure Processing of High Technology Ceramics . . . 765
E. Dow Whitney

Diamond Anvil Cell Technology for P,T Studies of Ceramics: ZrO_2 (8 mol% Y_2O_3) 783
R. G. Munro, S. Block, G. J. Piermarini, and F. A. Mauer

Effect of Strong Shock Compression on Covalent Materials and High Pressure Sintering 793
A. Sawaoka

A New Approach to the Reaction Sintering of Superhard Materials Under Very High Pressure. . . 809
M. Akaishi, T. Endo, O. Fukunaga, Y. Sato, and N. Setaka

ADVISORY COMMITTEE 821

CONTRIBUTORS 823

INDEX. 833

PART I

THE SCIENCE OF COLLOIDIAL PROCESSING

INTERFACIAL ELECTROCHEMISTRY OF DISPERSE SYSTEMS

J. Lyklema

Laboratory for Physical and Colloid Chemistry
of the Agricultural University
De Dreijen 6, 6703 BC Wageningen, Netherlands

INTRODUCTION

The present contribution describes the electrostatic properties of solid-liquid interfaces. Insight in such properties is a prerequisite for the understanding of the rheology and aggregative properties of suspensions and slurries which, in turn, are starting materials for the preparation of ceramic products.

Materials of common use in ceramic technology, such as inorganic oxides and clays, are not the most favorite model systems for electrical double layer studies. Electrochemists prefer systems like mercury because the mercury-solution interface is very smooth, it can be easily purified and high potentials can be applied across those interfaces (in electrochemical language: mercury has a high overpotential).

Over the past decades a wealth of information has been collected with the silver iodide system. Although much more difficult to handle than mercury (in particular, it is difficult to define the interfacial structure), this system has the advantage that suspensions and sols (colloidal solutions) can be made of it, enabling us to correlate double layer properties to colloid stability and rheology. Obviously, such a correlation is very relevant for the understanding of the behavior of ceramic materials.

Less understood than silver iodide is the group of insoluble oxides, but also for these systems considerable progress has recently been made. As a group, these oxides exhibit some common characteristic features. Because of their relevance for ceramics, these systems will receive some emphasis in this paper. For the same reason the interfacial electrochemistry of clays will be dealt with.

ORIGINS OF DOUBLE LAYERS AT SOLID-LIQUID INTERFACES

As suspensions of insoluble particles in aqueous solutions are as a whole electroneutral, a double layer at a phase boundary must consist of a charge on the particle (the surface charge, σ_o, usually defined per unit area and expressed in $\mu C\ cm^{-2}$) and an equal but opposite countercharge in the solution. The unequal distribution of charge over solution and particle is due to the specific accumulation of certain ions which have an affinity for the surface or bulk of the solid, that is high enough to overcome the counteracting electrical field. Let us call those ions that chemisorb to a surface so that they form a chemical unit with it and in this way give it a charge, charge-determining (c.d.) ions. Then σ_o can be defined as the charge attributed by the c.d. ions. (It is not always easy to discriminate between c.d. and other ions and therefore there can be cases where different definitions of the notion "surface charge" are open, a difficulty to which we shall return below.) For AgI in a solution of, say, KI and $AgNO_3$, Ag^+ and I^- are c.d. ions, these ions fit very well onto the solid lattice.[1] Denoting their surface excesses as Γ_{Ag+} and Γ_{I^-} (eq. cm^{-2}) respectively, we have for AgI

$$\sigma_o \doteqdot F(\Gamma_{Ag+} - \Gamma_{I^-}) \tag{1}$$

where F is the Faraday constant. For oxides, surface ROH groups (e.g., silanol groups on silica) can be protonated to become ROH_2^+ in acid solutions or they become negatively charged in alkaline solutions through the reaction $ROH + OH^- \rightarrow RO^- + H_2O$.[2] Hence, for oxides H^+ and OH^- are c.d. ions, so that, if other c.d. ions are absent,

$$\sigma_o \doteqdot F(\Gamma_{H^+} - \Gamma_{OH^-}) \tag{2}$$

For clays the situation is different.[3] Because of their structure, clays belong to the phyllosilicates. Morphologically, clay particles are flat plates or sheets, with a large fraction of the area on the platelet surface and a small part on the sides. The thickness varies from two to over 100 nm. Smectites and illites have the thinnest plates, kaolinites are usually thicker. The longest axis can be far over 1'000 nm. The platelets are built up of arrays of oxygen tetrahedra around Si^{4+} ions and oxygen or OH^- octahedra around Al^{3+} or Mg^{2+} ions. The precise architecture is different for the various types of clay minerals. In natural clays typically some cations are replaced by cations of lower valency without altering the structure, e.g., a number of Al^{3+} - or Mg^{2+} ions are found on sites where Si^{4+}

ought to be (isomorphic substitution). This process leads to a deficit of positive charges, rendering the clay platelet as a whole negative, with a compensating positive charge in the solution side. On the sides of the clay platelets there are some OH^- groups that can acquire a charge following (2).

It follows from the above that the greater part of the charges on clay particles (i.e., those on the plate surfaces) are typically bulk charges. Charges on AgI are essentially surface charges, although some so-called Frenkel-defects in the solid do occur. On oxides the surface- and counter-charge likely penetrates the solid over a number of lattice layers. Interaction of colloidal particles is not sensitive to the distribution of the charge inside the solid, but depends very strongly on that in the solution.

For clays the extent of isomorphic substitution is within narrow limits dictated by the nature of the clay or, for that matter, their cation exchange capacity (c.e.c., the number of compensating cations in the solution) can be established and it is not very sensitive to variables like pH, ionic strength ω or temperature. Hence, clays are, as far as the dominating surfaces of the plates are concerned, systems of constant charge. In contradistinction, oxides and AgI are systems of variable charge: in the former σ_o depends on pH, in the latter σ_o is a function of pAg and in both σ_o varies with T and ω. The surface at the sides of the clay platelets also bears a variable charge.

On oxides $\sigma_o > 0$ at low pH and $\sigma_o < 0$ at high pH. There is one pH value, the point of zero charge (p.z.c.), where $\sigma_o = 0$. For AgI, the p.z.c. is a special value of pAg. A p.z.c. can in principle also be assigned to the sides of clay platelets, but not to the plate surfaces. The p.z.c. is a property of the solid but depends also on the solution properties (ω, T, organic admixtures). For oxides, the p.z.c. is closely related to the relative acidity-basicity of the surface groups. Acidic oxides, like δ-MnO_2 or SiO_2 have a low p.z.c., basic oxides like hematite (α-Fe_2O_3) or gibbsite (γ-Al_2O_3) have a high p.z.c. It is straightforward to derive a simple approximative relation between the p.z.c. and the two constants.

$$K_a = \frac{[ROH_2^+]}{[ROH][H^+]} \qquad K_b = \frac{[RO^-][H_2O]}{[ROH][OH^-]} = \frac{1}{K_w}\frac{[RO^-][H^+]}{[ROH]} \tag{3a,b}$$

At the p.z.c. $[H^+]$ near the surface sites = $[H^+]$ in bulk and $[RO^-] = [ROH_2^+]$ so that

$$[H^+]_{pzc} = (K_b K_w / K_a)^{1/2} \quad \text{or} \quad \log(\text{p.z.c.}) = \tfrac{1}{2}(pK_a - pK_b - pK_w) \tag{4}$$

As the p.z.c. of an oxide is so closely related to pK_a and pK_b, it must be expected that surface treatments like preheating or laser induced modification do influence it. This indeed has been found. It is even possible to reverse the sign of σ_o at otherwise fixed conditions, and hence substantially alter the particle interaction. For homogeneous dispersions all particles bear the same charge so that this interaction is always repulsive, but for mixtures of oxides they can be partly repulsive and partly attractive, which may have great practical implications for the rheological properties of slurries. In this respect, clay minerals deserve particular attention: under conditions where the sides are positive they can aggregate among themselves, forming some card-house like structure.

Surface charges are measurable quantities. The differences $(\Gamma_{Ag+} - \Gamma_{I^-})$ for AgI and $(\Gamma_{H+} - \Gamma_{OH^-})$ for oxides can be obtained except for a constant provided the specific surface area of the suspension is known. Independent assessment of the p.z.c. is needed to obtain absolute values. In solutions of electrolytes, showing no non-electrostatic interaction with the surface (so-called <u>indifferent electrolytes</u>) this is found as the pAg or pH where σ_o does not depend on ω. The explanation is that the effect of ω is a screening of the charges, so that at high ω more changes can be adsorbed at given pH; if, however, the surface is uncharged, there is nothing to screen. Once the p.z.c. is established in indifferent electrolyte it can be obtained in other media.

Above, distinction has been made between c.d. - and indifferent ions, but it was stated that it is not always easy to discriminate, because there are sometimes species in the system that adsorb "intermediately" strong. We shall adopt the following classification.
(i) C.d. ions are chemically bound, they must have a very high affinity to the surface and binding energies of usually several tens of RT per mole. By way of example, in addition to the ions, mentioned in (2) for hematite, gibbsite and some other oxides, phosphate ions belong also to this category. If phosphate chemisorption occurs, the surface excess of the phosphate ions must be added to the r.h.s. of eq. (2).
(ii) Indifferent ions adsorb only through electrostatic interactions, their energy of binding is $zF\phi(x)$, if z is valency and $\phi(x)$ the potential at the locus of adsorption.
(iii) Ions with adsorption energies Φ of a few RT per mole in addition to $zF\phi(x)$ are called <u>specifically adsorbed</u>. The phenomenon is known as <u>specific adsorption</u> (s.a.). There are various sources for s.a., but the interaction is essentially of a physical nature. (By contrast, adsorption of c.d. ions might be called "specific chemisorption.")

It may be relevant for ceramic processing to have some insight in the various types and strengths of binding.

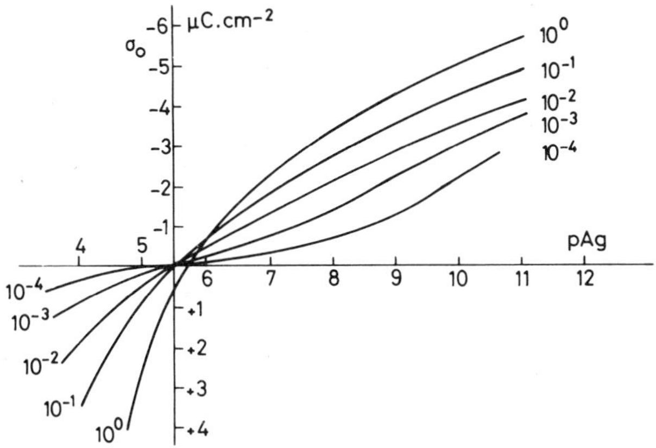

Fig. 1. Electrical double layer on AgI in the presence of KNO_3. T = 25° C. ω is indicated (reproduced from reference 1).

EXAMPLES OF DOUBLE LAYER STUDIES ON DISPERSE SYSTEMS

In Fig. 1 σ_o (pAg) curves for suspended AgI are presented. It is customary in electrochemistry to plot such curves with the negative side upward. The following features deserve attention. (1) Surface charges increase with ω at either side of the p.z.c. due to improved screening. (2) σ_o attains values of up to several μC cm^{-2}. Such values are comparable to those on mercury. Actually, the shapes of the curves of Fig. 1 are very similar to the corresponding ones on Hg if one unit of pAg is written as 59 mV (Nernst's law). This similarity indicates that the double layer structures resemble each other. (3) At low ω the p.z.c. is pAg = 5.65. The solubility product of AgI at 25° is 15.96, so that pI at the p.z.c. amounts to 10.30. Hence, the p.z.c. is very asymmetrical: the AgI surface has a much greater affinity for I^- than for Ag^+-ions. (4) The p.z.c. shifts slightly to the right with increasing ω. This points to weak s.a. of NO_3^-ions: accumulation of NO_3^-ions near the surface promotes sorption of Ag^+ over I^-, therefore in the presence of adsorbed NO_3^- a slightly higher bulk concentration of I^- is needed to reestablish the condition $\Gamma_{Ag+} = \Gamma_{I^-}$ than in the absence of NO_3^-. It may be added that many ions produce much greater shifts. For instance, on hematite, even traces of Ca^{2+} move the p.z.c. to the left (i.e., to lower pH) by about 2 units and SO_4^{2-} moves it strongly to the right.[4] Also organic adsorbates affect the p.z.c. because the molecules replace the oriented layer of adsorbed water dipoles on the surface.

Figure 2 is the corresponding picture for silica. It has in common with Fig. 1 that σ_o increases with ω at given pH and that the p.z.c. is very asymmetrical (for silica $K_a >> K_b$ in (3)). There are, however, two conspicuous differences: (1) the curves are not concave at high ω but convex, and (2) the absolute values of σ_o are much higher. Features (1) and (2) are common for all oxides, although between the various oxides (and even between differently pretreated but otherwise identical oxides) quantitative differences are observed. For instance, trends (1) and (2) have also been reported for cassiterite (SnO_2),[6] ludox,[7] quartz,[8] hematite (α-Fe_2O_3),[9] goethite (α-$FeOOH$),[10] rutile (TiO_2),[11] γ-Al_2O_3),[12,13] and MnO_2.[14] Qualitatively the explanation can be found in the fact that on oxides the c.d. ions tend to penetrate progressively deeper into the solid, the higher the pH.[15] In the example of Fig. 2, the maximum σ_o attainable would be ca. 75 $\mu C\ cm^{-2}$ if all surface silanols were charged. This maximum is clearly surpassed, so that it is impossible to accommodate all charges in the surface proper. The trend is that the more porous an oxide is for gases, also the deeper H^+, OH^- and counterions can penetrate[16] but there are exceptions. Double layers in which the surface side has some depth are sometimes called <u>porous double layers</u>.

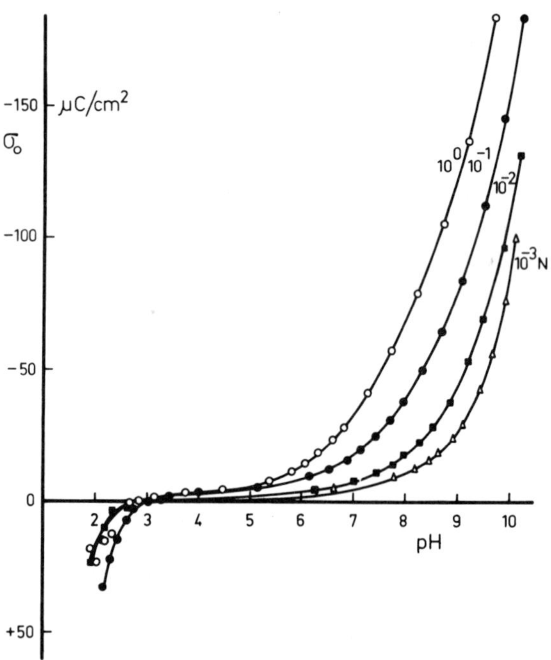

Fig. 2. Electrical double layer on precipitated silica. T = 25°C, ω is indicated. After[5]

Lyotropic sequences are of more than academic interest because they are reflected in the interaction energies and because they give insight into the nature of the binding of water to oxidic surfaces. Figure 3 shows that on SiO_2 Cs^+ screens better than Li^+, meaning that Cs^+ is better specifically adsorbed. However, on hematite it is the other way around. In the latter case, Li^+ adsorbs even so strongly that the p.z.c. is shifted. The observed order is correlated to the nature of the surface. SiO_2 is an example of a "low energy" surface whereas α-Fe_2O_3 has a "high energy" surface. Following Gurney,[18] the former group may be referred to as having a "structure breaking" surface whereas the latter has a "structure promoting" surface. Cs^+ ions in aqueous solution are structure breaking, whereas Li^+ ions are structure forming, therefore Cs^+ adsorbs more strongly on a structure breaking surface than Li^+ and conversely. The structure of water near surfaces can be important for the rheological properties of suspensions and lyotropic phenomena can be used to monitor this structure. For more details, see e.g.[2,19,20]

Some p.z.c. values of oxides are collected in Table 1. Their importance is twofold: (a) to establish the sign of σ_o ($\sigma_o \gtrless 0$ for pH $\lessgtr$ p.z.c.) and (b) to assess the nature of the interaction with water (the surface becomes gradually more high energy and more structure forming, the higher the p.z.c.).

Table 1. Points of Zero Charge of Some Oxides in the Absence of Specific Adsorption at Room Temperature

Oxide	p.z.c.	Reference
SiO_2 (precipitated)	2-3	5
SiO_2 (quartz)	2.7	21
SnO_2 (cassiterite)	5.6	6
TiO_2 (anatase)	6.2	22
TiO_2 (rutile)	7.3	22
TiO_2 (rutile)	6	23
TiO_2 (rutile)	5.3	6
β-MnO_2	7.3	24
Fe_3O_4 (magnetite)	6.5-7.3	25
α-Fe_2O_3 (hematite)	8.5	9,26
α-FeOOH (goethite)	8.4	27
α-FeOOH (goethite)	9.4	28
ZnO	8.5-9.5	29,30

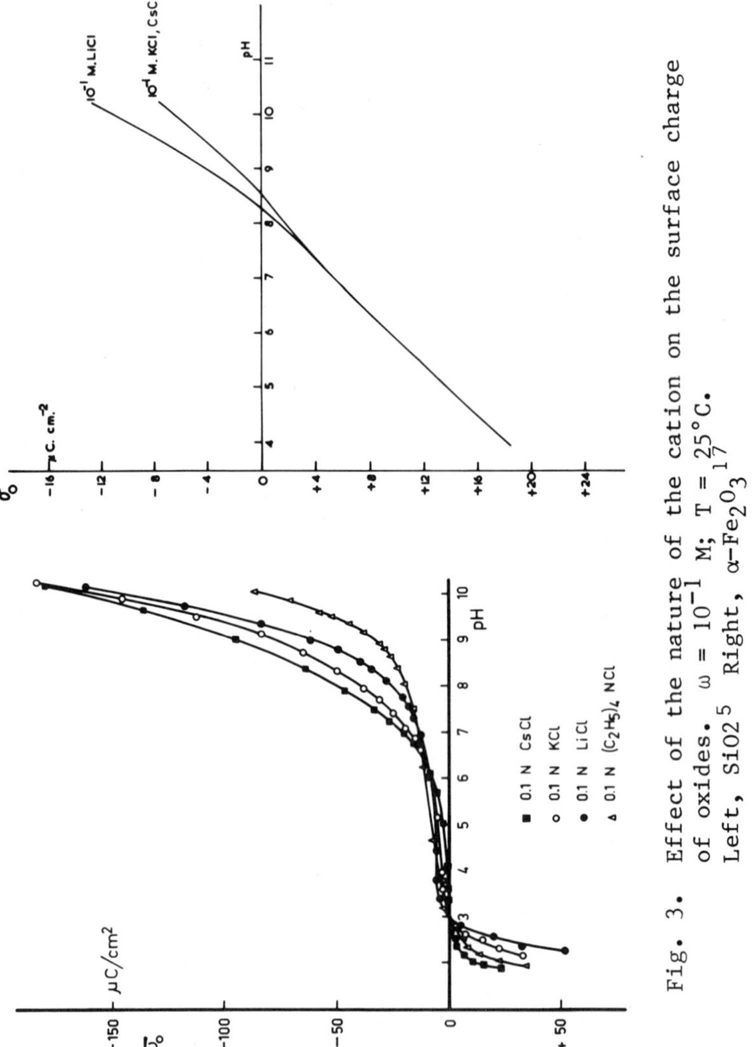

Fig. 3. Effect of the nature of the cation on the surface charge of oxides. $\omega = 10^{-1}$ M; T = 25°C. Left, SiO_2 [5] Right, α-Fe_2O_3 [17]

For clay minerals a p.z.c. can be assigned to the sides; usually it is about 6-7 so that the aforementioned card-house aggregates can occur below pH ~ 6. Plate charges can be obtained from c.e.c. values. Typical values are: montmorillonites ca. 10 $\mu C\ cm^{-2}$, illites up to 15 $\mu C\ cm^{-2}$ (if computed from the total number of cations and anions on exterior and interior surfaces, but its c.e.c. is lower than that of montmorillonite because only the external cations can exchange) and vermiculite 24 $\mu C\ cm^{-2}$. This matter has been reviewed and extensively tabulated by Bruggenwert and Kamphorst.[31]

DOUBLE LAYER THEORY

The purpose of double layer theory is to describe the potential and charge distribution around charged particles. For the present theme double layer (d.l.) pictures have to be related to the interaction between suspended particles and the quality of the various models can be judged by the extent to which experimental results, like those in Figs. 1-3 can be explained. As within the available space it is impossible to give full derivations, below restriction will be made to some important principles and results.

Unlike the treatment of surface charges so far, d.l. theory requires the introduction of the notion potential (ϕ), and this involves a basic problem. In electrostatics $\phi(x)$ is defined as the electrical work to transport a unit charge e from infinity ($x = \infty$, the reference point) to x. For ions near a surface, the work of transport, if done isothermally and reversibly, is a free energy and consists of a chemical and an electrical term: $\Phi_i + z_i F \phi(x)$ for ion i. Suppose that this sum could be measured, there would be no rigorous procedure to split it into its constituents, so that $\phi(x)$ is in principle unmeasurable. Some model assumption is needed to obtain it. Only at some distance from the surface, where $\Phi_i \to 0$ (because in contradistinction to electrical ones specific interactions are short range) is $\phi(x)$ fully defined.

Experience has shown that for practical purposes the <u>Gouy-Stern</u> (GS) picture works well. In this model the assumption is made that ϕ_i is finite only for the counterion that are very close (distances of one or two molecular cross-sections) to the surface. Beyond that, $\Phi_i \equiv 0$. The double layer part close to the surface is called the Stern layer and that further out the Gouy layer.[32,33] Alternative names for the former are <u>molecular condensor</u> or Helmholtz layer and for the latter <u>diffuse layer</u> or Gouy-Chapman layer. The Gouy part is the easier to describe rigorously. Here the ions distribute themselves according to the Boltzmann principle, finding a compromise between electrostatic interaction and the tendency to disperse because of thermal motion, very similar to the barometric density distribution in the earth's atmosphere. With increasing x, the space charge density $\rho(x)$ and $\phi(x)$ decay gradually to zero, hence, the name "diffuse" layer. Description fo the Stern part requires assumptions on the locus of adsorption of the various ions and on the dependence on σ_o. Two familiar approaches are the Langmuir treatment (Stern's own suggestion) and the Henderson-Hasselbalch (HH) method, stemming from the field of polyelectrolytes. Basically both approaches have the same principles, although the formulas look different.

The surface charge density in the diffuse part, σ_d, seldom exceeds ca. 2 $\mu C\ cm^{-2}$. Considering the values of σ_o (Figs. 1-3), it follows that the fraction σ_s/σ_o of charge compensated in the Stern layer is some tens of a percent for AgI, far over 90% for oxides and usually above 75% for clays. Notwithstanding the small fraction of charge in the Gouy layer, this part is of paramount physical significance since, because of its thickness, it is responsible for particle interaction. The DLVO theory of colloid stability is based on the overlap of diffuse double layers.[34-35] An important parameter is ϕ_d, the potential at the boundary between the Stern and the Gouy layer.

In Fig. 4 some typical cases of $\phi(x)$ distributions are drawn. Note that for oxides and clays the surface potential ϕ_o is not usually measurable. All distributions have in common that $\sigma_o + \sigma_s + \sigma_d = 0$, because of electroneutrality.

Figure 4(a) is the most simple. No s.a. is assumed but it is taken into account that counterions cannot approach the surface closer than some distance d, because of their finite sizes (hydration shells included; these are not shown in Fig. 4). The layer $0 \leq x \leq d$ is then charge-free, i.e., the space charge density $\rho(x)$ is zero in that layer. It follows from Poisson's law

$$\frac{d^2\phi(x)}{dx^2} = -\frac{\rho(x)}{\varepsilon_s\varepsilon_o} \tag{5}$$

where ε_s is the relative dielectric permittivity of the Stern layer and $\varepsilon_o = 8.854 \times 10^{-12}\ CV^{-1}\ m^{-1}$ that $\phi(x)$ is then linear. In this case ϕ_d follows simply from σ_o because $\sigma_o = \sigma_d$ and σ_d is related to ϕ_d according to GC theory (8). However, this simple situation has merely academic interest, it occurs only at very low ϕ_o and very low ω, much less than usually met in practice.

Case 4(b) is more common. Distinction is now made between the plane where anions adsorb specifically, the inner Helmholtz plane (iHp) and the slightly more remote outer Helmholtz plane (oHp), which is the distance of closest approach of non-specifically adsorbed

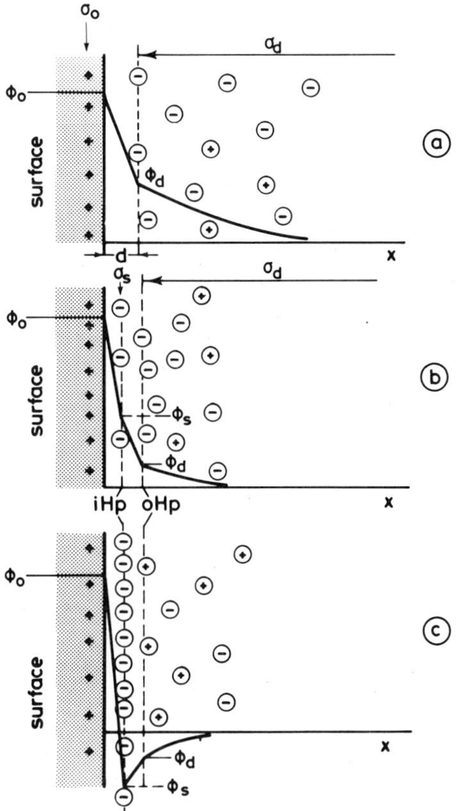

Fig. 4. Gouy-Stern double layer pictures. (a) No specific adsorption; (b) some specific adsorption; (c) super-equivalent specific adsorption (taken from 36). (N.B. σ_o increases if σ_s goes up, compare (b) with (a); however, to avoid crowding, this is not drawn in (c).)

ions. If ϕ_s is the potential at the iHp, the adsorption energy of an ion there is $\Phi_i + z_i F\phi_s$. In this case, d_d can only be found if σ_s is known. However, in practice the reverse path is often more passable because sometimes ϕ_d can be assessed with acceptable precision from stability and/or electrokinetics (see below).

In case 4(c) s.a. is so strong that at the iHp more negative charge is present than there is positive charge on the surface (superequivalent adsorption). It leads to a reversal of the signs of ϕ_d and σ_d. This possibility is of great importance for the preparation of suspensions with special properties: by choosing in mixed suspensions conditions under which some particles have a negative, and others a positive ϕ_d, an interaction balance can be set up to achieve certain desired properties. To the group of ions capable of superequivalent adsorption belong various surfactants and partially hydrolyzed higher-valency cations. It may be repeated that the distinction between very strongly superequivalently adsorbing and chemisorbing ions is not always sharp, nor is the ensuing definition of σ_o.

All three cases have in common that beyond x = d the d.l. is purely diffuse. It must, however, be realized that in practice surfaces are not so smooth as pictured in Fig. 4, so that the iHp and oHp are not always rigorously defined.

The distribution in the diffuse layer can be found from the Poisson-Boltzmann (PB) equation that is obtained by combining Boltzmann's law for each ionic species

$$c_i(x) = c_i(\infty)\exp.[-z_iF\phi(x)/RT] = c_i \exp.[-z_iy(x)] \qquad (6)$$

with (5). Here $y=F\phi/RT$ is a dimensionless potential; at 25°C one unit of y corresponds with 25.67 mV. From (6), at each x $\rho(x)$ follows as $F\sum_i z_ic_i(x)$ which is substituted in (5). Double integration gives $\phi(x)$ and one integration gives σ_d because

$$\sigma_d = \int_{x=d}^{\infty} \rho(x)dx = \varepsilon\varepsilon_o \int_{x=d}^{\infty} \frac{d\phi(x)}{dx} dx = -\varepsilon\varepsilon_o \left(\frac{d\phi}{dx}\right)_{x=d} \qquad (7)$$

Some results are represented in Figs. 5 and 6. It is assumed that the double layer is flat (spherical double layers give similar trends but are more difficult to handle). It is typical for the underlying premises that in the diffuse layer $\varepsilon \sim \varepsilon$ (bulk). The curves of Fig. 5 follow from this equation:

$$\sigma_d = (8\varepsilon\varepsilon_o cRT)^{1/2} \sinh.(zy_d/2) \qquad (8)$$

where c is the bulk concentration and $y_d=F\phi_d/RT$. The diagram applies to z=1. Because of the sinh. functionality σ_d increases progressively with ϕ_d. In practice σ_d tends to flatten off around ca. 2 μC cm^{-2}, additional countercharge being accumulated in the Stern part or inside the solid. This is also the reason for the flattening of σ_o in Fig. 1. The progressive increase of σ_o for oxides (Fig. 2) is not due to a high diffuse charge but to countercharge penetration inside the solid.

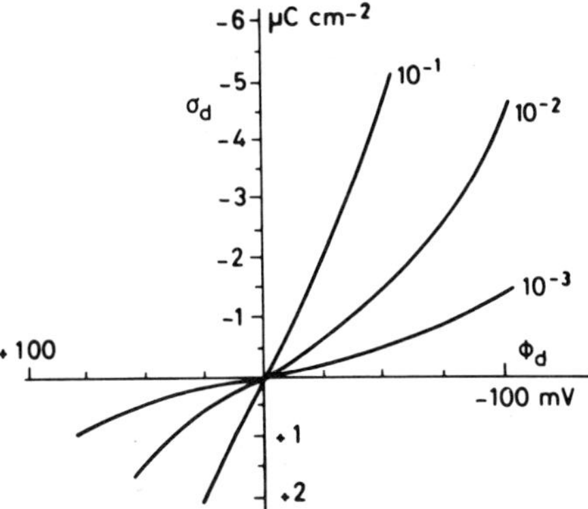

Fig. 5. Surface charge in a diffuse double layer. Monovalent electrolytes, T = 25°C.[36]

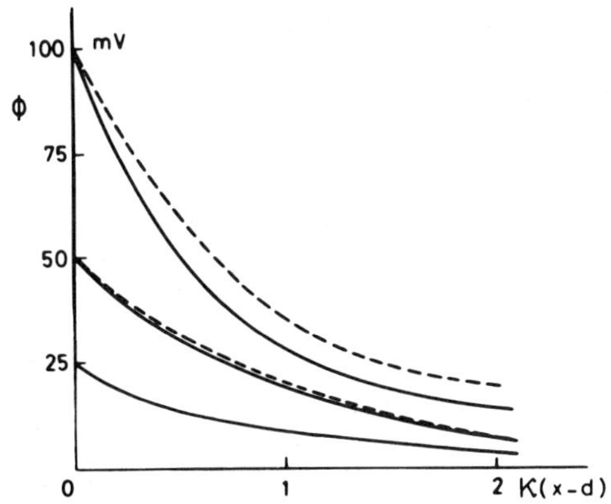

Fig. 6. Potential distribution in a diffuse double layer. Monovalent electrolytes, T = 25°C. Dashed curves: approximation (11).[36]

The potential distribution (Fig. 6) obeys

$$\tanh.[zy(x)/4] = \tanh.[zy_d/4] \exp.[-\kappa(x - d)] \tag{9}$$

where

$$\kappa = \left(\frac{2F^2cz^2}{\varepsilon\varepsilon_o RT}\right)^{1/2} \sim z\sqrt{10^{15}c} \text{ cm}^{-1} \text{ at } 25°\text{C if c in mol/l} \tag{10}$$

is the so-called Debye length, also occurring in the Debye-Huckel theory of strong electrolytes. Figure 6 is dimensionless, so that curves at various combinations of c and ϕ_d can be obtained from it by "scaling." In particular if at fixed ϕ_d c is increased κ goes up and this means that $\phi(x)$ decays more rapidly. In colloid-chemical terms: at higher κ the d.l. is compressed. This feature is again of great relevance for particle interaction: at elevated ω, particles can approach each other more closely without inhibition by electric repulsion so that it is easier for the attractive van der Waals forces to pull them together. This is basically the reason for the susceptibility of hydrophobic colloids to electrolytes. As the compression increases with z because of (10), and as s.a. tends to increase with z, leading to a lower ϕ_d at higher z, these two trends reinforce each other in making colloid stability very sensitive to the valency of the counterion (the Schulze-Hardy rule).

If the potentials are moderate, the hyperbolic tangents in (9) may be replaced by their arguments so that simply

$$\phi(x) = \phi d \exp.[-\kappa(x - d)]. \tag{11}$$

The dashed curves in Fig. 6 are based on this equation. From (11) it follows that over a distance κ^{-1} the potential has decayed to the e^{-1}th part of ϕ_d. Hence κ^{-1} is often called the (diffuse) double layer thickness. Double layers become thinner if c and z increase.

As stated above, there are two approaches to formulate ion adsorption in the molecular condensor. In the Langmuir-treatment it is assumed that on the surface there are adsorption sites on which an ion can adsorb. In the simplest case, with only one type of site and no superequivalent adsorption

$$\frac{\sigma_s}{\sigma_o - \sigma_s} = \frac{c_i}{55.5} \exp. -(\Phi_i + z_i y_s) \tag{12}$$

Here, $c_i/55.5$ is the mole fraction of ion i and the adsorption (free) energy contains the specific adsorption term Φ_i (often called "chemical" term, although no real chemical reaction occurs)[a] and the electrical contribution $z_i y_s = z_i F \phi_s/RT$. Equation (12) applies to the situation of Fig. 4(b). It is assumed that the charges on the surface act as the sites for counterions i. For the case of Fig. 4(c) there are more sites on the surface for counterions than there are for c.d. ions, or one surface site can bear more than one countercharge. It is not difficult to modify the equation accordingly. For

[a] Φ_i is counted in units of RT per mole and hence it is dimensionless.

instance, if N_{oi} is the number of sites for species i and N_{si} the number of ions i at the iHp, the LHS of (12) can be written as $N_{si}(N_{oi}-N_{si})$ and $\sigma_s = z_i e N_{si}$. Stern himself formulates an equation for two ionic species, a cation and an anion.[32] Consequently, his equation contains two specific adsorption energies.

The basic problem in applying (12) is to split the free energy in the RHS into its two constituents. Some assumption is needed. The equation gives an explicit expression for σ_s as a function of σ_o, but does not predict the relationship σ_o (pH).

Establishing this last prediction is attempted in the HH approach. Basically, the philosophy is as follows: σ_o is due to dissociation, obeying equations like (3a,b)

$$\sigma_o = e\{[ROH_2^+] - [RO^-]\} = E\ [ROH]\ \{K_a[H^+]_o - \frac{K_b K_w}{[H^+]_o}\} \qquad (13)$$

where $[H^+]_o$ is the proton concentration on the solid surface. If the potential of the surface is ϕ_o, $[H^+]_o = [H^+]_{bulk}$ exp. $(-y_o)$, so that σ_o can be related to pH. However, in elaborating (13) similar problems are encountered as in (12) because y_o is not known and the adsorption of c.d. ions is influenced by s.a. of other ions, whose influence must somehow be accounted for. It may be noted that since K_a = exp. $(-\Delta G^o/RT)$, the product $K_a[H^+]_o = K_a[H^+]_{bulk}\ e^{-y_o} \cong c_{H^+}$ exp. $-(\Delta G^o/RT - y_o)$ is identical to the same for cations in (12).

A feature worth mentioning is that (13) is a modification of the HH equation of polyelectrolyte theory. For a monomeric monabasic acid in solution, if α is the degree of dissociation and $K_{a,m}$ the dissociation constant one can write

$$pH = pK_{a,m} + \log \frac{\alpha}{1-\alpha} \qquad (14)$$

where $pK_{a,m} = 0.43\ \Delta G^o/RT$. If the same charges are found on a polymeric chain, the dissociation free energy of each group is influenced by its neighbors (because of the electrostatic long range action) and by ω (because of the screening). This can generally be accounted for by adding an additional term $\Delta G_{el}(\alpha,\omega)$, so that (14) becomes

$$pH = pK_{a,m} + \log \frac{\alpha}{1-\alpha} + \frac{0.43\ G_{el}(\alpha,\omega)}{RT} \qquad (15)$$

where it is immediately realized that the same "mixing up" of electrical and non electrical contributions to the free energy of binding of a proton is encountered as above. Plots of pH vs. $\log[\alpha/(1-\alpha)]$

are called HH-plots after Henderson, Hasselbalch et al.[37] For polyelectrolytes, it has been empirically established that often

$$pH = const. + n \log [\alpha/(1-\alpha)] \tag{16}$$

gives also a perfect straight line over a large pH-range.[38] Here the constant may be considered to be some apparent pK or, for that matter, some apparent binding free energy in units of RT and n is a measure for the interaction between groups: it is the higher the higher the charge density but it decreases with increasing ω and increasing counterion binding because of screening.

It is easily verified that (3) and (13) are virtually HH-type expressions by realizing that for only one surface group (eq. (36) only) $[\alpha/(1-\alpha)] = [RO^-]/[ROH]$. For oxides, therefore, the empirical equation (16) can also be advantageously used.

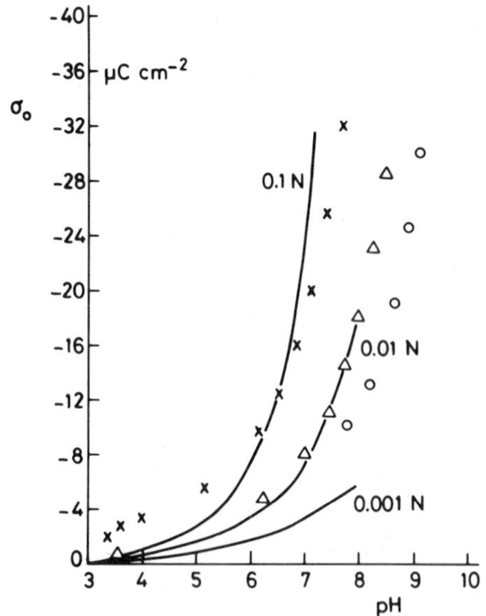

Fig. 7. σ_0 (pH) curves for silica. Experiments from Fig. 2. Theory: porous double layer with site dissociation picture. Specific adsorption of cations and anions, $\Phi_+ = \Phi_- = 1$. Penetration depth in solid 0.5 nm. $\varepsilon_{solid} = 15$, $p\overline{K}_a = 6.15$. The ionic strength is indicated.

Experience has shown how both treatments can be used to interpret experimental σ_o (pAg) or σ_o (pH) curves respectively.[1,2] By way of illustration, Fig. 7 gives an example of a more sophisticated computation that is characteristic for oxides in that charge penetration into the solid is allowed for.[39] In this particular example the double layer in the solid phase was considered to obey a Poisson-Langmuir distribution,[15] dissociation was accounted for according to (3) and (15), and two ions, a cation and an anion were supposed to adsorb specifically and to penetrate into the solid. Comparison with experiments show that perfect agreement has not yet been attained but that the characteristic shape and the electrolyte influence are well accounted for. Further elaboration is still under consideration.

For further information on double layers in general see references 36, 40-42, for AgI see 1, for clays see 3, 43, for oxides see 2, 44, 45. Much experimental information is collected in 20. The state of the art may be summarized as follows: much progress is being made, the basic phenomena are understood, but there is still much to do. Special attention is deserved by features involving the water structure near the phase boundary and the accounting for chemical phenomena. In particular the formation of oligomeric surface complexes and/or the reprecipitation of matter, dissolved in other parts of a suspension onto a certain surface is a complication deserving more study.

INFORMATION FROM ELECTROKINETICS

Electrokinetic phenomena, electrophoresis in particular, are the bases of a popular group of techniques to obtain rapidly information on double layers. For more than semiquantitative interpretation, however, great problems have to be overcome.

All electrokinetic phenomena have in common that they involve the tangential motion of the liquid with respect to the particle or conversely. In this motion, part of the countercharge (the <u>electrokinetic charge</u> σ_{ek}) is mobile, the remainder ($\sigma_o - \sigma_{ek}$) is fixed. The electrokinetic charge is related to the <u>electrokinetic potential</u> ζ, also known as the zeta potential, through (8) with σ_{ek} and $F\zeta/RT$ instead of σ_d and y_d, respectively. The assumption is that the charge beyond the slipping plane is fully mobile, it cannot be specifically bound and hence obeys Gouy theory.

This exposition indicates that in applying electrokinetic methods in principle two problems have to be solved. First, from the observed phenomenon σ_{ek} or ζ must be obtained. Next, σ_{ek} or ζ must be identified with the charge or potential respectively in some part of the double layer.

Measuring electrokinetic quantities is usually not so difficult with systems of interest for ceramics. Usually the particles are big enough and have sufficient contrast with the solution to be (ultra-) microscopically visible, so that the electrophoretic mobility in an electrical field is readily measured. Alternatively, the material under study can be compressed into a porous plug of which the electro-osmotic flow rate or streaming potential can be measured. The conversion of, say, the electrophoretic mobility into σ_{ek} or ζ is rigorously possible only under special conditions (spherical particles with electrostatically homogeneous and geometrically smooth surfaces, certain restrictions with respect to κa and the nature of the electrolyte). It is not sure if these conditions are always satisfied in ceramic suspensions. One of the unsolved theoretical problems is the proper accounting for relaxation phenomena. In plug experiments there are problems with surface conductance and double layer overlap. At high κa the simple Smoluchovski equation applies well, irrespective of the particle shape provided the surface is electrostatically homogeneous. For more information see references 46 and 47. Because of these problems, equations are often used that for the systems under study are only approximately valid. The consequence is that the computed ζ or σ_{ek} value is probably correct with respect to its sign and order of magnitude, but that the absolute value may be off by perhaps some tens of a percent.

The second problem, that of the identification with double layer properties, is also solved for more simple systems, but not yet entirely for oxides. If liquid flows tangentially to a surface, some molecules will stick to the solid; the further out from the surface, the higher the fluidity and at sufficiently large distance the mobility is the same as that in the bulk. The fluidity-distance relation is not usually known, hence it has become customary to replace it by a step function: within a so-called slipping plane, the viscosity $\eta = \infty$, (liquid does not move at all), beyond the slipping plane, $\eta = \eta$ (bulk). The ζ-potential is the potential of the slipping plane, and σ_{ek} is the charge at that plane. In this picture, the above problem is rephrased as "where to locate the slipping plane." For simple systems, like monolayers of surfactants and AgI it has been shown that within the limits of theoretical models available, the slipping plane and the oHp are identical ($\phi_d = \zeta$).[48] It is likely that this applies also to simple oxidic and clay surfaces, but the identification $\phi_d = \zeta$ becomes improbable if the surface contains oligomeric or polymeric adsorbates (they may be inadvertently present) or if the surface is very rough on a molecular scale.

The problem of the relation between ζ and ϕ_d is of crucial relevance for stability and rheology studies. As the diffuse part of the d.l. determines particle interaction, it is mandatory to establish ϕ_d properly. For instance, the DLVO equation for the coagulation concentration contains a factor ϕ_d^4. If ζ would be easily obtainable and if ζ would be equal to ϕ_d, a direct electrical characterization of the surfaces for stability purposes would be possible, but, as said, for oxides this alley is not yet without obstructions. More information on the relationship between ϕ_d or ζ and particle interaction follows in the chapter by Overbeek hereafter. In view of the close relationship, if not full identity, between ϕ_d and ζ, all features inherent in ϕ_d are also reflected in ζ. In particular, ζ reverses sign in the case of superequivalent adsorption (Fig. 4(c)) and σ_{ek} seldom exceeds a few $\mu C\ cm^{-2}$ even if σ_o amounts to several tens of $\mu C\ cm^{-2}$ (Figs. 2 and 3).

By way of example, Fig. 8 gives some electrophoretic mobilities for oxides. The left figure is typical for absence of specific adsorption. All curves pass through a common zero point, the <u>isoelectric point</u> (i.e.p.) which, in this case, is identical to the p.z.c. Increase of the $NaNO_3$ concentration reduces ϕ_d and ζ. In $Ca(NO_3)_2$ the situation is different because Ca^{2+} ions adsorb specifically, the more so, the higher the pH. The result is that ϕ_d tends to reverse sign at high pH. This trend becomes more pronounced if the Ca^{2+} concentration increases. At sufficiently high concentration $\phi_d > 0$ over the entire pH range. Not particularly high concentrations are needed to achieve ϕ_d-reversal. For practice, this implies that small amounts of ionic admixtures have a profound influence on the interaction. These admixtures may be intentionally added, but they may also be inadvertently present in the system, for instance as a result of leaching from other constituents in a mixed suspension.

If s.a. occurs, p.z.c. and i.e.p. are no longer identical. In fact, if s.a. of cations occurs at the p.z.c., this last quantity shifts toward lower pH.

Mobilities can also be measured in the presence of organic or inorganic admixtures. In combination with data on ϕ_o, such measurements are conducive to understand the inner layer properties of the double layers, and this information can in turn be used in the preparation of suspended materials of specific surface properties.

ACKNOWLEDGMENTS

The author appreciates permission to copy the following figures: Figs. 1-1, Elsevier Publ. Cy.; Figs. 4-6, The Royal Society of Chemistry.

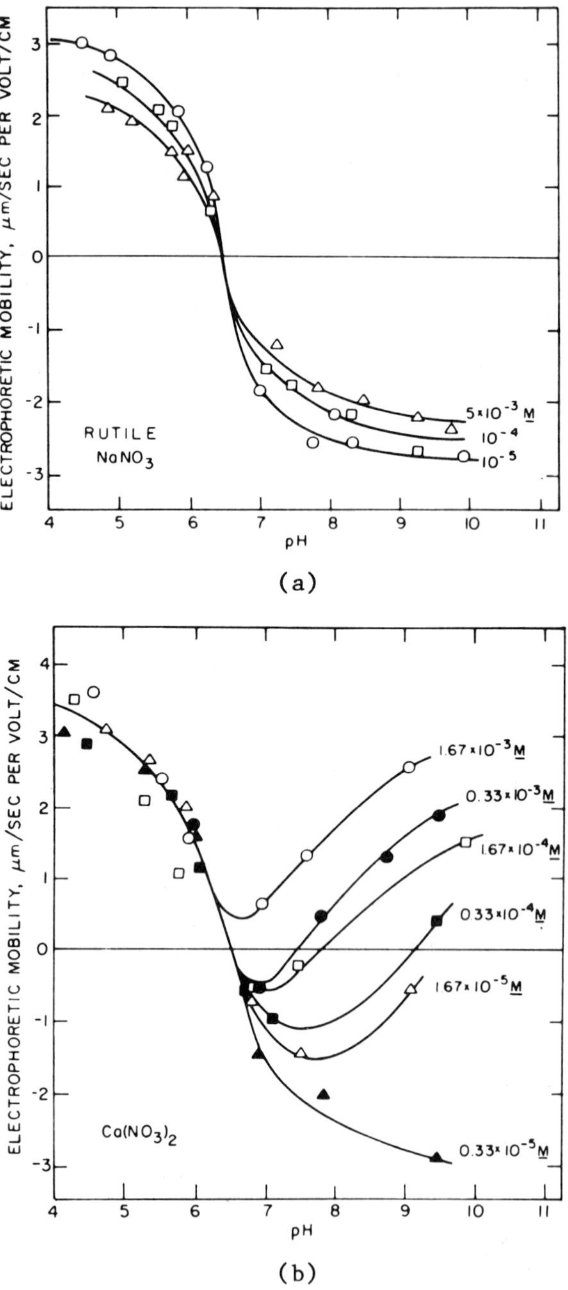

Fig. 8. Electrophoretic mobility of Rutile (TiO_2) in the presence of different concentrations of $NaNO_3$ (top) and $Ca(NO_3)_2$ (bottom).[49]

REFERENCES

1. B. H. Bijsterbosch and J. Lyklema, Advan. Colloid Interfac. Sci., 9, 147 (1978).
2. R. O. James and G. A. Parks, p. 119 in Surface and Colloid Science, Vol. 12, edited by E. Matijevic, Wiley-Interscience, 1982.
3. H. van Olpen, An Introduction to Clay Colloid Chemistry, John Wiley, 2nd ed., 1977.
4. A. Breeuwsma and J. Lyklema, J. Colloid Interfac. Sci., 43, 437 (1973).
5. Th. F. Tadros and J. Lyklema, J. Electroanal. Chem., 17, 267 (1968).
6. S. M. Ahmed and D. Maksimov, J. Colloid Interfac. Sci., 29, 97 (1969).
7. G. H. Bolt, J. Phys. Chem., 61, 1166 (1957).
8. H. C. Li and P. L. de Bruyn, Surface Sci., 5, 203 (1966).
9. A. Breeuwsma and J. Lyklema, discuss. Faraday Soc., 52, 324 (1971).
10. D. E. Yates and T. W. Healy, J. Colloid Interfac. Sci., 52, 232 (1975).
11. Y. G. Berube and P. L. de Bruyn, J. Colloid Interfac. Sci., 28, 92 (1968).
12. E. Herszynska, J. Inorg. Nucl. Chem., 26, 2127 (1964).
13. C. P. Huang and W. Stumm, J. Colloid Interfac. Sci., 55, 281 (1976).
14. J. J. Morgan and W. Stumm, J. Colloid Sci., 19, 48 (1964).
15. J. Lyklema, J. Electroanal. Chem., 18, 341 (1968).
16. J. Lyklema, Croat. Chem. Acfa., 43, 249 (1971).
17. A. Breeuwsma, (Thesis), Agricul. Univ. Wageningen, 1973.
18. R. W. Gurney, Ionic Processes in Solution, McGraw-Hill, 1953.
19. P. L. de Bruyn, Phys. Chem. Liq., 7, 181 (1978), see also ref. 49.
20. M. A. Anderson and A. J. Rubin, eds., Adsorption of Inorganics at Solid-Liquid Interfaces, Ann Arbor Science, 1981.
21. M. A. Malati, Discuss. Faraday Soc., 52, 377 (1971).
22. G. D. Parfitt, Progr. Surface Membr. Sci., 11, 181 (1976).
23. Y. G. Berube and P. L. de Bruyn, J. Colloid Interfac. Sci., 27, 305 (1968).
24. W. Strumm, C. P. Huang, and S. R. Jenkins, Croat. Chem. Acta, 42, 223 (1970).
25. S. Ardizzone, R. Biagiotto, and L. Formaro, J. Electroanal. Chem., acc. publ. (1982).
26. F. Dumont and A. Watillon, Discuss. Faraday Soc., 52, 352 (1971).
27. R. J. Atkinson, (Thesis), Univ. Western Australia, 1969.
28. J. D. Evans, (Thesis), Univ. of Newcastle Upon Tyne, 1976.
29. L. Blok and P. L. de Bruyn, J. Colloid Interfac. Sci., 32, 518, 527, 533 (1970).

30. H. F. A. Trimbos and H. N. Stein, J. Colloid Interfac. Sci., 77, 386 (1980).
31. M. G. M. Bruggenwert and A. Kamphorst, Vol. B, Ch. 5 in Soil Chemistry, edited by G. H. Bolt, Elsevier, 1979.
32. O. Stern, Z. Elektrochem., 30, 508 (1924).
33. G. Gouy, Ann. Chim. Phys., 8, 291 (1906); 9, 75 (1906).
34. E. J. W. Verwey and J. Th. G. Overbeek, Theory of the Stability of Lyophobic Colloids, Elsevier, 1948.
35. B. V. Deryagin and L. Landau, Acta Physicochim. URSS., 14, 633 (1941).
36. J. Lyklema, in Colloidal Dispersions, edited by J. W. Goodwin, The Roy. Soc. of Chem., 1982.
37. L. J. Henderson, Am. J. Physiol., 15, 257 (1906) and more recent papers; C. Bohr, K. Hasselbalch, and A. Kragh, Skand. Arch. Physiol., 16, 402 (1904).
38. A. Katchalsky and P. Spitnik, J. Polym. Sci., 2, 432 (1947).
39. M. Kleijn, A. de Keizer, and J. Lyklema, unpublished results (1981).
40. Comprehensive Treatise of Electrochemistry, Vol. 1 The Electrical Double Layer, edited by J. O'M. Bockris, B. E. Conway, and E. Yeager, Plenum Press, 1980.
41. M. J. Sparnaaij, The Electrical Double Layer, Pergamon Press, 1972.
42. D. C. Grahane, Chem. Revs., 41, 441 (1947).
43. Soil Chemistry B, Physicochemical Models, edited by G. H. Bolt, Elsevier, 1982.
44. T. W. Healy and L. R. White, Advan. Colloid Interfac. Sci., 9, 303 (1978).
45. G. R. Wiese, R. O. James, D. E. Yates, and T. W. Healy, Int. Rev. of Sci., Phys. Chem. Series Two, 6, 53 (1976).
46. J. Th. G. Overbeek, Vol. I, Ch. V in Colloid Science, edited by H. R. Kruijt, Elsevier, 1952.
47. R. J. Hunter, Zeta Potential in Colloid Science, Theory and Applications, Academic Press, 1981.
48. J. Lyklema, J. Colloid Interfac. Sci., 58, 242 (1977).
49. D. W. Feurstenau, D. Manmohan, and Ragharan, p. 93 in Adsorption from Aqueous Solutions, edited by P. H. Tewari, Plenum Press, 1981.

DISCUSSION

T. Wood (3M): Silver iodide and Hg surface structure and "surface chemistry" remain constant over the range of conditions of most investigations, and can be modeled and treated theoretically. In the case of metal oxides this is not true. For many metal oxides changes in the concentration of the potential determining ions (OH^-, H_3O^+) can result in changes of surface structure, alter surface-counterion interacting or produce surface complexes. Can this be the primary cause of the difficulty of the application of interfacial theory (electrical) to metal oxide systems?

Author: Provided the measurements are reversible and reproducible (i.e., no leaching, surface decomposition, dissolution, etc.), surface charges are well-defined and measurable, they can therefore be used as a basis for theoretical analyses. The problem is to define surface potentials since it is thermodynamically inoperational to split up the electrochemical potential of an adsorbed ion into a chemical and an electrical contribution. Because of this I disrecommend to use the notion "potential determining ions" for oxides.

R. Raj (Cornell University): Could you comment on the effect of particle size on the zeta potential?

Author: The zeta potential and the electrokinetic charge depend only on the surface properties and the charge distribution in a double layer, hence they are independent of particle size. There is, however, a substantial size effect in the relationship between ζ and the electrophoretic mobility u, from which ζ is calculated. For symmetrical electrolytes and spherical particles the theory of this has been elaborated (P. H. Wiersema, A. Loeb, J. Th. G. Overbeek, J. Colloid Interface Sci., 22, 78 (1966)).

C. Weissmantel (Tech. Hochschule): Recently a lot of information about surface adsorption and diffusion in zeolythe/liquid systems has been obtained by high resolution nuclear magnetic resonance (NMR). Has this method been applied to ceramic particles?

Author: Not to my knowledge, however, the specific kinetic conditions could pose problems.

H. Conrad (NCSU): What role does the density of the particles play in their agglomeration?

Author: As a rule, particles of higher density have a higher Hamaker constant and therefore attract each other more strongly.

J. H. Adair (Battelle Columbus Labs): Many different models have been used to compare "Stern potential" to zeta potential. However, calculated values are usually much higher than experimental at low ionic strengths (e.g., Healy & White, Adv. Coll. Int. Sci., Ionizable Surfaces etc.). What conjectures may explain this?

Author: If for various surface dissociation pictures always the same trend with respect to the properties of the zeta potential is observed, there must be a systematical defect in all these theories. One possibility is the neglect of charge accumulation inside the solid phase. Because of this simplification unrealistically high inner layer capacitances must be assumed and these tend to overestimate the potential drop in some parts of the double layer. However, more systematic analysis of these features is necessary before this matter can be definitely settled.

HOW COLLOID STABILITY AFFECTS THE BEHAVIOR OF SUSPENSIONS

J. Theodoor G. Overbeek

Van't Hoff Laboratory, University of Utrecht
Padualaan 8, 3584 CH Utrecht
The Netherlands

ABSTRACT

The stability of hydrophobic colloids toward added electrolytes, the valence rule of Schulze and Hardy, protective action and sensitization by large molecules are briefly treated. Sedimentation, the nature of the sediments, electrodeposition, and rheology are used to illustrate the difference in behavior between stable and unstable suspensions. The theoretical interpretation of stability is based on Van der Waals attraction, electrostatic repulsion and on the interaction of dissolved and adsorbed long chains ("hairy particles"). The rate of coagulation, repeptization and the possibility of thermodynamic stability are considered.

INTRODUCTION

The behavior of suspensions and emulsions during handling or even when left on the shelf is strongly affected by the interaction between the particles. These interactions are fairly well understood. If they are mainly repulsive, and if the suspended particles are small, the system does not change with time and is called colloidally stable. If, however, attraction between the particles prevails, the particles agglomerate, the suspension flocculates (= coagulates) and macroscopic phase separation results rapidly.

COLLOID STABILITY

These phenomena have been studied extensively with--usually aqueous--colloidal suspensions (called sols) in which the individual particles are so small that they show no sedimentation. Such particles may be formed by mixing two fairly dilute aqueous solutions so that a large number of nuclei are formed and consequently the particles remain small. The formation of a silver iodide sol by mixing $AgNO_3$ and KI solutions is an example.

$$AgNO_3 + KI \rightarrow AgI + KNO_3 \qquad (1)$$

Contaminating electrolytes, such as KNO_3 in the above example, may be removed by dialysis. The particles are electrically charged by the adsorption of ions, e.g., Ag^+ or I^- and the formation of an electric double layer as discussed by Lyklema.[1] Sols coagulate after addition of electrolyte, any electrolyte, to a critical coagulation concentration (c.c.c.), actually a narrow concentration range. The main factor determining the c.c.c. is the charge number of the ions, which are oppositely charged to the particles. This regularity has been known as the Schulze-Hardy rule.[2,3] The c.c.c.s are roughly 100 mM, 1 mM, 0.1 mM for counterions with charge number z = 1, 2, and 3 respectively, as shown in Table 1. Exceptions to the Schulze-Hardy rule occur with ions that are strongly adsorbed or that give rise to chemical reactions or precipitations.

Table 1. Critical Coagulation Concentrations[4] in m mole/liter

As_2S_3-sol Negatively Charged		AgI-sol Negatively Charged		Fe_2O_3-sol Positively Charged	
LiCl	58	$NaNO_3$	140	NaCl	9.25
NaCl	51	KNO_3	136	$1/2BaCl_2$	9.65
KNO_3	50	$RbNO_3$	126	KNO_3	12
$MgCl_2$	0.72	$Ca(NO_3)_2$	2.40	K_2SO_4	0.205
$MgSO_4$	0.81	$Ba(NO_3)_2$	2.26	$MgSo_4$	0.22
$ZnCl_2$	0.69	$UO_2(NO_3)_2$	3.15	$K_2Cr_2O_7$	0.195
$AlCl_3$	0.093	$Al(NO_3)_3$	0.067		
$1/2(Al)_2(SO_4)_3$	0.096	$La(NO_3)_3$	0.069		
$Ce(NO_3)_3$	0.080	$Ce(NO_3)_3$	0.069		

It further appears that coagulation can be prevented by the addition and adsorption of small amounts of large molecules, e.g., gelatin, gums. This is called protective action. A very old example is India ink, which is a suspension of soot in water, stabilized by gum. In many cases protective substances added in amounts too small to give protection sensitize the suspension, i.e., make it more easily flocculable.

Sedimentation

If the particles are larger (diameter > 1 μm) than in typical colloids (diameter < 0.1 μm) the difference in rate of sedimentation between stable and flocculating systems is less pronounced but there is a pronounced difference in the behavior of the sediments.

Stable suspensions sediment rather slowly, with a fuzzy boundary between supernatant and sedimenting suspension because the particles sediment individually with speeds varying according to their sizes. The sediment is very compact since the particles can glide along one another until the packing is as dense as possible. Such a sediment makes redispersion difficult and time consuming. It is a well known nuisance in a paint which is too well stabilized and has stood too long on the shelf.

If attraction prevails, the suspension coagulates while sedimenting. The sedimentation is faster. The boundary between supernatant and suspension is sharp, since the smaller particles are also caught in the flocs and sediment together with the larger ones. The final sediment is open. If the particles are not too small they can be easily redispersed by shaking or stirring. A little attraction leading to weak flocculation is good for shelf life. Figure 1, which illustrates this difference, also shows that a soil for agriculture must be flocculated, since it must allow easy passage of water and air.

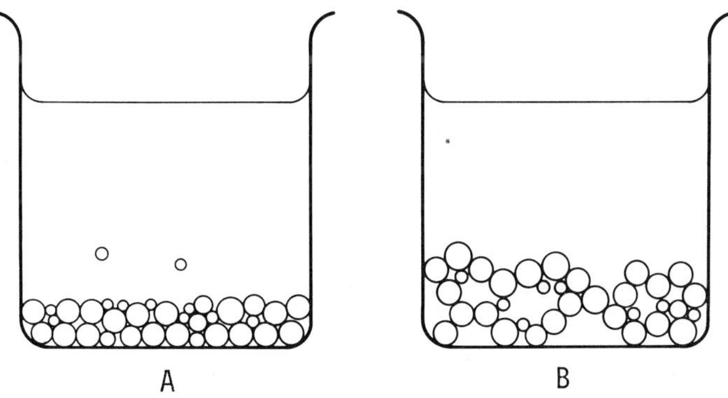

Fig. 1. Structure of sediments. (A) Compact sediment from stable suspension. (B) Loose sediment from flocculated suspension.

Electrodeposition

Since the particles are charged they can be transported towards an electrode by the application of an electric field, just as gravity transports them towards the bottom of the vessel. This process, called electrodeposition, has found important industrial applications, one of them being the painting of car bodies from a suspension. The particles which are in stable suspension are concentrated in a compact layer (c.f., Fig. 1A) at the electrode, and since the deposited layer has a high electrical resistance, open areas are covered preferentially. The layer becomes quite homogeneous, covers nooks and crevices and even the back side of the electrode. But, building this compact layer is not enough. The particles still repel each other, the layer, although viscous, is still fluid and tends to flow off the electrode. But now the electrode reaction, leading to electrolyte being formed comes to help. The electrolyte concentration increases just at the electrode and coagulates the concentrated suspension *in situ*.[5] Electrodeposition is a unique way to prepare in a single step a layer that is compact and at the same time coagulated.

Rheology

At *low rates of shear*, dilute *stable suspensions* show Newtonian behavior, i.e., their viscosity is independent of the shear rate. The viscosity is increased above that of the solvent with a term proportional to the concentration of particles, as shown in eq. (2);

$$\eta_s = \eta_o(1 + f.\phi_p) \qquad (2)$$

where η_s is the viscosity of the suspension, η_o that of the dispersion medium, ϕ_p the volume fraction of the particles and f a factor, which is 2.5 for spheres (Einstein)[6] and larger for non-spherical particles.[7,8] At higher concentrations the viscosity goes up faster than as described by eq. (2) and becomes extremely high when close packing is approached. Semi-empirical extensions of eq. (2) often contain a term $(1-\phi_p/\phi_{max})^{-1}$ where ϕ_{max} is the volume fraction at which the viscosity goes to infinity.[9,10] An extra complication is shown by suspensions of monodisperse spheres at high concentrations, where they form an ordered quasi crystalline state.[9,11,12]

If the particles are charged, an extra increase of the viscosity occurs. There are three of these *electroviscous effects*. The first one was already recognized by Smoluchowski.[13] The theory, improved by Booth,[14] is based upon the dissipation in the double layer in the field of shear. The second one[15,9,10] is due to the electrostatic repulsion between particles, when they pass close to each other in the shear field. The third effect occurs with flexible polyelectrolyte ions and is due to the stretching of these ions caused by the mutual repulsion of their charges.

Exact theoretical expressions for the viscosity of concentrated suspensions, even for monodispersed spherical particles are still lacking because the various interactions, such as the impenetrability of the particles (hard sphere effects), the mutual electrostatic repulsion and the hydrodynamic[16-19] interactions are difficult to evaluate. Moreover, when the particles are small, Brownian motion has to be taken into account, in particular for non-spherical particles and aggregates.

At high rates of shear two completely different effects are observed. At moderate concentrations small but elongated particles show shear thinning when the shear overcomes the randomizing effect of Brownian motion and the particles stay relatively longer in positions, where their contribution to the viscosity is small. At very high concentrations suspensions of more or less spherical particles show shear thickening and ultimately dilatancy.[20-22] In high shear the particles, instead of moving smoothly past each other, bump into each other and form a rigid network that blocks the motion.

Flocculated suspensions behave completely differently from stable ones. A concentrated flocculated system may form a continuous network, that stretches from wall to wall and turns the suspension into a gel. One may consider a gel as a suspension in which the sedimentation volume is larger than the total volume. Any influence (such as added salt) that flocculates a dilute suspension turns a concentrated one into a gel. This may also be the result of the addition of a small amount of a second liquid phase that wets the particles preferentially and forms capillary bridges between them (e.g., wet sand). Elongated particles (such as clays, that moreover carry plus and minus charges on each particle) form gels at a lower concentration than spherical or cubic particles.

At low shear stress a gel is elastically deformed, but does not flow. Above the yield stress a sufficient number of bonds are destroyed to allow the system to flow and at very high shear rates dispersion into primary particles or small flocs is complete.[10,23] At intermediate rates of shear flocs break up but are also regenerated, leading to a floc size distribution. Often the regeneration of the flocs is slow which causes the resistance against flow to decrease with time of shearing. The suspension may even lose its yield stress completely, become fluid, and gel only after more or less prolonged standing. This behavior is called thixotropy[24-25] or isothermal sol-gel transformation. In a shear rate against shear stress diagram hysteresis loops are formed as illustrated in Fig. 2.

Dispersed paints and drilling muds[26] illustrate the importance of rheological behavior. A dispersed paint should be thixotropic with a fairly short time constant. Then it does not settle in the container, it does not drip off a vertical surface, but during application it flows easily. The drilling mud should be fluid for easy pumping, but on the other hand, it should form a weak gel when the drilling is interrupted, so that the chips and the mud particles do not sediment.

Non-aqueous Media

Without mentioning this explicitly most of what I said so far was based on the behavior of suspensions in water, not only because water is cheap and omnipresent, and more data are available, but also because it is more difficult to prepare stable non-aqueous suspensions. Suspensions in polar organic solvents, such as the lower alcohols and acetone, behave similarly as aqueous suspensions, but they are much more sensitive to electrolytes.[27,28] This is not unexpected because the lower dielectric permittivity, ε, implies stronger electrostatic interactions between the surface charge and the counterions. With non-polar solvents electrostatic repulsion is usually absent, although it can be evoked by the use of large organic ions.[29] Protective action based upon the presence of oil soluble long chain molecules is the normal mechanism of stabilization.[30-32] Oil based paints are good examples of suspensions in non-polar media. Another example is engine oil, "doped" in order to keep carbon and other products of incomplete combustion suspended as small particles, that do not cause abrasion.

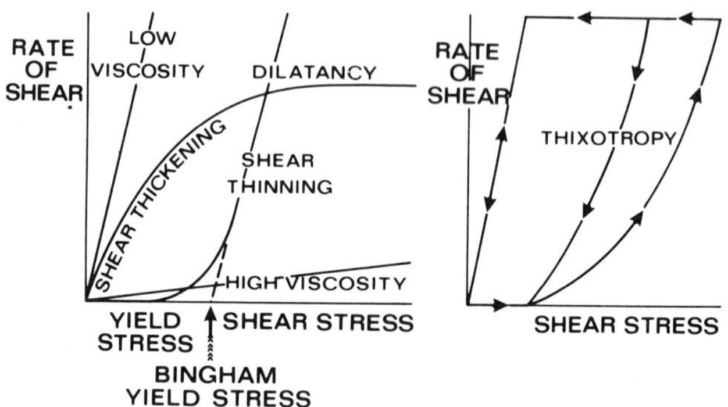

Fig. 2. Various types of rheological behavior. Left: time independent behavior. Right: thixotropy; structure decreases progressively during flow and is restored during rest.

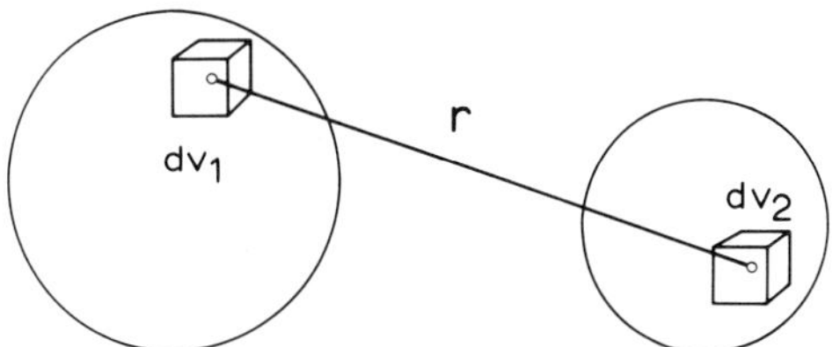

Fig. 3. Van der Waals attraction is built up from the attractions between all pairs of volume elements in the two particles.

THEORY OF COLLOID STABILITY

We mentioned that stable suspensions are obtained when the particles repel each other, and that suspensions flocculate when the interaction between particles is mainly attractive. Then we described a number of differences in behavior between stable and unstable suspensions. Now we shall analyze the several components of interaction more closely, first in static situations and then briefly also when the rate of changes plays a role.

Van der Waals Attraction

Kallmann and Willstaetter[33] were the first to suggest that Van der Waals forces are the main forces driving suspensions towards coagulation. This idea was worked out by de Boer[34] and Hamaker.[35] Since Van der Waals attraction between two atoms or molecules is inversely proportional to the sixth power of the distance and, to a first approximation, all Van der Waals energies are additive, the Van der Waals attraction between particles composed of many molecules has a fairly long range. The attractive energy between the two particles in Fig. 3 may be written

$$V_{att} = - \int_{v_1} \int_{v_2} \frac{\lambda q_1 q_2}{r^6} \, dv_1 dv_2 \qquad (3)$$

where λ is the London-Van der Waals constant for two molecules 1 and 2, q_1 and q_2 are the numbers of molecules per unit volume in particles 1 and 2 and v_1 and v_2 are their total volumes. V_{att} is independent of the scale of Fig. 3, since dv_1 and dv_2 are each proportional to the cube of the scale and r^6 is proportional to the sixth power of the scale. The attraction energy between two atoms is of the order of the thermal energy, kT, at a distance of an atomic radius. Thus, the energy between two particles is also of the order of kT at a particle radius distance between their surfaces. The energy of interaction is given as

$$V_{att}(\text{general}) = -f\frac{A}{H^n} \; ; \; V_{att}(2 \text{ spheres}) = -\frac{aA}{12H} \qquad (4)$$

where f is the n-th power of a length, $A = \pi^2 q^2 \lambda$ is the Hamaker constant, H is the distance of closest approach between the particles and n varies between 1 (spheres of radius a at small separations) to 6 (for large separations).

For particles suspended in a liquid the net Van der Waals attraction is smaller than in a vacuum, but it remains always an attraction.[36] The retardation[37] of the Van der Waals attraction, due to the finite speed of transmission of electromagnetic signals is rarely of practical importance in suspensions, because the attraction is already very small at distances where the retardation becomes significant.

Lifshitz and coworkers[38] have developed a more refined theory of the Van der Waals attraction, not based on the additivity of the interactions between pairs of atoms but on the dielectric properties of the solids and liquids involved. Numerical values of the Hamaker constants obtained in various ways have been compiled by Visser.[39] Cf. also Lyklema.[28] Typical values for the Hamaker constant A for particles in water are:

$$A(\text{metal-}H_2O\text{-metal}) = 5 - 30 \times 10^{-20} J = 12 - 75 kT$$

$$A(\text{oxide-}H_2O\text{-oxide}) = 0.5 - 5 \times 10^{-20} J = 1 - 12 kT$$

$$A(\text{hydrocarbon-}H_2O\text{-hydrocarbon}) = 0.3 - 1 \times 10^{-20} J = 1 - 2.5 \text{ kT}$$

Electrostatic Repulsion

As mentioned before and as discussed more fully by Lyklema[1] particles in suspension are electrically charged and thus repel each other. The repulsion is not simply Coulombic, but complicated due to the overlap of the diffuse parts of the two double layers. Just as the charge and the potential in the double layer fall off exponentially,[1] the energy and the force of repulsion decrease as $\exp(-\kappa H)$ both for parallel flat surfaces and for spheres. If the potential at the boundary between the molecular condenser and the diffuse layer is ϕ_d, a good approximation for the energy of repulsion, V_{rep}, between two equal spheres with radius a is given by

$$V_{rep} = 2\pi\varepsilon\varepsilon_0 a \left(\frac{4RT\gamma}{zF}\right)^2 e^{-\kappa H} \simeq 2\pi\varepsilon\varepsilon_0 a\phi_d^2 e^{-\kappa H} \tag{5}$$

where $\gamma = \tanh(zF\phi_d/4RT)$, $\kappa = (\Sigma z_i^2 c_i F^2/\varepsilon\varepsilon_0 RT)^{1/2}$ = the inverse Debye length, ε is the relative permittivity of the medium (the dielectric constant), ε_0 the permittivity of the vacuum, z_i and c_i the charge number and average concentration fo the ions in the solution, z the charge number of the counterions, H is the closest distance between the surfaces of the spheres where the potential is ϕ_d and R, T and F have their usual meaning. The second expression for V_{rep} in eq. (5) is a good enough approximation when $z\phi_d < 60$ mV.

Addition of electrolyte reduces the repulsion via $\kappa(\sim z\sqrt{c})$, compression of the double layer, and z in γ/z.

The difference in decay with the distance between attraction (exponentially) and repulsion (inverse power of the distance) has the very interesting consequence that at very large and at very small distances the attraction always prevails and that at intermediate distances the repulsion may prevail. Fig. 4 gives a number of combined repulsion and attraction curves, differing only in the steepness (value of κ) of the repulsion curve. If the maximum in V_{tot} is high enough (say 10 or 20kT), the suspension is stable, but if the maximum is low or absent particle encounters lead to entrapment in the deep, so called primary, minimum at contact. This minimum is not infinitely deep as might be inferred from Fig. 4 because the Born repulsion (= impenetrability of atoms) makes the curves swing up again near H = 0. Coagulation after the addition of electrolyte is now easily explained as a consequence of the steeper decay of the repulsion. Coagulation in the shallow "secondary" minimum (at the right in Fig. 4) may occur with large particles.

If the transition between stable and unstable is laid at curve 3 (V = dV/dH = 0), and the surface potential is so high ($z\phi_d > 150$ mV) that $\gamma \to 1$, the c.c.c. is inversely proportional to z^6, in agreement with the Schulze-Hardy rule. However, in most practical cases the ζ-potential ($\simeq \phi_d$) at coagulation is low and then the coagulation condition already proposed by Eilers and Korff[40] is found.

$$\frac{\phi_d^2}{\kappa} = \frac{\zeta^2}{\kappa} = \text{constant} = \frac{A}{24\pi\varepsilon\varepsilon_0\exp(-1)} \tag{6}$$

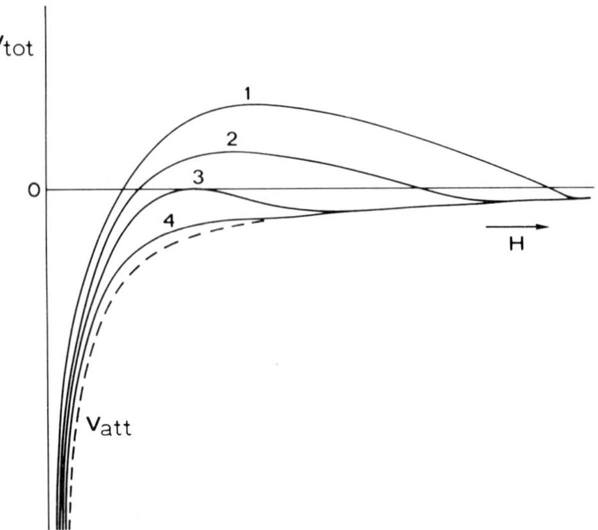

Fig. 4. Total interaction ($V_{tot} = V_{rep} + V_{att}$) between two spherical particles. All cases have the same attraction curve. The repulsion is progressively steeper (higher value of κ) in the direction 1, 2, 3, 4. Case 3 represents the borderline between stability and coagulation; $V_{max} = 0$. Schematic.

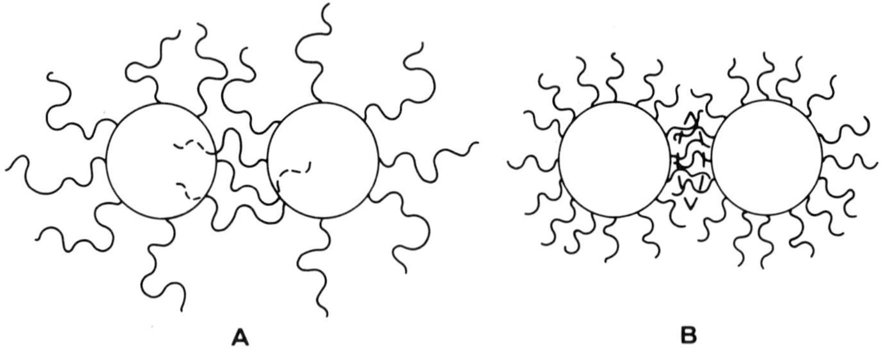

Fig. 5. Schematic illustration of the volume restriction effect (A) and the osmotic effect (B) in stabilization by adsorbed or chemically bound long chains.

In order to predict the c.c.c. from this equation a relation between ϕ_d, z and c is required and this again requires an adsorption isotherm for the counterions. The rule of Schulze and Hardy is then found when the charge of the diffuse double layer and the surface potential, ϕ_d, are decreased by adsorption of counterions, especially those with charge numbers above 1.[41]

In the condition V = dV/dH = 0 the c.c.c. is independent of the particle radius, a. But all energies involved are proportional to a, and thus large particles are more easily stabilized than small ones because a barrier of a given height (say 20kT) is more easily obtained with large a. When more accurate equations than eqs. (4) and (5) are used, larger particles may be less stable than smaller ones in exceptional cases.[42]

Effects of Large Molecules

A primitive, but essentially correct, explanation of the stabilizing effect of adsorbed large molecules is based upon their solubility in the medium, which prevents them from sticking together, whereas their bulk prevents the Van der Waals forces between the particles from coming into action.[43] More refined interpretations start from the theory of polymer solutions, as given by Flory[44] and Huggins.[45] From the many authors who have contributed to this field I mention Hesselink,[31] Napper[32] and more recently Fleer and Scheutjens.[46] When two surfaces, from which flexible long chains are sticking out into the solution, come close together two effects contribute to the repulsion. In the narrow gap between the surfaces the long chains lose some of their conformations (volume restriction effect). This results in a loss of entropy, in an increase in the free energy and thus in a repulsion. Furthermore the concentration of polymer segments in the gap increases and this so called osmotic effect results in another contribution to the repulsion. Figure 5 illustrates this schematically.

From these considerations it follows that stabilization occurs only on the solution side of the Θ-point and slightly below it in the case of the volume restriction effect. Changing the solvent to below the Θ-point results in flocculation. This flocculation is reversible and after changing the solvent again to above the Θ-point the flocs will redisperse.

The adsorption of the long molecules can be due to two different principles.

a. The stabilizing molecule may contain one part that is easily adsorbed (the anchor group) and one part, the chain, that is easily soluble.

b. The molecule is a homopolymer with a relatively weak adsorption per chain element, leading to the attachment of several segments in trains (one or a row of neighbouring segments in contact with the surface), but leaving two tails (ends of chains) and many loops (parts of the chain adsorbed at their two ends) sticking out in the solution. For the stabilizing interaction the tails are the most important.

Sensitization (destabilization) occurs when a molecule is attached with two anchors to two different particles. This occurs in particular at low polymer concentration, when the surfaces are incompletely covered with adsorbed molecules.

Free, i.e., not adsorbed, polymers also influence the stability. In a narrow gap between two surfaces polymer coils are forced out of the gap by the volume restriction effect. This leads to a concentration gradient, which pushes solvent out of the gap and thus causes flocculation. In high concentrations of polymer stabilization may result, because too much work is involved in pushing the polymers out of the gap against the now important concentration gradient. These effects have been studied experimentally and theoretically by Scheutjens and Fleer,[46] Vincent et al.,[47] Vrij,[48] Napper[49] and others.

Above a molecular weight of a few thousand the molecular weight of the polymer is not very critical, although at very high molecular weights the solutions may become too viscous and the coil extensions too large compared to particle sizes.

Most but not all water soluble polymers, gums, proteins, carboxymethylcellulose (CMC) are polyelectrolytes. They are good protective agents in which steric effects and electrostatic repulsion are combined. At low concentrations they lead to sensitization. The effects of electrolyte are complex, since they do not only decrease the electrostatic repulsion but often increase the adsorption of the polyelectrolyte. Polyelectrolyte counterions are of course powerful flocculants.

Structural Forces

The idea that hydration (more generally: solvation) at the surface may keep particles apart has recurred several times in colloid-chemical considerations, but until recently it has never been substantiated. Only in the last five years experimental evidence and model calculations have accumulated to show that the liquid structure near an interface is disturbed over a depth of several, up to ten, layers of molecules and that this disturbance may lead to a steep and strong repulsion. In a series of papers by Ninham, Israelachvili, Parsegian, Derjaguin and others in the proceedings of a recent symposium[50] this "structural force" is discussed. This effect, on which a great deal of work still has to be done, may be important for colloid stability. On the other hand, so far, no unambiguous evidence about its effect on suspensions of particles has been presented.

KINETIC EFFECTS IN COLLOID STABILITY

So far I have treated stability as a static, thermodynamic phenomenon. Kinetics, however, also play an important role. If the barrier is not infinitely high--and it never is--one should know at what rate coagulation occurs and even when a barrier is completely absent, coagulation requires a finite time. How much time? This question and related matters will be treated extensively in Dr. Gregory's lecture. But one aspect of kinetics has to be treated here.[51]

When two particles approach one another, be it in a Brownian encounter or in a systematic motion, as in shear or sedimentation, it is not correct, as we have tacitly assumed, that complete equilibrium reigns at every moment of the approach. Time is required for the rearrangement of the electric double layer, for the rearrangement of the conformations of macromolecules and for desorption and adsorption processes.

In the case of electrical stabilization we have to compare the time of a Brownian encounter with the relaxation time of the double layer and with the relaxation time involved in adsorption or desorption. The time of a Brownian encounter τ (Brown, double layer) can be defined as the time needed by a particle to diffuse through the thickness of the double layer, $1/\kappa$.

$$\tau(\text{Brown, d.l.}) = \frac{(1/\kappa)^2}{D(\text{particle})} \tag{7}$$

Similarly the relaxation time of the double layer is the time needed by an ion to diffuse through $1/\kappa$.

$$\tau(\text{relax. d.l.}) = \frac{(1/\kappa)^2}{D(\text{ion})} \simeq \frac{5 \times 10^{-11}\ \text{s}}{c/(\text{mol l}^{-1}} \tag{8}$$

Consequently, with $D_i = kT/6\pi\eta a_i$, where a_i is the radius of the diffusing entity,

$$\frac{\tau(\text{Brown, d.l.})}{\tau(\text{relax. d.l.})} = \frac{a(\text{particle})}{a(\text{ion})} \tag{9}$$

as this is usually 100 or larger. The relaxation time involved in adsorption or desorption is connected with the exchange current density of the electrode process involved and this varies over many orders of magnitude, but is at best of the order of time of a Brownian encounter and usually a great deal longer.[51]

Consequently in a Brownian encounter the double layer structure rearranges itself so rapidly that equilibrium may be assumed, but as a rule the surface charge does not adjust itself at all and remains constant. Calculations of double layer repulsion should be made for constant charge, not for constant surface potential.

Similar arguments, applied to a layer of adsorbed macromolecules, lead to the conclusion that here also desorption is too slow to occur during Brownian encounters, but the rearrangement of the conformations of loops and trains occurs easily within the time of the Brownian encounter. The slowness of desorption makes an adsorbed macromolecule a much better protective agent that it would be with complete adaptation.

REPEPTIZATION

One of the consequences of the theory of stability as presented so far is the conclusion that coagulation in the primary minimum is irreversible. The minimum is at a negative value of the free energy and spontaneous repeptization is impossible as illustrated in Fig. 6.

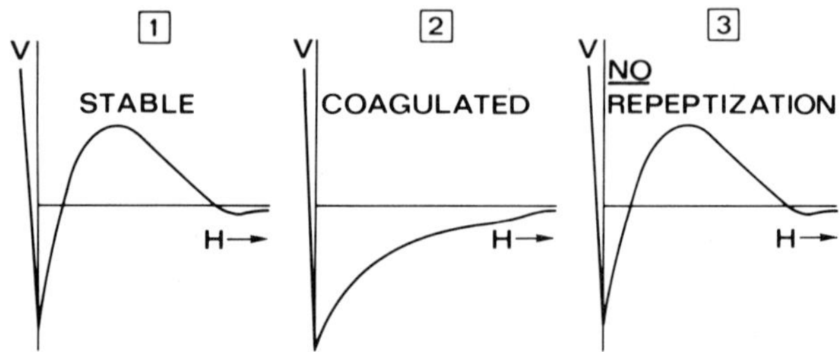

Fig. 6. Schematic energy vs distance diagrams. (1) Stable suspension; (2) electrolyte added, barrier removed, coagulation at H = 0; (3) electrolyte removed, but no repeptization, since this would require an increase in V not only to pass the barrier but also to reach the final dispersed stage (H → large).

This agrees with the observation that most coagulated colloids do not return spontaneously to the dispersed state when the coagulating electrolyte is removed. However, a number of cases are known in which repeptization does occur. For some sols[52,53] such as V_2O_5, HgS, Carey Lea's silver, this is even the preferred way of preparation. For many others, e.g., silver halides, $Fe(OH)_3$, other oxides and hydroxides repeptization can be obtained, if the electrolyte is washed out soon after flocculation. Therefore Fig. 6 cannot be correct. Our analysis of this figure shows that two essential elements in repeptization are:

a. The primary minimum must disappear or at least be situated at a positive value of V after removal of the coagulating electrolyte.

b. The energy barrier must be low or absent in the direction of increasing distance.

Both aims can be obtained if a layer of a few Ångstroms, i.e., one or two layers of solvent molecules keep the particles separated. Then instead of Fig. 6, we obtain the situation of Fig. 7.

It may well be that the layer δ disappears with time, or that the coagulated material recrystallizes through the layer or that the floccules rearrange themselves so as to have more extended contact. All such processes result in repeptization becoming more difficult and even disappearing on standing.

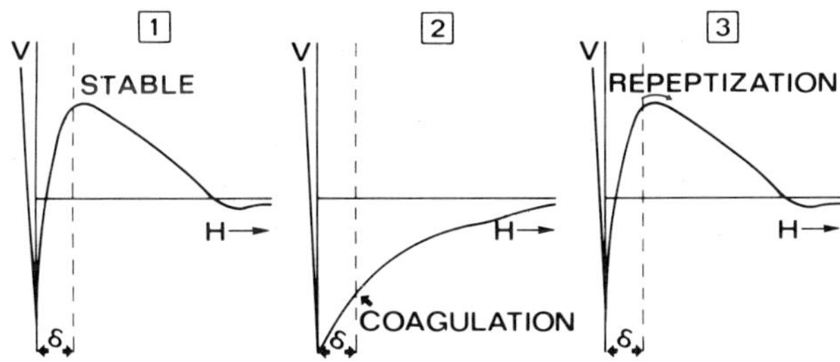

Fig. 7. Schematic diagram with a layer of thickness δ keeping the particles separated. (1) Stable; (2) coagulated, but at distance, δ, between the surfaces; (3) spontaneous repeptization if barrier at H = δ is not more than a few times kT. Numerical examples show that δ need not be more than a few Å.[53]

With coarse primary particles ($a \gg 1$ µm) matters seem to be different, because there aggregates can be redispersed by stirring or shaking, notwithstanding the fact that for the same distance, H or δ, the energies and forces of interaction are proportional to the particle radius, a, as can be read from eqs. (4) and (5). However, a shear field tends to separate the particles by a force proportional to a^2, because both the velocity difference at the centers of the two particles and their Stokes friction are proportional to a and for large enough particles this can overcome the Van der Waals attraction.

If stabilization is due to protective action, then, as mentioned earlier, repeptization after removing the flocculant is the normal course of affairs.

Repeptization is important for two reasons. It is often used in technical applications and moreover it produces information on details of the interactions at small separations.

THERMODYNAMIC STABILITY

Suspensions, emulsions and sols are only kinetically, not thermodynamically stable. They cannot be formed by simply mixing the macroscopic phases. The huge interface between particles and medium is a seat of free energy (interfacial tension) and this has to be brought into the system from the outside.

There are exceptions, however, to this too sweeping statement. A group of colloids, known as lyophilic, are thermodynamically stable. They are either solutions of large molecules, e.g., proteins in water, polystyrene in benzene, or solutions in which large but finite aggregates are formed, such as the soap micelles in water.

There are a few other cases in the borderland between lyophilic and lyophobic (= kinetically stable) systems. Microemulsions[54-56] belong here. Microemulsions are transparent mixtures of oil and water, stabilized by fairly large concentrations of one or two surfactants. Schulman[57] and coworkers, who first described these systems in the early forties, found already that the interfacial tension between the two phases becomes very low ($\ll 1$ mMm^{-1}) after the adsorption of the surfactants. When the interfacial tension is zero (or even, in passing, negative) spontaneous emulsification results, and if enough surfactant is available, the emulsion droplets become very small (~ 10 nm). Microemulsions had disappeared more or less from the scientific scene, but they have recently returned in people's interest, because the low interfacial tensions involved hold promise for applications in enhanced oil recovery.

Extremely low interfacial tensions need not be limited to the oil-water system. An electrical double layer, being formed spontaneously also causes a lowering of the interfacial tension and there may be cases in which suspensions are thermodynamically stabilized by their double layers. Cases in point are silica in alkaline solutions and aluminumoxide-hydroxide particles at a fairly low pH.[58] This is a new field that certainly deserves attention.

CONCLUSIONS

Knowledge about colloid stability does not only apply to true colloids, but also to dispersions of larger particles. For large particles gravity and hydrodynamic effects become relatively more important than Brownian motion. Colloid stability has a great influence on the properties of suspensions, especially on their packing and rheology.

Colloid stability is fairly well understood. The Van der Waals attraction is always there and not easily changed. The repulsion is either electrostatic or due to large molecules or possibly to structural forces. It is essential to realize that if one wants to increase or decrease the stability one has to manipulate the electrical double layer or the large molecules and thus one has to understand the theoretical background of these agents.

ACKNOWLEDGMENTS

I want to express my gratitude to Mr. N. M. van Galen for preparing the drawings, to Mr. M. Tollenaar for making the photographs and especially to Renske Kuipers for typing the manuscript in the short time available.

REFERENCES

1. J. Lyklema, Interfacial Electrochemistry of Disperse Systems, Proceedings of this meeting.
2. H. Schulze, J. Prakt. Chem., 25 [2], 431 (1882); 27, 320 (1883).
3. W. B. Hardy, Proc. Roy. Soc. London, 66, 110 (1900); Z. Physik. Chem., 33, 385 (1900).
4. J. Th. G. Overbeek, Stability of Hydrophobic Colloids and Emulsions, pp. 307-309 in Colloid Science, Vol. I, edited by H. R. Kruyt, 5th reprint, Elsevier, Amsterdam, 1969.
5. H. Koelmans, Philips Res. Rep., 10, 161 (1955); H. Koelmans and J. Th. G. Overbeek, Faraday Soc. Discussion, 18, 52 (1954).
6. A. Einstein, Ann. Physik., 19 [4], 289 (1906); 34, 591 (1911); Kolloid Z., 27, 137 (1920).

7. R. Simha, J. Phys. Chem., 44, 25 (1940); Proceedings of the International Congress on Rheology, Holland 1948, Amsterdam, 1949, p. II-68.
8. W. Kuhn and H. Kuhn, Helv. Chim. Acta, 38, 97 (1945).
9. J. W. Goodwin, T. Gregory and J. A. Stile, Adv. Colloid Interface Sci., 17, 185 (1982).
10. I. M. Krieger, Adv. Colloid Interface Sci., 3, 111 (1972).
11. S. Hachisu and K. Takano, Adv. Colloid Interface Sci., 16, 233 (1982).
12. E. A. Nieuwenhuis and A. Vrij, J. Colloid Interface Sci., 72, 321 (1982).
13. M. von Smoluchowski, Kolloid Z., 18, 194 (1916).
14. F. Booth, Nature, 161, 83 (1948); Proc. Roy. Soc. London, A203, 533 (1950).
15. G. J. Harmsen, J. van Schooten and J. Th. G. Overbeek, J. Colloid Sci., 8, 64, 72 (1953).
16. C. J. Lin, K. J. Lee and N. F. Sather, J. Fluid Mech., 43, 35 (1970).
17. G. K. Batchelor and J. T. Green, J. Fluid Mech., 56, 375 (1972).
18. T. G. M. van de Ven, Adv. Colloid Interface Sci., 17, 105 (1982).
19. W. R. Schowalter, Adv. Colloid Interface Sci., 17, 129 (1982).
20. Osborne Reynolds, Phil. Mag., 20 [5], 469 (1885); Nature, 33, 429 (1885).
21. H. Freundlich and H. L. Roder, Trans. Faraday Soc., 34, 308 (1938).
22. R. L. Hoffman, Adv. Colloid Interface Sci., 17, 161 (1982).
23. R. J. Hunter, Adv. Colloid Interface Sci., 17, 197 (1982).
24. H. Freundlich, "Thixotropy," Collect. Hermann, Paris (1935).
25. H. van Olphen, An Introduction to Clay Colloid Chemistry, 2nd ed., John Wiley and Sons, New York, 1977, p. 133.
26. Ref. 25, pp. 128, 140.
27. N. de Rooy, P. L. de Bruyn and J. Th. G. Overbeek, J. Colloid Interface Sci., 75, 542 (1980).
28. J. Lyklema, Adv. Colloid Interface Sci., 2, 65 (1968).
29. A. Klinkenberg and J. L. van der Minne, Electrostatics in the Petroleum Industry, Elsevier, Amsterdam (1958).
30. G. D. Parfitt, ed., Dispersion of Powders in Liquids, 2nd ed., Applied Science Editors, London, 1973.
31. F. Th. Hesselink, J. Phys. Chem., 73, 3488 (1969); 75, 65 (1971); F. Th. Hesselink, A. Vrij and J. Th. G. Overbeek, J. Phys. Chem., 75, 2094 (1971).
32. D. H. Napper, Trans. Faraday Soc., 64, 1701 (1968); J. Colloid Interface Sci., 32, 106 (1970); R. Evans and D. H. Napper, Kolloid Z. Z. Polym., 251, 329, 409 (1973).
33. H. Kallmann and M. Willstaetter, Naturwissenschaften, 20, 952 (1932).
34. J. H. de Boer, Trans. Faraday Soc., 32, 21 (1936).
35. H. C. Hamaker. Rec. Trav. Chim., 55, 1015 (1936); 56, 3, 727 (1937).

36. H. C. Hamaker, Physica, 4, 1058 (1937).
37. H. B. G. Casimir and D. Polder, Nature, 158, 787 (1946); Phys. Rev., 73, 360 (1948).
38. E. M. Lifshitz, Kokl. Akad. Nauk. SSSR, 97, 643 (1954); Sov. Phys. JETP, 2, 73 (1956); I. E. Dzyaloshinskii, E. M. Lifshitz and L. P. Pitaevskii, Advan. Phys., 10, 165 (1961).
39. J. Visser, Adv. Colloid Interface Sci., 3, 331 (1972).
40. H. Eilers and J. Korff, Chem. Weekblad, 33, 358 (1936); Trans. Faraday Soc., 36, 229 (1940).
41. J. Th. G. Overbeek, Pure and Appl. Chem., 52, 1151 (1980).
42. E. J. W. Verwey and J. Th. G. Overbeek, Theory of the Stability of Lyophobic Colloids, Elsevier, Amsterdam, 1948, p. 176 ff.
43. R. Zsigmondy, Z. Anal. Chem., 60, 697 (1901); Verh. Ges. Naturf. Arzte (Hamburg), 168 (1901).
44. P. J. Flory, J. Chem. Phys., 9, 660 (1941); 10, 51 (1942).
45. M. L. Huggins, J. Chem. Phys., 9, 440 (1941); Ann. N. Y. Acad. Sci., 43, 1 (1942); J. Chem. Phys., 46, 1 (1942).
46. J. M. H. M. Scheutjens and G. J. Fleer, Adv. Colloid Interface Sci., 16, 361 (1982).
47. B. Vincent, P. F. Luckham and F. A. Waite, J. Colloid Interface Sci., 73, 508 (1980).
48. A. Vrij, Pure Appl. Chem., 48, 471 (1976).
49. R. I. Feigin and D. H. Napper, J. Colloid Interface Sci., 74, 567 (1980); 75, 525 (1980).
50. IUTAM-IUPAC Symposium, "Interaction of Particles in Colloidal Dispersions," Canberra, March 1981, in Adv. Colloid Interface Sci., 16 (1982).
51. J. Th. G. Overbeek, J. Colloid Interface Sci., 58, 408 (1977).
52. H. Freundlich, Kapillarchemie, 4th ed., Vol. II, Akad. Verlagsges., Leipzig, 1932, pp. 201, 267, 273.
53. G. Frens and J. Th. G. Overbeek, Kolloid Z. Z. Polym., 233, 922 (1969); J. Colloid Interface Sci., 38, 376 (1972).
54. L. M. Prince, ed., Microemulsions, Academic Press, New York, 1977.
55. J. D. Robb, ed., Microemulsions, Plenum Press, New York, 1982.
56. J. Th. G. Overbeek, Faraday Discuss. Chem. Soc., 65, 7 (1978).
57. T. P. Hoar and J. H. Schulman, Nature, 152, 102 (1943); J. H. Schulman and J. B. Montagne, Ann. N. Y. Acad. Sci., 92, 366 (1961); J. H. Schulman and L. M. Prince, Kolloid Z., 169, 170 (1960).
58. R. J. Stol and P. L. de Bruyn, J. Colloid Interface Sci., 75, 185 (1980).

DISCUSSION

W. J. Lackey (Oak Ridge National Laboratory): Is colloidal science extrapolatable to suspensions of fibers or whiskers or is their geometry too different? For example, can stable suspensions be prepared?

Author: Colloidal suspensions of flat particles or rodlike particles are well known. Examples: V_2O_5 rods, tobacco mosaic virus rods, WO_3 flat plates, clays, etc. In these systems stability in the sense of coagulation concentrations of electrolytes are not different from those for more isodimensional particles. However, when these systems coagulate gels, often thisotropic gels, are formed at quite low concentrations by formation of very open networks. The viscosity of stable suspensions becomes non-Newtonian already at low rates of shear. I expect suspensions of fibers or whiskers to behave similarly.

G. L. Messing (Penn State University): Is it necessary to consider the Born repulsion when discussing the stability of concentrated (i.e., 50-60% solids) colloidal systems and if not why not?

Author: As long as the systems are colloidally stable, the particles are kept separated by electrostatic repulsion, or by a layer of large molecules, or possibly by a structured layer of the dispersion medium and the separation is large enough to prevent the Born repulsion to make itself felt. But if the system has lost or is losing its stability and coagulation in the primary minimum occurs, the Born repulsion is an essential factor in determining the depth of this minimum.

J. H. Adair (Battelle Columbus Labs): What effect does surface roughness have on interaction energy curves and coagulation kinetics in real systems?

Author: Since both attraction forces and repulsion forces have a fairly long range, the effect of small scale roughness is smoothed out. For roughness or mosaic like charges on a scale comparable to the double layer thickness ($1/\kappa$) one may apply equations for spherical particles and use the local radius of curvature instead of the particle radius. Only when mosaics of positive and negative charges occur on a scale comparable to or larger than $1/\kappa$, real deviations from simple behavior are to be expected (example: clay with negative surface charge and positive edge charge).

FORMATION AND STABILITY OF COLLOIDAL DISPERSIONS OF FINE PARTICLES IN WATER

Alan J. Rubin

Department of Civil Engineering
The Ohio State University
Columbus, Ohio 43210

INTRODUCTION

Colloid chemistry includes an extremely broad range of substrates and dispersion media. Colloids may be solutions that consist of very large solutes or of aggregates of smaller solutes or, in contrast, may be suspensions of very fine particles. the colloidal systems of greatest interest to ceramic science are the gels and discrete particle hydrosols. This discussion will be limited to the latter. Although there are many system-specific interactions between the surface of a particulate and the solution phase, enough is known if we also limit ourselves to aqueous systems to allow us to make several generalities. We will start with a few basic considerations and definitions pertinent to this paper.

A suspension of a finely divided solid that remains dispersed in a liquid for an extended period of time is said to be colloidally stable. In water, stability arises because the particles acquire a surface charge through any one of a number of mechanisms. Such a dispersion, a hydrosol, may be destabilized by electrostatic interaction between the electrical field surrounding the particles and oppositely charged ions. The latter are known as counterions and the process is called coagulation. Once destabilized, the particles may be restabilized, usually as a result of charge reversal following the adsorption of a counterion. It is apparent that stabilization and restabilization depend on the size and concentration of the particles, the nature of the particle surface, and the composition of the solution.

Probably the single most important parameter controlling the stability of hydrosols is the solution pH. This is because many sols have acid-base properties or because hydrogen and hydroxide ions are preferentially sorbed at the sol interface. As a result the minimum concentration of these ions for destabilization is much lower than for indifferent electrolytes of the same magnitude of charge. In fact, H^+ may coagulate dispersions at concentrations lower than observed for many 3+ species. Furthermore, because of the strong adsorption of hydrogen and hydroxide ions the sols may be reversed in charge, also frequently at relatively low concentrations. I wish again to emphasize the universality of this effect of pH. If unobserved with a particular sol it may be because the destabilization process is often quite slow and more patience is required.

An additional effect of pH is its role in determining the speciation of solutes. These interactions are particularly important with multivalent cations, especially the so-called hydrolyzable metals. In addition to its charge, the rate and extent of adsorption of a cation is altered upon hydrolysis.[2] In fact, the effect is so great that even neutral hydrolyzed species are strongly adsorbed.[18] Both of these aspects, the effects of pH on sol and solute, will be discussed.

EXPERIMENTAL

The preparation and experimental procedures are described in detail elsewhere.[4,13-17] Reagent grade chemicals were used and solutions and suspensions were made with carbonate-free deionized water. The titania suspensions were prepared with a water-dispersible powder of the anatase crystal type used in paper making.[15] The coal sols were prepared from a low-ash low-sulfur bituminous coal.[16] The oxidized coal was produced by heating a concentrated dispersion of unoxidized coal with hydrogen peroxide solution. The BET surface areas of the coal particles were 7.2 m^2/g prior to oxidation and 9.3 m^2/g afterward.[17] Turbidities initially and during settling were estimated by absorbance measurements at 400 and 550 nm for the coal and titania sols, respectively. Measurements of electrophoretic mobility were taken with a device manufactured by Zeta-Meter Inc., New York.

Two experimental approaches were used. With the first, a series of test samples were prepared in cuvettes at constant pH but with a systematic variation in metal salt concentration. Extrapolation of the steepest portions of the plots of the resultant turbidities for the series yielded the critical concentrations for coagulation and stabilization. The critical coagulation concentration (c.c.c.) is defined as the limiting or lowest concentration of counterion that just initiates aggregation. The critical stabilization concentration (c.s.c.) is the lowest concentration at which stability is just completed. The second approach involved the systematic variation of pH in the absence of metal ion or while holding its concentration constant in each series. From these results the critical pH values for coagulation and stabilization were obtained. The definitions of the critical pH for coagulation (pH_c) and stabilization (pH_s) are analogous to those of the corresponding critical concentrations in that a given sol is stable in the pH region greater than or equal to a pH_s and lower than a pH_c. These critical values, obtained graphically from the settling curves at their specified pH or metal concentration, are points along the boundary of a stability limit diagram. This diagram which is similar in construction to a phase diagram is a convenient device for summarizing large amounts of laboratory data.

THE CRITICAL pH AND RESTABILIZATION

It is the natural tendency of very small particles to adhere to one another. The forces of attraction are powerful but decrease rapidly with distance from the particle surface. Individual particles remain dispersed by acquiring a surface charge which has the net effect of being repulsive. Thus, to stabilize a particulate colloid means to develop this charge. This is easier than might be supposed. The real problem is maintaining the dispersion. The first requirement is that the particle concentration not be too high so as to minimize the probability of interparticule collision thereby preventing mutual coagulation. The second is the adjustment of the solute composition of the dispersion medium. Destabilization results when the repulsive force of the charge is reduced sufficiently by electrolytes to allow contact between particles.

The property of a given hydrosol most widely favored for predicting or relating the effect of hydrogen or hydroxide ion on its stability is the isoelectric point. This approach can be used with any particulate sol, including organic as well as inorganic systems. The isoelectric point (IEP) of a sol is the pH at which it has a net surface charge of zero and will not migrate in an electrical field. This is the point of zero electrophoretic mobility. It is difficult to relate the magnitude of the electrophoretic mobility to the magnitude of the surface charge of a particle, especially with large colloids. Nevertheless, the sign of the charge and the pH of zero charge are unambiguously determined by microelectrophoresis measurements. This is demonstrated in the first figure.

The upper curves in Fig. 1 are plots of the mobility of two coal sols as a function of pH. Each point is a separate experiment. "Negative mobility" means movement toward a positive electrode indicating that the particles are negatively charged except in acid solutions. The intersection with the line of zero mobility gives the IEP (or pH_i). As might be expected, the coal with the more highly oxidized surface has the lower IEP. Unfortunately, the IEP gives us only the point of minimum stability and does not define the minimum pH required to maintain the dispersion. A more practical and considerably easier approach is to observe the settling of the sol as a

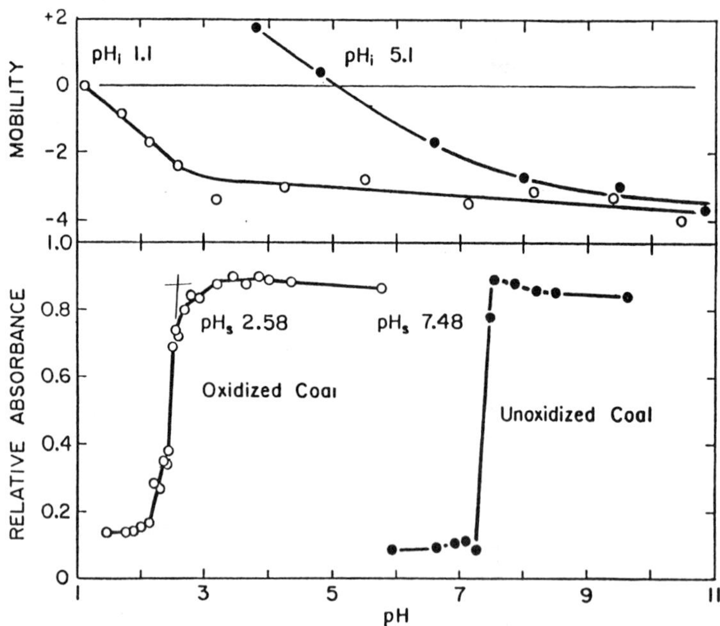

Fig. 1. Effect of pH on the stability and electrophoretic mobility of colloidal coals. 24 hour settling data.

function of pH. The lower curves in Fig. 1 are the corresponding turbidities of the two coal sols after 24 hours of quiescent settling. Extrapolation of the steep portions of the curves to the intersection with the horizontal part yields the critical pH. Above the pH_s the turbidity of the suspension is reduced by only about 10%, whereas in more acid solutions the reductions are 90%. Suspensions of the unoxidized coal would have to be buffered or otherwise protected from absorbing atmospheric carbon dioxide in order to remain dispersed for any extended period of time.

Figure 2 shows a similar set of data for colloidal titania. Above the pH_s the reduction in turbidity is only a few percent in 24 hours. This same sol may be restabilized below pH 6 by adsorbing a hydrolyzed metal cation. Figure 3 shows the turbidity of titania suspensions in the presence of varying concentrations of aluminum sulfate. The pH was 5.71 which is below the pH_s. Between Al(III) concentrations of 1 and 10 μM the turbidity increased quite sharply indicating redispersion of the sol. Destabiliation occurred once again at concentrations above 0.1 m$\underline{M}$ (the actual log molar values of the c.s.c. and c.c.c. were -5.20 and -3.82, respectively). This occurs because the sol, having been recharged by the adsorption of hydrolyzed aluminum, is now coagulated by the negative sulfate ion.[15]

This same effect is shown for the oxidized coal sol in Fig. 4. In this case, the solution pH is above the critical value and coagulation by aluminum(III) is followed by restabilization and, finally, coagulation by sulfate ion. The critical values for the coagulation of the two sols by sulfate ion are very close.

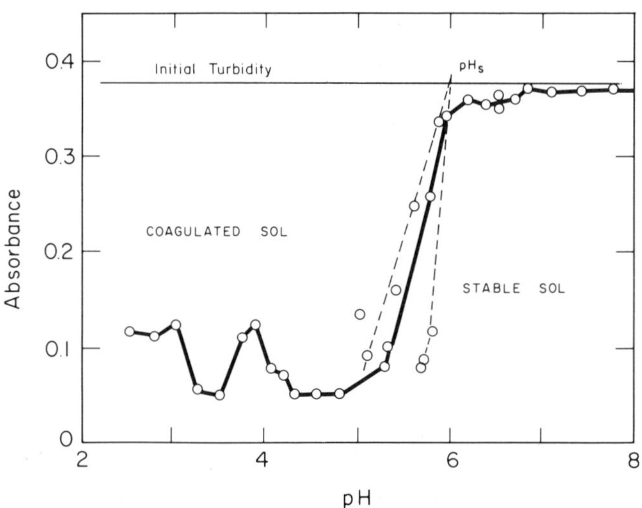

Fig. 2. Effect of pH on the stability of titania sol.

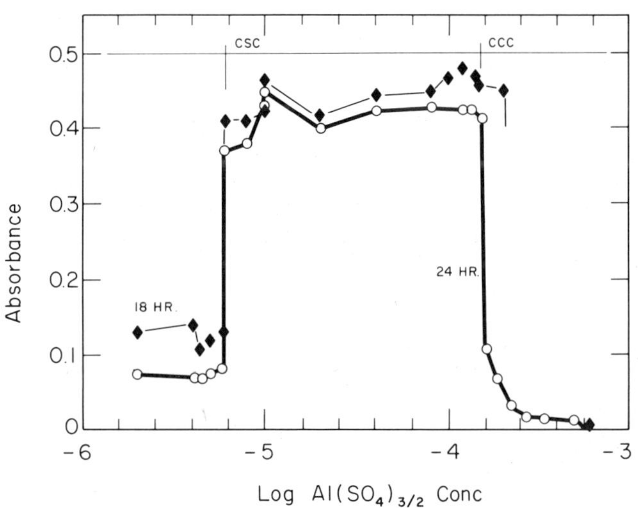

Fig. 3. Restabilization of titania sol with aluminum sulfate at pH 5.71.

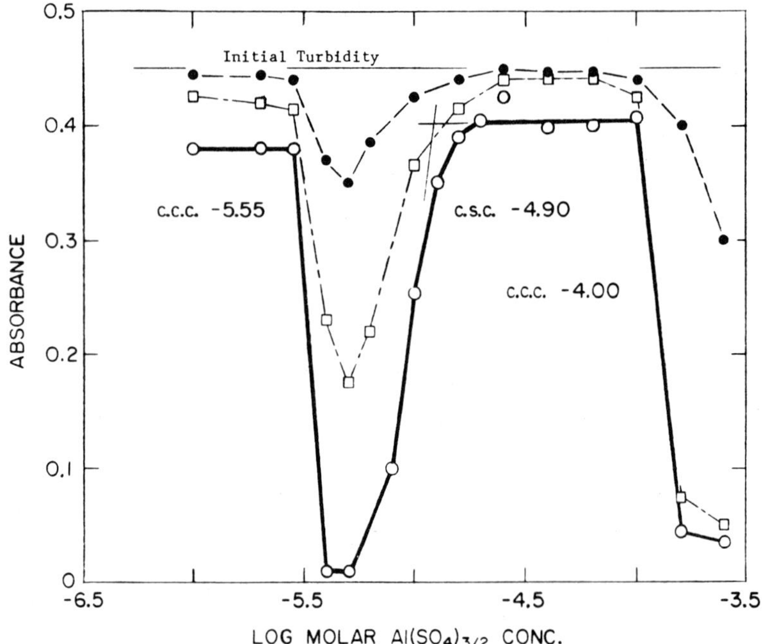

Fig. 4. The irregular series for colloidal coal. Aluminum sulfate at pH 6. 3, 10 and 20 hour settling data.

THE SCHULZE-HARDY RULE

Simple counterions which are not constituent ions cannot recharge a sol.[8] Such electrolytes coagulate hydrosols as predicted by the Schulze-Hardy rule which states that the coagulating power of a counterion increases exponentially with the magnitude of its charge. The coagulation of the oxidized coal sol by indifferent ions of different charge is shown by Fig. 5. The 1, 12 and 40 hour curves for calcium nitrate are in the middle of the figure. The curves to their right and left are for erbium chloride and sodium nitrate, respectively, after 40 hours of settling. The erbium ion is tripositive at the pH at which the data were obtained. Again, each point at a given time is a separate experiment. Notice the sharp break in the curves indicating that the critical concentrations of the electrolytes had been exceeded. Extrapolation of the steep portions of the settling curves, as before, gives the log c.c.c. values. There are roughly two orders of magnitude difference in the concentrations required as the charge is changed by one.

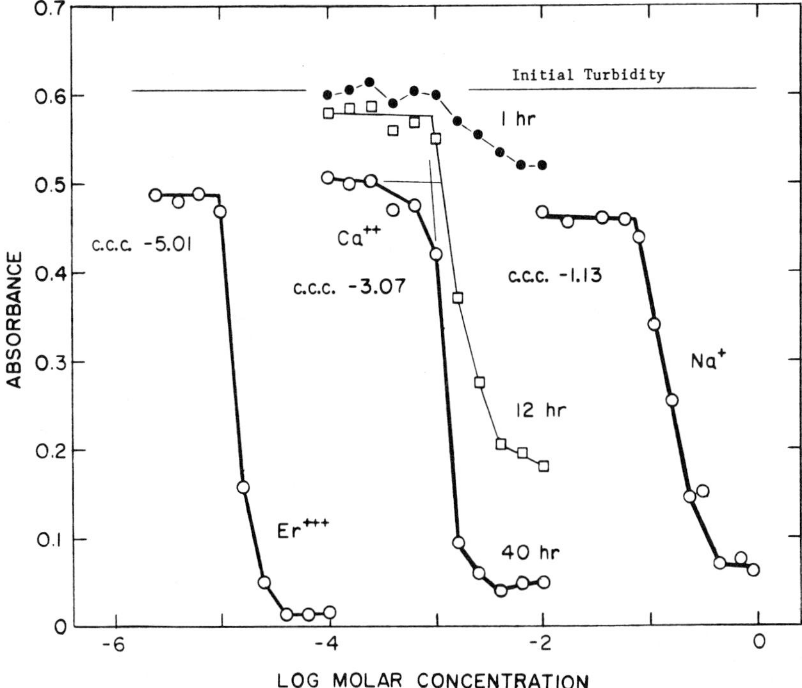

Fig. 5. Coagulation of colloidal coal with neutral salts.

A similar set of experiments were run with aluminum nitrate at pH 6. The log c.c.c. value was calculated from the data using the tetrapositive octameric structure proposed by Matijevic.[7] This species has been determined independently and its formation constant measured.[5] The octameric ion because of its very high charge coagulates at concentrations below which it is adsorbed.

Tezak and coworkers[19] have shown with silver halide sols that a plot of log c.c.c. values against the counterion charge is linear giving a quantitative expression of the Schulze-Hardy rule. Fig. 6 is such a plot for the coal and titania sols. Results for a latex sol,[3] for the negatively charged bacterium *Escherichia coli*,[14] and silver bromide[9] are also shown for comparison. Similar data have also been obtained for systems as diverse as gold sols, mineral oil emulsions and clays. The linearity of the plots implies that these sols have been destabilized by "simple coagulation." That is, at the c.c.c. the counterions coagulate without adsorption or other than electrostatic interaction with the sols.

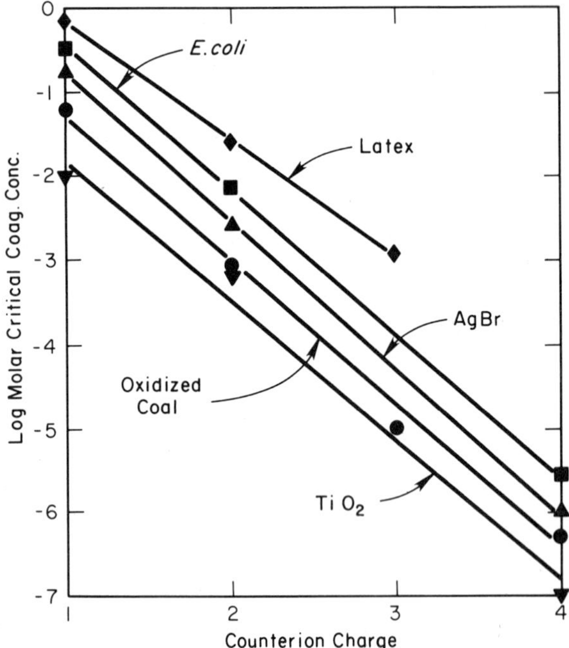

Fig. 6. Plot of the Schulze-Hardy rule for several sols.

Comparison of the c.c.c. for sodium ion with the pH_s for titania suggests that hydrogen ion, which coagulates the sol at about 1 μM, strongly interacts with its surface leading to "adsorptive coagulation." Adsorptive coagulation occurs at counterion concentrations much lower than predicted by the Schulze-Hardy rule. Ions which are capable of inducing adsorptive coagulation may also restabilize (peptize) the sol if the concentration of oppositely charged ions is sufficiently low. This sequence of the coagulation of a sol at a very low ion concentration followed by restabilization and subsequent coagulation at successively higher concentrations is known as the "irregular series." Figure 4 is a typical illustration of this effect.

INTERACTIONS WITH ALUMINUM(III)

The addition of an aluminum salt to water results in its dissolution and simultaneous hydration followed by lowering of the solution pH. This occurs because a series of stepwise hydrolysis reactions takes place in which Al(III) acts as a Bronsted-Lowry acid. In these reactions soluble hydroxo species are formed as hydrogen ions are released from the bound waters of hydration. Amorphous aluminum hydroxide may be precipitated as a consequence of hydrolysis, even in relatively acid solutions. Raising the pH to alkaline levels results in the dissolution of the precipitate. Thus, both positively and negatively charged hydroxo complexes may be formed. Because these hydrolyzed species interact so strongly with sols, a change in pH often has dramatic effects on its stability.

The interaction of aluminum sulfate with the titania sol was examined over a relatively broad concentration range. Most of the experiments were run in a series holding the metal concentration constant while systematically varying the pH. Critical pH values for coagulation and for stabilization were determined by extrapolating the steep portions of the settling curves up to the line of initial turbidity as described previously.

Figure 7 is typical of relatively high concentrations; 0.2 mM Al(III) in this example. Three distinct zones can be distinguished. The middle pH region is the zone of heavy floc formation and rapid settling of the sol. This is the "sweep zone" in which the solution is oversaturated with respect to aluminum hydroxide. On its right the sol is stable because the pH is high enough to dissolve the precipitate and the resulting negatively charged aluminate ion has the same sign as the sol. To the left of the sweep zone is a region of slow settling due to the destabilization of the sol by hydrogen ion or by positively charged aluminum species. In between this region and the sweep zone is a transient peak in stability. Similar "stability spikes" when using aluminum salts have been observed in past studies with colloidal coal and other sols and by Black and Hannah[1] in their work with kaolin, montmorillonite and Fuller's earth. They found that the spike corresponded to the point of greatest positive mobility of the sols.

With a decrease in Al(III) concentration to 0.1 mM as shown in Fig. 8, the stability spike has now expanded into a distinct zone of stability. The turbidity of the samples in this zone decreases in time but at a rate considerably less than in the slow coagulation region to its left. This is a zone of restabilization. Notice that the sweep zone has narrowed with the decrease in Al(III) concentration and that its boundaries are virtually time independent.

Critical pH and concentration values obtained from Figs. 3, 7 and 8 and similar data were plotted at their respective Al(III) concentrations or pH to establish the stability limit diagram shown in Fig. 9. The open symbols are critical data at the titania concentrations used in the figures shown here; the blackened symbols represent data for a lower sol concentration. Circles are critical pH values and diamonds are critical concentration values. The data at a higher sol concentration are represented by the dotted line and, except for the restabilization zone, were in perfect agreement with the results at the other concentrations. These data are not shown since they would only serve to clutter the figure.

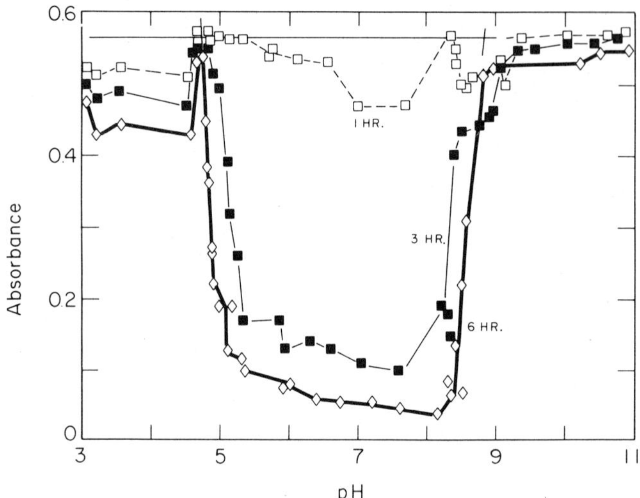

Fig. 7. Interaction of titania with aluminum sulfate. 0.2 mM Al(III). Typical three zones are shown.

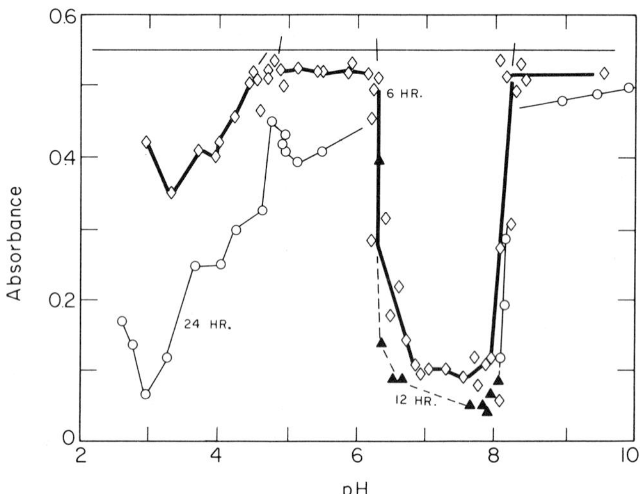

Fig. 8. Interaction of titania with aluminum sulfate. 0.1mM Al(III). Typical four zones are shown.

The stability limit diagram is characterized by four distinct zones. The central area is dominated by the sweep zone where heavy floc formation results in rapid settling of the titania. This zone surrounds the central stability zone and narrows with decreasing concentration reaching a minimum at the c.c.c. of 0.8 μM Al(III). This minimum is independent of sol concentration. The vertical line below the c.c.c. and the left boundary of the sweep zone form the upper limits of the slow coagulation zone. The lower vertical boundary was obtained by drawing the line through the critical point at 0.1 μM Al(III). This gives a pH_s of about 5.8, which is in good agreement with the value established in Fig. 2. In this zone the principal coagulating species is the hydrogen ion.

The boundary between the sweep zone and the region of stability on the right is determined by the equilibrium between aluminum hydroxide and the aluminate ion.[5,14] Thus, the right stability region is a zone of no coagulation due to the formation of the soluble aluminate ion which is negatively charged. The significance of the slopes and intercepts of the boundaries between these two zones and the slow and sweep zones on the left are described in detail elsewhere.[15]

It has been established that hydrolyzed species, polynuclear ions in particular, are responsible for the restabilization of sols.[11,12] These highly charged species adsorb to such an extent that charge reversal occurs, resulting in a positively charged sol. This effect was demonstrated in Fig. 3 and is the cause of the central stability zone in Fig. 9. Thus, the lower sol concentration independent c.c.c. and the subsequent c.s.c. and c.c.c. that define the lower and upper boundaries of the restabilization zone give rise to the phenomenon known as the irregular series. These same effects have also been demonstrated for iron(III)[6,10] as well as many other hydrolyzable metals.

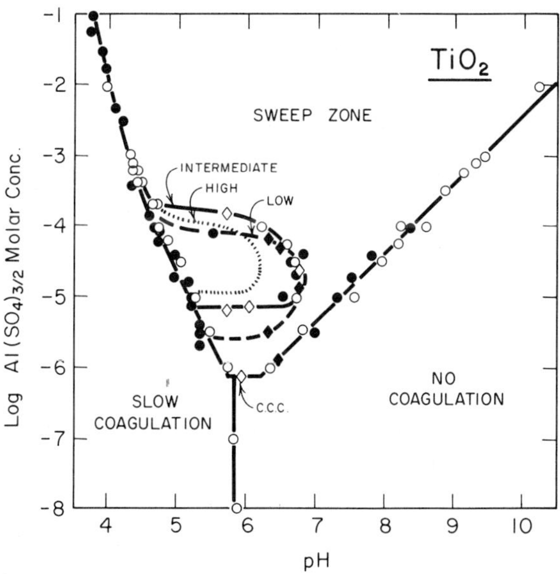

Fig. 9. Aluminum sulfate stability limit diagram for titania. The central region is the zone of restabilization. See the text for an explanation of the symbols.

It is apparent that aluminum(III), iron(III) and other metals may be used to stabilize a particulate in water. The concentration--pH requirements can be established quite readily by obtaining the stability limit diagram. Several effects not considered here, including solution age, temperature and the presence of complexing ions, can also be examined by constructing stability limit diagrams.

ACKNOWLEDGMENT

This research has been supported in part by grants from the Department of the Interior, Office of Water Research and Technology. The studies described in this paper were performed by Tom Kovac, Ralph Kramer and Paul Schroeder, whose contributions are also gratefully acknowledged.

REFERENCES

1. A. P. Black and S. A. Hannah, J. Am. Water Works Assoc., 53, 438 (1961).
2. L. Eriksson, E. Matijevic and S. Friberg, in Adsorption From Aqueous Solution, American Chemical Society, Washington, D.C. (1968).
3. C. G. Force and E. Matijevic, Kolloid-A., 224, 51 (1968).
4. G. P. Hanna and A. J. Rubin, J. Am. Water Works Assoc., 62, 315 (1970).
5. P. L. Hayden and A. J. Rubin, in Aqueous-Environmental Chemistry of Metals, edited by A. J. Rubin, Ann Arbor Science, Ann Arbor, Mich. (1974).
6. C. R. O'Melia and W. Stumm, J. Colloid Interface Sci., 23, 437 (1967).
7. E. Matijevic, J. Colloid Interface Sci., 43, 217 (1973).
8. E. Matijevic and L. H. Allen, Envir. Sci. Technol., 3, 264 (1969).
9. E. Matijevic, D. Broadhurst and M. Kerker, J. Phys. Chem., 63, 1552 (1959).
10. E. Matijevic and G. E. Janauer, J. Colloid Interface Sci., 21, 197 (1966).
11. E. Matijevic, G. E. Janauer and M. Kerker, J. Colloid Sci., 19, 333 (1964).
12. E. Matijevic and L. J. Stryker, J. Colloid Interface Sci., 22, 68 (1968).
13. A. J. Rubin and H. Blocksidge, J. Am. Water Works Assoc., 71, 102 (1979).
14. A. J. Rubin and G. P. Hanna, Envir. Sci. Technol. 2, 358 (1968).
15. A. J. Rubin and T. W. Kovac, in Chemistry of Water Supply, Treatment and Distribution, edited by A. J. Rubin, Ann Arbor Science, Ann Arbor, Mich. (1974).
16. A. J. Rubin and R. J. Kramer, Separation Sci., 17, 535 (1982).

17. A. J. Rubin, P. R. Schroeder and R. J. Kramer, Proc. Purdue Industrial Waste Conf., 35, 316 (1980).
18. L. J. Stryker and E. Matijevic, in Adsorption From Aqueous Solution, Am. Chem. Soc., Washington, D.C. (1968).
19. B. Tezak, E. Matijevic and K. F. Schultz, J. Phys. Chem., 59, 769 (1955).

DISCUSSION

M. F. Berard (Iowa State University, Ames Laboratory): Your first figure showed that the pH for coagulation of coal sols (in the absence of other added ions in solution) does not coincide with the pH for zero-point-of-charge. What factors are responsible for this deviation?

Author: The pH_s reflects the fact that there is a certain minimum in surface potential required for stability. This minimum is very similar for the two coal sols as is shown in Fig. 1.

E. A. Barringer (Massachusetts Institute of Technology): On the stability diagram for TiO_2, slow coagulation is shown for pH below 5. I would expect that at sufficiently low pH, the positive surface charge at the particle surface is large enough to stabilize suspensions. Why is this not shown on the stability diagram?

Author: Restabilization was observed at low pH but is not shown on the stability limit diagram since my interest was limited to the boundaries of the sweep zone.

K. Bridger (Martin Marietta Laboratory): I doubt that SO_4^{2-} is an indifferent electrolyte. Values for the equilibrium constant for the completion of Al^{3+} by SO_4^{2-}($Al^{3+} + SO_4^{2-} \rightleftharpoons AlSO_4^{+}$) have been reported as high as several hundred (M^{-1}). See for example "Handbook & Stablity Constants." Did you reproduce these data using fresh solutions and $Al(NO_3)_3$?

Author: Sulfate ion does form complexes with aluminum(III). However, the binding is not very great and the coagulation concentration is not much less than might be predicted by the Schulze-Hardy rule. There is other evidence of complexing in the data which is not discussed in the paper. An example is the "stability spike" at the left boundary of the sweep zone shown in Fig. 7. We believe this effect is due to a basic salt of aluminum sulfate. Several effects of complexing will be the subject of a paper that is planned for the Colloid and Surface Science meeting in Toronto next June. We have obtained a stability limit diagram with $Al(NO_3)_3$, but not for titania (see reference 15).

FLOCCULATION AND FILTRATION OF COLLOIDAL PARTICLES

John Gregory

Dept. of Civil Engineering, University College London
Gower Street
London WC1E 6BT UK

INTRODUCTION

Colloidal dispersions are said to be either stable or unstable depending on whether the particles remain dispersed or have a tendency to form aggregates. Colloid stability may arise from the fact that particles are charged and repel each other electrically or from the presence of adsorbed layers. These subjects and methods of reducing colloid stability have been discussed by Overbeek (this volume).

In many solid-liquid separation processes such as sedimentation, filtration and flotation, colloidal particles are too small to be easily removed and their size needs to be increased by the formation of aggregates. This involves destabilizing the particles (often by the addition of certain salts or polymers), followed by collisions of destabilized particles to form aggregates. The process is known either as <u>coagulation</u> or <u>flocculation</u> and only the latter term will be used here, although this is sometimes restricted to the special case of aggregation caused by long-chain polymers.[1]

Particle deposition onto surfaces is another process which can have great practical significance, for instance in certain types of filtration. The principles governing deposition are similar to those which are important in colloid stability and flocculation, although there are some complicating factors.

This chapter will be largely concerned with kinetic aspects of these processes.

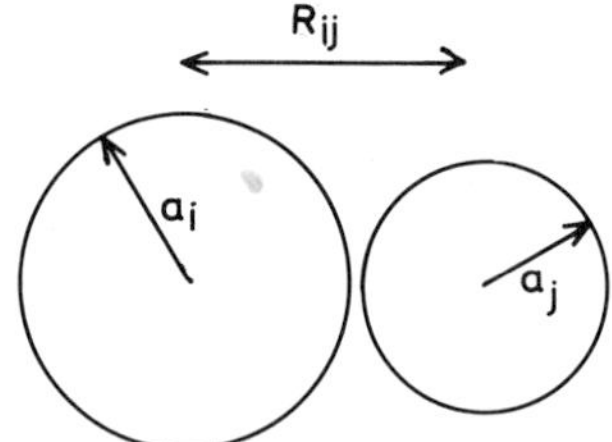

Fig. 1. Collision of unequal spherical particles.

FLOCCULATION KINETICS

The rate of flocculation of a suspension depends on the collision frequency of the particles. Collisions may be brought about in essentially three ways: (a) by Brownian motion (perikinetic flocculation); (b) by fluid motion (orthokinetic flocculation); (c) by differential settling.

The relative rates of these collision processes depend very much on the size of the particles and on the degree of shear applied to the suspension. After a brief survey of classical approaches to flocculation kinetics, we shall go on to consider hydrodynamic effects and the break-up of flocs.

Perikinetic Flocculation

The first quantitative treatment of flocculation kinetics was by Smoluchowski.[2] In a suspension containing spherical particles of type i and j, with radii a_i and a_j and number concentration N_i and N_j, the random Brownian motion of the particles causes collisions to occur between them at a frequency J_{ij} (per unit volume and unit time), given by:

$$J_{ij} = 2\pi(a_i + a_j)D_{ij}N_iN_j \tag{1}$$

In this expression, the collision radius R_{ij} has been replaced by the sum of the particle radii, which is an acceptable approximation in nearly all cases (see Fig. 1). D_{ij} is the mutual diffusion coefficient of the particles which is simply the sum of their individual diffusion coefficients. The Stokes-Einstein expression for the diffusion coefficient of a spherical particle radius a_i is:

$$D_i = kT/6\pi a_i\mu \tag{2}$$

where k is Boltzmann's constant, T the absolute temperature and μ the viscosity of the fluid.

With a similar expression for D_j and $D_{ij} = D_i + D_j$, Eq. (1) can be written:

$$J_{ij} = (kT/3\mu)N_iN_j(a_i + a_j)^2/a_ia_j \tag{3}$$

For an initially monodisperse suspension ($a_i = a_j$), the Smoluchowski treatment leads to the following expression for the change in total number concentration of particles, N, with time:

$$-dN/dt = (4\alpha kT/3\mu)N^2 = k_fN^2 \tag{4}$$

There is a negative sign since the number of particles decreases as aggregates are formed. The collision of two particles to form a doublet gives a net loss of one particle from the suspension. The term α has been introduced into Eq. (4) to allow for the possibility that not all collisions may be successful in producing aggregates. The fraction of successful collisions, α, is the collision efficiency factor and depends on the collidal stability of the particles. For a fully-destabilized suspension, we might expect every collision to be effective and $\alpha = 1$, but hydrodynamic effects (see later) can give a lower value.

Equation (4) is based on a number of simplifying assumptions. In particular, it is assumed that the aggregates are spherical and that collisions are mainly between aggregates of roughly equal size. Under these conditions, the size of the primary particles has no influence on the result. It is possible to use the Smoluchowski approach to calculate the concentration of different sizes of aggregate at various times and hence to predict the changing particle distribution of flocculating suspensions.[3] However, the predicted size distributions can be greatly affected by the shape of the aggregates, although the calculated total number concentration is relatively insensitive to the assumed aggregate shape.[4]

Equation (4) represents a standard, second-order rate process and can be integrated to give:

$$N = N_o/(1+k_fN_ot) \tag{5}$$

where N_o is the initial number concentration of particles and N is the concentration (including aggregates) remaining at time t. A plot of 1/N against t should give a straight line from which k_f can be calculated.

The concentration of particles is reduced to half the initial value ($N = N_o/2$) at a characteristic flocculation time, t_f, given by:

$$t_f = 1/k_fN_o \tag{6}$$

Assuming that $\alpha = 1$ (i.e., for a full-destabilized suspension and ignoring hydrodynamic effects), the rate constant k_f is a function only of temperature and viscosity. For water at 25°C, $k_f = 6.13 \times 10^{-16}\ m^3s^{-1}$ and the corresponding flocculation time is $t_f = 1.63 \times 10^{17}/N_o$.

So, for a suspension containing 10^{16} particles per m^3 (roughly 1% by weight of clay particles with an equivalent spherical diameter of 1 μm), t_f would be about 16 s. After this time, the average aggregate would consist of two primary particles. For more substantial aggregates, say 100-fold, about 27 minutes would be required if Brownian diffusion alone were operative. Perikinetic flocculation is thus a rather slow process, unless the initial particle concentration is very high. For a given suspension, the rate found when the particles are fully destabilized is called the <u>rapid</u> flocculation rate, even though this may be very slow in absolute terms.

When there is significant repulsion between particles, as between charged particles in dilute electrolyte solutions, the collision efficiency can take very low values and <u>slow</u> flocculation occurs. In colloid science it is usual to refer to the <u>stability</u> ratio, $W(= 1/\alpha)$, which is the ratio of the rapid rate to the slow rate for the given conditions. The stability ratio for charged particles increases markedly as the electrolyte concentration is increased and measurements of slow flocculation rates should be a sensitive test of current theories of colloidal interactions. Such measurements are not abundant[5] and the available results are not fully consistent with theoretical predictions.

Orthokinetic Flocculation

It is commonly observed that agitation of a destabilized dispersion gives a dramatic increase in flocculation rate, allowing much larger flocs to be formed in a reasonable time than would be possible by diffusion alone. This is a result of induced velocity gradients, which promote particle collisions. For spheres in a uniform shear field, Smoluchowski[6] showed that the collision rate between i and j particles (see Fig. 1) is given by:

$$J_{ij} = (4/3)N_iN_jG(a_i + a_j)^3 \qquad (7)$$

where G is the shear rate (s^{-1}).

For an initially monodisperse suspension of particles, radius a, the rate of decrease of particle concentration as a result of orthokinetic flocculation, is:

$$-dN/dt = (16/3)\alpha N^2Ga^3 \qquad (8)$$

Again, we have a second order rate process, but the presence of the a^3 term makes this a very different result from the equivalent perikinetic rate in Eq. (4). If it is assumed that, during a flocculation process, the total volume fraction, 0, of particles remains constant, then, since $\phi = (4/3)\pi a^3 N$, Eq. (8) reduces to a pseudo-first order form:

$$-dN/dt = 4\alpha G\phi N/\pi \tag{9}$$

or,

$$N/N_o = \exp(-4\alpha G\phi t/\pi \tag{10}$$

The term α in the orthokinetic rate expressions again indicates the fraction of successful collisions. However, for a partially-destabilized suspension, there is no reason to suppose that this will have the same value as the perikinetic collision efficiency since the two quantities are not physically equivalent. Nevertheless, the results of Swift and Friedlander[7] gave closely similar collision efficiencies for perikinetic and orthokinetic flocculation of the same suspension.

Equation (1) implies that the proportionate decrease in particle concentration, N/N_o, is exponential with time and independent of the initial concentration, in marked contrast to the perikinetic results. A plot of ln N/N_o against t should give a straight line which has been confirmed for the flocculation in Couette flow[7] and in laminar tube flow.[8]

Although Eq. (10) is based on several simplifying assumptions and cannot be expected to be accurate except in the early stages of a flocculation process, it remains a very useful result. It makes clear the importance of the volume fraction of suspended solids and of the dimensionless parameter Gt. In principle, the same degree of flocculation should be achieved for a given value of Gt, whether by applying high shear for a short time or low shear for a longer period. Essentially, Gt is a measure of the contact opportunity of particles as a result of orthokinetic collisions.

Laminar flow through a tube is a convenient method of studying orthokinetic flocculation, since it can easily be shown[9] that the average Gt value depends only on the tube dimensions and not on the flow rate.

Most practical examples of orthokinetic flocculation take place in shear conditions which are far from uniform, nearly always in some form of turbulent flow, for which calculations of collision frequencies are not easy. One approach, due to Camp and Stein,[10] is to derive an average shear rate $\overline{G}$, from the measured power input, P, to the flocculation vessel (e.g., a stirred tank):

$$\overline{G} = (P/\mu V)^{1/2} \tag{11}$$

where V is the volume of the vessel.

This effective value of $\overline{G}$ is then substituted into expressions for the uniform shear case, such as Eq. (10). Although this approach is incorrect for a number of reasons, the result is surprisingly close to that from a more rigorous treatment (see e.g., Spielman[11]).

Comparison of Flocculation Rates

Collision frequencies for the perikinetic and orthokinetic cases have been given in Eqs. (3) and (7). For the case of differential settling, the equivalent result is:

$$J_{ij} = (2\pi g/9\mu)(\rho_s + \rho)(a_i + a_j)^3(a_i - a_j)N_iN_j \tag{12}$$

where g is the acceleration due to gravity and ρ_s and ρ the densities of the particles and fluid respectively.

This collision mechanism depends on the fact that different particles have different sedimentation velocities (assumed to be given by Stokes law).

In all the collision processes considered, the frequency is given in terms of a second-order rate equation, which could be written:

$$J_{ij} = k_{ij}N_iN_j \tag{13}$$

where k_{ij} is a rate constant.

The rate constants for the perikinetic (diffusion), orthokinetic (shear) and differential settling collision mechanisms are shown in Fig. 2, for the case where one particle has a diameter of 1 μm and the size of the second particle is variable. The medium is assumed to be water at 25°C, the shear rate is taken as 50 s^{-1} (coresponding to quite mild agitation) and the specific gravity of the particles is 2.

The results show that orthokinetic collisions become significant relative to the perikinetic rate when the size of the second particle exceeds about 0.2 μm, a size at which diffusion is normally expected to predominate. Such calculations show that, for collisions between unequal particles, it is the size of the larger particle that mainly determines the relative importance of the orthokinetic mechanism. Even for the adsorption of macromolecules on suspended particles, it can be shown that shear-induced collisions may be much more important than diffusion.[12]

Differential settling is seen to become important when the diameter of the second particle is about 5 µm or greater. However, this value may be a little misleading for systems in turbulent flow, where the velocity fluctuations would tend to keep aggregates in suspension.[13]

Hydrodynamic Effects

When the particles approach each other in a liquid medium, the liquid must drain from the region of contact between the particles. As the gap between the particles narrows, this drainage becomes progressively slower because of the low velocity of liquids close to a solid surface. In the limit of zero separation, drainage ceases altogether and the implication is that, in the absence of other forces, no contact between particles could occur as a result of this hydrodynamic or viscous resistance. In practice, the hydrodynamic effect is counteracted by attractive forces between particles, notably van der Waals attraction.

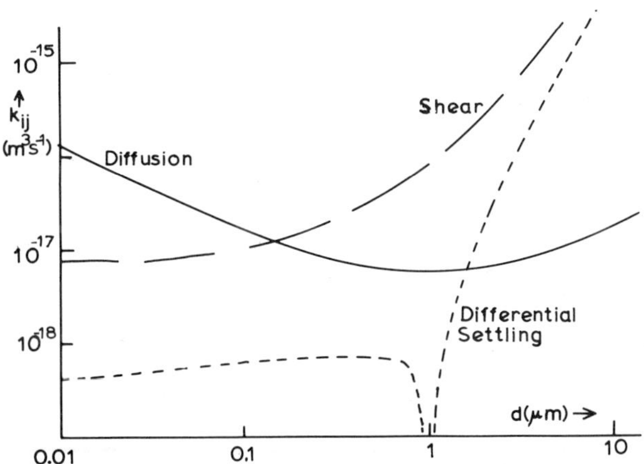

Fig. 2. Calculated collision rate constants for a particle of 1 µm diameter and a second particle of diameter d. For conditions, see text.

In the case of perikinetic flocculation, the hydrodynamic effect appears as a reduction in the diffusion coefficient of the particles as they approach close together. Spielman[14] and Honig et al.[15] showed that the rate of perikinetic flocculation would normally be reduced by a factor which depends on the Hamaker constant of the system, but usually would not exceed about two. For latex particles experimental flocculation rates about 50-60% of the Smoluchowski rate have been found,[3] in good agreement with theoretical predictions.

In the orthokinetic case, hydrodynamic effects have to be incorporated in trajectory computations for colliding particles in shear flow. Results of such computations[16] for equal particles show that the effect depends on particle size and shear rate and that flocculation rates may be up to a factor of 10 lower than that predicted by the Smoluchowski approach, Eq. (8). The magnitude of the effect will vary depending on a number of conditions, but will usually be rather more significant than the hydrodynamic effect in perikinetic flocculation.

When there is significant electrical repulsion between particles, the incorporation of viscous resistance leads to very complex behavior. Van de Ven and Mason[16] showed that partially-destabilized particles may flocculate at high and low shear rates, but show stability at intermediate values. The so-called "shear flocculation" examined by Warren[17] may be a case where particles colliding at high shear rates have sufficient energy to overcome the repulsion barrier.

For unequal size particles in shear flow, hydrodynamic effects become much more important.[18] It has been pointed out[19] that there is a limiting separation distance for interacting particles in shear flow and that fluid forces prevent closer approach. This distance is quite small for equal particles (but still sufficient to cause a marked decrease in flocculation rate). However, for particles of different size the limiting separation distance can be surprisingly large. For instance, for particles with radii of 10 μm and 1 μm it was shown[19] that the minimum distance of approach would be about 1.4 μm. This distance is far too large for colloid forces (electrical and van der Waals) to be operative and the implication is that such particles could not come into contact in a sheared suspension. This conclusion could be modified if Brownian motion of the particles were included and if surface roughness effects were important. Nevertheless, this "hydrodynamic gap" between unequal particles may be significant in many practical systems, which led van de Ven and Mason[19] to suggest that long-chain polymeric flocculants could be especially useful in such cases, since the chains may be long enough to span the gap.

Floc Break-up

When flocculation occurs in an agitated suspension, it is usually found that there is some limiting floc size, beyond which no further growth occurs. The more intense the agitation, the smaller is the limiting floc size.[20] Clearly, flocs must be subject to some break-up process and, ideally, this should be incorporated into a theoretical model of flocculation in sheared suspensions. Unfortunately, the subject of floc strength and break-up is an extremely difficult one, both theoretically and experimentally. Most approaches have been along empirical lines.

In laminar flow, Mason and co-workers[21] have shown that the type of shear can have a marked influence on the break-up of model aggregates, extensional shear being much more effective than rotational shear.

Most practical interest is in the break-up of flocs under turbulent conditions, where very complex behavior is found.[22] The nature of the break-up process depends on the size of the flocs relative to the turbulence microscale (which is of the order of 100 μm in typical agitated suspensions). For flocs of greater size, deformation and rupture may occur as a result of fluctuating dynamic pressure. For smaller flocs, viscous shear forces predominate and may cuase either an erosion of primary particles from the floc surface or fragmentation of the flocs. There is some evidence[23] that collisions between flocs may be important in the break-up process, although most observations can be explained in other ways.

From rheological studies of kaolin suspensions, Michaels and Bolger[24] drew a distinction between flocs of primary particles and aggregates composed of many flocs. Only aggregates were considered to break under shear, the flocs being virtually indestructable. The "elastic floc" model of Firth and Hunter[25] is based on similar ideas.

It is known that polymeric flocculants[26] can give very much stronger flocs than more conventional agents, presumably because of the formation of polymer bridges between particles with many points of attachment. It has been estimated that the force required to break a carbon-carbon bond in a polymer chain is of the order of 10^{-8} to 10^{-9} N, which is much smaller than the hydrodynamic force between particles in a suspension undergoing intense agitation.[19] Large numbers of chains would need to be adsorbed on the particles to give flocs capable of withstanding high shear.

O'Gorman and Kitchener[27] showed that very strong flocs could be formed in kimberlite clay slimes by a combination of salt destabilization (to form small aggregates) followed by the addition of a high molecular weight polymer. In the "pelleting flocculation" process[28] extremely tough flocs are produced by a combination of an aluminum salt and a long-chain polymer, together with "mechanical syneresis" brought about by uneven hydrodynamic forces.

When polymer flocs are disrupted by high shear, changes in configuration of the adsorbed polymer chains may occur, as well as chain scission, so that re-flocculation does not readily occur when the shear is reduced. It has been suggested that the reversibility or otherwise of floc breakage under shear may be an indication of whether polymer bridging or charge neutralization is the predominant flocculation mechanism.[29,30]

DEPOSITION AND FILTRATION

Particle deposition is an essential step in certain filtration processes, especially in _depth filtration_, which is commonly employed in water treatment. In this process, particles are removed by filtration through a porous medium in which the pores are very much larger than the particles, so that straining is not a significant removal mechanism. Particles are deposited on to _collectors_ (i.e., the grains or fibers making up the filter medium).

Deposition involves the _transport_ or _attachment_ of particles to the collector surface. Attachment is governed by the same colloidal forces which are important in flocculation (electrical, van der Waals, and steric interactions). These are of rather short range (usually less than 50 nm) whereas particles need to be transported over comparatively large distances from the bulk suspension to the collector surface. For this reason, it is usually justified to treat the transport and attachment steps separately.

Three main transport mechanisms are regarded as important in the filtration of liquid suspensions through porous media.[31] These are convective diffusion, interception and sedimentation. Theories have been developed which enable the capture efficiency of a collector to be calculated for each of the transport mechanisms. No details can be given here but the essential qualitative features of the results are summarized below:

(a) Capture by _convective diffusion_ decreases as the particle size and the collector size are increased and as the flow rate is increased.
(b) Capture by _interception_ increases as the particle size is increased and decreases as the collector size is increased. There is no effect of flow rate.
(c) Capture by _sedimentation_ increases as the size and density of the particles are increased and decreases as the collector size and flow rate are increased.

Because small particles are captured more effectively by diffusion and large particles by interception and sedimentation, there should be a certain particle size at which capture efficiency is minimal. The minimum usually occurs for particles around 1 μm in diameter,[32] which is of some practical significance. Such particles are not easily removed by filtration and may need to be flocculated in a pre-treatment step.

For attachment to occur the particle-collector interaction needs to be favorable, which means essentially no repulsion between them. If this condition is not met, some form of pre-treatment will be necessary, for instance by reducing the charge on particles or collectors or both. Attachment is certain to occur if particles and collectors have opposite charge but because of the rather limited range of the interaction little enhancement of the deposition rate is found under such conditions.[33]

In many filters, substantial (multilayer) deposits build up, so that, after an initial phase, deposition must occur on to previously deposited particles. For this to occur, particle-particle interaction must be favourable, i.e., the suspension must be colloidally unstable. In this case, filtration can be regarded as a special case of flocculation,[34] in which the filter is providing enhanced particle contact opportunity.

When suspensions are filtered through a porous medium with pore sizes comparable to the particle size, the medium quickly becomes blocked and particles are removed entirely by straining. After an initial period, all subsequent filtration occurs by straining through a layer of filtered particles (the _filter cake_). Cake filtration is a very important process which is widely used in the separation of solids from relatively concentrated suspensions. However, the subject is outside the scope of this review and reference should be made to standard works[35] for further details.

REFERENCES

1. V. K. La Mer, J. Colloid Interface Sci., 19, 291 (1964).
2. M. Smoluchowski, Physik. Z., 17, 557 (1916).
3. K. Higashitani and Y. Matsuno, J. Chem. Eng. Japan, 12, 460 (1979).
4. K. Okuyama, Y. Kausaka, and A. C. Payatakes, J. Colloid Interface Sci., 81, 21 (1981).
5. T. W. Healey, A. Homola, R. O. James, and R. J. Hunter, Faraday Disc. Chem. Soc., 65, 156 (1978).
6. M. Smoluchowski, Z. Phys. Chem., 92, 155 (1917).
7. D. L. Swift and S. K. Friedlander, J. Colloid Sci., 19, 621 (1964).

8. K. Higashitani, S. Miyafusa, T. Matsuda, and Y. Matsuno, J. Colloid Interface Sci., 77, 21 (1980).
9. J. Gregory, Chem. Eng. Sci., 36, 1789 (1981).
10. T. R. Camp and P. C. Stein, J. Boston Soc. Civ. Engrs., 30, 219 (1943).
11. L. A. Spielman, in The Scientific Basis of Flocculation, edited by K. J. Ives, Sijthoff and Noordhoff, Alphen aan den Rijn, 1978.
12. J. Gregory, in The Effect of Polymers on Dispersion Properties, edited by T. F. Tadros, Academic Press, London, 1982.
13. M. A. Delichatsios and R. F. Probstein, J. Wat. Poll. Control Fed., 47, 941 (1975).
14. L. A. Spielman, J. Colloid Interface Sci., 33, 562 (1970).
15. E. P. Honig, G. J. Roebersen, and P. H. Wierzerna, J. Colloid Interface Sci., 36, 97 (1971).
16. T. G. van de Ven and S. G. Mason, Colloid Polymer Sci., 255-468 (1977).
17. L. J. Warren, Chemtech, 11, 180 (1981).
18. P. M. Adler, J. Colloid Interface Sci., 83, 106 (1981).
19. T. G. van den Ven and S. G. Mason, Tappi, 64, 171 (1981).
20. N. Tambo and H. Hozumi, Water Research, 13, 421 (1979).
21. R. L. Powell and S. G. Mason, A.I.Ch.E.J., 28, 286 (1982).
22. L. A. Glasgow and R. H. Luecke, 1 & E.C. Fund., 19-148 (1980).
23. R. K. Ham and R. F. Christman, J. San. Eng. Div. ASCE, 95, 481 (1969).
24. A. S. Michaels and J. C. Bolger, Ind. Eng. Chem., 3, 14 (1964).
25. B. A. Firth and R. J. Hunter, J. Colloid Interface Sci., 57, 266 (1976).
26. J. Gregory, in Commercial Water Soluble Polymers, edited by C. A. Finch, Plenum Press, NY, 1983.
27. J. V. O'Gorman and J. A. Kitchener, Int. J. Min. Process, 1, 33 (1974).
28. M. Yusa, Int. J. Min. Process, 4, 293 (1977).
29. M. D. Sikora and R. A. Stratton, Tappi, 64, 97 (1981).
30. W. Ditter, J. Eisenlauer, and D. Horn, in The Effect of Polymers on Dispersion Properties, edited by T. F. Tadros, Academic Press, Longon, 1982.
31. L. A. Spielman, Ann. Rev. Fluid Mech., 9, 297 (1977).
32. K. Yao, M. T. Habibion, and C. R. O'Melia, Env. Sci. Technol., 5, 1105 (1971).
33. J. Gregory and A. J. Wishart, Colloids Surf., 1, 313 (1980).
34. W. Stumm, Environ. Sci. Tech., 11, 1066 (1977).
35. L. Svarovsky, ed., Solid-Liquid Separation, Butterworths, London, 1981.

THE SCIENCE OF THE INTERACTIONS OF COLLOIDAL PARTICLES AND CERAMICS PROCESSING

Alan Bleier

Massachusetts Institute of Technology
77 Massachusetts Avenue, Room 12-011
Cambridge, MA 02139

SYNOPSIS

The interactions among colloidal particulates, solvent, and solutes are discussed from the perspectives required to address current and near-future research needs in ceramics processing.

RESEARCH NEEDS IN CERAMICS

The general understanding of the fundamental, scientific and engineering requirements for the successful, continuous processing of high performance, technical ceramics is poor. This is especially evident when the desired specifications of technical ceramics are compared to the properties of components fabricated using current processing techniques.[1] For instance, many complex ceramic components are designed based on bulk material properties which may be optimized to provide a combination of electronic, magnetic, optical, structural, and chemical or refractory characteristics, with little regard for powder processing requirements. Specific examples include Si, Ge, PZT, Ba(Sr)TiO_3, SnO_2, GaAs, and ZrO_2 as electronic and energy-oriented materials[2-4] and ZrO_2, Al_2O_3, TiB_2, SiC, Si_3N_4, sialon, spinel, and mullite as structural and refractory ceramics.[5-8]

The problems which presently inhibit complete exploitation of ceramics for electronic and energy-oriented applications have been reviewed and summarized by Bowen.[1] Although his review principally addresses the energy industry (e.g., electronic, magnetic, and optical devices), the technological needs for structural and refractory ceramics parallel those outlined in reference (1). Basically, the generic guidelines which would ensure predictable fabrication of ceramic components by enabling the consistent development of specified fired-body components are currently either unknown or practically intractable given current processing designs. Since the concept of controllable, fired microstructures is predicated upon one of developing microstructurally desired (prefired) greenware, many of the problems focus specifically on the controlled generation, assembling, and compaction of ceramic powders.[1] Often, these are finely divided or colloidal powders.

The first step in the successful processing of such powders is the generation or attainment of "good" powder. A "good" powder may be considered as one which has the required chemical purity, phase composition, and desirable processing characteristics, including controllable packing density and acceptable sinterability.[9]

An understanding of particulate-liquid interactions, such as wetting and dispersibility phenomena must be developed in order to control the greenware density. This is particularly true with regard to powder agglomeration and its effects on ceramic slip properties (preparation, rheology, and mixing) and subsequent green-body uniformity. Additionally, though inadequate firing processes lead to serious problems, current limitations of components are often considered reflections of inadequate powder processing.[1,9] The ability to control particulate-liquid interactions and the potential to influence microstructural development in greenware are then of paramount importance in the successful fabrication of technical ceramics. Table 1 gives the principal processing needs as summarized in references (1) and (9). This table indicates that ceramics processing can be subjectively subdivided into the control of three distinct areas: powder formation, processing, and packing.

Paralleling these research needs are those related to powder characterization. This includes the determination of chemical and phase composition, purity of powder, size, shape, and roughness of primary particles, surface area of primary particles, number and size distribution of pores, degree of powder crystallinity, if any, nature of internal and surface stress in particulates, powder defect structure, and, finally, presence of surface impurities and adsorbed films on powders.[10] Many techniques for characterizing powders presently exist,[11] but require further development to analyze complex ceramic powders. Unfortunately, many of the advanced methods developed for solid surfaces[12-15] may not be easily adapted to the detailed characterization of powders.

Since the basic science and understanding of ceramics processing inherently involves many aspects which are traditionally encountered in colloid and surface science, it is natural that these two disciplines join forces in an attempt to develop concepts which would advance ceramics processing, while simultaneously improving our fundamental knowledge of physically and chemically complex interfaces.

Table 1. Current Processing Needs and Suggested Research Areas

Powder Formation
Nucleation and Growth Mechanisms and Effects on Powder Properties
Manifestation of Chemical Precursor Properties
Development of Amorphous, Crystalline, and Anisotropic Particulates
Powder Processing (Slip Preparation)
Inherent Powder Properties - Effects on Wetting, Dispersion Stabilization and Consolidation of Powder
Role of Surfactant, Polymeric, and Second Solvent Additives
Inter- and Intraparticulate Adhesional Forces
Role of Surface Roughness, Elasticity, and Plasticity
Rheology - Role of Agglomerate Structure
Powder Packing and Compaction Mechanics
Flow and Collapse of Agglomerated Particulate Structures
Drying

This paper endeavors to bridge the two disciplines by reviewing what is generally known about the interactions of colloidal particles with each other and their surroundings, while emphasizing contributions at specific surfaces. The goal is to relate the general state-of-the-art of ceramics processing to emerging processing approaches which are based on colloidal interactions.

Interactions of Colloidal Particles

Status - What is Known. The following discussion illustrates the range of phenomena for which current understanding provides general principles by which colloids are governed. This knowledge is sufficient for many current technologies, including ceramics. However, considerably more remains unknown so that the full potential to control colloidal powders is not realized presently on a wide scale.

Synthesis of Uniform Powder. The fundamentals of preparing and controlling many desirable colloidal solids are known.[16-18] Technical aspects of their preparation either involve nucleation and growth of a new phase or are related to the state of subdivision, as for instance, in the comminution of coarse solids. Three examples of processes, by which colloids have been prepared via nucleation and growth, are: condensation of vapor to yield solid directly,[10] condensation of vapor to form colloidal liquid droplets which are subsequently solidified during chemical reaction,[19] and precipitation from liquid solution.[20] Beddow[21] comprehensively reviews the formation of colloidal solids by subdivision.

Table 2 contains a partial list of the numerous "monodispersed" sols of elements, inorganic compounds, and organic polymers that colloid scientists have prepared. Recently, the underlying principles in the preparation of metal (hydrous) oxide sols related to high performance ceramics have been elucidated, permitting the aqueous and nonaqueous synthesis of model particulates which represent a relatively large range of compositions. Matijevic[34] has reviewed the mechanisms by which uniform oxidic particulates can be synthesized: homogeneous precipitation, forced hydrolysis, thermal decomposition of complex species, calcination, and chemical reactions of aerosols. See Table 3.

Table 2. Examples of Monodispersed Sols

Elements[22-26]
Gold, Sulfur, Silver, Selenium
Inorganic Compounds[27-30]
Silica, ZrO2, Iron Hydrous Oxide, α-Fe_2O_3, Silver Chloride, Barium Sulfate, Chromium Hydroxide, TiO_2, CdS, $FePO_4$
Organic Polymers[31-33]
Polyvinyl Acetate, Polyvinyl Chloride, Polystyrene, Polyacrylic Lattices

Powder Characterization. The "proper" characterization of colloids, particularly that of their surfaces, involves various analyses, depending upon the purpose for which the information is sought. Physical properties typically evaluated include gas and solution adsorption, heat of immersion, permeability or porosity, and wetting phenomena (e.g., contact angle). The colloid's surface charge may be important and this is usually deduced from electrokinetic measurements or surface titration studies.[35-37] Chemical characterization of colloidal surfaces is often accomplished using IR,[15,38-42] ESR,[43-45] NMR,[46] and Raman[46,47] spectroscopies and electrochemical methods.[24]

Table 3. Controlled Powder-Generating Mechanisms and Some Important Parameters

Mechanism	Parameters
Homogeneous Precipitation Forced Hydrolysis	Salt Type, Nucleating Agent, Temperature, pH, Mechanical Peptization
Thermal Decomposition Calcination	Chelant Type, Degree of Hydration, Temperature
Aerosol Reaction	Temperature, P_{H_2O}, Metal Alkoxide Type

Dispersibility Phenomena. Parfitt[48] noted that a successful dispersion process consists of three stages: wetting of powder, disintegration of large agglomerates into colloidal particles or small clusters of primary particles, and stabilization of powder against reagglomeration. Taking the first two stages collectively as powder dispersibility and treating the third stage separately, it follows that powder dispersibility consists of, firstly, replacing a solid-vapor (air) interface by a solid-liquid one and, secondly, providing sufficient energy to mechanically break agglomerate structures. The change in free energy per unit area associated with dispersibility (ΔG_d), given by Eq. (1), may provide sufficient compensation to separate agglomerate

$$\Delta G_d = \gamma_{SL} - \gamma_{SV} = - \gamma_{LV} \cos \Theta_e \tag{1}$$

components into primary particles, provided thermal energy is relatively large. Usually, the reverse is true, that is, mechanical action is required to separate particles comprising large agglomerates to permit the complete wetting or engulfment of particulates by liquid. Bleier[49] has recently demonstrated that the Girifalco-Good theory[50-52] and Fowkes' modification of it[53] adequately describe the observed dispersibility data obtained previously[54] with one ceramic powder, silicon. It was demonstrated for this powder that ΔG_d could be calculated using contact angles and physical data of silicon and various liquids. It was also shown that equilibrated, microscopic contact angles and contributions to the surface energy of the powder can be evaluated using the Girifalco-Good model.

Such calculations and related ones are possible if optical, dipole moment, and ionization data are available for the solid and liquid being considered. As these properties become more widely measured or estimated for other ceramic powders, the wetting phenomena associated with chemically complex powders can be treated theoretically. The practical outcome of analyses, such as those given in reference (49), is the development of a fundamental foundation from which complicated slips can be studied. For instance, the role of surfactants and second solvents on various wetting and drying phenomena in powder processing can be evaluated quantitatively and, in principle, optimal choices can be made from the myriad of potential chemical additives.[55] Alternate approaches[56–65] to wetting focus on the energetics of the phenomenon via adsorption isotherms,[30] calorimetry,[30–32] and specific behavior of surface active agents. Examples of the latter class are adsorption phenomena and formation of micelles and hemimicelles.

Stabilization of Colloidal Suspensions. Stability refers to the ability of a colloidal suspension to resist reagglomeration brought about by particulate collisions. Collisions arise from diffusion, as in Brownian motion, velocity gradients (shear) in the liquid, and gravitational settling.[66] The collisional frequency in the first case depends on the liquid's viscosity, concentration of particles, temperature, and the fraction of collisions that lead to irreversible agglomeration. If fluid velocity gradients are present and are sufficiently large, the collisional frequency depends on the volume fraction of solids and the mean velocity gradient, in addition to the parameters already mentioned. Assuming that sedimentation is slow compared to the first two mechansims, the overall rate of agglomeration, expressed in terms of the decrease in particle concentration, N, is given by Eq. (2).

$$-dN/dt = k_d N^2 + k_s N \qquad (2)$$

The rate constants, k_d and k_s, correspond to the diffusion-controlled process and shearing or velocity gradient process, respectively. These constants depend on particle properties such as chemical composition (including dielectric and dipolar properties), size, size distribution, shape, surface charge, solid phase distribution within particles, and anisotropy. Important liquid properties which have not been mentioned thus far are dielectric constant, dipole moment, and ability to support electrolytes. These physical properties combine to determine the relative magnitudes of repulsive electrical double layer and attractive van der Waals forces which are responsible for the electrostatic stabilization of powder dispersions in aqueous and nonaqueous media.[16,18,23] The k_d-term usually dominates in quiescent systems for submicron particles. The full expression in Eq. (2) should be used for other systems. A more detailed description on the application of Eq. (2) can be found in reference (66).

The following chemical reactions decisively affect the rate constants in Eq. (2) and, therefore, in suspension stability: dissolution, reprecipitation, hydrolysis, precipitation, and chemical complexing.[66-68] The last reaction may involve either simple solutes (e.g., $Al^{3+} + SO_4 \rightarrow AlSO_4{}^{+}$) or complicated species, e.g., $Al_8(OH)_{20}{}^{4+}$ and chelated metal species.[68] The use of natural and synthetic polymers to stabilize aqueous and nonaqueous colloidal suspensions is technologically important. Research in this area has focused on polymer adsorption and steric stabilization. Of course, electrolyte and bridging flocculation by polymers have been addressed experimentally and theoretically. References (69) and (70) are comprehensive reviews on suspension phenomena related to polymers.

Periodic Colloidal Structures and Powder Consolidation. Recent work on the structure of regularly arranged colloidal suspensions has indicated that ordering of particles in a fluid occurs under restricted solution conditions and solids concentrations exceeding ~ 50% v/v.[71-75] Experimental advances have spurred theoretical modeling of the liquid-like - to - solid-like phase transition.[76,77] Regular patterns corresponding to hexagonal or cubic packing have been identified, along with associated lattice defects, dislocations, grain boundaries, and segregation phenomena.[73] The interesting feature of these systems is that the liquid separates particles and there are virtually no particle-particle contacts.

Stable ordered suspensions can, therefore, be considered precursor structures for ordered greenware components. Potential applications include casting (slip, tape, freeze, pressure, centrifugal, ultrasonic) isostatic and hot pressing.[78] The development of novel techniques based on periodic structures are in the offing.[79,80]

EMERGING CERAMIC PROCESSES

Many new processes for preparing ceramic powders of desirable powders are already available to the ceramics community.[78,81,82] Examples are decomposition of salts and organometallics, aqueous and nonaqueous precipitation, cryochemical methods, and vapor-phase and molten-salt reactions. Some commercially viable applications appear economically limited[83] because synthetic powders are typically highly agglomerated and require milling,[84] a process which does not always produce optimal particle size distributions and may even introduce detrimental contaminants.

CLOSING REMARKS: CERAMIC-COLLOID RESEARCH NEEDS

The discussions and examples given earlier highlight some important and timely research frontiers in the science of colloidal interactions that are pertinent to existing ceramics and their current limitations.[1,9,78] Table IV contains a short list of general problems in colloid science that impact powder processing and which must be addressed if the potential of high performance, technical ceramics is to be realized. Simply, much is known about colloidal interactions, but considerable work remains before these fields mutually develop to the extent that homogeneous, complex ceramic powders can be predictably and controllably prepared and processed. Some advances in this regard have been made.[49,86,87] Reduced sintering times and temperatures have been achieved.[86] Similarly, the ability to control fired microstructural development for projected ceramic needs is being elucidated.[88]

Table 4. Research Needs in Colloid Science

Design and Optimize Processing Methods to Control Powders
Extend Current Understanding to Concentrated Suspensions (i.e., Extensive Overlap of Electrical Double Layers)
Test Predictions Experimentally, Using Model Ceramic Powders
Examine Relaxation Phenomena Exhibited by Periodic Colloidal Suspensions
Understand Semiconductor-Liquid Interfaces
Determine Role of Space Charge Layer on Electrical Double Layer (e.g., Consult Pleskov[85])
Test Predictions for Wetting and Stability Phenomena

ACKNOWLEDGMENTS

Corning Glass Works funded this project through the Corning Assistant Professorship. The help provided by K. S. Henchey and W. C. Hasz in preparing this manuscript is greatly appreciated.

REFERENCES

1. H. K. Bowen, Mat. Sci. Eng., 44, 1 (1980).
2. S. Iwato, Am. Ceram. Soc. Bull., 61, 926 (1982).
3. L. M. Levinson, ed., in Advances in Ceramics, Vol. 1, Amer. Ceram. Soc., Inc., Columbus, OH (1981).
4. A. N. Frumkin, Surface Properties of Semiconductors, Consultants Bureau, NY, 1964.
5. R. M. Spriggs, Ceram. Eng. Sci. Proc., 3, 1 (1982).

6. I. M. Lachman and R. N. McNally, Ceram. Eng. Sci. Proc., 2, 337 (1981).
7. R. W. Rice, Ceram. Eng. Sci. Proc., 2, 493 (1981).
8. V. J. Tennery, G. C. Wei, and M. K. Ferber, Ceram. Eng. Sci. Proc., 2, 1171 (1981).
9. G. Y. Onoda and H. L. Hench, eds., Ceramic Processing Before Firing, John Wiley & Sons, Inc., NY (1972).
10. C. R. Veale, Fine Powders, Halsted Pr., NY, 1972.
11. G. D. Parfitt and K. S. W. Sing, eds., Characterization of Powder Surfaces, Academic Press, NY (1976).
12. R. J. Blattner and C. A. Evans, Crystal Growth: A Tutorial Approach, 2, 269 (1979).
13. H. H. Brongersma, F. Meijer, and H. W. Werner, Philips Tech. Rev., 34, 357 (1974).
14. G. A. Somorjai, Chemistry in Two Dimensions: Surfaces, Cornell Univ. Press, Ithaca, 1981.
15. A. T. Bell and M. L. Hair, eds., Symp. Ser., No. 137 (1980).
16. H. R. Kruyt, Colloid Science, Elsevier, Amsterdam, 1952.
17. E. Matijevic, Chem. Technol., 3, 656 (1973).
18. G. D. Parfitt, ed., Dispersion of Powders in Liquids, Applied Science Publ., London, 1981.
19. B. J. Ingebrethsen, (Ph.D. Thesis), Clarkson College of Technology, Potsdam, 1982.
20. H. Furedi-Milhofer and A. G. Walton, p. 203 in reference 18.
21. J. K. Beddow, Particulate Science and Technology, Chemical Publishing, NY, 1980.
22. R. Zsigmondy, Z. Physik Chem., 56, 65 (1906).
23. V. K. LaMer and M. D. Barnes, J. Colloid Sci., 1, 71 (1946).
24. V. K. LaMer and R. Dinegar, J. Am. Chem. Soc., 72, 4847 (1950).
25. E. Wiegel, Kolloidchem. Beihefte, 25, 176 (1927).
26. H. R. Kruyt and A. E. van Arkel, Rec. Trav. Chem. Pays-Bas, 39, 656 (1920).
27. W. Stober, A. Fink, and E. Bohn, J. Colloid Interface Sci., 26, 62 (1968).
28. D. Sinclair and V. K. LaMer, Chem. Rev., 44, 245 (1949).
29. J. H. L. Watson, W. Heller, and W. Wojtowicz, J. Chem. Phys., 16, 997 (1948).
30. S. Hamada and E. Matijevic, J. Colloid Interface Sci., 84, 274 (1982).
31. J. W. Vanderhoff, H. J. van den Hul, R. J. M. Tausk, and J. Th. G. Overbeek, in Clean Surfaces, Marcel Dekker, Inc., NY, 1970.
32. A. Homola and R. O. James, J. Colloid Interface Sci., 59, 123 (1977).
33. V. I. Eliseeva, S. S. Ivanchev, S. I. Kuchanov, and A. V. Lebedev, Emulsion Polymerization and its Applications, Consultants Bureau, NY, 1976.
34. E. Matijevic, Acc. Chem. Res., 14, 22 (1981).
35. R. J. Hunter, Zeta Potential in Colloid Science, Academic Press, NY, 1981.

36. P. C. Hiemenz, Principles of Colloid and Surface Chemistry, Marcel Dekkar, Inc., NY, 1977.
37. R. O. James and G. A. Parks, in Surface and Colloid Science, Vol. 12, edited by E. Matijevic, Plenum Press, NY, 1982.
38. L. H. Little, Infrared Spectra of Adsorbed Molecules, Academic Press, NY, 1966.
39. M. L. Hair, Infrared Spectroscopy in Surface Chemistry, Marcel Dekker, NY, 1967.
40. C. H. Rochester, Progr. Colloid & Polymer Sci., 67, 7 (1980).
41. C. H. Rochester, Powder Technology, 13, 157 (1976).
42. C. H. Rochester, Adv. Colloid Interface Sci., 12, 43 (1980).
43. G. P. Luzos and B. M. Hoffman, J. Phys. Chem., 78, 200 (1974).
44. S. Abdo, R. B. Clarkson, and W. K. Hall, J. Phys. Chem., 80, 2431 (1976).
45. M. F. Ottaviani and G. Martini, J. Phys. Chem., 84, 2310 (1980).
46. A. W. Adamson, Physical Chemistry of Surfaces, John Wiley & Sons, Inc., NY, 1982.
47. J. F. Rabolt, R. Santo, and J. D. Swalen, Appl. Spectrosc., 13, 549 (1979).
48. G. D. Parfitt, p. 1 in reference 18.
49. A. Bleier, Am. Chem. Soc., submitted for publication.
50. L. A. Girifalco and R. J. Good, J. Phys. Chem., 61, 904 (1957).
51. R. J. Good and L. A. Girifalco, J. Phys. Chem., 64, 561 (1960).
52. R. J. Good, in Surface and Colloid Science, Vol. 11, edited by R. J. Good and R. R. Stromberg, Plenum Press, NY, 1979.
53. F. M. Fowkes, Ind. Eng. Chem., 56, 40 (1964).
54. S. Mizuta, W. R. Cannon, A. Bleier, and J. S. Haggerty, Am. Ceram. Bull., 61, 872 (1982).
55. McCutcheon's Detergents and Emulsifiers, MC Publishing Co., Glen Rock, NJ, 1980.
56. Reference 46, pp. 369 and 517.
57. G. C. Benson, H. P. Schreiber, and F. van Zeggeren, Can. J. Chem., 34, 1553 (1956).
58. A. J. Tyler, J. A. G. Taylor, B. A. Pethica, and J. A. Hockey, Trans. Faraday Soc., 67, 483 (1971).
59. W. C. Preston, J. Phys. Colloid Chem., 52, 84 (1948).
60. A. Ray, Nature (London), 231, 313 (1971).
61. C. R. Singleterry, J. Am. Oil Soc., 32, 446 (1955).
62. T. Wakamutsu and D. W. Fuerstenau, Adv. Chem. Ser., 79, 161 (1968).
63. C. H. Giles, A. P. D'Silva, and I. A. Easton, J. Colloid Interface Sci., 47, 766 (1974).
64. M. J. Rosen, Surfactants and Interfacial Phenomena, John Wiley & Sons, Inc., NY, 1978.
65. W. Black, p. 149 in reference 18.
66. W. Stumm and J. J. Morgan, Aquatic Chemistry, John Wiley & Sons, Inc., NY, 1981.

67. L. G. Sillen and A. E. Martell, Stability Constants of Metal-Ion Complexes, Spec. Publ. Nos. 17 and 25, The Chemical Society, London, 1964 and 1971.
68. E. Matijevic, J. Colloid Interface Sci., 43, 217 (1973).
69. Th. F. Tadros, ed., The Effect of Polymers on Dispersion Properties, Academic Press, London, 1982.
70. T. Sato and R. Ruch, Stabilization of Colloidal Dispersions by Polymer Adsorption, Dekker, NY, 1980.
71. P. A. Hiltner and I. M. Krieger, J. Phys. Chem., 73, 2386 (1969).
72. A. Kose and S. Hachisu, J. Colloid Interface Sci., 46, 460 (1974).
73. S. Okamuto and S. Hachisu, J. Colloid Interface Sci., 61, 172 (1977).
74. K. Takano and S. Hachisu, J. Colloid Interface Sci., 66, 124 (1978).
75. I. F. Efremov, in Colloid and Surface Science, Vol. 8, edited by E. Matijevic, Plenum Press, NY, 1976.
76. W. van Meegan and I. Snook, Faraday Disc., 65, 92 (1978).
77. E. Dickenson, J. Chem. Soc., Faraday Trans. II, 75, 466 (1979).
78. F. Y. Wang, ed., Ceramic Fabrication Processes, Academic Press, NY, 1976.
79. E. A. Barringer, Disorder-Order Transition in Monodisperse Titania Sols, Ceram. Proc. Res. Lab. Report No. 8, M.I.T., Cambridge, MA, December 1980.
80. (a) T. C. Huynh, A. Bleier, and H. K. Bowen, (b) E. Barringer and H. K. Bowen, presented at the 84th Annual Meeting of the Am. Ceram. Soc., Cincinnati, May 4, 1982, Paper Nos. 53-B-82 and 54-B-82; also Am. Ceram. Soc. Bull., 61, 336 (1982).
81. R. H. Arendt, J. H. Rosolowski, and J. W. Szymaszek, Mat. Res. Bull., 14, 703 (1979).
82. W. R. Cannon, S. C. Danforth, J. H. Flint, J. S. Haggerty, and R. A. Marra, in Laser Applications in Materials Processing, Soc. Photo-Opt. Inst. Eng., Bellingham, WA, 1980.
83. J. L. Pentecost, p. 1 in reference 18.
84. G. Lowrison, Crushing and Grinding, Chem. Rubber Co., Boca Raton, 1974.
85. Y. U. Pleskov, in Progress in Surface Membrane Science, edited by J. F. Danielli, M. D. Rosenberg, and D. A. Cadenhead, Academic Press, NY, 1973.
86. R. L. Pober, E. A. Barringer, M. V. Parish, N. Levoy, and H. K. Bowen, in this volume.
87. A. Bleier, Am. Ceram. Soc., submitted for publication.
88. U. Chowdhry and R. M. Cannon, in Processing of Crystalline Ceramics, edited by H. Palmour III, R. F. Davis, and T. M. Hare, Plenum Press, NY, 1978.

PREPARATION OF SHAPED GLASSES THROUGH THE SOL-GEL METHOD

Sumio Sakka and Kanichi Kamiya

Faculty of Engineering, Mie University

Kamihama-cho, Tsu, Mie, 514, Japan

ABSTRACT

In the sol-gel preparation of oxide glasses using metal alkoxides as raw materials, the content of water added for hydrolysis of the alkoxides must be controlled depending on the desired shape of the resulting glass. Larger water contents of the starting solution are preferred for the production of bulk glasses, while smaller water contents lead to the formation of fiber glasses. The cross-section of glass fibers made by the sol-gel method is circular or non-circular, depending on the composition of the starting solution. This effect is examined, and it is shown that the volume change of gelling fibers occurring during the solidification process is related to the shape of the cross-section of the resulting glass fibers.

INTRODUCTION

The sol-gel method using metal alkoxides as raw materials is one of the new techniques for making oxide glasses without melting.[1,2] In this method, metal alkoxides undergo hydrolysis and polycondensation to yield gels which are converted to oxide glasses by heating at relatively low temperatures. Oxide glasses, including those of CaO-SiO_2, ZrO_2-SiO_2 and SrO-SiO_2 systems[3-5] which are difficult to obtain by conventional melting techniques, have been prepared by this method in the form of fiber, film and bulk. Papers discussisng the recent development in the science and technology of this field are collected in the special issue of the Journal of Non-Crystalline Solids (Vol. 48, No. 1, 1982).

In this paper, the factors to be considered in preparing two types of shaped glasses are reviewed; bulk glasses and fiber glasses. The water content of the starting solution is of prime importance for both types of shaped glasses. Factors in the preparation of bulk glasses other than the water content will be discussed. Finally, the factors affecting the shape of the cross-section of fibers made by the present method will be described on the basis of our recent experimental results.

FACTORS TO BE CONSIDERED IN PREPARATION OF SHAPED GLASSES

The sol-gel method for making oxide glasses consists of three processing stages.[1] In the first stage, metal alkoxides are diluted with alcohols and intimately mixed. In the second stage, the addition of water causes hydrolysis and polycondensation of the alkoxide, which gives a gel as equations (1) - (4) illustrate.

$$M(OR)_n + H_2O \rightarrow M(OR)_{n-1}OH + ROH \quad (1)$$

$$M(OR)_{n-1}OH + M(OR)_n \rightarrow (OR)_{n-1}MOM(OR)_{n-1} + ROH \quad (2)$$

$$2M(OR)_{n-1}OH \rightarrow (OR)_{n-1}MOM(OR)_{n-1} + H_2O \quad (3)$$

$$(OR)_{n-1}MOM(OR)_{n-1} + M(OR)_{n-1}OH \rightarrow (OR)_{n-1}MOMOM(OR)_{n-1} + ROH \quad (4)$$

In the final stage, the gel is converted to an oxide glass by heating. The heating temperature is usually lower than the glass transition temperature of the resulting glass.

Conventionally, glass melts are formed into desired shapes such as fiber, film and plate in the viscous state at higher temperatures than their glass transition temperatures. In the sol-gel method, however, the processing temperature is lower than the temperatures corresponding to the viscous state and the success in obtaining glasses of the desired shape depends on whether the gel of the corresponding shape is obtainable.

It can be assumed that a sol containing linear or chain-like polymers may be favorable for drawing fibers, while a sol containing three-dimensional network polymers may give a bulk gel. It has been reported that the hydrolysis and polycondensation reactions of metal alkoxides are affected by the temperature, the concentration of materials, the water content and the catalyst.[6] Among those factors, the water content is most important in determining the form of polymers. A smaller water content leads to linear polymers which are favorable for drawing fibers, while a larger water content leads to three-dimensional network polymers which give an elastic bulk gel.[7] The experimental results concerning the possibility of forming fibers and bulk gels are shown in Table 1. In Table 1, the composition of

Table 1. Compositions and properties of $Si(OC_2H_5)_4$-H_2O-C_2H_5OH-HCl solutions for SiO_2 glass.

$Si(OC_2H_5)_4$ (g)	H_2O[a]	C_2H_5OH (ml)	HCl[a]	Temperature(°C)	Fiber drawing	Gel
25	1.0	25	0.003	30	Yes	Particulate
25	1.0	25	0.03	30	Yes	"
25	2.0	25	0.03	30	Yes	"
25	2.0	25	0.3	30	Yes	"
62	2.0	90	0.01	25	Yes	"
62	2.0	90	0.01	80	Yes	"
25	4.0	25	0.03	30	Yes	"
25	4.0	25	0.3	30	Yes	"
62.5	4.0	18.4	0.01	80	Yes	"
25	6.0	25	0.003	30	No	"
25	6.0	25	0.3	30	No	"
62.5	6.0	36.4	0.01	80	No	"
25	10.0	30	0.03	30	No	Bulk
25	29.0	30	0.3	30	No	"

[a]Contents of H_2O and HCl are represented by their molar ratio to $Si(OC_2H_5)_4$.

starting $Si(OC_2H_5)_4$ solutions and the reaction temperature are correlated to the fiber drawing ability and the shape of gel. The solutions in which the molar ratio of water to $Si(OC_2H_5)_4$ is less than 4 becomes viscous as the reaction progresses, showing spinnability just before gelling, irrespective of the reaction temperature and the HCl content. The ease of drawing fibers from such a viscous solution increases with decreasing water content. The solutions of larger water content do not show spinnability but are solidified into bulk gels, like jelly.

The isolated thin films are made by pulling the viscous and spinnable solutions through a thin slit.[2] The $Si(OC_2H)_4$ solution with the mole ratio [water]/[alkoxide] ranging from about 10 to 20 is preferred as the starting solution for SiO_2 coating films.[8]

The vaporization of organic matters and water from gel occurs in the course of heating to convert the gel to oxide glass. The volatile materials are easily removed from gel fibers and films. However, in the case of bulk bodies, the internal stress due to different evaporation rates between surface and interior may cause serious fragmentation of gels. This difficulty must be overcome in order to produce bulk glasses. Several requirements for making bulk oxide glasses by the sol-gel method will be described.

Requirement 1. The content of water added for hydrolysis of metal alkoxides in the starting solution should be larger than that theoretically required for the reaction to be completed, thus yielding corresponding oxide. Figure 1 shows the variation of the size of fragments of gel for the $10TiO_2$-$90SiO_2$ (wt%) glass with the water content of the starting solution. The solutions of $Ti(OC_3H_7)_4$ and $Si(OC_2H_5)_4$ with different water contents were prepared, stored in glass container and kept open in air to be gelled. The results shown in Fig. 1 were obtained after drying the gels in the laboratory for a few months. It is clear that the size of the fragment increases with increasing molar ratio of water to alkoxide.

Requirement 2. The use of containers made from non-hydrophilic materials such as teflon and polystyrene is preferred to minimize the effect of adherence and friction between gels and container wall. Lining the glass containers with paraffin may also be effective. Fractures may be caused by the adherence of the gel surface to the glass container wall and by the friction between the gel and container surfaces on contraction of the gel. In Fig. 2, several bulk gels thus obtained are shown. The gels are easily machined into plates, rectangular bars and other shapes.

Requirement 3. Gels must be heated slowly to minimize fracture of the resulting glasses. Several heating schedules leading to successful production of bulk glasses from gels have been reported. Relatively slow heating is favorable in order for the possible internal stress to be relaxed. The internal stress would arise in the gels due to non-uniform vaporization of volatile materials.

Requirement 4. Hydrolysis of alkoxides at relatively high temperatures (below the boiling point of the solution) is preferred to make bulk gels which are easily converted to bulk glasses with less fracture.[9] Larger pores are found in such gels, which makes vaporization of volatile materials easy.

Requirements 1 to 4 described above are not independent from each other. Therefore, it is necessary to seek for an optimum combination of the requirements in order to make bulk glasses by the sol-gel method.

TiO_2-SiO_2 glasses[10] shown in Fig. 3 have been made in our laboratory by following the above requirements. The solution consisting of $Ti(OC_3H_7)_4$ and $Si(OC_2H_5)_4$ with the molar ratio of water to the alkoxides equaling 50 was used as the starting solution. Hydrolysis was conducted at 40°C. The resultant gel was heated to 900°C with a slow heating rate of 6°C/h.

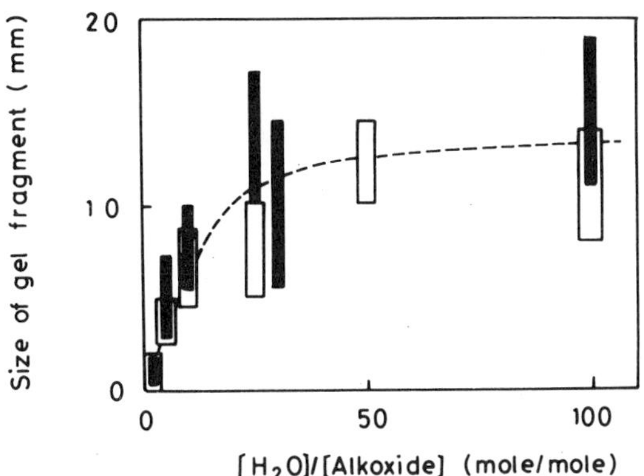

Fig. 1. Variation of the size of gel fragments for the $10TiO_2 \cdot 90SiO_2$ (wt%) glass with the water content expressed by the molar ratio to alkoxide.

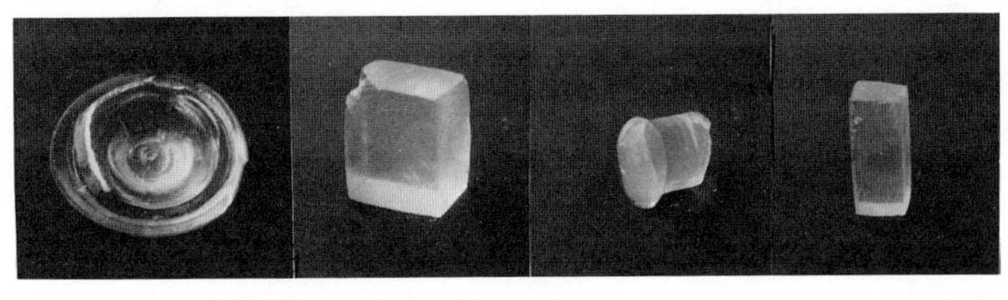

Fig. 2. Different shapes of bulk gels of the SiO_2 composition.

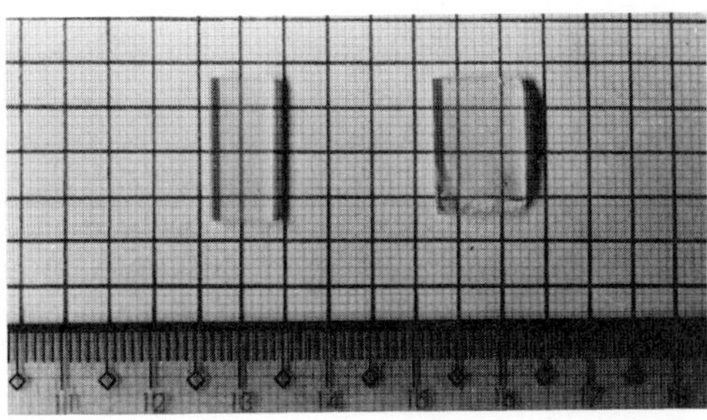

Fig. 3. TiO_2-SiO_2 glasses made by sol-gel method.

PREPARATION OF GLASS FIBERS BY SOL-GEL METHOD AND SHAPE OF THEIR CROSS-SECTION

Figure 4 shows variations of the viscosity of an alcoholic solution of $Si(OC_2H_5)_4$, in which the molar ratio of water to the alkoxide is 2, with reaction time at three different temperatures. Hydrochloric acid was used as a catalyst for hydrolysis. Details of preparing mixed solutions of $Si(OC_2H_5)_4$-H_2O-C_2H_5OH-HCl have been reported elsewhere.[7] The viscosity of the solution increases with increasing time, or with the progression of the hydrolysis reaction. After viscosities greater than 10 poises were reached, the solution became sticky and spinnable. Gel fibers could then be drawn by dipping a glass rod in the solution and pulling it up quickly. It is seen from Fig. 4 that the spinnable state is attained in shorter times as the reaction temperature is raised. At 80°C, fibers can be drawn within two hours.[11] Gel fibers for SiO_2 glass fibers are shown in Fig. 5.

Fig. 6 is the composition diagram showing the relation between fiber drawing behavior and composition of the starting $Si(OC_2H_5)_4$-H_2O-C_2H_5OH solution with $[HCl]/[Si(OC_2H_5)_4] = 0.01$ hydrolyzed at 80°C. The diagram is roughly divided into four areas. In area I where the C_2H_5OH content is low, materials are not miscible with each other. In area II where the molar ratio of water to $Si(OC_2H_5)_4$ is larger than 5, the solution gels into a jelly like solid without showing spinnability as mentioned in the previous section. In area III, where the molar ratio of water to alkoxide is less than 1.5 neither gelling nor spinnability is observed.[11] This is possibly due to the vaporization of water while keeping the solution open in air. The solutions show spinnability and fiber drawing is possible in area IV where the $[H_2O]/[Si(OC_2H_5)_4]$ ratio ranges from 1.5 to 4. The drawn fibers can be converted to SiO_2 glass fibers by heating to 800 ~ 1000°C.

It has been noticed that the cross-section of fibers thus obtained is not always circular unlike conventional glass fibers prepared by drawing through nozzles from the melt. The shape of the cross-section of the fibers is related to the composition of the starting solutions as shown in Fig. 6. The solutions of compositions marked by solid circles yields fibers with circular cross-sections. This area is very limited and non-circular cross-sections are produced in most part of area IV. Dependence of the shape of cross-section of fibers on the solution composition is more clearly shown in Fig. 7. Fig. 7a shows the change of cross-section of fibers with the water content of solution. The water content is changed along tie-line (1) in Fig. 6, so that the molar ratio of C_2H_5OH to $Si(OC_2H_5)_4$ is equal to one. Figure 7b shows the change of cross-section of fibers with the C_2H_5OH content. It is changed along tie-line (2) in Fig. 6, so that the molar ratio of water to $Si(OC_2H_5)_4$ is equal to four. Hydrolysis temperature, viscosity of the solution from which fibers are drawn, as well as the tools and methods adopted to draw fibers scarcely affected the cross-sectional shape of the fibers.

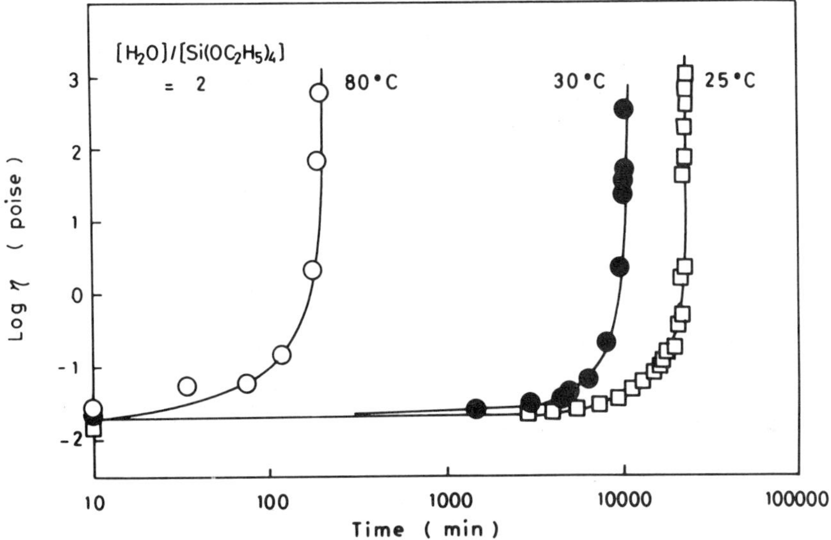

Fig. 4. Variation of the viscosity of a $Si(OC_2H_5)_4$ solution with the molar ratio $[H_2O]/[Si(OC_2H_5)_4] = 2$ as a function of time at 25°, 30° and 80°C.

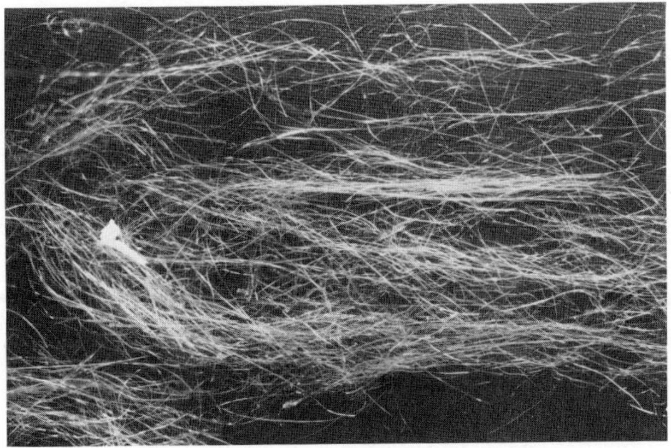

Fig. 5. SiO_2 gel fibers.

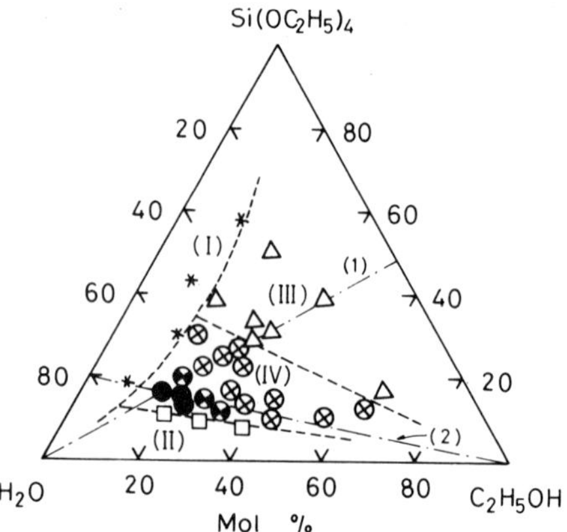

Fig. 6. Relation between fiber drawing behavior and composition of $Si(OC_2H_5)_4$-H_2O-C_2H_5OH solution with $[HCl]/[Si(OC_2H_5)_4]$ = 0.03 hydrolyzed at 80°C.
✱immiscible (area I); □ not spinnable (area II); △ not-gelling (area III); ● circular cross-section (area IV); ⊗ non-circular cross-section (area IV); ⊗ circular and non-circular cross-sections (area IV).

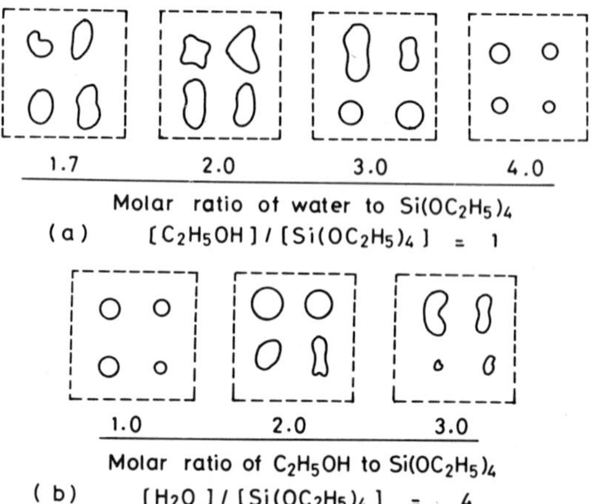

Fig. 7. Schematic representation showing the change of the shape of cross-section of SiO_2 fibers with the composition of the starting $Si(OC_2H_5)_4$ solution.
(a) Change with the water content at a molar ratio of C_2H_5OH to $Si(OC_2H_5)_4$ being unity.
(b) Change with the C_2H_5OH content at a molar ratio of water to $Si(OC_2H_5)_4$ being four.

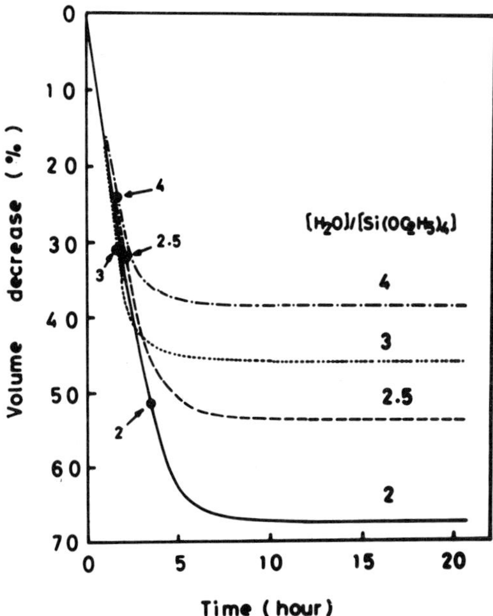

Fig. 8. Change of the volume of $Si(OC_2H_5)_4$ solutions with different water contents with holding time at 80°C. The molar ratio of C_2H_5OH to $Si(OC_2H_5)_4$ is kept unity.

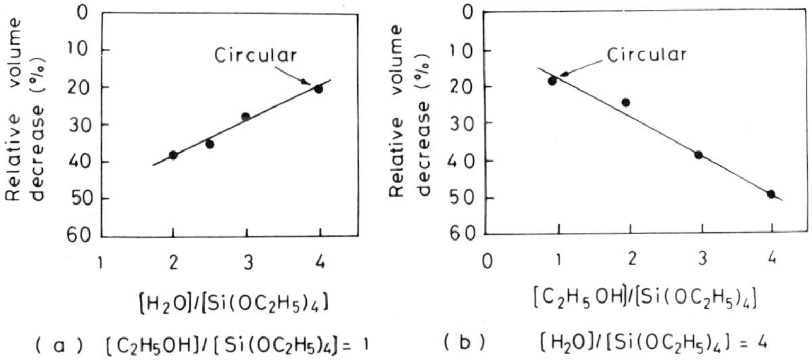

Fig. 9. Relative volume decrease of $Si(OC_2H_5)_4$ solutions on solidification of fibers as functions of the content of water (a) and C_2H_5OH (b). Relative volume decrease (%) = [(Volume at fiber drawing shown in Fig. 8) - (Volume observed at constant volume after drawing)]/(Volume at fiber drawing) x 100.

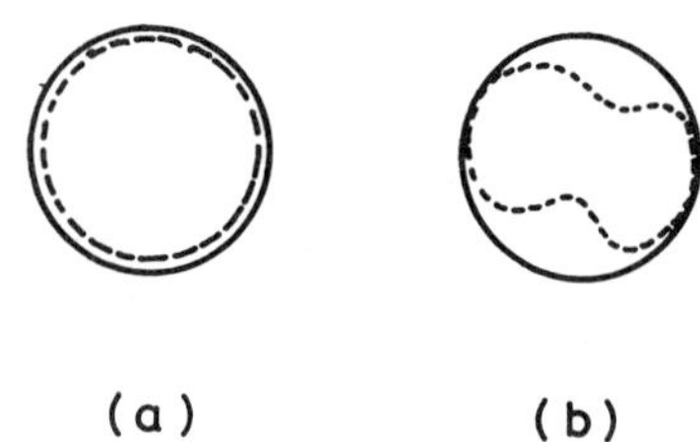

Fig. 10. Illustration of change of the cross-section of fibers (a) when the volume change is small and (b) when the volume change is large on solidification.
At spinning, ----- After solidification.

In order to discuss the cause for occurrence of different cross-section shapes, the volume change of the $Si(OC_2H_5)_4$ solution in the course of sol-gel transition was measured. For that purpose, the solutions of compositions on tie-line (1) of Fig. 6 were prepared. The 23 cm^3 of the solution was transferred to a glass container of 2.5 cm diameter and 6 cm in height, and kept at 80°C until gel. The height of the meniscus of the solution was measured at some time intervals and the volume of the solution calculated. After gelling, the gel was kept at room temperature. Fig. 8 shows the volume changes of the solutions on tie-line (1) as a function of time. Fibers were drawn at times indicated by solid circles on the curves. In Fig. 9, relative volume decrease of a solution occurring during the period from the time of fiber drawing to the time when constant volume is achieved is shown as a function of the content of water (a) and C_2H_5OH (b). This volume decrease corresponds to that of fibers occurring during solidification or the drying process. It is seen from Fig. 9a that gel fibers drawn from the solution with a molar ratio of water of 2 show 40% volume decrease on solidification, while those obtained from the solution with a molar ratio of water of 4 show only 20% volume decrease. It should be noted that the latter solutions yield fibers with circular cross-section. Fig. 9 shows that gel fibers drawn from the solution with smaller C_2H_5OH contents have smaller volume changes. It should be noted that solutions with smaller C_2H_5OH contents yield fibers with circular cross sections.

The above observations indicate that the cross-section of drawn fibers is circular if the volume change occurring in the solidification process is small and is non-circular if the volume change is large. This is explained as follows. The cross-section of drawn gel fibers must be circular due to surface tension just at the time of drawing. The surface of the fibers is solidified more quickly than the interior. The circumference length of the fiber, therefore, would scarcely change on further solidification, while the interior would gradually contract with vaporization of the included volatile materials. Thus, the circular circumference would be constant when the volume change of the fiber on solidification is small, and the circumference could not be circular and thus would be deformed as illustrated in Fig. 10 when the volume change is large.

The difference in volume change in the sol-gel transition between alkoxide solutions with different compositions may reflect the type of polymers produced in the solution. The present authors have suggested[12] that linear polymers and three-dimensional network polymers are produced in the $Si(OC_2H_5)_4$ solutions with molar ratio of water of 1 and 20, respectively. The latter solution showed a smaller volume change in the sol-gel transition. On the basis of above results, it is concluded that the complexity or degree of branching of the alkoxide polymers is increased with increasing water content of the starting solution. The solution with molar ratio of water of 4 which yields fibers of circular cross-section probably contains polymers of intermediate type between linear and network, showing smaller volume change than other solutions with smaller water content in the sol-gel transition.

ACKNOWLEDGMENT

The authors wish to thank T. Kato and H. Asao who assisted in the experiments. This work was partly supported by Grant-in-Aid for Co-operative Research (1982) of the Ministry of Education, Science and Culture, Japan.

REFERENCES

1. S. Sakka and K. Kamiya, J. Non-Crystal. Solids, 42 [1-3], 403-422 (1980).
2. S. Sakka, pp. 129-167 in Treatise on Materials Science and Technology, Vol. 22, Glass III, edited by M. Tomozawa and R. Doremus, Academic Press, New York (1982).
3. T. Hayashi and H. Saito, J. Mat. Sci., 15 [8], 1971-77 (1980).
4. K. Kamiya, S. Sakka, and Y. Tatemichi, J. Mat. Sci., 15 [7], 1765-71 (1980).
5. M. Yamane and T. Kojima, J. Non-Crystal. Solids, 44 [1], 181-190 (1980).
6. D. C. Bradley, pp. 410-446 in Inorganic Polymers, edited by F. G. A. Stone and W. A. G. Graham, Academic Press, New York (1962).
7. K. Kamiya, S. Sakka and M. Mizutani, Yogyo-Kyokai-Shi, 86 [11], 552-59 (1978).
8. Y. Yamamoto, K. Kamiya and S. Sakka, Yogyo-Kyokai-Shi, 90 [6], 328-33 (1982).
9. M. Yamane and S. Okano, Yogyo-Kyokai-Shi, 87 [8], 434-39 (1979).

10. K. Kamiya and S. Sakka, J. Mat. Sci., 15 [11], 2937-39 (1980).
11. S. Sakka, K. Kamiya and T. Kato, Yogyo-Kyokai-Shi, 90 [9], 555-56 (1982).
12. S. Sakka and K. Kamiya, J. Non-Crystal. Solids, 48 [1], 31-46 (1982).

INORGANIC OXIDE GELS AND GEL-MONOLITHS: THEIR CRYSTALLIZATION BEHAVIOR

Shyama P. Mukherjee[a]

Battelle's Columbus Laboratories
Columbus, Ohio

ABSTRACT

The significance of kinetics in the ordering process in metal oxide gels and gel-monoliths during densification is discussed. Crystallinity of TiO_2 gels after different thermal treatments were examined by x-ray diffraction and electron diffraction techniques. The sequence of phase transformation was noncrystalline gel → anatase → rutile. Crystallinity of gel powders and gel-monoliths of two compositions in the Na_2O-Ba_2O_3-SiO_2 system were examined after thermal treatment at different temperatures above the Tg. Results indicate that the crystallization rates are controlled by the composition and the catalytic activity of Na^+ ions associated with hydroxyl ions in rupturing the Si-O-Si bonds. Porous gel-monoliths are transformed into transparent monocrystalline glass by appropriate heat treatment just above Tg.

INTRODUCTION

Preparation of noncrystalline silicate gels and gel-monoliths and subsequent conversion to glasses and glass-ceramics has been reported[1-5] as have noncrystalline metal oxide gels containing no silica.[6-9] Forming a noncrystalline phase by rapid quenching of melts or from evaporated solids is not unusual. In these cases, the

[a]Now at Jet Propulsion Laboratory, California Institute of Technology, Mail Stop 67-201, 4800 Oak Grove Drive, Pasadena, CA 91109.

quenching of the disordered melt or vapor phase structures plays a major role in developing noncrystallinity. The kinetic forces such as cooling rate do not appear to play a major role in forming noncrystalline solids (from solutions) by chemical processes. Chemical process conditions during polymerization/gelling produce certain stable disordered polymeric molecular configurations that may play major roles in developing noncrystalline inorganic oxide gels.

Some questions might be raised: What is the nature of noncrystallinity of metal oxide gel structures? How does it differ from the structure derived by quenching melts? What is the stability of the gel's structure during subsequent thermal treatment? Answers are important, not only for understanding the fundamentals of noncrystalline solids, but also in applying gel-derived dielectric materials in electro-optics/fiber optics and electronic technologies.

The study of the gel structures becomes more complex because of (a) the presence of submicroporosity, structural hydroxyl groups, and chemically bonded organic groups in gels; and (b) subsequent changing of pore structures and formation of new oxy-bridges on heat treatment during densification.

Protons can participate in forming structures in different ways and can strongly influence both structure-sensitive properties and stability toward crystallization. High specific surface areas of gels could increase the heterogeneous nucleation sites. Also, noncrystalline materials formed at low temperatures might have fewer impurities and structural defects than those prepared at high temperatures.[10]

In discussing crystallization kinetics in the gel-glass transformation, Zarzycki[11] pointed out that glass formation results from competition between phenomena that lead to densification and those that promote crystallization.

Published work on gels and gel-derived glasses indicate different crystallization rates from those for vitreous solids produced from melts. The effects of minor components and impurities on the crystallization behavior of gels are found to be different.[3,11,17] Gel preparation procedures influence the crystallization rates of multicomponent gels.[15-17] Critical point drying of gel-monoliths in autoclaves introduces a new process parameter influencing the crystallization. Hence, study of the crystallinity and crystallization behavior of gels is becoming more important and more complex as the sol-gel process for making glasses and microcrystalline ceramics is developed.

This paper discusses work on crystallization behavior of gels and gel-monoliths prepared from metal-alkoxide systems: TiO_2 and $Na_2O-B_2O_3-SiO_2$.

Noncrystalline gels and gel-monoliths in TiO_2 systems are prepared by the controlled hydrolytic polycondensation of titanium alkoxides.[8] Structural complexity is altered by such processing conditions as the proportion of water to the alkoxide and the chemical nature of the alkoxide.[18] Gels prepared by different procedures might crystallize at different rates. Moreover, TiO_2 exhibits three polymorphic forms: anatase, brookite, and rutile. Relating the development of any particular phase(s) from gels to preparation procedures and thermal history might throw some light on the structural characteristic of noncrystalline metal oxide gels.

Gels in the sodium borosilicate system are interesting for crystallization studies because they contain two distinctly different oxides as minor components: B_2O_3, a strong glass former, enters the silicate network, thus inhibiting crystallization rates; whereas, Na_2O, a modifier, causes crystallization of silica gels even when present in a trace amount.[12,13,16]

We want to determine the influence of composition, thermal history, and structure on the crystallization behavior of gels and gel-monoliths during gel-to-glass transformation. The work presented here is preliminary.

EXPERIMENTAL WORK AND RESULTS

TiO_2 System

TiO_2 gels and gel-monoliths were prepared by controlled hydrolytic polycondensation of titanium isopropoxide, $Ti(OC_3H_7)_4$, in the presence of an acid catalyst.[8] The gels treated at 353K for 7 days remained transparent and noncrystalline, and were used for subsequent thermal treatment.

The gel samples were analyzed by differential thermal analysis (DTA) and thermogravimetric analysis (TGA) at temperatures up to 1273K. Two exothermic peaks were observed in the DTA curve; the first, at 653K, might be attributed to the combined effects of exodation of organic groups and transformation of noncrystalline TiO_2 gels to the anatase phase. The second, at 1013K, is due to the transformation from the anatase to the rutile phase.

The TGA curve of wet gels indicates a total weight loss of about 65 percent--a sharp loss of 50 percent from 473 to 710K, a gradual loss of about 15 percent up to 873K, and no weight loss above 873K. The first loss covers removal of free alcohols, water, and other free organic reaction products; the second, elimination of carbonaceous products formed during thermal cracking of organics and structural hydroxyl groups. Gel samples were examined by x-ray powder diffraction to detect crystallinity and to identify the crystalline phases developed. The results of various thermal treatments are shown in Table 1 and x-ray diffraction patterns of a noncrystalline gel sample and progressively heat treated gel samples are shown in Fig. 1. The evolution of the crystalline anatase phase from the noncrystalline gels is evident.

Table 1. Thermal Treatment Effects on Crystallization of TiO_2 Gels

Gel	Thermal Treatment Temp., K	Thermal Treatment Time	Crystallinity	Crystalline Phases
1	343	5-7 days	Noncrystalline, transparent	---
2	423	24 hours	Noncrystalline	(Sample 1)*
3	Raised slowly to 523	No holding	Noncrystalline	(Figure 3b)
4	Raised slowly to 623	No holding	Crystallinity detected	Broad peak due to anatase
5	Raised slowly to 773	No holding	Crystallinity detected	Broad peak due to anatase (Sample 2)*
6	Introduced to furnace at 773	30 minutes	Crystalline	Anatase (Sample 3)*
7	1173	30 minutes	Crystalline	Rutile (Sample 4)*

* See Fig. 1.

Scanning electron microscopy could not resolve the microstructure of the air-dried gel. Microstructure of a gel heat treated at 773K for 30 minutes is shown in Fig. 2.

The crystallinities of two gel samples were examined by the electron diffraction technique. These samples after air drying at 343K for 7 days were heat treated at two different temperatures. One sample was heat treated at 423K for 24 hours. The other was heated at 50K/hr up to 523K with no hold period. Then, finely ground gel particles were sprayed onto a carbon substrate supported by a 200 mesh copper grid. Electron diffraction studies on 25 particles on each grid detected no crystallinity. A typical transmission electron micrograph sample heated to 523K is shown in Fig. 3a and the electron diffraction pattern of the particles is shown in Fig. 3b. Extremely fine particles and pores are observed. Also note the broad diffuse ring patterns which preclude crystallinity.

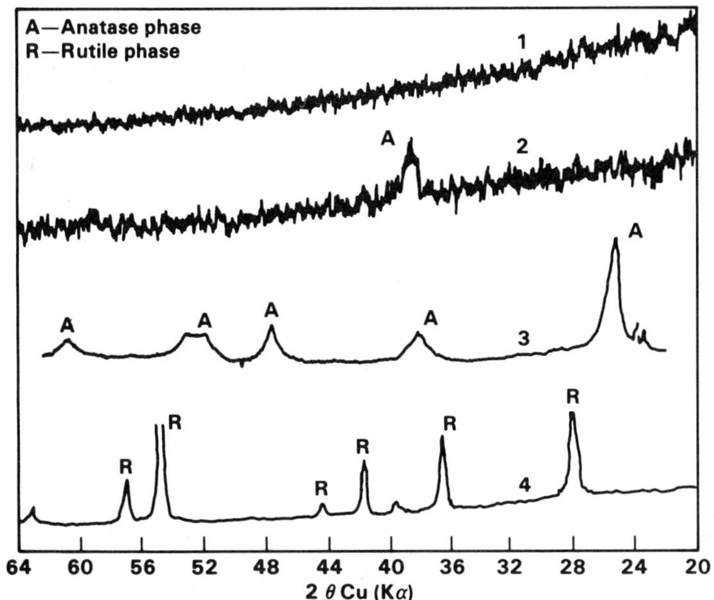

Fig. 1. X-ray diffraction patterns of TiO_2 gels after different thermal treatments: (1) after 423K; (2) raised to 773K; (3) 773K for 30 minutes; and (4) 1173K for 30 minutes.

Na_2O-B_2O_3-SiO_2 System

Gels investigated in the Na_2O-B_2O_3-SiO_2 system are shown in Table 2. Glasses of like composition prepared by melting conventional batches show Tgs of 520°C for Composition 1 and 873K for Composition 2.

In preparing gels and gel-monoliths,[17] $Si(OCH_3)$ was partially hydrolyzed in methanol. Solutions of HBO_3 and $NaOCH_3$ dissolved in methanol were added. Final pH was ~9 before gelation at room temperature.

Fig. 2. Scanning electron micrograph of TiO_2 gel after heat treatment at 773K for 30 minutes. (Magnification 51,000X)

Table 2. Composition of Na_2O-B_2O_3-SiO_2 Gels

Composition	Weight Percent			Mole Percent		
	SiO_2	B_2O_3	Na_2O	SiO_2	B_2O_3	Na_2O
1	60	15	25	61.78	13.30	24.91
2	84	12	4	85.53	10.52	3.95

(a) micrograph (96,000X)

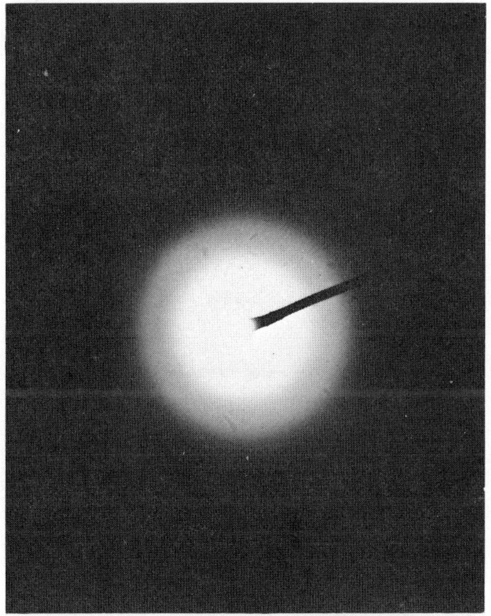

(b) diffraction pattern

Fig. 3. Transmission electron micrograph of a gel sample dried at 243K and heated at the rate of 500K/hr up to 523K; (no holding at 523K) and electron diffraction pattern from specimen particles.

Table 3. Heat Treatment Effects on Crystallization of Gels

Thermal History K	Thermal History Hr	Heat Treatment K	Heat Treatment Hr	Crystallinity	Physical Appearance
				Composition 1	
Slow heat to 773 C		773	13	Noncrystalline	Partially sintered mass
Ditto		773	24	Ditto	Ditto
"		873	2	"	"
"		873	40	"	"
"		1273	2	"	Transparent, grey colored glass
				Composition 2	
423 623	72 1.25	773	4	Noncrystalline	Transparent, porous
423 623 773	72 1.25 4	873	1/4	Noncrystalline	Transparent, porous
773	4	873	1/2	Noncrystalline	Transparent, porosity exists
773	4	923	1/4	Noncrystalline	Transparent, densified
773	4	923	1/2	Noncrystalline	Transparent, densified
773	4	973	1/2	Noncrystalline	Transparent, densified
773	4	1023	1/2	Crystallinity detected, β-quartz	Transparent to translucent, densified
773	4	1073	1/4	Crystalline, α-crystobalite	Completely white and opaque
773	4	1123	1/4	Crystalline, α-crystobalite	Completely white and opaque
773	4	1473	1	Crystalline	Completely opaque and white

Table 3. Continued

Thermal History K	Hr	Heat Treatment K	Hr	Crystallinity	Physical Appearance
423 623 773	16 1.5 4	973	1/2	Partially crystallized, α-crystobalite	Opalescent
773	4	973	1/4	Crystallinity detected	Translucent
423 623 773	96 1.5 4	973	1/2	Extremely broad peak; initiation of ordering	Fairly translucent

Table 4. Influence of Leaching on the Crystallization of Composition 2 Porous Gels

Thermal History K	Hr	Leaching Treatment	Heat Treatment K	Hr	Crystallinity	Physical Appearance
423 623 773	16 1.5 4	None	973	1/2	Partially crystallized, β-crystobalite	Opalescent to opaque
773	4	Dilute HCl, 1/2 hr at RT	973	1/2	Noncrystalline	Completely transparent, colorless
773	4	Dilute HCl, 1/2 hr at RT	1023	1/2	Noncrystalline	Completely transparent, colorless
773	4	Dilute HCl, 1/2 hr at RT	973	1/2	Noncrystalline	Completely transparent, colorless

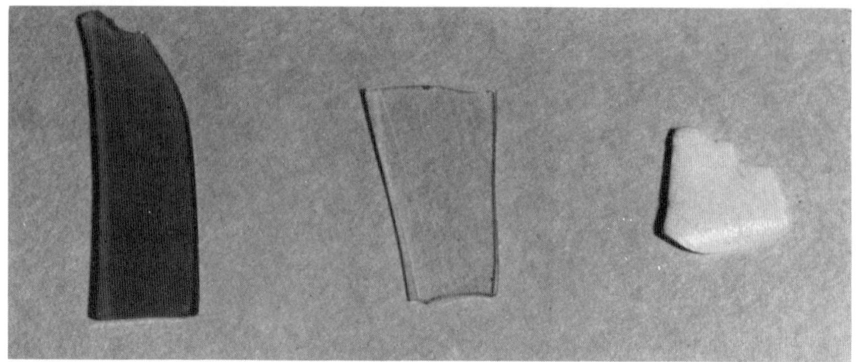

Fig. 4. Composition 2, gel-monoliths at various stages of thermal treatment. (Left to right--after 423K, after 773K, after crystallization at 1123K.)

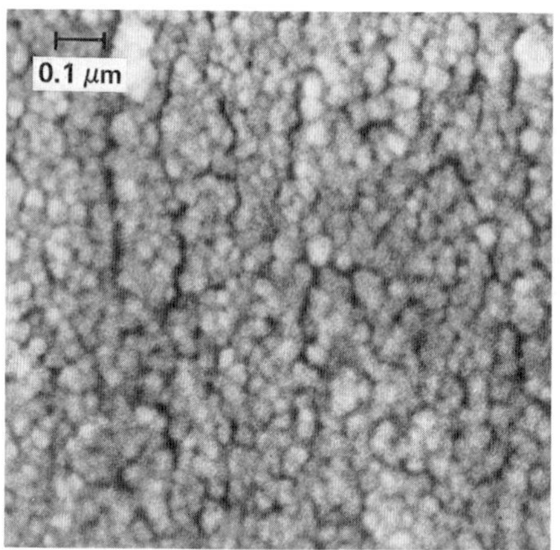

Fig. 5. Scanning electron micrograph of a porous transparent Composition 2, gel-monolith after thermal treatment at 773K for 2 hours. (Magnification 50,000X)

Particulate gels of Composition 1 and monolithic gels of Composition 2 were prepared. The thermal histories and heat treatments for both compositions are given in Table 3 as are DTA and x-ray powder diffraction results. To determine the influence of leaching on crystallization behavior, gel-monoliths leached with 5N HCl for different periods and subsequently heat treated at 973K, were compared with unleached samples. Table 4 shows that the chemical leaching of Na^+ ions reduced the crystallization rate significantly. Heat treated monoliths at different stages of transformation are shown in Fig. 4 and a porous gel-monolith of Composition 2 after thermal treatment at 773K is shown in Fig. 5.

To examine the influence of Na_2O on the crystallization rate, silica gels doped with Na_2O were prepared by hydrolytic polycondensation of $Si(OC_2H_5)_4$ containing Na^+ ions added as an aqueous solution of sodium acetate. The pH of the solution during gelation was ~ 8. After drying in air at 70°C for several days, gels were examined by differential thermal analysis. Results are given in Table 5.

Table 5. Differential Thermal Analysis Results

Composition of Gels, wt percent			Heating Rate During DTA, K/min	Exothermic Peak Position Indicating Crystallization, K
SiO_2	B_2O_3	Na_2O		
60	15	25	10	No peak
84	12	4	10	1073
100	--		10	No peak up to 1673
100			2	Peak at 1623
99.09	0.01		10	No peak up to 1573
99.9	0.1		10	1382
99.9	0.1		2	1223
99.5	0.5		10	1303
99.0	1.0		10	1258

DISCUSSION

Titanium Dioxide Gel System

Our preliminary results indicate that titanium dioxide gels prepared at room temperature are completely noncrystalline and transparent. Thermal treatment at temperatures up to 523K alter the chemical nature of gels due to removal of organics but do not develop crystallinity (Fig. 3b). Crystallinity develops with heat treatment up to 623K (Table 1). The first ordered phase is the anatase phase (Fig. 1, Sample 2). At 773K (Fig. 1, Sample 3), anatase is the only phase evolving. At 1173K, the rutile crystallizing phase evolves (Fig. 1, Sample 4). Hence, the sequence of the solid-state phase transformation is noncrystalline titanium dioxide gel → anatase → rutile. Observing a similar phase transformation sequence during the hydrothermal crystallization of amorphous titanium dioxide prepared by the hydrolysis of $TiCl_4$, Mathews contends the anatase phase nucleates first because it is structurally closer to the reactant and the rutile phase forms later when a sufficient surface anatase phase is available for nucleation.[19]

Mathews attempted to interpret the sequence with the help of the Ostwald Step Rule (based on the molar volume of the reactant and product) in which the preferred reaction gives the product with the highest molar volume with respect to the reactant. In this system, the anatase phase has a considerably higher molar volume than its polymorphs.[19,21]

The product sequence observed in the present study is explained by the Ostwald Step Rule. Consequently, one might anticipate that when noncrystalline titanium dioxide gels having a molar volume higher than that of the anatase phase are heat treated, the first step in the rearrangement will be to overcome a nucleation barrier and to form an intermediate metastable phase, such as anatase, by the structural reorganization of the titanium-oxygen network of disordered structures. Presumably, the removal of structural hydroxyl groups and organic groups chemically bonded to the network during thermal treatment plays a role in the structural reorganization of the network.

It is speculated that the polymeric species in the gels are structurally related to the anatase phase. Consequently, the reorganization induced by thermal treatment forms the nuclei of the anatase phase. However, systematic nucleation and crystallization kinetic studies of gels of different structural types are needed to establish the suggested mechanism.

Na_2O-B_2O_3-SiO_2 System

Results of the crystallization studies of these gels at temperatures above Tg can be summarized as follows:

- Composition 1 gels, having practically no tendency toward crystallization on heat treatment above 773K (Tg), can be transformed easily without crystallization.
- Composition 2 gel-monoliths, having a strong crystallization tendency during transformation, form glass with appropriate thermal treatment above Tg.

- The crystallization rate of Composition 2 gels is controlled by heat treatments above Tg and by aging treatments below Tg.
- The concentration of leachable sodium ions has a strong influence on the crystallization rate.
- DTA of Na_2O-doped SiO_2 gels shows increased crystallization rate with increased Na_2O concentration.

Small quantities of cations adsorbed on silica gels cause crystallization at lower temperatures.[12,13,15] Adsorbed cations entering the network cause the Si-O-Si bridges to rupture, thereby collapsing the structure and lowering the crystallization temperature.

The crystallization of Composition 2 gels is presumably based on the same principle of rupturing the Si-O-Si bonds due to the presence of Na^+ ions. Strong bases, such as caustic soda and caustic potash, are excellent catalysts for crystallizing amorphous silica in hydrothermal conditions.[22] Thus, the mineralizing action on the crystallization of silica gels might be due to the combined effect of the Na^+ ions and hydroxyl ions in the vicinity rather than the Na^+ ions alone. If so, the improved stability toward crystallization by prolonged heat treatment at lower temperatures might be attributed to the removal of hydroxyl groups and the formation of more Si-O-Si bridges.

The effects of acid leaching on crystallization rates of porous gels might be due to the removal of Na^+ ions that are more reactive in the rupture of Si-O-Si bonds, thus leaving those less reactive in rupturing of Si-O-Si bonds. The absence of a crystallization tendency in Composition 1 gels may be a compositional effect; in this system, a viscous, low-melting, soda borosilicate glassy phase might form at lower temperatures. Na^+ ions, being structurally incorporated into the glassy phase, cannot act as a mineralizer.

CONCLUSIONS

Titanium dioxide gels prepared by the hydrolytic polycondensation of titanium iso-propoxide are found to be noncrystalline at the wet stage. Gels remain noncrystalline after thermal treatment up to 523K. Crystallinity develops on thermal treatment at 623K. The sequence of phase transformation is: noncrystalline gels → anatase → rutile. The metastable development of the anatase phase can be explained by the Ostwald Step Rule based on molar volumes of the phases. Presumably, in the first step, titanium dioxide gels having a molar volume greater than that of the anatase phase and being structurally closer to the anatase phase transform to that phase.

The crystallization rates of gels and gel-monoliths in the $Na_2O-B_2O_3SiO_2$ system are controlled by the composition and catalytic activity of Na^+ ions (presumably) associated with OH^- ions in rupturing the Si-O-Si bonds. With appropriate heat treatments above Tg, porous gel-monoliths can be transformed into transparent noncrystalline "glass" without melting.

ACKNOWLEDGMENTS

The author acknowledges the laboratory assistance of J. C. Debsikdar in gel preparation experiments and of A. Skidmore in the electron diffraction experiments.

REFERENCES

1. H. Dislich, Angew. Chem. Int. Ed. (Engl.), 10 [6], 363-70 (1971).
2. K. Kamiya, S. Sakka, and I. Yamanaka, pp. 13-14 in Tenth Inter. Congress on Glass, Part II, edited by M. Kunugi, M. Tashiro, and N. Saga, Ceramic Soc. of Japan, Kyoto, 1974.
3. S. P. Mukherjee, J. Zarzycki, and J. P. Traverse, J. Mater. Sci., 11 [2], 341-55 (1976).
4. S. P. Mukherjee, J. Non-Cryst. Solids, 42, 477-88 (1980).
5. C. J. Brinker and S. P. Mukherjee, J. Mater. Sci., 16, 1980-88 (1981).
6. B. E. Yoldas, J. Appl. Chem. Biotechnol., 23, 803-09 (1973).
7. I. M. Brown and K. S. Mazdiyasni, J. Amer. Ceram. Soc., 53 [11], 590-94 (1970).
8. S. P. Mukherjee and J. C. Debsikdar, presented at the 84th Annual Meeting of Amer. Ceram. Soc., Cincinnati, OH, May 2-5, 1981.
9. H. Murayawa, K. Kobayashi, M. Koishi, and K. Meguro, J. Colloid Interface Sci., 32 [3], 470-76 (1970).
10. S. P. Mukherjee, J. Non-Cryst. Solids, 48, 177-84 (1982).
11. J. Zarzycki, J. Non-Cryst. Solids, 48, 105-16 (1982).
12. C. Mougey, J. Francois-Rossetti, and B. Imelik, pp. 266-84 in Structure and Properties of Porous Materials, edited by D. H. Everett and F. S. Stone, Academic Press, NY, 1958.
13. S. Kondo, F. Fujiwara, and M. Muroya, J. Colloid Interface Sci., 55 [2], 421-30 (1976).
14. P. Cloos, A. J. Leonard, J. P. Moreau, H. Herbillon, and J. J. Fripiat, Clay and Clay Minerals, 17, 279-87 (1969).
15. S. P. Mukherjee and J. Zarzycki, J. Amer. Ceram. Soc., 62, 1-2 (1979).
16. J. Phalippou, M. Prassas, and J. Zarzycki, J. Non-Cryst. Solids, 48, 17-30 (1982).
17. S. P. Mukherjee, pp. 321-31 in Materials Processing in the Reduced Gravity Environment of Space, edited by G. E. Rindone, North-Holland, NY, 1982.

18. D. C. Bradley, R. Gaze, and W. Wardlaw, J. Chem. Soc., 3977 (1955).
19. A. Mathews, Am. Mineral., 61, 419-24 (1976).
20. W. S. Fyfe, F. J. Turner, and J. Verhoogen, Geol. Soc. Am. Mem., 73, 259 (1958).
21. R. A. Robie and D. R. Waldbaum, U.S. Geol. Surv. Bull., 1259, 256 (1968).
22. A. S. Campbell and W. S. Fyfe, Am. Mineral., 45, 464-68 (1960).

BORON NITRIDE FIBER SYNTHESIS FROM BORIC OXIDE PRECURSORS

Arthur E. Lindemanis[a]

Combustion Engineering, Inc.

Stamford, CT

ABSTRACT

The effective application of boron nitride fibers as battery separators or as high modulus fibers requires improving the economics significantly, particularly in the conversion of B_2O_3 fibers to BN with ammonia. The kinetics were studied as a function of fiber diameter and temperature and found to fit a shrinking core model. Low melting point surface phases and a high temperature B-O-N phase also influence the bulk fiber nitriding process.

BORON NITRIDE FIBER APPLICATIONS

Boron nitride fiber exhibits chemical and physical properties that make it uniquely suitable for a number of advanced ceramic applications. In inert atmospheres the material demonstrates thermal stability at 2000°C. In air, extended operation is possible at 800°C. The resistance to attack from acidic or basic media, including molten salts and metals, led to boron nitride felts being developed as electrode separators for the high temperature Li-Al/LiCl-KCl/FeS battery.[1] The two other possible ceramics, MgO and BeO, are not presently available in a fibrous form.[2] With fiber felts porosities exceeding 80-90% can be generated to achieve the required lower cell resistance and higher current densities.

---

[a]Formerly with Kennecott Corp. where this work was supported in part by Argonne National Laboratories under contract No. 31-108-38-5747.

High-strength, high-modulus continuous boron nitride fibers are being developed for radar window composites, utilizing its low electrical conductivity and dielectric constant.[3] Individual samples have exhibited ultimate tensile strengths exceeding 3100 MPa and elastic moduli up to 800 million.[4] Other applications to plastic or metal matrix composites may be possible provided competitively priced BN fiber is available.

The present second generation pilot facility for BN felt produces 26 Kg/week at $550/Kg, partially due to the bench scale being labor intensive. The projected cost for the third generation process is $31/Kg for 100,000 Kg/yr semi-works and $9/Kg for a large scale commercial operation.[4]

MANUFACTURING PROCESS

Since BN vaporizes rather than melts above 2500°C and no suitable solvents have been identified, BN fibers cannot be prepared by traditional ceramic processing techniques. The preparation of BN fibers through the chemical reaction of a precursor B_2O_3 filament with ammonia was disclosed in 1966 by Economy and Anderson.[5] Subsequently, they used the chemical conversion of precursor fibers to also prepare B_4C, NbCN, TiN and other fibers.[6] The process flow sheets for manufacturing BN felt are given in Fig. 1.

The second generation process for manufacturing BN felt (Fig. 1(b)) replaced the earlier version (Fig. 1(a)) for pilot production in 1980. Both utilized 5-20 micron diameter B_2O_3 filaments drawn continuously from a platinum bushing at 800-1000°C. The wound fiber undergoes first stage nitriding with ammonia at temperatures below 1000°C. Temperatures exceeding 1800°C are then used to stabilize the fiber by removing residual oxygen and modifying the crystal structure. For the high modulus fibers, the "hot stretch" stabilization step is analogous to that utilized for carbon fibers. Felt is formed from chopped BN fiber mixed with approximately 15% B_2O_3 fiber which provides bonding nodes by fusion at 500°C. The B_2O_3 nodes require repeating the nitriding step. The use of continuous Fourdrinier paper molding machine and hot calender rolls represent the most important change. The improved physical properties for the Fourdrinier felt will be reported in a future paper.

The third generation process (Fig. 1(c)) represents a significant simplification into three unit operations:
(1) Felt Formation--Spun B_2O_3 fiber is air laid directly into a fiber matte.
(2) Felt Bonding--Needling the B_2O_3 (or BN) felt with optional hot rolling provides bonding.
(3) Nitriding--B_2O_3 fiber is converted to BN in a single nitriding and stabilization step.

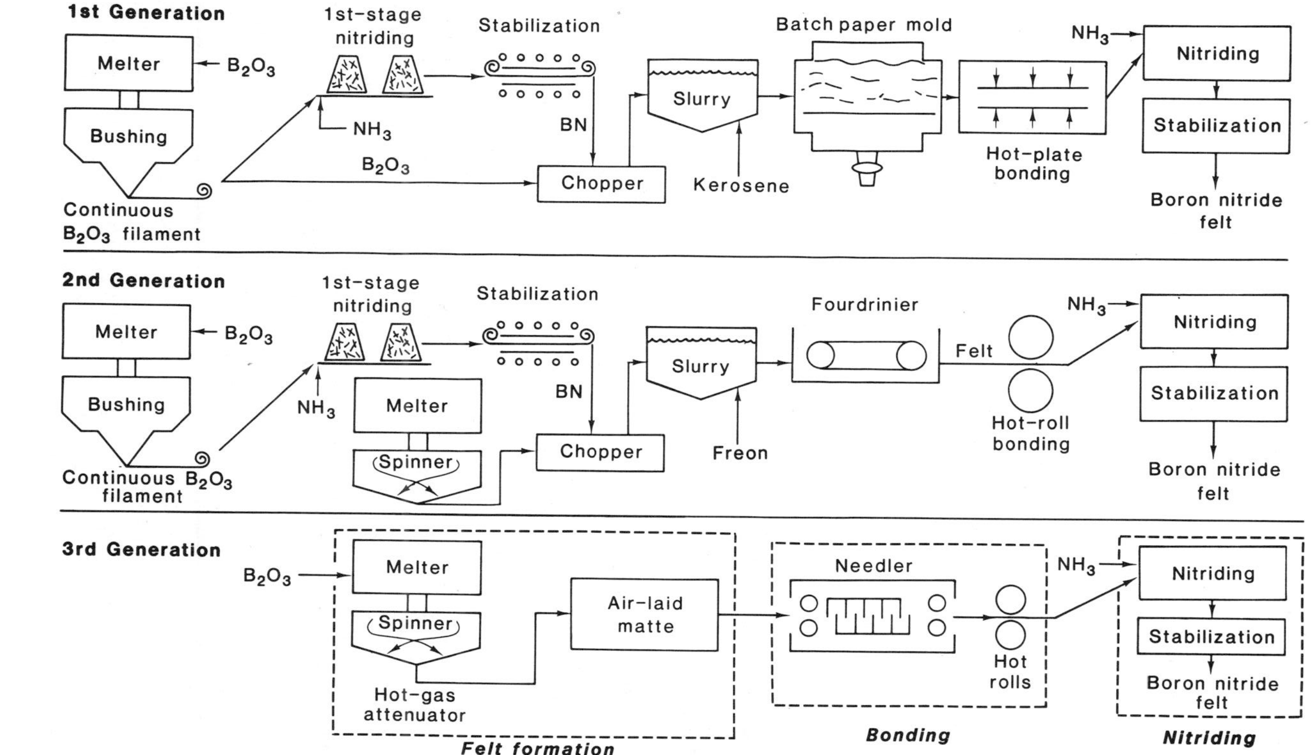

Fig. 1. Manufacturing schematics for BN fiber felts: (A) first; (B) second; and (C) third generation process flowsheets.

While the parameters needed further optimization, the rotary fiberizer, fiber collection train, and needler were demonstrated during 1980-81 at the semi-works scale (100,000 Kg/yr). The nitriding operation became the critical step in achieving targeted cost reductions. Significant improvements were required over previous nitriding practices: eliminating multiple nitriding steps, reducing ammonia consumption 3 orders of magnitude through recycling and more efficient reactor design, decreasing the 72 hour nitriding time 1 order of magnitude and increasing the reactor volume utilization efficiency 2 orders of magnitude. A formal analysis of previous nitriding process data and additional laboratory experiments showed significantly improved nitriding performance is feasible.

NITRIDING THERMODYNAMICS

Conditions favorable for the conversion of B_2O_3 to BN can be evaluated in terms of the following equations:

$$B_2O_3\ (s,l) + 3H_2(g) + N_2(g) \rightarrow 2BN(s) + 3H_2O(g) \qquad (1)$$

$$B_2O_3\ (s,l) + 2NH_3(g) \rightarrow 2BN(s) + 3H_2O(g) \qquad (2)$$

$$2NH_3(g) \rightarrow N_2(g) + 3H_2(g) \qquad (3)$$

using available thermodynamic values.[7]

The reaction for B_2O_3 with molecular nitrigen and hydrogen (Eq. 1) is unfavorable given positive standard free energies which require that the H_2O partial pressure be kept below 5×10^{-3} atm in order to maintain a driving force for the nitriding reaction. This also argues against mechanisms in which H_2 and N_2 from the dissociation of NH_3 (Eq. 3) are the reactive species.

The reaction of NH_3 directly with B_2O_3 (Eq. 2) is much more favorable given a standard free energy which decreases from +88.3 Kj/mole at 25°C to -40.2 Kj/mole at 1000°C. However, ammonia possesses a positive free energy above 181°C and tends to dissociate (Eq. 3). In a closed system, assuming both reactions 2 and 3 go to completion, the pNH_3 becomes less than 0.002 atm above 400°C and the equilibrium pH_2O is less than 0.003-0.005 atm. Figure 2 presents the calculated equilibrium H_2O pressures and the stable phase regions.

Fortunately, as discussed later, the NH_3 dissociation rate is significantly lower than the conversion of B_2O_3 to BN. One can therefore assume that only the nitriding reaction (Eq. 2) proceeds to equilibrium and that the initial pNH_3 is set by the bulk gas composition. The resulting partial equilibrium calculations show that considerably higher pH_2O levels can be tolerated than previously calculated for complete equilibrium. The pH_2O increases from 0.0053 to

0.390 atm with no NH_3 dissociation at 527°C. As also seen in Fig. 2, at 0.1 atm pNH_3, which represents 82% dissociation, the maximum tolerable pH_2O is still 1 order of magnitude higher. Off-gas analyses found H_2O exceeding several percent as predicted for both laboratory packed fiber bed experiments and commercial boron nitride powder manufacturing operations. This impacts the reactor design by reducing the flow rates required and increasing the fraction of dissociation NH_3 which can be recycled.

Appreciable NH_3 losses will not occur without catalytic surfaces. Homogeneous gas phase decomposition is negligible at 1000°C. The rate extrapolated from shock tube data[8,9] would be 8.9×10^{-5} $\%NH_3$/sec for a 10 cm/sec velocity. While ammonia reactions have been studied extensively for metal-promoted catalyst systems, limited information is available for ceramic surfaces. Stable oxide coatings such as Al_2O_3 and SiO_2 do inhibit decomposition on metals.[10] The specific surface reactivity cannot be determined from existing studies[11,12] due to experimental limitations.

Preliminary laboratory experiments found BN surfaces the least catalytic. Utilizing a perpendicular impinging jet geometry to achieve high mass transfer coefficients, the rate constant measured at 800°C was less than 5×10^{-6} mol/cm^2/sec. This is consistent with the low H_2, N_2 concentrations in the off-gas from packed bed experiments in BN-coated quartz reactors in comparison to earlier Inconel reactors.

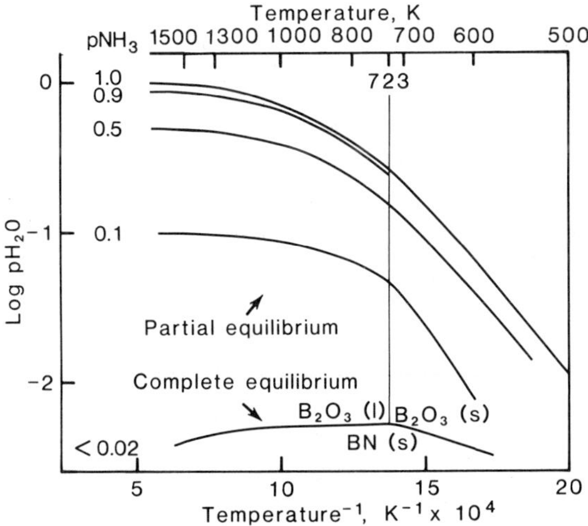

Fig. 2. Partial water pressure for B_2O_3-BN phase equilibria at various levels of NH_3 dissociation.

NITRIDING KINETICS

Individual Fibers

The role of fiber diameter, retorting temperature and time can be established for the reactor design by analyzing the individual fiber nitriding kinetics. The present analysis of existing thermogravimetric analysis data[13] for drawn B_2O_3 fiber in pure ammonia encompasses 3 fiber diameters, 5.3, 8.0, and 10.5 microns, and temperatures from 380° to 860°C. The isothermal data from 380° to 420°C show 30 to 50% conversions within 30 minutes. The two higher temperatures, 610° and 850°C, were done in stages to provide a stable BN sheath which contains the molten B_2O_3 core above 450°C. Conversions exceeding 90% were observed in less than one hour, contrasting markedly with the present 72 hour nitriding schedule. This justified further examination using models based on a shrinking unreacted core with a porous product layer. Relationships for the fraction of B_2O_3 converted to BN, X, to time, t, for a cylindrical particle with radius, r, have been derived.[13] Ignoring mixed control, two rate equations are considered:

Mass Transport Control

$$k_m r^{-2} t = X + (1 - X)\ln(1 - X) \qquad (4)$$

Chemical Reaction

$$k_r r^{-1} t = 1 - (1 - X)^{1/2} \qquad (5)$$

Over the entire nitriding interval, including the initial rates where X is small, quite different functional dependences are seen for the two cases. One can therefore proceed to evaluate graphically which mechanism is rate controlling. The experimental data for three different fiber diameters at 400°C are presented in Fig. 2. The functional fit is good for mass transport but not for chemical reaction control. The rate constants for mass transport control, k_m, calculated from the slopes are in good agreement at 0.170 ± 0.011 μm^2/min. Figure 3 shows the good linearity observed with the mass transport model from 420° to 860°C for both 5.3 and 8.0 micron fibers. The rate of nitriding at the higher temperatures was assumed to be unaffected by any initial conversion at lower temperatures. Although not shown, chemical control did not give a good fit. The temperature dependence for k_m is shown in Fig. 5 is given by:

$$k_m = -4.25 \times 10^3 \, T^{-1} + 4.375 \qquad (\mu m^2/\text{min}) \qquad (6)$$

Using this the calculated nitriding times for 90% conversion at 800°C for 3 and 10 micron fibers are 0.07 and 0.74 hours respectively.

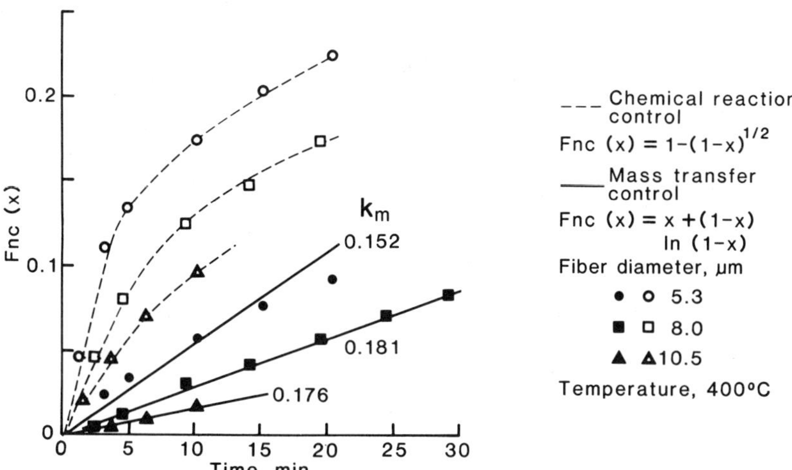

Fig. 3. Comparison between mass transport and chemical reaction control for shrinking core analysis of B_2O_3 fiber conversion to BN.

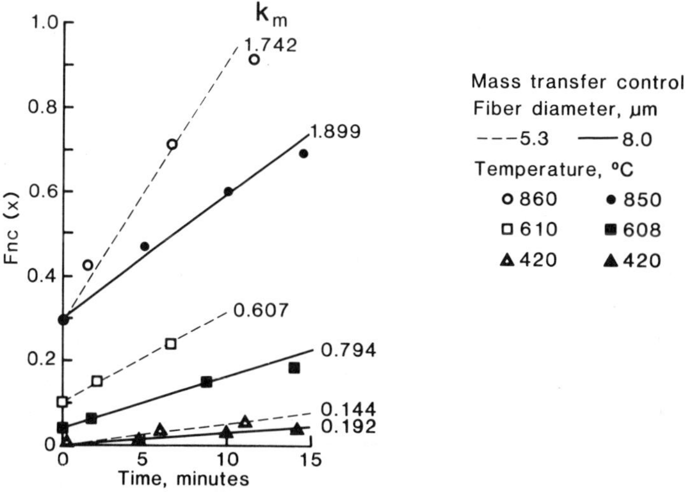

Fig. 4. Conversion analysis with the shrinking core model for different fiber diameters and temperatures.

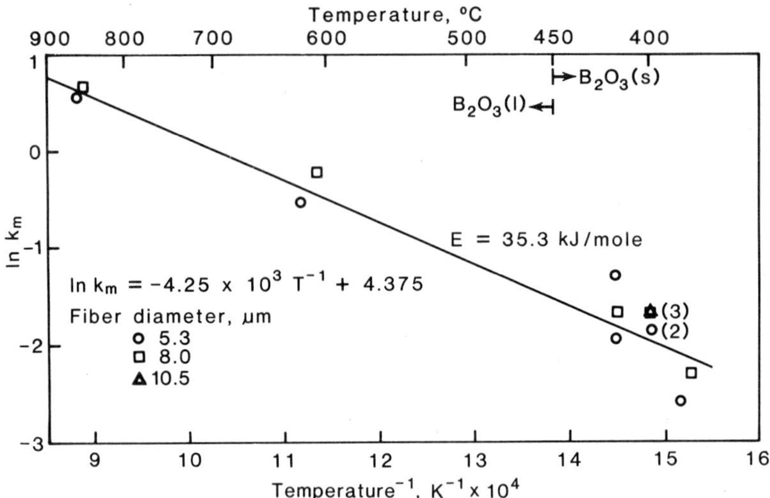

Fig. 5. Rate constant for the shrinking core analysis as a function of temperature.

The 35.3 Kj/mole activation energy, E, is consistent with a mass transfer mechanism. It is too low for either an interfacial reaction or solid state diffusion. The value is slightly high for gaseous diffusion. Possible diffusion mechanisms can be indicated by calculating the effective diffusivity D_e from the measured k_m using the expanded version of Eq. 4:

$$\frac{K_2}{1+K_2}\left\{C^o_{NH_3} - \frac{C^o_{H_2O}}{K_2}\right\}\frac{4D_e}{\rho r^2}\, t = X + (1 - X)\ln(1 - X) \qquad (7)$$

where K_2 is the reaction equilibrium constant; C^o, initial molar concentration; ρ, B_2O_3 molar specific volume. One obtains approximately 10^{-6} cm^2/sec. This D_e is orders of magnitude smaller than the 10^{-2} and 10^{-3} cm^2/sec calculated for large pore and Knudson diffusion using available relationships[14] with reasonable assumptions. Although high resolution work was not attempted, earlier scanning and transmission electron microscopy had not shown pores in the fibers except as gross defects. Given the several hundred angstrom crystallite sizes and two dimensional graphite-like layered structure reported for BN fibers,[5,15] "interactive" diffusion parallel to these layers or between crystallites would explain the low D_e and high E. For fine pore (4-5A) xeolites, Satterfield[16] reported "interactive" diffusivities as low as 10^{-12} cm^2/sec with 12-46 Kj/mole activation energies. Additional microstructural analysis seems worthwhile especially for conversions above 90% where B-O-N phases formed resist "stabilization" at 2200°C.

Gas Phase Mass Transfer

In nitriding either individual fibers or homogeneous packed fiber beds, gas mass transport resistance does not appear to be a significant factor. For a 3 micron fiber at 400°C with 1 cm/sec average space velocity, the estimated gas mass transfer coefficient, k_c, is 7.1×10^3 cm/sec from established correlations to the Peclet number for spherical geometries.[17] Since the effective boundary layer thickness can be shown to approach the particle radius for small Peclet numbers, one can conclude that a packed fiber bed will behave as independent fibers for porosities greater than approximately 75%. The effect of relative packing densities is reflected in the greater ease in nitriding felts 80-95% porous compared to continuous B_2O_3 fiber "bundles" typically 40-50% porous. Bushing drawn sliver, roving, high modulus fiber, and felt fibers displaced by needling perpendicular to the web approach one dimensional packing. Unlike spheres with a 0.741 close packing density, cylinders can exhibit densities as high as 0.907 in parallel packing.

While limited correlations are available for packed beds with particles less than 500 microns or porosities greater than 0.7,[17] one estimates the k_c to be substantially greater than 1 cm/sec. At flow rates avoiding starvation or stagnation, this is sufficient to prevent local accumulation of gaseous products in a homogeneously packed bed. Fiber clusters would present difficulties only for reaction times much less than 72 hours.

FIBER FUSION

The rapid nitriding predicted by individual fiber kinetics can be inhibited by premature fiber fusion and sintering. Hot stage microscope experiments confirmed that partially hydrolyzed fibers are covered with a low melting point surface phase which causes premature fiber agglomeration. Node formation between adjacent fibers was observed at 150°C for drawn B_2O_3 sliver. This temperature is substantially below the 450°C melting point for pure B_2O_3. With less fiber-fiber contact, the nodes for felts were observed at 230°C. Above 400°C, softening of the fibers resulted in additional nodes. Heating under a 4.0×10^{-3} kPa vacuum prevented fiber fusion until about 420°C. X-ray analysis confirmed the presence of orthoboric acid, H_3BO_3. SEM micrographs of fiber cross sections showed a distinct surface layer. Assuming H_3BO_3, the thicknesses correlated reasonably with the measured water contents of several percent.

These observations are consistent with the literature[18-20] where H_3BO_3 is reported to lose water and transform at around 130°C to phases of metaboric acid, HBO_2, having 176-236°C melting points. Investigators have found that heating above 150°C can yield metastable viscous liquid phases. The details of the H_3BO_3 dehydration behavior depend on the heating rates, gas composition, and other experimental conditions. Directly dissociating the acid into amorphous B_2O_3 and steam in a one step process was reported for a 1.3 kPa vacuum at 96-98°C.[21]

Given that the water pressure above H_3BO_3 is 0.048 kPa at 25°C, normal countermeasures such as controlling the humidity or applying protective sizing coatings were not as effective as desired. Concepts for the commercial nitriding process needed to be designed to deal with partially hydrolyzed fiber. This is especially true for the more densely packed drawn fibers. Prolonged prenitriding with hot dry gas or vacuum dehydration are possibilities.

CONCLUSIONS

In the next generation process for BN fiber felts, the nitriding operation represents the greatest opportunity to achieve cost reductions and productivity increases. The kinetic analysis demonstrates that significant reductions are possible in the presently practiced nitriding schedule. The critical step is low temperature dehydration of partially hydrolyzed fibers to prevent surface fusion and agglomeration. Several options are available to do this. Except for this initial step, the reaction sensitivity to H_2O was shown to be considerably less than previously believed. Ammonia losses can be minimized by selecting appropriate reactor materials and flow conditions. Sufficient data exists to design a semi-works scale reactor with reasonable confidence.

ACKNOWLEDGMENTS

Assistance with the nitriding and fiber fusion experiments from Drs. Michael Redman and George Rainville and other staff at Kennecott Corp. is gratefully acknowledged. Technical discussions with Prof. John Elliott, MIT, and Dr. Jacob Mu, Occidental Research Corp., were very helpful.

REFERENCES

1. Development of Boron Nitride Felt Separators for Lithium/Metal Sulfide Batteries, Argonne National Laboratories Contract No. 31-109-38-5747 with Kennecott Corp., 1980.
2. R. B. Swaroop, J. E. Battles, and R. S. Hamilton, 6th Inter-American Conf. on Mat. Tech., San Francisco, CA, August 1980.
3. Development of High Strength, High Modulus Continuous Boron Nitride Fiber, Naval Surface Weapons Center/NASA Langley Contract N60921-82-C0015 with Kennecott Corp., 1981.
4. Priv. communication, F. Tompkins, Carborundum Co., Niagara Falls, NY.
5. J. Economy and R. V. Anderson, Inorg. Chem., 5 [6], 989-92 (1966).
6. J. Economy, Chem. Tech., 240-6 (April 1980).
7. JNANAF Thermochemical Tables, 2nd Ed., U.S. National Bureau of Standards, Washington, D.C., 1972.
8. M. C. Branch and R. F. Sawyer, AFOSR Scientific Report AFOSR-TR-72-0200, June 1971.
9. D. L. Johnson, Disser. Abstr., B27, 1898 (1966).
10. H. Carube, Trans. Steel Inst. Jap., 14 [6], 404-10 (1974).
11. N. Kelvin, (Ph.D. Thesis), Yale University, 1969.
12. J. Voelter and M. Schoen, Z. Anorg. Allgem. Chem., 322, 213-14 (1963).
13. J. Economy and R. Anderson, U.S. Patent No. 3,429,722 (1969) and U.S. Patent No. 3,62,780 (1971).
14. J. Szekely, J. W. Evans, and H. Y. Sohu, Gas-Solid Reactions, Academic Press, NY, 1976.
15. J. Thomas, N. E. Weston, and T. E. O'Connor, J. Am. Chem. Soc., 84, 4619-22 (1962).
16. C. N. Satterfield and W. G. Margetts, A.I.Ch.E. J., 17, 295-99 (1971).
17. T. K. Sherwood, R. L. Pigford, and C. R. Wilke, Mass Transfer, McGraw-Hill, NY, 1975.
18. N. J. Croissant and G. Garanud, J. Therm. Anal., 5 [5-6], 577-97 (1973).
19. J. Haladjain and G. Carpeni, Bull. Soc. Chim. France, 1679-82 (1956).
20. M. J. Lecomte, C. R. Acad. Sc. Paris Ser. A, B, 264B [3], 235-38 (1967).
21. V. V. Urusov, Doklady Akad. Nauk vs. S.S.R., 116, 97-100 (1957).

PART II

NOVEL POWDER-FORMING AND POWDER-PROCESSING METHODS

SOME COMMON ASPECTS OF THE FORMATION OF NONOXIDE POWDERS BY THE VAPOR REACTION METHOD

Akio Kato, Junichi Hojo and Takanori Watari

Department of Applied Chemistry, Faculty of Engineering
Kyushu University 36
Fukuoka, 812 Japan

INTRODUCTION

As technology has advanced, needs for fine homogeneous powders have increased. Among many new powder preparation techniques, the vapor reaction method is receiving increasing attention as an important technique for producing powders of refractory oxides and nonoxides. Vapor-phase techniques are noted for their: (1) high purity powder products; (2) highly discrete particles or unaggregated powders; (3) ease of preparation of ultrafine powders with narrow particle size distribution; and (4) versatility for the direct preparation of metals, oxides and nonoxides.

The present authors have investigated the formation of ultrafine powders of several refractory carbides and nitrides using the vapor reaction method. In the present paper, the formation processes of the powders of several carbides and nitrides are discussed with emphasis on the similarities and general trends existing among them.

POWDER FORMATION BY VAPOR PHASE REACTION

The type of solid deposit produced from the vapor phase varies, depending on the reaction systems and the reaction conditions, from particles formed by homogeneous nucleation in the gas phase to films, whiskers or bulk crystals grown on a solid surface via heterogeneous nucleation. Homogeneous nucleation requires a higher degree of supersaturation than heterogeneous nucleation. The nucleation rate of crystals from the vapor phase may be treated formally by the same equations as the nucleation of liquid drops from a supersaturated vapor. The nucleation rate of liquid drops (I) is given by Eq. 1.[1]

$$I = I_0\sigma^{1/2}(P/T)^2\exp(-16\pi\sigma^3\nu^2/3kT(kT\ln P/P_0)^2) \quad (1)$$

where σ and P are the surface tension and the vapor pressure of the liquid, respectively, ν the volume per molecule in the liquid and P_0 the equilibrium vapor pressure of the liquid at temperature T. The nucleation rate is an exceedingly steep function of the supersaturation ratio (SS). SS can be derived in terms of thermodynamic quantities. For a general deposition reaction, Eq. 2, SS is given by Eq. 3.[2]

$$aA(g) + bB(g) = cC(s) + dD(g) \quad (2)$$

$$SS = (P_A^a\, P_B^b/P_D^d)_{react}/(P_A^a\, P_B^b/P_D^d)_{equil} = K_2(P_A^a\, P_B^b/P_D^d)_{react} \quad (3)$$

K_2 is the equilibrium constant for the reaction in Eq. 2. SS given by Eq. 3 is the overall supersaturation ratio for the deposition. The deposition of a solid by a vapor phase reaction consists of several consecutive elementary processes: reaction in a vapor phase, vapor phase diffusion, adsorption, surface reaction, desorption, etc. The driving force for the deposition, i.e., SS, is generally divided into each elementary process. When homogeneous nucleation is the rate determining step for the deposition, SS given by Eq. 3 has a maximum value.

Equilibrium Constant and Powder Formation

The initial requirement for the production of a fine powder is the achievement of a high degree of supersaturation in order to produce a large number of nuclei by homogeneous nucleation. Since the supersaturation ratio is proportional to the equilibrium constant for the deposition reaction as shown above, the value of the equilibrium constant should be related to the formation of powder by the vapor reaction method. The equilibrium constants for the formation of several nonoxides by vapor phase reactions and the results of the formation of the powders observed by the present authors are summarized in Table 1. In accordance with the above considerations, one can observe that the formation of powders occurs only when the equilibrium constants are large. The same results were observed for the formation of oxide powders by vapor phase reactions.[3] In those cases where the equilibrium constants are small and powders were not formed, the depositions occurred on the reactor wall. Such reaction systems with a low degree of supersaturation are suitable for whisker growth.[4]

Table 1. Equilibrium constants for the formation of nonoxides from gaseous systems and powder formation.

No.	Reaction	Equilibrium constant ($\log K_p$) 1000°C	1500°C	Formation of powder at ≤1500°C [a)]
1	$SiCl_4 + 2H_2 + \frac{4}{6}N_2 = \frac{1}{3}Si_3N_4 + 4HCl$	1.1	1.4	×
2	$SiCl_4 + \frac{4}{3}NH_3 = \frac{1}{3}Si_3N_4 + 4HCl$	6.3	7.5	○
3	$SiH_4 + \frac{4}{3}NH_3 = \frac{1}{3}Si_3N_4 + 4H_2$	15.7	13.5	○
4	$SiCl_4 + CH_4 = SiC + 4HCl$	1.3	4.7	×
5	$CH_3SiCl_3 = SiC + 3HCl$	4.5	(6.3)	×
6	$SiH_4 + CH_4 = SiC + 4H_2$	10.7	10.7	○
7	$(CH_3)_4Si = SiC + 3CH_4$	11.1	10.8	○
8	$TiCl_4 + 2H_2 + \frac{1}{2}N_2 = TiN + 4HCl$	0.7	1.2	×
9	$TiCl_4 + NH_3 + \frac{1}{2}H_2 = TiN + 4HCl$	4.5	5.8	○
10	$TiCl_4 + CH_4 = TiC + 4HCl$	0.7	4.1	×
11	$TiI_4 + CH_4 = TiC + 4HI$	0.8	4.2	○
12	$TiI_4 + \frac{1}{2}C_2H_4 + H_2 = TiC + 4HI$	1.6	3.8	○
13	$ZrCl_4 + 2H_2 + \frac{1}{2}N_2 = ZrN + 4HCl$	-2.7	-1.2	×
14	$ZrCl_4 + NH_3 + \frac{1}{2}H_2 = ZrN + 4HCl$	1.2	3.3	○
15	$ZrCl_4 + CH_4 = ZrC + 4HCl$	-3.3	1.2	×
16	$NbCl_4 + NH_3 + \frac{1}{2}H_2 = NbN + 4HCl$	(8.2)		○
17	$NbCl_4 + \frac{1}{3}SnCl_4 + \frac{8}{3}H_2 = \frac{1}{3}Nb_3Sn + \frac{16}{3}HCl$	>0.3		○
18	$MoCl_5 + \frac{1}{2}CH_4 + \frac{3}{2}H_2 = \frac{1}{2}Mo_2C + 5HCl$	19.7	(18.1)	○
19	$MoO_3 + \frac{1}{2}CH_4 + 2H_2 = \frac{1}{2}Mo_2C + 3H_2O$	11.0	(8.0)	○
20	$WCl_6 + CH_4 + H_2 = WC + 6HCl$	22.5	22.0	○

a) ○ : Powders were formed; × : Powders were not formed.

Control of Particle Size

The control of particle size is of prime importance in the production of powders. As the system which produces powders by vapor phase reaction has a large equilibrium constant, the metal-carrying reactant can in practice be completely convertedinto the products by adjusting the other reactant until it is in excess. In this case, the following relation holds.

$$d = \left(\frac{6}{\pi}\frac{C_0 M}{N\rho}\right)^{1/3} \qquad (4)$$

where d is the diameter of particle produced, C_0 and N the number of moles of metal source and the number of particles formed per unit volume of reacting gases, respectively, M the molar weight of product per one mole of metal and ρ the density of the product. The particle size is determined by the ratio of the concentration of metal-carrying reactant to the number of nuclei formed. The homogeneous nucleation rate (R_N) is a function of temperature and composition of reacting gases as seen in Eq. 1. In the simple condensation of a vapor, the change of R_N with temperature arises from the exponential term in Eq. 1. On the other hand, in the vapor reaction process in which the growth species or the precursors are formed by chemical reactions the temperature dependence of P^2-term in the pre-exponential factor in Eq. 1 should also be considered in addition to the exponential term. The effect of the consumption of reactants may be exceedingly larger on R_N than on growth rate, because the former is very sensitive to the supersaturation ratio which decreases rapidly as the reaction proceeds. In addition, when particles are once formed in a vapor phase, the deposition on the surfaces is more favorable thermodynamically than the formation of new nuclei. The reason why a vapor reaction method gives a powder with a uniform particle size may be that the nucleation takes place predominantly in the early stage of the reaction and stops in practice in the later stage.

As mentioned above, the size of the particles produced by a vapor phase reaction can be controlled by the reaction temperature and the composition of the reacting gases. This is illustrated later for the formation of the powders of carbides and nitrides. An illustration of particle-size control during the formation of oxide powders is given elsewhere.[5]

PREPARATION AND PROPERTIES OF POWDERS

Reactor for Powder Formation

The present authors have studied the formations of powders of about ten nonoxides using the reaction systems shown in Table 1.

Each reaction was carried out by a flow method using a horizontal alumina reactor. Two typical reactors used are illustrated in Fig. 1. Metal-carrying reactant which is liquid and has a considerable vapor pressure below 100°C was carried into the reactor (type I) from the evaporator by using a carrier gas. When metal-carrying reactants were less volatile, they were sublimed or evaporated, or prepared by the halogenation of metal in the reactor using that of type II.

Formation Processes and Properties of Powders

The reaction temperatures and properties of the nitride and carbide powders produced are summarized in Table 2. The particle formation processes can be classified into the following three types:

(A) Formation of particles of reactants and their thermal decomposition into nitrides or carbides.
(B) Formation of nuclei of nitrides or carbides and their growth into particles.
(C) Formation of metal particles and their nitridation or carburization.

The formation process changes with reaction conditions, mainly with reaction temperature, even in the same system. Furthermore, the method of introduction of the reactant gases into the reaction zone significantly influences the formation process and properties of the resulting powders, as described below.

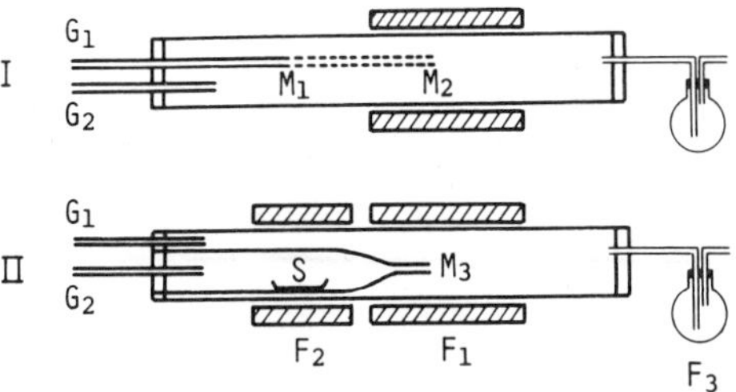

Fig. 1. Reactor for powder preparation by vapor reaction method G_1, G_2: inlet tube for reactant gases; S: metal chloride or metal source; F_1, F_2: electric furnace; F_3: powder collecting flask. Reactor: alumina tube (I.D. 22 mm).

Table 2. Preparation of Fine Powders of Nitrides and Carbides

	Reaction system	Reaction temp. (°C)	Product	Particle size (μm)
Nitrides	$SiCl_4$-NH_3	1000-1500	SiN_aH_b[a)]	0.01-0.1
	$TiCl_4$-NH_3	600-1500	TiN	0.01-0.4
	$ZrCl_4$-NH_3	1000-1500	ZrN	<0.1
	VCl_4 -NH_3	700-1200	VN	0.01-0.1
Carbides	$(CH_3)_4Si$	900-1400	SiC	0.01-0.2
	CH_3SiCl_3	plasma	SiC	<0.03
	SiH_4 -CH_4	1300-1400	SiC	0.01-0.1
	$TiCl_4$-CH_4	plasma	TiC	0.01-0.2
	TiI_4 -CH_4	1200-1400	TiC	0.01-0.15
	$MoCl_4$-CH_4	1200-1400	Mo_2C	0.02-0.4
	WCl_6 -CH_4	1300-1400	WC	0.02-0.3

a) Non-crystalline compound containing excess N and H.

Nitrides. The formation of nitrides from gaseous metal halides and ammonia have large equilibrium constants, and fine nitride powders can be produced by their vapor phase reaction at relatively low temperatures. TiN powders are obtained from the $TiCl_4$-NH_3 system at 700-1500°C.[6] The reactor used is type I shown in Fig. 1. In this case, the particle formation process changes with the mixing temperature of the $TiCl_4$-N_2 and NH_3-H_2 streams (M_1 or M_2 in Fig 1) as follows:

Mixing temp. $\leq$ 250°C ------ Process (A)

$$TiCl_4(g) + nHN_3(g) \xrightarrow{< 250^\circ C} TiCl_4 \cdot nNH_3(\text{particle})$$

$$TiCl_4 \cdot nNH_3(\text{particle}) \xrightarrow{> \sim 500^\circ C} TiN(\text{particle})$$
(n = 2 at 200°C)

Mixing temp. $\geq$ 600°C -------- Process(B)

$$TiCl_4(g) + NH_3(g) + H_2(g) \xrightarrow{> 600^\circ C} TiN \text{ nucleus} \longrightarrow TiN \text{ particle}$$

When the mixing temperature is below 250°C, the process consists of the initial formation of $TiCl_4$-NH_3 particles and their subsequent thermal decomposition into TiN in the high temperature zone. On the other hand, when the mixing temperature is above 600°C, TiN nuclei are formed by the vapor phase reaction of $TiCl_4$ with NH_3 and grow into TiN particles. Since metal halides are Lewis acids and NH_3 is a

Lewis base, their compounds are easily formed. The change in the formation process with the mixing temperature of the reactants may be a common feature in the formation of fine nitride powders in the metal halide-NH_3 system. Electron micrographs of TiN powders produced from $TiCl_4$-NH_3 system are shown in Fig. 2. TiN powders produced by process (A) consist of spherical polycrystalline particles and have a wide distribution of particle size from 0.01 to 0.4 μm. A surface area measurement showed their high porosity. On the other hand, TiN powders produced by process (B) consist of single-crystal particles finer than those by process (A). In the case of process (B), the particle size can be controlled via the reaction conditions (mainly with reaction temperature) as seen in Fig. 3.

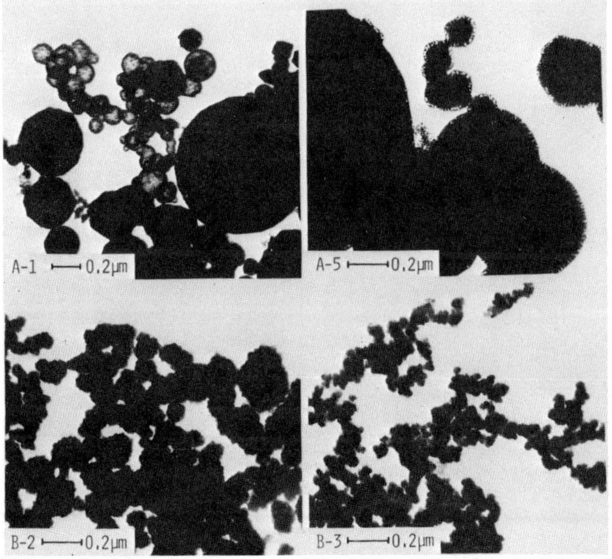

Fig. 2. TiN particles. Reaction temperature: A-1, 1100°C; A-5, 1500°C; B-2, 900°C; B-3, 1100°C. Mixing temperature of $TiCl_4$ and NH_3: A series, ca 200°C; B series, equal to reaction temperature. Gas composition: $TiCl_4$ = 2.1%; NH_3 = 19%; H_2 = 19% (balance: N_2).

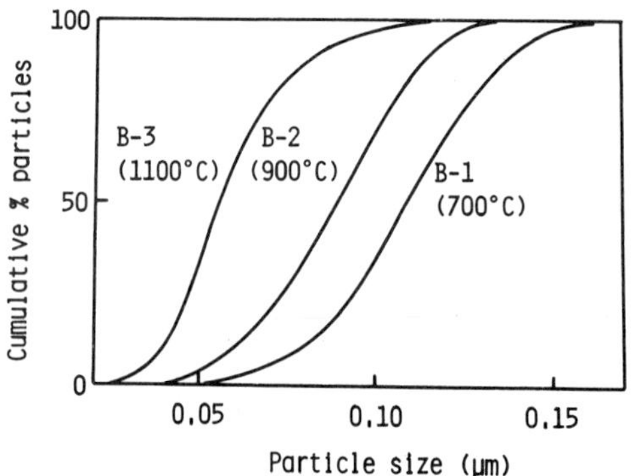

Fig. 3. Particle size distributions of TiN powders. Reaction conditions, see Fig. 2.

The reaction occurring in the $ZrCl_4$-NH_3 system at 1000-1500°C gives powder products consisting of ZrN with particle size below 0.1 μm and hydrolyzable intermediate compounds.[7] The content of ZrN increases remarkably with an increase in the mixing temperature of $ZrCl_4$ and NH_3 streams beyond 1000°C or with an increase in the reaction temperature. This effect of the mixing temperature is also interpreted by the change in the formation process of ZrN particles from (A) to (B) with increasing mixing temperature. The vapor phase reaction occurring in the VCl_4-NH_3 system at 700-1200°C gives VN powders with a particle size below 0.1 μm. When the mixing temperature of the VCl_4 and NH_3 streams is above 400°C, VN particles are formed by process (B). The vapor phase reaction in the $SiCl_4$-NH_3 system at 1000-1500°C gives powder products with a particle size below 0.1 μm.[9] However, the products are not pure Si_3N_4 but non-crystalline compounds containing excess N and H. The non-crystalline products crystallize into α-Si_3N_4 by heat treatment at 1400°C. Prochazka et al. have obtained similar results in the vapor phase reaction of the SiH_4-NH_3 system at 500-900°C.[10] The particle formation process in these reaction systems may be classified into process (A).

Carbides. Although most of the reactions which form carbides from gaseous metal compounds and hydrocarbons have small equilibrium constants at low temperatures, these constants increase sharply with increasing temperature, as seen in Table 1. This indicates that the formation of carbide powders from these systems are more favorable at higher temperatures. Therefore, the application of plasma or arc processes to the powder production is widespread.[11] Although these reaction systems have large equilibrium constants, powders of carbides are obtained by the vapor reaction method at temperatures below 1500°C.

Fine powders of β-SiC are obtained by the vapor phase reaction of SiH_4-CH_4 system at about 1400°C[12] and by the thermal decomposition of $(CH_3)_4Si$ in hydrogen above 1000°C.[13] The formation processes of SiC particles in these reaction systems are summarized as follows:

a. $SiH_4(g) + CH_4(g) \longrightarrow SiC(s) + 4H_2(g)$

$$\text{Process(C)} \begin{cases} SiH_4(g) \xrightarrow{500\sim700^\circ C} Si(\text{particle}) \\ Si(\text{particle}) + CH_4(g) \xrightarrow{\geq 1000^\circ C} SiC(\text{particle}) \end{cases}$$

b. $(CH_3)_4Si(g) \longrightarrow SiC(s) + 3CH_4(g)$

$$\text{Process(A)} \begin{cases} (CH_3)_4Si(g) \xrightarrow{>700^\circ C} \text{Polymer particle } (\ell \text{ or } s) \\ \text{Polymer particle} \xrightarrow{>900^\circ C} SiC(\text{particle}) \end{cases}$$

In the SiH_4-CH_4 system, the carburization of Si with CH_4 becomes observable above 1000°C and gives almost pure SiC at about 1400°C. The Si particles are single crystals, but SiC particles are polycrystalline. The structure of the SiC particles produced changes with the flow rate of the reactant gas mixture, viz., the heating rate of reactants. When the flow rate is small, hollow SiC particles are formed. The formation of hollow SiC particles shows that the carburization of Si particles by CH_4 occurs by the out diffusion of Si through the SiC layer formed.[14] When the gas flow rate is large, solid SiC particles are formed. This may be due to the rapid carburization of Si particles which induces cracks in the SiC layers and the inward diffusion of CH_4 becomes possible. SiC particles produced by the pyrolysis of $(CH_3)_4Si$ are spherical and polycrystalline. In the two reaction systems described above, the size of the SiC particles can be controlled by the reaction temperature, the concentrations of reactants and the gas flow rate, as illustrated in Fig. 4.

The formation of WC the in WCl_6-CH_4-H_2 system has a large equilibrium constant and the carbide powders are easily formed by vapor phase reaction at about 1400°C.[15] This reaction system is characterized by a large equilibrium constant for the reduction of WCl_6 to W by H_2 and the formation process of WC particles changes with the mixing temperature of WCl_6 and H_2 streams as follows. When the mixing temperature is low (~400°C), carbide particles are formed by the following two-step process via W particles (Process (C)).

$$WCl_6 + H_2(g) \xrightarrow{<1000°C} W(particle)$$

$$W(particle) + CH_4(g) \xrightarrow{>1000°C} W_2C,\ WC(particle)$$

On the other hand, when WCl_6 is mixed with H_2 at high temperatures (>1000°C) where the carburization rate of W with CH_4 becomes large, the formation of carbide particles occurs by the direct deposition process consisting of the formation of carbide nuclei and their growth (Process (B)). As the reaction temperature in both processes increases the carburization ratio increases and powders consisting of almost pure WC are obtained at 1400°C. Electron micrographs of these WC powders are shown in Fig. 5. The two processes give markedly different particle sizes: 0.05-0.3 μm in process (B); 0.02-0.1 μm in process (C). The sizes of the carbide particles produced by process (C) are nearly equal to those of W particles produced by the reaction in the WCl_6-H_2 system as the carbide particles are formed via W particles. This two-step process of WC formation, process (C), is also confirmed by the kinetic study on the carburization of W metal. The formation of Mo_2C particles in the $MoCl_4$(or MoO_3)-CH_4-H_2 system also proceeds by similar processes to the formation of WC particles.[16] The WC and Mo_2C particles obtained are spherical in both processes (C) and (B), but their particle sizes are smaller in process (C) than in process (B). The sizes of the carbide particles can be controlled by controlling the reaction temperature and the concentrations of reactants in addition to the mixing temperature of the metal halide and hydrogen.

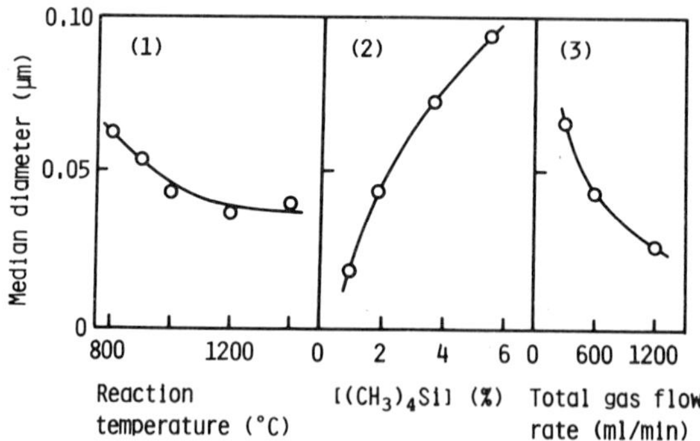

Fig. 4. Variations of the particle size of SiC powders formed from the $(CH_3)_4Si$-H_2 system under the following reaction conditions. (1) $(CH_3)_4Si$ = 2%, total gas flow rate (TFR) = 600 ml/min; (2) 1000°C, TFR = 600 ml/min; (3) 1000°C, $(CH_3)_4Si$ = 2%.

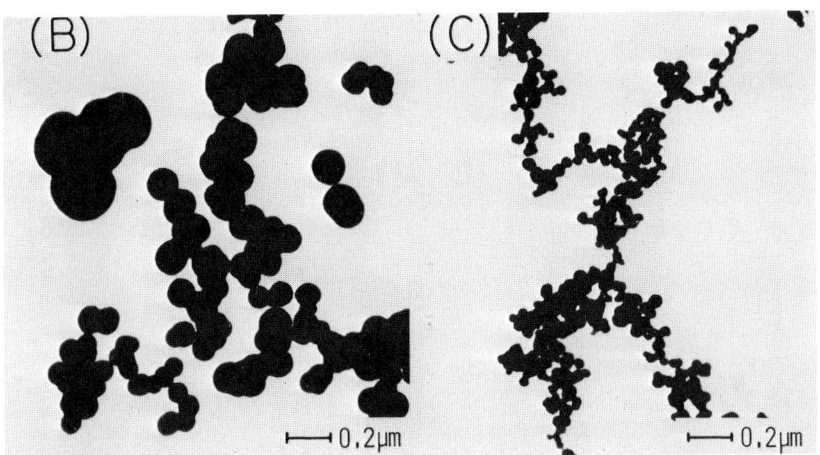

Fig. 5. WC particles. Reaction temperature = 1400°C. Mixing temperature of WCl_6 and H_2: (B), 1400°C; (C), ca 400°C. Gas composition: WCl_6 = 1%; CH_4 = 4.8%; H_2 = 48% (balance: N_2).

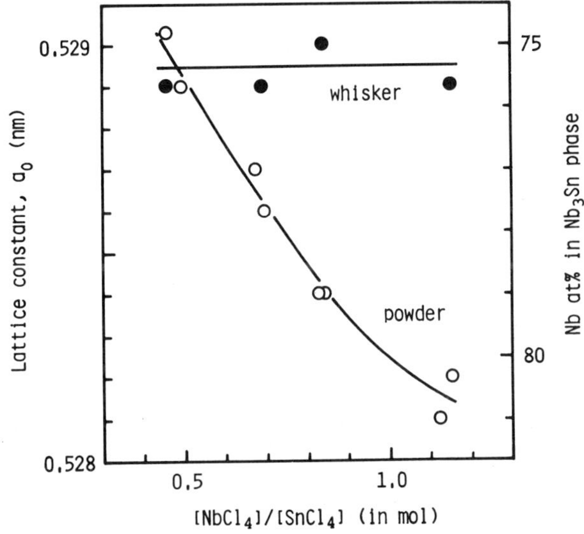

Fig. 6. Lattice parameter of Nb_3Sn phase and gas composition. Maximum temperature of reactor: 1400°C. Whisker: grown at 820-920°C.

Nonstoichiometry of Powders

When a compound can have a nonstoichiometric composition, the powders produced by vapor phase reaction show a wide range of composition depending on the reaction conditions. An example is shown in Fig. 6.[17] This figure shows the lattice parameters of the powders and whiskers of Nb_3Sn produced from the $NbCl_4$-$SnCl_4$-H_2 system. With the increase in $NbCl_4/SnCl_4$ ratio in the vapor phase, the lattice constant of the powders decreases; whereas, the whisker shows a constant value close to the lattice parameter of stoichiometric Nb_3Sn(cubic), a_o = 0.5290 nm. The decrease in the lattice constant of the Nb_3Sn powders shows the increase of the Nb content in the Nb_3Sn phase. The change of lattice constant, i.e., the change of chemical composition, of powder products with vapor-phase composition or reaction temperature was observed on TiN powders from the $TiCl_4$-NH_3-H_2 system,[6] ZrN powders from the $ZrCl_4$-NH_3-H_2 system,[7] and TiC powders from the $TiCl_4$-CH_4, $TiCl_4$-C_2H_4 and $TiCl_4$-CH_4 systems.[18,19] In the growth of whiskers of TiN, TiC and ZrC from the MCl_4-N_2-H_2 or MCl_4-CH_4-H_2 systems, on the other hand, the lattice parameters changed little with the reaction conditions such as vapor phase composition, growth temperature and type of substrates and were very close to that corresponding to the stoichiometric composition.[4]

The reason why the chemical composition of powders varies widely with reaction conditions may be that the particles are formed by rapid deposition of metal and nonmetal atoms and leave the hot zone in so short a period that atomic rearrangement to the stable structure cannot occur. In the growth of whiskers, the growth occurs more slowly over a long period at a given temperature so that the most stable structure results.

REFERENCES

1. R. F. Strickland-Constable, Kinetics and Mechanism of Crystallization, Academic Press, London and New York, 1968.
2. A. P. Levitt, ed., Whisker Technology, Wiley-Interscience, New York, 1970.
3. A. Kato, Seramikkusu, 13, 625 (1978).
4. A. Kato, N. Tamari, et al., J. Crystal Growth, 29, 55 (1975); 37, 293 (1977); 46, 221 (1979); 49, 199 (1980); Nippon Kagaku Kaishi, 1973, 1608; 1977, 650; 1979, 882, 1771; J. Less-Common Met., 58, 147 (1978).
5. Y. Suyama and A. Kato, J. Am. Ceram. Soc., 59, 146 (1976).
6. A. Kato, M. Iwata, J. Hojo, and M. Nagano, Yogyo Kyokai Shi, 83, 453 (1975); J. Hojo and A. Kato, Yogyo Kyokai Shi, 89, 227 (1981).
7. Y. Okabe, J. Hojo, and A. Kato, Yogyo Kyokai Shi, 85, 173 (1977).

8. J. Hojo, O. Iwamoto, and A. Kato, Nippon Kagaku Kaishi, 1975, 820.
9. A. Kato, Y. Ono, S. Kawazoe, and I. Mochida, Yogyo Kyokai Shi, 80, 114 (1972).
10. S. Prochazka and C. Greskovich, Ceram. Bull., 57, 579 (1978).
11. C. R. Veale, Fine Powders, Preparation, Properties and Uses, Applied Science Publ., Essex, England, 1972.
12. Y. Okabe, J. Hojo, and A. Kato, Nippon Kagaku Kaishi, 1980, 188.
13. Y. Okabe, J. Hojo, and A. Kato, J. Less-Common Met., 68, 29 (1979).
14. A. Kato and Y. Okabe, J. Am. Ceram. Soc., 63, 236 (1980).
15. J. Hojo, T. Oku, and A. Kato, J. Less-Common Met., 59, 85 (1978).
16. N. Tamari and A. Kato, Yogyo Kyokai Shi, 84, 409 (1976); J. Hojo, M. Tajika, and A. Kato, J. Less-Common Met., 66, 151 (1979).
17. T. Nakamatsu, T. Watari, and A. Kato, unpublished.
18. Y. Okabe, J. Hojo, and A. Kato, Yogyo Kyokai Shi, 86, 518 (1978).
19. J. Hojo and A. Kato, unpublished.

DISCUSSION

R. Rice (Naval Research Laboratory): You showed variations in lattice parameters for powder produced by various methods and at various temperatures. Did the lattice parameters of these powders change with various subsequent heat treatment conditions? If so, how?

Author: TiN_{1+x} and VN_{1+y} powders evolve excess nitrogen above ~600°C on heating under vacuum. The details are described in J. Less-Common Met., 53, 265-76 (1977).

R. R. Kirk (Pennsylvania State University): Have you measured the efficiencies of this process for the different powders? If so, what are they?

Author: The yields of powders of nitrides and carbides range from 10 to 60% on the basis of metal-carrying compounds. The yields are generally high in the reactions with large equilibrium constants. In a given reaction, the yield increases when the reacting gas is made to flow faster over the low temperature zone of reactor by using small inlet tube. At low temperature zone, the deposition on the wall occurs mainly.

SYNTHESIS OF POWDERS AND THIN FILMS BY LASER INDUCED GAS PHASE REACTIONS

John S. Haggerty

Senior Research Scientist
Massachusetts Institute of Technology
Cambridge, MA 02139

ABSTRACT

Laser induced gas phase reactions have been used to produce highly controlled powder and thin film reaction products. Powders of Si, Si_3N_4 and SiC exhibit presumed ideal characteristics for consolidation into dense ceramic parts. Individual powder characteristics can be manipulated with this process while being highly efficient in terms of materials utilization and energy consumption. The CVD process permits physical, electronic and crystallographic properties to be controlled. Resulting amorphous Si films exhibit superior properties.

INTRODUCTION

This paper will review the principal features of two laser driven gas phase chemical reactions that are being investigated at M.I.T. Powders are formed[1] in one and thin films in the other.[2] Both processes yield materials having superior, technologically important properties on a laboratory scale and we believe that they can be developed to operate in economically viable manners.

Reactant gases are heated by absorbing IR photons in these processes. They differ from photochemical reactions induced by ultraviolet or visible wavelength photons because absorption of single or multiple IR photons generally leaves the molecules in vibrationally or rotationally excited electronic ground states where they will ultimately dissociate when sufficient energy has been adsorbed. With the high gas pressures used, and the combination of the anharmonicity, the sparseness of the rotational structure and high dissociation energy of the SiH_4 molecules, it is unlikely that collisionless unimolecular multiphoton reactions can be induced. Thus, these reactions basically proceed as thermal reactions with a Boltzmann distribution of energies. Even if collisions and other relaxation processes occur, these reactions may differ from conventional reactions because the short times and rapid heating rates may result in different intermediate states. Our investigation has proceeded on the premise that these laser induced reactions will differ from conventional reactions in the level of process control and the process conditions that can be achieved uniquely with them and that they will have value because of the superior properties of the resulting powders and films.

We have modelled[3,4] these processes simply as thermal processes described by the laser energy absorbed by the reactant gas stream, the heat capacity of all species, and latent heats of reaction. The absorption of laser energy is generally presumed to terminate progressively as the reaction proceeds. There are in fact important differences between this simple model and actuality.

The powder and CVD laser driven reactions permit important process attributes to be achieved. For both, these include:

(1) highly controlled atmospheric conditions;
(2) cold wall reaction vessels;
(3) unusually rapid heating and cooling rates;
(4) precise control of process variables;
(5) accessibility to the reaction with process diagnostics.

For the laser CVD process, the ability to establish the substrate temperature at a level that is different from the gas reaction temperature is an important feature.

LASER INDUCED POWDER SYNTHESIS

Process Description

Powders have been synthesized with crossflow and counterflow gas stream-laser beam configurations and with both static and flowing gases. Most of the process research has been conducted with the reaction cell shown schematically in Fig. 1.

In a crossflow configuration, the laser beam having a Gaussian shaped intensity profile orthogonally intersects the reactant gas stream possessing a parabolic velocity profile. The laser beam enters the cell through a KCl window. The premixed reactant gases, under some conditions diluted with an inert gas, enter through a 1.5 mm ID stainless steel nozzle located 2-3 mm below the laser beam. A coaxial stream of argon is used to entrain the particles in the gas stream. Argon is also passed across the inlet KCl window to prevent powder build-up and possible window breakage. Cell pressures are maintained between 0.08 to 0.90 atm with a mechanical pump and throttling valve. At present, the powder is captured in a microfiber filter located between the reaction cell and vacuum pump. In the future, the particles will be collected by electrostatic precipitation or by separation with a fluid used for dispersion.

The powders of interest, Si, Si_3N_4, and SiC, have been synthesized from reactions based on electronic grade SiH_4, NH_3, C_2H_4, and CH_4 gases. Prepurified argon is used as the carrier gas. Preliminary studies indicated that the CH_4/SiH_4 reaction did not yield a SiC product, so we have concentrated our efforts on SiH_4 decomposition and both the NH_3/SiH_4 and C_2H_4/SiH_4 reactions.

Under normal synthesis condition only about 2-15% of the laser energy is absorbed by the reaction zones. In this experimental configuration, the reactant gas stream was made optically thin purposely to insure uniform thermal histories and to simplify analyses. In a production scale process, different optics would be employed with optically dense gas streams to avoid wasting 85-98% of the light energy. For instance, the counterflow configuration achieves efficient utilization of the laser light while maintaining other important attributes; it was not used extensively because it is not well suited to the wide range of process conditions employed in this experimental program. One of the important features of this process is the efficient use of reactants. Between 90-100% of the reactant gas is utilized with a single pass through the laser beam. It appears that Si, Si_3N_4 and SiC powders can be produced with as little as 2 kW/hr of energy per kilogram of powder.

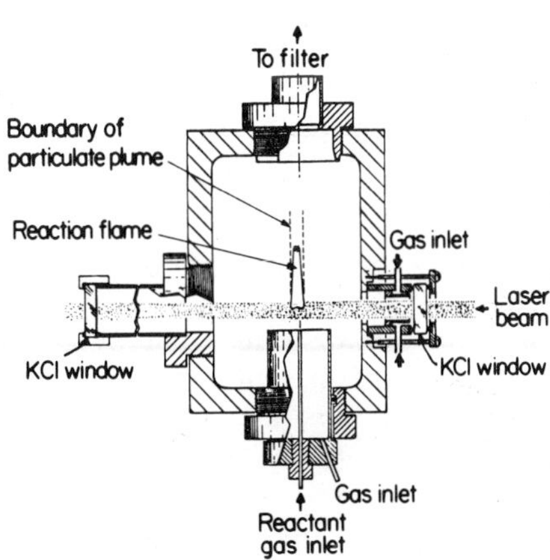

Fig. 1. Powder synthesis cell.

Rather simple analyses of the reaction zone permit heating rates, nucleation rates, growth rates and temperatures to be estimated surprisingly accurately based on measured zone and particle dimensions, calculated as velocities and uncorrected pyrometrically determined temperatures. Typical reaction conditions, Table 1, employ a total reactant gas flow rate of approximately 100 cm^3/min and a pressure of 0.2 atm. The velocity of the gas decreases from approximately 500 cm/sec at the nozzle to 350 cm/sec at the center of the laser beam. With a NH_3/SiH_4 flame, the reaction commences approximately 3 to 5 mm into the laser beam. Thus, the exposure time needed to initiate the reaction is nominally 10^{-3} sec. For a reaction temperature of approximately 1000°C, this indicates a gas heating rate of approximately 10^6°C/sec. The radial particle growth rates are approximately 10^4–10^5 Å/sec. Time-temperature profiles have been computed[3,4] as a function of varied process parameters using calculated velocities, actual laser intensities and absorptivities that we have measured. Qualitatively, the calculations agree with observed effects.

Table 1. Typical Processing Conditions for Si, Si_3N_4, and SiC Powders

Powder Type	Si	Si_3N_4	SiC
Cell Pressure (atm)	0.2	0.2	0.2
SiH_4 Flow Rate (cc/min)	11	11	11
NH_3 Flow Rate (cc/min)	0	110	0
C_2H_4 Flow Rate (cc/min)	0	0	9
Argon Flow Rate (cc/min) (to window)	600	600	600
Argon Flow Rate (cc/min) (to annulus)	400	400	400
Laser Intensity (W/cm_2)	760	760	760

Normally the reactants are usually completely consumed; thus, particle growth is limited by depletion of the reactants rather than elapsed time. The narrowness of the resulting particle size distribution indicates that nucleation and growth processes occur largely at different times. Process variables probably affect particle size mainly through the nucleation step.

Newly developed light scattering and transmittance measurements permit nucleation and growth processes to be analyzed in terms of temperature and reactant partial pressures with approximately 1 mm spacial resolution throughout the reaction zone and particulate plume. They also permit important observations about the crystalline state of the reaction product and the agglomeration process. Using an experimental technique developed by Marra[5] and Flint[6] and modifications[6] of computational techniques reported by Sarofim,[7] it is possible to calculate the local emissivity, particle diameter and number density from the extinction and scatter-extinction ratio results. These calculations employ the complex index of the particulate materials which are strongly dependent on crystallinity, temperature, stoichiometry and wavelength. This variance causes some uncertainty in the computed values; but, when calculated results are compared with other direct measurements, the variance can be used to gain insights about the reaction process.

Typical results[6,8] are given in Figs. 2, 3 and 4 for Si powders made from SiH_4. The temperatures shown in Fig. 2 are calculated from the pyrometrically determined brightness temperatures and the calculated emissivities. Measured heating rates (10^5–10^6°C/sec) agree with estimated heating rates and the cooling rates (~ 10^5°C/sec), that agree with radiant cooling models, are fast enough to quench in non-equilibrium structures. The particles exhibit a region of rapid growth followed by constant size when the reactions are complete. Both the absolute sizes and estimated growth rates agree with values calculated on the basis of elevated temperature, polycrystalline indices of refraction. The temperature begins to decrease as soon as the growth process terminates indicating that the Si particles are not heated by the laser beam. Under some synthesis conditions, the reaction is initiated before the gases reach the laser beam showing that a significant amount of heat is transferred to incoming gas molecules from sources other than the laser.

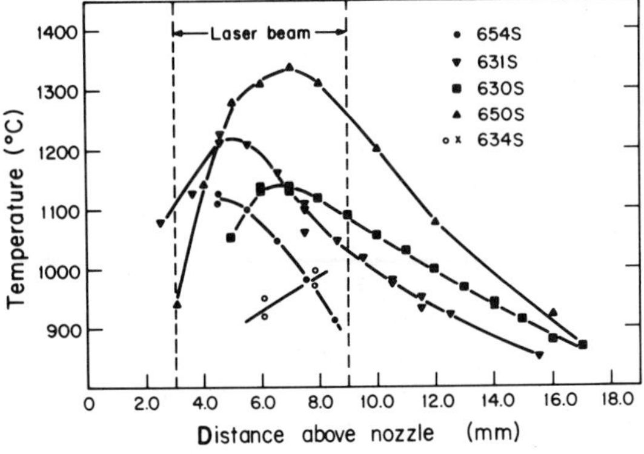

Fig. 2. Temperature versus distance from nozzle.

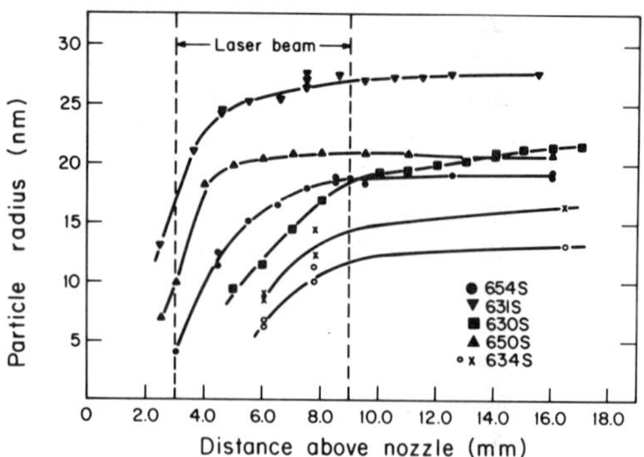

Fig. 3. Particle size versus distance from nozzle.

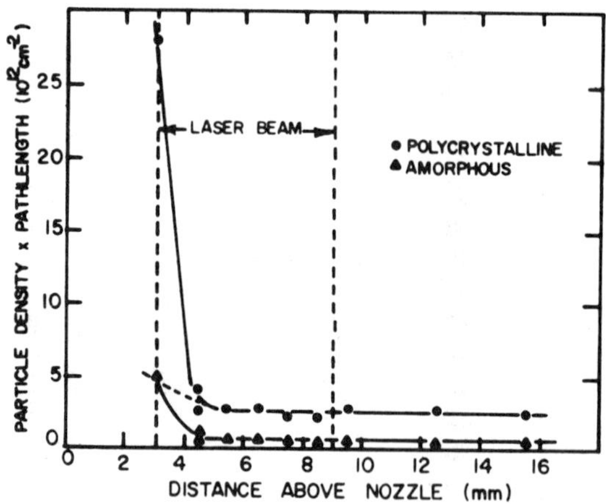

Fig. 4. Particle number density versus distance from nozzle.

The particle number density calculations in Fig. 4 show their extreme sensitivity to the presumed index. For either amorphous or polycrystalline indices, the calculated number density decreases rapidly over a distance of approximately 1 mm ($\sim 3 \times 10^{-4}$ sec) and then remains constant. Supported by direct STEM observations and annealing studies,[8] these results show that the Si particles form with an amorphous structure then crystallize when the gases become hotter as they penetrate into the laser beam.[6] The actual number density history follows the dotted line from the amorphous curve ($\sim$ 3 mm) to the polycrystalline curve ($\sim$ 4.5 mm) and remains essentially constant ($N \approx 4 \times 10^{12}$ cm^{-3}) throughout the growth process. The cessation of nucleation is an essential requirement for achieving the desired uniform particle size. Equally important, this result shows that agglomeration does not occur within the hot region where hard agglomerates could form in agreement with the constant particle diameter regions in Fig. 3.

Mass balance calculations, based on local number density and particle diameter, permit the local reactant concentrations to be calculated. These, combined with local temperatures, permit reaction kinetics to be interpreted in terms of rate controlling models. Marra demonstrated[8] that Si particle growth is rate controlled by molecular bombardment of Si vapor species probably formed by the unimolecular dissociation of SiH_4. The description of the nucleation process is less complete because the theory is inherently more complex and the measurement techniques employed do not permit observations with particle diameters less than 80 Å. There is reasonable agreement with nucleation rates calculated by the Volmer-Weber formulism assuming collisions between silicon atoms at an equilibrium concentration based on unimolecular dissociation of silane. The cessation of nucleation apparently results from the very sharp decrease in Si(v) supersaturation with a small extent of reaction.

More limited studies of this type have been done for SiC synthesized from laser heated SiH_4 + C_2H_4. Suyama et al.[9] concluded that these gases react to form an equilibrium concentration of SiC(g) molecules which participate in the nucleation and growth processes. Growth was probably controlled by molecular bombardment rates because the particles are much smaller than the mean free path. Unlike the silicon synthesis process, the nucleation and growth processes were generally completed by the time the gases reached the laser beam. For reasons that are not completely understood, these reactant gases are heated to a greater extent before interacting directly with the laser beam than occurs with pure SiH_4 or NH_3/SiH_4 mixtures. Also, the SiC particles are heated by the CO_2 laser beam causing the temperature distribution in the laser effected region to reflect the Gaussian intensity distribution of the laser beam. Despite these detailed differences, it is possible to control SiC particle diameter and other important characteristics by means of process parameters; laser intensity and reactant gas pressure are the most important. B_2H_6 had a marked effect on the crystallinity of the SiC particles. The undoped SiC particles generally had the same polycrystalline structure observed in Si particles while the B doped particles were faceted single crystals.

Virtually all of the Si_3N_4 synthesis research was conducted before the reaction diagnostics were developed. Consequently, the synthesis of Si_3N_4 powders from SiH_4 + NH_3 gases has not been analyzed to determine nucleation and growth kinetics and active gas species. However, the effect of process variables have been investigated systematically to determine their effect on particle characteristics as well as the volume of gas depleted around each growing nuclei which provides a direct measure of the nucleation rate • nucleation time product.[1] Reaction temperature had a strong effect on the gas depletion volume whether the temperature change was caused by laser intensity, gas velocity, gas mixture or gas pressure. An increased temperature caused an increased nucleation rate • nucleation time product (reduced depletion volume) for the Si_3N_4 process which is opposite to the results with the Si and SiC synthesis processes. These reactions apparently occur entirely within the laser beam region; thus, the heating rates during nucleation are directly affected by the laser. The Si_3N_4 particles are not heated by the laser beam.

Powder Characteristics

In general, the Si, Si_3N_4 and SiC powders are all similar in character. They fulfill many of the requirements which are sought. Table 2 lists the general characteristics of these powders and the ranges of properties that have been achieved.[1,4]

A bright field TEM photomicrograph of a typical SiC powder is shown in Fig. 5. The particles are spherical and uniform in size. In general, the Si_3N_4 and SiC powders are smaller and have a narrower Si_3N_4 and SiC powders. Dispersion results[1] have shown primary bonding does not exist between particles.

Table 2. Summary of Powder Characteristics

Powder Type	Si	Si_3N_4	SiC
Mean Diameters (Å)	190–1700	75–500	200–500
Standard Deviation of Diameters (% of Mean)	46	23	~ 25
Impurities (ppm by wt.)			
O_2	680–7000	3000	3300–13,000
Total Others	≤ 200	≤ 100	NA
Major Elements	Ca,Cu,Fe	Al,Ca	NA
Stoichiometry	---	0–60% (excess Si)	0–10% (excess C or Si)
Crystallinity	crystalline-amorphous	amorphous-crystalline	crystalline Si and SiC
Grain Size:Mean Diameter	1/5–1/3	~ 1/2	1/2–1

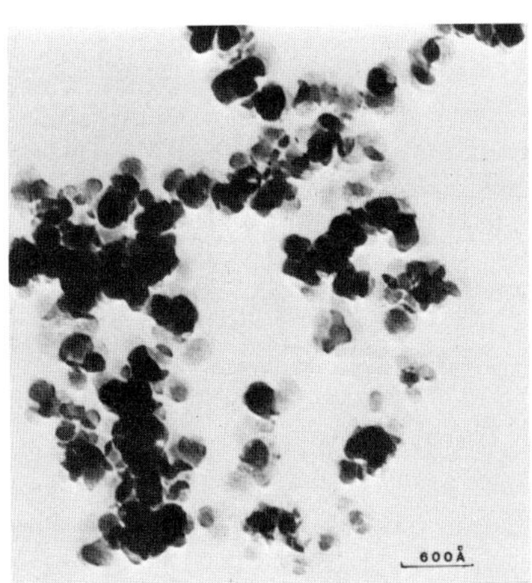

Fig. 5. TEM photomicrograph of SiC powder.

The BET equivalent spherical diameter and the diameter measured from TEM micrographs have always been nearly equal. This indicates that the particles have smooth surfaces, no porosity accessible to the surface, relatively narrow size distributions, and nearly spherical shapes. Powder densities, measured by He pycnometry, indicated the particles had no internal porosity.

Chemical analyses have shown that the oxygen content is generally less than 1.0% by weight and some powders are as low as 0.05 wt.%. The total cation impurities are typically less than 200 ppm. For Si_3N_4 and SiC powders, the stoichiometry varied substantially depending on the process conditions. For Si_3N_4 powder, the stoichiometry ranged from nearly pure Si_3N_4 (< 1.0 wt.% excess Si) to Si_3N_4 + 60 wt.% excess Si. Near stoichiometric values are caused by increased laser intensity, increased pressure, and lower gas stream velocities. C_2H_4/SiH_4 runs produced SiC powders with up to 10 wt.% excess C or Si. The more stoichiometric powders were produced with increased laser intensity and close to stoichiometric reactant gas ratios. In both the Si_3N_4 and SiC powders, it appears the excess Si and C are distributed uniformly throughout the individual particles.

Except when synthesized under very low laser intensities, the Si powders are crystalline. In all cases, the crystallite size was a fraction (1/5 -1/3) of the BET equivalent diameter, indicating the individual particles are polycrystalline. It was found that the particle size ratio reflected the nucleation and growth of crystals in the amorphous particles. Virtually all process conditions produced Si_3N_4 powder which was amorphous. High laser intensities and high reactant pressure resulted in stoichiometric, crystalline Si_3N_4 powders. All SiC powders were crystalline. Most process parameters produced polycrystalline SiC particles with the BET equivalent spherical diameter approximately twice as large as the crystallite size measured by X-ray line broadening.

LASER INDUCED CHEMICAL VAPOR DEPOSITION

Process Description

Our initial research[2,10] has demonstrated the feasibility of producing silicon films by laser induced pyrolysis of SiH_4. It is apparent that the process can be applied to other materials; we selected silicon chemistries because we already have measured the optical characteristics of SiH_4 gas and there is such an enormous market for Si, as well as related SiO_2 and Si_3N_4 films, in various electronics applications. Also, the unique ability to operate with low substrate temperatures is particularly important for inducing high levels of H_2 in amorphous silicon.

This process differs from other laser CVD processes that induce localized pyrolysis reactions[11-13] with laser heated substrates. In this process, the gases are heated directly by absorbing light emitted from an IR laser until the reactant temperatures reach a level where the reaction proceeds. Because of the heating method, the reaction zone can be precisely controlled. Also, the laser beam does not provide a heated surface that typically dominates heterogeneous nucleation and other reaction effects.

Recently, Bilenchi and Musci[14] reported successfully depositing amorphous silicon onto substrates by a technique very similar to the one described in this paper. Others[15-16] have induced ultraviolet photochemical CVD reactions using similar geometries to those we are using. Each laser CVD process configuration claims specific advantages. We believe these processes will ultimately find extensive use, probably in configurations incorporating important features from each.

The extreme sensitivity of the IR optical absorption coefficient (α) to the gas pressure, gas mixture and probably temperature is an important difference between the powder and CVD processes. The low pressure IR absorption spectra of gases such as silane are made up of many narrow absorption lines; it is the difference between these line positions and the emission lines of the laser and their respective line widths that determines α. The SiH_4 absorption line nearest the P(20) CO_2 line[17] is at 944.213 cm^{-1}. Natural and Doppler line broadening are not sufficient to cause adequate coupling.

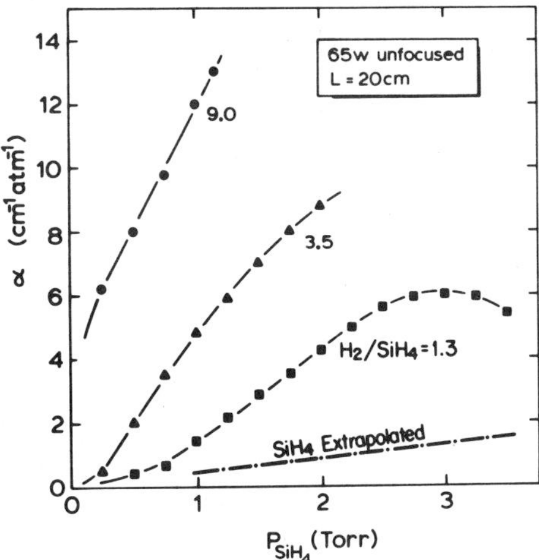

Fig. 6. The absorption coefficient of SiH_4 as a function of silane partial pressure and the hydrogen to silane ratio for the P(20) CO_2 laser line.

At system pressures above a few Torr, collisions between like or unlike gas molecules cause the absorption line widths to broaden; for many gases, this line broadening is proportional to the system pressure.[18] Lorentz broadening results from the temporary formation of a quasi-molecule[19] and is explained by transitions that occur between the potential energy curves of the colliding molecules. Figure 6 shows the dramatic effect of hydrogen, which is transparent, on silane absorptivity.[2-10] At a silane partial pressure of 1 Torr the absorptivity increases by over two orders of magnitude as the H_2:SiH_4 ratio is increased from 0 to 9. In contrast, the effective absorptivities are essentially insensitive to gas pressure and mixture at the high pressures used for powder synthesis.

It is evident that gas temperature and the intense laser radiation can also effect the absorptivity of the gas. Although important, these effects have not been quantified. The extreme sensitivity of the energy transfer to these process variables requires precise control of process variables.

A modification of the reaction cell shown in Fig. 1 has been used in the CVD process research. Substrates are mounted on a temperature controlled stage positioned parallel to the beam axis. Both vertical and horizontal substrate orientations have been used. The distance between the axis of the laser beam and the substrate plane was varied, but was usually set approximately equal to one laser beam diameter ($d \leq 6$ mm). Substrate temperatures up to 400°C were used to investigate their effect on process kinetics and film properties. Aluminum, aluminum oxide, single crystal silicon, borosilicate glass and borosilicate glass coated with platinum, molybdenum, and indium-tin oxide have been used as substrates.

Both constant volume (static cell) and constant pressure (flowing gas) conditions have been used. Pure silane at pressures between 4 and 10 Torr are generally used. Mixtures of He/SiH_4 and H_2/SiH_4 have been used. With flowing gases, the inlet stream is located at the same position used with powders but it enters through a diffusor.

A Coherent Radiation Everlase 150 untuned CO_2 laser and an Advanced Kinetics MIRL 50 grating-tuned CO_2 laser have been used in cw mode. The unfocused beam diameters are 6 mm; a lens with a focal length of 12.7 cm was used in the experiments involving a focused beam. A thermopile was used to monitor the laser power transmitted through the cell. Incident laser energy intensities have been in the 100 to 1000 watts/cm^2 range.

Figure 7 shows a plot of thickness versus time constructed from the positions of the maxima and minima of a typical reflectance versus time record.[2] Figure 7 also shows the instantaneous deposition rate versus time. The general curve shapes in Fig. 7 are typical; the deposition rate increases continuously until a steady state, constant rate is reached. The long time required to reach steady state deposition rates has not been explained. Deposition rates in excess of 160 Å/min have been achieved which is comparable to normal CVD deposition rates for a-Si.[20]

Deposition rates in the laser CVD process increase with increasing substrate temperatures, laser power, silane flow rate and silane pressure. Figure 8 shows the effect of SiH_4 flow rate on deposition rate. The maximum rate is limited by the onset of powder production. It is expected that the range of film production can be extended by using mixtures of silane with either H_2 or HCl to suppress the onset of homogeneous reactions.

The modelling of the laser CVD process is not as complete as for the powder process. Most effects can be understood qualitatively on the basis of heat transfer. Increased gas temperatures within the envelope of the laser beam result from increased substrate temperature, laser power and silane pressure. Simplified analyses predict no effect of varied intensity with constant laser power, in general agreement with observations. The effect of varied flow rate appears related to early cessation of the reaction but the factors that cause the CVD reaction to terminate with only 1-5% of the SiH_4 are still not completely understood. The molecular mechanisms for decomposition of SiH_4 under CVD rates and process conditions require further study. Others[14,21,22] have reported SiH_2 and SiH_3 vapor species and have suggested these decompose to Si on the substrate surface. While we have not addressed this issue, these conclusions differ from ours[8] based on Si particle growth.

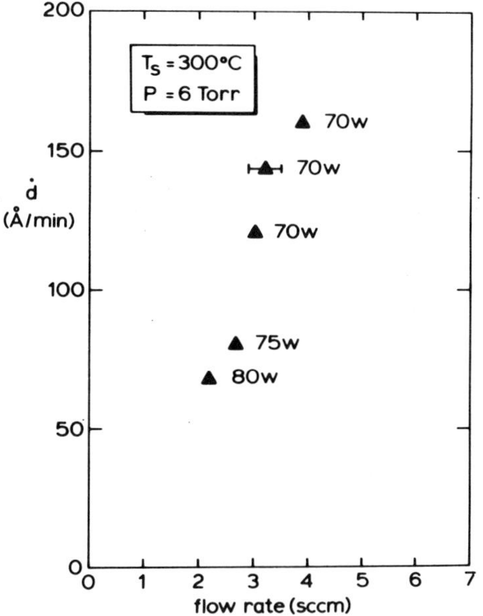

Fig. 7. Thickness and deposition rate versus time for $P(SiH_4)$ = 6 Torr, T_s = 300°C and 70 W of laser power.

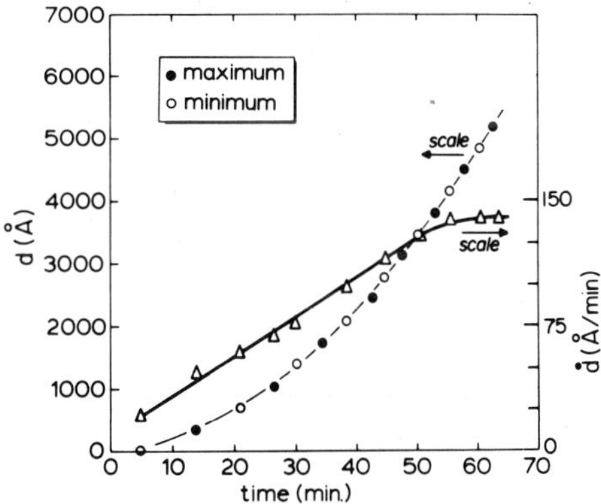

Fig. 8. Steady state deposition rate versus silane flow rate for constant substrate temperature, cell pressure and (nominally) laser power.

Film Characteristics

Characterization of the resulting a-Si films[2,10] shows that they exhibit high quality levels and that properties can be manipulated through altered process conditions. Films produced by this technique are amorphous in terms of electron and X-ray diffraction criteria. With clean substrates, film adhesion was found to be very good. Pealing, from points of contamination or rough edges, is observed occasionally. The curvature of detached films indicates that their surfaces are in tension.

One of the most important a-Si film properties for devices is the concentration of unpaired spins. Incorporation of hydrogen can decrease the defect concentration sufficiently to unpin the Fermi level and allow substitutional doping. Figure 9 shows the spin concentration (N_s) as a function of substrate temperature (T_s) for these a-Si films. The minimum concentration is about a factor of 100 lower than in conventional CVD films[23-24] and only about a factor of 10 higher than in the best glow-discharge films.[25] N_s appears to be independent of the laser power, but it decreases with decreasing T_s, presumably due to the increased hydrogen which saturates dangling bonds.

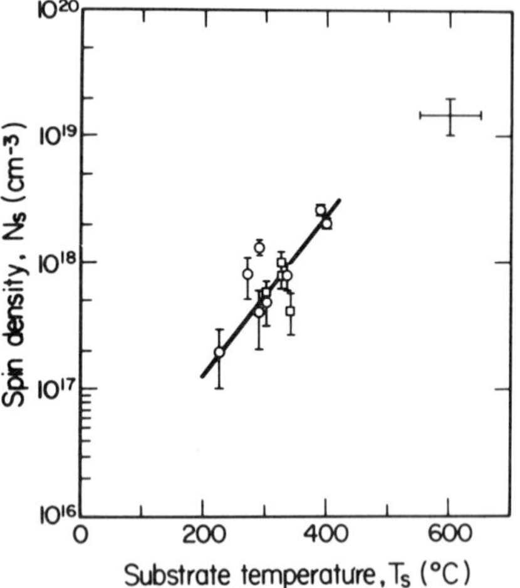

Fig. 9. Spin concentration N_s versus substrate temperature T_s for films made with (o) laser power = 40W, SiH_4 = 8.8 Torr; (□) laser power = 60W, SiH_4 = 70 Torr; (+) normal CVD.

Optical absorption, optical band gaps (E_{opt}) and electrical conductivities of these films have also been measured. They agree with those cited for films having the high H_2 contents found in films produced by glow-discharge or sputtering processes. The substrate temperatures' effect on E_{opt} and the high temperature activation energy for electrical conductivity (E_{act}) both follow expectations for varying hydrogen contents. Increasing T_s (200-400°C), resulted in decreasing H_2, decreased E_{opt} (2.2-1.6 eV) and decreased E_{act} (0.44-0.27 eV). These electronic characteristics are indications of high quality films and they can be manipulated through optimization of process variables.

SUMMARY

Two new processes have been developed which permit high quality ceramic powders and semiconductor thin films to be synthesized from laser heated gas phase reactants. The unusually high qualities result from the ability to manipulate time-temperature-position histories very precisely and the consequence this has on our ability to control the reactions. High purity, agglomerate-free Si, Si_3N_4 and SiC powders have been synthesized; size, crystallinity and stoichiometry can be adjusted while maintaining conversion efficiencies in excess of 95%. Hydrogenated amorphous Si films were deposited at high rates which exhibited excellent physical and electronic characteristics. These processes can be applied to many other materials including oxides, carbides, borides and nitrides.

ACKNOWLEDGMENTS

This research has been sponsored by the Department of Defense (ARPA, ONR and ARO), the Standard Oil Company of Indiana and the 3M Corporation. Many cited and uncited students and staff have participated in these programs. All contributions are gratefully acknowledged.

REFERENCES

1. J. S. Haggerty, "Sinterable Powders from Laser Driven Reactions," Energy Laboratory Report MIT-EL 82-002, Massachusetts Institute of Technology, Cambridge, MA, 1981.
2. T. R. Gattuso, M. Meunier, and J. S. Haggerty, "Laser Induced Deposition of Thin Films," Energy Laboratory Report MIT-EL 82-002, May 1982.
3. W. R. Cannon, S. C. Danforth, J. H. Flint, J. S. Haggerty, and R. A. Marra, J. Am. Ceram. Soc., 65, 7, 324-30 (1982).
4. W. R. Cannon, S. C. Danforth, J. S. Haggerty, and R. A. Marra, J. Am. Ceram. Soc., 65, 7, 330-35 (1982).

5. R. A. Marra and J. S. Haggerty, Ceram. Eng. and Sci. Proc., 3, 1-2, 3-19 (1982).
6. J. H. Flint, (M.S. Thesis), M.I.T., February 1982.
7. A. D'Alessio, A. Dilorenzo, A. F. Sarofim, F. Beretta, S. Masi, and C. Venitozzi, p. 1427 in Fifteenth Symposium (International) on Combustion, The Combustion Institute, Pittsburg, PA, 1975.
8. R. A. Marra, (Ph.D. Thesis), M.I.T., February 1983.
9. Y. Suyama, R. A. Marra, J. S. Haggerty, and H. K. Bowen, submitted for publication to the J. Am. Ceram. Soc., October (1982).
10. T. R. Gattuso, M. Meunier, D. Adler, and J. S. Haggerty, Proceedings of the Symposium on Laser Diagnostics and Photochemical Processing for Semiconductor Devices, Materials Research Society, November 3-4, 1982, Boston, MA.
11. C. P. Christensen and K. M. Lakin, Appl. Phys. Lett., 32, 254-55 (1978).
12. V. Baranauskas, C. I. Z. Mammana, R. E. Klinger, and J. E. Greene, Appl. Phys. Lett., 36, 930-32 (1980).
13. S. D. Allen, J. Appl. Phys., 52, 6501-05 (1981).
14. R. Bilenchi and M. Musci, Electrochemical Soc., 275-83 (1981).
15. P. K. Boyer, G. A. Roche, W. H. Ritchie, and G. J. Collins, Appl. Phys. Lett., 40, 716-19 (1982).
16. T. F. Deutsch, D. J. Ehrlich, and R. M. Osgood, Jr., Appl. Phys. Lett., 35, 175-77 (1979).
17. J. W. C. Johns and W. A. Kreiner, J. Mol. Spec., 60, 400-11 (1976).
18. A. C. G. Mitchell and M. W. Zemansky, Resonance Radiation and Excited Atoms, Cambridge University Press, Cambridge, 1961.
19. H. Okabe, Photochemistry of Small Molecules, Wiley, NY, 1978.
20. J. Bloem and L. J. Giling, Current Topics in Materials Science, edited by E. Kaldis, 1, 147-342 (1978).
21. B. A. Scott, R. M. Plecenik, and E. E. Simonyi, Appl. Phys. Lett., 39, 73-75 (1981).
22. T. F. Deutsch, J. Chem. Phys., 70, 1187-92 (1979).
23. H. Hirose, J. de Physique, C4, 705-14 (1981).
24. S. Hasegawa et al., Phil. Mag. B, 43, 149-56 (1981).
25. H. Fritzche, Sol. En. Mat., 5, 229-316 (1981).

DISCUSSION

D. Nath (Aetna Telecom Labs): What are the (1) deposition rates of the powders; (2) the film thicknesses vs. time; and (3) the collection efficiencies of the powders.

Author: (1) The laser powder process appears scalable to commercially interesting levels provided the reactions are exothermic. Using a 1.5 mm ID nozzle, we synthesize up to 20 grams of Si, Si_3N_4 and SiC powders per hour; others have achieved synthesis rates that are 10 times higher using a more powerful laser. As a first approximation, we estimate that a 20 mm ID nozzle, which would capture 100% of the laser light, operated at the presently used conditions of 1 atm pressure and 500 cm/sec gas velocity, would produce 4.0-15 kg/hr of powder using a 2.4-9 kW CO_2 laser.

(2) We have achieved amorphous Si deposition rates from SiH_4 in excess of 160 Å/min with the laser CVD process which is comparable to conventional CVD rates for amorphous silicon. Higher rates (~ 900 Å/min) were reported by Bilenchi et al. (Materials Research Society, 1982 Annual Meeting, November 1-4, Boston, MA). We are confident that even higher rates can be achieved by controlling the supersaturation and the unique ability to increase the temperature difference between the gas and the substrate.

(3) Powder collection efficiencies are essentially 100%.

J. P. Skalny (Martin Marietta Labs): Did you consider scale-up of the processes and what is the estimated energy efficiency?

Author: The feasibility for scale up looks excellent provided the reactions are exothermic and the reactants are strongly absorbing at wavelengths accessable to a CO_2 laser or other high efficiency lasers. For these conditions, the laser is used only to heat the gases through the reaction temperature range; other more efficient heat sources are used to bring the reactants up to the spontaneous reaction temperature and no latent heats are required. With proper design, all of the laser light is converted to sensible heat. We estimate the total heat requirement to be less than 2 kWhr/kg of powder for the chemistries that we have investigated (see response to first question, above).

D. Biswas (ITT/EOPD, Roanoke, VA): (1) If the reactions are complete (i.e., 100% conversion), what are the deposition rates of Si, Si_3N_4 and SiC powders and how can they be improved? (2) During the film depositions, how can one maintain the preferential temperature gradient between the substrate and the chemicals when the laser beam is operating at a particular (i.e., fixed) wavelength?

Author: (1) Answered in responses to the questions asked by Nath and by Skalny.

(2) The temperature difference between the reactant gas and the substrate is determined by the laser power absorbed and heat transfer to cold surfaces. The laser power absorbed per unit path length is related to the laser intensity, the optical absorptivity and gas pressure. The substrate temperature can be established with active heating or cooling or can be allowed to reach an equilibrium value

with respect to the laser heated gases. We have generally heated the substrates to predetermined levels (~ 200-400°C). CVD reactant gas temperatures with SiH_4 have not been measured but are presumed to be 550-700°C.

S. Dutta (NASA): Would you comment on how you would improve the stoichiometry in SiC and Si_3N_4 powder; and whether that approach might have any effects on the resultant powder characteristics?

Author: By adjusting process variables (gas mixture, laser intensity, gas pressure, gas velocity, etc.), the stoichiometry of the compound powders can be set at precise levels. Silicon nitride powders can be made with varying quantities of excess silicon (including zero excess), and silicon carbide powders can be made with compositions ranging from excess Si to C. Thus, stoichiometry is a powder characteristic that can be established to optimize subsequent processing into parts or the properties of the finished parts. Stoichiometry can usually be adjusted independent of other powder characteristics (size, purity, crystallinity, agglomeration, etc.).

PREPARATION OF ZIRCONIA-ALUMINA FINE POWDERS BY HYDROTHERMAL OXIDATION OF Zr-Al ALLOYS

Shigeyuki Sōmiya, Masahiro Yoshimura and Shinya Kikugawa

Laboratory for Hydrothermal Syntheses
Research Laboratory of Engineering Materials
Tokyo Institute of Technology
4259 Nagatsuta, Midori, Yokohama 227 Japan

ABSTRACT

Fine powders of pure metal oxides can be prepared by hydrothermal oxidation method. This technique can provide the fine powders of homogeneous ZrO_2-Al_2O_3 mixtures from Zr-Al alloys at temperatures 600-700 °C under 100 MPa pressure for 3 h, according to the reactions,

$2\ ZrAl_3 + 13\ H_2O = 3\ Al_2O_3 + 2\ ZrO_2 + 13\ H_2$, *and*

$2\ Zr_5Al_3 + 29\ H_2O = 10\ ZrO_2 + 3\ Al_2O_3 + 29\ H_2$.

The average crystallite sizes were 27-35 nm for a-Al_2O_3, 13-24 nm for nonoclinic ZrO_2 and 9-12 nm for tetragonal ZrO_2 in the products.

INTRODUCTION

High purity fine powders have become important in modern ceramic science and technology. Some new techniques[1-5] have been developed to prepare high purity ceramics fine powders with submicron to hundreds angstroms size. They include thermal decomposition of metal salts and/or hydroxides derived from inorganic or organometallic compounds, by hydrolysis, precipitation and/or drying, and chemical vapor deposition or physical vapor deposition, etc. Hydrothermal crystallization[6] and spark discharge processes[7] under water have also been used.

Hydrothermal Oxidation Method[8-13]

We have developed a method to prepare fine oxide powders from metal by the reaction with high temperature high pressure water, and called it "hydrothermal oxidation method." Fine grains can be prepared by chemical dividing of metal grains during reactions to form oxide and hydride (in the cases of Hf[11,12] and Zr[8-10]), or hydroxide and oxide (in the case of Al[13]) as follows,

$$Zr + 2\ H_2O = ZrO_2 + 2\ H_2 \quad (1)$$

$$Zr + H_2 = ZrH_2 \quad (2)$$

$$ZrH_2 + 2\ H_2O = ZrO_2 + 3\ H_2 \quad (3)$$

$$Al + 4\ H_2O = Al_2O_3 \cdot H_2O\ (\text{Boehmite}) + 3\ H_2 \quad (4)$$

$$Al_2O_3 \cdot H_2O = Al_2O_3 + H_2O \quad (5)$$

This method can provide fine grains of ~24 nm for ZrO_2[9], ~ 33 nm for HfO_2[11] and ~ 110 nm for Al_2O_3,[13] respectively under 100 MPa at temperatures 200 ~ 700°C. The reaction process and grain growth in the hydrothermal oxidation method have been studied in detail in the case of hafnia.[12] It is noteworthy that this method has the advantage to introduce no impurities into the products during reactions because of the reaction between high purity metal and pure water.

Fine-grained Mixed Oxide Powder

Even if fine powders of pure oxide could be prepared, homogeneously mixed (that is, uniformly dispersed) high-purity fine powders are not so easy to prepare. Because impurities might be incorporated into the products during grinding and mixing processes mianly from containers and/or grinding medium if mechanical mixing or blending techniques were used. Some chemical procedures, therefore, such as co-precipitation,[14] co-pyrolysis[15] or co-deposition[16] have been employed to prepare mixed powders from the starting materials of homogeneous mixtures as solutions or gases.

We have tried to prepare homogeneously mixed oxides by the hydrothermal oxidation of alloys. Since an alloy contains different atoms homogeneously mixed in atomic scale, it is expected that the direct oxidation of alloys can produce homogeneously mixed oxides at relatively low temperatures under hydrothermal conditions. Here the hydrothermal oxidation process of Zr-Al alloys is studied to prepare ZrO_2-Al_2O_3 mixed oxides, which are of interest in toughened ceramics.

EXPERIMENTAL PROCEDURE

Preparation of Alloys

Starting materials were Zr powder[a] (99.9% purity involving ~ 2% Hf, < 300 mesh) and Al powder[a] (99.9% purity, 200 ~ 300 mesh). They were weighed in appropriate ratios and mixed thoroughly using a boron carbide mortar and pestle in a glove box under 1 atm nitrogen atmosphere. The mixed metal powder was fused on a cold hearth of water cooled copper in flowing high-purity argon atmosphere by the radiation of a 10 kw Xe arc image furnace.[17] The fused alloy globule about 3 mm in diameter was solidified quickly by removing from the focus of the image furnace.

Hydrothermal Oxidation

The alloy sample was placed in a gold crucible and heated in pressurized (100±5 MPa) water in a Tuttle-Roy type pressure vessel at the rate of 15°C min. After the heat treatment at a certain temperature between 200 and 700°C for 0 ~ 6 h, the pressure vessel was quenched by immersion in cold water. The temperature was measured at the outside wall of the vessel by a Platinel thermocouple[b] (Type K, IEC) calibrated against m.p. of Zn (419.5°C).

Characterization of the Products

All reaction products of the hydrothermal treatments were weighed to check the weight change after drying in air at 120°C for 24 h. Then powder X-ray diffraction[c] was performed for the phase identification and the determination of the average crystallite size (D) of the produced phases. The Scherrer equation,

$$D = 0.9\ \lambda/(\beta \cdot \cos\Theta) \tag{6}$$

was used in this study to calculate the D values based on the diffraction lines, (111) of monoclinic ZrO_2, (111) of tetragonal ZrO_2 and (113) of α-Al_2O_3. The peak width (β) at the half maximum of these peaks were corrected by Warren equation,

$$\beta^2 = \beta^2_{obs} - \beta^2_{inst} \tag{7}$$

using β_{inst} for the (101) line of low quartz as an internal standard.

---

[a] Soekawa Chemicals, Co., Ltd., Tokyo, Japan.
[b] Engelhard Ind. Div. East Newark, NJ, U.S.A.
[c] RU-200 Rigaku Denki Co., Ltd., Tokyo, Japan.

The quantity of each phase in the products was determined by X-ray diffraction using a calibration curve derived from the mixture of reagent grade α-Al_2O_3 and m-ZrO_2 powders.

The products were also examined by scanning electron microscope (SEM[d]) and transmission electron microscope (TEM[e]) with energy dispersive spectroscopy (EDS). The TEM samples were made by the immersion of a collodion film supported on a copper grid in the suspension of sample powder ultrasonically dispersed in water.

RESULTS AND DISCUSSION

Arc image fusion yielded Zr_5Al_3, Zr_2Al_3, $ZrAl_2$ and $ZrAl_3$ phases in the system Zr-Al. The alloy samples of Zr_5Al_3 and $ZrAl_3$ (including trace amounts of free Al) were subjected to hydrothermal oxidation. The weight gain of the fused spherical sample of Zr_5Al_3 was observed above 400°C but it scarcely proceeded at this temperature (Fig. 1). The observed weight gain approached the calculated one, 43.2% according to the reaction,

$$2\ Zr_5Al_3 + 29\ H_2O = 10\ ZrO_2 + 3\ Al_2O_3 + 29\ H_2 \qquad (8)$$

after 6 h at 500°C, 3 h at 600°C or 1 h at 700°C.

Scanning electron microscope revealed the formation of boehmite crystals on the surface of the spherical sample (Fig. 2(b)) after 3 h at 400°C. It caused cracking of the sample. The formation of boehmite and zirconia, and the change of boehmite to corundum at higher temperature brought about the pulverization of the sample into the powder, Fig. 2(c), of the aggregate of the mixed oxide of ZrO_2 and Al_2O_3.

Figure 3 shows that the hydrothermal oxidation of $ZrAl_3$ samples did not proceed below 500°C but became significant at 600°C. The oxidation of samples was almost completed after 1 h at 600°C or 0.5 h at 700°C. The microstructure of the surface and inner part of the samples changed similarly to those for Zr_5Al_3 samples.

A higher temperature was needed for the reaction of the fused alloy samples than the constituent metals, Zr and Al powders, for which a temperature as low as 250°C[10] and 200°C[13], respectively, were sufficient to start the reaction with water under the same conditions. This appeared to be caused by the inactivation of the surface of the samples due to the fusion. When the fused alloy samples were crushed into powder, even though they were such coarse grains as 100

---

[d]JSM T-200, JEOL, Co., Ltd., Tokyo, Japan.
[e]JEM 200CX, JEOL, Co., Ltd., Tokyo, Japan.

μm in diameter, they started to react above 300°C as shown in Fig. 4. This fact suggests that the alloy samples having active surfaces would react quickly with water at relatively low temperatures and yield finer powders.

The reaction products of alloys with water were boehmite, small amounts of Kl-Al_2O_3 and monoclinic and tetragonal ZrO_2 below 400°C, and α-Al_2O_3 and monoclinic and tetragonal ZrO_2 above 500°C as seen in Fig. 5. The zirconium hydrides were not formed in the products. The tetragonal ZrO_2 has never been produced in the hydrothermal oxidation of Zr metal alone under the same conditions, so that the formation of t-ZrO_2 is characteristic in the hydrothermal oxidation of Zr-Al alloys. Particularly, the fraction of t-ZrO_2 was higher in early stage of the reaction and at lower temperatures, then it decreased gradually in the prolonged treating time as seen in Fig. 6.

The average crystallite size of each phase increased with the treating time at 600°C and 700°C (Table 1) due to grain growth. The average crystallite size of t-ZrO_2, however, is always smaller than that of m-ZrO_2. This would be one of the reasons why t-ZrO_2 was formed (or stabilized) in the products of the alloy oxidation. In the case of the Zr_5Al_3 samples, the crystallite size of the produced phases was a little larger than those in the case of the $ZrAl_3$ samples, and had similar time dependency.

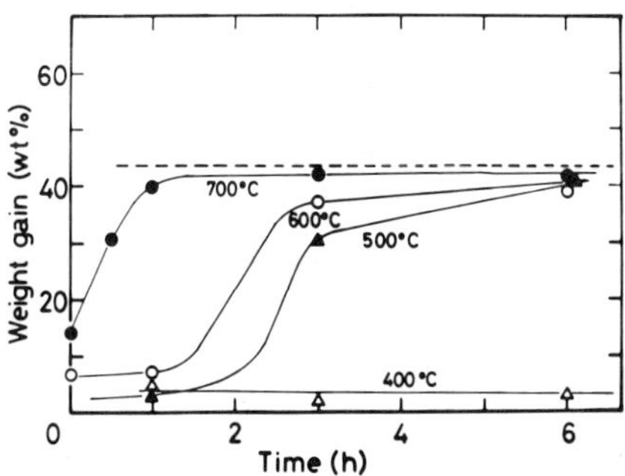

Fig. 1. Weight gain of fused Zr_5Al_3 sample during reactions with 100 MPa water at various temperatures.

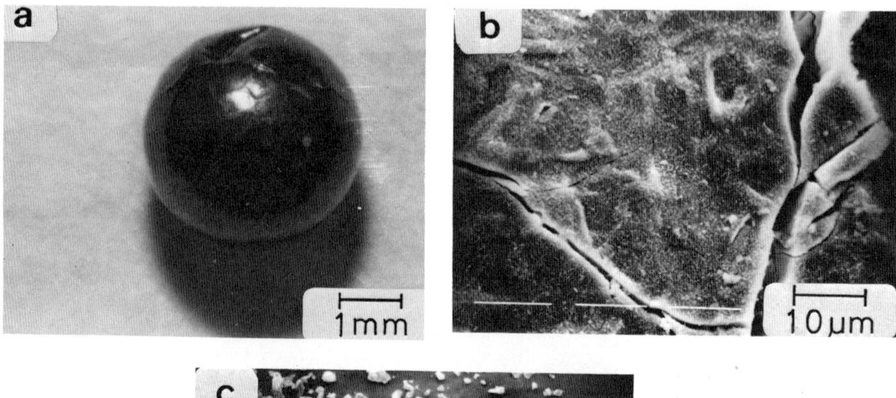

Fig. 2. Scanning electron microphotograph of the Zr_5Al_3 samples under hydrothermal oxidation. (a) Fused spherical alloy sample; (b) surface of partially reacted sample (400°C, 3 h); (c) completely oxidized sample, α-Al_2O_3 + ZrO_2 powder (700°C, 3 h).

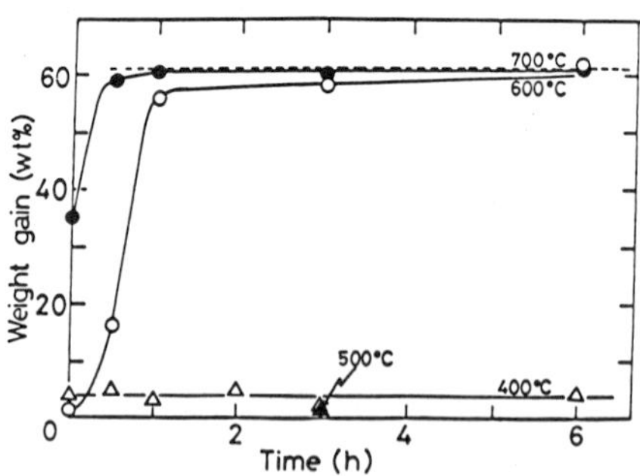

Fig. 3. Weight gain of fused $ZrAl_3$ sample during reactions with 100 MPa water at various temperatures.

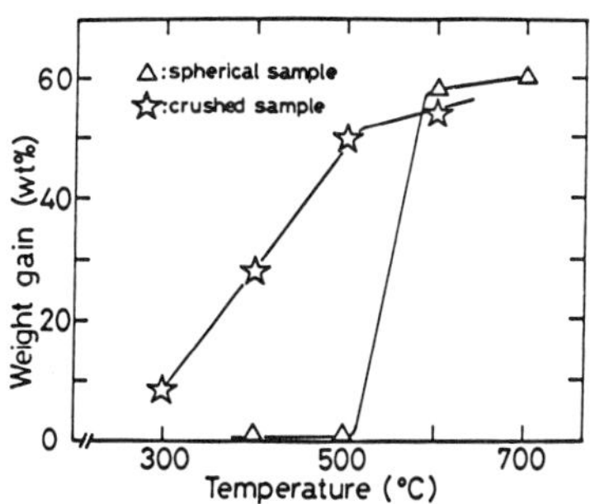

Fig. 4. Weight gain of spherical and crushed $ZrAl_3$ samples by hydrothermal oxidation under 100 MPa for 3 h.

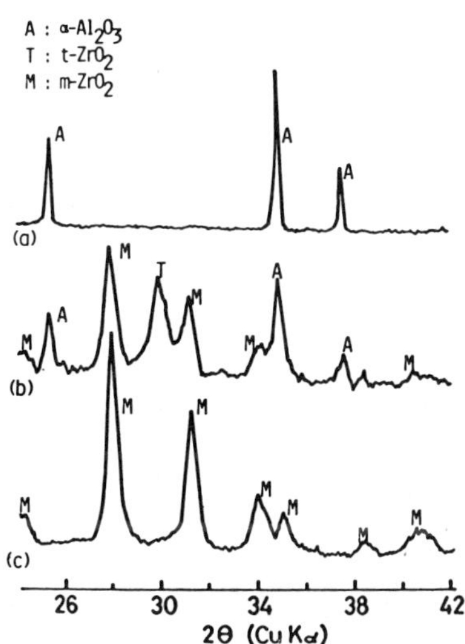

Fig. 5. X-ray powder diffraction patterns of the products prepared at 700°C under 100 MPa for 3 h from: (a) Al and H_2O; (b) $ZrAl_3$ and H_2O; or (c) Zr and H_2O.

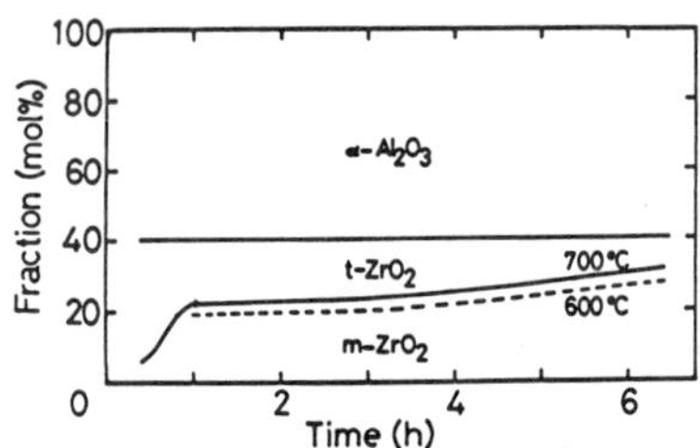

Fig. 6. Change of the contents of ZrO_2 and Al_2O_3 in the products prepared from $ZrAl_3$ and 100 MPa H_2O.

Table 1. The Average Crystallite Size of the Produced Phases with Treating Time of Hydrothermal Oxidation for Fused $ZrAl_3$ Samples.

		Crystallite Size (nm)		
Phase	Temp. (°C)	1 h	3 h	6 h
α-Al_2O_3	600	26.6	27.1	28.3
	700	27.2	30.0	31.2
m-ZrO_2	600	12.6	12.9	15.6
	700	16.9	20.2	21.8
t-ZrO_2	600	9.0	9.0	9.8
	700	11.5	11.7	13.2

The crystallite sizes of ZrO_2 and Al_2O_3 are greatly reduced in the case of the alloy oxidation in comparison with the cases of constituent metal as demonstrated in Table 2. In particular, the crystallite of α-Al_2O_3 is about 5 times less in the alloy oxidation than that in pure Al oxidation. These facts suggest that the growth of the crystallites of ZrO_2 and Al_2O_3 interfered with each other, because Zr and Al atoms mixed in atomic scale in the alloys had to segregate from each other to form ZrO_2 and Al_2O_3 phases separately. Such segregation and crystallite growth processes are difficult at low temperatures, therefore the fine grains of ZrO_2 and Al_2O_3 mix with each other in the products. This is one of the favorable features in the hydrothermal oxidation of alloys to prepare mixed fine powders with homogeneous dispersion.

Table 2. Products and Their Average Crystallite Size of Hydrothermal Oxidation of Metals and Alloys with 100 MPa Water.

Starting Material	Temperature (°C)	Products	Crystallite Size (nm)
Zr	250 - 700	m-ZrO_2	23 - 27
Al	100 - 500	γ-AlOOH	---
	500 - 700	α-Al_2O_3	110 - 200
$ZrAl_3$	600 - 700	m-ZrO_2	13 - 20
		t-ZtO_2	9 - 12
		α-Al_2O_3	27 - 35
Zr_5Al_3	500 - 700	m-ZrO_2	15 - 24
		t-ZtO_2	10 - 12
		α-Al_2O_3	28 - 35

Figure 7 shows the TEM photograph and EDS analysis of the products composed of m-ZrO_2, t-ZrO_2 and α-Al_2O_3. The observed crystallite sizes ranged 9 ~ 35 nm are in good agreement with those calculated by powder X-ray diffraction. The EDS analysis indicated that the area (b) in Fig. 7 contained ZrO_2 and Al_2O_3 particles and that an individual particle was either ZrO_2 or Al_2O_3; for example, the particle (a) in this figure was ZrO_2. These results clearly demonstrate the homogeneous dispersion of the fine particles of ZrO_2 and Al_2O_3 in the products. The EDS analysis showed no detectable Zr contents in Al_2O_3 particles, whereas a small amount of Al (estimated to be ~ 3.5 atom%) was often detected in the ZrO_2 particles as illustrated in Fig. 7(a). Although it is not clear whether this Al_2O_3 is present as a solid solution with ZrO_2 or as a separate phase on the surface and/or in the ZrO_2 particle, this presence of Al_2O_3 would help the formation (or stabilization) of t-ZrO_2 phase. The solid solubility of Al_2O_3 in ZrO_2 phase has not been established yet, but it is suggested[18] at high temperatures around 1900°C. The formation of the t-ZrO_2 phase by the presence of Al_2O_3 was also observed when the mixtures were melted and chilled by a plasma flame.[19] Additional strain (or surface) energy imparted to ZrO_2 grains by Al_2O_3 grains contacting with ZrO_2 seems to be the probable reason for forming (or stabilize) the tetragonal ZrO_2 phase.

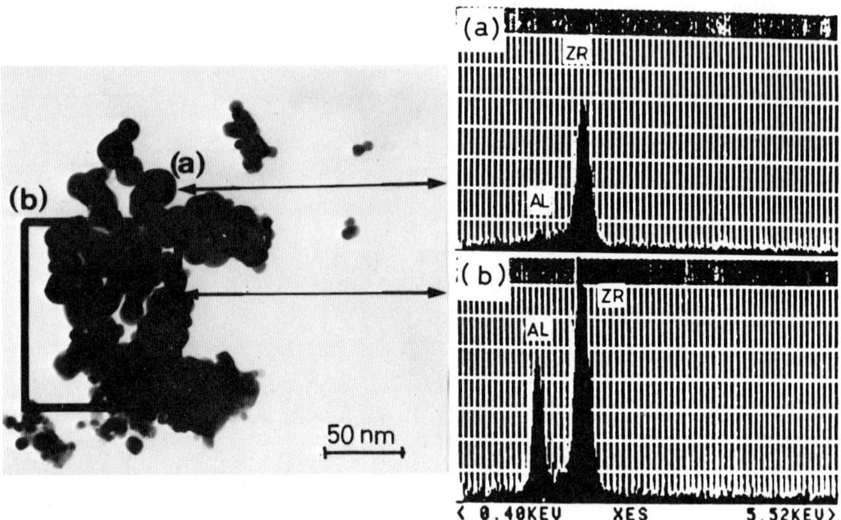

Fig. 7. TEM photograph of the products (m-ZrO_2 + t-ZrO_2 + α-Al_2O_3) by hydrothermal oxidation (700°C, 100 MPa, 3 h) of $ZrAl_3$ sample, and EDS analyses of the particle (a), and the area (b).

In addition to the homogeneous mixtures of fine grained ZrO_2 and Al_2O_3 as shown in Fig. 7, larger Al_2O_3 particles have also been formed in the hydrothermal oxidation of Zr-Al alloys. These Al_2O_3 particles have the morphology of hexagonal plate 1 to 5 μm diameter. They appeared to be grown by hydrothermal dissolution/precipitation due to high temperature and high pressure water. The elimination of these larger Al_2O_3 particles is an important study needed in further application of the hydrothermal oxidation method.

CONCLUSION

The hydrothermal oxidation of Zr-Al alloys lead to the following conclusion:

(1) Fused Zr_5Al_3 sample reacted with 100 MPa water above 400°C to form boehmite, then formed α-Al_2O_3, m-ZrO_2 and t-ZrO_2 above 500°C.

(2) Fused $ZrAl_3$ samples reacted similarly but needed above 600°C for the complete oxidation.

(3) The reaction temperature could be reduced below 300°C by crushing the fused alloy samples.

(4) The fine powders consisted of α-Al_2O_3, m-ZrO_2 and t-ZrO_2 having the average crystallite size of 27-35 nm, 13-24 nm and 9-12 nm, respectively, could be prepared at 600-700°C.
(5) The hydrothermal oxidation of alloys could produce the finer oxide powders than that of single component metal could.
(6) The possibility of this method was demonstrated to produce homogeneously dispersed mixed fine powders.

ACKNOWLEDGMENTS

The authors wish to thank JEOL, Co., Ltd. and Hitachi Co., Ltd. for use of TEM and EDS apparatus.

This work was partially supported by Grant-in-Aid for Scientific Research No. 443015, 1980-81, and the Asahi Glass Foundation for Industrial Technology.

REFERENCES

1. D. W. Johnson, Jr., Am. Ceram. Soc. Bull., 60 [2], 221-24, 243 (1981).
2. C. R. Veale, Fine Powders, Preparation, Properties and Uses, Halsted Press Div., Wiley & Sons, NY, 1972.
3. J. L. Pentecost, pp. 1-14 in Ceramic Fabrication Processes, edited by F. F. Y. Wang, Academic Press, NY, 1976.
4. M. Hoch, pp. 13-28 in Science of Ceramics, 9, edited by K. J. DeVries, Nederlandse Keramische Verenging, Noordwijkerhout, The Netherlands, 1977.
5. K. S. Mazdiyasni, Ceram. Int., 8 [2], 42-56 (1982).
6. E. Tani, M. Yoshimura, and S. Somiya, J. Am. Ceram. Soc., 66 [2], in press (1983).
7. T. Shibuya, Yogyo Kyokai Shi, 90 [10], 603-609 (1982).
8. M. Yoshimura and S. Somiya, Am. Ceram. Soc. Bull., 59 [2], 246 (1980).
9. M. Yoshimura and S. Somiya, pp. 453-63 in Science and Technology of Zirconia, Advances in Ceramics, Vol. 3, edited by A. H. Heuer and L. W. Hobbs, The American Ceramics Society, Columbus, 1981.
10. M. Yoshimura and S. Somiya, in Proc. of 5th Int. Round Table Conf. on Sintering, Sept. 7-10, 1981, Portoroz, Yugoslavia, in press.
11. H. Toraya, M. Yoshimura, and S. Somiya, J. Am. Ceram. Cos., 65 [5], C-72 (1982).
12. H. Toraya, M. Yoshimura, and S. Somiya, J. Am. Ceram. Soc., 66 [2], in press (1983).
13. M. Yoshimura, S. Kikugawa, and S. Somiya, submitted to Funtai oyobi Funmatsu Yakin (J. Japan Soc. Powder Metallurgy).

14. P. E. D. Morgan, pp. 67-76 in Processing of Crystalline Ceramics, edited by H. Palmour III, R. F. Davis, and T. M. Hare, Plenum Press, NY, 1978.
15. P. K. Gallagher and F. Schrey, Thermochimica Acta, 1, 465-76 (1970).
16. C. Sheer, S. Korman, D. J. Angier, and R. P. Cahn, pp. 133-52 in Fine Particles, 2nd International Conference, edited by W. E. Kuhn, Electrochemical Society, Princeton, NJ, 1974.
17. M. Yoshimura and S. Somiya, pp. 23-26 in Proc. of the 4th Int. Conf. on Rapidly Quenched Metals, Vol. I, edited by T. Masumoto and K. Suzuki, Japan Institute of Metals, Sendai, 1982.
18. A. M. Alper, pp. 339 in Science of Ceramics, 3, edited by G. H. Steward, Academic Press Inc., London, 1967.
19. A. Krath and H. Meyer, Ber. Deut. Keram. Ges., UU, 61-72 (1965).

COMBUSTION SYNTHESIS OF TRANSITION METAL NITRIDES[a]

J. B. Holt and D. D. Kingman

University of California
Lawrence Livermore National Laboratory
Livermore, CA 94550

INTRODUCTION

The solid-state synthesis of refractory compounds, such as transition metal carbides, borides and nitrides etc., usually requires high temperature furnaces and long reaction times. More innovative methods for the formation of refractory powders include the gaseous reaction of metal and non-metal ions in high temperature environments. Disadvantages of the former method are impure products and inefficient operation, while the latter method deals with complex reactions and expensive equipment.

Recently, Russian scientists led by A. G. Merzhanov introduced another process for the formation of refractory materials.[1-3] Based on the high exothermic heat of chemical reactions between many elements and/or compounds, the process is both simple and efficient. For example, metal powders may be mixed with non-metal components, such as boron and carbon, and cold-pressed into porous cylindrical compacts. If a compact is ignited at one end with a hot tungsten coil or pulsed laser, then the high temperature reaction front rapidly self-propagates through the sample in a few seconds, producing a metal boride or carbide. Whether a powder or porous sintered part is formed depends on several factors, the most decisive being the combustion temperature. The advantages of this process, described as Self-Propagating High Temperature Synthesis (SHS), are

[a]This work was sponsored by DARPA and performed under the auspices of the U.S. Department of Energy by the Lawrence Livermore National Laboratory under contract number W-7405-ENG-48.

high purity products, high rates of synthesis, energy efficient (no furnaces), potential for the formation of new metastable compounds and enhancement of physical and chemical properties of the refractory products.

One important class of refractory materials is the transition metal nitrides. According to several reports[4-6] TiN,ZrN, and HfN have been synthesized by the SHS process. Cold-pressed compacts of the metal powder are ignited in either liquid nitrogen or nitrogen gas. The combustion product is a highly porous (40-50%) nitride material. One serious drawback of this process for the formation of nitrides is the relatively low percent of conversion of metal to nitride. For example, in one experiment we ignited a loosely packed -325 mesh titanium powder in nitrogen gas at one atmosphere. By weight gain measurements there was only 35% conversion, assuminmg the reaction to proceed according to equation (1)

$$Ti + 1/2N_2 = TiN \tag{1}$$

The weight gain measurements were confirmed by metallographic examination of the reacted product which revealed large regions of unreacted metal. To circumvent this problem the Russian investigators conducted their combustion experiments in high nitrogen pressures (10-4,000 atmo).[7] With these high nitrogen densities they claim to have achieved full conversion. Unfortunately, the requirement of high pressure complicates the experimental procedure and significantly reduces the advantage of simplicity and efficiency.

The reason for the low degree of conversion of Ti to TiN may be the high combustion temperature. Table 1, provides a list of the calculated adiabatic temperatures for many of the technically important nitrides. Although the actual combustion temperature for reaction (1) is below the adiabatic temperature, it is higher than the melting point of titanium metal (1660°C). Consequently, the titanium metal in front of the rapidly moving combustion wave melts and forms a barrier for the inward transport of nitrogen gas. Russian workers[7] suggest that reducing the combustion temperature by the addition of an inert diluent such as TiN will promote higher conversion. Figure 1 shows the experimental data for the percent of conversion of a loosely packed titanium powder versus the weight percent of TiN additive. An increase in conversion is observed but much lower (55%) than full reaction.

A new process developed at LLNL, utilizing a solid source of nitrogen, that provides 100% conversion of metal to the nitride will be discussed in this paper. This process retains all of the favorable advantages of the SHS or Combustion Synthesis process. In addition to the synthesis of single phase compounds, the process has been extended to the formation of composite materials.

Table 1. Calculated Adiabatic Temperatures and Corresponding Melting Temperatures of a Group of Technologically Important Nitrides

Nitride	Adiabatic Temp. (°C)	Fusion Temp. (°C)	Heat of Formation (Cal/g)
HfN	4820	3310	458
ZrN	4630	2980	830
TiN	4630	2950	1300
Si_3N_4	4030	1900 (sublimes)	1260
BN	3430	3000 (sublimes)	2448
VN	3230	2230	924
TaN	3090	3090	308
AlN	2620	2230	1560

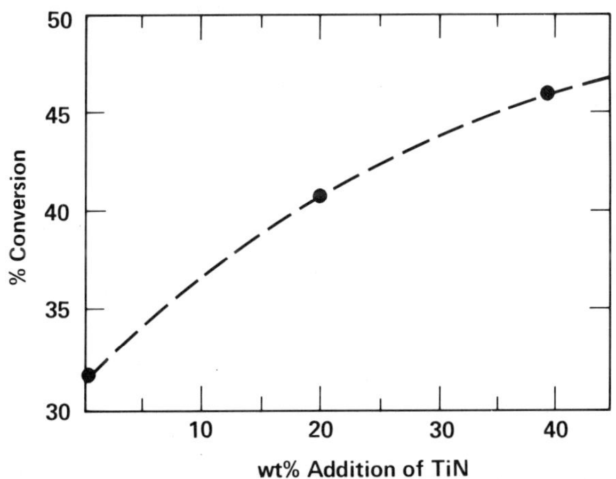

Fig. 1. Percent conversion of titanium to titanium nitride is plotted versus the weight percent of titanium nitride added to the starting powder as a diluent.

EXPERIMENTAL PROCEDURE

The search for a material that would provide the necessary nitrogen for complete conversion in the combustion reaction identified Sodium Azide (NaN_3) as a potential solid source of nitrogen. Since NaN_3 is non-explosive and decomposes into nitrogen and sodium metal vapor, then it becomes the ideal material for the combustion synthesis of the refractory transition metal nitrides. Mixtures of Zr, Ti, and Hf powder with NaN_3 readily ignite producing 100% conversion of the metal to the nitride.

For example, a -325 mesh Ti powder with an average particle size of 22 μm was mixed with NaN_3 powder in a glass jar on a vibratory mill for 5 minutes. The powder was cold-pressed into cylinders or poured into quartz crucibles ~ 2.5 cm in diameter and 2.5 cm high. The powder composition was formulated according to equation (2)

$$3Ti + NaN_3 = 3TiN + Na\uparrow \qquad (2)$$

The quartz crucible containing the powder was placed in a glove box which was pumped down to ~ 200μm. After back-filling with nitrogen to one atmosphere the powder was ignited by a tungsten coil which was in contact with the surface of the powder. The combustion wave moved down the crucible in an irregular mode, i.e., the front was not a smooth plane. When the combustion wave reached the bottom of the crucible the nitride product glowed even brighter than the combustion front. This observation confirms that chemical reaction continues (several seconds) after the combustion front has passed. The combustion temperature was estimated to be at least 2200°C at the maximum point. A thick cloud of smoke, consisting of Na vapor and other volatile gases from the initial powders, evolved during reaction. The reaction product is a loose, friable powder that is easily removed from the quartz crucible. Figure 2 shows a photograph which demonstrates the actual movement of the combustion front through the reactant materials. The average grain size of the TiN powder is 10μm as shown in a SEM micrograph in Figure 3. X-ray diffraction anlysis verified the powder to be TiN with no trace of Ti metal. Since the stoichiometry of TiN can vary over a large range there was a question as to the N/Ti ratio of the combustion product. The value of this ratio is dirctly related to the lattice parameter. From previous experimental measurements[8] the lattice parameter of $TiN_{0.9}$ has been determined to be 4.234A. The experimental lattice parameter from our x-ray diffraction data was 4.238A which indicates that the product is near stoichiometric TiN.

Emission Spectrochemical Analysis of the powders in Table 2 shows TiN to be as pure as the Ti reactant powder. The increased amount of Si, B and Fe in the TiN are a result of sampling procedure and not any contamination during reaction. Na was not detected in

Fig. 2 Photograph of the combustion of titanium and sodium azide powders as the combustion front has moved nearly halfway down the powder charge. The reaction is completed in several seconds.

Fig. 3 An SEM micrograph of the titanium nitride grains after being produced by the combustion of titanium powder with sodium azide.

Table 2. A list of cation impurities in the reactant titanium powder and the combustion-formed titanium nitride as determined by emission spectroscopy.

	Ti	(ppm by wt.) TiN
Al	3000-4000	4000-5000
Ca	2000	1000
Zr	1000	600
Mg	100	6
Si	⩽100	2000
Ba	Nd <30	100
Mn	20	100
Sr	10	20
Fe	⩽5	100
B	Nd <3	100
Cu	Nd <3	3
Cr	Nd <1	20
Ni	Nd <1	3

Na was not detected <300

this analysis which confirms our assumption that Na is volatile and will not be a contaminant in the product phase. ESCA examination of the first few atomic layers of the surface of the grains reveal O, C, Al and a small amount of Na to be present. The O and Al were probably impurities in the initial Ti powder. The overall percentage of oxygen in the TiN has not been determined but is believed to be quite low ($< 0.1\%$).

HfN and ZrN were synthesized by the same procedure as described for TiN. In both instances the metal powders were completely converted to the nitride. To prove that the nitrogen was supplied by the NaN_3 powder, several experimental runs were conducted in argon gas at one atmosphere pressure. The nitride products were near full conversion but there was still some unreacted metal. For this reason all of the subsequent experiments were carried out in one atmosphere of nitrogen.

The versatility of this new synthesis process was demonstrated further by several other reactions shown by the following equations:

$$3TiO_2 + 4Al + NaN_3 = 3TiN + Al_2O_3 + Na\uparrow \qquad (3)$$

$$Ti + 0.5C + 0.167NaN_3 = TiC_{.5}N_{.5} + 0.167Na\uparrow \qquad (4)$$

Equation (3) shows a composite material of TiN and Al_2O_3 while equation (4) is a solid solution of TiC and TiN. (Titanium Carbonitide). These examples illustrate the potential for the production of a wide variety of products. A full description of these other systems will be reported in another paper.

DISCUSSION

The mechanism of the combustion process for the synthesis of the transition metal nitrides was not investigated and no attempt was made to refine the product phase by variation of experimental parameters such as metal particle size and distribution, combustion temperature, amount of NaN_3, initial porosity of reactants, etc. As noted above, the combustion front moved in an irregular, undulating fashion. This is much different from the smooth steady-state combustion wave observed in the combustion of a mixture of Ti and C powders. The theory of the combustion of metal powders in nitrogen has been studied[8,9] but is still not well understood.

This new process, utilizing NaN_3 as a solid source of nitrogen, has many advantages over more conventional methods for the production of transition metal nitride materials. First, the process is very simple in concept and procedure. The metal powders are fully converted to the nitride in a matter of seconds. Second, the process forms high purity products. Usually, the nitrides are purer than the original metal because of the volatilization of impurities at the high temperatures achieved in the reaction (similar to zone refining) and the absence of contaminants routinely found in high temperature furnace atmospheres. Third, the process is energy efficient since no furnaces are required and the rate of synthesis is fast.

Further study of the mechanism should provide optimum conditions for the process. Full exploitation of the combustion process will be realized when synthesis and densification are combined in one step.

REFERENCES

1. A. G. Merzhanov and I. P. Borovinskaya, Doklady Akad. Nauk SSSR, 204 [2], 366-69 (1972).
2. E. I. Maksimov, A. G. Merzhanov, and V. M. Shkiro, Fizika Goreniya i Vzryra, 1 [4], 24-30 (1965).
3. A. G. Merzhanov, A. K. Filonenko, and I. P. Borovinskaya, Doklady Akademii Nauk SSSR, 208 [4], 892-894 (1973).
4. I. P. Borovinskaya and A. N. Pityulin, Fizika Goreniya i Vzryva, 14 [1], 137-140 (2978).
5. I. P. Borovinskaya and V. E. Loryan, Doklady Akad. Nauk. USSR, 231 [4], 911-914 (1976).

6. Yu. M. Shul'ga, V. E. Loryan, V. K. Yatsimirskii, I. P. Borovinskaya, and Yu. C. Borod'ko, Poroshkovaya Metallurgiya, [11] (203), 1-5 (1979).
7. I. P. Borovinskaya and V. E. Loryan, Poroshkovaya Metallurgiya, [11], 42-45 (1978).
8. A. G. Merzhanov, I. P. Borovinskaya and Yu. E. Volodin, Doklady Akad. Nauk SSSR, 206 [4], 905-908 (1972).
9. A. G. Merzhanov, and I. P. Borovinskaya, 10, 195-201 (1975).

DISCUSSION

A. Pasto (GTE Labs): I would suggest the application of uniaxial pressure during reaction to obtain an enhanced densification, yielding a high density product. The residual gases (Na↑) would have to be drawn off, e.g. *in vacuo*.

Author: Going from the initial elements to a final dense product would be the optimum condition for this process. The method you suggest might be used for the in situ densification.

J. Ings (U.S. Army): The Army would like to investigate potential processes for making dense, large shapes of carbides, borides and nitrides (6 inch X 6 inch X 1 inch). A comment by someone in the audience suggested applying pressure while combusting your mixtures. Does this have any merit in your opinion?

Author: The in situ densification of certain carbides and borides formed by the combustion process is a realistic goal. Although the densification possibility also exists with the nitrides much more work is required. Specifically, we need to know the ignition temperatures of the transition metals with nitrogen from the sodium azide.

R. H. Baney (Dow Corning): Is the process really energy efficient since pyrophoric materials and electrolytic processes are needed?

Author: We understand that sponge titanium is a relatively inexpensive form of the metal. If that can be used with the sodium azide, we still think that the combustion process is energy efficient.

R. Rice (NRL): (1) The density of the products is typically higher than the average densities of the constituents. How much of a problem does this present, e.g., in resultant porosity? (2) How scalable is the process in view of the outgassing and heat dissipation being very dependent on size and shape of the compact?

Author: The first question relates to the possibility of in situ densification of the TiN, ZrN and HfN material that we have discussed all are powders. Whether in situ densification can be realized with the nitrides is still an open question.

As the process is scaled up, the outgassing may present problems. We are not sure of the difficulty right now. The heat dissipation should not be a problem. Although the reaction time will be fast, more time will be required before removing the products from the reactor.

S. Dutta (NASA): Chemical analysis indicated a significant increase in Al, Si and Fe impurities in TiN powder. How could you claim that the "combustion synthesis" method results in "high purity" products?

Author: Si, B and Fe impurities were introduced into the TiN by the handling procedures after combustion. The material was taken from the quartz crucible with a steel spatula and then ground in a boron carbide crucible. The large amount of Al comes from the original Ti metal powder. If properly sampled, the TiN should be as pure or purer than the reactants.

THE INFLUENCE OF POWDER SYNTHESIS TECHNIQUES ON PROCESSES OCCURRING DURING COMPACT FORMATION AND ITS SINTERING

Max Paulus

Laboratoire d'Etude et de Synthese des Microstructures
C.N.R.S.-E.S.P.C.I.
10, rue Vauquelin 75231 Paris cedex 05 France

INTRODUCTION

Usual equations for sintering assume a regular arrangement of spherical particles in a compact system and uniform small grain size, composition and structure. In fact, these conditions are seldom fulfilled. This explains most of the limitations of the sintering process.

Prime causes responsible for sintering defects may be classified in four interdependent groups:
(1) particle arrangement in the compact;
(2) chemical homogenization;
(3) particle and green compact density;
(4) heating thermocycle for multicomponent powder pellets.

These four topics and their interactions will be discussed in this paper. Furthermore, an attempt will be made to derive from this analysis the most suitable preparation processes to obtain the ideal compact and thermocycle.

PARTICLE ARRANGEMENT IN THE COMPACT

If we consider a compact containing an undeformed stacking of spherical particles, the theoretical interstitial porosity of this "ideal" compact is ≃ 25%. However, experimentally, one obtains we get after compression (with eventual deformation and fracture) a green porosity ranging from 40 to 60%. This means that porosity is twofold: (1) the interstitial porosity ≃ 25%, and (2) the unoccupied site porosity which is normally between 15 and 35%.

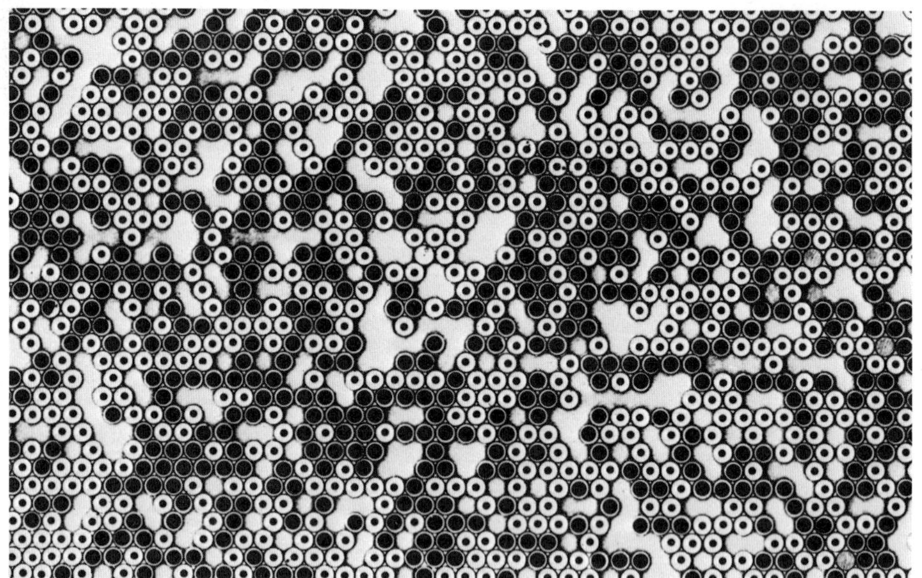

Fig. 1. Statistical distribution of two kinds of particles in a honeycomb net. 2/3 of the sites are occupied by the particles (circles); 1/3 ◉, 1/3 • and 1/3 unoccupied. These values assume a porosity of the sample of 50% due to the interstitial voids. This aleatory picture, from a Montecarlo method, points out that clusters of specific particles or unoccupied sites are not due to insufficient mixing but to statistical distribution.[1]

This author believes this unoccupied porosity is due to insufficient or bad pressing conditions. In contradiction to this statement, Lacour[1] indicates that unoccupied site porosity is normal when pressing a powder. In a statistical study he has established a random distribution of spherical particles with the same diameter and located these particles at the nodes of a compact hexagonal lattice. The location of the particles was obtained by computer generation of uniformly distributed random numbers. Comparison with pseudo-spherical, well separated powders of Fe_2O_3 having the same total porosity indicates a similarity between the unoccupied site porosity in the experimental green compact and the calculated random distribution. Figure 1 shows the statistical distribution of two kinds of particles and of the unoccupied sites. It is clear that both singular unoccupied sites and clusters of these sites may be observed. The number of clusters of each size (given in numbers of unoccupied sites) is drawn in Fig. 2. The existence of unoccupied sites drastically increases the pore size. For example, the interstitial pore size is multiplied by 2 - 3 for one unoccupied site. It is also important to keep in mind that for a porosity of 50%, one percent of the clusters reaches a size of 9 sites. These big pores reduce and limit the sintering process, not only by increasing the radius of curvature of the pores and the length of the diffusion path, but also by migration of vacancies from small pores interstitial to big ones (unoccupied site clusters) resulting in a pore coalescence. Therefore, one must use every endeavour to reduce the formation of unoccupied site clusters by using appropriate synthesis and consolidation processes.

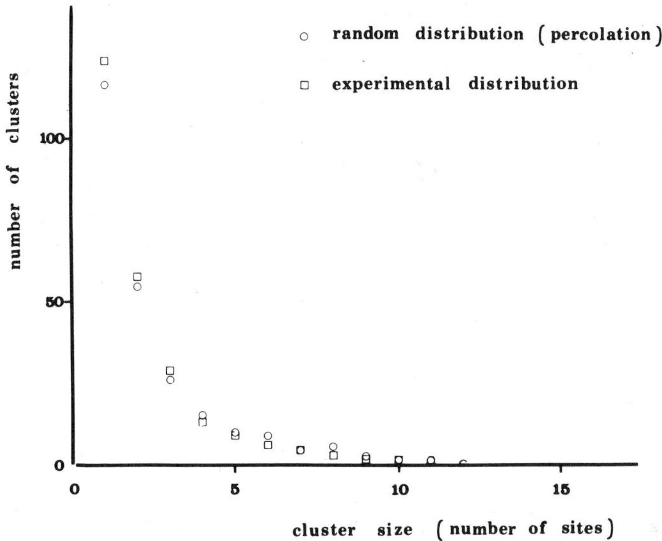

Fig. 2. Random (O) and experimental (□) distribution of unoccupied site clusters as a function of their size (expressed in number of sites).[1]

CHEMICAL HOMOGENIZATION

This parameter has already been discussed many times[2,3] but due to its fundamental role it seems necessary to review the main points.

It is an experimental fact that solid mixtures are chemically heterogeneous. Local electron beam analysis and X-ray pictures (Fig. 3) reveal this heterogeneity. Even a perfect mixing of particles does not lead to a homogeneous distribution of species. A statistical approach to this problem shows that the random distribution of particles or agglomerates gives rise to clusters of the different species (Fig. 1). This distribution is the best we can expect after a theoretically perfect mixing. It will be shown momentarily that reduction of the particle or agglomerate size to that of the molecule or ion is the only way to increase the chemical homogeneity.

These heterogeneities in composition oppose complete densification of the material. Since the diffusion coefficients of the different elements are not identical, a Kirkendall effect with migration of initial porosity and pore coalescence is generally observed during homogenization.[5] This decreases the rate of sintering drastically because big pores from Kirkendall effects (like those from non-uniform initial porosity distribution) are almost impossible to eliminate. The diffusion path is greater and the saturation in vacancies much lower in the vicinity of big pores in comparison with small ones. Therefore, chemically homogeneous powders are necessary for a rapid and nearly complete sintering (to theoretical density).

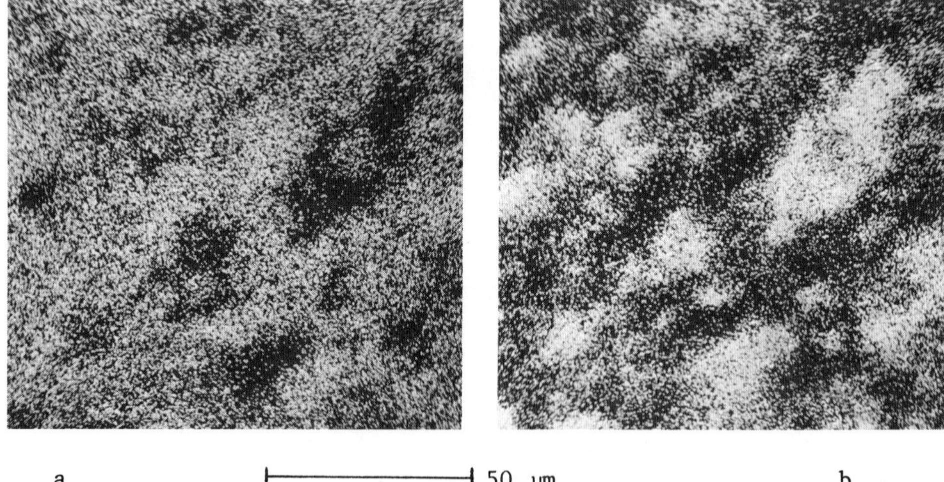

Fig. 3. Green pressed mixture of oxide particles (1-2 µm) having an overall cordierite composition[4] $5SiO_2$, $2Al_2O_3$, 2MgO. (a) X-ray image of Al; (b) X-ray image of Si.

PARTICLE AND GREEN COMPACT DENSITY

In order to increase the chemical homogeneity of the powder particles and to reduce their size, different chemical preparation processes have been proposed. Usually these methods yield powders with a grain size of 50 to 200 Å, resulting in an increased sintering rate. In fact during the heat treatment a small amount of grain growth was primarily observed, but also (and it is the most significant restraining feature) the formation of necks between loose particles leading to a foam powder with a rigid skeleton (Fig. 4). This foam powder is very difficult to press to a low porosity compact. Porosity is far from the optimum interstitial one and may reach 50 to 60%.

Even an increase in the applied pressure during compaction does not solve the problem. The green density increases with the applied pressure but the sinted density decreases for a given applied pressure during the compaction. This phenomenon is clearly observed for yttria stabilized zirconia prepared by the freeze-drying technique from sulfate solutions (Fig. 5). Micrographic observation shows that this decrease in sintered density originates from cracks. Above a given pressure the cracks form during the pellet formation, and the two edges separate during the sintering. The overall density is decreased although local density increases.

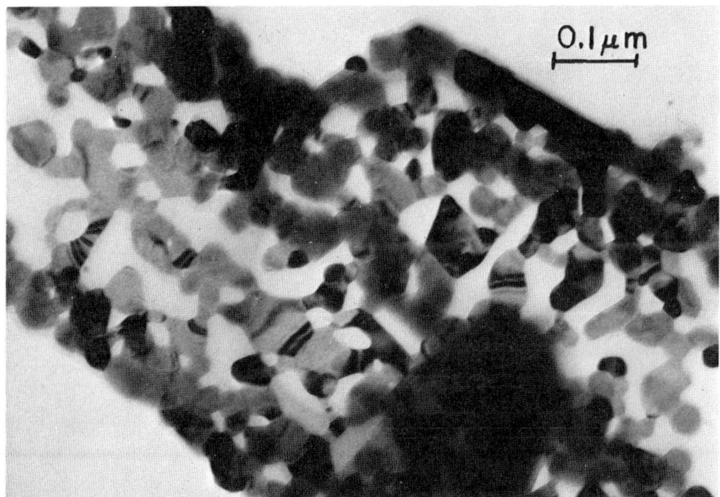

Fig. 4. S.E.M. micrograph of Yttria stabilized Zirconia powder prepared by freeze-drying of a sulfate solution followed by heat treatment at 1100°C in oxygen during 2 hours. The formation of a rigid skeleton in the loose powder appears clearly.[6]

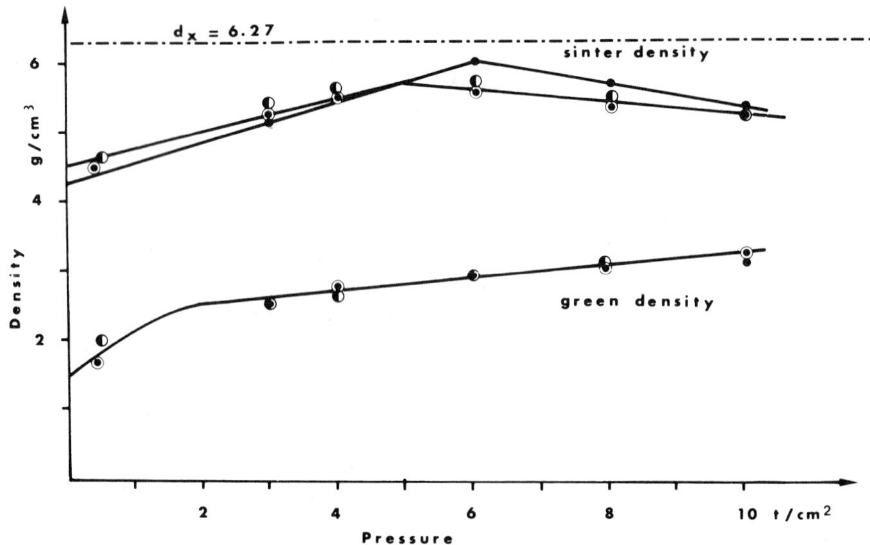

Fig. 5. Green and sinterdensity as a function of compaction pressure for yttria stabilized Zirconia powder prepared by the freeze-drying technique. Sintering at 1400°C during 2 h--heating rate 300°C/h--heat treatment at 900°C during 2 h--liquid solution concentration expressed in oxide per liter.[6] O 65 g/l; ⊙ 100 g/l; ◑ 160 g/l.

When the pressed density decreases the number of bonds between particles (12 per particle in a hexagonal compact lattice) decreases drastically. Therefore the local sintering is no longer even locally isotropic. Figure 6 shows the particle arrangement of transition Al_2O_3 after pressing and a very light sintering at 1100°C in order to stabilize the compact for observation after impregnation. It is clear that for a total porosity of ≃ 60% (this means 35% of unoccupied sites) one obtains a stacking of particle chains (2 bonds or necks in a plane). These chains are contorted and the sintering (neck growth and closer approach of particle center) results not so much in densification but in straightening of particle chains.[7] This phenomena is enhanced when a volume-altering phase transformation occurs, as in the transition to α-Al_2O_3, because it allows the breaking up of some necks leading to chain-like particle arrangement.

Depending on the initial state (density, particle shape and size, heat treatment parameters, neck formation in loose powder, pressing of powder, etc.) through the straightening of the particle chain one may obtain a reticulation of the compact structure. Figure 7 shows this phenomena which results in a relatively dense grain boundary-like structure with an inner radial low density structure. Figure 8 gives evidence of neck vanishing or breaking up which occurs during the straightening of particle chains. This phenomena, also results in low density and low bonding rate, which clearly acts as a deterrent to complete sintering.

HEATING SCHEDULE FOR MULTICOMPONENT POWDER PELLETS

We have seen in the second topic that, due to the Kirkendall effect associated with homogenization, the chemical heterogeneity limits the sintering process. But the influence of the heterogeneity may be radically different if some of the elements can form a mixture with a low melting point, e.g., eutectic. As the elements are distributed in analeatory manner, we can find in the pellet small volumes of each composition of the phase diagram formed by the elements present in the pellet.

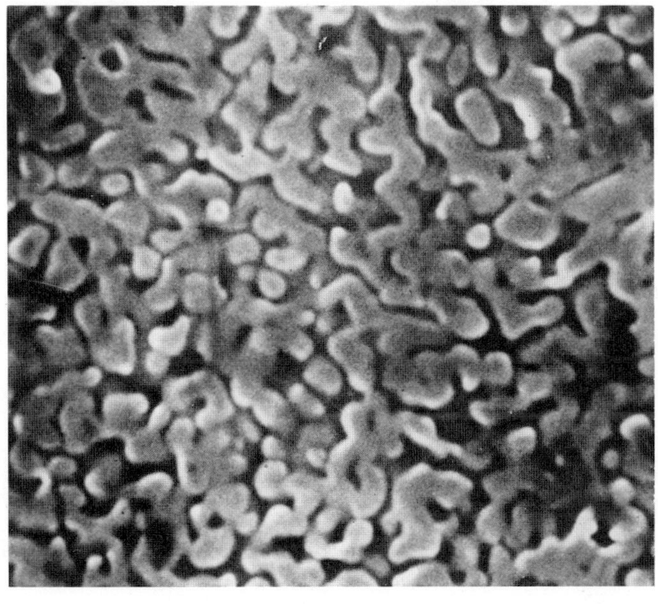

1μm

Fig. 6. Contorted particle chains. S.E.M. section of a transition Al_2O_3 pellet heated up to 1100°C. Impregnated with araldite before the metallographic preparation. Al_2O_3 powder elaborated by the freeze drying technique from sulfate solution.[7]

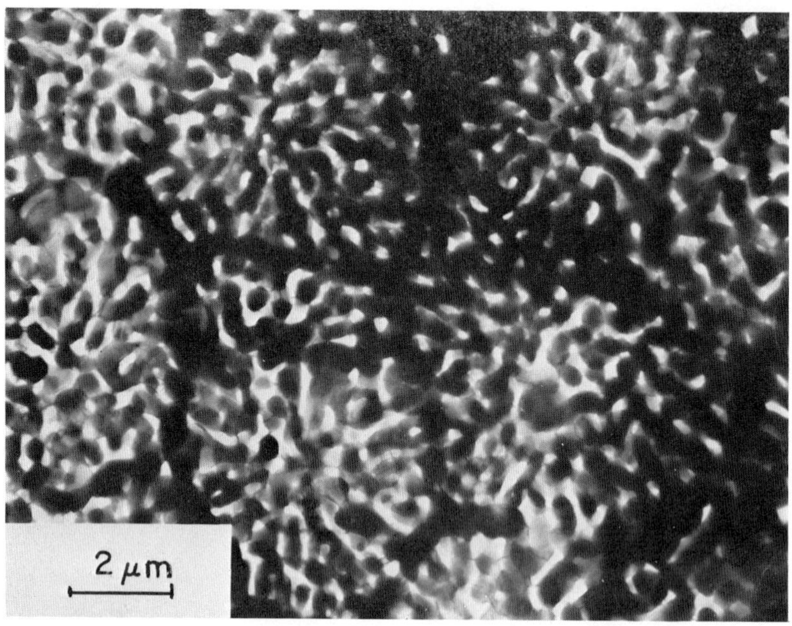

Fig. 7. Dense grain boundary-like structure with an inner radial low density structure[7] (negative grain). Same sample as for Fig. 6 but heated up to 1500°C (T.E.M.). Sample thinned by ionic bombardment.

Therefore, a mixture which gives rise to an overall compound which does not melt at a given temperature may be melting transitorily and at least locally. On the other hand, a compound normally in equilibrium with a liquid phase may show some part of the pellet in the solid state. This effect is valid whatever the sintering process (pure solid state or liquid phase sintering) and may be very important.[4] The main conditions for its observation are both the existence of a low melting point in the phase diagram and the limitation of the diffusion between the different species before the pellet reaches the maximum sintering temperature.

In order to simplify the discussion, we consider only two kinds of particle groups: the melting ones and the non-melting ones. The melting particles are in fact arrangements of different grains forming the melting compositions.

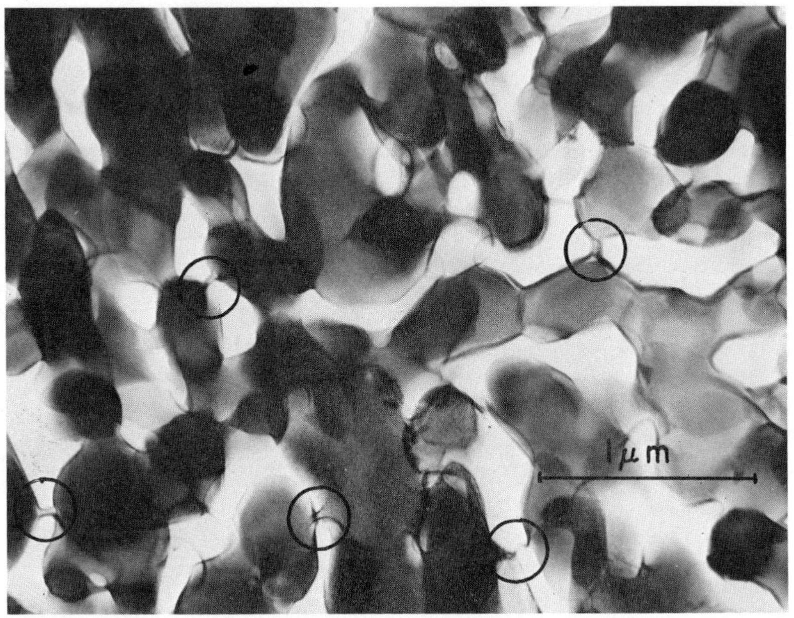

Fig. 8. Micrograph giving evidence (circle) of the vanishing or breaking up of neck during the straightening of the particle chain[7] (T.E.M.). Same sample as for Fig. 7.

Four different cases may be analysed separately depending on the amount of melting phases, their locations, and the sizes of grains and agglomerates. The cases A, B and C, containing decreasing amounts of transitory liquid phase, will be discussed. However, for convenience in explanation case C is presented before B.

A. <u>The melting particle exceeds a critical value</u>

The volume of the melting particles exceeds a critical value. The resulting liquid phase penetrates and separates the different grains, dissolving locally each oxide in accordance with the extent of the liquid phase area on the phase diagram (Fig. 9). The pellet deforms to a shiny drop and air bubbles are trapped in the liquid phase.

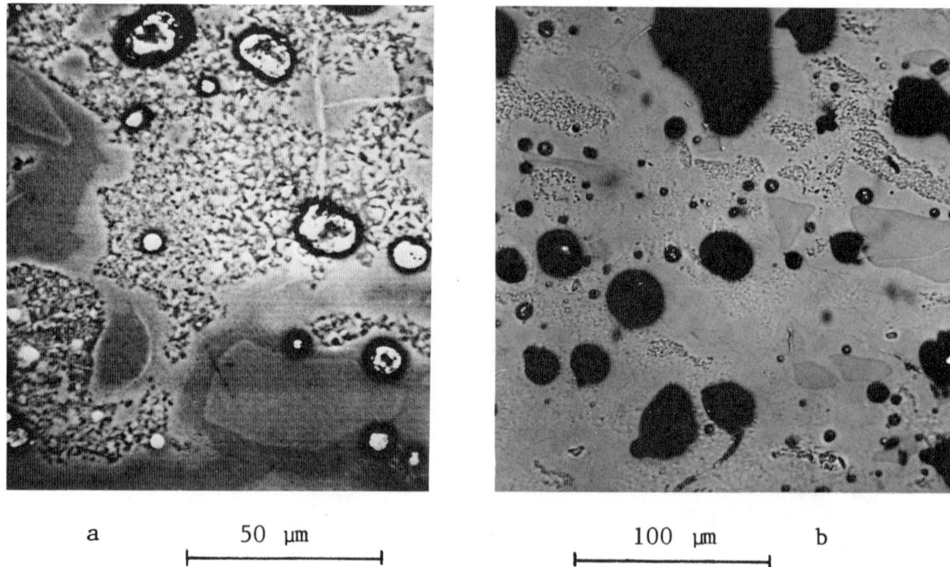

Fig. 9. Sample having the overall composition of cordierite which has been flash heated from room temperature to 80°C above the eutectic temperature and air quenched after 2 min at this temperature (1430°C). Oxide grain size ≈ 1-2 μm.[7] (a) S.E.M. (adsorbed electrons) (600X); SiO_2, Al_2O_3, glassy phase and cordierite precipitates--no pores in this area. (b) Optical micrograph (300X), same phases as found in (a). Pores (dark) are clearly seen. The small dark circles with the white structure in the middle are Al_2O_3 agglomerates.

C. The melting particle volume is below a critical value

In this case, we do not observe the large deformation as in the former case. Furthermore, two kinds of arrangement must be considered depending on the distribution type of the melting particles.

C-1 Very localized melting particles. If we have in the pellet very localized melting of particles with their volume remaining below the critical value, as in Fig. 10a, the sample remains apparently solid. Due to capillarity transport of the melt between the non-melting particles, the melted area leaves a large cavity (Fig. 10b). This suction of the melt into the interstices between the non-melting particles can be noted by the formation of pot-holes at the surface of the sample.

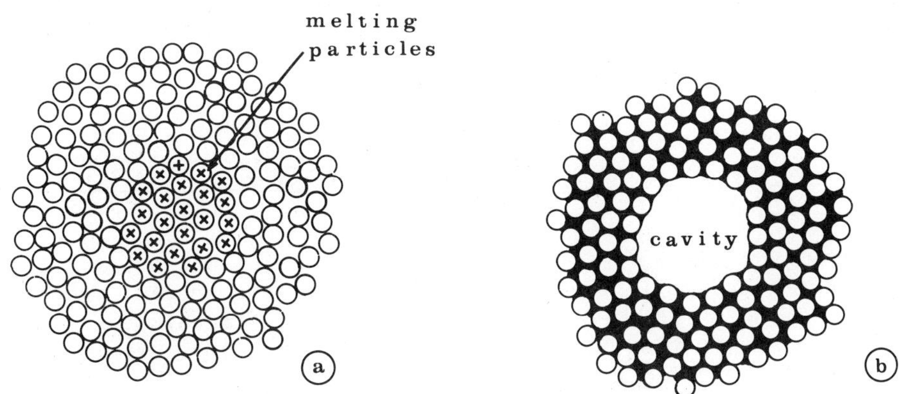

Fig. 10. Evolution of very localized melting particles.[4]
(a) Before melting; (b) after melting, the liquid phase diffuses by capillarity in the surrounding particles.

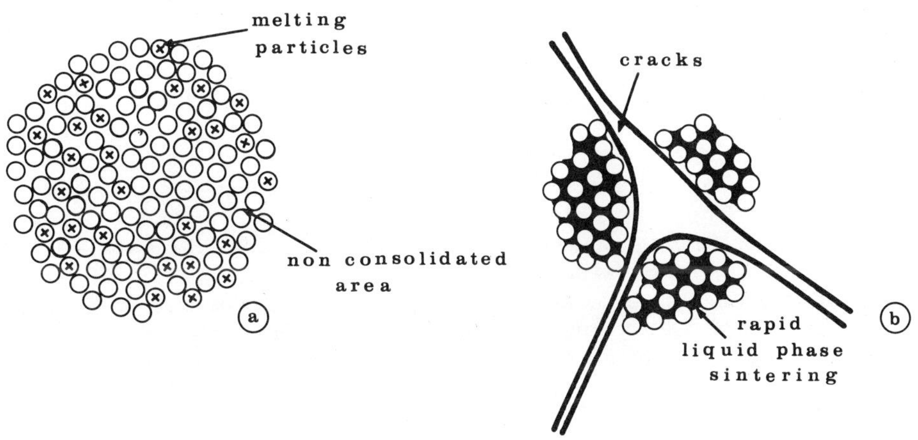

Fig. 11. Evolution of roughly well distributed particles.[4]
(a) Before melting; (b) after melting, the liquid phase in some areas allows a rapid liquid phase sintering and formation of cracks in non consolidated areas.

C-2 Moderately well distributed melting particles. In this case we have not enough liquid phase to get the shiny drop but the melting particles are roughly well distributed as in Fig. 11a. In the parts of the pellet with melting particles we observe rapid liquid phase sintering with formation of cracks in parts almost free of liquid phase which have not yet been consolidated by solid state reaction (which occurs at higher temperature) as shown on the schematic of Fig. 11b.

B. The volume of melting particles is roughly equivalent to the critical value.

At the limit between the cases A and C the sample does not deform easily and rapid liquid phase sintering is observed. However, as the sample is not yet homogenized some parts contain an excess of liquid phase. During sintering the surrounding almost rigid skeleton squeezes the liquid phase out of the sample if channels to the surface exist. The schematic representations of Fig. 12a and 12b show the process and the micrograph.

Parameters Controlling the Non-Equilibrium Melting of a Sample and its Microstructures

The above discussion assumes little or no compound formation before reaching the melting temperature. The parameters controlling diffusion and formation are essentially temperature, time, grain size, and the degree of heterogeneity. Therefore, for one overall composition, as well as a given initial grain size of the starting oxides and a mixing or other process procedure for the powder which defines the degree of heterogeneity, an isothermal diagram can be established. For each isothermal heat treatment below the lower melting temperature of the phase diagram, the isothermal diagram will show the features of the sample after a heat treatment of a few minutes above the said melting temperature but below the melting temperature of the compound or solid solution. Figure 13 gives an example[4] for the sintering of cordierite ($2MgO$, $2Al_2O_3$, $5SiO_2$) which melts at 1465°C but is contained in a phase diagram which has a eutectic at 1350°C. Changes in the starting oxides or of the processing procedure for of the powders moves the different areas on the map.

OPTIMUM POWDERS AND HEAT TREATMENT SCHEDULES

From this survey of the parameters which affect the processes occurring during compact formation and its subsequent sintering, four main clues appear.

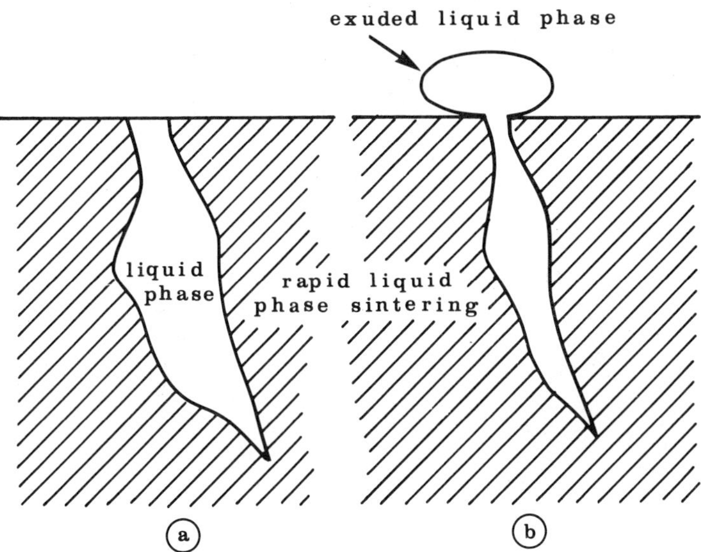

Fig. 12. Schematic representation of the formation of a pimple by the process of exuding of a liquid phase from a semi-rigid skeleton.[4] (a) At the beginning of melting; (b) after 2 min. of rapid liquid sintering around the liquid ladle.

1. One must start with very fine homogeneous powders but statistical and experimental results indicate that mixing per se cannot be the right way to achieve this objective. The use of stabilized liquid solutions (sol-gel as employed in the freeze-drying processes) or stabilized gas phases (as used in laser photochemical reaction) appear to be more appropriate approaches.

2. In addition, powders must not form chains or foams, otherwise the sintering process goes astray and "negative grains" (Fig. 7) may appear. Electrical or ultrasonic milling of powder would be a way to solve the problem but the best approach (if possible) is to avoid the neck formation between loose particles by using a very dilute medium for the formation of starting powders (e.g., gas phase photochemical reaction induced by a laser).

3. The random packing of particles leads to clusters of unoccupied sites limiting the final density. Processes controlling the stacking of particles such as electrophoretic or colloidal deposition, may reduce the number of these clusters.

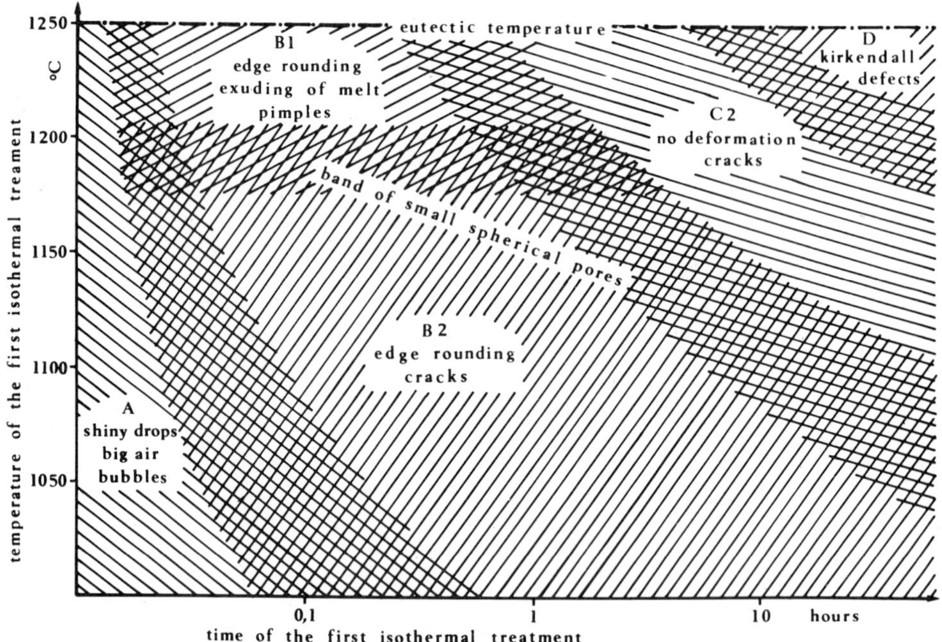

Fig. 13. Features map for cordierite composition versus diffusion parameter in the sample. The time and temperature axis are connected with the low temperature isothermal treatment (diffusion below the lower melting point, with these oxides: 1250°C). The given features have been observed after a second heat treatment (2 mn at 1430°C) identical for all samples. For the two successive treatments, the samples were flash heated.[4]

4. A mixture having a controlled heterogeneity may be chosen in order to sinter a non-melting compound with a transitory liquid phase by an optimum, two steps, fast firing cycle.[4]

REFERENCES

1. C. Lacour, Phys. Stat. Sol., (b), 96 (1979); Phys. Stat. Sol., (a) (to be published).
2. M. Paulus, Seminaires Chim. Etat Sol, t. 10 Masson edit, Paris, 1976.
3. M. Paulus, in "Processing of Crystalline Ceramics," Materials Science Research Vol. 11, H. Palmour IIIm R. F. Davis, T. M. Hare, eds., Plenum Press, New York, 1978, p. 17.

4. M. Paulus, F. Laher-Lacour, P. Dugleux, and A. Dubon, British Ceramic Society (to be published).
5. M. Paulus and P. Eveno, 6th International Symposium on the Reactivity of Solids, Shenectady (U.S.A.), 25-30 August 1968.
6. L. Rakotoson, These de 3eme cycle de Science des Materiaux, 16 janvier 1981, Universite Pierre et Marie Curie, Paris; L. Rakotoson and M. Paulus, Annales de Chimie Science des Materiaux (to be published).
7. M. Paulus, J. P. Fitoussi, and A. Dubon, British Ceramic Society (to be published).

DISPERSION AND PACKING OF NARROW SIZE DISTRIBUTION CERAMIC POWDERS

R. L. Pober, E. A. Barringer, M. V. Parish, N. Levoy, and H. K. Bowen

Materials Processing Center
Massachusetts Institute of Technology
Cambridge, Massachusetts 02139

INTRODUCTION

A major problem in ceramics processing is that improved and reliable ceramic materials having the required properties, critical elements in the development of high technology systems, cannot be reproducibly manufactured. This lack of reproducibility in properties lies in the inability to control the development of specified microstructures, which for sintered materials depends on the characteristics of the starting powder, the green compact microstructure, and the sintering and coarsening processes.[1] Although much significant work has been performed on the processes which occur during firing, many researchers now believe that the microstructures developed during sintering are essentially determined by the powder characteristics and packing.[2] Recent studies have demonstrated improved stability of the microstructure against exaggerated grain growth with a monodisperse powder,[3] and a significant reduction of sintering time and temperature with a submicron, narrow particle size distribution and with uniform, but not necessarily high, green density.[4,5]

More control of sintered microstructures is possible through the control of raw material properties (particulate size, size distribution, shape, and state of agglomeration) and of the packing process. Ideally, equiaxed, narrow size distribution powders should be uniformly packed so that the interparticle pores are no larger than a single particle; this type of green microstructure leads to shorter sintering times and more uniform shrinkages.

A class of processing techniques which produces greenware with uniform microstructures is based on the packing of uniform size particles from fully dispersed suspensions. Important considerations for the application of this technique are: a maximum size ratio of about 2 or 3 for the largest to smallest particles, a mean size of about 0.5 μm, and a liquid medium which will support a fully stabilized dispersion of the smallest particles present. After these conditions have been obtained, the particles may then be formed into dense, uniformly-packed greenware by various techniques such as: sedimentation, slip casting, tape casting (doctor blade), and electrophoretic deposition. Such microstructures sinter to full density and exhibit good stability against exaggerated grain growth due to the narrow size distribution of the pores and grains.

PARTICLE PREPARATION

The first step toward attaining uniform particle packing in greenware is to obtain or produce particles of narrow size distribution. Fortunately for the dedicated practitioner, there are a number of routes available to this end. A simple but tedious method is the fractionation (classification) of ground (comminuted) materials which have a wide size distribution. This can be accomplished by continuous air classification, which is not too tedious; however, the lower size limit which can be handled effectively is about 1 μm. In order to get to the submicron particle range, it is necessary to resort to gravitational or centrifugal sedimentation techniques. Sedimentation is conceptually quite simple, but the successful application of the technique requires an understanding of the surface and colloid chemistry of the powder and solvent systems of interest. This understanding is necessary in order to prepare fully dispersed suspensions from which the particles may be sedimented into various size fractions.

An alternative route to classification is the direct synthesis of a narrow size distribution powders; two basic techniques are solution precipitation reactions and gas phase reactions. The controlled liquid phase hydrolysis of metal alkoxides and organometallics is a promising technique for the synthesis of high-purity, uniform-size powders. Monosize, spherical TiO_2,[5] doped TiO_2,[6] and SiO_2[7] have been successfully synthesized in our laboratory. Initial experiments on $SrTiO_3$ have not yielded uniform size particles; however, concentration homogeneity has been achieved. The synthesis of TiO_2 from titanium tetraethoxide is discussed below in more detail to illustrate the potential of controlled precipitation reactions in raw materials fabrication.

A variety of techniques exist for powder synthesis via gas phase decomposition or oxidation of reactants. In our laboratory, high-purity, uniform-size powders are produced by controlled decomposition of gaseous reactants; a CO_2 laser is used as an intense, localized heat source.[8,9] This technique has been successful for silicon, silicon carbide, silicon nitride, and recently for aluminum oxide.

Classification of Barium Titanate

Classification of powders into narrow size increments is accomplished through the sedimentation of a dispersion. The rate of sedimentation is estimated by Stokes' law, a hydrodynamic relationship based on a balance of the drag and gravitational forces acting on a spherical particle:

$$\nu = \frac{2}{9}\,\frac{R^2(\rho_2 - \rho_1)g}{\eta} \qquad (1)$$

where ν is the steady state sedimentation velocity, R is the particle radius, ρ_1 and ρ_2 are the liquid and particle densities, respectively, g is the effective gravity, and η is the absolute viscosity of the liquid. Since the terminal velocity of a particle is proportional to the square of the particle size, a large particle or floc falls at a considerably greater rate than a smaller, individual particle.

Increments of particle sizes may be extracted by allowing a desired particle size to settle out of the dispersion of a known height in a calculated time. For example, a 0.4-0.6 µm powder may be obtained by allowing all particles larger than 0.6 µm to fall out of the dispersion. The remaining dispersion, which contains only particles less than 0.6 µm, is removed and the particles greater than 0.4 µm are allowed to settle. This sediment now consists of particles in the 0.4-0.6 µm size range.

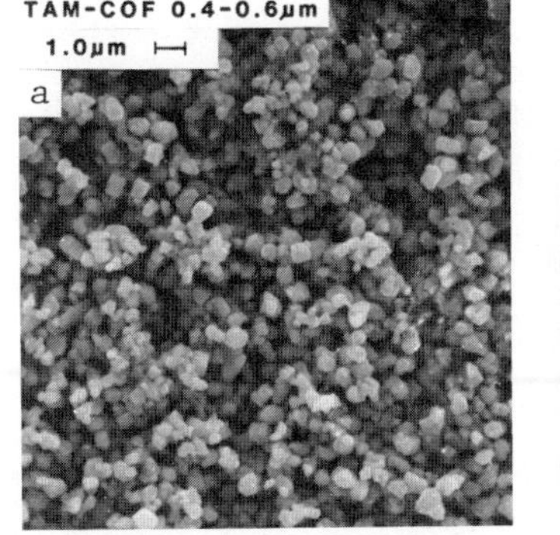

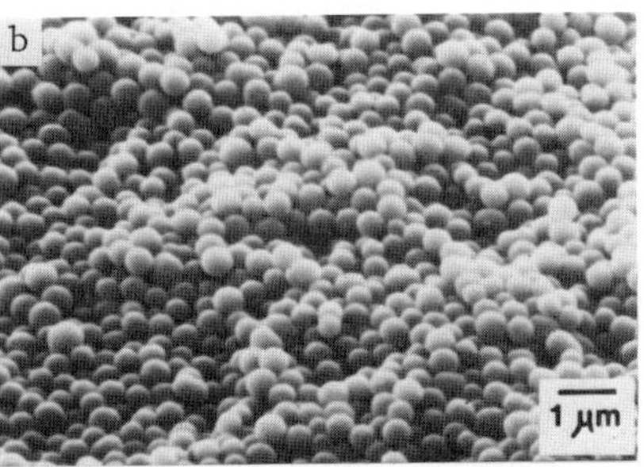

Fig. 1. (a) Classified $BaTiO_3$ (0.4-0.6 µm cut); (b) Monosize TiO_2 from $Ti(OC_2H_5)_4$ hydrolysis.

An SEM photomicrograph of a sample from a 0.4-0.6 μm classified powder is shown in Fig. 1a.[10] Note that the particle size is close to that predicted by Stokes' Law. A closer look at Fig. 1a shows a significant percentage of particles less than 0.4 μm. This is unavoidable since the sedimentation time for the second cut is determined for the removal of 0.4 μm particles from the dispersion. During this time, smaller particles settle from lower heights into the sediment as the larger size particles (> 0.4 μm) settle from the top to the bottom of the dispersion.

Solution Synthesis of Titania

High-purity, monosize TiO_2 powders are prepared by the controlled hydrolysis of dilute alcoholic solutions of titanium tetraethoxide, $Ti(OC_2H)_4$.[5] Since homogeneous nucleation of particulates is desired, all liquids (absolute ethanol, tetraethoxide, and deionized water) are ultrafiltered through 0.22 μm pore-size filters to minimize the level of insoluble impurities. For a given batch reaction (synthesis performed in a dry box), the tetraethoxide is dissolved in ethanol, and water is dissolved in a separate portion of ethanol. The two solutions are mixed using a magnetic stirrer, giving a solution with concentrations of 0.1 to 0.2 M ethoxide and 0.3 to 1.0 M water; the initial molar ratio of water to ethoxide is always 3 or greater. Subsequent precipitation of amorphous, hydrated TiO_2 occurs uniformly throughout the solution in 2 to 120 seconds (at room temperature); the time decreases as the initial concentration of either reagent is increased. The powder is repeatedly washed in deionized water or alcohol, depending on the choice of the final dispersion medium, and ultrasonically dispersed in the desired solution.

Figure 1b shows the fracture surface of a sediment cake consisting of monosize, spherical TiO_2 particles. The average particle size for a batch reaction is controlled by the initial reagent concentrations; powders have been produced with the average size ranging from 0.3 to 0.6 μm. The size distribution is very narrow for these powders. As a rule, the size ratio of the largest particles to the smallest particles is < 3, and the ratio of the largest particles to the mean size is ≤ 2. Semiquantitative spectroscopic results for the TiO_2 powder show a low cation impurity level (< 10 ppm Al, Ca, Cu, and Mg and < 100 ppm Si).

DISPERSION

The state of aggregation of a dispersed powder, which strongly influences powder packing and subsequent sinterability, depends on the stability of the dispersion against coagulation. When two particles approach each other, due to Brownian motion, to sedimentation, or to processing forces, they flocculate or repel depending on the sign and magnitude of the interparticle forces. The general equation describing these forces consists of attractive and repulsive terms,

$$V_T = V_A(\text{van der Waals}) + V_R(\text{electrostatic}) + V_R(\text{steric}) + V_R(\text{solvation}). \quad (2)$$

The van der Waals attraction between two spherical particles of the same nature with radius a and center-to-center distance r is given by[11]

$$V_A = -\frac{A}{6}\left(\frac{2a^2}{r^2 - 4a^2} + \frac{2a^2}{r^2} + \ln\frac{r^2 + 4a^2}{r^2}\right), \quad (3)$$

where A, the Hamaker constant, depends on the properties of the particles and the dispersion medium ($A \sim 10^{-12} - 10^{-13}$ ergs).

Generally for a given dispersion system, either electrostatic or steric forces dominate the repulsive term, while solvation forces contribute to both cases. Electrostatic repulsion arises from the interaction of the electrical double layers surrounding dispersed particles (described below), whereas steric repulsion results from the interaction of macromolecules which are adsorbed onto particle surfaces. In most cases, solvation forces are small; however, in some nonaqueous solvents of moderate to low dielectric constant, solvation forces contribute significantly to dispersion stability. In these systems electrostatic repulsion is small, and thus coagulation is retarded by the energy required to alter and remove solvation layers surrounding each particle.

The total two-body interaction energy curve, V_T, has the general form shown in Fig. 2. The height and position of the maximum repulsion, V_{max}, which controls the coagulation rate [$\propto \exp(-V_{max}/-RT)$], is primarily determined by the chemistry of the solid-liquid interface. Particle size and size distribution also affect dispersion stability; narrow size distribution powders are more stable than wide distribution (typical) powders.[12,13] In order to control and manipulate these interparticle forces during greenware fabrication, the surface and colloid chemistry of powder dispersions must be understood.

Electrostatic Stabilization

An electrostatic force results from the interaction of diffuse double layers surrounding two approaching particles; this force is repulsive for particles of the same chemical nature which have an electrostatic surface charge of the same sign and similar magnitude. The surface charge is formed when an oxide powder is put into a solvent that has a sufficiently high dielectric constant to support electrostatic charge development (especially water); the magnitude and sign of the surface charge depend on the solution chemistry. In response to the surface charge, counter ions in the solution are distributed at the solid/liquid interface to form an electric double layer.

The theory describing the total interaction energy is the Derjaguin-Landau-Verwey-Overbeek treatments[14] which incorporate a van der Waals attraction term and an electrostatic repulsive term of the form

$$V_R(\text{electrostatic}) \propto \varepsilon\varepsilon_0\psi_\delta^2\exp(-\kappa H) \tag{4}$$

where ε = medium dielectric constant, H = particle separation, ψ_δ = the Stern (diffuse layer) potential. The quantity $1/\kappa$ is the Debye length given by

$$1/\kappa = \left[\frac{\varepsilon\varepsilon_0 RT}{F^2 \sum_i c_i z_i}\right]^{1/2} \tag{5}$$

where F = Faraday constant, and c_i and z_i are the counter ion concentration and valence, respectively.[15]

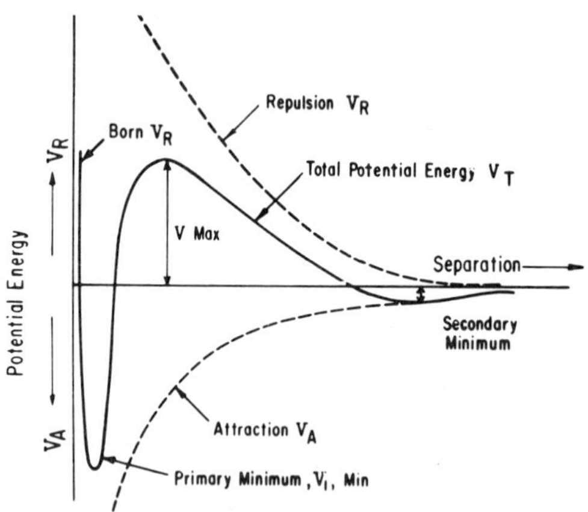

Fig. 2. Schematic of the two-body interaction energies.

The two dominant solution parameters that control the interaction energy are the electrolyte concentration, c_i, through κ, and $\psi\delta$, which is often approximated by the zeta (ζ) potential. The ζ-potential is a strong function of solution pH and a weak function of c_i, as shown in Fig. 3 for a crystalline TiO_2.[16] This figure shows that the isoelectric point (IEP), the pH at which no net charge exists in the solid-liquid interface region, is about 5.9. Thus, in order to develop a sufficient electrostatic repulsion to prevent coagulation (i.e., a large V_{max}), both solution pH and electrolyte concentration must be controlled. Specifically, the pH must be several pH units above or below the IEP and c_i must be low.

The principles of electrolyte and pH control have been successfully applied to the dispersion of many ceramic oxides. For example, aqueous dispersions of monodisperse TiO_2, with up to 35 volume percent solids, remain stable at a pH of 8 to 9. Although these dispersions may become concentrated (up to 60 volume percent solids) due to gravitational settling to form sediments, the particles remain nonagglomerated; the particles are readily resuspended by hand agitation. However, dispersions with much lower powder concentration rapidly coagulate at pH values between 5 and 7. Rapid coagulation also occurs at a pH of 9 when the electrolyte concentration is raised above 10^{-2} M. Thus, solution chemistry plays a major role on dispersion stability; the influence of stability on particle packing in green compacts will be demonstrated in a following section.

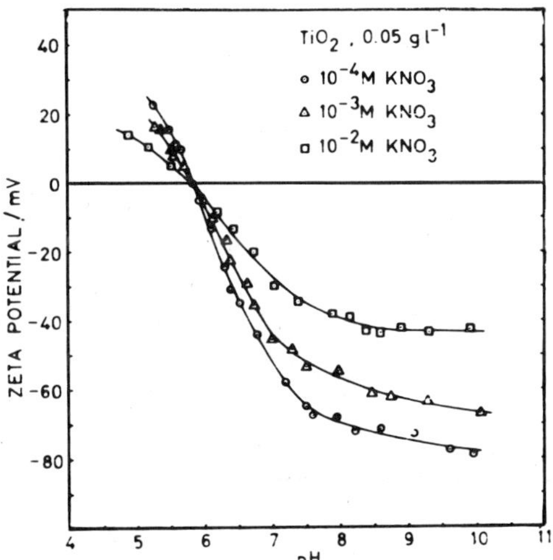

Fig. 3. Zeta potential versus pH for a crystalline TiO_2 (after Wiese and Healy[14]).

The electrostatic and solvation contributions to the stability of nonaqueous dispersions were evaluated in experiments performed on a commercial $BaTiO_3$ in a variety of pure organic solvents (alcohols, aldehydes, carboxylic acids, esters, ketones, ethers, and hydrocarbons), and solutions of hydrogen bonding solutes (e.g., benzoic acid, p-hydroxybenzaldehyde, and p-hydroxybenzoic acid) in ethanol.[9] Although stable dispersions were formed in all solvents, except the hydrocarbons, under dry conditions, moisture caused flocculation for the ethers, ketones, and esters studied. The mechanism for moisture sensitivity in these solvents, and the insensitivity in most aldehydes, alcohols, and acids is not well understood. However, competitive adsorption processes and changes in particle solvation are thought to be important. Benzaldehyde was found to be the best dispersing medium among the pure solvents.

Of the organic solutes in ethanol studied, p-hydroxybenzoic acid produced the most stable dispersions (much more stable than for the other solutes); the combination of acid and para alcohol functional groups on the aromatic ring resulted in extremely favorable solid-liquid interface interactions. The precise interaction mechanism is not known; however, since pure solvents which form stable dispersions interact with the $BaTiO_3$ surface through hydrogen bonding, those solutes which significantly enhance surface solvation through extensive hydrogen bonding should increase stability. In addition, surface charge is developed by the ionization of amphoteric $BaTiO_3$ surface sites through acid-base reactions with the organic molecules. The addition of an acidic solute increased the surface charge, thus increasing the electrostatic contribution to interparticle repulsion and dispersion stability. These experiments with $BaTiO_3$ indicate that both electrostatic and solvation forces are important in nonaqueous dispersions.

Steric Stabilization

In nonpolar organic solvents, electrostatic charge development and solvation are insufficient to obtain stable dispersions; thus, steric stabilization, arising from the repulsive interactions of adsorbed macromolecules as particles approach each other, must be employed. The magnitude of the repulsive force depends on macromolecule size, adsorption density, and conformation. (See review by Sato and Ruch.[17])

The steric stability of high-purity α-alumina powders in toluene has been studied for both menhaden fish oil and glycerol trioleate (used in the substrate industry).[18] The results show that adsorption, which occurs via hydrogen bonding, was much greater for the fish oil than for glycerol trioleate and that the fish oil imparted stability to the alumina dispersions, whereas the adsorbed trioleate did not. The degree of stability increased with an increase in the fish oil adsorption density.

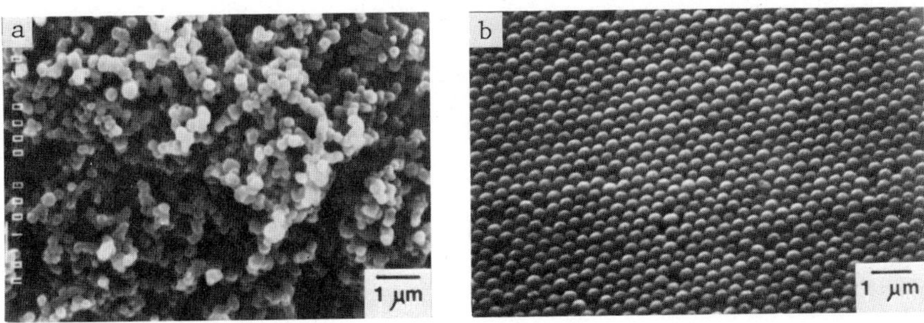

Fig. 4. TiO_2 sediment compacts formed from (a) an agglomerated dispersion, and (b) a stable dispersion.

PARTICLE PACKING

To achieve the goal of reliability and reproducibility in ceramics processing, uniformly-packed green microstructures must be obtained. Uniform particle packing, the consequence of a well dispersed, narrow size distribution powder, not only influences sinterability, but also results in uniform, controlled shrinkages. In contrast, wide size distribution powders cannot be packed to a high degree of uniformity (i.e., uniform pore-size distribution).

Particles may be packed from stable dispersions by various techniques which are suitable for both experimental and technological purposes. We have had success with such techniques as sedimentation and centrifugation, slip/dip casting, tape casting (doctor blade), and electrophoretic deposition; the powders investigated were TiO_2, $BaTiO_3$, ZnO, and Al_2O_3.

Sedimentation

The monosize TiO_2 powders, discussed previously, were dispersed and gravitationally settled into green compacts for sintering. The adverse effects of improperly chosen dispersion conditions on agglomeration and powder packing is shown in Fig. 4a. In this case, the powder was dispersed in deionized water at a pH of 5.9 (slightly acidic due to the absorption of atmospheric CO_2). Since the pH was near the IEP, the particles flocculated during sedimentation and subsequently formed a powder compact with nonuniform distribution of porosity. However, similar powder, when dispersed at a pH of 9, remained stable. Upon settling the powder from this dispersion a uniformly packed, dense green body was formed, as shown in Figs. 1b and 4b. These compacts, with approximate bulk densities of 65 percent of powder density, have a uniform distribution of pore volume and contain few voids larger than one particle diameter. The average number of particle-particle contacts appears to be greater than 11. In fact, the spherical particles were often packed into ordered arrays (Fig. 4b) which exhibited irridescence when illuminated with white light.

Several classified $BaTiO_3$ powders were centrifugally cast, dried, and observed by SEM. Narrow size distribution powders, 0.1-0.2 µm and 0.4-0.5 µm cuts, packed to uniform and high densities (up to 70% of theoretical). In contrast, 0.5-0.9 µm and unclassified powders packed to much lower and less uniform densities. Similar results were obtained for sedimentation cast thin films; uniform powders packed densely (Fig. 5a), whereas the unclassified powder segregated during the casting process (Fig. 5b).

Other Techniques

To test the general applicability of the aforementioned precepts for attaining very uniform green microstructures, experiments were performed to simulate commercial fabrication techniques. The techniques investigated included slip/dip casting, tape casting (doctor blade), and electrophoretic deposition. Uniform, dense packing was obtained for all casting techniques when stable dispersions of narrow size distribution powders were used; however, nonuniform, and often porous, packing resulted for wide size distribution powders.

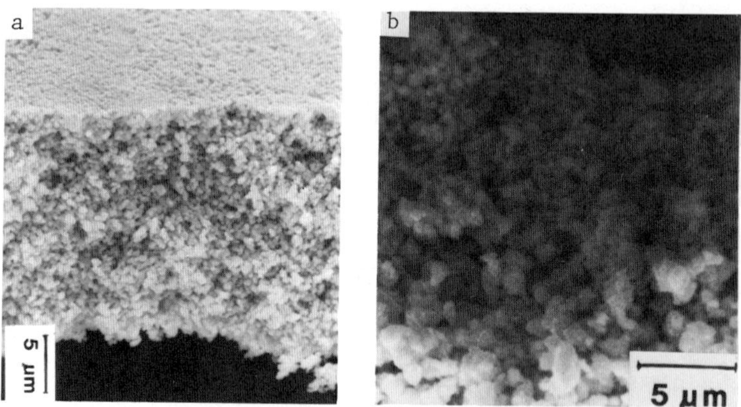

Fig. 5. Fracture surface of sedimentation cast thin films of (a) 0.4-0.5 µm $BaTiO_3$, and (b) as-received $BaTiO_3$.

Slip/dip cast thin films, formed by dipping a porous substrate into a stable slurry, were made using ZnO, TiO_2, and $BaTiO_3$. The casts for classified 0.4-0.5 μm and unclassified $BaTiO_3$ dispersed in isopropanol are shown in Fig. 6. Tapes were cast using classified (< 0.4 μm XA139) Al_2O_3 dispersions in isopropanol with p-hydroxybenzoic acid; uniform packing was observed for both the bottom and top (Fig. 7) surfaces of the film.[19]

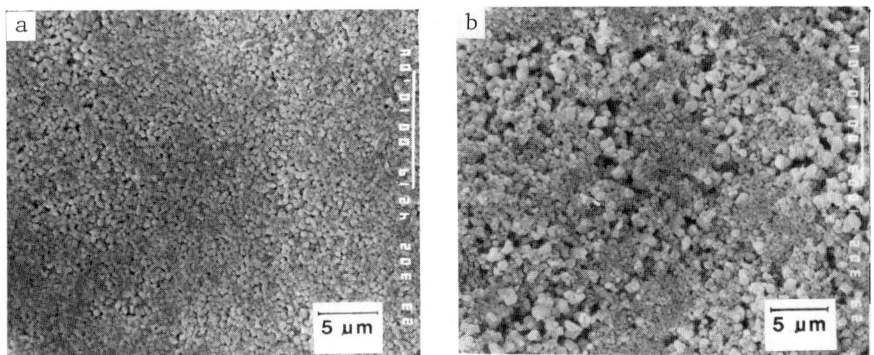

Fig. 6. Slip/dip cast thin films of (a) 0.4-0.5 μm $BaTiO_3$, and (b) 0.5-0.9 μm $BaTiO_3$ in isopropanol.

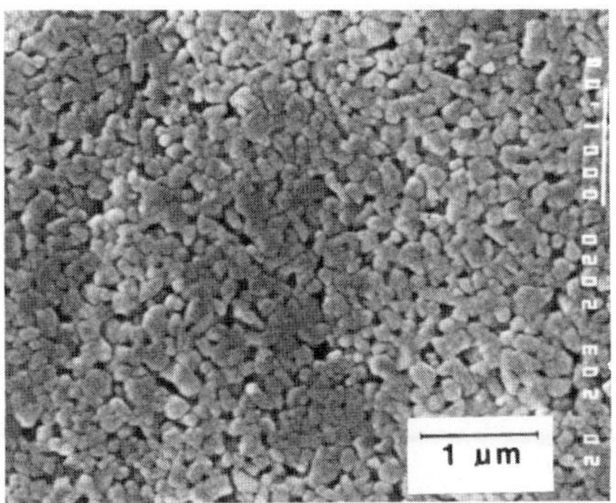

Fig. 7. Tape cast, classified XA139 Al_2O_3 in isopropanol with p-hydroxybenzoic acid.

Electrophoretic deposition experiments were performed using 0.5-0.9 μm $BaTiO_3$ dispersions in isopropanol for a variety of currents, voltages, and deposition times.[20] Figure 8 shows that deposit packing densities increase current; high currents (high deposition velocities) caused particles to impact the deposit and adhere, rather than allowing them to migrate to lower potential positions on the deposit surface, and thus to pack more densely.

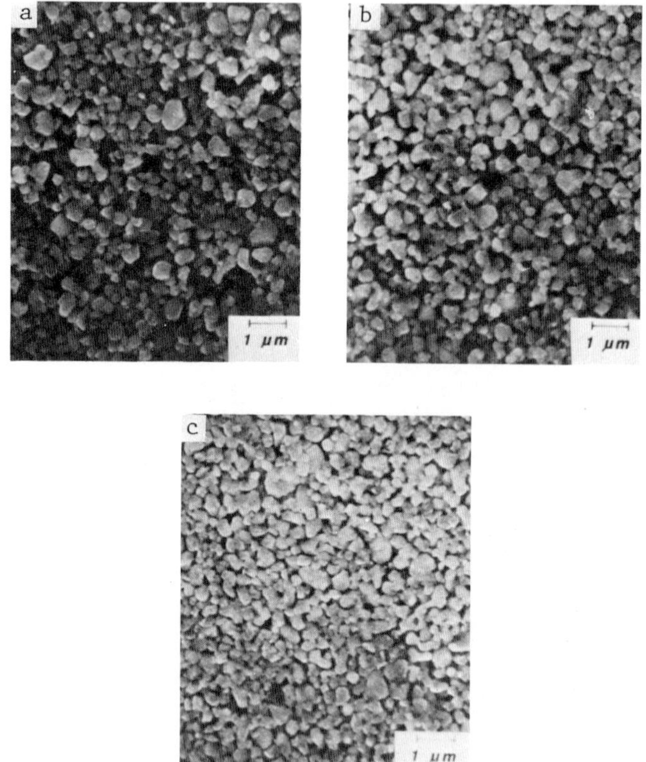

Fig. 8. Electrophoretically deposited $BaTiO_3$ (0.5-0.9 μm films showing the effect of current density (deposition rate); (a) 0.6 mA, (b) 0.4 mA, and (c) 0.1 mA.

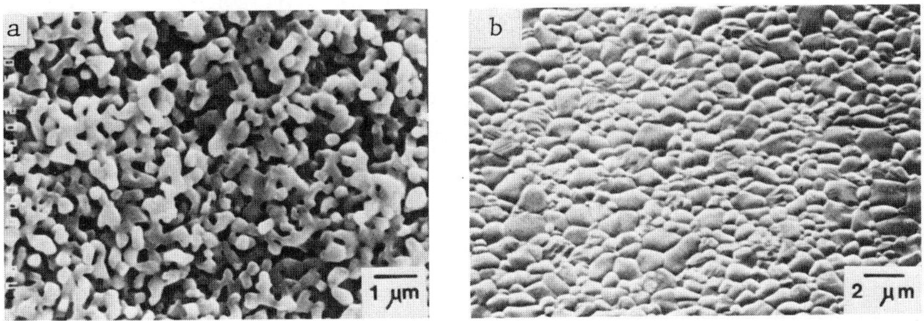

Fig. 9. Sintered microstructures 1100°C, 2 h of TiO_2 sedimentation cast compacts consisting of (a) agglomerated powder (Fig. 4a), and (b) nonagglomerated powder (Fig. 4b).

SUMMARY AND CONCLUSIONS

The preparation, dispersion, and packing of narrow size distribution powders has been discussed and demonstrated for several powder systems and processing techniques. The formation of uniform microstructures in greenware, which requires an understanding of the surface and colloid chemical principles, results in lower sintering temperatures and more uniform shrinkages.

The sintered microstructures for the monosize TiO_2, shown in Fig. 9, clearly demonstrate the effects of powder packing uniformity on the final microstructure. Attempts to sinter a porous compact consisting of agglomerated powder (Fig. 4a) resulted in low density microstructures (Fig. 9a). Conversely, dense, uniform compacts (Fig. 4b) sintered to greater than 99% of theoretical density at 1050-1100°C (Fig. 9b), a temperature much lower than the 1300-1400°C reported to sinter conventional TiO_2 powders to 97% of theoretical density. For these compacts, little grain growth was observed and uniform, fine-grained microstructures were obtained without using sintering aids; thus demonstrating that a uniform particle size provides a measure of stability against exaggerated grain growth. The observed enhanced sinterability was the result of the small particle size and the uniform pore-size distribution in the powder compacts.

REFERENCES

1. H. K. Bowen et al., Mat. Sci. and Eng., 44, 1056 (1980).
2. G. Y. Onoda, Jr. and L. L. Hench, eds., Ceramics Processing Before Firing, John Wiley and Sons, NY, 1978.
3. M. F. Yan, R. M. Cannon, U. Chowdhry, and H. K. Bowen, Bull. Am. Ceram. Soc., 64, 19 (1981).

4. W. H. Rhodes, J. Am. Ceram. Soc., 64, 19 (1981).
5. E. A. Barringer and H. K. Bowen, J. Am. Ceram. Soc., 65 [12] (1982).
6. B. Fegley, M.I.T., personal communication.
7. T. C. Huynh, (Ph.D. Thesis), M.I.T., 1983.
8. W. R. Cannon, S. C. Danforth, J. H. Flint, J. S. Haggerty, and R. A. Marra, J. Am. Ceram. Soc., 65, 324 (1982).
9. W. R. Cannon, S. C. Danforth, J. S. Haggerty, and R. A. Marra, J. Am. Ceram. Soc., 65, 330 (1982).
10. M. V. Parish, (M.S. Thesis), M.I.T., 1982.
11. H. C. Hamaker, Physica, IV [10], 1058 (1937).
12. R. J. Pugh and J. A. Kitchner, J. Coll. and Interface Sci., 35, 656 (1971).
13. M. T. Strauss, to be published in J. Am. Ceram. Soc.
14. E. J. W. Verwey and J. Th. G. Overbeek, Theory of the Stability of Lyophobic Colloids, Elsevier, Amsterdam, 1948.
15. J. Th. G. Overbeek, J. Coll. and Interface Sci., 58, 408 (1977).
16. G. R. Weise and T. W. Healy, J. Coll. and Interface Sci., 51, 427 (1975).
17. T. Sato and R. Ruch, Stabilization of Colloidal Dispersions by Polymer Adsorption, Marcel Dekker, Inc., NY, 1980.
18. E. S. Tormey, (Ph.D. Thesis), M.I.T., 1982.
19. T. R. Gattuso, M.I.T., personal communication.
20. C. Foss, (M.S. Thesis), M.I.T., 1982.

PLASMA SINTERING OF CERAMICS

D. Lynn Johnson, Victoria A. Kramb, and Donald C. Lynch

Department of Materials Science and Engineering
Northwestern University
Evanston, Illinois 60201

INTRODUCTION

Although the first plasma sintering of ceramics was reported a number of years ago,[1,2] there has been relatively little follow-up work.[3-8] Bennett et al.[1] inserted a quartz glass or alumina tube through a waveguide, excited plasmas at 2450 MHz at gas pressures of 100-7000 Pa, and sintered Al_2O_3 and a few other ceramic powders. They heated small pressed compacts for various lengths of time at various temperatures in the plasma and at the same times and temperatures in a conventional furnace. At all temperatures, that is, 1300°-1700° for Linde A alumina in air, the densities of the plasma sintered pellets were significantly higher than those of the conventionally sintered pellets. Moreover, the grain size was smaller and the strength was higher for the plasma sintered specimens. They observed that various gases gave the same densities at the same specimen temperatures and times.

Bennett and McKinnon[2] reported another experiment which demonstrated that the plasma was producing effects other than simply heating the specimen. Specimens sintered for 20 minutes in a plasma, 100 minutes in a conventional furnace, and 100 minutes in a plasma showed about the same final density as specimens sintered 120 minutes without interruption in the plasma. Densification during the intervening time in the conventional furnace was negligible, but the densification resumed upon return to the plasma.

Cordone and Martinsen[3,4] and co-workers[5,6] sintered small cylindrical pressed compacts of Al_2O_3 and UO_2 in a plasma generated in a cylindrical hollow cathode. Sintering rates were reported to be greater than those in a conventional furnace.

Johnson and Rizzo[7] used still a third type of plasma generator, namely an RF induction coil operating at 5 MHz to plasma sinter thin-wall tubes of β"-alumina using a rapid pass-through technique in which the tube was lowered through the plasma. They observed unusually rapid densification rates and small grain size.

Kim and Johnson,[8] using the same apparatus employed by Johnson and Rizzo, sintered rods and tubes of "pure" and MgO-doped α-Al_2O_3 in an argon plasma, again using the rapid pass-through technique. At translation velocities of 2 and 3 cm/min, they observed a final density of 96.1% of the x-ray density for undoped rods and about 98.5% for doped rods. Thin-wall tubes of doped alumina passed through at 6 cm/min exhibited a final density of about 99.5%.

Thus, all three groups of workers, each using a different type of excitation of the plasma, have reported enhanced densification rates. Bennett et al.[1] advanced three hypotheses to explain the rapid sintering. They are (1) an increased surface tension and, therefore, driving force through cleansing of the particle surfaces by ionic bombardment from the plasma, (2) an increase in diffusion rate by the creation of vacancies and other phenomena caused by ionic bombardment, and (3) an increase in diffusion rate through local increases in temperature inside the pores of the compact. They presumed, as these three hypotheses require, that the plasma existed within the pores of the compacts. No firm data are available with which to evaluate the first two hypotheses. However, the third one can be ruled out because local temperature variations could only be sustained for very short times since the entire compact would rapidly reach the local temperature. While it is conceivable that the plasma could be excited within the pores at microwave and RF frequencies, since the sample resides in the alternating electromagnetic fields, it is not clear that such would be the case in the DC hollow cathode. Here, the plasma is generated by bombardment of atoms by electrons accelerated under the influence of the field, and the plasma would presumably only exist as far as it could diffuse into the compact. Insufficient data exist to explain the observed enhanced densification with confidence. (The β"-alumina sintered by Johnson and Rizzo may have sintered by a transient liquid phase process.)

EXPERIMENTAL METHODS

Further work was conducted by sintering specimens of Al_2O_3 both in the 5 MHz RF plasma furnace of Johnson and Rizzo, and in a 2450 MHz microwave system. The latter included a 13 mm quartz plasma tube which passed through a reduced height slotted waveguide. In both cases the quartz plasma tubes were partially evacuated by mechanical pumps. All studies in the RF plasma were conducted in argon, while specimens were sintered in argon and nitrogen in the microwave plasma. Both thin-wall tubes (15 mm O.D.) and rods (5 mm diameter) were sintered in the RF plasma, while rods (3-5 mm diameter) were sintered in the microwave plasma after presintering at 600°C to burn out binders. In all cases, specimens were translated through the plasmas at velocities ranging from 1-4 cm/min in the RF plasma, and up to 11 cm/min in the microwave plasma. Magnesium oxide was added to some powders using magnesium nitrate solutions.

RESULTS AND DISCUSSION

In general, the undoped Al_2O_3 powders could be sintered to about 98.6% of theoretical density under optimal conditions. The MgO-doped powders could be sintered to 99.1-99.5% of the theoretical x-ray density under the best conditions. Densification was very rapid and appeared to be limited only by the rate at which specimens could be inserted into the plasmas without thermal shocking.

Overall linear densification rates, defined as the total percent linear shrinkage divided by the time required for any given point on a specimen to densify, naturally increased with increasing translation rate and exceeded 150%/min at 11 cm/min translation rate. The maximum densification rate was not determined but would be somewhat higher than these overall values.

Some surprising phenomena were observed. For instance, it was found that pre-sintered specimens of undoped Al_2O_3 could be heated to the melting temperature quite readily in an Ar plasma, whereas the maximum temperature achievable upon insertion of sintered specimens into the plasma was approximately 800°C less. If the translation of a sintering specimen was stopped, a temperature would spontaneously drop the same amount. The maximum temperature observed increased as the rate of translation of the specimen through the plasma increased, quite the opposite to expectations for conventional heating. The maximum density achievable increased with increasing translation rate to a broad maximum, falling off only slightly at 11 cm/min. The surface of undoped alumina specimens could be melted by increasing the translation rate, whereas the surface of the MgO-doped specimens could not be melted at the same power levels at any speed. At the lowest translation rate, 1 cm/min, the specimen temperature oscillated several hundred degrees. A section of specimen would become hot, densify rapidly, and then spontaneously cool down. As a new section of specimen moved into the plasma, it would subsequently and suddenly heat up, densify, and then cool down, and the cycle was repeated.

It was obvious that the effects were related to the porosity. The role of porosity in these heating effects was investigated using specimens which remained porous throughout the firing cycle. One type of specimen was prepared by adding 40% by volume of organic binder, while the other employed about the same amount of organic particulates on the order of 0.5 mm in diameter. The sintered specimens contained interconnected porosity with a normally dense matrix, but the porosity of the former specimens was finer. In both cases, the transient heating effects were not observed, i.e., the specimens did not spontaneously cool down when held stationary within the plasma, and sintered specimens could be reheated to the sintering temperature. It was therefore concluded that the transient heating effects are related to the porosity.

The other transient heating effects can now be understood. At higher translation rates, the specimens would be carried further into the plasma before the porosity shrank to small values, thus permitting the attainment of a higher temperature relative to more slowly translated specimens. The greater ease in melting the undoped powders can also be explained, since the MgO doping causes more rapid densification and, therefore, densification in the cooler regions of the plasma. It is still not clear whether these effects are due to penetration of the plasma into the pores or some other effect related to the porosity.

CONCLUSIONS

Aluminum oxide and some other ceramics have been sintered to high densities at high rates in gas plasmas generated by microwave and RF and DC hollow cathode plasma devices. The universal result is rapid densification rates. Rapid pass-through sintering produces the highest densification rates. Transient heating effects are related to the porosity of the specimens, i.e., porous specimens can be heated to higher temperatures than can non-porous specimens.

ACKNOWLEDGMENTS

Supported by the National Science Foundation Grant No. DMR-7918403. This research was carried out in the facilities of Northwestern University's Materials Research Center, supported in part under the NSF-MRL Program (Grant No. DMR-7923573).

REFERENCES

1. C. E. G. Bennett, N. A. McKinnon, and L. S. Williams, Nature, 217, 1287 (1968).
2. C. E. G. Bennett and N. A. McKinnon, in Kinetics of Reactions in Ionic Systems, edited by T. J. Gray and V. D. Frechette, Plenum Press, NY, 1969.
3. L. G. Cordone and W. E. Martinsen, J. Am. Ceram. Soc., 55, 380 (1972).

4. L. G. Cordone and W. E. Martinsen, Nature, 241PS, 86 (1973).
5. G. Thomas, J. Freim, and W. E. Martinsen, Trans. Am. Nucl. Soc., 17, 177 (1973).
6. G. Thomas and J. Freim, Trans. Am. Nucl. Soc., 21, 182 (1975).
7. D. L. Johnson and R. R. Rizzo, Am. Ceram. Soc. Bull., 59, 467 (1980).
8. J. S. Kim and D. L. Johnson, J. Am. Ceram. Soc. Bull., to be published.

DISCUSSION

W. A. Kaysser (Max-Planck-Institut, Stuttgart; Massachusetts Institute of Technology): During plasma sintering internal heat vanishes quite rapidly after the alimination of major parts of the porosity. Did this effect cause any thermal shock damage to the samples?

Author: Thermal shock problems arose during insertion of the rods into the plasma. No thermal shock damages were observed during cooling.

H. Palmour III (North Carolina State University): Recent dilatometric data indicates that the role of magnesia doping in alumina may be to alter densification rates in the early stages of densification--near D ~ 0.70--thereby introducing changes in the pore-grain morphologies which continue to influence densification all the way through to final stage. Question: Does the starting density of the rod influence sinterability in the plasma?

Author: This has not been investigated yet.

R. Rice (Naval Research Laboratory): Have you investigated other more volatile additives, e.g., LiF in MgO, to evaluate effects of vaporization of the additives and the effect of this on the plasma?

Author: We have observed Mg lines in the spectrum of the plasma surrounding MgO-doped Al_2O_3 tubes. Other doping studies are planned.

P. R. Johnson (Alfred University): Have you found any relationship between the frequency at which your plasma was generated and the size of the resulting pores in your sintered product?

Author: This has not been evaluated yet.

PLASMA MELTING OF SELECTED COMPOSITIONS IN THE Al_2O_3-ZrO_2-SiO_2 SYSTEM

James V. Portugal and L. David Pye

New York State College of Ceramics
Alfred University
Alfred, NY 14802

INTRODUCTION

The application of plasma spraying for coating substrates with refractory materials is a well established discipline.[1,2] The use of a plasma to prepare non-crystalline solids from refractory compositions, however, has received far less attention.[3,4] There are many reasons why plasma melting is attractive for the preparation of non-crystalline solids. Among these are high melting temperatures (> 5,000°K), availability of strong oxidizing or reducing conditions, the attainment of high quench rates for single particles (> 10^6 °C/s), the maintenance of high purity (for RF plasma torches) and its potential for inducing gas-phase reactions, e.g., vapor phase oxidation of chloride solutions. These potential benefits offer the possibility of preparing refractory glasses of unusual compositions. It was for this reason primarily that the present work was undertaken.

The following is a progress report on our attempts to study the melting behavior of selected compositions in the Al_2O_3-ZrO_2-SiO_2 system. In this work, we report the effects of particle size, melting temperature and other parameters on the melting efficiency of an RF plasma torch. The characterization of the glasses prepared and the region of glass formation in this system will be the subject of a future communication.

EXPERIMENTAL PROCEDURE

Materials

A summary of the materials investigated is presented in Table 1. Except for the standard glass (see below), the powders were obtained through chemical preparation techniques. The single component oxides, Al_2O_3 and ZrO_2, were prepared via precipitation of their hydroxides from reagent grade sulfate solutions. The various A-Z mixtures were prepared by co-precipitation, while for the A-Z-S composition, colloidal silica (ludox) was added to the A-Z solution prior to co-precipitation. The resulting precipitates were then filtered, repeatedly washed with hot 2% NH_4Cl solution (to remove any excess sulfate ions) and dried. Analysis of the prepared powders by energy dispersive spectrometry (EDS), showed a very homogeneous distribution of the cationic species on a spatial resolution scale of 0.5 μm.

Table 1. Materials Investigated

System	Composition (mol%)			Melting Point (°C)
	Al_2O_3	ZrO_2	SiO_2	
Al_2O_3	100	---	---	2045
ZrO_2	---	100	---	2715
A-Z (1)	30	70	---	2340*
A-Z (2)	50	50	---	2040*
A-Z (3)	70	30	---	1790*
A-Z-S	5	5	90	1700**
NaO-CaO-SiO_2 Glass†				~ 1100

*Based upon phase diagram from reference 7.
**Based upon phase diagram from reference 8.
†N.B.S. Standard Reference Material No. 710.

Since particle size of the powders was thought to have an appreciable influence on melting efficiency, a relatively low melting soda-lime-silica glass was first investigated in an attempt to determine optimum particle size. To this end, the glass was first ground and then separated into nine discrete size intervals (from less than 38 to over 150 μm), using an Allen-Bradley sonic sifter. About three grams from each size interval was passed through the torch and the percent of melting assessed. Based on the results of this preliminary study (see below), a particle size of 45-63 μm was used for all other compositions given in Table 1. Choosing this particle size range was consistent with those selected by other investigators.[5,6]

Equipment

An oxygen radio-frequency plasma unit was employed for this study.[a] A general schematic diagram of the system is depicted in Fig. 1. The water cooled plasma torch was inductively coupled to a Lepel 25 KW power supply. The frequency of the resulting RF field was 2.5-5.0 MHz.

The feeder depicted in Fig. 1 operated by vibrating powder up an inclined ramp along the edge of a conical bowl. When the powder reached an orifice at the top of the bowl, it was transported through a tube connecting the feeder to the torch where it was discharged into the plasma. An oxygen carrier gas was also used for maintaining steady flow rates of powder into the torch. A powder feed rate of approximately 1 gm/min. was found to be optimal for the particle size used.

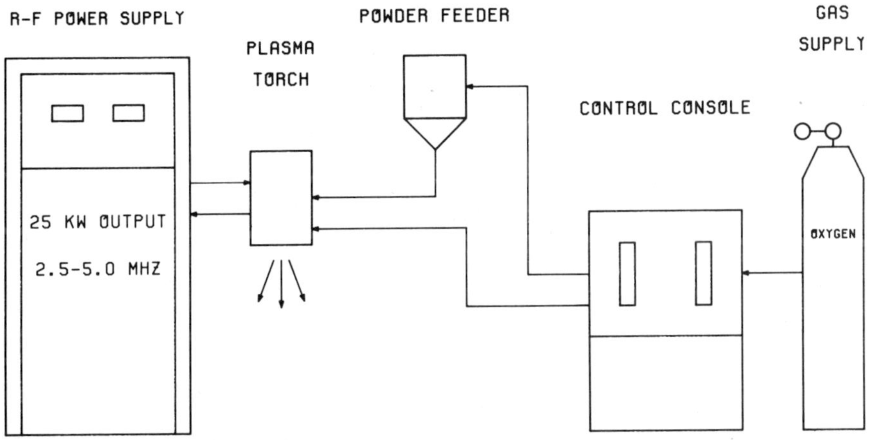

Fig. 1. Block diagram of induction plasma system.

[a]TAFA Corp., Dow Rd., Bow (Concord), NH 03301.

Fig. 2. Plasma melting of zirconia.

A 35 mm camera (Canon AE-1) affixed with a zoom lens (Vivitar 90-230 mm) was used for a photographic analysis of the plasma during melting of pure zirconia. Film speeds of ASA 100, 200 and 400 were used at shutter speeds ranging from 1/30 to 1/1,000 sec., all under incident light. Another camera (Canon F-1) was also used to record exposures of 1/2,000 sec.

Photomicrographs of the prepared powders (before and after melting) were taken using an ETEC scanning electron microscope. Standard light and polarizing microscopes were used for low magnification examinations.

RESULTS AND DISCUSSION

Photographic Analysis

An enlarged color photograph (8 x 12 in.) was obtained, similar to that shown in Fig. 2. From this photograph, an estimation of the quench rate was possible by using a luminosity technique. This method makes use of the fact that a heated particle becomes incandescent at a temperature above 600°C. Therefore, a particle previously heated above its melting temperature (T_m) will radiate light continuously as it is cooled to ~ 600°C. The streaks shown in Fig. 2 (zirconia particles) are a manifestation of this phenomenon. Knowing the minimum shutter speed of the camera (t_s) necessary to reveal the full length of the streaks, it was possible to estimate the quench rate ($\Delta T/\Delta t$) through the following equation:

$$\Delta T/\Delta t = (T_m - 600)/t_s. \quad (1)$$

Using a shutter speed of 1/1,000 sec. an estimated quench rate of ~ 2 x 10^6 °C/sec. was obtained. This shutter speed was chosen on the basis that it represented experimentally the time limit of detection for the luminous streaks of melted ZrO_2. A faster shutter speed, i.e., 1/2,000 sec., did not reveal a full luminous streak such as that shown in Fig. 2.

An approximate particle velocity was also determined by dividing the distance the particle travelled by the shutter speed. A velocity on the order of 10^4 cm/sec. was determined. Finally, a melting time of ~ 10^{-4} sec. was calculated by knowing the minimum distance in the plasma required for melting to occur (estimated at 1 cm) and dividing this quantity by the particle velocity. These values are in excellent agreement with other estimations.[1,9]

Photomicrographic Analysis

The effect of particle size on melting efficiency is evident by the photomicrographs of the soda-lime-silica glass shown in Figs. 3 a-f. These three sets of photos (before and after melting) typify the results for the nine size intervals tested. Figs. 3-a and b show particles of large size range (125-150 μm) and reveal incomplete melting. The angular corners of the glass are only partially rounded, and thus the particles have not totally responded to surface tensional forces as they would have if their viscosity had been sufficiently reduced (i.e., the fluid state had been achieved). Figs. 3-e and f represent the melting of the smallest size range (less than 38 μm). Although melting is quite complete, substantial agglomeration is evident. Powders of this size also introduced problems in obtaining uniform powder flow, particularly at the point of injection into the plasma. As a result, material entered the plasma stream not as discrete particles, but rather as larger agglomerates which fused together upon melting. This may explain why the melted spherules (Fig. 3-f) show a larger average size than that of the pre-melted glass (Fig. 3-e). An optimum range of 63-75 μm (Figs. 3-c and d) therefore gives the best compromise between particle size necessary for melting, and particle size necessary to insure steady flow rates thereby avoiding agglomeration.

Fig. 3. SEM photomicrographs of glass of various particle size, before and after plasma melting.

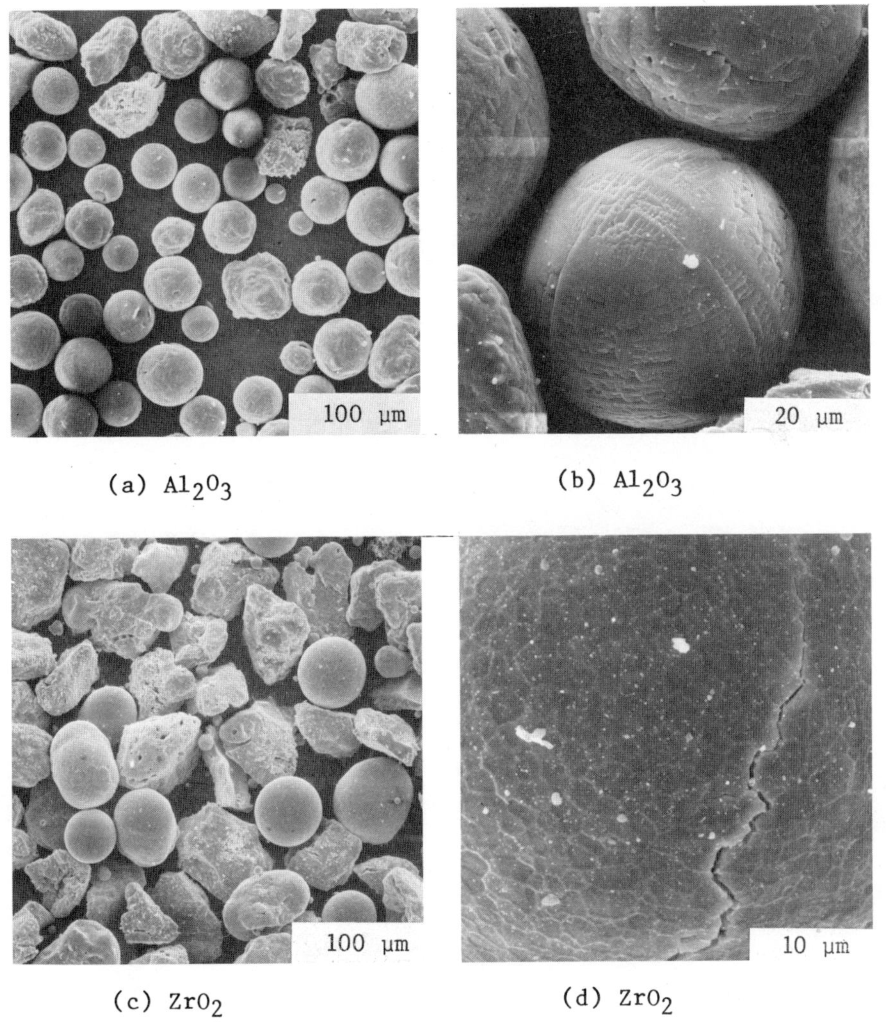

Fig. 4. SEM photomicrographs of alumina and zirconia after passing through the plasma torch. All particles are 45-63 µm.

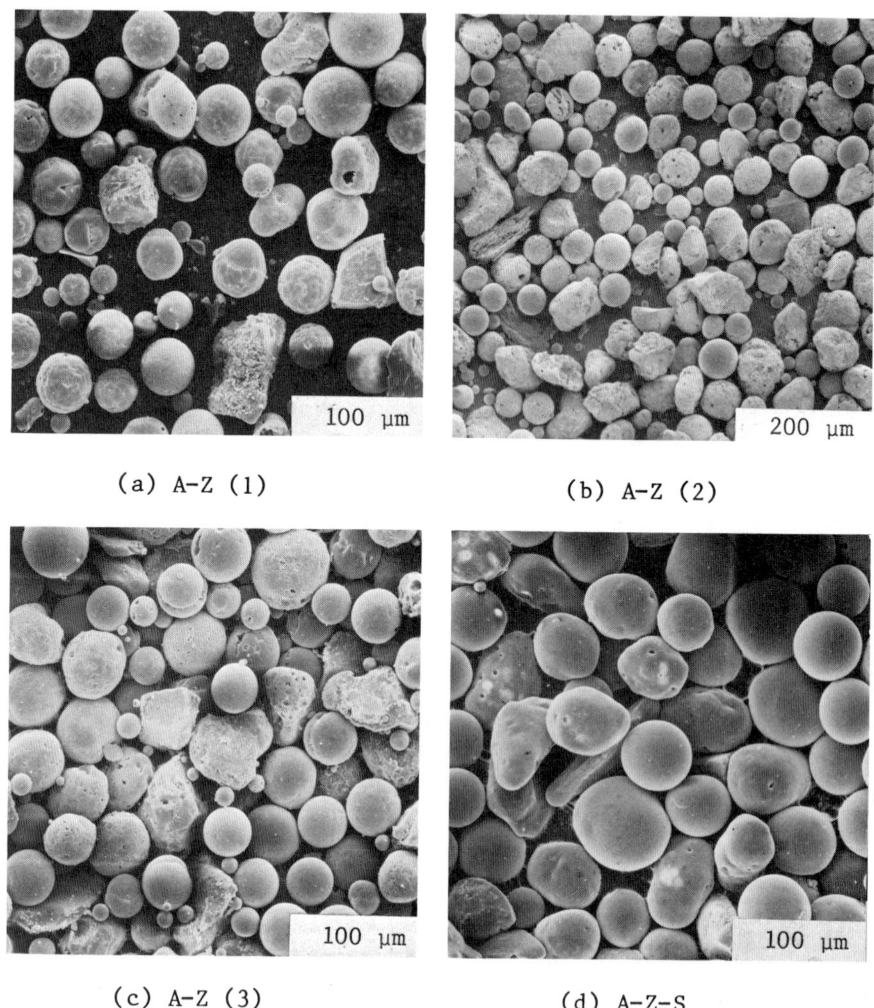

(a) A-Z (1)

(b) A-Z (2)

(c) A-Z (3)

(d) A-Z-S

Fig. 5. SEM photomicrographs of various Al_2O_3-ZrO_2-SiO_2 plasma sprayed powders. See Table 1 for compositions.

The succeeding photomicrographs show the various compositions introduced in Table 1. A variety of microstructures are evident on the surfaces of the alumina and zirconia (Figs. 4 a-d). In particular, the alumina shows both a platy-type of microstructure, as well as a possible dendritic surface (see Fig. 4-b). The zirconia (Fig. 4-d) exhibits a crack propagating along an intergranular route, which could be evidence for a thermal shock effect upon rapid cooling.

The various A-Z and A-Z-S compositions (Figs. 5 a-d) exhibit much more uniform and smooth surface morphologies. Most are transparent under transmitted light and become extinct under crossed polarizers and 360° rotation. On this basis the presence of a non-crystalline phase may be inferred. We note that others have reported glass formation for compositions around the eutectic in the Al_2O_3-ZrO_2 system.[5,3,11]

Many of the spherules were found to be hollow, which may be explained by the cooling of the particle from the outside first and cooling radially inward. Surface tension would draw the still molten center toward the cooler outer shell producing thin walled spherules.

Melting Kinetics

A major interest in plasma studies is a determination of melting efficiency of a given torch when various powders are fed through it. Melting efficiency (defined here as the ratio of particles actually melted to the total number of particles examined), is influenced by various torch and particle related variables. Among the former are power input, type of gas (and flow rate), and plasma geometry; the latter include particle size, composition, thermal diffusivity and melting temperature. There are a number of theoretical models available which discuss the relationship of melting efficiency to these variables.[1,6,10] One discussed by Gerdeman and Hecht (1972), utilized a heat transfer approach.[1] It was assumed that the surface of a solid particle with radius R is instantaneously brought to a constant temperature T_s and for the duration of the residence time in the plasma, the interior of the particle is heated radially and approaches the surface temperature. This rate of temperature rise is expressed according to:

$$T(r,t) = T_s \left\{ 1 + \frac{2R}{\pi r} \sum_{n=1}^{\infty} \frac{(-1)^n}{n} \sin \frac{n\pi r}{R} \exp - \frac{\alpha t n^2 \pi^2}{R^2} \right\} \qquad (2)$$

where:

T = temperature of particle at radius r, time t,

α = thermal diffusivity,

t = residence time in the plasma.

Thus for a constant particle size and residence time, an estimation of melting efficiency using this model would be primarily a function of the thermal diffusivity.

The above heat transfer model, however, only comprises half of the necessary conditions for arriving at a parameter indicative of the ease of melting of a material. It is also important to consider the total amount of energy required to melt the material, i.e., one must take into account respective heat capacities (C_p) and all heats of transition (ΔH_{tr}) necessary to attain the liquid state. This total enthalpy change (ΔH) is expressed by the equation:

$$\Delta H = \Sigma \int_{T_1}^{T_2} C_p \, dT + \Sigma \, \Delta H_{tr} \qquad (3)$$

For the present purpose, then, one may define an ease of melting parameter as the ratio of the particle diffusivity (h) to the total heat content of the liquid <u>at the melting temperature</u>, expressed on a volume basis (L). This definition assumes, then, that particle size, feed rates and plasma variables of a given unit are held constant for purposes of comparing different materials. Table 2 shows calculated values of this parameter for the soda-lime-silica glass, Al_2O_3 and ZrO_2. In the seventh column, the values have been normalized with respect to the glass. The glass was chosen as a reference base since 100% melting was attained for the same particle size used for the Al_2O_3 and ZrO_2. The melting efficiency for each material is given in the last column and was estimated by examining SEM photomicrographs of the powders after they had been passed through the torch (see Figs. 3-d, 4-a, and 4-c). It is seen that as the ease of melting parameter decreases, the melting efficiency falls as well. Thus, our results are in qualitative agreement with the predicted melting efficiencies of these materials. An examination of other materials with varying combinations of h and L values would be of additional benefit in confirming the usefulness of this approach.

Table 2. Ease of Melting Parameters and Melting Efficiencies

Material	ρ (g/cm^3)	h (cm^2/a)	ΔH (kcal/mol)	L ($kcal/cm^3$)	Parameter x 10^{-4} (h/L)	Rel. to glass	Melt Eff. (%)
Glass	2.50	0.0600	18.7	0.723	830.0	1	100
Al_2O_3	3.97	0.0076	87.6	3.41	22.3	37	~50
ZrO_2	6.27	0.0025	69.0	3.52	7.10	117	~20

SUMMARY/CONCLUSIONS

This work has examined the plasma melting of selected compositions in the Al_2O_3-ZrO_2-SiO_2 system. Photographic analyses of the plasma used in this work indicated that high quench and melting rates were achieved, i.e., > 10^6 °C/sec. It was found that a particle size of 63-75 μm of a soda-lime-silica glass produced the best combination for good flow rate and melting behavior. A newly defined ease of melting parameter correlated well with observed melting efficiencies of materials studied in this work.

ACKNOWLEDGMENTS

The financial support of the N.Y. State Science and Technology foundation is greatly appreciated. Other technical support from W. Votava, J. Cerdan, G. Cartledge, V. Frechette, L. Hanks, P. Gignac, P. Johnson and D. Snowden was both highly appreciated and essential.

REFERENCES

1. D. A. Gerdeman and N. L. Hecht, Arc Plasma Technology in Materials Science, Springer-Verlag, NY, 1972.
2. R. F. Smart and J. A. Catherall, Plasma Spraying, Butler & Tanner Ltd., London, 1972.
3. T. Suzuki and A. M. Anthony, Mat. Res. Bull., 9, 745 (1974).
4. A. Revcolevschi, A. Rouanet, F. Sibieude, and T. Suzuki, High Temp. High Press., 7, 209 (1975).
5. V. A. Krauth and H. Meyer, Ber. Deut. Keram. Ges., 42, 61 (1965).
6. D. R. Mash, N. E. Weare, and D. L. Walker, J. Metals, 13, 473 (1961).
7. G. Cevales, Ber Deut. Keram. Ges., 45, 217 (1968).
8. P. P. Budnikov and A. A. Litvakovskii, Dok. Akad, Nauk. S.S.S.R., 106, 268 (1956).
9. P. T. Sarjeant and R. Roy, Mat. Res. Bull., 3, 265 (1968).
10. C. W. Marynowski, F. A. Halden, and E. P. Farley, Electrochem. Tech., 3, 110 (1965).
11. S. L. Dahar, (B.S. Thesis), Alfred University, Alfred, NY, 1980.

LIQUID PHASE SINTERING OF CERAMICS

W. A. Kaysser and G. Petzow

Max-Planck-Institut fur Metallforschung
Institut fur Werkstoffwissenschaften
Heisenbergstr. 5, D-7000 Stuttgart 80, FRG

INTRODUCTION

Liquid phase sintering is widely used for consolidation of ceramics as well as metallic powders into final shapes. Main advantages of this production method are low sintering temperatures, fast densification, high final densities and resulting microstructures often providing mechanical or physical material properties superior to solid state sintered materials. Densification during liquid phase sintering is based on rearrangement and shape change of solid constituents. Rearrangement of larger particles (<10 μm) is caused by short range movements due to capillary forces between a few adjacent particles.[1] In systems containing particles and pores of a smaller scale or larger content of melt, pores may be eliminated by cooperative movements of particles and liquid, comparable to viscous flow of a dispersion.[2] Shape change of particles is due to matter transport from solid areas near solid/liquid interfaces to similar areas of lower chemical potential by diffusion either in the solid or in the melt. The latter is connected to the dissolution and reprecipitation of solid material in and from the liquid. This paper is intended to discuss some features of liquid phase sintering which may be of relevance to the processing of ceramics. Ceramic powders are usually characterized by small average grain sizes (often <1 μm) and broad ranges of reduced particle sizes. To circumvent the poor flow properties of fine powders preagglomeration is usual causing frequently inhomogeneous pore distributions after compaction. Areas of small approximately equal sized pores alternate with areas of lower density including pores much larger than the average particle and pore size. The elimination of large pores by liquid flow and shape accommodation will be discussed for zero and nonzero degree wetting conditions. Since the small average grain size of ceramic powders causes rapid grain coarsening during the early stages of sintering, relations between grain growth and shape accommodation will be considered. In a variety of systems chemical composition and/or phases of dissolved and reprecipitated material are different. The virtually large free energies released by alloying and phase transformations cause changes of the particle geometry which are described. Shrinkage of Si_3N_4 in the presence of small amounts of liquid phase will be discussed on the basis of the foregoing considerations.

LARGE PORES

In the case of complete wetting, i.e., $\gamma_{sl} + \gamma_{lv} = \gamma_{sl}, \gamma_{lv}$, where γ_{sv} is the specific energies of the interfaces solid/liquid, liquid/gas and solid/gas) all pores will be filled by liquid phase if a compact is brought into contact with an external unlimited reservoir of melt.[a] For wetting angles $\Theta < 0$ deg. a plane monolayer of equal sized spheres will not cover itself completely with liquid phase, i.e., a complete immersion of the spheres in the melt is possible through applied external forces only. The critical wetting angle Θ_{crit} which yields pore elimination by intrinsic capillary forces increases with decreasing pore size (Fig. 1). For calculation, pores were modelled as spherical space enclosed by spheres of radius R_s whose centers are situated on a sphere of radius $R_s + R_p$. The centers of three adjacent spheres were assumed to form triangles of equal sides. Simple geometrical considerations yield

$$\Theta_{crit1} = \text{arcsine}\ \left\{\frac{2R_s}{3(R_s + R_p)}\right\} \tag{1}$$

shown as curve I in Fig. 1. Curve II was obtained from energy considerations for the same pore geometry. The sum of all interface energies was calculated. The critical wetting angle Θ_{crit2} was determined as the maximum wetting angle which yielded no energy minimum between $h = 0$ and $h(\Theta_{crit1})$. The latter value was calculated from relation (1). As mentioned in the introduction, pores with $R_p/R_s >$ 10 may frequently occur in fine grain ceramics. For wetting angles > 25 deg. results shown in Fig. 1 would predict a considerable number of pores which do not exert forces for their elimination, i.e., for shrinkage. Even small perturbations of the pores from sphericity, however, result in much higher critical wetting angles. When a small area of a pore with a local radius of curvature $R_1 < R_p$ is covered by liquid phase, complete filling of the residual pore requires only a very modest wetting behavior. Thus no problems of pore elimination should arise in respect to intrinsic capillary forces for wetting

---

[a]Effects of gas inclusions and gravity were excluded throughout this chapter.

angles < 50 deg. For larger wetting angles, as often observed in cermets, the formation of larger spherical pores during processing should be carefully omitted.

A second case where the capillary forces exerted by pores may be insufficient to induce further shrinkage is connected with shape accommodation. Starting a with roundish particle shape accommodation results in both densification and an increasing curvature of particular areas of the solid/liquid interface. Elimination of large pores by filling with liquid will be directly related to shape accommodation if the melt is withdrawn from melt channels situated along junctions of three adjacent grains. Smaller melt channels require smaller radii of curvature of the solid/liquid interface. An analysis of the change in interfacial area in a periodic spatial arrangement of equal sized truncated dodecahedrons when liquid phase is taken from prism shaped channels at the junctions of three grains results in a maximum pore size which could be eliminated

$$r_{pc} = C^{-1}\{\frac{2\gamma_{1v}}{\gamma_{1s}}\}R\{\frac{\Phi_L}{(1 - \Phi_L)^{1/3}}\} \quad (2)$$

where R is the radius of a sphere with a volume equivalent to the truncated dodecahedron, Φ_L the volume fraction of prism shaped channels and C an explicitly known geometry factor related to the dihedral angle. The critical pore radius, r_{pc}, increases with R, i.e., grain growth should increase the possibility of successfully eliminating larger and larger pores. Using a series expansion for the right hand term of Eq. (2) this relation may be approximated by

$$\Phi_L = (r_{pc}K_o)^2R^{-2} \quad (3)$$

for $\Phi_L < .25$ with an error less than 1%. The time derivation of Eq. (3) and the assumption of an Ostwald ripening type of diffusion controlled grain growth with $R^3 = R_o{}^3 + kt$ yields

$$\frac{\delta\Phi_L}{\delta t} = -.66(r_{pc}K_o)^2k(R_o{}^3 + kt)^{-5/3} \quad (4)$$

(curve II in Fig. 2). It is obvious from Eq. (4) that a pore free state in systems where Φ_L is small may be achieved by liquid phase sintering only if either very fine pores are present throughout sintering or after grains have grown very large.

It may be mentioned that once a pore is filled with melt, grains may grow excessively into the liquid reservoir replacing melt by solid phase, thus increasing Φ_L again. The latter effect was observed in W-Ni.

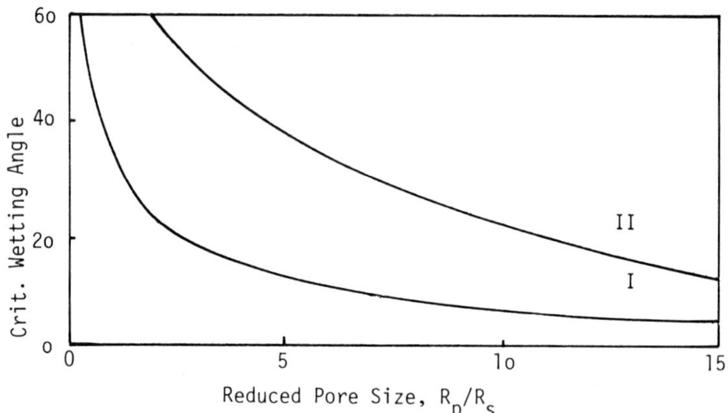

Fig. 1. Critical wetting angle for filling of spherical pores.

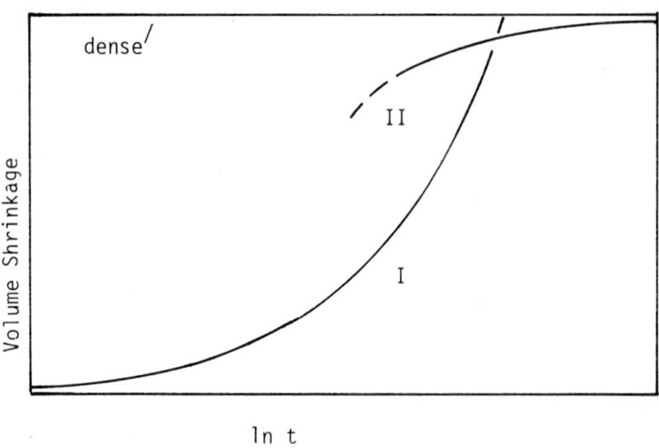

Fig. 2. Shrinkage during liquid phase sintering (schematic). (I) Calculated for contact flattening of dissolving particles. (II) Calculated for final stage Eq. (4).

SHRINKAGE AND GRAIN GROWTH

In the preceding section, grain growth during liquid phase sintering was considered with respect to the maximum curvature of solid/liquid interfaces. In this section the mechanism of shrinkage by shape accommodation will be discussed. Growth rate coefficients, k, measured during liquid phase sintering are usually of the order 10^{-18} m^3 s^{-1}. Assuming a cubic time law, the average particle radius will be doubled for the average initial radii of .5, 1.5, 5.0 and 10.0 μm after .1, 2.5, 90 and 715 s ($k = 1.4\ 10^{-18}$ m^3 s^{-1}). These short time periods indicate the importance of grain growth in all considerations of relations between dissolution and reprecipitation and shrinkage. In the classical approach shrinkage was treated as contact flattening in a system of equal sized spheres where grain growth was absent. Taking grain growth of a diffusion controlled type into consideration,[4]

$$\frac{\Delta l}{l} = \left[\frac{C_1}{2k}\ \{1 - (1 + ktR_o^{-3})^{-3/2}\}\right]^{1/3} \qquad (5)$$

is obtained with the constant C_1 in the same order of magnitude as k. Equation (5) should hold if shrinkage is determined by contact flattening of particles larger than the average particle size. The discrepancy between the prediction of Eq. (5) and experimental data indicates the major contribution of dissolving grains to shrinkage.

Space, i.e., the possibility of a center to center approach of adjacent particles will open up if contact flattening occurs during dissolution of small particles (curve I in Fig. 2), if small particles are pushed from between larger particles or if the surface of reprecipitated material is smoothened by diffusion. It is not clear at present which one of these mechanisms is actually dominating shrinkage, but all are linked to the rearrangement of larger particles, i.e., larger particles have to move to take up at least partially the space occupied by the smaller particles before. The reprecipitation of material evoded from contact regions and dissolving particles on nonshrinking particles must occur nonuniformly, i.e., less material should be deposited in contact areas than on other areas of the solid/liquid interface. This was demonstrated for W-Ni by Yoon and Huppmann.[5] Mixtures of 48 wt% large W spheres (200 um), 48 wt% fine W powder (10 um) and 4 wt% Ni were sintered at 1670°C. After 20 min most of the initially small grains had disappeared. Material had reprecipitated nonuniformly on large particles and grain coarsening had occurred in clusters of fine particles. Although the shrinkage mechanism could not be revealed from these experiments, shape accommodation during growth of larger particles became obvious.

DIRECTIONAL GRAIN GROWTH

When single crystal W spheres of uniform large size are sintered in the presence of liquid Ni, growth of one sphere at the expense of its immediate neighbors occurs.[6] This directional grain growth has also been found to occur in Fe-Cu, Fe-Cu-C and Mo-Ni systems. During directional grain growth additional interface area may be created, i.e., energy may be consumed. It is believed that compositional differences between dissolving and reprecipitating material measured in all systems provide a decrease in free energy. Typical free energies were on the order of 100 $Jmole^{-1}$. Thus directional grain growth may be defined as growth of grains at the expense of their neighbor grains regardless of the corresponding change of interface area and energy. In mixtures of solid particles and liquid phase, initially growing particles may be covered by material of a lower free energy. These particles may then all grow further by directional grain growth until all initially shrinking particles are completely dissolved.

DENSIFICATION OF Si_3N_4 IN PRESENCE OF LIQUID PHASE

An attempt will be made to correlate several of the mechanisms mentioned above to some experimental observations during liquid phase sintering of Si_3N_4. The high viscosity of the liquid phase containing MgO, CaO, SiO_2 or other additions is linked to low diffusivity, i.e., grain growth and shape accommodation were expected to be rather sluggish. In contrast, however, a certain period of liquid phase sintering of this material is characterized by rapid grain growth and considerable shrinkage.[7] Driving forces for this acceleration may be provided by the phase change when α-Si_3N_4 particles dissolve and β-Si_3N_4 particles grow. The enhanced dissolution of α particles may include both purely directional grain growth when the β-particles are still small and a certain contribution of Ostwald ripening when the β grains were grown large. The assumption that directional grain growth is essential for coarsening and shrinkage of Si_3N_4 is supported by the observation that the rate of shrinkage is proportional to the initial fraction of β particles.[7] In addition, the amount of shrinkage after various sintering times was found to relate well to the amount of β-Si_3N_4 already formed.[8] It may be deduced from the foregoing sections that small α-grains dissolve in the neighborhood of β-Si_3N_4 providing space for rearrangement of both β-Si_3N_4 and residual α-Si_3N_4 particles. The increased particle mobility due to the dissolution of small α grains may yield even more shrinkage if an external pressure is simultaneously applied. During hot pressing Si_3N_4 shows an increasing shrinkage during the transformation period with increasing applied pressure. Rearrangement and shrinkage cessate as more and more β-Si_3N_4 particles impinge each other. When a pseudo "rigid skeleton" is formed, subsequent dissolution of residual α-particles provides no further shrinkage.[9]

REFERENCES

1. V. Smolej, S. Pejovnik, and W. A. Kaysser, Powder Met. Int., 14, 34 (1982).
2. W. D. Kingery, ed., in Ceramic Fabrication Processes, Wiley, NY.
3. W. A. Kaysser, to be published.
4. W. A. Kaysser, O. J. Kwon, and G. Petzow, Proc. P/M 82, Firence, June 1982.
5. D. N. Yoon and W. J. Huppmann, Acta Met., 27, 973 (1979).
6. W. J. Huppmann, W. A. Kaysser, D. N. Yoon, and G. Petzow, Powder Met. Int., 11, 50 (1979).
7. H. Knoch and E. E. Gazza, Ceramurgia Int., 6, 5 (1980).
8. L. J. Bowen, R. J. Weston, T. G. Carruthers, and R. J. Brook, J. Mat. Sci., 13, 341 (1978).
9. J. Weiss and W. A. Kaysser, in Nitrogen Ceramics, edited by F. L. Riley, Noordhoft, Leiden, Brighton, UK, 1981.

PRECISION DIGITAL DILATOMETRY:

A MICROCOMPUTER-BASED APPROACH TO SINTERING STUDIES

A. D. Batchelor, M. J. Paisley, T. M. Hare, and
H. Palmour III

North Carolina State University
Raleigh, North Carolina 27650

ABSTRACT

Representing significant advances in precision, atmosphere control, programming flexibility, experimental convenience and rate-controlled sintering capability, a new University-constructed instrument designed for operation to 1873K incorporates evolutionary thermal and mechanical features with novel microcomputer-based functions for measurement, control, data logging, data reduction, and graphical data representation. Principal design features, hardware selections, and performance data are reviewed; rationales for software development are discussed and typical sintering applications shown.

INSTRUMENT DESCRIPTION

The instrument is a vertical tube-type dilatometer[1] that includes time-proven thermal and mechanical design elements[2,3] and novel digital control and data management features and capabilities. Among its special features are a low thermal mass, high temperature (1873K) furnace operating in air and capable of rapid heating rates (>100K/min at 1273K, and >20K/min at 1873K); provision for regulated flow of controlled atmosphere (or optional air) within the hermetically sealed specimen chamber; and a fully digital, microcomputer-based controller/recorder/data analyzer system, strongly supported by appropriate software. It was initially developed and used for sintering studies on SYNROC,[4,5] a polyphase, titanate-based ceramic intended to serve as the crystalline, geologically stable host matrix for very complex mixtures of high level radioactive wastes, many of which required reducing atmospheres (e.g., Ar, 4% H_2) to maintain proper valence states for incorporation of waste ions into the matrix phases.

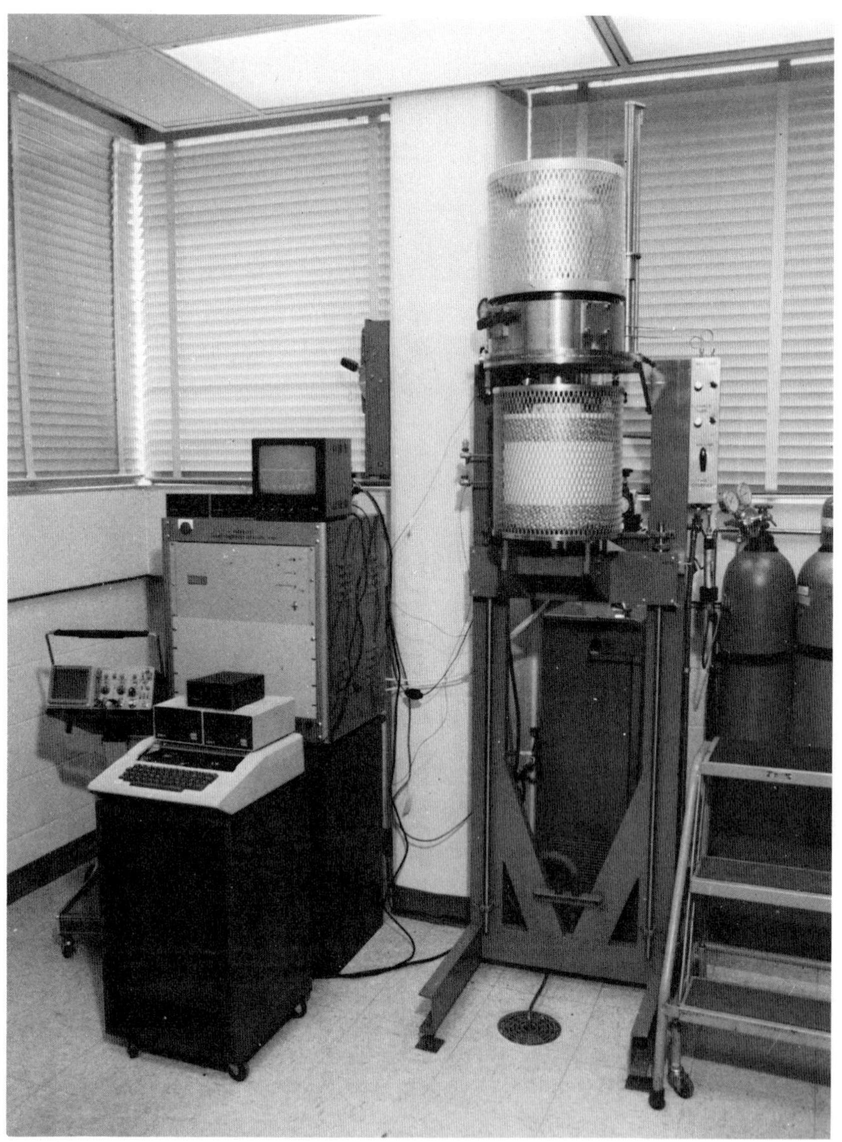

Fig. 1. Overall view of precision digital dilatometer.

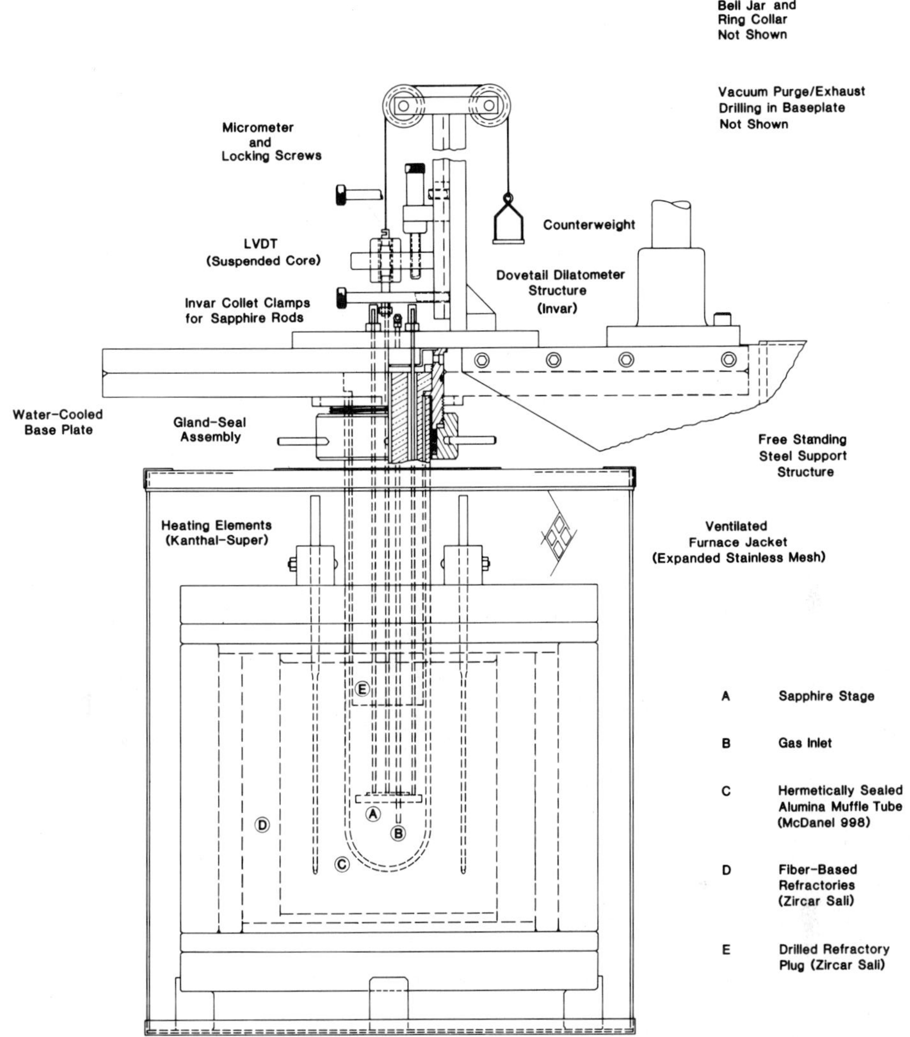

Fig. 2. Cross-section view of the dilatometer and furnace.

Figure 1 shows the overall configuration of the dilatometer and its peripherals, and Fig. 2 shows its principal components in cross-section. They include: (1) the Invar dilatometer mechanism supporting the shrinkage-sensing linear variable differential transformer (LVDT) within the "cold" upper chamber; (2) the precision machined, water-cooled base plate and seal gland; the "hot" dilatometer extension rods and specimen stage (Fig. 3), all made of C-axis oriented sapphire; (4) the lower specimen chamber of dense, inpermeable alumina, in the form of a closed-end tube (Fig. 4), hermetically sealed at the base plate by O-rings within the gland nut; (5) the flowing atmosphere inlets and outlets; and (6) the pneumatically-lifted, top loading furnace.

A design similar to those employed in earlier NCSU-built dilatometers[2,3] was chosen for the heart of the dilatometer, which utilizes sapphire extension rods connecting with a "cold" linear variable differential transformer (LVDT) for sensing and measuring changes in "hot" specimen dimensions during firing.

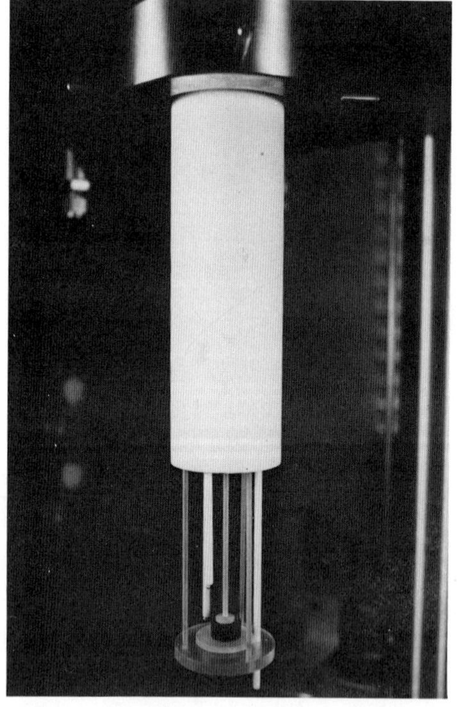

Fig. 3. Sapphire stage, rods, thermocouples, gas inlet tube, shims and specimen in situ.

Fig. 4. Dense alumina tube enclosing and sealing the "hot" lower chamber.

Fig. 5. Dilatometer/LVDT assembly in situ on the base plate.

Fig. 6. Base plate halves before welding and final machining.

The LVDT is supported and pre-adjusted by a simple but sturdy vertical slide structure (Fig. 5) made of Invar, a nickel-based alloy chosen for its extremely low coefficient of thermal expansion under near-ambient conditions. When secured in prealigned position on the internally channeled, water cooled stainless steel base plate (Fig. 6), within the ring collar and bell jar[a] which complete the "cold" upper chamber, the entire dilatometer/LVDT assembly is provided with both firm support and reliable protection against thermal aberrations.

The three sapphire extension rods which align and support the hot stage are referenced at the principal horizontal plane of the Invar dilatometer structure, and are collet-clamped with Invar fittings. A sapphire probe rod is attached with an Invar collet clamp to the moving core of the LVDT. The core/probe rod assembly is counterweighted to minimize the load on the specimen (typically $<5g$), thereby reducing the possibility of stress-induced creep. Sapphire shims above and below the specimen (Fig. 3) also act as load spreaders, eliminating point load problems, thus reducing external stresses to negligible levels.

From the base of the Invar dilatometer, the sapphire probe and support rods extend downward through a close-fitting stainless steel radiation shield/gas baffle positioned within the base plate, and then through the precision machined inner refractory plug of fibrous alumina to the "hot" specimen stage (Fig. 3). To minimize creep and thermal drift effects, the c-axis of all sapphire components is oriented parallel to the direction of shrinkage measurement.

With the present atmosphere control system, samples can be fired selectively in oxidizing, neutral, or reducing atmospheres (for safety reasons, however, gas atmospheres are restricted to those which are nontoxic and nonexplosive). Containment of the particular atmosphere is provided by a dense (impervious) alumina tube[b] which encloses the c-axis sapphire sample stage and support rods (see Figs. 2 and 4). A high temperature (473K) silicone rubber cement is used to bond and seal a stainless steel band at the top of the alumina tube, which is then finished machined to establish axiality, concentricity and proper diameter for O-ring sealing. Mechanical support for the suspended alumina tube, as well as the gas tight mechanical seal, is accomplished with three high temperature (473K) O-rings with stainless steel spacers, situated between the cemented ring and the inner surface of the well-tightened (threaded) gland nut. [With the

---

[a]Conventional vacuum equipment components (products of Denton Vacuum, Inc., Cherry Hill, N.J.).

[b]Type 998 Alumina, a product of McDanel Refractory Porcelain Company, Beaver Falls, PA.

existing combination of fan-assisted convective air and water cooling, the silicone sealed joint has remained unaffected for furnace operating temperatures to at least 1923K. The original alumina tube has successfully survived more than 100 firings!]

The atmosphere control system is a simple, manually operated one which provides for (a) vacuum outgassing, or (b) optional popoff release to atmosphere, both accomplished through a side drilling and small annular plenum within the base plate; (c) parallel, separately valved and metered gas flow paths entering (1) the "cold" upper chamber (bell jar) and (2) the "hot" specimen chamber (preheated while passing downward through the small bore alumina tube shown in Figs. 2 and 3); and (d) optional inputs to the gas handling system of either of two selected, pressure regulated, prefiltered gases (including line air).

By introducing gas into the upper (LVDT) chamber at a higher flow rate than that entering the "hot" lower specimen chamber, the hot gases rising from the lower chamber become buffered, diluted and cooled in the vicinity of the radiation shield/baffle plate, from whence they are readily exhausted through the inner plenum and side drilling of the base plate. The higher flow rate downward from the cold upper chamber tends to prevent unwanted upward flow of hot (or contaminated) gases from the lower specimen chamber, thereby assisting in maintaining closely controlled ambient conditions for the temperature-sensitive LVDT/dilatometer assembly. Flow rates are adjustable, typically being 6.0 (upper) and 2.0 liters/hour, respectively.

The electrically heated, air atmosphere furnace is designed around a hexagonal configuration of six small $MoSi_2$ heating elements.[c] Each element is offset from and backed by a hot face wall of 2.54 cm thick, high alumina refractory fiberboard.[d] Between the hot face wall and the outer shell (a 2.54 cm thick cylinder of less-refractory insulating fiberboard[e]) is an air gap. The 3.81 cm thick hearth is machined from refractory alumina fiberboard,[d] and is backed by an additional 2.54 cm layer of less-refractory material. The machined flat crown is of similar thickness, material and construction, and is additionally reinforced against sagging by radially inserted support rods of dense alumina. Finally, a protective expanded mesh jacket of stainless steel encloses and supports the furnace, and also provides for fan-assisted convective cooling.

---

[c]Super Kanthal 33, a product of Kanthal Corporation, Bethel, CN.

[d]Zircar SALI, a product of Zircar Corporation, Florida, NY.

[e]Zircar ZAL-15, a product of Zircar Corporation.

PRECISION DATA MANAGEMENT SYSTEM

In keeping with the highly precise thermal and mechanical components described above, major improvements have been called for in establishing the accuracy, reproducibility and flexibility of the chosen dilatometric data management system. Typically in the past, such devices have been of the analog controlled, strip chart recorded type, often capable of yielding at best only manually transcribed comparative data for supposedly systematic studies.

From the outset, the instrumentation for this dilatometer has been planned as being fully digital and microcomputer-based. This not only provides for precise digital control, but also assures real time acquisition (and frequent-interval magnetic disc storage) of digitally precise data. Inputs are obtained directly (via BCD circuits) from stable, reliable digital meters which continuously monitor both the temperature (via thermocouple) and the specimen displacement (via LVDT), as well as time (from an internal digital clock). Once magnetically stored (on diskette), the pertinent data base (whether recalled from a single experiment or from several comparable ones) can subsequently be re-entered into this (or another) microcomputer, and thus be corrected, reduced, analyzed and/or plotted in detail, and in a variety of ways.[5-7]

After considering compatibility of components, installation costs, ease of programming and operation and other pertinent factors, the system chosen for dilatometer control/data logging functions was based on a small "personal" microcomputer,[f] conventionally equipped with 48 kilobytes of memory, two disk drives, and a CRT monitor. For this application, it was additionally equipped with (a) a digital clock accurate to ±0.001 sec, (b) an interface board for two BCD circuits, and a special interface board digitally controlling the furnace power supply. The 0-10 volt output of the LVDT system[g] is read directly--and displayed--by a digital voltmeter[h] to ±0.001 volt, corresponding to a displacement sensitivity $<0.5\times10^{-6}$ m. The emf of the Pt, Pt,10%Rh thermocouple adjacent to the specimen is converted and read directly--and displayed--by a digital thermometer[h] to ±0.1K. [Furnace temperature also is monitored (displayed) to ±1K by another (nonreading) digital thermometer.[h]]

---

[f]Apple II Computer, a product of Apple Computer Co., Cupertino, CA.

[g]HR-050 LVDT and CAS-025 LVDT Signal Conditioner, products of Schaevitz Engineering, Pennsauken, NJ.

[h]Series 410-A Digital Trendicator(s), product(s) of Doric Scientific Division, San Diego, CA.

Given proper software (discussed in the following section), this powerful combination of digital meters and computer programming enables the experimenter to establish and maintain (and importantly, reliably to reproduce!) accurate control of the dilatometer system throughout the firing schedule, including cooling. One may choose between several different operating modes, including--but not limited to--(a) constant power level; (b) constant temperature [the classic isothermal mode]; (c) constant rate of heating [CRH mode]; (d) ramp at preset rate to preset temperature, with preset soak period [CTS mode]; and (e) densification rate control [RCS mode; see ref. 8 for further comparisons of CTS and RCS modes].

The flexibility of the microcomputer allows a variety of profiles to be entered with high precision. Different types of control can be combined in practically any arrangement, as may be dictated by the needs of a particular experiment. Once such a control profile is set up the system operates independently, without operator intervention. Data taken during the operation is recorded with high digital accuracy, allowing even small effects to be detected and analyzed. Since all the data are conveniently and routinely stored on diskettes, subsequent data analyses become almost automatic, and are not subject to human error during transcription.

DILATOMETER CONTROL PROGRAM

Figure 7 schematically illustrates the main steps involved in setting up and running the Dilatometer Control Program. The first four stages involve the operator in preprogrammed option selections and data input sequences (via keyboard). Operation under computer control begins at the "initialize" step: from that point on, the "looped" parts of the program repeat continuously, in part under clock control. Typically, the inner "control" loop is completed in about 1 sec, and data is recorded on the diskette at about 30 sec intervals.

A computation cycle this rapid (unusual for such a complex program in a small, nominally "slow" microcomputer) is facilitated by the use of the external digital meters for the temperature and displacement data inputs. With them, all of the signal conditioning (e.g., TC reference junction compensation and linearization) is external to the microcomputer itself, thus avoiding the "overhead" computation time which would have otherwise been involved in doing things such as averaging or smoothing over many readings, other filtering steps, etc.

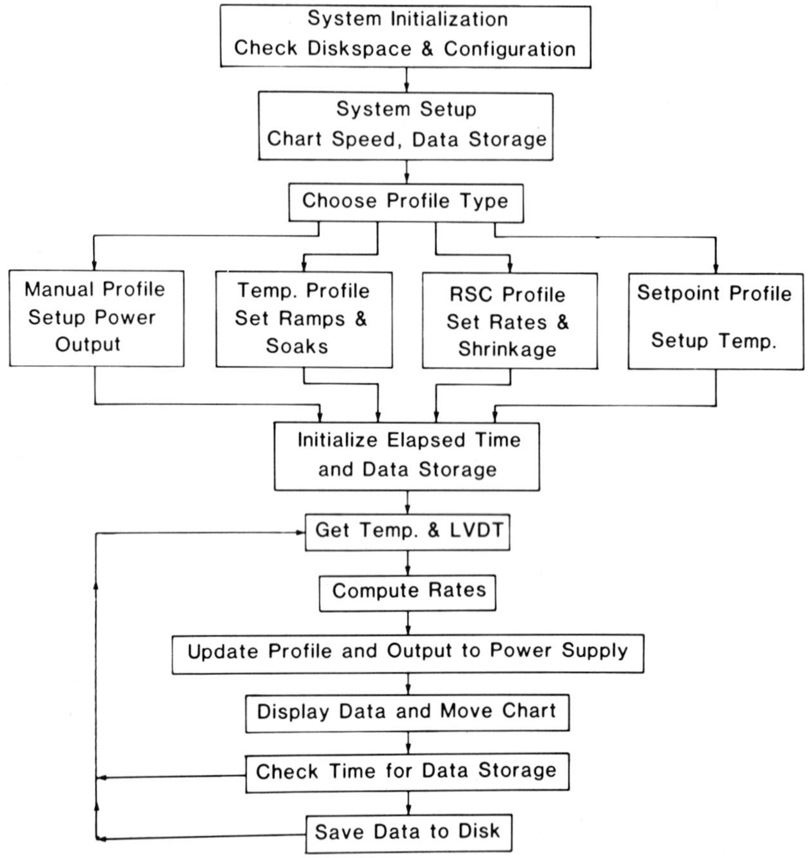

Fig. 7. Functions of the dilatometer control program.

The CRT display format (not illustrated) consists of two parts. The lower part shows digital data representing current conditions (including clock time, control set point, temperature, LVDT output in volts, calculated heating and densification rates, power level, etc.). Above it, a pseudo two-pen, strip-chart trace of thermocouple and LVDT output is displayed, indexing (scrolling) to provide the operator with a time-history view of recent trends at an operator-selected "chart" speed (typically spanning 1-3 h depending on dT/dt).

Many of the functions being undertaken sequentially within the inner control loop are more-or-less standard computational and/or data management tasks common to many kinds fo microcomputer applications. However, those that directly involve the programming and control of the furnace power supply, and thereby of the independent experimental variable of interest (T, dT/dt or dD/dt), must selectively be made sensitive to the response time(s) of that particular variable (within the overall context of the operating "hot" system). For the conventional temperature control modes, they may be considered to involve primarily the characteristic thermal response time(s) of the hot parts of the system (e.g., the furnace/ceramic envelope assembly, Figs. 2 and 4). For the RCS control mode,[2,3,7-9] the generally dissimilar time constant(s) of the shrinking specimen itself must also be reflected.

One means for "tuning" the microcomputer/controller with respect to actual system response is the familiar three mode control combination (proportional band, reset and rate time), long known in analog, and now in digital, versions. It is used quite widely as a process controller; in most situations, it does a very good job of handling the process (and meeting the needs of the operator).[10] More complex control algorithms are available[11] but are generally not used because more sophisticated equipment would be required, and special process analyses would also be needed before they could be implemented. In addition, these complex methods require many more calculations, which would pose a problem for real-time control by a typical small microcomputer.

For these reasons, algorithms representing a digital three mode controller were chosen for incorporation in the dilatometer control program. If needed, real time changes in the preset values for proportional band, rate time and reset can be implemented during a run by operator intervention through the microcomputer keyboard.

The shrinkage rate variable responds to control in a quite different manner than does temperature. There is, for example, no direct relationship between shrinkage rate and power level, yet the power required to maintain a specified shrinkage rate must change with time as the sample shrinks. Furthermore, the effective RCS profiles[2,3,7-9,12] call for both fast and slow densification rates, deliberately established, with rather abrupt transitions occurring between them at preselected fractional densities.[12] These relatively sharp transitions further complicate the control issue, and also introduce some time-dependent calibration correction effects, discussed below.

For these reasons, the rate controlled sintering (RCS) mode involves a different and fairly complex set of program/control objectives. Significant difficulties were encountered with earlier analog RCS controllers, in part because the control variable then in use (shrinkage, $\Delta l/l$) was one-sided (noncyclic; shrinkage can only increase!). Two types of control had to be manually selected and/or retuned at critical times, because conventional (CRH) temperature control was also required, both before the RCS segment (up to the onset of shrinkage) and after (during cooling). These issues now are easily handled by the microcomputer: in the RCS mode, it is preprogrammed to change the control logic and the respective settings smoothly and automatically from (a) heating rate to (b) densification rate, and (c) back again. It does so at well defined transition points, which in principle can be expressed in terms of <u>any</u> of the pertinent variables (e.g., at some given level of dD/dt, or D, or dT/dt, or T, etc.).

With existing software, modified three mode control algorithms are being employed with reasonable success for early experiments involving RCS control of alumina and other ceramic materials. Satisfactory (though not yet ideal) control of the RCS mode is being obtained for realistic shrinkage times, e.g., from D_o to D_f in >2 h. Tuning the three mode control system for critical damping of dD/dt still is not an easy task, especially at and near major rate transitions, and at the higher densification rates (>>0.01/min). [As more operating experience in the RCS mode accumulates on this new and precise digital control system, it is likely that better (and probably more complex) algorithms can be identified and developed specifically for this unique and useful firing mode.]

DATA CORRECTION, REDUCTION AND MANIPULATION

The limiting precision of the dilatometric data is governed in part by adequate knowledge of the appropriate temperature calibrations, machine corrections, thermal expansivities, and specimen shrinkage anisotropies. The particular material and/or sintering objective determines the extent to which the correction details must be explored and/or applied. However, once worked out in detail, their routine applications become simple (though perhaps time consuming) because even the most complex correction procedure can be implemented easily by the computer.

Figure 8 describes schematically the overall sequence of corrections and data reduction steps. The temperature calibration is applied first, from an empirical equation which gives the true temperature (T*) as a function of the thermocouple emf and the heating rate. This relationship is easily established experimentally by placing a small loop of fine wire of a noble metal (Au or Pd) between two alumina shims and, at a given dT/dt observing its melting, with sudden LVDT displacement serving as the marker for the melting temperature. In this machine, the true temperature was found to be a strong function of heating rate near the melting point of gold (1337K), resulting in an offset of 70K at 20K/min, but only 40K at 5K/min. At the palladium point (182K), there was no observable heating rate effect, and the offset error did not exceed 5K.

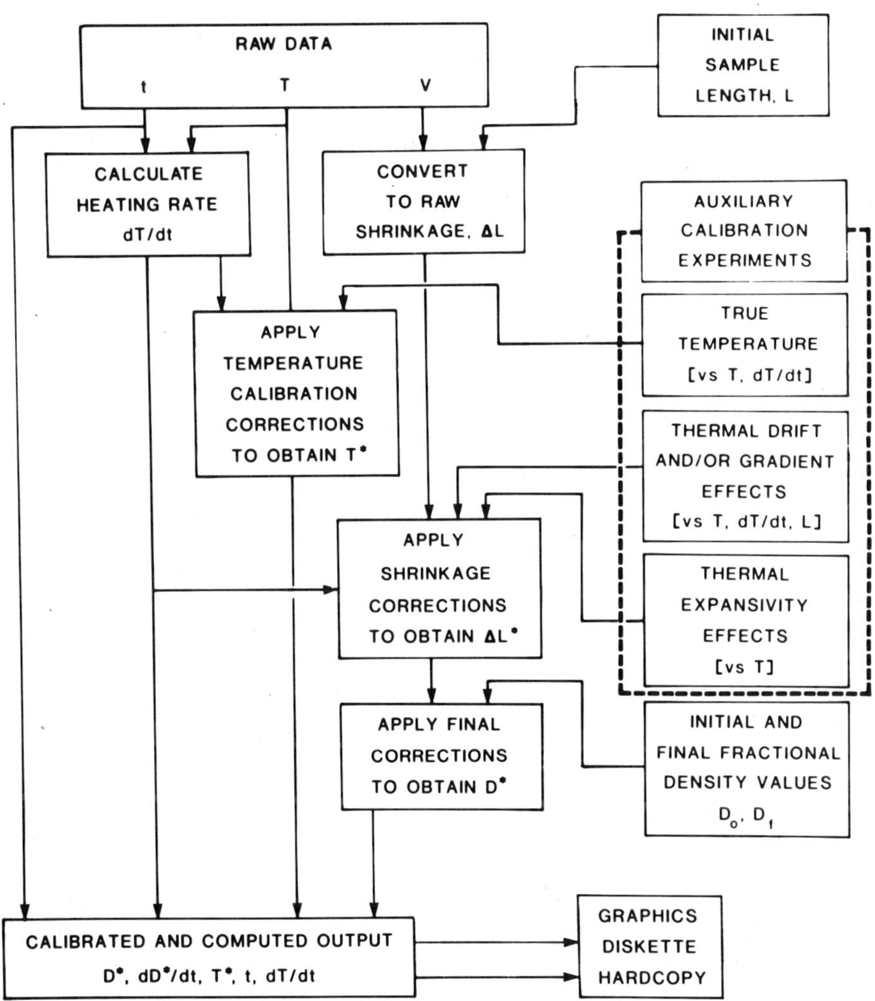

Fig. 8. Schematic diagram for correction and reduction of dilatometric data.

The heating rate is either read directly from the raw data or by regression of the data. Simple linear or parabolic fits can be employed, depending on the type of control program in use.

Because the data to be processed includes carefully measured starting and final densities, D_o and D_f, the calculation of fractional density at all intervening levels as a function of time and temperature is force fitted to include both D_o and D_f. For materials having isotropic shrinkage, the total linear shrinkage thus becomes fixed (this should be the case for isostatically pressed materials, but not for uniaxially pressed compacts). However, as long as the ratio of the shrinkages in each direction remains constant during densification, there will be no error in fixing the starting and final densities and scaling the shrinkages at each point in time by a factor calculated by dividing the expected shrinkage (based on D_o and D_f) by the observed shrinkage (from the LVDT). If the shrinkage anisotropy varies (e.g., due to high uniaxial pressures and/or low densities), then that nonlinear function must be determined from auxiliary experiments.

The raw shrinkage data is converted to length from the LVDT output voltage, then scaled to the expected shrinkage based on D_o and D_f.

If all calibration effects were linear with respect to temperature, then the density could be computed without further need for correction. For most ceramic materials, corrections for thermal expansivities (relative to sapphire) are small to begin with, and are nearly linear over normal ranges of sintering temperatures. Unless a phase change or reaction is known to occur within that sintering temperature range, it usually is not necessary to know the actual thermal expansivities in order to reduce the data properly.

The "baseline" machine correction is a quite different issue, and cannot be ignored, especially during non-linear heating, including the isothermal part of the conventional up-and-hold (CTS) firing schedule. The machine correction curve shown in Fig. 9 is known to be a strong function of heating rate and rate history, particularly at the higher temperatures. When the heating rate is changed suddenly, as in going on temperature hold, it can take as long as 40 minutes to come to "equilibrium."

Experimentally, a series of calibration runs with various heating rate changes over a range of temperatures can be used to determine a set of time constant coefficients.[13] That baseline data has now been established for use in the correction software. By summing backwards in time, the contribution of all recent heating rate changes can be usefully approximated. Correct application of the exponentials is essential, especially for the determination of late-stage sintering behavior, as illustrated in Fig. 10.

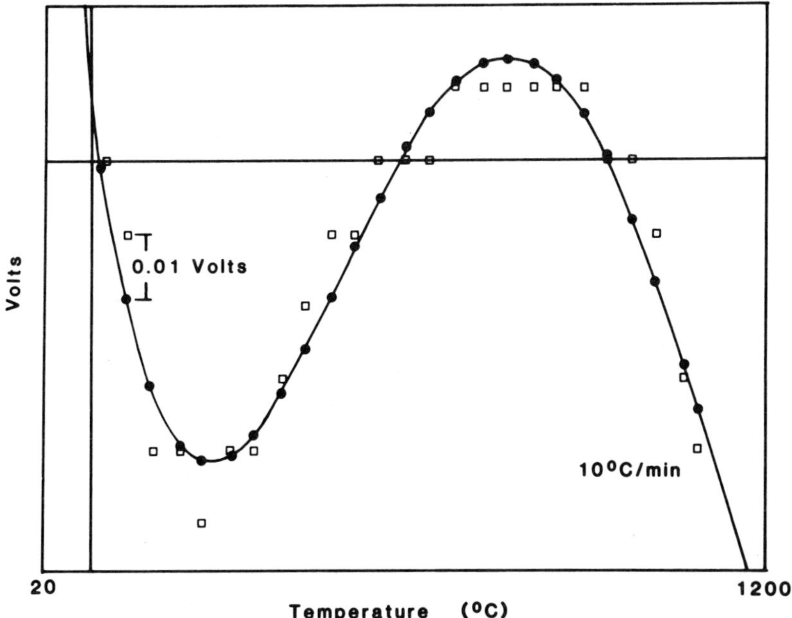

Fig. 9. Experimentally determined machine correction curve.

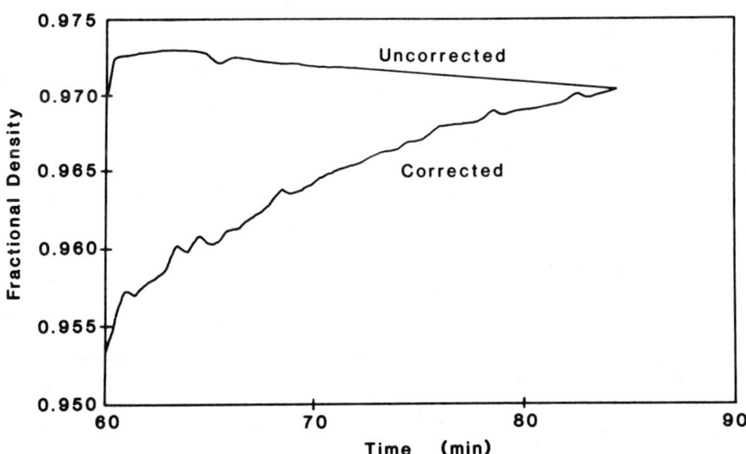

Fig. 10. Comparison of uncorrected and corrected density data for the case of late-stage isothermal sintering.

The complete shrinkage correction procedure, then, utilizes corrected temperature, LVDT displacement data, heating rate data over the recent past (≤60 min.), experimentally derived time constants, and the independently measured initial and final densities to produce the best possible estimates of all the pertinent densification parameters.

Though diagramed together here, in practice it has been found desirable to treat corrections (of original data) in a separate software package, so that the data reduction software ("Krunch" program) can be used routinely to process either corrected data (where warranted), or uncorrected (where not), in keeping with specific experimental needs. In many cases, particularly for those firings carried out under constant-rate (CRH) conditions, the uncorrected but reduced data are all that is required, since they differ only very slightly from the fully corrected and reduced values.

In the case of rate controlled sintering, where the final density is not known during the run, both the heating rate and final temperature must be estimated in advance, since part of the apparent shrinkage at the end (near D_f) will be due to machine response. In practice, it is easiest to make a trial run in the RCS mode to establish approximate ranges, followed by any necessary adjustments, either in another shrinkage rate-controlled (RCS) run, or in an equivalent, adapted (derived) temperature-time schedule. In either case, the workup after the final density has been determined is the same, regardless of the control method.

The data-handling computer routines produce a corrected data file which includes time, temperature, heating rate, density and densification rate. Raw data files can be saved for recalibration as and/or if improved methods are developed. Finally, graphics displays are generated, for the monitor (CRT) and for hardcopy as needed. A wide variety of interesting plots can then be examined.[7]

TYPICAL APPLICATIONS

The dilatometer thus far has been used for two principal purposes: (a) comparisons of firing schedules and (b) comparisons of materials. Figure 11 compares uncorrected but reduced data from two different firing schedules for alumina; shrinkage rate control (RCS) and conventional temperature control (CTS). Final stage microstructural features are optimized from a series of such runs.[8,9]

Figure 12 shows an uncorrected but reduced comparison of two similar complex titanate ceramics (SYNROC),[5] with and without simulated radwaste dopants. The effect of the dopant on sintering behavior can readily be seen with the aid of the densification rate plots.

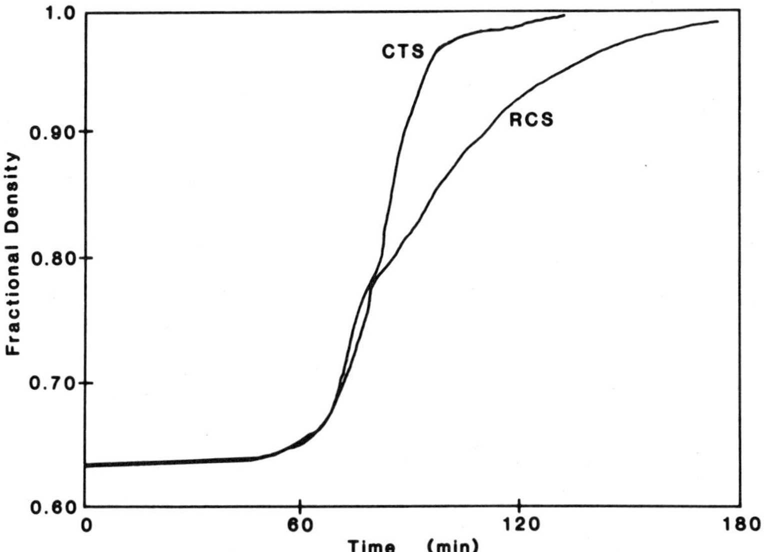

Fig. 11. Comparisons of RCS and CTS profiles for sintered alumina.

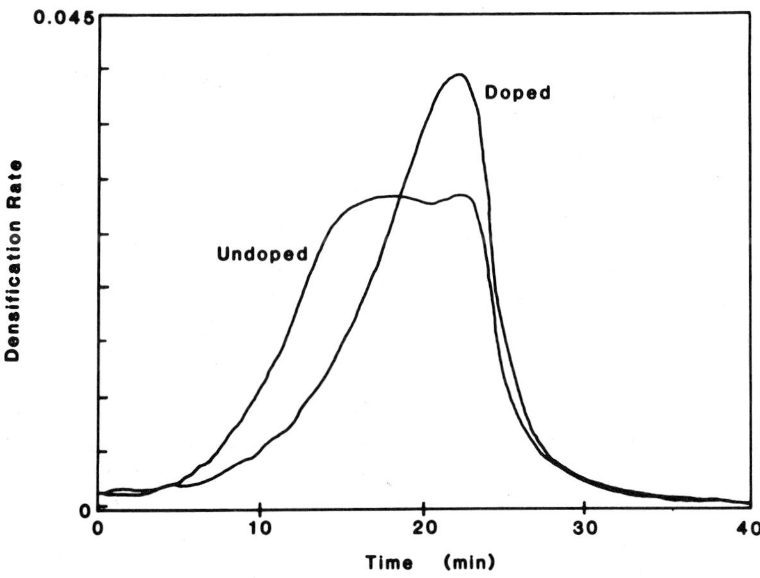

Fig. 12. Comparisons of sinterability of doped and undoped SYNROC.

SUMMARY

A new controlled atmoshpere, digitally precise, computer-controlled dilatometer system has been designed, constructed, programmed and tested. It is capable of attaining 1873K at high heating rates. The availability of densification rate-controlled and conventional temperature-controlled sintering modes, precise and accurate calibration methods for length change and temperature, and data handling programs which can conveniently process, utilize and display the data base created, combine to provide an apparently unique capability for investigating the sintering of ceramic materials (even unusual or "difficult" ones) under hitherto largely unexplored conditions.

ACKNOWLEDGEMENTS

Supported by the U.S. Department of Energy under Contracts DE-AC09-80ET-41902 and DE-AC09-81SR-10957, with W. Bruce Wilson serving as Technical Monitor; and under LLNL Subcontract UCRL 31-109-38-5843, with Clarence L. Hoenig serving as Technical Monitor.

Additional support, particularly in the development and utilization of detailed calibrations and computer-based correction procedures, has come from a DARPA-supported program under LLNL Subcontract UCRL 6853501, with Carl F. Cline serving as Technical Monitor.

Further cost-sharing support has been provided by the School of Engineering's Engineering Research Program. The personal interest and encouragement evidenced by F. D. Hart, Associate Dean of Engineering for Research, is gratefully acknowledged.

Among those who have made significant contributions to the design and/or construction of the facility have been: David Fike (preliminary design sketches); Craig and Bernie Hamling, of Zircar (furnace design concepts); J. C. Russ (microcomputer application concepts); K. R. Brose and Wade Howell (power supply and interface); designer-draftsman Ralph Stecker (detailed working drawings); R. L. Cochrane (gas-handling system, project engineering); and Alex Russell, David Strickland, Marvin Strickland and Tony Mason (instrument makers).

E. M. Gregory, Jim Mehaffy, Michelle Bridges, Mark Engelhardt, Laura Freed, Karren More, Ellice Luh, Tom Goudey and Betty Randall assisted in the preparation of material for the poster presentation, and for this paper.

REFERENCES

1. J. Valentich, Tube Type Dilatometers, Instrument Society of America, Research Triangle Park, NC, 1980.
2. M. L. Huckabee and H. Palmour III, Am. Ceram. Soc. Bull., 51 [7], 574-76 (1972).
3. M. L. Huckabee, T. M. Hare, and H. Palmour III, pp. 205-15 in Processing of Crystalline Ceramics, edited by H. Palmour III, R. F. Davis, and T. M. Hare, Mat. Sci. Res., Vol. 11, Pelnum Press, New York-London, 1978.
4. H. Palmour III, T. M. Hare, J. C. Russ, C. B. Boss, A. G. Solomah, and A. D. Batchelor, Topical Report, Contract DE-AC09-81SR-10957, July, 1981.
5. H. Palmour III, T. M. Hare, J. C. Russ, A. D. Batchelor, M. J. Paisley, and L. E. Freed, Final Technical Report, LLNL Subcontract UCRL 6548901, September, 1982.
6. H. Palmour III, K. Y. Kim, K. L. More, and R. C. Motley, pp. 331-72, LLNL Report UCID 19663, edited by C. F. Cline, August, 1982.
7. K. Y. Kim, A. D. Batchelor, K. L. More, and H. Palmour III, "Rate Controlled Sintering of Explosively Shock-Conditioned Alumina Powders," this volume.
8. H. Palmour, M. L. Huckabee, and T. M. Hare, pp. 46-56 in Sintering - New Developments, edited by M. M. Ristic, Elsevier Science Publishers, Amsterdam, 1979.
9. H. Palmour and M. L. Huckabee, pp. 278-97 in Factors in Densification and Sintering of Oxide and Non-Oxide Ceramics, Tokyo Institute of Technology, Tokyo, Japan, 1978.
10. F. J. Hale, Introduction to Control System Analysis and Design, Prentice-Hall, Inc., Englewood Cliffs, NJ, 1973.
11. J. Hogan, Measurements and Control Applications, 2nd edition, Instrument Society of America, Pittsburgh, PA, 1979.
12. H. Palmour III and M. L. Huckabee, U.S. Patent No. 3,900,266, 1975.
13. Y. Horie, H. Palmour III, and J. K. Whitfield, PP. 130-216 in C. F. Cline, ed., Second Quarterly Report, DARPA Dynamic Synthesis and Consolidation Program, Lawrence Livermore National Laboratory Report UCID-19663-83-1, March, 1983.

PART III

CERAMICS DERIVED BY POLYMER PROCESSING

THE CONVERSION OF METHYLCHLOROPOLYSILANES AND POLYDISILYLAZANES TO SILICON CARBIDE AND SILICON CARBIDE/SILICON NITRIDE CERAMICS, RESPECTIVELY

Ronald H. Baney, John H. Gaul, Jr., and Terrance K. Hilty

Dow Corning Corporation
Midland, Michigan 48640

BACKGROUND

Engineering or structural ceramics have received attention by materials scientists in recent years. Much of the effort has been prompted by a need for energy conservation. This has translated into the search for materials with higher specific strengths for weight savings and material for higher temperature Carnot efficiencies for heat engine applications. The super alloys have been asymptotically approaching a limit of around 1373K; thus design engineers have turned to ceramics. The use of ceramic materials presents several technical problems one of which is their formation into complex shapes.

Hot pressing usually yields billets which must be further machined to produce complex shapes. Pressureless sintering allows for near net shapes but useful temperatures are limited to the softening temperature of the glassy phase used as sintering aids. Reaction-bonding gives near net shapes but is a very slow process requiring high temperatures. Hot isostatic processing gives near net shape but with complex relatively expensive processing. Ceramics, in contrast to most metals, usually fail through catastrophic brittle fracture. Organometallic processing of ceramics offers potential solutions to both of these technical problems.

Organometallic routes to ceramics have advantages over the more traditional routes such as the Acheson process for the production of silicon carbide which involves a high temperature solid/gas reaction. Distillation and other traditional purification methods of the organometallic intermediates allow for higher purity ceramics. Molecular level mixing of organometallic compounds and the molecular level architectures of the starting compounds yields ceramic materials with a variety of elemental compositions.

Over the past few years numerous patents and publications have revealed the use of simple organometallic compounds and polymers for routes to ceramics. Essentially three distinct types of technologies are described. These are chemical vapor depositions, CVD; polymer pyrolysis; and sol/gel processing. The CVD process involves vapor phase decomposition of, for example, methyltrichlorosilane, to give beta silicon carbide. The latter two routes involve densification and/or pyrolysis of organometallic polymers to ceramics.

The high Gibbs free energy of formation of organometallic compounds frequently allows for lower temperature formations of ceramics. Thus, for example, the pyrolysis of methyltrichlorosilane to silicon carbide becomes thermodynamically favorable about 900K while the formation of silicon carbide by the reduction of silica with carbon is not favorable until 1800K. Likewise, inorganic colloidal sols are prepared through hydrolysis of metal alkoxides. The high surface free energy found in the pores of gelled colloidal sols allows for densification at lower temperatures and thus, formation of materials not possible by higher temperature processing.[1]

The polymer routes also allow for formation of complex shapes as continuous filaments or yarns before pyrolysis. Molecular level mixing in the organometallic state can facilitate faster reaction rates and give ceramic material not possible by high temperature solid or solid/gas phase reactions. Variable atomic ratio in starting organometallic compounds also can yield novel ceramic compositions.

Distillation of volatile organometallic intermediates allows for simple purification so that resulting ceramics have a higher level of purity than ceramics obtained through high temperature solid reactions. This is completely analogous to the purification of electronic grade silicon through formation of volatile, trichlorosilane from hydrogen chloride and silicon followed by pyrolysis back to silicon.

The conversion of polyacrylonitrile fibers to graphite fibers has been known for a long time.[2] Within recent years analogous work has been published describing the conversion of silicon-containing polymers to silicon carbide-containing ceramic yarns[3] and silicon carbide/silicon nitride containing ceramic yarns.[4]

Certain polymer characteristics are desirable for optimum polymer-to-ceramic systems. If a shaping operation such as spinning is required, the polymer should obviously be tractable. This means that the polymer cannot be crosslinked into an infinite network before spinning. Pyrolysis requires some crosslinking of polymer chains for two reasons. If after forming (for example melt spinning) the polymer is to undergo a subsequent pyrolysis step it must not melt down during that step. It is also desirable to maintain as high a char or ceramic yield as possible both for economic reasons and to minimize shrinkage and avoid dealing with large amounts of pyrolysis gases. High crosslink densities promote free radial homolytic pyrolysis bond cleavage reactions which generally increase char yield. The requirement of high char yield and corresponding high crosslink density and also tractability is a dichotomy which can be met by utilization of polycyclic structures.

The very property of high covalent bond density which gives silicon carbide and silicon nitride ceramics their desirable properties must be avoided to produce tractable polymers. This is best accomplished by employing low molecular weight mono-functional fragments and/or catenated bonds. Another obvious consideration is the cost of raw materials and processing to obtain the polymers. Methylchlorodisilanes are produced as byproducts in the direct synthesis of methylchlorosilanes which are intermediates in the silicone industry. These disilanes, described in this report, consist of about 10% $[(CH_3)_2ClSi]_2$; 55% $(CH_3Cl_2Si)_2$; and 35% $Ch_3Cl_2SiSiCl(CH_3)_2$ and were employed to prepare the polymers systems.

METHYLCHLOROPOLYSILANE AND ITS DERIVATIVES

Yajima[3] used sodium reduction dimethyldichlorosilane to prepare his preceramic polymer.

$$X(CH_3)_2SiCl_2 + 2XNa \dashrightarrow [(CH_3)_2Si]_x + 2XNaCl \quad (1)$$

The polysilane was then subsequently rearranged in a high temperature autoclave step to a carbosilane.

$$2[(CH_3)_2Si]_x \xrightarrow{\Delta} [(CH_3)_2SiCH_2SiH(CH_3)]_x \quad (2)$$

The reduction with sodium and subsequent autoclave rearrangement are expensive process steps. Bond redistribution reactions have also been reported to form catenated silicon structures. Bond redistributions in silicon chemistry has been extensively reviewed.

$$\overset{|}{Si}X + \overset{|}{Si}Y \rightleftarrows \overset{|}{Si}Y + \overset{|}{Si}X \quad (3)$$

In order to avoid a sodium reduction step, rearrangement of methylchlorodisilanes was chosen.

$$3X(CH_3Cl_2Si)_2 \xrightarrow{\text{Catalyst}} 2(CH_3Si)_x + 4XCH_3SiCl_3 \quad (4)$$

The catalyst employed was tetrabutylphosphonium chloride at ca. 1% concentration. The mixed methylchlorodisilane obtained in the direct synthesis of methylchlorosilanes were heated with the catalyst.[7] Methyltrichlorosilane and dimethyldichlorosilane distilled from the reaction. As the pot residue increased in temperature and became more and more yellow, it became more viscous. Subsequent studies on samples which had been methylated by reaction with methyl Grignard reagents as a function of pot temperature showed increasing molecular weight with increasing pot temperature. By careful mass balance analyses with the aid of gas chromatography, the following empirical formula was determined for a mixed disilanes polymerized to 523K:

$$(CH_3Si)_{1.0}[(CH_3)_2Si]_{5.7}Cl_{1.7}$$

The assumption was made that no methyl/silicon bond redistribution reactions took place during polymerization. Studies on the polymerization of $[(CH_3)Cl_2Si]_2$ and $CH_3Cl_2SiSiCl(CH_3)_2$ verified that assumption. The molecular weight of the polymer polymerized to 523K was about 1200 g/mole. From assumptions about ring size (i.e., none smaller than 5), a knowledge about functionality around silicon, and the molecular weight, it must be concluded that polycyclic structures are the most likely structures present. Figure 1 represents one such structure.

Silicon-chlorine functionally reacts with numerous reagents. Thus, the pre-ceramic polymer described above reacted with water to replace the chlorine with oxygen and crosslink the system. The SiCl bond also reacted with alcohols to form alkoxy derivatives,[8] with amines and ammonia to form amino silanes,[9] with reducing agents such as lithium aluminum hydride to form hydrosilanes,[10] with alkylating agents such as Grignard reagents to form fully alkylated polymers;[11] and with hexamethyldisiloxane to form trimethylsiloxy derivatives.[12] When these derivatized polymers were pyrolyzed, a variety of ceramic compositions resulted. In the case of the amino derivatives, Si_3N_4 was detected only when the chloro-containing polymers were pyrolyzed in ammonia. The hydride polymer gave a high ceramic yield of over ninety percent when the heavy Cl-fragment lost in pyrolysis was replaced with a light H-fragment.

Fig. 2 shows a thermogram of a polymer obtained by polymerizing mixed disilanes to 523K. A significant rate of pyrolysis begins at around 523K. Evolved gases were analyzed by mass spectrometry and the results of these studies are also included in Fig. 2. Methyltrichlorosilane and dimethyldichlorosilane are initially evolved. As polymerization continued, these products give way to methane and hydrogen chloride, the products expected from free radical homolytic bond cleavage reactions. Si-H moieties may arise from methyl group insertion reactions similar to those observed for the Yajima autoclave step.[3]

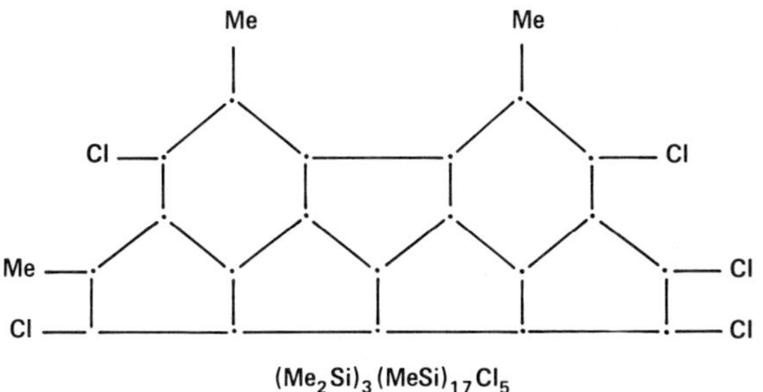

Fig. 1. Representative structure of methylchloropolysilane polymerized to 523K; "." = CH_3Si moieties.

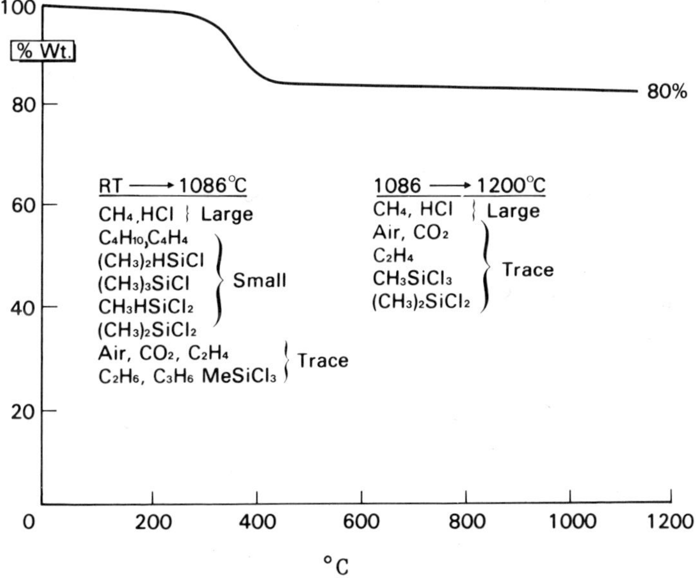

Fig. 2. TGA of methylchloropolysilane polymerized to 523K.

Samples of the polymer prepared at 523K, and its derivatives, were fired to 1473K. X-ray diffraction patterns revealed only very small grain β-SiC (6-20 nm) similar to the results reported by Yajima.[3] Precise elemental analysis is frequently difficult to obtain for ceramic materials. Table 1 lists the elemental composition on several derivatives and compares these data with similar data for the Yajima polymer.[3] Silicon and carbon were determined by the Leco[13] method. Oxygen was determined by the Phoenix laboratories by neutron activation.[14] The data, though not precise, illustrate the wide variety of ceramic compositions possible by varying the preceramic polymer composition.

Table 1. Composition of Ceramics

Polymer* Precursor	Si:C Ratio in Polymer	1473K	1873K
PCP-Cl-250	1:1	$Si_{1.0}C_{1.0}O_{.05}$	--
PCP-O-250	1:1	$Si_{1.0}C_{.62}O_{.42}$	$SiC_{.74}O_{.004}$
PCP-Me-250	1:1.3	$Si_{1.0}C_{.53}O_{.15}$	$SiC_{.63}O_{.02}$
Yajima's Carbosilane	1:2	$Si_{1.0}C_{1.1}O_{.6}$ (1300°C)	--

*PCP-Cl-250 is the designation used for the methylchloropolysilane polymerized to 523K. O and Me stand for oxygen and methyl derivatives, respectively.

The methylchloropolysilane system had some handling problems. The extended catenated polysilane structure was very reactive toward oxygen and in the case of the hydride derivatives was pyrophoric at room temperature. This promoted the search for another more stable system which is the subject of the next section of this report.

POLYDISILYLAZANES

Like the methylchloropolysilanes, the polydisilylazanes were prepared from the disilanes obtained as by products in the direct synthesis of methylchlorosilanes. Hexamethyldisilazane reacted with the disilanes through silicon-chlorine/silicone-nitrogen bond redistribution reactions to produce trimethylchlorosilane and polydisilylazane.[15]

Excess hexamethyldisilazane reagent was generally employed. When hexamethyldisilazane is introduced into a pot containing the mixed disilanes a small amount of precipitate consisting of ammonium chloride is formed. Excess hexamethyldisilazane was generally employed. As the reaction is heated, trimethylchlorosilane distills and as the boiling point of hexamethyldisilazanes is approached it is co-distilled from the reaction. The reaction clears as the small amount of ammonium chloride initially formed sublimes from the reaction mixture.

The reaction can be represented by the following equation:

$$XCH_3Cl_2SiSiCl(CH_3)_2 + 2X[(CH_3)_3Si]_2NH \dashrightarrow (CH_3\overset{|}{\underset{|}{Si}}Si(CH_3)_2NH-)_x$$

$$+ 3X(CH_3)_3SiCl \quad NHSi(CH_3)_3 \tag{5}$$

The reaction is more complex than is shown in Eq. (5). After thirty seconds a reaction of 1:1 ratio of starting silazanes to mixed disilanes reaction was examined by gas chromotography. All of the starting disilanes had reacted. After 30 seconds the volatiles consisted of 0.62 mole ratio of $(CH_3)_3SiCl$ and 0.55 mole ratio of $[(CH_3)_3Si]_2NH$ with 0.45 mole ratio being consumed. The 463K distillate composition consisted of 84 mole % $[(CH_3)_3Si]_2NH$ and 16 mole % $(CH_3)_3SiCl$. These observations are consistent with the following set of reactions:

$$\equiv SiCl + [(CH_3)_3Si]_2NH \rightleftarrows \equiv SiNHSi(CH_3)_3 + (CH_3)_3SiCl \tag{6}$$

$$\equiv SiCl + SiNHSi(CH_3)_3 \rightleftarrows \equiv SiNHSi + (CH_3)_2SiCl \tag{7}$$

$$2 \equiv SiNHSi(CH_3)_3 \rightleftarrows \equiv SiNHSi\equiv + [(CH_3)_3Si]_2NH \tag{8}$$

The weight of residue and amounts of volatile materials determined by gas chromatography were used to determine mass balance and deduce structural features of the polymer. A polymerization carried to 523K was found to have an empirical formula of

$$\left[(CH_3)_2Si_2]_{0.6}[(CH_3)_3Si_2]_{0.4}[(CH_3)_4Si_2]_{0.1}(NH)_{1.5}[NHSi(CH_3)_3]_{0.4}\right]$$

and a molecular weight of 3100. The number of different ways the structure units can be put together makes speculation about specific structures difficult. It can be concluded, however, that the polymer structure must be polycyclic.

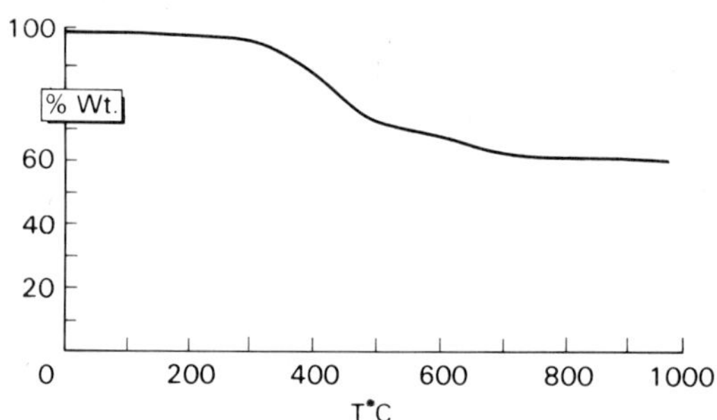

Fig. 3. TGA of polydisilyazane polymerized to 548K.

A thermogram of the polymer polymerized to 548K is shown in Fig. 3. A significant rate of pyrolysis is observed to begin about 593K. The ceramic char yield was about 61 percent but it is dependent on the extent of polymerization.

The pyrolysis gases were examined by mass spectrometry as a function of temperature in helium. Table 2 lists the compounds observed. Small amounts of oxygen-containing species were observed which probably arise from small amounts of oxidation or hydrolysis reaction inadvertently occurring during workup and handling. Polymerization continues at the lower temperature with evolution of $[(CH_3)_3Si]_2NH$. The SiH species may arise from methyl group insertion into silicon-silicon bonds.[3] Only free radical reactions leading to formation of methane are observed at elevated temperatures.

Infrared spectra of the polymer and its pyrolysis product fired to 1473K are shown in Fig. 4. Bands associated with NH and CH at 2800-3500 cm^{-1} have disappeared at 1473K. The band associated with a methyl group on silicon found between 1200 and 1500 cm^{-1} is also gone at 1473K. Only a broad band associated with silicon nitrogen and silicon oxygen stretching frequencies are observed for material heated to 1473K. The density increased from 1.9 cm^{-3} to 2.2 gcm^{-3}. X-ray diffraction patterns revealed only a broad peak in the region of β-SiC for the material fired at 1473K. Broad peaks indicating about 12 nm β-SiC were also observed for polymer fired to 1600°C. No silicon nitride peaks were observed even when the material was fired to 2000°C; however, β-SiC and α and β-Si_3N_4 occurred with the addition of nucleation agents at as low as 1273K. Elemental analysis by the Leco method[13] and neutron activation,[14] though not totally reliable, give the following composition: $Si_{1.0}C_{0.76}N_{0.16}O_{0.03}$. The polymer was melt spun into yarn and allowed to cure in moist air. The elemental composition for the fiber fired to 1473K was shown to be $Si_{1.0}C_{0.92}N_{0.22}O_{0.59}$.

Table 2. Pyrolysis of Mixed Disilane Polymer in Helium

T K	
623	Trace - $Me_3SiOSiMe_2H$, $(Me_3Si)_2NH$
673	Trace - Me_2SiH_2, Me_4Si, Me_3SiOH, $Me_3SiOSiMe_2H$, $(Me_3Si)_2NH$
723	Trace - $Me_3SiOSiMe_2H$, $(Me_3Si)_2NH + CH_4$
773	Trace - Me_3SiH_2, Me_3Si, $Me_3SiOH + CH_4$
823	CH_4
873	CH_4
923	CH_4
973	CH_4
1073	CH_4

Both polymer systems have been melt spun and fired to give ceramic yarns. They both have been employed as binders for ceramic powders. Very low density ceramic foams can be prepared from each polymer by pyrolyzing the polymer under vacuum or in the bulk in argon and allowing the material to foam during firing.

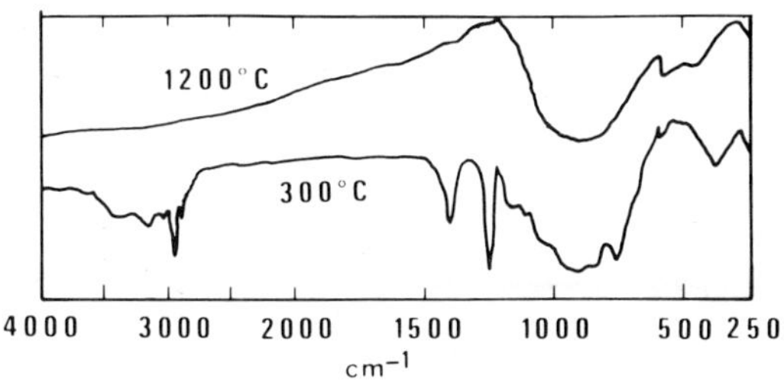

Fig. 4. Infrared spectra of polydisilazane polymerized to 523K after heating.

REFERENCES

1. R. Roy, J. Am. Cer. Soc., 52, 344 (1969).
2. A. Shindo, Rept. Govt. Ind. Res. Inst., Osaka, No. 317 (1961).
3. S. Yajima, J. Hayashi, M. Omori, and K. Okamura, Nature, 261, 683-5 (1976); S. Yajima, J. Hayashi, and M. Omori, U.S. Patent 4,100,233, July 11, 1978; S. Yajima, J. Hayashi, and K. Okamura, Nature, 2661, 521 (1977).
4. W. Verbeek, U.S. Patent 3,853,567, November 8, 1973; W. Verbeek and G. Winter, Ger. Offen. 2,236,078, March 21, 1974; G. Winter, W. Verbeek, and M. Mansmann, Ger., May 16, 1974, U.S. Patent 3,892,583.
5. G. D. Cooper and A. R. Gilbert, J. Am. Chem. Soc., 82, 5042 (1960).
6. J. C. Lockhart, Chem. Rev., 65, 131-51 (1965); K. Moedritzer, J. Organometal. Chem. Rev., 6, 179-278 (1968); D. Weyenberg, L. Mahone, and W. Atwell, Am. NY Acad. Sci., 159, 38-55 (1969).
7. R. H. Baney and J. H. Gaul, Jr., U.S. Patent 4,310,651, Jan. 12, 1982.
8. R. W. Baney and J. H. Gaul, Jr., U.S. Patent 4,298,558, Nov. 3, 1981.
9. R. H. Baney and J. H. Gaul, Jr., U.S. Patent 4,314,956, Feb. 9, 1982.
10. R. H. Baney, U.S. Patent 4,310,482, Jan. 12, 1982.
11. R. H. Baney and J. H. Gaul, Jr., U.S. Patent 4,298,559, Nov. 3, 1981.
12. R. H. Baney, U.S. Patent 4,310,481, Jan. 12, 1982.
13. Analysis performed by the Leco Corp., St. Joseph, MI.
14. Analysis performed by Pheonix Memorial Labs, Ann Arbor, MI.
15. J. H. Gaul, Jr., U.S. Patent 4,340,619, July 20, 1982.

DISCUSSION

R. F. Davis (North Carolina State University): (1) Would it not be possible to work in evacuated or non-oxygen containing atmospheres and therefore eliminate the oxygen incorporation in the material? (2) Is the only reason for going to 1873K to remove oxygen via SiO evaporation?

Author: (1) The polymers described in the paper "cure" after fiber formation and before pyrolysis through chemical reactions which introduce oxygen. (2) It is not necessary to go to 1873K for conversion to ceramic materials.

SILICON-NITROGEN POLYMERS AND CERAMICS DERIVED FROM REACTIONS OF DICHLOROSILANE, H_2SiCl_2

Dietmar Seyferth, Gary H. Wiseman and
Christian Prud'homme
Department of Chemistry
Massachusetts Institute of Technology
Cambridge, Massachusetts 02139

INTRODUCTION

Silicon nitride, Si_3N_4, is of great current interest as a material of high thermal and oxidative stability and high hardness for ceramics, electronic and other applications.[1] Silicon nitride may be prepared by the reaction of elemental silicon with gaseous nitrogen or ammonia at elevated temperatures. Ceramic bodies may be produced directly (reaction sintering) or, alternatively, Si_3N_4 powder may be obtained first and converted to solid bodies by subsequent processing.[1] Chemical vapor synthesis also has served in the preparation of Si_3N_4 (films or powders), using high temperature reactions of ammonia with chlorosilanes (H_2SiCl_2, $HSiCl_3$, $SiCl_4$) for this purpose.[2] Solution-phase chemistry also has found application in the synthesis of Si_3N_4. In particular, the reaction of ammonia with silicon tetrachloride, carried out in an inert solvent, has been the subject of investigations by a number of groups of workers.[3] This ammonolysis reaction initially produces a solid, insoluble, cross-linked product, $[Si(NH)_2]_x$, together with four molar equivalents of ammonium chloride. Pyrolysis of this product at 1250°C gives α-Si_3N_4, but the fact that $[Si(NH)_2]_x$ is a nonvolatile, insoluble solid brings some limitations to its practical applications.

For some purposes, a liquid, polymeric precursor to silicon nitride could be very useful, e.g., to serve as a binder for Si_3N_4 and other ceramic powders or fibers, as an infiltration agent in porous ceramic bodies, and as a coating for diverse solid bodies. In all of these applications, the final process step would be the pyrolysis of the Si-N precursor to Si_3N_4. Organosilazane polymers with methyl substituents on silicon and nitrogen have been used as silicon nitride precursors.[4] However, if such organic, i.e., carbon-containing, silicon-nitrogen precursors are used, silicon carbide as well as silicon nitride will be formed upon pyrolysis. The resulting Si_3N_4/SiC ceramic product may find useful application. However, a route to SiC-free silicon nitride would be desirable.

We have chosen to investigate the products which are obtained in the ammonolysis of dichlorosilane, H_2SiCl_2, which should contain only silicon, nitrogen and hydrogen, as potential precursors of silicon nitride. The reaction of ammonia with dichlorosilane was examined sixty years ago by Stock and Somieski.[5] In the gas phase, the reaction of H_2SiCl_2 and ammonia gave only intractable solids. However, Stock and Somieski found that in benzene solution ammonia and dichlorosilane react to yield a soluble silicon-nitrogen (silazane) product. Benzene-insoluble ammonium chloride is formed as well. The initial process is shown in eq. 1.

$$H_2SiCl_2 + 4\ NH_3 \xleftrightarrow{C_6H_6} \underset{\mathrm{I}}{H_2Si(NH_2)_2} + 2NH_4Cl_{(s)} \quad (1)$$

Diaminosilane, in the context of known silicon-nitrogen chemistry,[6] is not expected to be stable with respect to condensation (eq. 2).

$$x\ H_2Si(NH_2)_2 \longleftrightarrow \underset{\mathrm{II}}{[H_2SiNH]_x} + xNH_3 \quad (2)$$

In any case, filtration of the precipitated ammonium chloride and removal of solvent from the filtrate by distillation at reduced pressure left a viscous oil which was not stable at room temperature. On standing for one day (absence of moisture) it changed to a clear, hard glass. Cryoscopic molecular weight measurements showed that the oil immediately after its preparation had a molecular weight of about 350. Chemical quantitative analysis of the oil for Si, N and H established a composition $\sim H_3SiN$. Thus, an oligomer of type II ($x \sim$ 7-8) appeared to have been formed. This work was not followed up by Stock or by later workers.

At the time of Stock's investigation, dichlorosilane was prepared by the gas-phase reaction of hydrogen chloride with monosilane, SiH_4, which was at that time difficult to prepare and purify[7] and thus was available only in small quantities. Today, dichlorosilane is a commercially available material, and the Linde Division of Union Carbide Corporation offers dichlorosilane (b.p. 8.2°C) in gas cylinders in amounts of up to 250 pounds. (There are, it is importnat to note, potential hazards associated with the handling of dichlorosilane, especially in large quantities.[8]) In view of the availability of dichlorosilane and the potentially interesting product formed in its ammonolysis, we have investigated this reaction as an alternate entry to silicon nitride.

RESULTS AND DISCUSSION

As noted above, Stock and Somieski carried out the reaction of H_2SiCl_2 and ammonia in benzene solution. We have found that this reaction proceeds better in polar solvents such as dichloromethane and diethyl ether. The products are polysilazane oils whose composition appears to be more complex than the expected $[H_2SiNH]_x$.

The addition of gaseous ammonia to an ether solution of H_2SiCl_2 (at 0°C with exclusion of air and moisture) resulted in a vigorous reaction and formation of copious amounts of a white precipitate. The latter was filtered and the filtrate was evaporated at reduced pressure (0.1 mm Hg) to leave a fairly mobile, air-sensitive, nonvolatile, colorless oil, generally a 60-70% yield. No volatile products (i.e., $[H_2SiNH]_n$ cyclics) appeared to have been formed. The white solid formed in the reaction reacted vigorously with water to give SiO_2, showing that some insoluble, apparently cross-linked polymeric Si-N product had been formed in addition to the soluble oil. When the H_2SiCl_2/NH_3 reaction was carried out in dichloromethane, the initial precipitate was almost completely soluble in water and the yield of organic-soluble oil was 75%.

Our studies of the silazane oil produced in these reactions are not yet completed. Since it contains Si-N bonds, it is very reactive toward moisture and must be handled and stored under an inert atmosphere. At -30°C, under a nitrogen atmosphere, this oil appears to be indefinitely stable. However, when it is kept at room temperature under nitrogen, the viscosity of the oil increases gradually and, after 3-5 days a glassy solid has formed, without significant weight loss. Very likely, cross-linking processes are operative.

Spectroscopic studies of this oil have, thus far, failed to give a conclusive indication of structure. The infrared spectrum of the oil showed only one peak in the N-H stretching region at 3390 cm^{-1}, which suggests that -NH- groups but no $-NH_2-$ groups are present. Other bands were observed at 2172 (νSi-H), 1180 (δN-H) and 1020-840 cm^{-1} (δSi-N-Si). There were no bands in the 1150-1000 cm^{-1} region, so no Si-O-Si linkages are present. The 250 MHz proton NMR spectrum of the oil showed two complex multiplets at δ4.3 and 4.7 due to Si-$\underline{H}$ protons and also a broad resonance between 1.0 and 1.7 ppm. The latter is assigned to the N-$\underline{H}$ protons. The integrated signal area ratio, Si-H/N-H, was 3.3.

Mass spectroscopy was not applicable to the study of the silazane oil. Its introduction into the mass spectrometer ion source invariably resulted in fouling of the focusing plates. Extended bakeout or complete ion source cleaning was required after every experiment. It is likely that further polymerization of the oil at the higher ion source temperature, perhaps on contact with the metal surface, is responsible for this behavior.

Thermogravimetric analysis of the silazane oil (quartz sample boat, dry nitrogen or argon blanket gas) at a constant heating rate of 1°C per min. (room temperature to 1200°C) showed that the thermolysis proceeded smoothly as shown in Fig. 1. The onset of weight loss was observed at about 50°C and the weight loss was virtually complete at 450°C. The final ceramic yield was 69% of the original charge. Prior thermal treatment at 100°C or 200°C had essentially no effect on the final ceramic yield. The final product was a black solid. In theory, the "ideal" H_2SiCl_2 ammonolysis product, $[H_2SiNH]_x$, should lose only hydrogen on pyrolysis as shown in eq. 3 and on this basis, a 6.7% maximum weight loss would be expected. The

$$4[H_2SiNH] \text{------} Si_3N_4(s) + Si(s) + 6H_2(g) \tag{3}$$

observed weight loss of 31% indicates that more deep-seated decomposition processes are occurring. Indeed, the exit gases from the pyrolysis contain ammonia, although this product has not yet been quantified. The chemistry of the decomposition process remains to be elucidated.

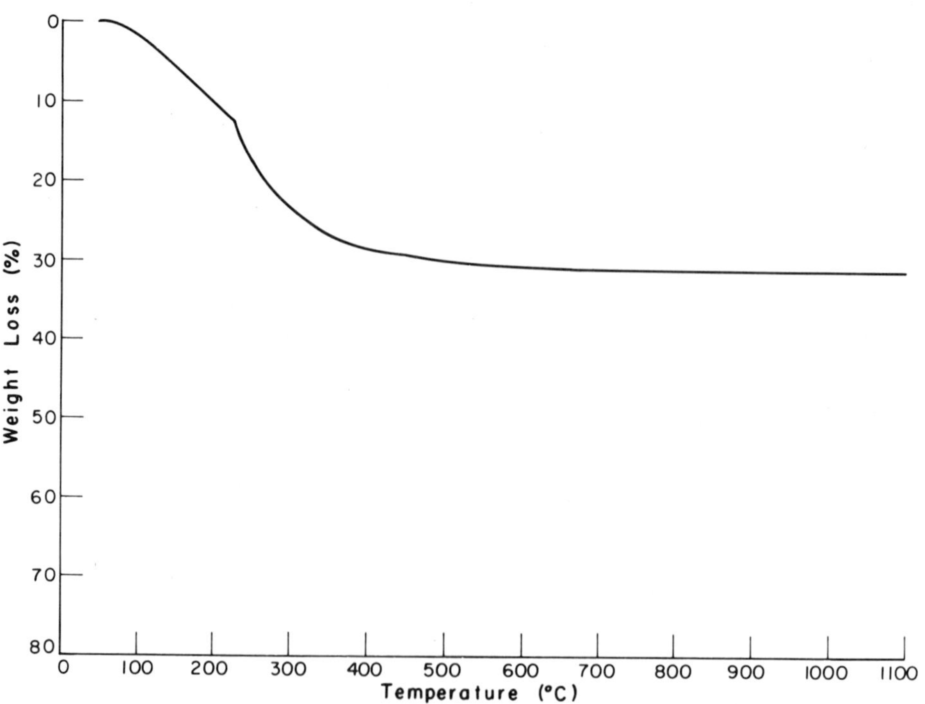

Fig. 1. Thermoanalysis of silazane.

Preparative-scale experiments were carried out with the silazane oil sample in a fused silica boat in a horizontal quartz pyrolysis tube in a tube furnace (always under an inert atmosphere). Initial heating for one hour at 200°C under a slow flow of argon served to solidify the sample. The temperature was then increased slowly to 1150°C and maintained there for 5 hours while the argon flow was maintained. The brown, crystalline solid which was produced was examined by x-ray diffraction. The observed powder pattern showed the lines characteristic of αSi_3N_4[9], β-Si_3N_4[9], and elemental silicon.[10] This indicated that the initial silazane oil was deficient in nitrogen (with respect to the Si_3N_4 formulation). In an attempt to convert the elemental silicon to Si_3N_4, the pyrolysis was carried out in a stream of nitrogen for 12 hours at 1150°C (rather than argon). This approach was successful and the brown solid which was obtained contained only a trace of elemental silicon as evidenced by the x-ray powder pattern.

Ceramics studies of the silicon nitride obtained by this procedure are reported in more detail in the paper by R. W. Rice, G. H. Wiseman, P. B. Davis and W. S. Coblenz of the Naval Research Laboratory. However, we note that solid pieces with dimensions of one to a few mm with immersion densities of 2.5 g/cc and open porosities > 10% could be prepared. Examination of these pieces by scanning electron microscopy showed some cracking and blistering of the solid (Fig. 2). This is not surprising since gas evolution and, most probably, substrate shrinkage occur during the pyrolysis. A fine grain structure (~ 0.2 μm) is apparent which suggests that fibers or bodies from silicon nitride produced in this manner could show good mechanical properties.

We are extending our investigations to include organic group-containing silazane polymers that should, as mentioned above, give Si_3N_4/SiC mixtures on pyrolysis. The reactions of H_2SiCl_2 with monomethylamine and other monoalkylamines and the polymerization of the cyclic oligomers which are obtained (e.g., $[H_2SiNCH_3]_3$) are under active study.

We have also examined the ammonolysis of a number of alkyldichlorosilanes, $RSiHCl_2$ (R = CH_3, $(CH_3)_2CH$, $(CH_3)_3C$, $C_6H_5CH_2$) and have isolated and characterized the cyclic silazanes, $[RSiHNH]_n$ (n = 3 and/or 4) which are the major products.

In the case of the methyldichlorosilane ammonolysis product, the pyrolysis has been studied in a preliminary manner by TGA. When the crude, unfractionated, liquid ammonolysis product was pyrolyzed (heating rate, 2°C/min., maximum temperature 100°C), a 20% yield of black, noncrystalline solid was obtained. When the ammonolysis product was heated under nitrogen with 6% by weight of NH_4Cl for 14 hours at 120°C before pyrolysis, the solid product yield was increased to 39%. This pretreatment no doubt caused partial ring-opening polymerization of the $[CH_3SiHNH]_n$ cyclics to nonvolatile linear polymer (a process demonstrated by Kruger and Rochow[11] for the $[(CH_3)_2SiNH]_n$ cyclics) and so a lesser amount of volatiles was lost during the subsequent heating period. The benzyldichlorosilane ammonolysis product also could be pyrolyzed to give, at 400°C, a hard, white solid and, at 1200°C, a black, noncrystalline solid. Characterization of these pyrolysis products is in progress.

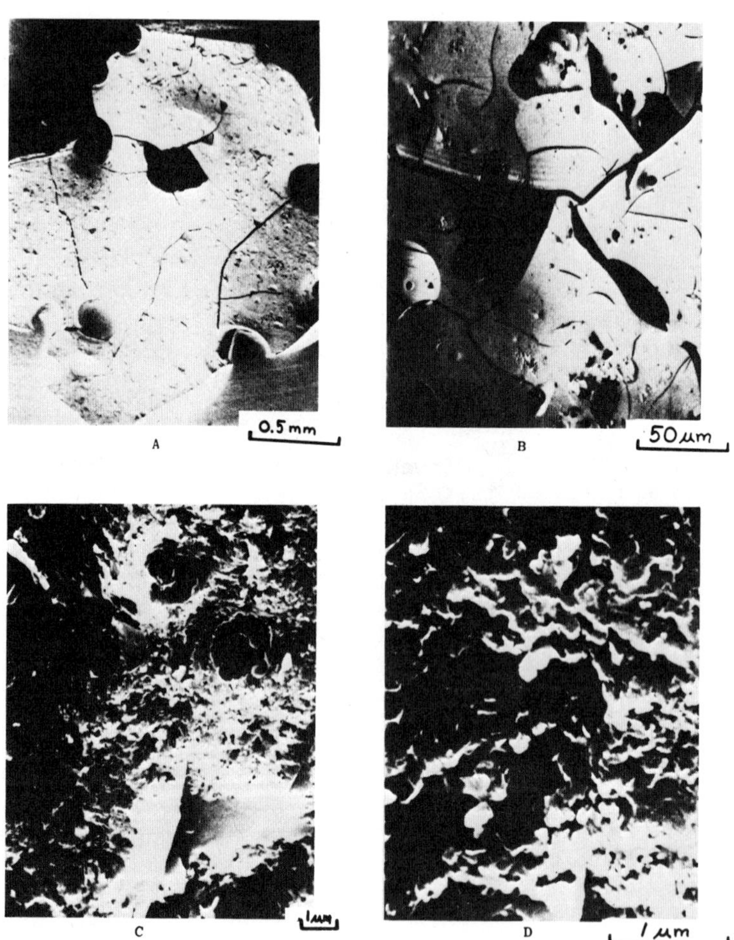

Fig. 2. SEM micrographs of sample II-21 $(SiH_2NH)_x$ 3 hours, 1200°C, under N_2.

ACKNOWLEDGMENTS

This work was supported in part by the Office of Naval Research. We are grateful to Roy W. Rice and William S. Coblenz of the Naval Research Laboratory, Washington, D.C., for helpful discussions and advice concerning the ceramics aspects of this work. Thanks are due to Barry Bender of the Naval Research Laboratory for the SEM pictures.

REFERENCES

1. D. R. Messier and W. J. Croft, Preparation and Properties of Solid State Materials, Vol. 7, edited by W. R. Wilcox, Marcel Dekker, New York, 1982, pp. 131-212.
2. T. Hirai and S. Hayashi, J. Am. Ceram. Soc., 64, C88-89 (1981), and references cited therein.
3. O. Glemser and P. Naumann, Z. Anorg. Allg. Chem., 298, 134-41 (1959); M. Billy, Ann. Chim. (Paris), 4, 795-851 (1959); K. S. Mazdiyasni and C. M. Cooke, J. Am. Ceram. Soc., 56, 628-33 (1973).
4. W. Verbeek, U. S. patent 3,853,567 (Dec. 10, 1974); W. Verbeek and G. Winter, Ger. Offen. 2,236,078 (March 27, 1974); G. Winter, W. Verbeek and M. Marsmann, U.S. patent 3,892,583 (July 1, 1975); R. A. Markle, D. L. Hill, and W. E. Hauth, XV Organosilicon Symposium, Durham, NC, March 27-28, 1981; R. H. Baney, J. H. Gaul, Jr., and T. K. Hilty, XVI Organosilicon Symposium, Midland, MI, June 16-18, 1982.
5. A. Stock and K. Somieski, Ber. dt. chem. Ges., 54, 740-58 (1921).
6. R. Fessenden and J. S. Fessenden, Chem. Rev., 61, 361-388 (1961).
7. A. Stock and C. Somieski, Ber. dt. chem. Ges., 52, 695-725 (1919).
8. Material Safety Data Sheet, Union Carbide® Silane A-199; Linde Specialty Gases Safety Precautions and Emergency Procedures, F-12-237.
9. S. N. Ruddleson and P. Popper, Acta Cryst., 11, 465-68 (1958).
10. H. E. Swanson and R. K. Fuyat, NBS Circular 539, 2, 6 (1953).
11. C. R. Kruger and E. G. Rochow, J. Polymer Sci. [A], 2, 3179 (1964); E. G. Rochow, Monatsh. Chem., 95, 750-65 (1964).

FORMATION OF CERAMIC COMPOSITES AND COATINGS UTILIZING POLYMER PYROLYSIS

W. S. Coblenz, G. H. Wiseman, P. B. Davis, and R. W. Rice

Naval Research Laboratory
Washington, D.C. 20375

ABSTRACT

This paper reports the utilization of polymer pyrolysis for the fabrication of ceramic composites and for the improvement of the oxidation resistance of carbon materials by infiltration and coating. The dominant features of the pyrolysis process are: The evolution of pyrolysis gases which may result in pore formation and growth; and shrinkage which may result in micro- and macro-cracking.

Modest improvements in strength and oxidation resistance have been achieved using either a carborane siloxane polymer or a silazane oligomer. Preliminary results on the fabrication of fiber composites are also reported.

INTRODUCTION

The pyrolysis of organometallics to produce ceramics is of interest for two reasons. First, it is a method of forming materials such as covalently bonded glasses which would be difficult or impossible to form by any other method. Second, the shape forming methods used with polymers (casting, injection molding, fiber drawing, or painting may be applied to form refractory ceramic materials. Glassy carbon, carbon-carbon composites and high modulus graphite fibers are all formed by pyrolytic processing and indicate the possibilities for polymers yielding more oxidation resistant materials such as SiC, Si_3N_4, BN and mixtures of these materials. The evolution of pyrolysis gases which may result in pore formation and growth, and shrinkage which often results in micro- or macro-cracking, are both key features of the pyrolysis process. Both of these problems are minimized in forming fibers or powders by polymer pyrolysis since diffusion distances are small. It is the objective of this study to explore the use the polymer pyrolysis process to form composites and coatings.

EXPERIMENTAL

Two polymers have been used in this study. Dexsil[a] 202 polymer is a dimethylsiloxydiphenlysiloxycarborane. Its composition is $B_{10}Si_4C_{30}O_3H_{42}$ and its molecular weight is 1000 to 2000. It is a viscous liquid at room temperature (like honey) and becomes very fluid at ~ 100°C. Crosslinking with elimination of benezene occurs at ~ 300°C resulting in a rubbery solid. The pyrolytic yield is ~ 60 wt% (at heating rates of 1 to 10°C/min in Argon) with the major weight loss occurring between 100 and 500°C. B_4C and β-SiC are observed by x-ray diffraction on heating to 1200°C in one atmosphere of argon. One hot pressing experiment was conducted with the Dexsil polymer. It was crosslinked at 300°C to produce a solid and then vacuum hot pressed in a bed of BN powder to 1200°C and ~ 0.49 MPa. The BN powder remains porous during hot pressing and thus does not trap volatile pyrolysis products from the sample while allowing the transmitted force to aid densification.

A second polymer used in this study is a silazane oligomer. It is synthesized by the ammonolysis of dichlorsilane. The reaction is:

$$3NH_3 + SiCl_2H_2 \xrightarrow[(CH_3CH_2)_2O]{0^\circ C} (SiH_2NH)_x + 2\ NH_4Cl$$

Soluble products are separated from the NH_4Cl by-products by filtration. Details of the chemistry and synthesis conditions are given by Seyferth.[1] The silazane is unstable at room temperature, probably reacting with water vapor or oxygen. It is converted from a fluid liquid first to a gel and then to a glassy solid. Freshly prepared silazane was stored below 0°C and transferred by use of N_2-filled gas-tight syringes for this study. N_2 filled glove bags were also used to minimize exposure to air. Pyrolysis in N_2 results in mixtures of Si_3N_4 and Si with ~ 70 wt% yield.

---

[a]Olin Corporation, Chemical Division, New Haven, CT 06504.

Four types of carbon materials were used for coating experiments hereafter referred to as Nuclear, Carbitex,[b] Poco[c] and French.[d]

The Nuclear carbon was a reactor grade carbon of unknown origin. It contained large grains (up to 300 μm), considerable uneven porosity and was of low strength. The Carbitex was a 2-D graphite fiber carbon composite. The Poco graphite was a fine grain (~ 4 μm) material with uniform porosity. The French material was a 3-D woven graphite fiber carbon composite formed by the multiple impregnation and high pressure pyrolysis of a carbon woven graphite. Samples were machined into nominally 2.54 mm x 6.35 mm x 25.4 m test bars (ground along the length of the bar and in a fiber direction for the composites) and washed in acetone and dried at 100°C for several hours prior to impregnation.

Limited experiments were conducted on the impregnation of a reaction sintered Si_3N_4[e] material with the silazane oligomer. A variety of techniques were used for the impregnation and coating of carbon materials. A control experiment was run in which one of each type of carbon was heat treated under the standard pyrolysis conditions. No changes in mass, density, open porosity or strength were observed for the controls compared to the as-received material. Two bars of each material were impregnated so that one could be used for density, porosity and strength measurements while the other could be used intact for oxidation experiments. The impregnation methods listed in Tables 1 and 2 were performed as follows:

Vacuum Impregnation. Samples were loaded into a holding flask and evacuated to 80 mtorr. Polymer was next added while still under vacuum, covering the samples. The pressure was then raised to 1 atmosphere (N_2 for silazane polymer) and soaked for 30 minutes. The Dexsil polymer was heated to reduce its viscosity.

Vacuum Impregnation + Isostatic Pressing. After vacuum impregnation these samples were sealed in rubber bags and held at 173 MPa for 15 minutes in an isostatic press.

---

[b]Carbitex 500, The Carborundum Company, Graphite Products Div., Niagra Falls, NY 14302.

[c]Standard Grade, (HPD-1) Poco Graphite, Inc., Decatur, TX 76234.

[d]Societe Nationale Industrielle Aerospatiale, 33160 Saint-Medard-En-Jalles, France.

[e]NC-350, Norton Company, Worcester, MA.

"Puddle" Pyrolysis. Samples were covered with liquid polymer and pyrolyzed. Excess pyrolyzed polymer was rubbed off samples after pyrolysis.

Multiple Dip. A vertical quartz tube furnace was assembled such that samples, suspended with nichrome wire, could be transported from a container with polymer at room temperature to the hot zone of the furnace, all under a continuous flow of nitrogen. These experiments were carried out with both the neat silazane polymer and a 50% solution of the silazane polymer in dry ether. Samples were alternately soaked in the polymer for 15 minutes, then pyrolized at 550°C for 15 minutes. On the third soak/pyrolyze cycle, samples were held for 1 hour at 550°C and cooled to room temperature. The samples were then pyrolized under the standard conditions (to 1000°C) to complete the pyrolysis.

Coated. Samples were dip-coated first in 20 wt% solution of Dexsil in benzene and a second coating, after pyrolysis under the standard conditions to 1000°C, in neat Dexsil, heated (~ 100°C) to lower its viscosity.

Composites were prepared by copyrolysis of rayon[f] fabric or polycarbosilane[g] fibers as well as the graphite[h] woven fabric, graphite[i] fibers and SiC (Nicalon)[g] fibers. In each case, polymer was added to the fibers or fabrics cut in approximately 5 cm lengths and pyrolized in fused silica boats.

Pyrolysis of both coatings and composites were carried out in a vacuum tight alumina tube furnace. The heating rate was 2°C/min from room temperature to 500°C and 10°C/min from 500 to 1000°C with a 12 hour hold at temperature. An argon atmosphere was used for the Dexsil samples while a nitrogen atmosphere was used for pyrolysis of the silazane.

In one experiment, Thornel graphite fibers were used with the silazane polymer and the composite was partially pyrolized to 500°C followed quasi-isostatic pressing in BN powder to 1200°C with 34.5 MPa applied pressure.

Densities and open porosities were measured using the Archimedes method in water. Three-point bend strengths were measured with a 12.7 mm span at a crosshead speed of 12.7 mm/min. Strength values reported in Tables 1 and 2 are the average of two measurements.

---

[f] Burlington Industries, Rockleigh, NY 07647.
[g] Nippon Carbon Co., Ltd., Tokyo, Japan.
[h] Pyron KF, Knit Fabric, Stackpole Fibers Co., Lowell, MA 01852.
[i] Thornel-P, Union Carbide Company, Niagara Falls, NY.

RESULTS AND DISCUSSION

While pyrolysis yield was not greatly sensitive to heating rate for the two polymers used in this study, the quality of the product (micro- and macro-cracks and pores) was found to be very sensitive. The microstructure shown in Fig. 1 illustrates some of the problems which were encountered in early pyrolysis experiments with the silazane polymer. Cracking is related to shrinkage and amplified in this third soak/pyrolyze cycle by poor temperature control. Pore formation and growth is related to pyrolysis gas and is also amplified by poor temperature control. X-ray analysis indicates the presence of a α-Si_3N_4 and silicon metal.

The Dexsil hot pressing experiment gave a fine grain product (Fig. 2). X-ray analysis gave broad β-SiC peaks indicative of a very fine grain size (~ 100 Å). The absence of large pores is particularly encouraging. The application of an external pressure during pyrolysis counteracts the gas pressure within pores which would tend to inflate them while the matrix is still plastic. Gas pressure should also increase the solubility of pyrolysis gases in the matrix which should improve the yield.

The as-received reaction sintered Si_3N_4 (NC-350) had a density of 2.50 gm/cc and an open porosity of 8.12%. After impregnation and pyrolysis with the silazane polymer the density increased to 2.62 gm/cc and the open porosity decreased to 2.0%. Mazdiyasni et al.[2] have reported similar increases in density by infiltration and pyrolysis of NC-350 with hexaphenylcyclotrisilazane and with methylphenyl polysilane. They also report strength increases of 35 and 21% respectively after pyrolysis.

The poor quality of coatings with both silazane and Dexsil on either Si_3N_4 or carbon materials is illustrated in Fig. 3. Shrinkage during pyrolysis causes the coating to break up into ~ 10 µm plates giving the surface an alligator skin texture. Spalling of these plates then depends upon the thermal expansion mismatch and adhesion to the substrate as illustrated by the example of the Dexsil coating (Fig. 4) on a carbon-carbon fiber composite. The coating has spalled off preferentially where fibers are parallel to the surface.

The results listed in Table 1 for silazane impregnations and/or coating on carbon bodies indicate weight gains of 1 to 10%, with the greatest increases occurring in the more porous Nuclear and Poco graphites after vacuum and pressure impregnation. Given the cracked character of the coating, we feel that much of the improvement in properties (open porosity, density, strength) is due to the impregnation. Complete oxidation of the impregnated fiber composites leaves a skeletal structure which replicates the fiber structure of the starting material. The coatings are porous and probably contribute to the open porosity. The accelerating oxidation kinetics indicate

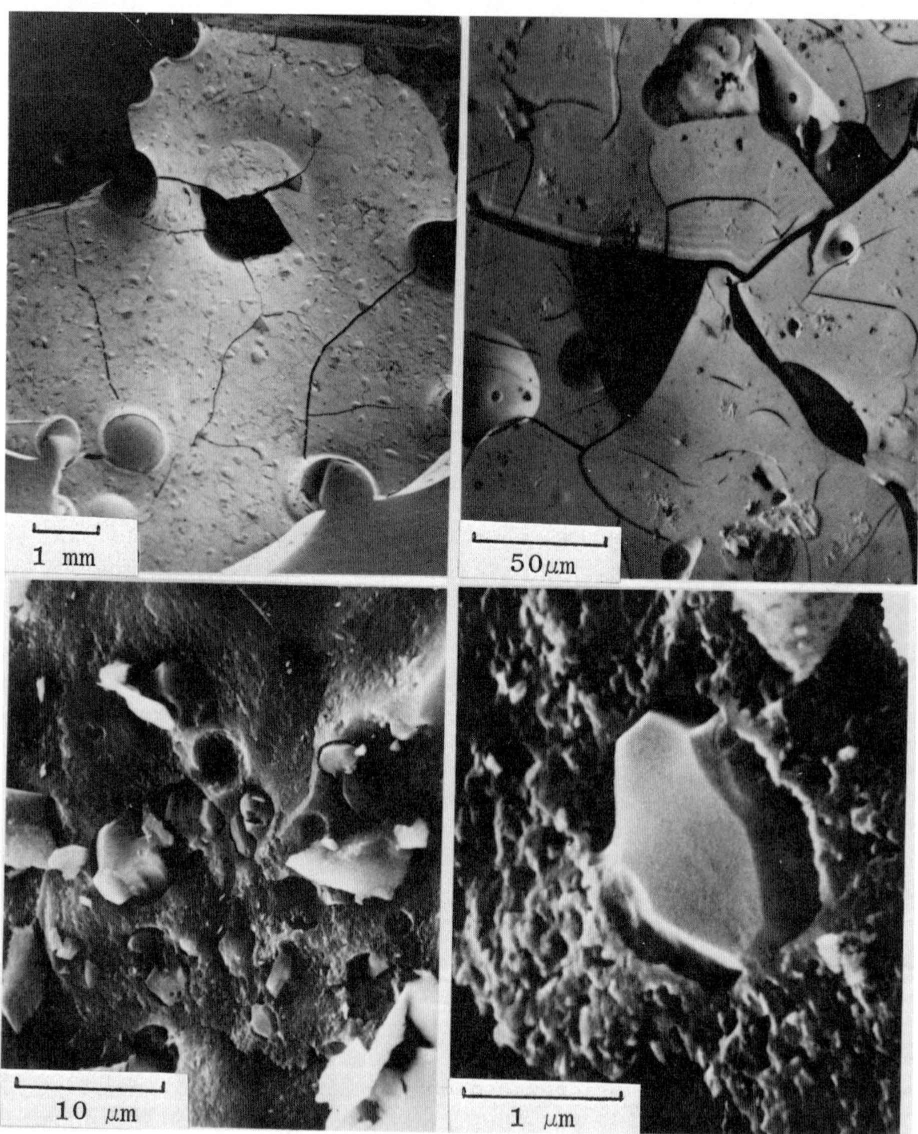

Fig. 1. Microstructure of pyrolyzed silazane $(SiH_2NH)_x$, 3 hours, 1200°C, under N_2.

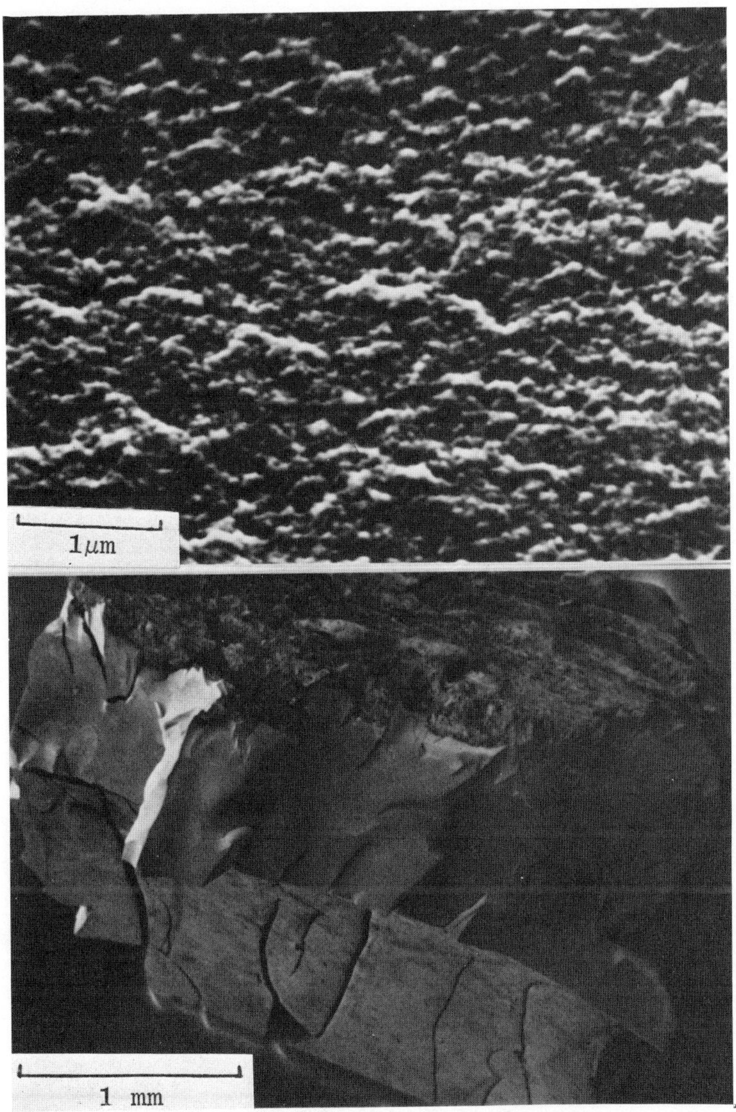

Fig. 2. Dexsil polymerized at ~ 300°C and hot-pressed at 1200°C and 0.49 MPa in BN powder.

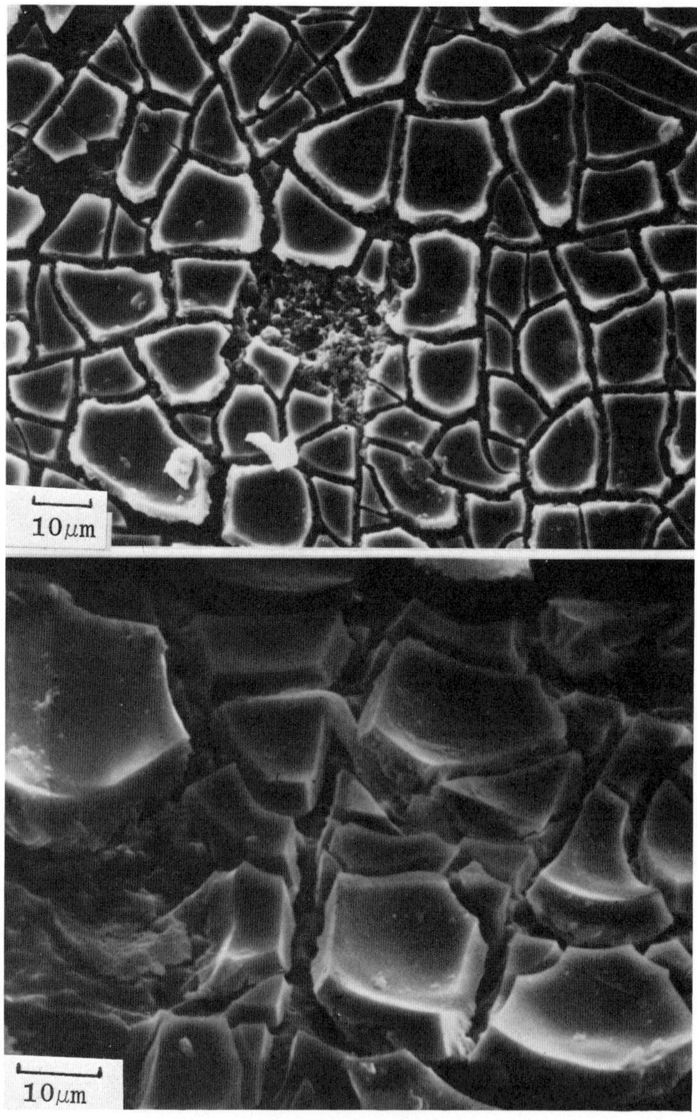

Fig. 3. Silazane derived Si_3N_4 coatings on Poco graphite (top) and reaction sintered Si_3N_4 (bottom).

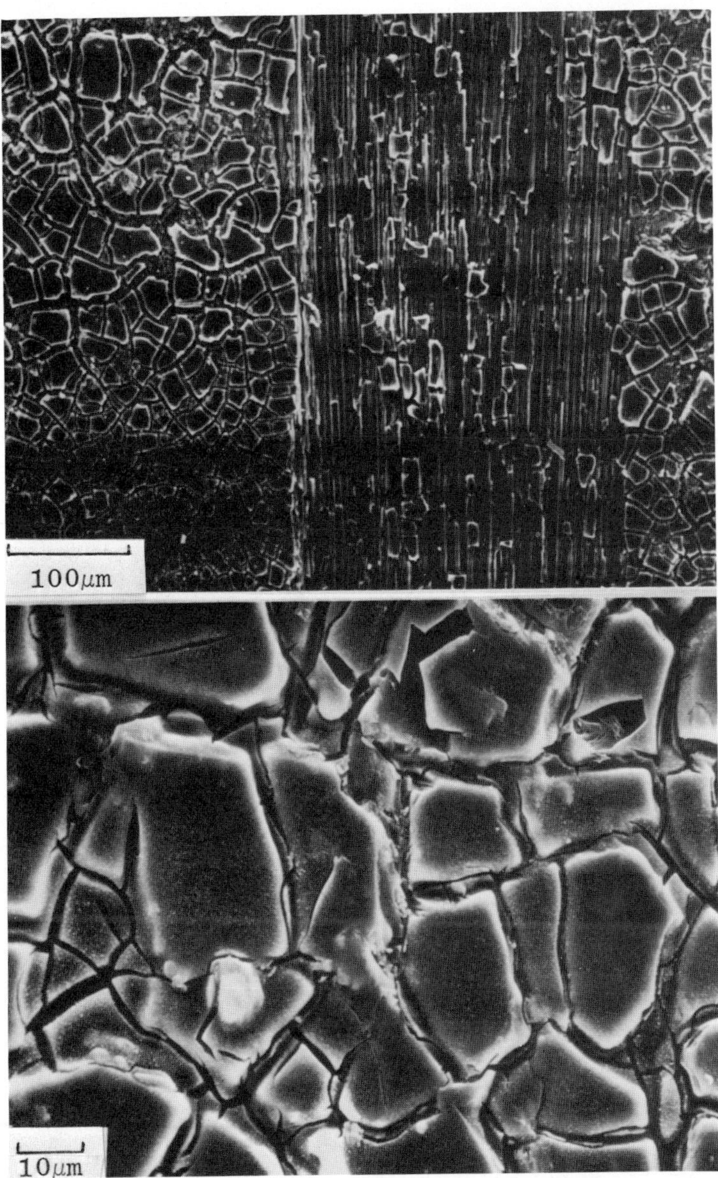

Fig. 4. Dexsil coating on French (Aerospasitiale) carbon-carbon composite.

Table 1. Impregnation and Coating of Carbon Materials with Silazane Polymer

Impregnation Method	% Weight Increase	Density gm/cc (% Change)	Open Porosity (% Change)	Strength, MPa (% Change)	% Weight Loss on Oxidation at 600°C 1 Hr.	4 Hr.	18 Hr.
MATERIAL: NUCLEAR CARBON							
Control	–	1.74 (–)	10.4% (–)	38.9 (–)	81	100	–
Vacuum Impregnation	2.1	1.83 (+5.2)	9.7% (–6.7)	37.58 (–4.2)	–	–	87
Vacuum Impregnation + Isostatic Pressing	8.4	1.89 (+8.6)	9.4% (–9.6)	37.54 (–4.2)	2.6	17	88
'Puddle' Pyrolysis	0.44	1.76 (+1.1)	12.9% (+24)	36.36 (–7.4)	5.5	45	99
MATERIAL: POCO GRAPHITE							
Control	–	1.82 (–)	14.7 (–)	104.6	80	100	–
Vacuum Impregnation	2.2	1.86 (+2.2)	15.0 (+2.0)	115.8 (+11)	–	–	96
Vacuum Impregnation + Isostatic Pressing	8.1	1.94 (+6.6)	6.4 (–56.9)	120.6 (+15)	9.7	74	91
'Puddle' Pyrolysis	(no change)	1.83 (+0.5)	16.7 (+13.6)	112.7 (+8.0)	24	94	100
MATERIAL: FRENCH CARBON							
Control	–	1.90 (–)	5.4% (–)	104.1 (–)	62	100	–
Vacuum Impregnation	1.9	1.92 (+1.1)	6.5% (+17)	113.4 (+9)	–	–	62
Vacuum Impregnation + Isostatic Pressing	3.7	1.94 (+2.1)	6.1% (+13)	138.7 (+33)	3	30	96
'Puddle' Pyrolysis	1.3	1.92 (+1.1)	6.8% (+26)	129.81 (+25)	8	63	98
Multiple Dip Polymer Coatings	3.7	1.96 (+3.2)	5.2% (+3.7)	119.9 (+15)	5	39	96
Multiple Dip from 50 wt% Polymer in Ether Solution	3.2	1.95 (+2.6)	5.7% (+5.6)	126.5 (+22)	7	66	96
MATERIAL: CARBITEX							
Control	–	1.34 (–)	6.92% (–)	90.17 (–)	94	100	–
Vacuum Impregnation	3.5	1.50 (+12)	6.62% (–4)	90.13 (no change)	–	–	97
Vacuum Impregnation + Isostatic Pressing	6.7	1.55 (+16)	4.1% (–41%)	86.67 (–4)	12	74	93
'Puddle' Pyrolysis	1.0	1.48 (+10)	6.6% (–4.0)	113.99 (+26)	26	98	98

that the coatings do offer short term oxidation protection at 600° which is surprising given their cracked nature. In control experiments with untreated bars, catastrophic weight losses (62-94%) were observed after one hour and all the bars had completely oxidized after four hours. The best results were obtained with the vacuum impregnated and isostatic pressed impregnated French material which lost only 3% of its weight after one hour.

Dexsil coated/impregnated samples (Table 2) had, in general, smaller changes in density and open porosity than the samples treated with silazane. Oxidation of a set of bars was carried out progressively at 400°C for 4 hours, 600°C for 16 hours, and 800°C for 6 hours. Oxidation resistance at 400 and 600°C was greatly improved, particularly for the French and Nuclear material. This improvement may be due to formation of liquid B_2O_3 which can flow and fill in pores. A conditioning of this liquid oxide layer may be necessary since impregnated bars oxidized by placement in a furnace at temperature (800°C) show greater weight losses after two hours (French and Nuclear carbons) than the samples heated at progressive temperatures, including 6 hours at 800°C.

Several ceramic precursor fibers were used with the silazane polymer in hopes that co-pyrolysis might reduce cracking due to differential shrinkage and also enhance fiber-matrix bonding. An immediate reaction at room temperature was observed with rayon fabric releasing NH_3 and converting the fabric to a white solid. This is probably a reaction of the polymer with the hydroxyl group of the rayon since the Si-N bonds are easily hydrolyzed. In another experiment, strips of Pyron KF cloth (a polyacrylonitrile-derived, prepyrolyzed fabric) were soaked in the silazane. The cloth seemed to stiffen and layers pulled away from each other. Pyrolysis of this sample produced a semi-hollow black composite (ρ = 1.38 gm/cc open porosity = 17%) with very low strength. In a third experiment, polycarbosilane fibers (precursors to Nicalon SiC fibers) were used. These fibers were wetted nicely by the silazane, however pyrolysis seriously degraded the structure of the fibers resulting in a weak composite of low density (ρ = 1.41 gm/cc and 14% open porosity).

Composites were also formed using SiC fibers (Nicalon, II). These fibers were wetted by the silazane and drawn together into a bundle by capillary action. Pyrolized bodies held their shape but were of low density (~ 1.4 gm/cc), had considerable open porosity (30 to 40%), and low strength (10 to 15 MPa). Axial and radial cracks in the matrix (Fig. 5) were observed around the fibers. These shrinkage cracks are due to a lack of plasticity in the matrix during pyrolysis.

Thornel graphite fibers were also wetted nicely by the silazane polymer. Partial pyrolysis (to 500°C) followed by hot pressing (1200°C and 34.5 MPa) produced a composite which was still quite porous (31%) with low density (1.44 gm/cc) and low strength (13.8 MPa). Apparently there is little plasticity in the partially pyrolyzed silazane.

Composites formed by the co-pyrolysis of polycarbosilane fibers with the Dexsil polymer produced bodies with large pores. Examination of a fracture surface (Fig. 6) indicated that the fibers had reacted with the matrix and bonding between fiber and matrix was so strong that little interaction of the crack with the fibers was observed.

Table 2. Impregnation and Coating of Carbon Materials with Dexsil Polymer

Material (treatment)	% Weight Increase	Density, gm/cc (% Change)	Open Porosity (% Change)	Strength, MPa (% Change)	% Weight Loss on Oxidation 4 hrs. at 400°C	+ 16 hrs at 600°C	+ 6 hrs. at 800°C	2 hrs. at 800°C
Poco Graphite (control)		1.82 (–)	14.7 (–)	104.6 (–)	72	100	100	
(coated)		1.86 (+2.2)	15.0 (+2.04)	106.8 (+2.1)	0.27	73.0	93.0	
(Vacuum + Isostatic Pressing)	1.6	1.87 (+2.7)	11.87 (–19.3)	(–) (–)				70.0
Nuclear (control)		1.74 (–)	10.4 (–)	39.25	12.2	100	100	
(coated)		1.83 (+5.17)	9.73 (–6.44)	39.05 (–0.5)	0.17	2.8	9.6	
(Vacuum + Isostatic Pressing)	2.0	1.81 (+4.0)	9.9 (–4.8)	36.4 (–7.3)				56.0
French (control)		1.90 (–)	5.73 (–)	104.1 (–)	18.8	100	100	
(coated)		1.92 (+1.05)	6.48 (+13.1)	137.6 (+32.2)	0.0	2.0	12.4	
(Vacuum + Isostatic Pressing)	1.0	1.92 (+0.95)	6.31 (+10.1)	124.9 (+19.9)				48.0
Carbotex (control)		1.34 (–)	6.92 (–)	90.17 (–)	85.6	100	100	
(coated)		1.50 (+11.9)	6.62 (–4.34)	97.43 (+8.1)	0.35	73.0	95.9	
(Vacuum + Isostatic Pressing)	1.7	1.483 (+10.7)	5.61 (–19.0)	99.8 (+10.7)				79.0

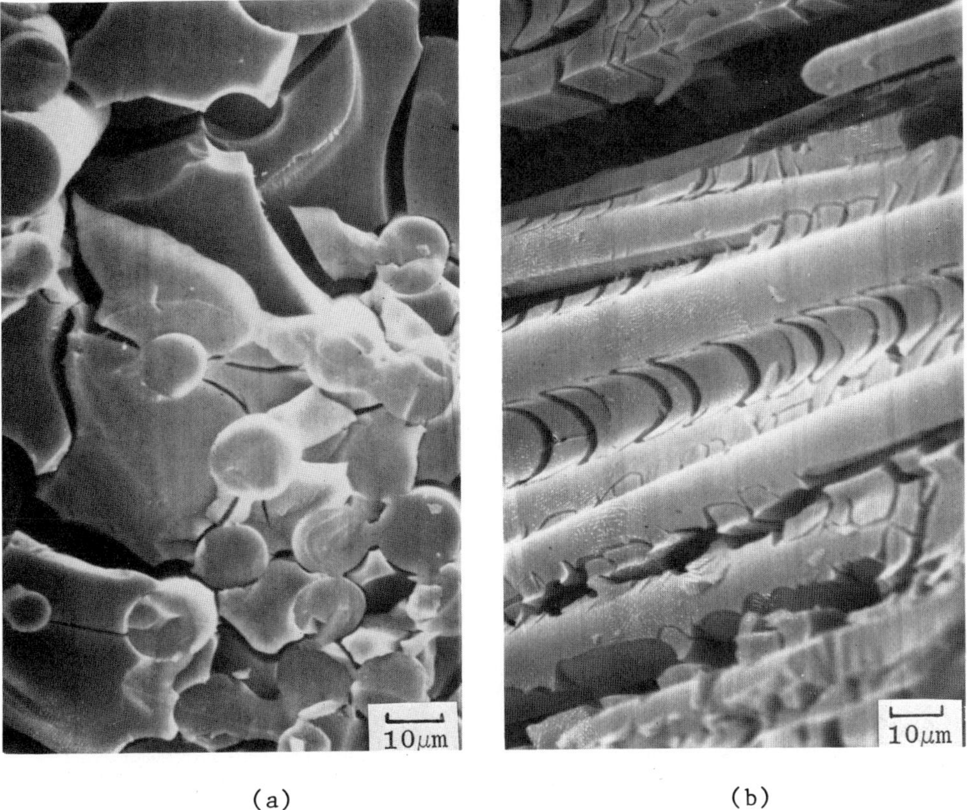

(a) (b)

Fig. 5. Fracture surface of a composite formed with SiC fibers and silazane derived Si_3N_4 matrix.

CONCLUSIONS

Cracks associated with shrinkage during pyrolysis present a major problem to be overcome in the use of polymer pyrolysis for coatings and composites. Powder fillers are used in paints and in carbon-carbon composites to inhibit the formation of shrinkage cracks. The use of fillers in ceramic precursor polymers is suggested both for this purpose and to modify thermal expansion and to modify oxidation behavior.

Bubble formation and growth results from evolution of pyrolysis gases. This problem may be minimized by either slow heating rates or by the application of pressure during pyrolysis.

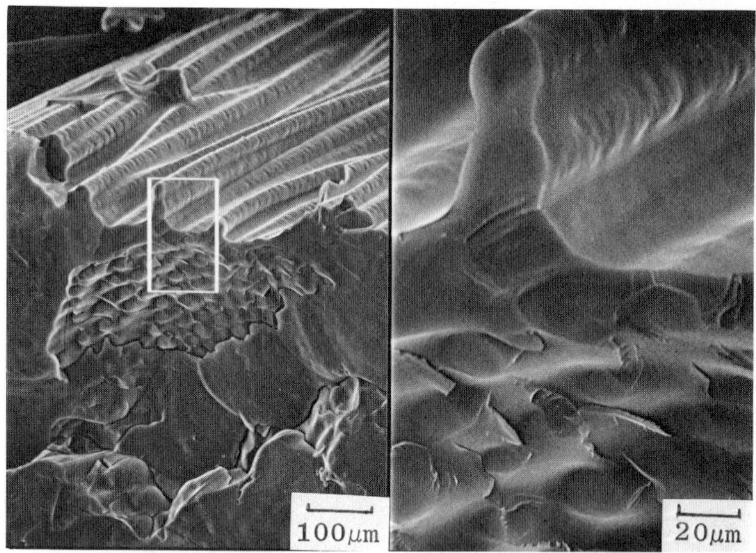

Fig. 6. Composite formed by co-pyrolysis of polycarbosilane fibers and Dexsil matrix.

The oxidation resistance of carbon materials can be improved by coating and impregnating with both Dexsil and Silazane polymers. Modest increases in strength were also noted for these materials. The poor quality of coatings suggests that greater improvements are possible if the cracking can be eliminated.

Preliminary results on the fabrication of composites by the polymer pyrolysis have been reported. While the processing of such composites is simple, acceptable mechanical properties can be expected only after the shrinkage cracking and fiber-matrix interface bonding problems are addressed.

ACKNOWLEDGMENTS

The authors wish to acknowledge Ms. Lyda Green and Mr. Charles Taylor for SEM examination of coatings and composites. Discussions with Dr. Jean Jamet of ONERA during preparation of this paper are also greatly appreciated. We would also like to thank Monsieur Grenie of Societe Nationale Industrielle Aerospatiale for supplying the 3-D woven fiber carbon composites.

REFERENCES

1. D. Seyferth, these proceedings.
2. K. S. Mazdiyasni, R. West, and L. D. David, J. Am. Ceram. Soc., 61 [11-12], 504-08 (1978).

DISCUSSION

R. Baney (Dow Corning): Have you tried multistep infiltration on your materials?

Author: We have used multiple impregnations on graphite samples coated with Dexsil. This helps fill in some of the shrinkage cracks from the first impregnation. Multiple impregnation should have similar advantages for composites while also increasing the density of matrix.

GAS ANALYSIS DURING THE PYROLYSIS OF CARBOSILANE

J. J. Poupeau, D. Abbe, and J. Jamet[a]

O.N.E.R.A. (Office National d'Etudes et de Recherches Aerospatiales)
92320 Chatillon-sous-Bagneux, France

ABSTRACT

Research and development of ceramic materials formed by the pyrolytic transformation of organometallics needs detailed information on the behavior of the polymer (weight, rheology, chemical reactivity) as a function of temperature, time and pressure.

This paper describes the behavior of polycarbosilane (Nippon Carbon: Mark I), particularly the analysis of gaseous byproducts during the pyrolysis by gas chromotography and mass spectrometry as a function of temperature and oxidation state. A correlation with the results of other thermal analysis methods (DTA, TGA, TMA) indicates 3 stages to the transformation, based on the nature of the gaseous byproducts.

Finally, these results suggest an analogy with carbon-carbon processing, as well as ways to use such organometallics in ceramic composite development.

INTRODUCTION

Polymers able to generate ceramics by pyrolytic transformations similar to those used in the carbon sciences are of increasing interest, especially since the fruitful works of Yajima[1] on carbosilane to produce SiC. This new ceramic processing perspective opened jointly

---

[a]Currently working for one year as a visiting scientist at the Naval Research Laboratory, Ceramics Branch, Washington, DC 20375.

for chemists and ceramists[2] makes knowledge of the pyrolytic evolution particularly important; especially in the early stages of assessing and developing this technology.

Pyrolytic evolution involves several physicochemical phenomena, including distillation,; cross linking, degradation, and crystallization. Their existence and their importance depend upon parameters such as initial chemical structure, temperature, pressure, and time. These phenomena imply modifications of the polymer, e.g., its mass, enthalpy, rheology, and density. Thus, the thermal analysis forms an important and much needed tool for study of this evolution by ceramists.

The present work concerns the carbosilane MARK I from NIPPON CARBON: its oxidation behavior (thermorheological analysis, oxidation treatment, IR analysis); its gas analysis during pyrolysis [distillation, main production (chromotography), secondary production (mass spectrometry)]; and its correlations between TGA and DTA.

Finally, these results suggest an analogy with carbon-carbon processing, as well as ways to use such organometallics in ceramic composite development.

SAMPLE PREPARATION AND OXIDATION BEHAVIOR

The carbosilane Mark I was treated at 553K under vacuum (700 Pa) for 1 hour to eliminate accumulated gas before use. The PCS was then reduced to powder at room temperature under nitrogen to yield an average size of 160 μm.

To determine the oxidation treatment of the PCS its thermorheological evolutions under air and nitrogen were compared. These measurements were performed with a torque pendulum using a PCS impregnated carbon tow. The modulus and damping were measured at the same time between room temperature and 573K. The evolution of the damping under these two conditions is presented in Fig. 1.

In air, the increase of the damping appears clearly at 473K. At the same temperature, the PCS under nitrogen continued softening until 503 to 513K. At higher temperatures a cross-linking seemed to appear.

Two oxidized samples were prepared at 463K, one for 30 min, the second for 4 hours using the previous powders.

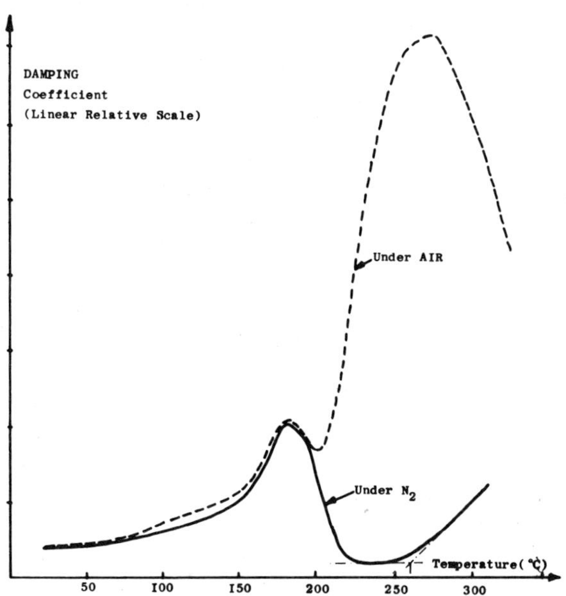

Fig. 1. Rheological thermal analysis of the carbosilane.

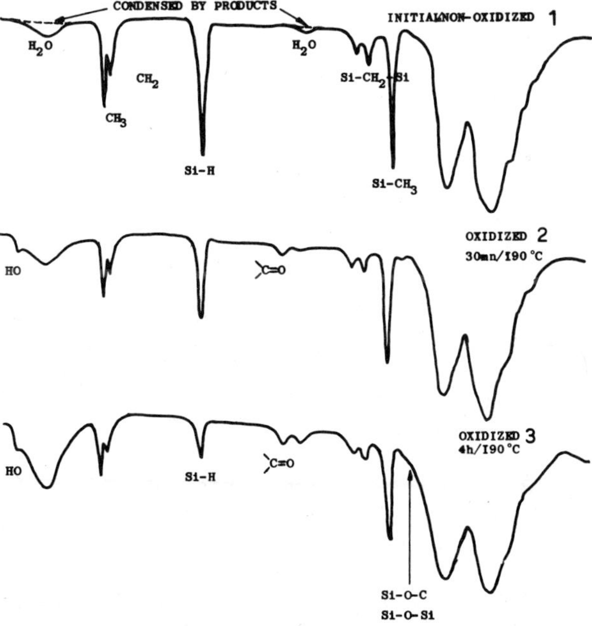

Fig. 2. Infrared spectrograms of carbosilanes before gas analysis.

The three states of the PCS so obtained have been compared by IR analysis. The three spectrums presented in Fig. 2 show:

(1) Lateral bonds such as Si-H are the first attacked by oxygen compared with the second $Si-CH_3$ and cross-linking takes place so creating new bonds like Si-O-Si.

(2) Two other bonds appear; C = O and -OH.

GAS ANALYSIS DURING PYROLYSIS

Apparatus (Fig. 3)

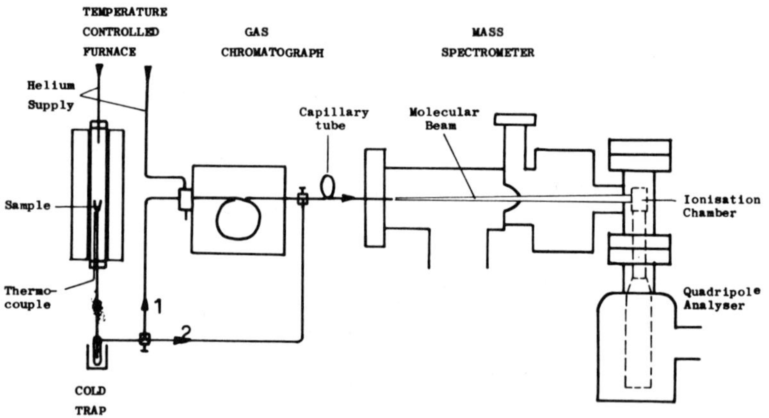

Fig. 3. Gas analysis apparatus

The samples were placed in an alumina crucible in the center of a silica tube and were heated at a rate of 423K/h. Helium gas swept the gaseous species to a cold trap in order to remove heavy species. Afterwards the gas was analyzed: either with the chromatograph alone or with the chromatograph followed by the mass spectrometer, or with the mas spectrometer alone.

The chromotograph used a 1.5 m Porapak Q column (80/100 mesh) at +323K. The injector takes 0.3 ml TPS into the stream each 5 seconds. A catharometer is used as output transducer.

The mass spectrometer is a RIBER quadripole analyzer recording 12 spectrums per second in a 120 mass range. The gas-input uses a capillary tube inducing a molecular beam with an aperture reduced by 2 diaphragms.[3]

Coarse Gas Analysis (Fig. 4)

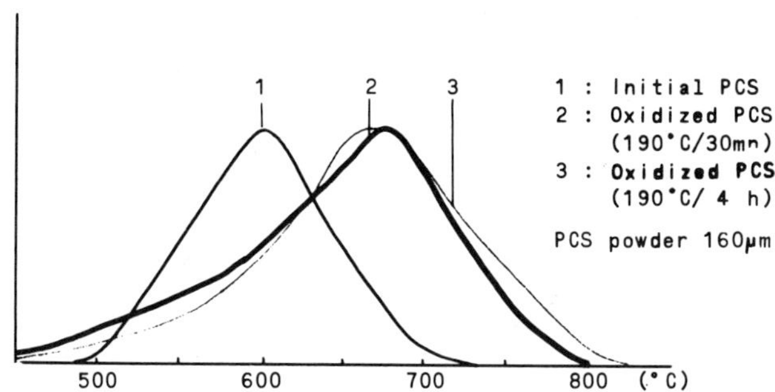

Fig. 4. Methane production analyzed by gas chromotography.

The first analysis with the chromatograph provided: the temperature range of the gas production from 673 to 973K with initial PCS, from 673 to 1073K with oxidized PCS; the maximal temperature of the production rate (873K with initial PCS, 953K with oxidized PCS); the identification of the most abundant species with the mass spectrometer (methane, hydrogen).

On the other hand, these first runs have clearly shown that the heavy species are produced essentially with the initial PCS and only before 673K. These species which have been analyzed on IR spectrometer, present exactly the same spectrum as the initial PCS. So, before 673K, the nonoxidized PCS goes through an intense distillation.

High Sensibility Gas Analysis

To detect the very low concentration species produced during the pyrolytic transformation between 673 and 1173K and condensed with the cold trap in liquid nitrogen, the trap was reheated very slowly with the mass spectrometer running during this time.

In a first step, during this fractional distillation we measured a large release of pure methane, bearing out its majority presence with hydrogen. In a second step, we obtained the typical spectrum presented in Fig. 5. It shows, in addition to the presence of methane: the silanes, with mass 31 identifying the major abundance as SiH_3; the ethylenes, with the 26, 27 and 28 masses which don't rule out traces of ethane; and, finally, the methylsilanes with the 41 to 45 masses. These results are summarized in Fig. 6.

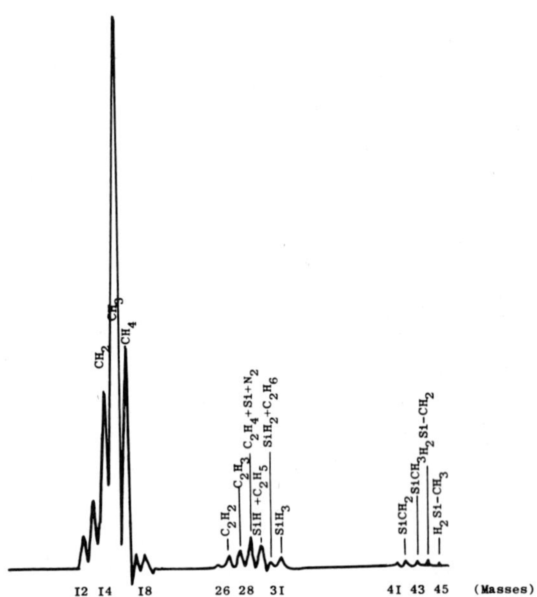

Fig. 5. Typical mass spectrogram of gas species from the reheated cold trap.

1st Importance	1	METHANE	CH_4
	2	HYDROGEN	H_2
2d Importance	3	SILANE	SiH_4
	4	ETHYLENE	C_2H_4
3d Importance	5	MONOMETHYLSILANE	H_3SiCH_3
	6	ETHANE(presumed)	C_2H_6

-Similar for every states of the P.C.S

-No oxygenated species with crosslinked P.C.S

-Max.Rate Temperature higher with crosslinked P.C.S

Fig. 6. Analyzed gas species during the pyrolytic transformation.

Moreover, the analysis performed with oxidized PCS provided similar gas compositions and no oxidized gas species have been detected. So, it seems that the oxygen reacted fully with PCS at low temperature and is conserved at high temperature in the structure of the new product.

CORRELATION WITH TGA AND DTA

In addition to the gas analysis during pyrolysis, TGA and DTA of these three states of the PCS were performed in order to correlate the various informations. The results are presented in Fig. 7.

First, the TGA of the initial PCS indicated a 30% weight loss before 673K, serving as an important distillation. With oxidized PCS, weight losses are, respectively: 4% with the 30 min oxidized; 1% with the 4 hours oxidized. Above 673K, if weight losses are very similar for every PCS (15%), weight evolutions are quite different: temperatures of the maximum rate losses are higher with oxidized PCS, bearing out the gas analysis results and the effect of cross-linking.

As for DTA results, in addition to the degradation evolution, they yield some information in connection with crystallization occurring by an exotherm after degradation. We can remark that they don't occur at the same temperature for each PCS: 393K with the initial PCS; 1203K with the oxidized PCS. This result may be explained by the inhibiting effect of the carbon in the silicon carbide since the free carbon concentration in initial PCS product is higher than in the oxidized one.

IMPLICATIONS ON COMPOSITE PROCESSING

The whole physicochemical results are summarized in Fig. 8. They suggest some remarks regarding the use of such precursors in new ceramic composite processes. First we must reduce, indeed suppress, the distillation phase before pyrolytic transformation during which a large amount of product can be saved (~ 30% for PCS). Several methods can be followed.

(1) Standard pressure cross linking. (a) By heterogeneous oxidation. This is largely used in SiC fibers process but is inapplicable in composite process.[1] (b) By acid catalyst. (d) By reaction in situ; Yajima recently showed the interest of the alkoxides.[4]

100
90
80
70
60
3
2
T.G.A
1
1
D.T.A
2/3
1
2/3
M.S.A
0
200
400
600
800
1000
1200 (°C)

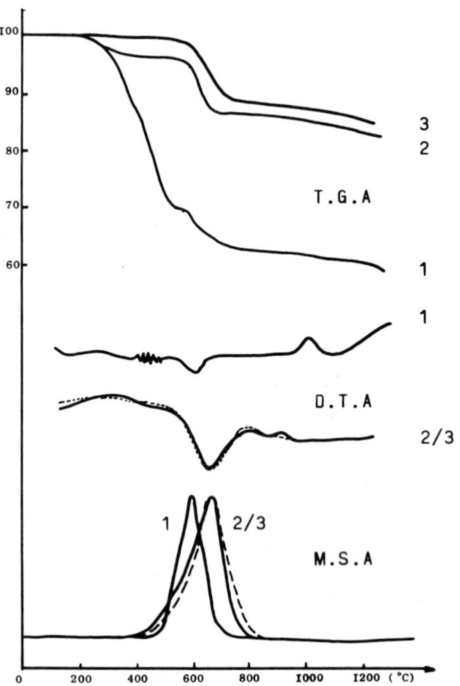

Fig. 7. Thermal analysis correlations.

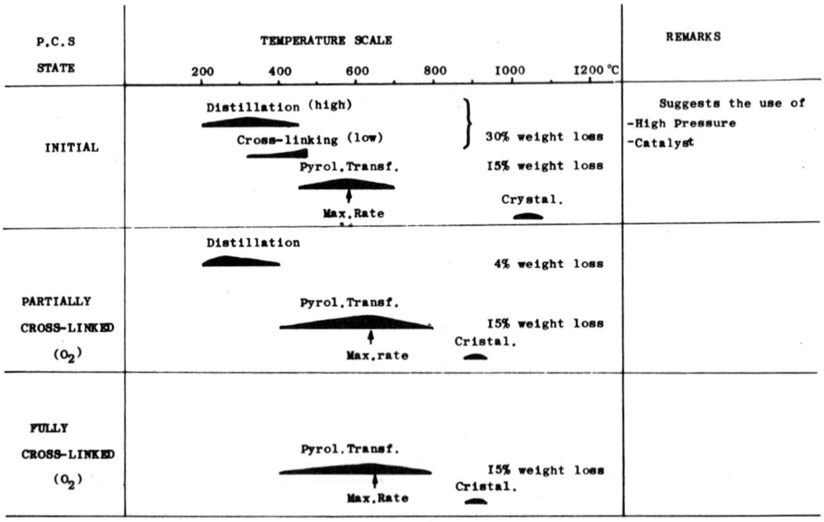

P.C.S
STATE
TEMPERATURE SCALE
200
400
600
800
1000
1200 °C
REMARKS
INITIAL
Distillation (high)
Cross-linking (low)
30% weight loss
Pyrol.Transf.
15% weight loss
Max.Rate
Crystal.
Suggests the use of
-High Pressure
-Catalyst
PARTIALLY
CROSS-LINKED
(O_2)
Distillation
4% weight loss
Pyrol.Transf.
15% weight loss
Max.rate
Cristal.
FULLY
CROSS-LINKED
(O_2)
Pyrol.Transf.
15% weight loss
Max.Rate
Cristal.

Fig. 8. PCS analysis of the influence of crosslinking on the physicochemical evolution between 20 and 1100°C.

(2) Transformation under high pressure. This allows reaching the pyrolytic transformation without distillation. It is used in carbon-carbon processing. Such an experiment has been performed in our laboratory under 70 MPa pressure up to 973K. Results presented in Fig. 9 show that the yield can reach 76% with a high content of free carbon but with a good oxidation resistance.[5]

(3) Finally, these two methods could be favorably used together by a high pressure cross linking before the pyrolytic transformation. This intermediate composite could be tested at high temperature under standard pressure.

Secondly, the gas analysis has shown that the main part of the transformation occurs in a narrow temperature range (673-873K) and the main part of the gaseous byproducts have a low molecular weight (CH_4-H_2). So, with the simultaneous shrinkage of the matrix, they generate internal stresses in a vitreous and weak solid producing intense cracks as shown in Fig. 10. This suggests, as in carbon-carbon processing, not only a treatment at very slow heating rate or hot pressing but also the need of a second phase in addition to the precursor to obtain: a relaxation of stresses during shrinkage; a better diffusion of the gaseous species; a toughness of the intermediate matrix. It used in high temperature fiber composite, this second phase must be: a refractory compound intimately mixed in a colloidal state with the precursor; able to impregnate fibers in a homogeneous manner; able to reduce the Young's modulus of the matrix. This has been used previously with success in carbon-carbon processing with thermosetting precursor (phenolic-furanic) using a colloidal graphite.[6]

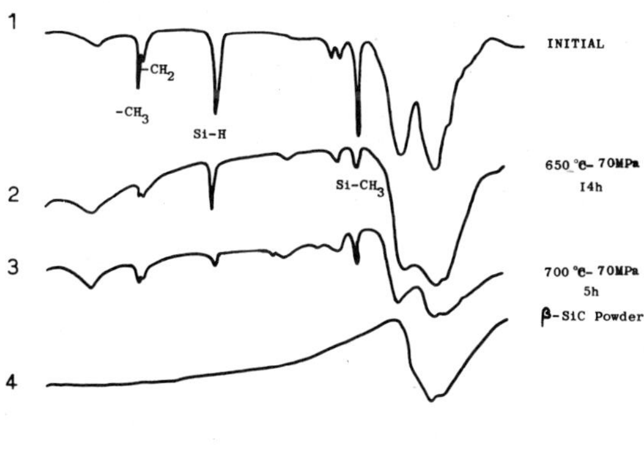

Fig. 9. Infrared spectrograms PCS evolution under high pressure (70 MPa).

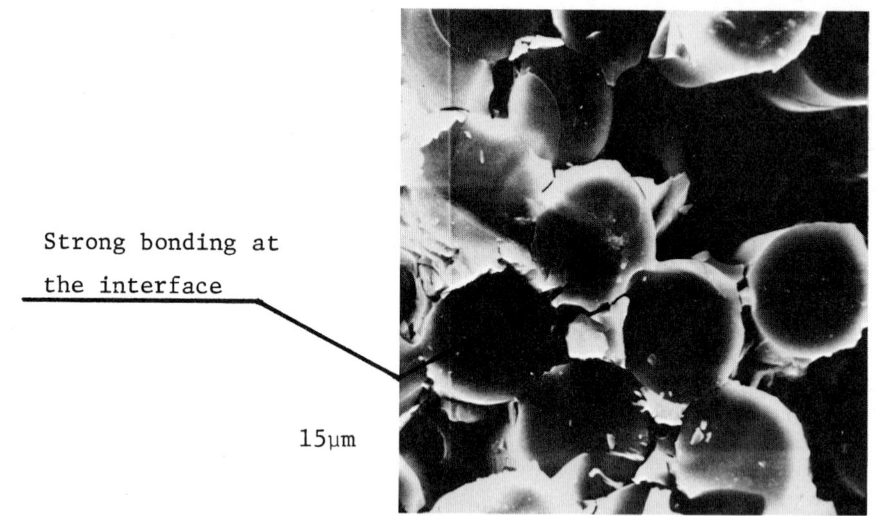

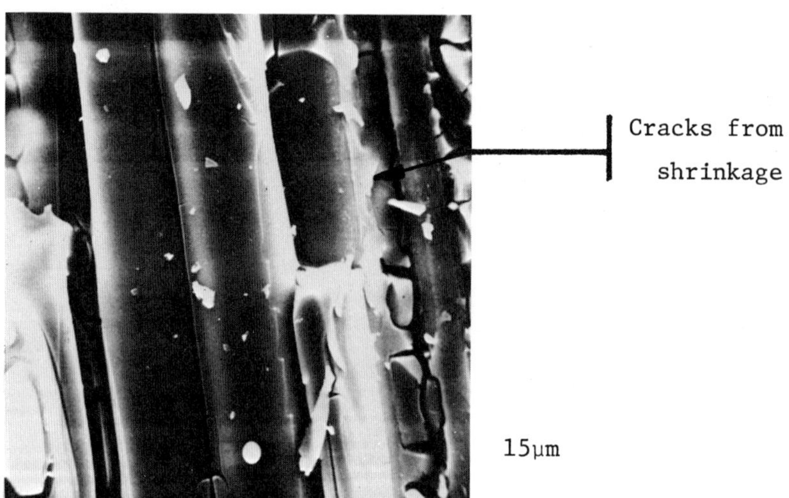

Fig. 10. SEM photographs of SiC/SiC from polycarbosilane.

Finally, the very strong bonding obtained particularly between SiC fibers and the matrix decrease considerably the strength and the toughness of the composite. So, the control of the interface will be another challenge during developments of such processing.

ACKNOWLEDGMENTS

The authors wish to acknowledge the support of the French Agency (Direction des Recherches, Etudes et Techniques) for this work which is being continued as the last author visits the Ceramic Branch of the Naval Research Laboratory in Washington, DC.

REFERENCES

1. S. Yajima, K. Okamura, and J. Hayashi, J. Mat. Sci., 13, 2569-76 (1978).
2. R. W. Rice, 81st Annual Meeting of the Am. Cer. Soc. May 81 (paper 13-B-81 Bull. of Am. Cer. Soc., 60 [3], 374 (1981)).
3. J. Jamet, R. Jalin, and C. Letourneur, Recherches Aerospatiales, 233-44 (1973-4).
4. S. Yajima, T. Iwai, T. Yamamura, K. Okamura, and Y. Hasegawa, J. Mat. Sci., 16, 1349-55 (1981).
5. J. Jamet and J. P. Barret, RT ONERA n° 2/3548 M. April 1982.
6. J. Jamet, J. Loubeau, and J. Omnes, RT ONERA n° 1/1784 M. (1972).

PART IV

CHEMICAL VAPOR DEPOSITION

CHEMICAL VAPOR DEPOSITION OF CERAMIC MATERIALS

John M. Blocher, Jr., Melvin F. Browning, and
David M. Barrett
Battelle's Columbus Laboratories
505 King Avenue
Columbus, Ohio 43201

ABSTRACT

The chemical vapor deposition (CVD) of ceramic materials such as pyrolytic carbon, silicon carbide, boron nitride, and silicon nitride is finding increased application. Factors involved in the control of uniformity and morphology of vapor-deposited structures, as well as the use of plasma and laser technology in extending the scope of CVD, are discussed.

INTRODUCTION

When Battelle became interested in chemical vapor deposition about fifty years ago, it was essentially a laboratory curiosity, but with a good potential for the preparation of high-purity materials, the application of coatings, and the vapor-forming of complex shapes. Today it has permeated all areas of technology. It is particularly important in the manufacture of microelectronic integrated circuits, from the preparation of ultra-pure semiconductor silicon for the wafers on which the circuits are based, to the silicon nitride passivating coating used to protect the resultant chips.

TYPES OF CHEMICAL VAPOR DEPOSITION (CVD)

Broadly defined, chemical vapor deposition, or CVD, is the formation of solid products by chemical reaction of gaseous precursors. The reaction is usually thermally activated at or near a heated surface. However, activation by plasma discharge or by radiation is possible. Thus, we have plasma-activated CVD (P-CVD) and laser-induced CVD (L-CVD), both of which are receiving considerable attention currently. Plasma-CVD is not new. It has been known since the turn of the century. However, it has been applied extensively only in the last five to ten years. The planar configuration of silicon wafers for microelectronic circuitry is admirably suited to P-CVD, where the advantage of lowered processing temperature for the required thin films is exploited.

By contrast, L-CVD is new and the potential is only currently being explored. With a laser, high concentrations of photons may be obtained within a narrowly limited energy range and within a narrowly defined area. The laser beam can be directed perpendicular to the substrate surface, in which case it is usually a means of highly localized thermal activation [1-9] and accordingly can be used to form patterned deposits. When the laser is directed so as not to intersect a radiation-absorbing surface, its action is limited to the gas phase.[10-21] There, free radicals or other dissociation products of the reactants may form either directly or indirectly through a secondary species. These reaction products can either diffuse to a nearby surface to give a deposited film, or nucleate to form a finely divided gas-phase precipitate.

Similar free radicals and other dissociation products are formed in P-CVD, but at the lower pressures usually used to sustain the electrical discharge, the potential for gas-phase precipitation is reduced. No such upper pressure limit exists for laser activation, so it has a higher potential for the preparation of gas-phase-precipitated materials of very fine particle size.

It is not surprising that more is known about thermally-activated CVD in terms of the mechanisms of deposition and the process/structure/property relationships. In this presentation some of the fundamentals of thermally-activated CVD will be reviewed briefly and then the results of some recent work at Battelle's Columbus Laboratories (BCL) will be discussed. In what follows, the term CVD will refer to the thermally-activated case, unless otherwise qualified.

CVD is important to ceramics in three areas, depending upon whether or not a substrate is used and whether the substrate is functional in the final application, or is expendable, or ultimately nonfunctional. The products are powders, coatings, and free-standing shapes.

Powders are made from CVD reactions having a large driving force or negative free-energy change under the conditions used. Coatings are applied by less favorable, preferably endothermic, reactions so that the reaction is predominantely heterogeneous at the surface, and gas-phase precipitation is suppressed. Vapor forming is basically a matter of prolonging the coating until the desired thickness of the deposit is obtained. However, as will be evident later, this extension of the coating process is by no means simple.

Table 1 shows some of the most useful reactions for the deposition of ceramic materials. The first reaction, representing the thermal decomposition of hydrocarbons, is used for the formation of pyrolytic carbon and pyrolytic graphite. The latter is the highly anisotropic form in which the graphitic planes are essentially parallel to the deposition surface and the thermal conductivity parallel to the surface is many times that perpendicular. The next five reactions are representative of the use of the chlorides as precursors.

Table 1. Reactions for CVD of Ceramic Materials

Reaction	Temperature Range
$C_xH_y(g) \rightarrow xC(s) + \frac{Y}{2} H_2(g)$	1100 - 2700 K
$SiCl_4(g) + 2H_2O(g) \rightarrow SiO_2(s) + 4HCl(g)$	700 - 1300 K
$SiCl_4(g) + 2CO_2(g) + 2H_2(g) \rightarrow SiO_2(s) + 4HCl(g) + 2CO(g)$	1100 - 1300 K
$TiCl_4(g) + O_2(g) \rightarrow TiO_2(s) + 2Cl_2(g)$	1200 - 1500 K
$SiCl_4(g) + CH_4(g) \rightarrow SiC(s) + 4HCl(g)$	1300 - 1700 K
$3SiCl_4(g) + 4NH_3(g) \rightarrow Si_3N_4(s) + 12HCl(g)$	1200 - 1800 K
$3SiH_4(g) + 4NH_3(g) \rightarrow Si_3N_4(s) + 6H_2(g)$	1000 - 1800 K
$WF_6(g) + CH_4(g) + H_2(g) \rightarrow WC(s) + 6HF(g)$	1500 - 1800 K
$W(CO)_6(g) \rightarrow WC(s) + CO_2(g) + 4CO(g)$	600 - 1100 K
$W(CO)_6(g) \rightarrow W(s) + 6(CO)(g)$	1000 - 1300 K
$(CH_3)SiCl_3(g) \rightarrow SiC(s) + 3HCl(g)$	> 1200 K
$(C_3H_5O)_4\ Si(g) \rightarrow SiO_2(s) + \ldots$	1300 - 1400 K

The second and third represent the formation of an oxide by hydrolysis of a halide. However, in the third, the reaction of hydrogen with CO_2 is used for hydrolysis, a reaction that itself is surface catalysed, thus limiting the oxide formation to the substrate surface, suppressing the gas-phase precipitation of finely divided oxide "smoke" that would tend to form in the direct hydrolysis depicted by the second reaction.

The last six reactions represent the use of hydrides, carbonyls, and organometallic compounds as precursors.

Apparatus

The apparatus used for CVD depends upon the deposition reaction being used, its reaction temperature requirements, and the configuration of the substrate. Figure 1 shows examples of laboratory reactors for the deposition of coatings on planar substrates, such as silicon wafers. In one case the resistively heated support plate is rotated to average out exposure to the reactant gases. Fluidized-bed reactors such as that shown in Fig. 2 are particularly well suited for the coating of particles 0.01 to 1 mm in diameter because of the inherently high rates of heat and mass transfer in the fluidized bed. Exposure to the reactant gases is averaged out by circulation of particles through the bed, and good uniformity of coating results. Incidentally, the reactor shown in Fig. 2 is an example of a highly reliable high-temperature reactor (to 2700K) that can be easily assembled from readily available materials.

The objective with any CVD reactor is to provide uniform exposure of the substrate to the coating atmosphere. Since the rate of deposition is a function of reactant concentration and the reactant concentration decreases downstream as the result of reactant depletion upstream, some compensation is necessary if efficient use of the reactants is to be attained along with good uniformity. Such compensation can take the form of rotating axially symmetrical substrates, translating the substrate relative to the reactant flow, increasing the substrate temperature downstream, or constricting the flow channel downstream so as to increase the rate of gas flow and decrease the boundary layer thickness through which gas-phase mass transfer must occur.

Reaction Conditions

The pressure at which a CVD reactor is operated influences first the concentration of reactants in the gas phase and second the diffusivity of reactants toward the substrate and reaction products away. Where satisfactory results can be obtained at atmospheric pressure, it is most convenient to operate in that mode. However, due to the increased uniformity of deposition that results from increased diffusivity at decreased pressure, many CVD reactors are operated at pressures in the 650 to 13,300 (5-100 Torr) range.

The substrate temperature influences not only the rate of deposition, it is the major factor in determining the structure of the deposit, mainly because of its effect on the relative importance of surface diffusion and bulk diffusion in the grain growth mechanism.

Relation of Kinetics to Deposit Uniformity

In general, as the substrate temperature is raised, the rate of deposition goes through two stages. The first, an exponential increase of the rate with reciprocal temperature, follows the typical Arrhenius relation from which a characteristic activation energy for a kinetically controlled reaction can be derived. The second stage evolves when the interdiffusion of reactants and reaction products can no longer keep up with the rate of reaction at the surface of increasing temperature. In this latter stage the rise in deposition rate at the substrate drops off, becoming much less dependent on temperature and much more dependent on the rate of diffusion through the boundary layer at the surface, which is, in turn, sensitive to the gas flow dynamics of the system.

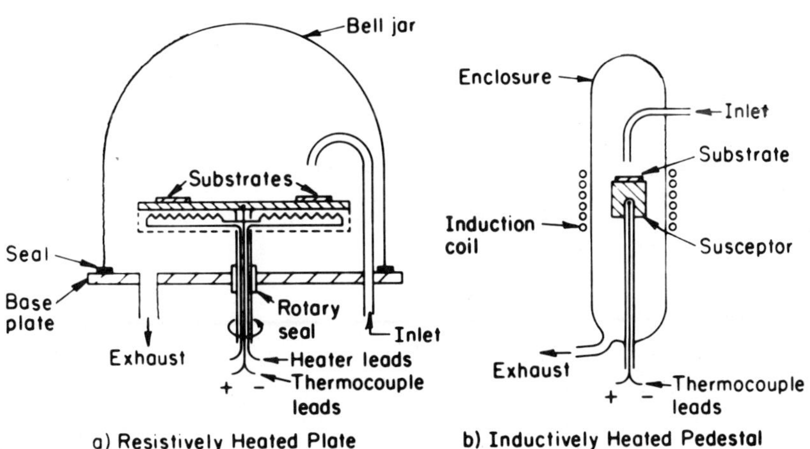

Fig. 1. Experimental CVD reactors.

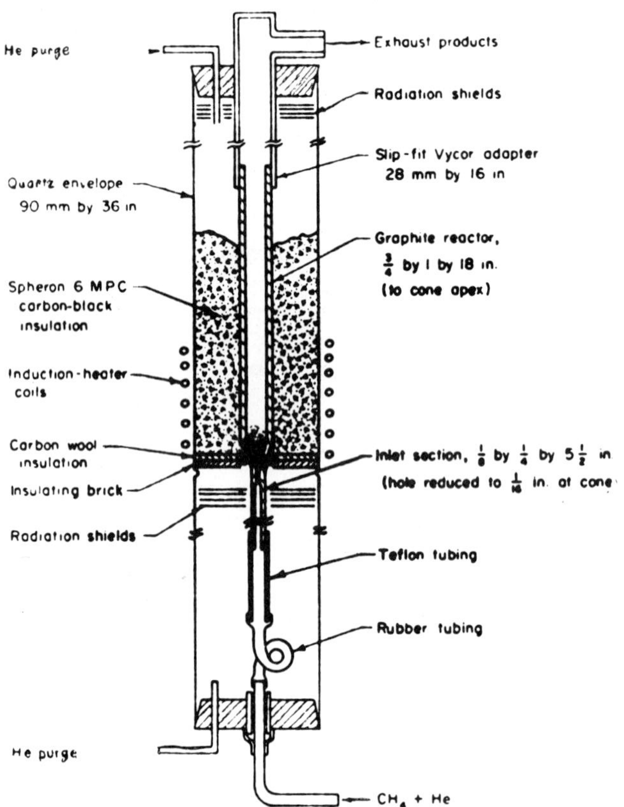

Fig. 2. Carbon-coating reactor for operation above 1773K.

As the operating pressure is decreased and diffusivity is increased, the transition from kinetic control to diffusion limited control takes place at higher temperatures and higher deposition rates.

In the region of diffusion limitation, a reactant concentration gradient develops in the gas phase adjacent to the surface. Incipient protrusions penetrating such a gradient grow faster at the tips than at the root and become exaggerated, leading eventually to so-called "berries" of the type shown in Fig. 3 for a chromium carbide deposit from dicumene chromium (bis-isopropylbenzene chromium, $C_{18}H_{24}Cr$).

For maximum uniformity and the suppression of surface protrusion growth, it is desirable to operate within the kinetically controlled range. Thus, in order to optimize conditions it is desirable to characterize the CVD process being used as to the relative importance of the diffusional and kinetic factors. Fortunately, it is no longer necessary to run a series of deposition rates versus temperature to determine where and in which regime one is operating.

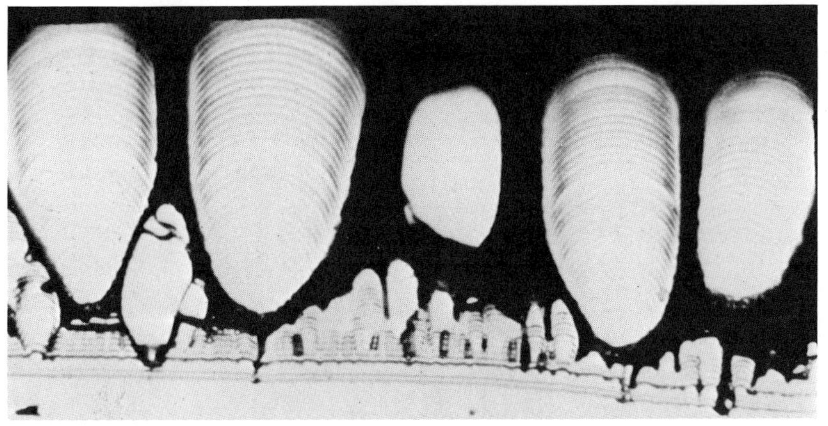

Fig. 3. Chromium "berries" caused by deposition in the region of diffusion limitation.

Characterization of CVD Processes

Van den Brekel[22] has analysed quantitatively the uniformity of deposition and the growth of surface irregularities as a function of deposition regime, and has shown that the deposition regime can be determined from the deposit profile in a blind hole ($L/D \geqslant 30/1$) adjacent to the point of interest. One simply plots the logarithm of the ratio of the deposit thickness along the hole to that at the entrance versus the position/radius. The slope gives the desired characterization. Its true physical significance depends upon the kinetics of the reaction in question. However, the slope is a useful number, in any case, because its variation with conditions and from one reaction to another is greater than the differences in the kinetics by which it needs to be corrected to be physically significant.

For a first-order reaction, the basic equation is given in Fig. 4. The quantity K, determined from the plot, when divided by the radius R of the blind hole gives the ratio of the first order reaction rate constant to the reactant diffusivity. When this is multiplied by an estimated boundary layer thickness, it gives the dimensionless Sherwood number, the ratio of diffusional impedance to kinetic impedance. The transition from kinetic control to diffusion limitation occurs at a Sherwood number of unity. It is predicted that this simple means of characterizing CVD systems will be used universally in the future.

Precipitation of gas-phase-nucleated material can occur if the gas-temperature/composition exceeds a critical value near the substrate. This material can be entrained and carried downstream and out of the reactor, in which case the rate of deposition on the

$$d_z/d_o = \exp\left(-\frac{z}{R}\sqrt{2K}\right)$$

where

d = the deposit thickness at position z from the entrance of a deep blind hole of radius R

plot = $\ln d_z/d_o$ vs. z/R

slope = $-\sqrt{2K}$, $K = (\text{slope})^2/2$

K/R = k/D, the ratio of the first-order reaction rate constant, k, to the reactant diffusivity, D

$k\delta/D$ = Sh, the Sherwood number or ratio of diffusional to kinetic impedances

δ = boundary layer thickness

Sh > 1 = Diffusion limited

Sh < 1 = Kinetically controlled

Fig. 4. Van den Brekel characterization of a first-order CVD reaction.

substrate actually decreases with increasing temperature. The gas-phase-precipitated material can also be incorporated in the deposit, modifying its characteristics. Such gas-phase precipitation is highly detrimental in thin films for microelectronics, but is beneficial in decreasing the anisotropy of some desirable forms of pyrolytic graphite.[23] Gas-phase precipitation can result in grain refinement, porosity, or surface protrusions. These effects may be desirable or undesirable, depending upon the requirements.

Table 2 summarizes the conditions for promoting uniformity of deposition and Table 3 the avoidance of surface protrusions and attainment of surface conformity. It is obvious that some sacrifice in deposition rate may be required to attain these goals.

VAPOR FORMING

The promotion of uniformity and suppression of surface protrusions are particularly important in the vapor forming of free-standing

Table 2. Conditions for Promoting Uniform Deposition Rate in CVD (Uniform Deposit Thickness)

Increase Diffusivity (decrease pressure) or Decrease Temperature to Assure Operation in the Kinetically Controlled Regime

Offset Effect of Reactant Depletion Downstream by:

- Rotation or Translation of Substrate Relative to Gas Stream
- Limited Reactant Conversion per Pass
- Compensating Temperature Gradient
- Increased Gas Velocity Downstream

Table 3. Avoidance of Surface Protrusions in CVD and Attainment of Surface Conformity

---

Operate in the Kinetically Controlled Regime

Avoid Gas-phase Precipitation by:

- Reducing Pressure
- Decreasing Temperature
- Decreasing Reactant Concentration

Eliminate Dust Particles

Minimize Substrate Surface Irregularities

---

shapes by CVD. However, in such cases one is often faced with the problem of columnar deposits with relatively low intergranular strength such as in the tungsten deposit shown in Fig. 5. This calls for some means of grain refinement during deposition, which is best accomplished by introducing a competing growth mechanism as summarized in Table 4. One of the possibilities already mentioned is the judicious allowance for gas-phase precipitation as in the case of some pyrolytic graphites. Other possibilities are listed below it. Holman and his associates[24] have mechanically burnished or "brushed" tungsten and tungsten alloy surfaces to promote renucleation of new grains. The same effect is observed in the inherent self peening of particles being coated in a fluidized bed.

One can use additives to the reactant stream to produce a second-phase grain refiner, such as the carbon addition to a beryllium oxide nuclear fuel particle coating developed at Battelle[25] to suppress anisotropic deposition. Some CVD systems inherently provide for grain refinement. For example, from the $SiCl_4/CH_4/H_2$ system it is possible to obtain silicon, silicon carbide, carbon, or mixtures of silicon carbide with either element, depending upon conditions. Thus, the choice of conditions can lead to a fine-grained silicon carbide, the grain growth being suppressed by the competing second phase.

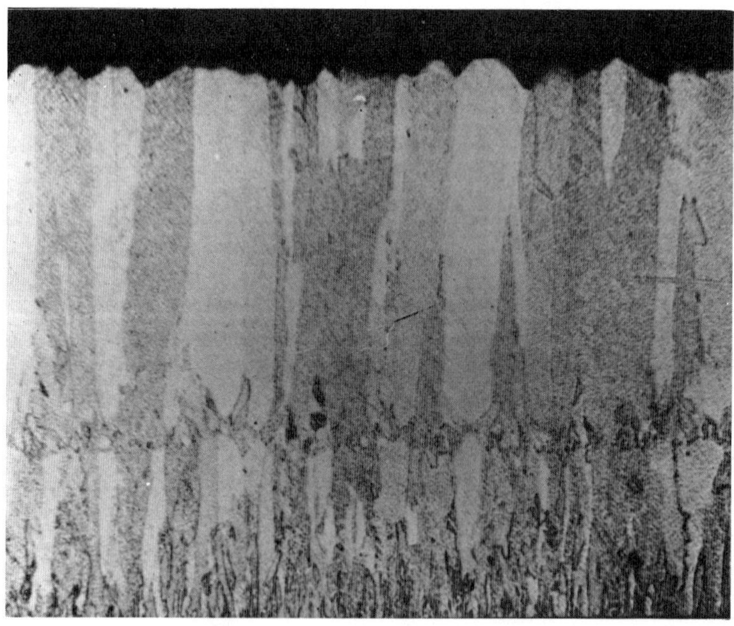

Fig. 5. Columnar tungsten deposit.

Table 4. Grain Refinement in CVD by Providing Competition to Grain Growth

Allow Controlled Gas-Phase Precipitation
Mechanically Distort Surface Structure (Brushing or Peening)
Add Competing CVD Reaction to Precipitate Second Phase
Utilize Inherent Competing Processes:
Co-deposition of second phase
Condensation and surface decomposition of polymer droplets

Another grain-growth suppression possibility is inherent in deposits for which liquid droplets of intermediate polymeric materials can form in the vapor phase, condense on the surface and undergo further decomposition independent of the main heterogeneous deposition reaction. Such possibilities are observed with silicon compounds, e.g., SiC, and Si_3N_4, and in depositions from organometallic compounds.

Attention to these fundamentals of CVD can increase the already long list of vapor-formed ceramic products shown in Table 5.

ACTIVATED CVD

As indicated earlier, there is currently a surge of interest in activated CVD technology such as plasma and laser-induced processes. This interest spawns primarily from a practical desire to apply CVD technology to the coating of substrate materials that cannot withstand the relatively high temperatures usually associated with CVD. Low-temperature CVD has always been a dream of people working in the area but only in the last few years has there been a real need for this technology. In fact, Battelle is currently attempting to establish a group research program with the primary objective of addressing all the main known approaches to low-temperature CVD--catalysis, less stable precursors, plasma-activated and photon-induced (laser).

Laser-Induced CVD (L-CVD)

Laser-induced CVD has reportedly[1-21] been used to produce many materials (e.g., aluminum, bismuth, boron, carbon, cadmium, cobalt oxide, chromium, germanium, iron, lead, manganese, molybdenum, nickel, silicon, silicon carbide, silicon nitride, silicon dioxide,

Table 5. Applications of CVD Vapor-Formed Ceramics

Pyrolytic Graphite	Nose Cones, Atmosphere Re-entry Bodies, Smoking Pipe Bowls	Commercial
Pyrolytic Carbon	Medical Prostheses	Commercial
Silicon	Furnace Tubes, Hardware	Commercial
Silicon Carbide	Various Shapes	Commercial
Silicon Nitride	Various Shapes, Turbine Parts	Developmental
Anisotropic Boron Nitride	Various Shapes	Commercial
Isotropic Boron Nitride	Various Shapes	Commercial
Zinc Selenide	Radomes	Commercial
Zinc Sulfide	Radomes	Developmental
Silicon Carbide/ Tungsten	Thermionic Emitter Shells	Developmental
Boron Fibers	Fiber Reinforcement	Commercial
Silicon Carbide Fibers	Fiber Reinforcement	Commercial
Doped Silica Fibers	Fiber Optics	Commercial

tin, tin oxide, titanium carbide, titanium dioxide, zinc and tungsten) by either pyrolytic or photolytic mechanisms on substrates or as powders. In the reported work, pyrolytic decomposition of the reactant species has been carried out by heating either the substrate or the reactant gases or both with laser radiation. Those materials deposited by a photolytic mechanism generally utilize matching of the laser frequency with either an absorption band of the reactant of interest, or with the absorption band of an intermediate species which would attain an excited state. The energy of the excited state would, in turn, be transfered to the reactant of interest, similar to the decomposition of reactant species by an argon plasma via excited electrons.

Although some L-CVD reactions have been conducted at atmospheric pressure at other sites, most have been performed at lower pressures since there are inherent potential advantages in regards to gas-phase nucleation and coating quality, as discussed earlier. However, systems operating at atmospheric pressure are generally more desirable from an economic and production adaptability point-of-view. Accordingly, Battelle elected to focus its L-CVD studies, at least initially, on producing materials at atmospheric pressures. In addition, whenever possible, reactants that have been commonly used in traditional CVD reactions were used. Much of the work done at other sites has been with precursors with which experience is limited.

It has been demonstrated at Battelle that coatings, such as aluminum, nickel, and carbon, can be deposited by L-CVD at atmospheric pressure and in some cases at temperatures somewhat lower than those used in thermally activated systems. Incident laser power, duration of laser incidence (and therefore substrate temperature), deposition rates, and quality have not been optimized. The parameters used in these studies are tabulated in Table 6 for review, but for the sake of brevity only generalities of each system will be discussed.

Table 6. Experimental Data for Selected L-CVD Experiments

Reactant	Diluent	Total Flow $m^3\ sec^{-1}$ (cc/min)	Mole Percent Reactant	Substrate Material	Incident Laser Power (Watts)	Nominal Substrate Temperature, K	Comments
Triisobutyl Aluminum	Argon	55×10^{-6} (3300)	0.3	Quartz	25	∿560	Carbon noted in aluminum deposit
"	"	"	"	"	23	∿520	Reflective aluminum deposit
"	"	"	"	Borosilicate	25	∿400	Very thin, partially oxidized coating
Nickel Tetracarbonyl	"	7.8×10^{-6} (470)	1.4	Quartz	15	∿350	Reflective nickel coating except for region adjacent to quartz which appeared to have carbon in coating
"	"	"	"	Borosilicate	16	∿320	Reflective nickel coating on substrate: substrate moved relative to quartz for large area coverage
Acetylene*	"	2.3×10^{-6}	5.2	Quartz	14	∿373	Substrate temperature kept at ≤373 K by periodically removing incident laser beam. Gas-phase nucleation downstream of substrate due to high concentration of C_2H_2 in reactant stream

* Note reaction was catalyzed with $Ni(CO)_4$, concentration ∿0.2 mole percent.

The laser-CVD system currently devoted to the Battelle study is a continuous wave CO_2 laser rated at 50 w with a horizontally positioned 27-mm ID quartz reactor ~ 1 meter in length equipped with an argon purged potassium bromide window. A neon laser is used for alignment purposes. The apparatus has been assembled with the appropriate reactant flow, temperature, and pressure instrumentation usually associated with CVD systems.

Aluminum. In the Battelle studies on L-CVD of aluminum, argon-diluted, triisobutyl aluminum (TIBA) was used at atmospheric pressure. TIBA is normally decomposed thermally to aluminum at temperatures of ~ 550K (~ 275°C) at atmospheric pressure and should not be a strong absorber of the CO_2 laser radiation (10.6×10^{-3} mm). If temperatures are exceptionally high, the reaction by-products may be thermally cracked resulting in a carbon-contaminated aluminum.

In the initial experiments an unfocused beam was directed against a quartz substrate, a strong absorber of CO_2 laser radiation. The results essentially confirmed that the mechanism of decomposition with this precursor is pyrolytic in that reflective metallic coatings were obtained at temperatures estimated to be ~ 520K (~ 250°C) and that black-appearing carbon-contaminated films were obtained at temperatures greater than ~ 560K (~ 290°C). A very thin coating was deposited on borosilicate glass in an area estimated to be in the range of ~ 400K (~ 130°C). Such results are similar to those obtained when TIBA is used in a thermally activated system. In addition, there was no obvious difference in deposition rate between laser and thermal activation. However, the L-CVD parameters were not optimized and the coating rates were highly uncertain because of thinness and lack of uniformity.

Nickel. In the Battelle laser study, nickel carbonyl [$Ni(CO)_4$] was used since it is the most common precursor for applying this metal in a thermally activated system. A reflective nickel was deposited from $Ni(CO)_4$ on quartz and borosilicate glass at temperatures as low as ~ 320K (~ 50°C) and at one atmosphere pressure using an unfocused CO_2 laser. Since $Ni(CO)_4$ is normally deposited at 390K (~ 120°C) in thermally activated CVD, Battelle's results would imply a photolytic mechanism which is consistent with the fact that $Ni(CO)_4$ could be expected to absorb CO_2 laser radiation. However, no significant gas-phase nucleation was noted which suggests that parameters used (reactant concentration) favored surface activation. There were surfaces in the system which attained higher temperatures that were coated with a poorer quality coating. This was thought to be the result of a combination of pyrolytic and photolytic deposition mechanisms. The deposition rate (2×10^{-7} mm sec^{-1}) was lower than that typically attainable with thermally activated systems (2×10^{-3} mm sec^{-1}), but has the probability of being improved with optimization.

Carbon. In the Battelle studies on L-CVD of carbon, acetylene gas, one of the most common precursors, was used to deposit carbon on quartz at atmospheric pressure and at an estimated temperature of ~ 370K (~ 100°C). A small quantity of $Ni(CO)_4$ was used to catalyze the reaction. The catalyzing effect of nickel on carbon had been demonstrated earlier at Battelle in a thermally activated system. The carbon coatings were black, hard and dense, but did contain some nickel. Deposition rate was 4×10^{-7} mm sec^{-1}. Gas-phase nucleated material was found downstream of the substrate, clinging to the reactor wall. It is believed that nickel incorporation in the deposit and gas-phase nucleation can be minimized through process optimization.

In conventional CVD, carbon coatings are usually applied from acetylene at temperatures of ~ 1370K (~ 1100°C), but coatings have been obtained at temperatures as low as ~ 1120K (~ 850°C). As indicated above, nickel additions have been shown to reduce the minimum temperatures at which coatings can be applied at a practical rate to about 670K (~ 400°C). Consequently, it is believed that the L-CVD carbon coatings were applied primarily by a photolytic mechanism since they were formed at temperatures of about 370K (~ 100°C).

The rate of deposition for the L-CVD carbon was significantly lower than that typically obtained in thermal systems but was sufficient to be practical and could probably be increased with optimization.

This preliminary investigation into laser-activated processes confirm the work of others that the potential exists for developing this technique to (1) pyrolytically deposit substances in localized areas, and (2) apply selected materials at lower substrate temperatures than that typically associated with thermally activated CVD.

Plasma-Enhanced CVD (P-CVD)

Although much has been reported in the literature over the last twenty years or so in P-CVD[26-36], very little has been done in combining P-CVD technology with fluidized-bed processing. The use of fluidized-bed systems with thermally activated CVD for the coating of particulates was pioneered at Battelle approximately thirty years ago, and some work has been done in the intervening years with Plasma/fluidized bed. More recently this interest has been reactivated in view of the current thrust to lower the deposition temperature of typical CVD systems. An attractive approach is to convert the molecular hydrogen normally used to reduce typical CVD precursors to atomic hydrogen by passing the hydrogen through an argon RF plasma. This technique is ideally suited for use in coating particulates in a fluidized bed.

Titanium. In a conventional system titanium is deposited at temperatures of >1370 K (~ 1100 C) by the hydrogen reduction of titanium tetrabromide. Utilizing a plasma activated fluidized-bed system, preliminary EDAX results would indicate that titanium was deposited on silica sand (0.44 mm) in the temperature range of ~ 770K (~ 500 C) to 870K (~ 600 C). This was accomplished in a typical 25-mm diameter reactor in which a RF plasma (2.5 MHz) was maintained below the distributor plate. A hydrogen/argon mixture was used as the fluidizing gas and titanium tetrachloride ($TiCl_4$) as the titanium source, instead of the less stable $TiBr_4$. $TiCl_4$ was introduced above the distributor plate. Pressure in the plasma section was ~ 1660 Pascal (~ 20 Torr).

The deposition temperature given for the P-CVD system is the minimum that could be achieved with the particular plasma used since all the heating resulted from the plasma. However, even in the temperature range of this work, which is roughly 500 K below that typically used in thermal systems using a less stable precursor (TiBr), it would appear that the reduction was enhanced by atomic hydrogen formed in the plasma below the distributor plate.

Carbon. In the Battelle studies on deposition of carbon onto a fluidized bed of hollow carbon particles (~ 2 mm), by the hydrogen reduction of carbon tetrachloride (CCl_4) utilizing P-CVD, a deposit was obtained at a bed temperature in the range of 770K (500°C) to 870K (600°C). This was done in the same equipment at a pressure similar to that described earlier with the titanium P-CVD studies. When using thermal activation, carbon is typically deposited from CCl_4 at ~ 1070 K (~ 800°C). The results of this work also suggest that atomic hydrogen is playing a role in lowering the deposition temperature. As mentioned earlier, the temperature range given for deposition in P-CVD has thus far been controlled by the plasma generated heat and not by the minimum required to form carbon.

Niobium. As in the two previous systems, niobium was also applied to a fluidized bed (hollow carbon spheres) in the same equipment and at the same conditions of temperature and pressure utilizing niobium pentachloride ($NbCl_5$) as the precursor. The qualifications as to temperature also apply. In thermal-CVD, niobium is typically deposited at temperatures above ~ 1070K (~ 800°C) from $NbCl_5$.

In general the results of this preliminary investigation of P-CVD in a fluidized bed have indicated a strong potential for the establishment of fluidized-bed CVD systems capable of applying coatings at lower temperatures than those used in a conventional system.

REFERENCES

1. S. D. Allen, J. Appl. Phys., 52 [11], 6501 (1981).
2. R. S. Berg and D. M. Mattox, Proceedings of the Fourth International Conference on Chemical Vapor Deposition, 1973, The Electrochemical Society, Inc., Pennington, NJ.
3. J. Mazumder and S. D. Allen, Proceedings of the SPIE, 198, 73 (1979), San Diego, CA.
4. G. Leyendecker, D. Bauerle, P. Geittner, and H. Lydtin, Appl. Phys. Lett., 39 [11], 921 (1981).
5. C. P. Christensen and K. M. Lakin, Appl. Phys. Lett., 32 [4], 256 (1978).
6. W. M. Steen, International Conference on Advances in Surface Coating Techniques, London, 1978.
7. V. Baranauskas and C. I. Z. Mammana, Appl. Phys. Lett., 36 [11], 930 (1980).
8. O. Tabata et al., Proceedings of the Eighth International Conference on Chemical Vapor Deposition, 1981, The Electrochemical Society, Inc., Pennington, NJ.
9. R. Bilenchi and M. Musci, Proceedings of the Eighth International Conference on Chemical Vapor Deposition, 1981, The Electrochemical Society, Inc., Pennington, NJ.
10. T. F. Deutsh, D. J. Ehrlich, and R. M. Osgood, Appl. Phys. Lett., 35 [2], 175 (1979).
11. D. J. Ehrlich, R. M. Osgood, and T. F. Deutsch, Appl. Phys. Lett., 38 [11], 946 (1981).
12. D. J. Ehrlich and R. M. Osgood, Thin Solid Films, 90, 287 (1982).
13. W. R. Cannon et al., Proceedings of the SPIE, 198, 65 (1979).
14. W. R. Cannon and J. S. Haggerty, Laser Induced Chemical Processes, edited by J. I. Steinfield, Plenum Press, New York, NY (1981).
15. J. S. Haggerty and W. R. Cannon, Energy Laboratory Report, MIT-EL, 81-003 (1980).
16. P. K. Boyer, G. A. Roche, W. H. Ritchie, and G. J. Collins, Appl. Phys. Lett., 40 [8], 716 (1982).
17. R. W. Andreatta et al., Appl. Phys. Lett., 40 [2], 183 (1982).
18. M. Hanabusu, A. Namiki, and K. Yoshihara, Appl. Phys. Lett., 35 [8], 626 (1979).
19. R. Solanki et al., Appl. Phys. Lett., 38 [7], 572 (1981).
20. C. W. Draper, Metallurgical Trans. A., 11A, 349 (1980).
21. A. M. Ronn, Chem. Phys. Lett., 42 [2], 202 (1976).
22. C. H. J. Van den Brekel et al., Proceedings of the Eighth International Conference on Chemical Vapor Deposition, 1981, The Electrochemical Society Inc., Pennington, NJ.

23. W. H. Smith and D. H. Leeds, Modern Materials, 7, 139-218 (1970).
24. W. R. Holman and F. J. Huegel, Proceedings of the Conference on Chemical Vapor Deposition of Refractory Metals, Alloys, and Compounds, 1967, The American Nuclear Society, Hinsdale, IL.
25. W. J. Wilson, M. F. Browning, V. M. Secrest, and J. M. Blocher, Jr., USAEC Report BMI-1718 (1965).
26. J. R. Hollahan and A. T. Bell, Techniques and Applications of Plasma Chemistry (1974).
27. A. T. Bell, Proceedings of the Eighth International Conference on Chemical Vapor Deposition, 1981, The Electrochemical Society, Pennington, NJ.
28. D. Kuppers, Proceedings of the Seventh International Conference on Chemical Vapor Deposition, 1979, The Electrochemical Society, Pennington, NJ.
29. K. R. Sarma et al., Proceedings of the Seventh International Conference on Chemical Vapor Deposition, 1979, The Electrochemical Society, Pennington, NJ.
30. J. R. Hollahan et al., Proceedings of the Sixth International Conference on Chemical Vapor Deposition, 1977, The Electrochemical Society, Pennington, NJ.
31. B. Bourbon et al., Proceedings of the Sixth International Conference on Chemical Vapor Deposition, 1977, The Electrochemical Society, Pennington, NJ.
32. K. Wahawili and F. J. Weinberg, AIChe Symposium Series 186, 75, 11 (1979).
33. D. E. Carlson, J. Electrochem. Soc., Solid State Science & Tech., 122 [10], 1334 (1975).
34. A. K. Sinha et al., J. Electrochem. Soc., Solid State Science & Tech., 124 [4], 601 (1978).
35. S. B. Hyder and T. O. Yep, J. Electrochem. Soc., Solid State Science & Tech., 123 [11], 1721 (1976).
36. D. R. Secrist and J. D. Mackenzie, J. Electrochem. Soc., 113 [9], 914 (1966).

THE APPLICATION OF THERMODYNAMIC CALCULATIONS TO CHEMICAL VAPOR DEPOSITION PROCESSES

Angus I. Kingon* and Robert F. Davis**

*National Physical Research Laboratory, CSIR
P. O. Box 395, Pretoria 0001, South Africa
**Department of Materials Engineering, North Carolina State University, NC 27650, USA

ABSTRACT

Computer codes may be used to calculate CVD phase diagrams. These predict the condensed phases deposited as a function of the thermodynamic conditions, and are particularly important in complex systems, such as the Si-C-H-Cl system. The phase diagrams, combined with the calculated compositions of the gas phase, allow an understanding of the process. The applications and the limitations of these calculations are discussed.

INTRODUCTION

Chemical vapor deposition (CVD) is a processing technique which is growing in importance. It is the reaction of one or more gaseous precursors to form a condensed phase (or phases) and gaseous products. The condensed phase is normally formed on a heated substrate.

The CVD technique is currently used for the fabrication of the following groups of materials:
(1) Solid state devices, typically based on Si and SiO_2;[1]
(2) Coatings, typically for wear resistant applications;[2]
(3) Monolithic hard materials, for structural and wear-resistant applications;[3,4]
(4) Precursor ceramic powders, such as SiC and Si_3N_4.

The CVD technique has a number of advantages which include:

(1) The high purity and quality of the deposits which can be attained. In particular this implies an absence of impurities on the grain boundaries, which can significantly improve high temperature strength and creep resistance.[5]
(2) The ability to deposit homogeneous materials of complex composition,[6] as well as multiple coatings.[7]
(3) The deposition temperature can be low in comparison with sintering temperatures.
(4) Controllable crystallinity and morphology of the deposits. This includes the deposition of coatings with very small grain size, which can show a marked increase in strength.[8]

The chemical vapor deposition process is complex, with many parameters requiring optimization in practice. Thermodynamic calculations can be used to assist in the evaluation of a particular CVD system, to optimize conditions, and to gain an understanding of the system. In this paper the role of thermodynamic calculations are discussed, with particular emphasis on the prediction of condensed phases deposited at equilibrium in complex systems by the use of CVD phase diagrams. The limitations and the range of applicability of the calculations are discussed in a separate section.

METHOD OF CALCULATION

The computer code used in the present work is a modification of SOLGASMIX,[9] which has been described by Spear[10] for use in CVD systems. The procedure, originally developed by White et al.[11] involves the calculation of that composition which results in a minimum for the free energy of the system, under given conditions of temperature, total pressure, and input gas concentration. The calculated composition includes condensed and gaseous phases. Up to 36 species are included in the calculations. In most cases the results are presented as CVD phase fields as a function of temperature and input gas concentration ratios.

The necessary thermodynamic values for most of the species were taken from the JANAF tables.[12]

RESULTS

The SiC System

In this paper, only deposition from the $SiCl_4/CCl_4/H_2$ system is discussed. The thermodynamics of other systems of commercial importance for the deposition of SiC have been discussed previously.[13]

One of the problems in the deposition of SiC is the occurrence of mixed phases, typically SiC + C or SiC + Si.[13] The CVD phase diagrams calculated in the present work indicate that the range of equilibrium conditions resulting in SiC as the only condensed phase, are relatively small, as discussed below.

Reaction of $SiCl_4$ and CCl_4 without the use of a carrier gas such as H_2 results in the deposition of C(s) (graphite) alone over a wide range of temperatures, system pressures, and $SiCl_4/CCl_4$ input gas ratios.

CVD phase diagrams for the $SiCl_4/CCl_4$ reaction in the presence of H_2 are shown in Figs. 1-3 for a wide range of system pressures and H_2 concentrations. In these diagrams, phase fields of C(s); β-SiC; β-SiC + C(s); Si(s) + β-SiC; Si(ℓ) + β-SiC; Si(s); and Si(ℓ) are observed. Under certain conditions, especially low temperatures, no condensed phases are deposited. Referring to the conditions for the deposition of single phase β-SiC in this system, there are several implications for CVD processors. If one desires to work at low system pressures (LPCVD) to minimize contamination, one should avoid high H_2 carrier gas concentrations, where the conditions for single phase β-SiC deposition are limited (compare Figs. 1(d) and 3(d)). Rather, one should work at lower carrier gas concentrations, and at $Si/(Si + C) \geqslant 0.6$ in the input gas. At system pressures around 10^5 Pa, conditions are less limited, with high H_2 carrier gas concentrations allowing a wide choice of temperatures and Si/Si + C parameters (Fig. 3).

By studying the calculated gas phase compositions, it is possible to deduce that the following are the principal controlling equilibria at lower temperatures:

$$SiCl_4 + CCl_4 + 4H_2 \quad \beta\text{-}SiC + 8HCl \qquad \text{equ. (1)}$$
$$CCl_4 + 2H_2 \quad C(s) + 4HCl \qquad \text{equ. (2)}$$
$$SiCl_4 + 2H_2 \quad Si(s) + 4HCl \qquad \text{equ. (3)}$$

In the $SiCl_4/CCl_4/H_2$ system, the CVD phase diagrams show that it is possible to establish conditions such that any one of the equilibria is predominantly to the right, while the remaining two are completely to the left. At low H_2 gas concentrations, competition for the reactant gases is primarily between equilibria 1 and 2 (Fig. 1), while at higher H_2 concentrations it is also necessary to consider equilibrium (3) (see Figs. 2 and 3) as condensed Si phases also occur.

The Si_3N_4 System

The CVD technique is used to process Si_3N_4 for structural applications[14] and electronic devices.[15] The literature on the CVD of Si_3N_4 has been discussed previously.[16] The co-deposition of Si with Si_3N_4 is recognized as a problem area.

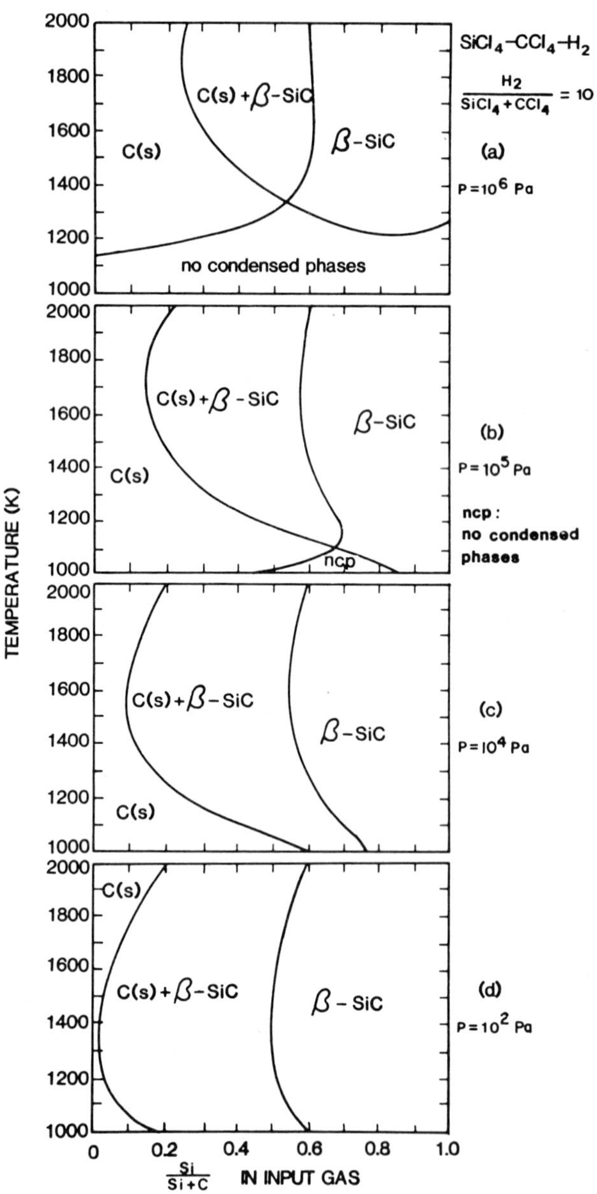

Fig. 1. The $SiCl_4/CCl_4$ system with $H_2/(SiCl_4+CCl_4)$ = 10.

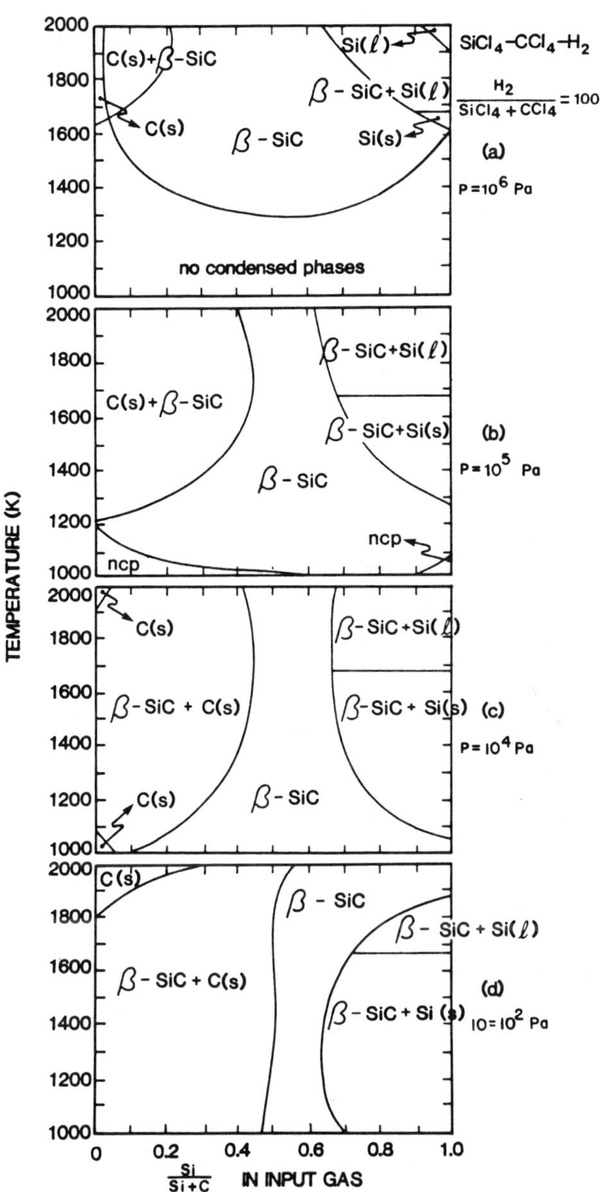

Fig. 2. The $SiCl_4/CCl_4$ system with $H_2/(SiCl_4+CCl_4) = 100$.

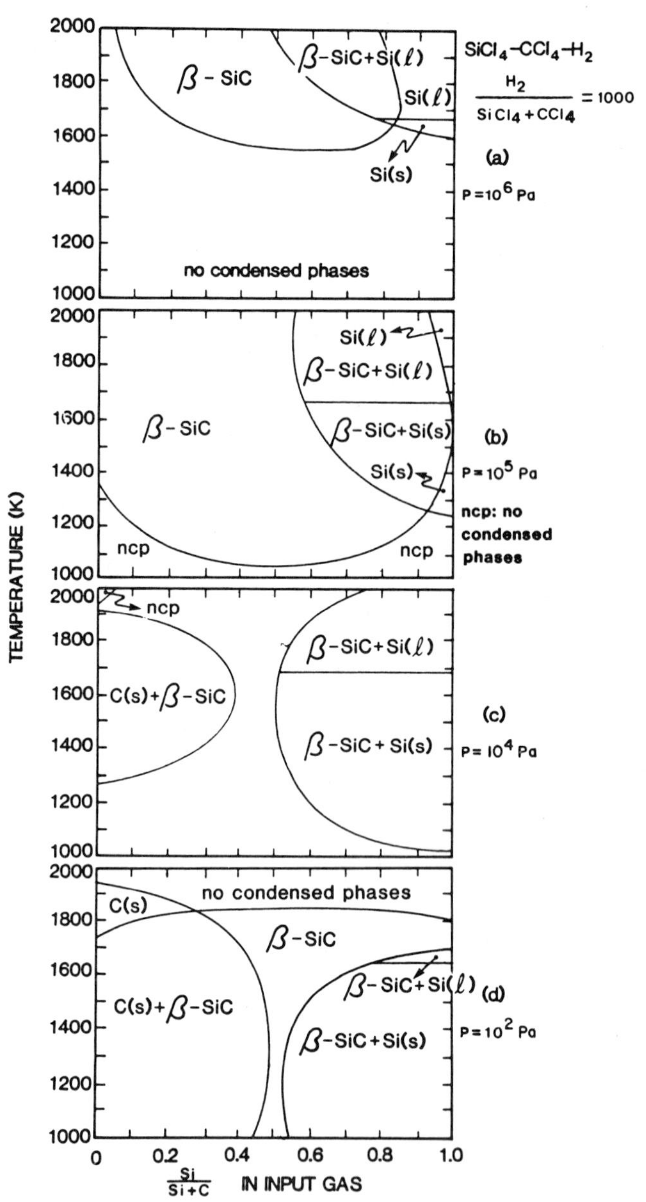

Fig. 3. The $SiCl_4/CCl_4$ system with $H_2/(SiCl_4+CCl_4)$ = 1000.

CVD phase diagrams for the NH_3/SiH_4 system without H_2 carrier gas are shown in Fig. 4. It can be seen that single phase Si_3N_4 is predicted to deposit under equilibrium conditions for $Si/(Si + N) \leqslant 0.43$ in the input gas. As the total system pressure is decreased, it can be seen that the maximum temperature at which Si_3N_4 can be deposited is lowered. This is a result of the high stability of N_2 in the gas phase at high temperatures and reduced pressures.[16]

The addition of H_2 to the NH_3/SiH_4 system does not markedly change the CVD phase diagrams (in contrast to the effect of the H_2 additions to the $SiCl_4/CCl_4$ system). Rather than substantially affecting the controlling equilibria, the effect is similar to lowering the system pressure. This is due to the lowered partial pressures of SiH_4 and NH_3 in the input gas.

The numerals shown on the CVD phase diagrams in Fig. 4 represent the "theoretical deposition efficiencies." This has been defined[10] as the amount of condensed product phase which can be deposited under equilibrium conditions expressed as a percentage of the amount which could be deposited if no thermodynamic or kinetic limitations existed. It therefore represents a "maximum" or "limiting" deposition.

Condensed phases in the $SiCl_4/NH_3$ system are shown in Fig. 5, at a system pressure of 10^6 Pa. The composition of the gas phase, at a $Si/(Si + N)$ ratio in the input gas of 0.5, is shown as a function of temperature in Fig. 6. This figure shows that at low temperatures the major equilibrium controlling the system is:

$$3SiCl_4 + 2N_2 + 6H_2 \rightarrow Si_3N_4(c) + 12HCl.$$

Figure 5 indicates that the deposition efficiency is greatest at low $SiCl_4$ concentrations in the input gas.

The SiF_4/NH_3 system is shown in Fig. 7. Of note are the low theoretical deposition efficiencies, as well as the implication that one must utilize low SiF_4 input gas concentrations if one wishes to deposit at relatively high temperatures.

APPLICATIONS AND LIMITATIONS

One limitation to the validity of the calculated equilibrium CVD phase diagrams results from the accuracy of the thermodynamic data used. Bernard,[18] among others, has pointed out the influence of uncertainties in the data on the results.

Secondly, the validity of the assumption that equilibrium is achieved, and that it is the equilibrium conditions which are controlling the process should also be of concern. This has been discussed recently by Spear.[18]

In modelling the CVD process as shown in Fig. 8, Spear has listed the following important steps:[19]

1. Forced flow of reactant gases into the system.
2. Diffusion and Stefan flow of reactant gases through the gaseous boundary layer to the substrate.
3. Adsorption of gases onto the substrate.
4. Chemical reactions of the adsorbed species, or of adsorbed and gaseous species.
5. Desorption of adsorbed species from the substrate.
6. Diffusion and Stefan flow of product gases through the boundary layer to the bulk gas.
7. Forced exit of gases from the system.

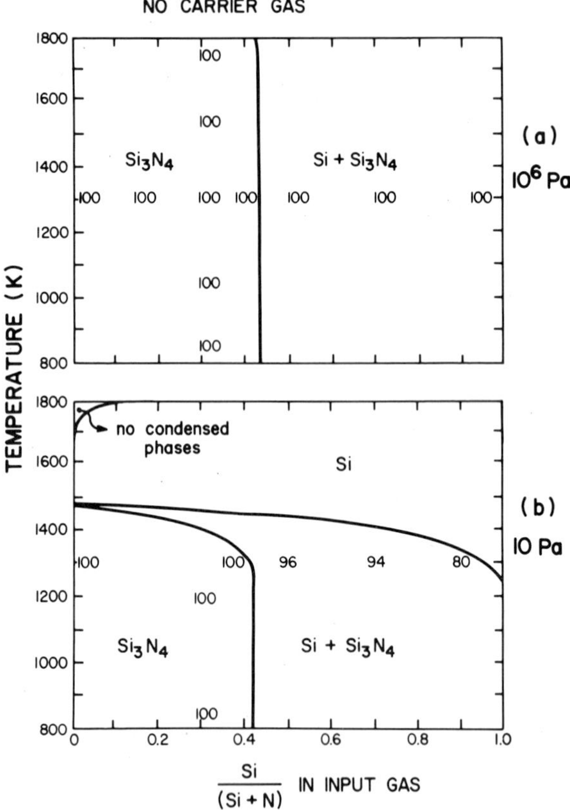

Fig. 4. Condensed phases in the SiH_4/NH_3 system showing the effect of pressure.

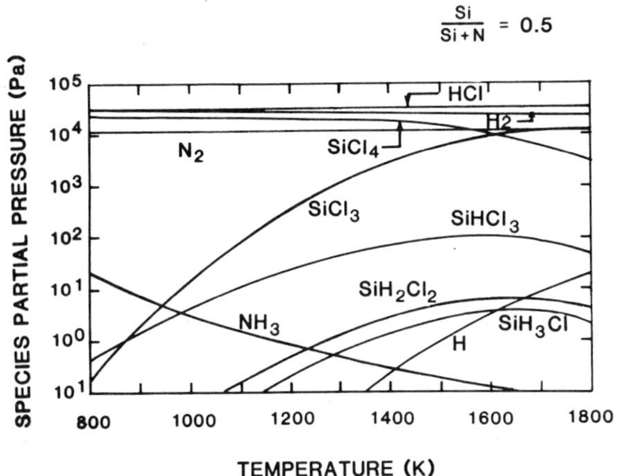

Fig. 5. Condensed phases in the $SiCl_4/NH_3$ system.

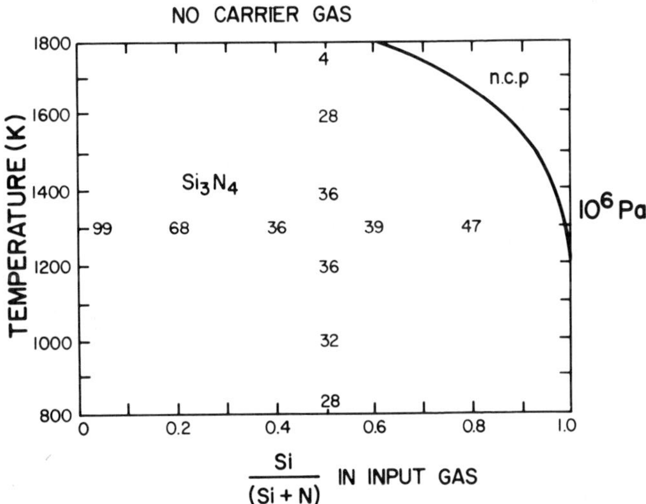

Fig. 6. Composition of the gas phase in the $SiCl_4/NH_3$ system as a function of the temperature.

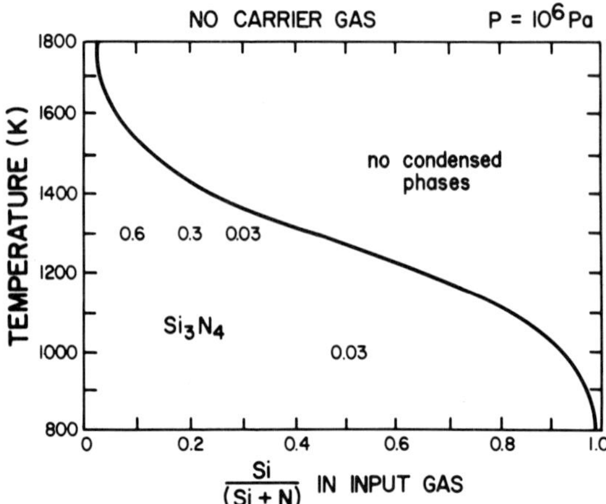

Fig. 7. Condensed phases in the SiF_4/NH_3 system.

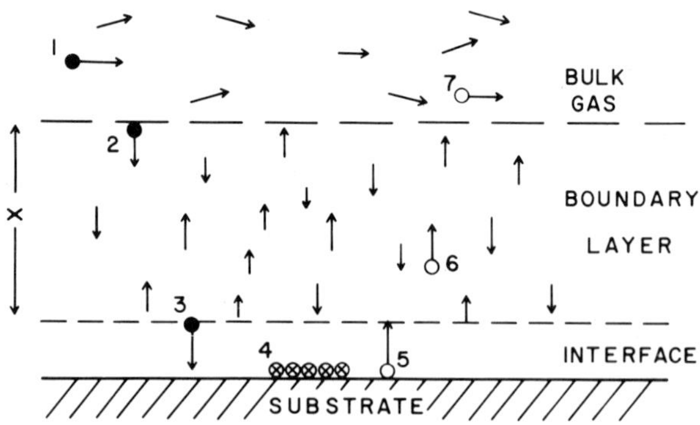

Fig. 8. Schematic showing mechanistic steps occurring during the CVD process (after Spear[18]).

In the case where the rate limiting step in the process is the input gas flow rate, then the thermodynamic properties will be controlling the CVD process. The thermodynamic calculations described above will therefore be valid.

It should also be noted, however, that in the other two cases described by Spear,[18] where the residence time of the reactant gases in the reaction zone of the system is short in comparison with gaseous diffusion and chemical reaction rates, then equilibrium gas phase concentrations are necessary in order to model the CVD process.[17,18]

REFERENCES

1. R. S. Rosler, Solid State Technol., 20 [4], 63-70 (1977).
2. E. Fitzer and D. Kehr, Thin Solid Films, 39, 55-67 (1967).
3. E. Fitzer, D. Hegen and H. Strohmeier, pp. 525-535 in Proceedings of the Seventh International Conference on Chemical Vapor Deposition, 1979, edited by T. O. Sedgwick and H. Lydtin, The Electrochemical Society, Inc., Princeton, NJ (1979).
4. A. C. Airey, S. Clarke and P. Popper, Proc. Br. Ceram. Soc., 22, 305-320 (1973).
5. W. J. McDonough, C. Cm. Wui and P. E. D. Morgan, Commun. Am. Ceram. Soc., 64 [3], C45-C47 (1981).
6. H. O. Pierson, Thin Solid Films, 40, 41-47 (1977).
7. R. W. Kidd, M. F. Browning and J. M. Rusin, pp. 563-577 in Proceedings of the Seventh International Conference on Chemical Vapor Deposition, 1979, edited by T. O. Sedgwick and H. Lydtin, The Electrochemical Society, Princeton, NJ (1979).
8. S. Dutta, R. W. Rice, H. C. Graham and M. C. Mendiratta, J. Mater. Sci., 15, 2183-2191 (1980).
9. G. Eriksson, Chemica Scripta, 8 [3], 100-103 (1975).
10. K. E. Spear, pp. 1-16 in Reference 7.
11. W. B. White, W. M. Johnson and G. B. Dantzig, J. Chem. Phys., 28 [5], 751-755 (1958).
12. JANAF Thermochemical Tables (a) Second Edition, NSRDS-NBS 37, National Bureau of Standards, USA (1971), (b) 1974 Supplement, J. Phys. Chem. Ref. Data, 3, 311-480 (1974), (c) 1975 Supplement, J. Phys. Chem. Ref. Data, 4 [1], 1-175 (1975).
13. A. I. Kingon, L. J. Lutz, P. Liaw and R. F. Davis, "Thermodynamic Calculations for the Chemical Vapor Deposition of Silicon Carbide," accepted for publication by J. Am. Ceram. Soc.
14. R. A. Tanzilli, J. J. Gebhardt and J. D'Andrea, Processing Research on Chemically Vapour Deposited Silicon Nitride; Final Technical Report, General Electric Co., Philadelphia, PA, Dec. 1981. Government Accession No. AD-A109 204/8.
15. C. E. Morosanu, Thin Solid Films, 65 [2], 171-208 (1980).
16. A. I. Kingon, L. J. Lutz and R. F. Davis, "Thermodynamic Calculations for the Chemical Vapour Deposition of Silicon Nitride," submitted to J. Am. Ceram. Soc.

17. C. Bernard, pp. 3-16 in Proceedings of the Eighth International Conference on Chemical Vapor Deposition, 1981, edited by J. M. Blocher, Jr., G. E. Vuillard and G. Wahl, The Electrochemical Society, Inc., Pennington, NJ (1981).
18. K. E. Spear, Pure and Appl. Chem., 54 [7], 1297-1311 (1982).

CVD OF Si_3N_4 AND ITS COMPOSITES

Toshio Hirai

The Research Institute for Iron, Steel and Other Metals
Tohoku University
Katahira, Sendai 980, Japan

INTRODUCTION

It has been widely recognized that the development of new materials is indispensable to the establishment of original science and technology. There are two approaches to the development of new materials. One is to develop quite new materials which have never existed, and the other is to add new functions to existing materials. The research and development of composites include both approaches simultaneously.

Generally ceramics/ceramics composites are prepared by mixing, then sintering raw powders. However, non-oxide ceramics/ceramics composites which are composed of semi-metal elements and non-metal elements, such as SiC, Si_3N_4, B_4C and BN, are difficult to sinter. In comparison with the sintering method, the CVD method provides the following characteristics: (1) possible to synthesize a non-oxide ceramics/ceramics composite at a temperature lower than the sintering temperatures of composites; (2) possible to produce the composites with high density without including sintering additives; and (3) possible to obtain the composites whose fillers are dispersed in ultra fine size.

In the author's laboratory, research for developing non-oxide ceramics/ceramics composites by the CVD method is being conducted. In the present report, the author reviews the synthesis, structure and some properties of the Si_3N_4 base ceramics/ceramics composites prepared by the CVD method.

GENERAL PREPARATION METHOD FOR COMPOSITES

The preparation methods for composites vary greatly, but are classified into the phase-joining method and the phase-separating method. The former is a method of combining separately prepared material A and material B into a composite:

$$A + B \rightarrow \text{Composite } (A + B)$$

The latter is a method of preparing composites from a homogeneous material (X) containing both A and B by proper treatment:

$$X_{(A+B)} \rightarrow \text{Composite } (A + B)$$

There are two CVD techniques for synthesizing ceramics/ceramics composites: Continuous CVD technique for which the CVD parameters (deposition temperature, gas pressure, gas flow rate, gas mixing ratio, etc.) are kept at a steady state, and intermittent CVD technique for which one of the CVD parameters is varied intermittently. Table 1 shows the types of ceramics/ceramics composites synthesized by these CVD methods. Various research has been made on the synthesis of the phase-joining composites by the CVD methods. There are innumerable examples: TiC, TiN or Al_2O_3 coating to cemented carbides as tools, carbide coating to graphites for improving the oxidation resistance, and so on. However, only a little investigation has been done on the synthesis of the ceramics/ceramics composites by the phase-separating CVD methods.

CVD OF THE Si_3N_4 BASE CERAMICS/CERAMICS COMPOSITES

The Si_3N_4 base ceramics/ceramics composites are listed in Table 2 which are prepared by the phase-separating methods. Most of them deal with their films, therefore their structures and bulk properties have not been well investigated. On the other hand, the Si_3N_4(C), Si_3N_4(TiN) and Si_3N_4(BN) composites prepared in the author's laboratory are in the thickness of mm order. In the following sections the preparation method, structure and some properties of CVD-Si_3N_4 and its composites will be described.

Preparation

Various raw gases are considered from preparing Si_3N_4 by the CVD method, but $SiCl_4$ and NH_3 have little been used as the raw gases. The reason is that two kinds of gases react with each other as soon as they make contact with each other to produce NH_4Cl and complicated amino-compounds. Nevertheless, since it is necessary for obtaining deposits at a high deposition speed to use raw gases which are high in reactivity, we have been using $SiCl_4$ and NH_3.

Table 1. CVD Methods for Each Composite

Type of Composite	Preparation Method	CVD Technique	Example
Spherical particle dispersed type	Phase-separating	Continuous	PyC(TiC)[2], PyC (BeO)[3], Am-Si_3N_4 (C)[16], Am-Si_3N_4 (AlN)[4], Am-Si_3N_4 (TiN)[5,6], α-Si_3N_4 (TiN)[5,6]
Flake-like particle dispersed type	"	"	PyC(SiC)[7]
Thin layer dispersed type	"	"	Ti_3SiC_2(TiC)[8]
Fiber dispersed type	"	"	β-Si_3N_4(TiN)[5,6]
Laminated type (different materials)	Phase-joining	"	Coatings
"	"	Intermittent	Multi-coatings
Laminated type (same materials with different structures)	"	"	Porous PyC/ dense PyC[9]
"	Phase-separating	"	rutile/anatase[10], Am-Si_3N_4/Cry-Si_3N_4[11], α-Si_3N_4/ β-Si_3N_4

Table 2. CVD of Si_3N_4(X) Composite by the Phase-separating Method

X	Raw Gas	Deposition Temperature (°C)	Researcher
Si	$SiCl_4$-N_2-H_2	1100~1300	Nickl and Braunmuhl[12]
Si	SiH_4-NH_3	700	Dong et al.[13]
Si+SiC	$SiCl_4$-CCl_4-H_2-N_2	1100~1300	Nickl and Braunmuhl[12]
Ge	SiH_4-Ar-GeH_4-N2-NH_3	650~850	Tamaki et al.[14]
AlN	SiH_4-$AlCl_3$-NH_3-H_2	600~1100	Zirinsky and Irene[5]
AlN	$SiCl_4$-NH_3-H_2-$AlCl_3$-O_2		Landingham and Taylor[15]
C	$SiCl_4$-NH_3-H_2-C_3H_8	1100~1300	Hirai and Goto[16]
TiN	$SiCl_4$-Nh_3-H_2-$TiCl_4$	1050~1450	Hirai and Hayashi[11,17]
BN	$SiCl_4$-NH_3-H_2-B_2H_6	1100~1300	Hirai et al.[18]

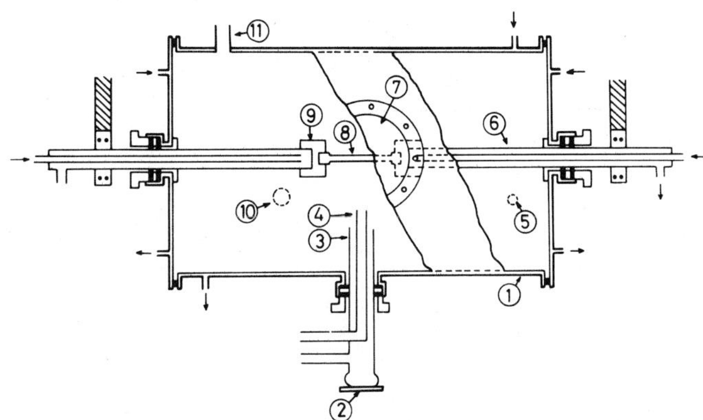

Fig. 1. CVD-furnace for synthesizing Si_3N_4 and its composites. (1) water-cooled vacuum chamber; (2), (7) quartz glass windows; (3) gas inlet for $SiCl_4+H_2$, $SiCl_4+H_2+C_3H_8$, $SiCl_4+H_2+TiCl_4$ or $SiCl_4+H_2+B_2H_6$; (4) gas inlet for NH_3; (5) pressure gauge; (6) water-cooled copper electrode; (8) graphite heater (substrate); (9) graphite socket; (10), (11) gas outlets.

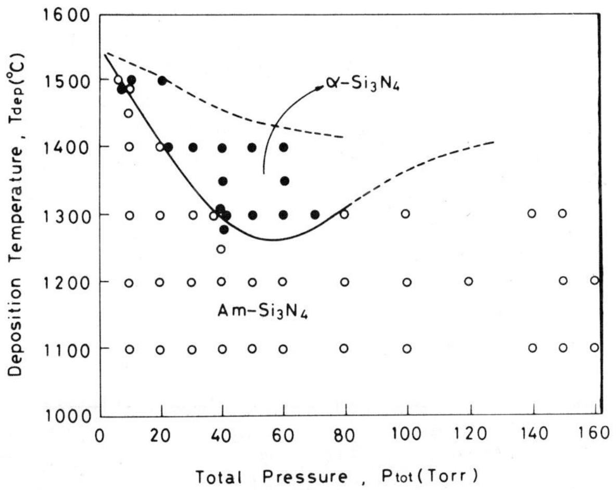

Fig. 2. Effect of CVD conditions on the structure of CVD-Si_3N_4.

Table 3. CVD Conditions of Si_3N_4 and its Composites

	Si_3N_4	$Si_3N_4(C)$	$Si_3N_4(TiN)$	$Si_3N_4(BN)$
Raw Gas	$SiCl_4$-NH_3 -H_2	$SiCl_4$-NH_3 -H_2-C_3H_8	$SiCl_4$-NH_3 -H_2-$TiCl_4$	$SiCl_4$-NH_3 -H_2-B_2H_6
Gas Flow Rate (FR)				
$SiCl_4$	100~260	170	136	170
NH_3	60	60	120	60
H_2	400~1000	700	2720	700
X		25~1000	18	1~12
Deposition Temperature T_{dep}(°C)	1100~1500	1100~1300	1050~1450	1100~1300
Total Gas Pressure P_{tot}(Torr)	5~300	30~70	10~80	30~70
Reference	19, 20	16	11, 17	18

Figure 1 shows a CVD-furnace with which we prepared CVD-Si_3N_4 and its composites. The main body of the CVD-furnace is a water-cooled vacuum chamber; the diameter of the chamber is about 30 cm for the smaller furnace and about 50 cm for the larger. The graphite substrate on which deposit is formed is directly heated by flow of electric current. Then the raw gases are introduced into the furnace through a double nozzle tube. The reason why we succeeded in the high-speed synthesis is that we used the nozzle tubes as parts 3 and 4 in Fig. 1.[19] CVD conditions are included in Table 3.

For CVD-Si_3N_4, as shown in Fig. 2, two types of CVD-Si_3N_4 plates, the amorphous and crystalline (α-type) deposits, were obtained, mainly depending on T_{dep} and P_{tot}.[19]

Figures 3 and 4 show the results of Si-N-C deposits obtained under various CVD conditions. From the $SiCl_4$-NH_3-H_2-C_3H_8 system, Am-Si_3N_4, α-Si_3N_4, carbon and β-SiC were locally co-deposited with the Am-Si_3N_4(C) composites on the substrate. It is necessary to find the optimum condition in which only the Am-Si_3N_4(C) composites are obtained. Uniform plate-like deposits, the Am-Si_3N_4(C) composites, were obtained at T_{dep} of 1100° to 1300°C, as shown in Figs. 3 and 4, but were not obtained at T_{dep} of 1400° to 1600°C.[16] The carbon content of the Am-Si_3N_4(C) composites increased with an increase in the flow rate of C_3H_8. It was possible to add 10 wt% of carbon to the Am-Si_3N_4 matrix, but too much carbon caused many visible micro-cracks in the composite to be obtained, so the maximum carbon content was found to be about 5 wt%.

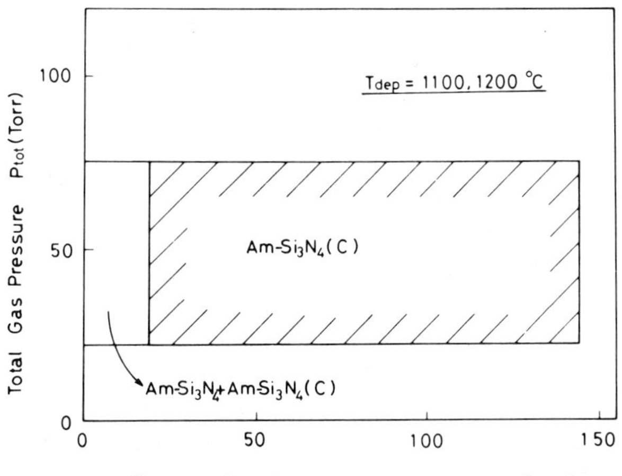

Fig. 3. CVD conditions for synthesizing the Am-Si_3N_4(C) composites at T_{dep} of 1100° and 1200°C.

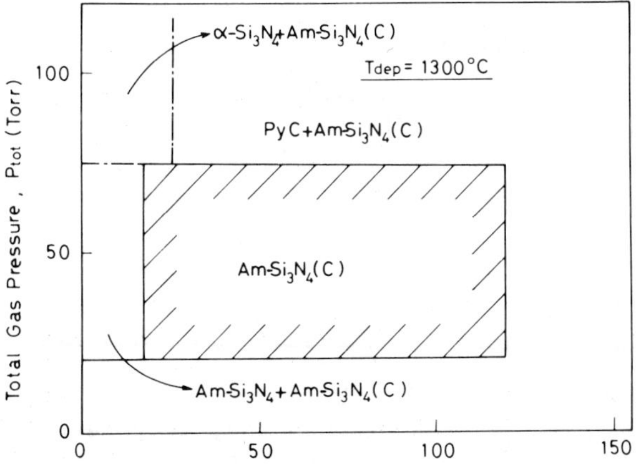

Fig. 4. CVD conditions for synthesizing the Am-Si_3N_4(C) composites at T_{dep} of 1300°C.

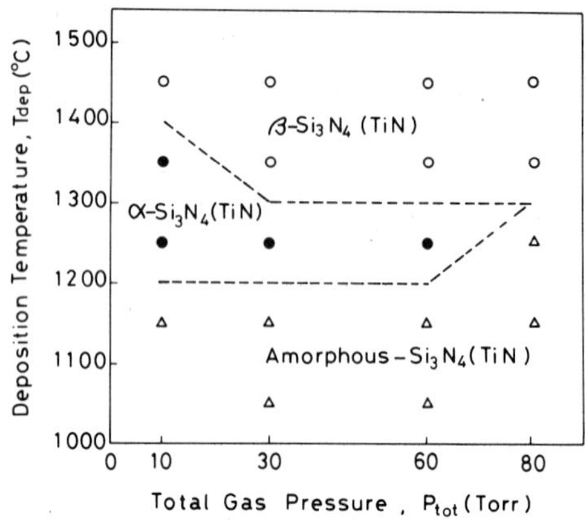

Fig. 5. Effect of CVD conditions on the structure of the Si_3N_4 matrix for the Si_3N_4(TiN) composites prepared at FR($TiCl_4$) of 18 $cm^3 \cdot min^{-1}$.

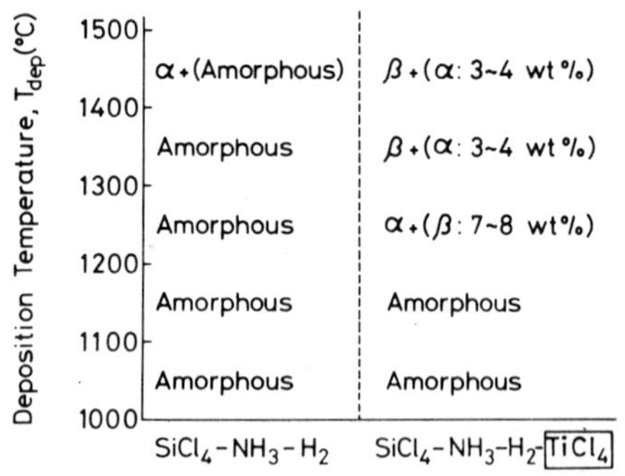

Fig. 6. Effect of the $TiCl_4$ addition on the structure of the Si_3N_4 matrix for the Si_3N_4(TiN) composites prepared at P_{tot} of 30 Torr.

Figure 5 shows the structures of the Si_3N_4 matrix in the Si_3N_4 (TiN) composites prepared at FR($TiCl_4$) of 18 $cm^3 \cdot min^{-1}$ and various T_{dep} and P_{tot}.[11] The Am, α- and β-Si_3N_4 matrices were formed depending on the CVD conditions. The former CVD experiments on Si_3N_4 produced only an α-type or amorphous one, in some cases a small amount of β-type was included in the α-type. In our laboratory, however, β-Si_3N_4 was obtained by the CVD process. An example of the effect of the $TiCl_4$ addition on the structure of Si_3N_4 is clearly demonstrated in Fig. 6. The addition of $TiCl_4$ to the $SiCl_4$-NH_3-H_2 system had the effect of lowering the deposition temperature of

α-Si_3N_4 and promoting the formation of β-Si_3N_4. The TiN content of the Si_3N_4(TiN) composites decreased with an increase in T_{dep}.[22]

The Am-Si_3N_4(BN) composites were obtained in any CVD conditions shown in Table 3. The color of CVD·Am-Si_3N_4 was white, and CVD·α-Si_3N_4 and the Am-Si_3N_4(C) and Si_3N_4(TiN) composites were black. On the other hand, the color of the Am-Si_3N_4(BN) composites varied with an increase in the flow rate of B_2H_6 from white to black via yellow and brown. The variation of the color of the Am-Si_3N_4(BN) composites with the CVD conditions are shown in Fig. 7.[18] The B content of the Am-Si_3N_4(BN) composites increased with an increase in the flow rate of B_2H_6.

Some data on CVD-Si_3N_4 and its composites are included in Table 4.

Table 4. Some Data on CVD-Si_3N_4 and its Composites

	Si_3N_4	Si_3N_4(C)	Si_3N_4(TiN)	Si_3N_4(BN)
Max. Deposition Speed ($mm \cdot h^{-1}$)	1.2	0.6	0.47	0.52
Content of X (wt%)		C < 10	TiN < 32	BN < 83
Density ($g \cdot cm^{-3}$)	2.60~3.18	2.75~3.00	3.02~3.37	1.51~2.96
Structure of the Si_3N_4	Am, α	Am	Am, α, β	Am
Reference	19, 20, 23	16, 21	11, 22	18

Structure

The structure of CVD·Am-Si_3N_4 was investigated by Suzuki et al. The electronic structure of CVD·Am-Si_3N_4 was found to be β-type by the Compton scattering measurement.[24] Also the neutron scattering measurement revealed that the major defects existing in CVD·Am-Si_3N_4 were voids with an average diameter of about 10 Å and a volume fraction of about 4% and that there were no N-N and Si-Si bonds.[25] The X-ray small angle scattering measurement indicated that the short range structure of CVD·Am-Si_3N_4 was similar to that of β-type and that clusters in sizes of 8~10 Å were included in CVD·Am-Si_3N_4 .[26] However, recent investigations using recent data on structural parameters of Si_3N_4 indicate that the basic structure of CVD·Am-Si_3N_4 may be close to that of α-Si_3N_4.

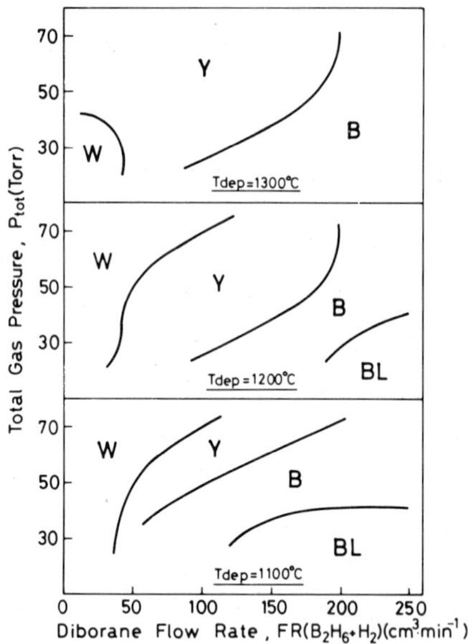

Fig. 7. Variation of color of the Am-Si_3N_4(BN) composites with CVD conditions. (W) white; (Y) yellow; (B) brown; (BL) black.

The preferred orientation of the crystal is the structural characteristics of the CVD crystalline materials. As a result, anisotropy appears in various properties. This preferred orientation was also observed in CVD·α-Si_3N_4, which depended on T_{dep} and P_{tot}.[27] The (110, (210), or (222) planes can be arranged in parallel on the substrate surface by selecting the CVD conditions. The lattice parameter "a" was independent of P_{tot}, approximately 7.75 Å, while the parameter "c" increased with an increase in P_{tot} and with a decrease in oxygen content, 5.618~5.627 Å.[27]

The Am-Si_3N_4(C) composites were found to be composed of the Am-Si_3N_4 matrix and the free carbon particles in the size of about 1000 Å. No bondings between Si and C were confirmed. These were examined by ESCA, positron annihilation,[28] neutron diffraction, heat-treatment, and HF-treatment methods. Figure 8 shows the surface of the Am-Si_3N_4(C) composite (including 6 wt% of carbon) heat-treated at 1400°C and for 3 hr. in vacuum. By this heat-treatment, dispersed carbon particles reacted with the Am-Si_3N_4 matrix to form SiC. Figure 8 indicates that the carbon particles are considered to contact with each other in the Am-Si_3N_4 matrix. The Am-Si_3N_4(c) composite is a typical example of the particle dispersed type.

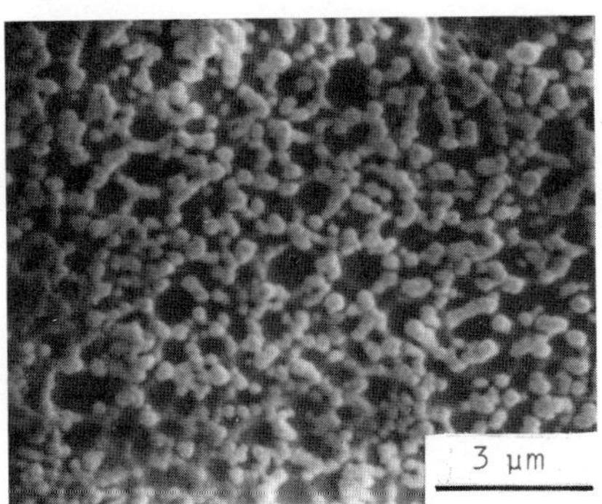

Fig. 8. The surface of the Am-Si_3N_4(C) composite heat-treated at 1400°C and for 3 hr. in vacuum.

The state of TiN in the Si_3N_4(TiN) composites was mainly examined by a high resolution electron microscope. Spherical particles of TiN dispersed in the Am- and α-Si_3N_4 matrices and the size of the dispersions were found to be about 30 and 100 Å, respectively. Figures 9(a) and (b) show electron micrographs of the β-Si_3N_4(TiN) composite prepared at T_{dep} of 1350°C and P_{tot} of 30 Torr. Figures 9(a) and (b) were taken with the incident electron beam parallel and perpendicular, respectively, to the substrate surface. Figure 9(b) reveals clearly that these TiN dispersions were fibers extending their axis in parallel to the c-axis of the β-Si_3N_4 crystal. The diameter of the TiN fiber was several tens Å. The crystal lattice relationship determined by a 1 MV electron microscopy is exhibited in Fig. 10, that is, $(110)_{TiN}//(001)_{\beta\text{-}Si3N4}$. The detailed relationships and structure defects in the β-Si_3N_4 matrix will be reported elsewhere.[29,30] The α- and β-Si_3N_4(TiN) composites also had preferred orientations which mainly depended on T_{dep} and P_{tot}.[11] Lattice parameters a_o and c_o of the β-Si_3N_4 matrix were 7.608 and 2.909 Å, respectively.[17] The Am- and α-Si_3N_4(TiN) composites are the particle dispersed type, while the β-Si_3N_4(TiN) composite is the fiber dispersed type.

The structure of the Si_3N_4(BN) composites are not well understood at the present time. The relationship between density and boron content of the Si_3N_4(BN) composites indicates that the Si_3N_4(BN) composite is the mixture of Am-Si_3N_4 and BN.[18]

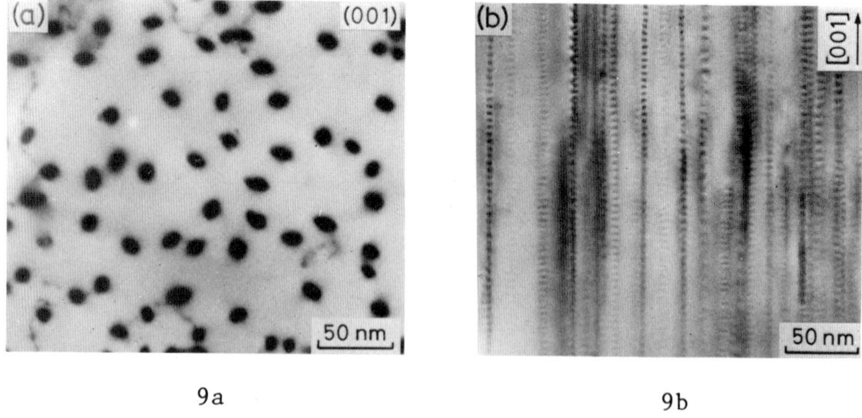

9a 9b

Fig. 9. Electron micrographs of the β-Si_3N_4(TiN) composite prepared at T_{dep} of 1350°C and P_{tot} of 30 Torr.

Fig. 10. The crystal lattice relationship between the TiN fiber and the β-Si_3N_4 matrix.

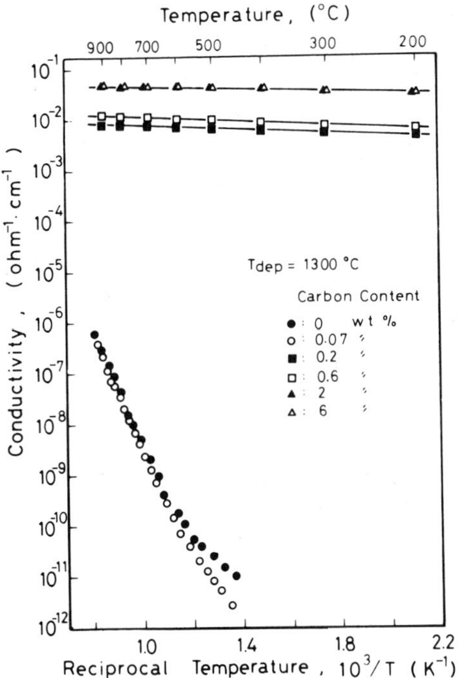

Fig. 11. Temperature dependence of the d.c. electrical conductivity of the Am-Si_3N_4(C) composites prepared at T_{dep} of 1300°C.

Properties

Various properties have been measured; however, only the most remarkable properties of CVD-Si_3N_4 and its composites are described here.

Generally, the hardness of ceramics decreases with an increase in temperature.[31] However, CVD-Si_3N_4 does not show excessive deterioration of hardness at high temperatures in comparison to other ceramics, and it still maintains high hot-hardness even at 1400°C. The oxidation behavior of CVD-Si_3N_4 was investigated in dry oxygen at 1550° to 1650°C.[32] It was concluded that CVD-Si_3N_4 had extremely good oxidation resistance.

The most remarkable characteristic of the Am-Si_3N_4(C) composite was found in its d.c. conductivity.[33] Figure 11 shows temperature dependence of d.c. conductivity in the direction perpendicular to the substrate surface for the Am-Si_3N_4(C) composites prepared at T_{dep} of 1300°C. CVD-Si_3N_4 is an insulator, however, for the

Am-Si_3N_4(C) composite, electrical conductivity increased abruptly at about 0.2 wt% of carbon. The slope which represents the temperature coefficient of the electrical conductivity became small. The activation energy for electrical conduction for the Am-Si_3N_4(C) composites was 0.02 to 0.06 eV. The electrical conduction may be assumed to take place through the carbon, as shown in Fig. 8.

The thermal conductivity in the direction perpendicular to the substrate surface for the α-Si_3N_4(TiN) composite was measured at 20°C by a laser flash method. As shown by (a) in Fig. 12, the experimental thermal conductivity decreased drastically with an increase in TiN volume fraction. On the other hand, a theoretical equation on the thermal conductivity of a composite gives the curve (b) which is markedly different from the curve (a). As previously described, the TiN particles dispersed in the composite were as small as about 100 Å and were included in α-Si_3N_4 grains. The area of the matrix-dispersed particle boundary surface, which acts as a source of thermal resistance, increases with a decrease in the size of the dispersed particle. Moreover, particles dispersed in a grain distort the crystal lattice of α-Si_3N_4 and generate dislocations and sub-boundaries. These imperfections act as another source of thermal resistance. Thus the experimental thermal conductivity seems to be lower than the theoretical thermal conductivity.

The properties of the Am-Si_3N_4(BN) composite have not been fully examined yet. The most notable characteristic is that the Am-Si_3N_4(BN) composites obtained in the region shown by B in Fig. 7 are transparent.[18]

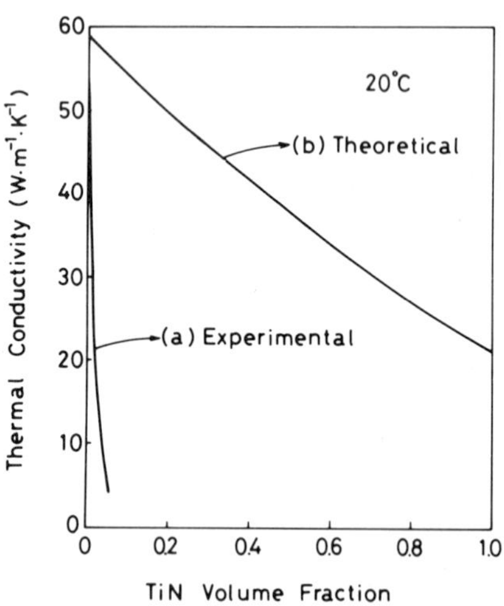

Fig. 12. Thermal conductivity of the α-Si_3N_4(TiN) composites.

CONCLUSION

For developing original materials, it is necessary to design original processes and devices. In the field of CVD, the research on the synthesis of composites of ternary or multi-systems has been initiated, and many interesting results have been obtained and more are forthcoming. The phase-separating CVD method, together with the phase-joining CVD method, will be useful for the future development of new materials.

ACKNOWLEDGMENTS

The author wishes to express his appreciation to Drs. K. Niihara and S. Hayashi and Messrs. T. Goto and A. Ohkubo for their contributions to the preparation and characterization of the CVD materials.

REFERENCES

1. T. Hirai and S. Hayashi, Bull. Japan Inst. Met., 16, 205 (1977).
2. A. S. Schwartz and J. C. Bokros, Carbon, 5, 325 (1967).
3. J. J. Gebhardt, Decompos. Organometal. Compound Refract. Ceram., Metals, Metal Alloys, Proc. Inst. Symp., 1967 (Pub. 1968), p. 319.
4. S. Zirinsky and E. A. Irene, J. Electrochem. Soc., 125, 305 (1978).
5. T. Hirai and S. Hayashi, 8th Inter. Conf. CVD 1981, edited by J. M. Blocher, Jr., G. E. Vuillard and G. Wahl, Electrochem. Soc., Inc., Pennington (1981), p. 790.
6. S. Hayashi, T. Hirai, K. Hiraga and M. Hirabayashi, J. Mater. Sci., in press.
7. S. Yajima and T. Hirai, J. Mater. Sci., 4, 424 (1969).
8. J. J. Nickl, K. K. Schweitzer and P. Luxenberg, J. Less-Common Met., 26, 335 (1972).
9. J. M. Blocher, Jr., M. F. Browning, W. J. Wilson, V. M. Secrest, A. C. Secrest, R. B. Landrigan and J. H. Oxley, Nucl. Sci. Eng., 10, 153 (1964).
10. S. Hayashi and T. Hirai, J. Crystal Growth, 41, 41 (1977).
11. T. Hirai and S. Hayashi, J. Mater. Sci., 17, 1320 (1982).
12. J. J. Nickl and C. von Braunmuhl, J. Less-Common Met., 37, 317 (1974).
13. D. Dong, E. A. Irene and D. R. Young, J. Electrochem. Soc., 125, 819 (1978).
14. Y. Tamaki, S. Isomae, A. Shintani and M. Maki, J. Electrochem. Soc., 126, 2271 (1979).

15. R. L. Landingham and R. W. Taylor, Energy and Ceramics, Materials Science Monographs 6, edited by P. Vincenzini, Elsevier Scientific Publishing Co., New York (1980), p. 494.
16. T. Hirai and T. Goto, J. Mater. Sci., 16, 17 (1981).
17. T. Hirai and S. Hayashi, J. Amer. Ceram. Soc., 64, c-88 (1981).
18. T. Hirai, T. Goto and T. Sakai, in this issue.
19. K. Niihara and T. Hirai, J. Mater. Sci., 11, 593 (1976).
20. T. Hirai, K. Niihara and T. Goto, J. Japan Inst. Met., 41, 358 (1977) and J. Mater. Sci., 12, 631 (1977).
21. T. Hirai and T. Goto, J. Mater. Sci., 16, 2877 (1981).
22. T. Hirai and S. Hayashi, J. Mater. Sci., in press.
23. K. Niihara and T. Hirai, J. Mater. Sci., 11, 604 (1976).
24. F. Itoh, T. Honda, K. Niihara, T. Hirai and K. Suzuki, J. Phys. Soc. Japan, 48, 561 (1980).
25. M. Misawa, T. Fukunaga, K. Niihara, T. Hirai and K. Suzuki, J. Non-Crystalline Solids, 34, 313 (1979).
26. T. Aiyama, T. Fukunaga, K. Niihara, T. Hirai and K. Suzuki, J. Non-Crystalline Solids, 33, 131 (1979).
27. K. Niihara and T. Hirai, J. Mater. Sci., 12, 1233 (1977).
28. F. Itoh, T. Honda, T. Goto, T. Hirai and K. Suzuki, Proc. 8th Inter. Conf. CVD 1981, edited by J. M. Blocher, Jr., G. E. Vuillard and G. Wahl, Electrochem. Soc., Inc., Pennington (1981), p. 227.
29. K. Hiraga, M. Hirabayashi, S. Hayashi and T. Hirai, J. Amer. Ceram. Soc., in contribution.
30. K. Hiraga, K. Tsuno, D. Shindo, M. Hirabayashi, S. Hayashi and T. Hirai, Phil. Mag., in contribution.
31. R. D. Koester and D. P. Moak, J. Amer. Ceram. Soc., 50, 290 (1967).
32. T. Hirai, K. Niihara and T. Goto, J. Amer. Ceram. Soc., 63, 419 (1980).
33. T. Goto and T. Hirai, J. Mater. Sci., in press.

DISCUSSION

Skalny (Martin Marietta Labs): You showed the formation of various forms of Si_3N_4, composites, etc. as a function of temperature-pressure. By what method(s) did you identify the amorphous phases?

Author: I identified the existence of crystalline phase by normal x-ray technique. Therefore the amorphous deposits shown in my talk do not contain crystalline phase. However, crystalline deposits prepared in some CVD conditions might contain amorphous phase. I do not examine the exact amorphous/crystalline ratio in deposits. Measurements of x-ray diffraction intensity of a sample roughly show that the sample is a mixture of crystalline and amorphous or not.

J. J. Petrovic (Los Alamos National Lab): How controllable is the volume fraction of the second dispersed phase in the Si_3N_4 matrix and what is the current range of volume fraction that can be produced?

Author:

For Si_3N_4-C, $C < 10$ wt%;
for Si_3N_4-TiN, $TiN < 32$ wt%; and
for Si_3N_4-BN, $BN < 83$ wt%.

PREPARATION OF AMORPHOUS Si_3N_4-BN COMPOSITES BY CHEMICAL VAPOR DEPOSITION

Toshio Hirai, Takashi Goto and Tadashi Sakai[a]

The Research Institute for Iron, Steel and Other Metals
Tohoku University
Katahira, Sendai 980, Japan

ABSTRACT

Chemical vapor deposition of an Si-N-B system has been studied by using $SiCl_4$, NH_3, H_2 and B_2H_6 as source gases at deposition temperatures of 1100° to 1300°C and total gas pressures of 30 to 70 Torr. The chemical composition and density of the deposits were measured. The structure of the deposits was investigated by X-ray diffraction and IR absorption techniques. The deposits were composed of amorphous Si_3N_4 and turbostratic BN.

INTRODUCTION

The authors have studied the chemical vapor deposition of Si-N-C and Si-N-Ti systems, and successfully prepared the electrically conductive amorphous Si_3N_4(C) composites[1,2] and the Si_3N_4(TiN) composites.[3,4] In the present work, the CVD of a Si-N-B system was carried out by adding B_2H_6 gas to the $SiCl_4+NH_3+H_2$ gases. This paper describes the preparation and some properties of the deposits of the Si-N-B system.

---

[a]Currently with the Toshiba Research and Development Center.

EXPERIMENTAL

Sample Preparation

The experimental set-up is illustrated in Fig. 1. The graphite substrate (40 mm x 12 mm x 2 mm) was heated by transmitting an electric current. The deposition temperature (T_{dep}) was measured using a two-color pyrometer. The T_{dep} was varied from 1100° to 1300°C. The total gas pressure (P_{tot}) was regulated in the range of 30 to 70 Torr by needle valves. The source gases were Si_3Cl_4 vapor, NH_3, H_2 and a mixture of 5 vol% B_2H_6 in 95 vol% H_2. The Si_3Cl_4 container was held at 20°C and the saturated vapor of Si_3Cl_4 was carried with H_2. The Si_3Cl_4 vapor and NH_3 were separately introduced through a double tube nozzle into the CVD chamber to prevent premature vapor reactions.[5] In the present experiment, NH_3 and the mixture Si_3Cl_4+B_2H_6+H_2 were introduced through the inner and outer tubes, respectively. In reference to previous work,[6] the flow rates of H_2 and NH_3 [FR(H_2) and FR(NH_3)] were fixed at 700 and 60 $cm^3 \cdot min^{-1}$, respectively. The flow rate of diborane [FR(B_2H_6)] was controlled in the range of 1 to 12 $cm^3 \cdot min^{-1}$. The deposition conditions are summarized in Table 1.

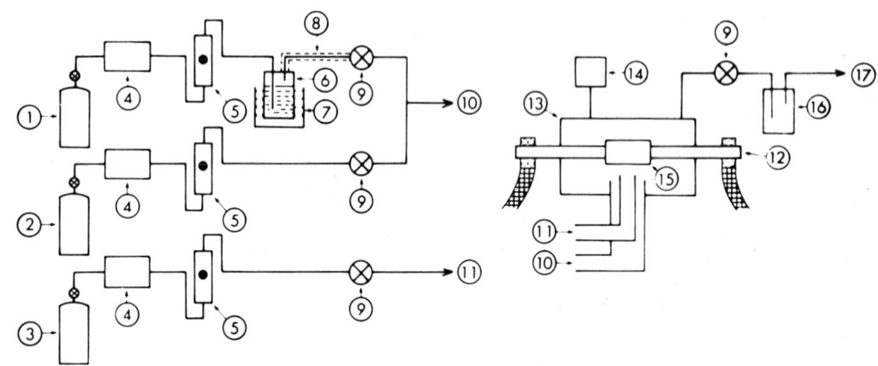

Fig. 1. Schematic diagram of the deposition apparatus. (1) H_2 gas; (2) B_2H_6+H_2 gases; (3) NH_3 gas; (4) gas purifier; (5) flow meter; (6) Si_3Cl_4 reservoir; (7) constant temperature bath; (8) ribbon heater; (9) pressure regulator; (10) Si_3Cl_4+B_2H_6+H_2 gas inlet; (11) NH_3 gas inlet; (12) copper electrode; (13) reaction chamber; (14) manometer; (15) graphite heater (substrate); (16) cold trap; (17) rotary pump.

Table 1. Deposition Conditions

Deposition Temperature, T_{dep}(°C)	:	1100 to 1300
Total Gas Pressure, P_{tot}(Torr)	:	30 to 70
Gas Flow Rate, FR($cm^3 \cdot min^{-1}$)		
FR(H_2)	:	700
FR(Si_3Cl_4)	:	170
FR(NH_3)	:	60
FR(B_2H_6)	:	1 to 12
Deposition Time, t(h)	:	1 to 8

Characterization

The boron and silicon contents of the deposits were determined by chemical analysis and electron probe microanalysis. The density of the deposits was measured in toluene using the Archimedean method. The structure of the deposits was examined by IR absorption, and electron and X-ray diffractions (Ni filtered CuKα).

RESULTS AND DISCUSSION

Figure 2 shows the color of Si-N-B deposits obtained under various CVD conditions. The color of the deposits varied from white (opaque) to yellow (opaque), to brown (transparent) and to black (opaque) with increasing FR(B_2H_6) at each T_{dep} and P_{tot}. A typical transparent deposit obtained in the present work is shown in Fig. 3. The energy band gap of amorphous Si_3N_4 is 4.4 to 5.2 eV,[7-10] so that electron transitions from the valence band to the conduction band do not occur in the visible range.[11] Therefore a and completely amorphous Si_3N_4 (a-Si_3N_4) is considered to be transparent. However, the present CVD·a-Si_3N_4 is white (opaque). This suggests that the CVD·a-Si_3N_4 may contain impurities such as O and Cl or structural defects such as microcracks and voids. In the present work, such impurities or defects may not be included in the deposits by adding the diborane gas to the $SiCl_4$+HN_3+H_2 gases.

Figure 4 indicates the effects of FR(B_2H_6) on the boron content of the deposits prepared at T_{dep} of 1100° to 1300°C. The boron content increased with an increase in $FR(B_2H_6)$ This tendency is more remarkable at lower T_{dep} and P_{tot}. The maximum boron content of the Si-N-B deposits was 36 wt%. Figure 5 represents the relationship between silicon content and boron content. The silicon content decreased with an increase in boron content. The solid line in Fig. 5 indicates the calculated values assuming the deposits are a mixture of Si_3N_4 and BN. The experimental values are in good agreement with the calculated values.

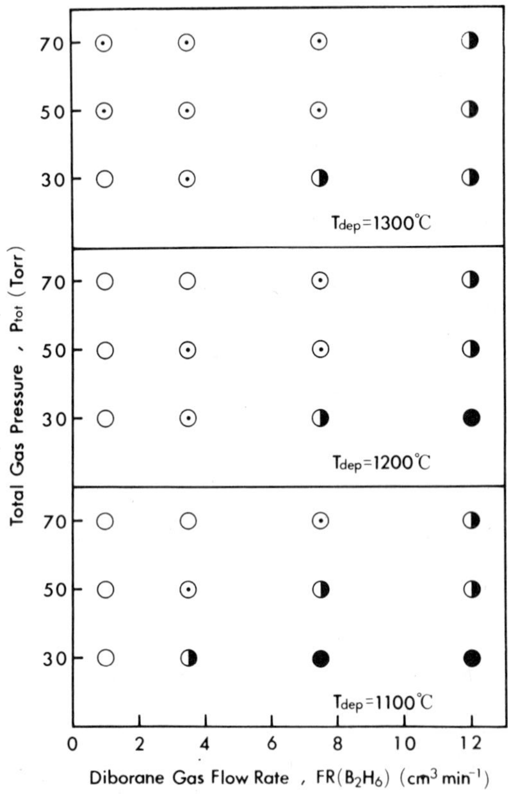

Fig. 2. Effect of the diborane gas flow rate $[FR(B_2H_6)]$ and the total gas pressure (P_{tot}) on the color of the deposits prepared at T_{dep} of 1100° to 1300°C.
○: white (opaque); ⊙: yellow (opaque); ◑: brown (transparent); ●: black (opaque).

Fig. 3. Transparent Si-N-B deposit.
T_{dep} = 1300°C, P_{tot} = 30 Torr and $FR(B_2H_6)$ = 12 cm^3/min.

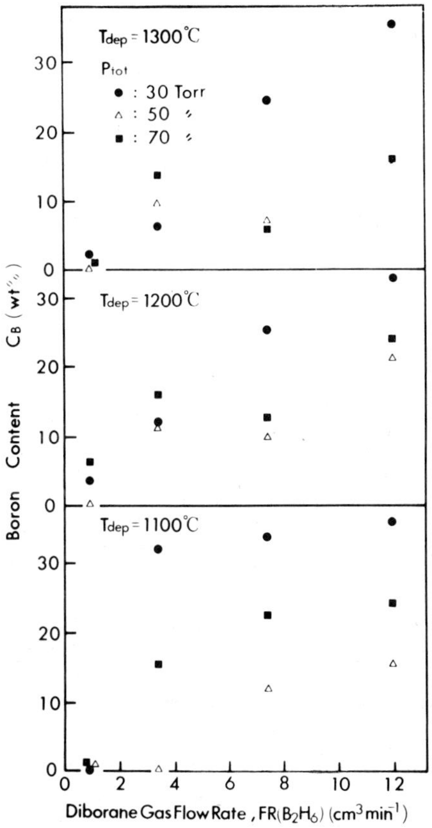

Fig. 4. Effect of the diborane gas flow rate [$FR(B_2H_6)$] on the boron content.

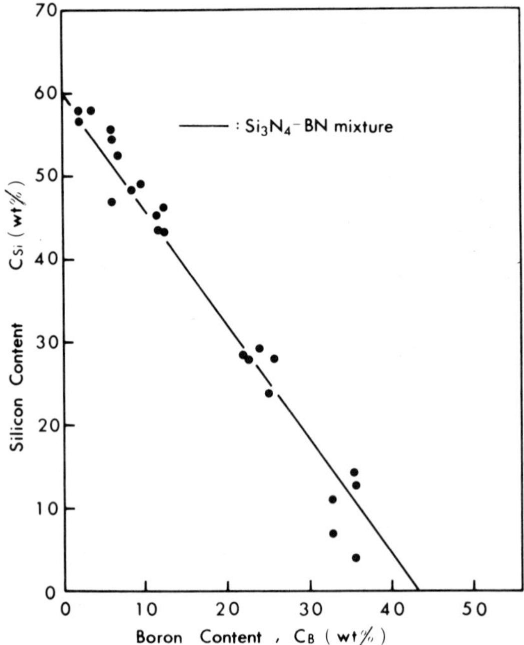

Fig. 5. The relationship between silicon content and boron content.

Figure 6 shows the effect of $FR(B_2H_6)$ on the density of the deposits prepared at T_{dep} of 1100° to 1300°C. The density decreased with an increase in $FR(B_2H_6)$. Fig. 7 represents the relationship between density and boron content. The density decreased with an increase in boron content. The solid line in Fig. 7 indicates the calculated values of the deposits assuming that the density of the deposits is represented by the linear combination of the densities of the CVD·a-Si_3N_4 and BN. This implies that the deposits are the mixture of the CVD·a-Si_3N_4 and BN. The densities of the CVD·a-Si_3N_4 and BN were assumed to be 3.0 and 1.6 $g \cdot cm^{-3}$, respectively.

Figure 8 shows the effect of the boron content on the IR absorption spectra of the deposits prepared at T_{dep} of 1200°C and P_{tot} of 30 Torr. At a low boron content, the IR absorption bands resulting from Si-N bonding are found at 850 to 1000 cm^{-1}. The antisymmetrical Si-N stretching frequency is in the range of 900 to 1000 cm^{-1}.[5] At an intermediate boron content, the IR absorption bands resulting from B-N bonding are found at 1300 to 1500 cm^{-1} as well as the Si-N bands. At a high boron content, the IR absorption bands resulting from B-N bonding are found at 1300 to 1500 cm^{-1} and at about 800 cm^{-1} which are due to the B-N stretching and B-N-B bending frequencies, respectively.[12,13] The IR absorption bands resulting from Si-B, Si-O and B-O bonding were not found in the present work.

Fig. 9 represents X-ray diffraction patterns of the deposits prepared at T_{dep} of 1200°C and P_{tot} of 30 Torr, and that of a commercial hexagonal BN. In Fig. 9(c), broad diffraction peaks are observed at about 2Θ = 25 and 42°. These peaks correspond to those of turbostatic BN.[14] The lattice constant (c_o) and the particle dimension (L_c) of the present turbostatic BN estimated from the (002) reflection are 3.56 A and 20 A, respectively. These values are in good agreement with the previously published data on turbostatic BN.[14] The intensity of the peaks decreased with a decrease in boron content. All the deposits obtained in the present work are amorphous according to the results of the X-ray and electron diffraction patterns.

CONCLUSIONS

Si-N-B deposits were prepared by chemical vapor deposition on graphite substrates at T_{dep} of 1100 to 1300°C using $SiCl_4$+NH_3+B_2H_6+H_2 gases. All the deposits were amorphous. The silicon content decreased with an increase in boron content. The silicon and boron contents were varied from 4.1 to 58 wt% and 1.9 to 36 wt%, respectively. These boron contents correspond to 4.4 to 83 wt% BN. The density decreased from 2.96 to 1.51 $g \cdot cm^{-3}$ with an increase in boron content. The IR absorption bands in the deposits were due to the Si-N and B-N bonds. The deposits obtained in the present work are thought to be composed of amorphous Si_3N_4 and turbostatic BN.

ACKNOWLEDGMENTS

This research was supported in part by the Grant-in-Aid for Special Project Research from the Ministry of Education, Science and Culture under Contract No. 56211005.

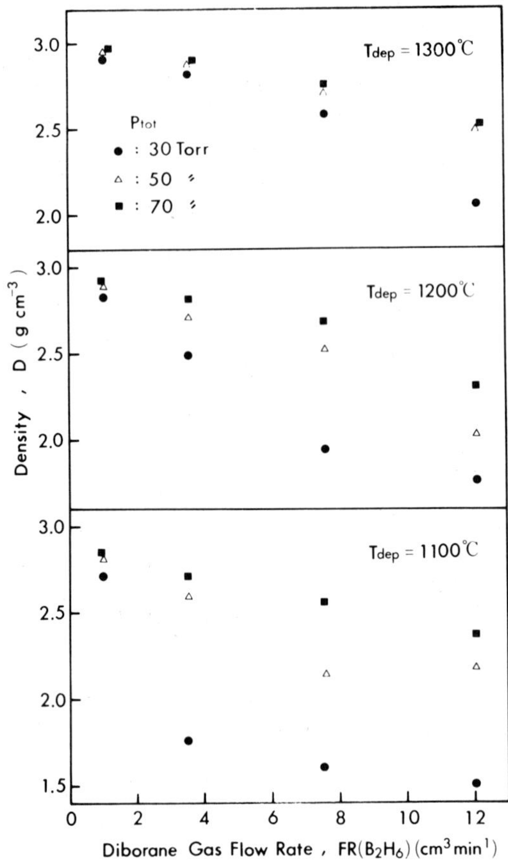

Fig. 6. Effect of the diborane gas flow rate [FR(B_2H_6)] on the density.

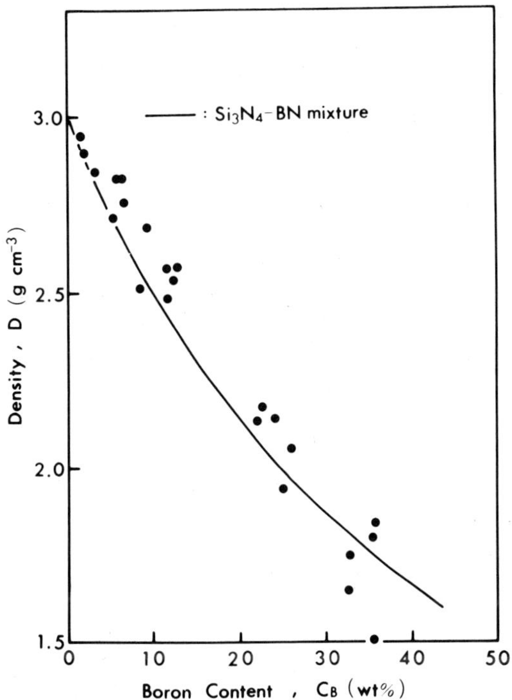

Fig. 7. The relationship between density and boron content.

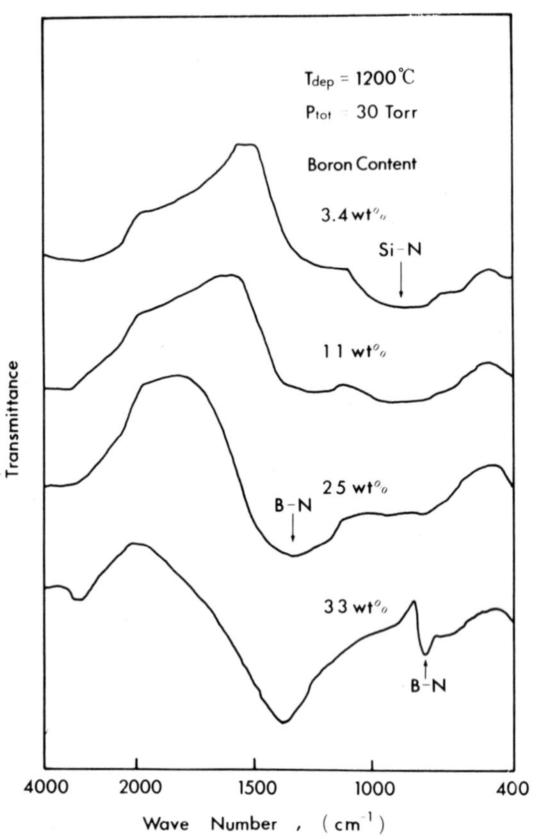

Fig. 8. Effect of the boron content on IR absorption spectra.

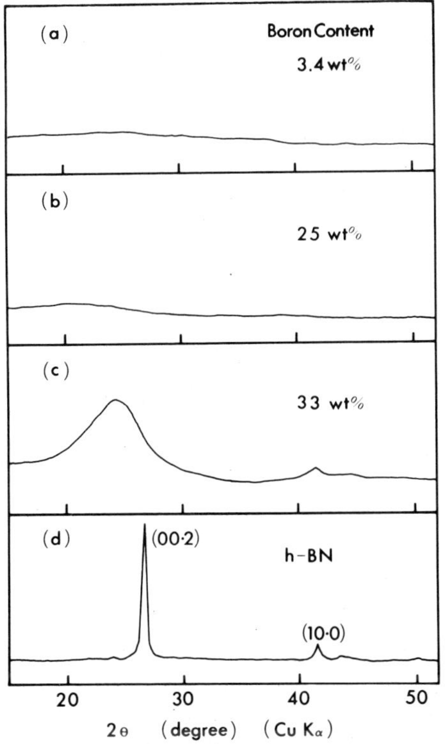

Fig. 9. X-ray diffraction patterns of the deposits prepared at T_{dep} of 1200°C and P_{tot} of 30 Torr.
(a) : $FR(B_2H_6)$ = 1 $cm^3 \cdot min^{-1}$; (b) : $FR(B_2H_6)$ = 7.5 $cm^3 \cdot min^{-1}$
(c) : $FR(B_2H_6)$ = 12 $cm^3 \cdot min^{-1}$; (d) : commercial hexagonal BN.

REFERENCES

1. T. Hirai and T. Goto, J. Mater. Sci., 16 [1], 17-23 (1981).
2. T. Goto and T. Hirai, J. Mater. Sci., in press.
3. T. Hirai and S. Hayashi, pp. 790-797 in Proceedings of the 8th International Conference on CVD, edited by J. M. Blocher, Jr., G. E. Vuillard and G. Wahl, Electrochemical Society, Pennington, NJ (1981).
4. T. Hirai and S. Hayashi, J. Mater. Sci., 17 [5], 1320-1328 (1982).
5. K. Mazdiyasni and C. M. Cooke, J. Amer. Ceram. Soc., 56 [12], 628-633 (1973).
6. K. Niihara and T. Hirai, J. Mater. Sci., 11 [4], 593-603 (1976).
7. K. E. Bean, P. S. Gleim, R. L. Yeakley and W. R. Runyan, J. Electrochem. Soc., 114 [7], 733-737 (1967).
8. D. M. Brown, P. V. Gray, F. K. Heumann, H. R. Philipp and E. A. Taft, J. Electrochem. Soc., 115 [3], 311-317 (1968).
9. V. V. Voskoboynicov, V. A. Gritsenco, N. D. Dikovskaya, B. N. Zaitsev, K. P. Mogilnicov, V. M. Osadchii, S. P. Sinitsa and F. L. Edelman, Thin Solid Films, 32 [2], 339-342 (1976).
10. J. Bauer, Phys. Stat. Sol. (a), 39 [2], 411-418 (1977).
11. C. Kittel, pp. 611 in Introduction to Solid State Physics, 4th edition, John Wiley, New York (1971).
12. M. J. Rand and J. F. Roberts, J. Electrochem. Soc., 115 [4], 423-429 (1968).
13. T. Takahashi, T. Itoh and T. Takeuchi, J. Crystal Growth, 47 [2], 245-250 (1979).
14. J. Thomas, Jr., N. E. Weston and T. E. O'Conner, J. Amer. Chem. Soc., 84 [24], 4619-4622 (1963).

DISCUSSION

R. W. Rice (NRL): Did I understand that your Si_3N_4 was black in "pure" form and yellow with BN? The reason I inquire, we have commonly seen yellow Si_3N_4 (crystalline) with no additives, and occasionally see brown or more red material.

Author: The color of amorphous CVD Si_3N_4 is white and varies to black via yellow and brown with increasing the amount of BN. Color of bulk crystalline Si_3N_4 is black on the deposition surface, but red-brown on the polished surface. Single crystal of Si_3N_4 was translucent light yellow.

A MORPHOLOGICAL STUDY OF SILICON BORIDES PREPARED BY CVD[a]

R. R. Dirkx and K. E. Spear

Materials Research Laboratory
The Pennsylvania State University
University Park, PA 16802

ABSTRACT

Compounds in the silicon-boron system have been prepared by chemical vapor deposition in the form of refractory protective coatings. Morphological changes which take place as a function of substrate temperature, total pressure, reactant gas composition and carrier gas composition were investigated. Marked changes in film morphology were observed as these CVD parameters were systematically varied.

INTRODUCTION

The silicon borides investigated in the present research have the potential of meeting today's technology demands for materials which will perform under extreme conditions, either in the bulk form or as protective coatings. Nicoll et al.[1] have shown that compounds formed in the silicon-boron system exhibit high resistance to cyclic oxidation, sulfidation, and thermal shock. These are also lightweight compounds which are very hard[1,2] and exhibit a high elastic modulus.[1,3]

The present paper is concerned with the formation of these refractory compounds by chemical vapor deposition (CVD), a technique which can be used at relatively low temperatures. Only a few reports on their CVD preparation have appeared in the literature. Silicon

---

[a]This research was supported by the U.S. National Science Foundation, Division of Materials Research, Grant No. DMR-8109260.

boride coatings of different compositions have been previously deposited by hydrogen reduction of silicon tetrachloride and boron halides.[4-6] Armas and coworkers[7] deposited a silicon boride compound characterized as SiB_{14} by the pyrolysis of the respective silicon and boron bromides. Silane and a boron halide were reacted on a thin, hot wire in the manufacture of high strength monofilaments by McCandless and coworkers.[3,8]

This paper reports investigations of the CVD of SiB_4 coatings from the hydrogen reduction and reactions of boron trichloride-silane mixtures on a heated graphite substrate. The morphology of these films would be of extreme importance if they were to be used as protective coatings. The microstructure would have to be tailored for a designated service. In this light, this paper reports investigations of the morphological changes which take place as experimental parameters are systematically varied.

The nomenclature "SiB_4" is used in this paper with the understanding that it does not necessarily represent the true stoichiometry. Most of the literature indicates that "SiB_3" more closely represents the stoichiometry of this compound.[9,10] However, this phase was identified by comparison of its x-ray diffraction pattern with the JCPDS pattern which lists the chemical formula as "SiB_4."[11a]

EXPERIMENTAL PROCEDURE

The CVD apparatus employed is a simple horizontal tube reactor designed for this preliminary study of the silicon-boron system. A schematic of the substrate zone of the reactor is shown in Fig. 1. The overall reaction studied was

$$SiH_{4(g)} + 4BCl_{3(g)} + 4H_{2(g)} = SiB_{4(s)} + 12HCl_{(g)},$$

where the tetraboride deposited on an rf inductively heated graphite substrate. The graphite substrates were cut from 1/4-in. rods of ultra purity spectroscopic electrodes from Ultra Carbon Corporation. The SEM micrograph shown in Fig. 2 characterizes the nature of the graphite surface prior to deposition. The reactant gases were ultra high purity grade and were metered into the reactor through type FM360 Tylan mass flowmeters. Additional cleaning and drying of the hydrogen and/or argon was accomplished prior to metering using an Alltech oxygen trap and dryer. The exhaust gases were removed from the reactor by a Nash vacuum pump.

The variable experimental parameters in this study include substrate temperature, total pressure, gas composition and total flow rate. The substrate chemistry and the geometry of the reactor and substrate were maintained constant. The total pressure of the system was 50 Torr for most experiments, but ranged as high as 600 Torr.

Substrate temperatures were measured with an optional pyrometer and ranged from 1073 to 1673K, but were typically 1473K. The total flow rate was usually 1040 standard cubic centimeters per minute (sccm), where the flow ratio $BCl_3:SiH_4$ = 3:1 and the ratio of hydrogen flow to the total flow of the reactant gases was 25.

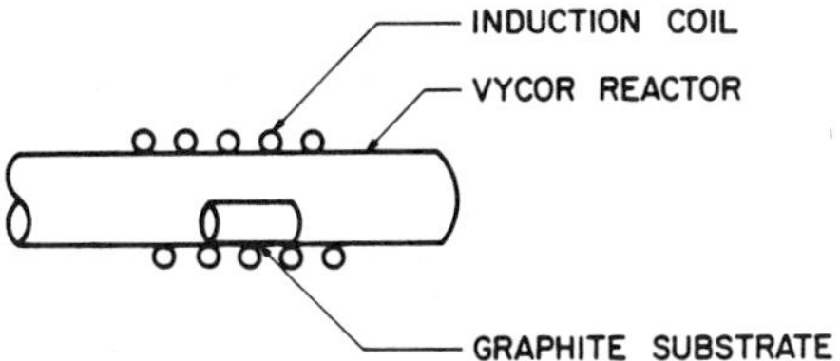

Fig. 1. Schematic of the substrate zone of the reactor drawn to scale.

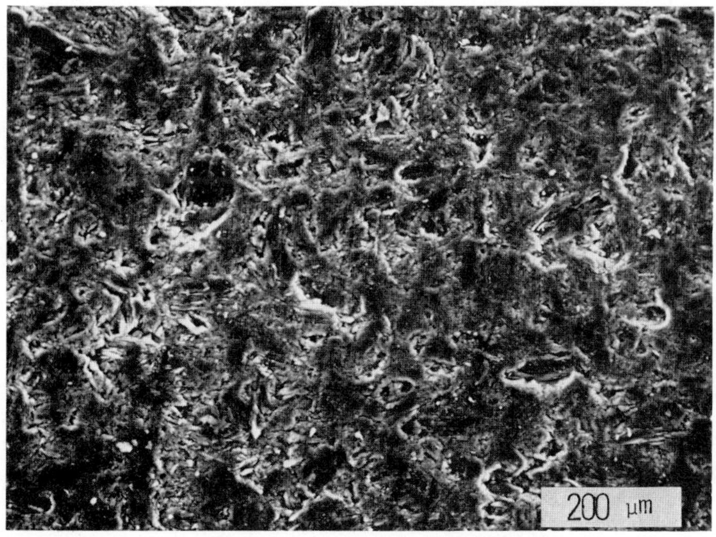

Fig. 2. The surface of the spectroscopic grade graphite substrate prior to deposition.

The film thickness and deposition rate were measured for each experiment. Film thicknesses varied with experimental conditions from 5 to 300 μm. The time of deposition for this study was 15 minutes. However, in experiments not reported in this paper, deposition rates were observed to be linear with respect to time. X-ray diffraction was used to identify the phases present in the deposits, and the morphologies were examined with a scanning electron microscope. Additional characterization of the samples is continuing.

RESULTS AND DISCUSSION

Effects of Substrate Temperature

In any CVD process, substrate temperature always plays an important role in determining the resulting morphology of the deposit. Both the nucleation and growth stages of deposition are affected. As substrate temperature is increased, the mobility of atomic species on the surface also increases. The gas temperature and therefore the glass reactor wall temperature will also increase. All these considerations make the interpretation of results most complex. However, the observed microstructural changes shown in Fig. 3 are quite clear.

The following experimental conditions were fixed as the substrate temperature was changed;

Total pressure = 50 Torr,
B:Si = 3:1,
$H_2:(BCl_3 + SiH_4) = 25$,
Total flow rate = 312 sccm.

The temperatures ranged from 1073 to 1673K in the twelve deposition experiments performed. From a practical point of view, this temperature study is quite valuable since morphologies observed from other studies (in which other parameters are varied) often are similar to the morphologies observed in this temperature study.

Figure 3, on the following page, summarizes the morphologies which were observed in this temperature study. As the substrate temperature is increased from 1073 to 1423K (Fig. 3a-c), there is a tendency to form larger, equiaxed, nodules. The intra-nodular structure also becomes more pronounced, and between 1423 and 1448K, sharply defined crystallites make up the structure of the nodules. The increasing surface mobility of atomic species with increasing surface temperature is strongly supported by these observations. At 1473K the large, equiaxed nodular structure is no longer visible and the sharply defined crystallites dominate the microstructure. Further increases in temperature result in the deposition of a second phase (SiB_6). The micrograph of the deposit formed at 1573K (Fig. 3) shows clearly the presence of two phases. At 1673K, the hexaboride is the only phase present.

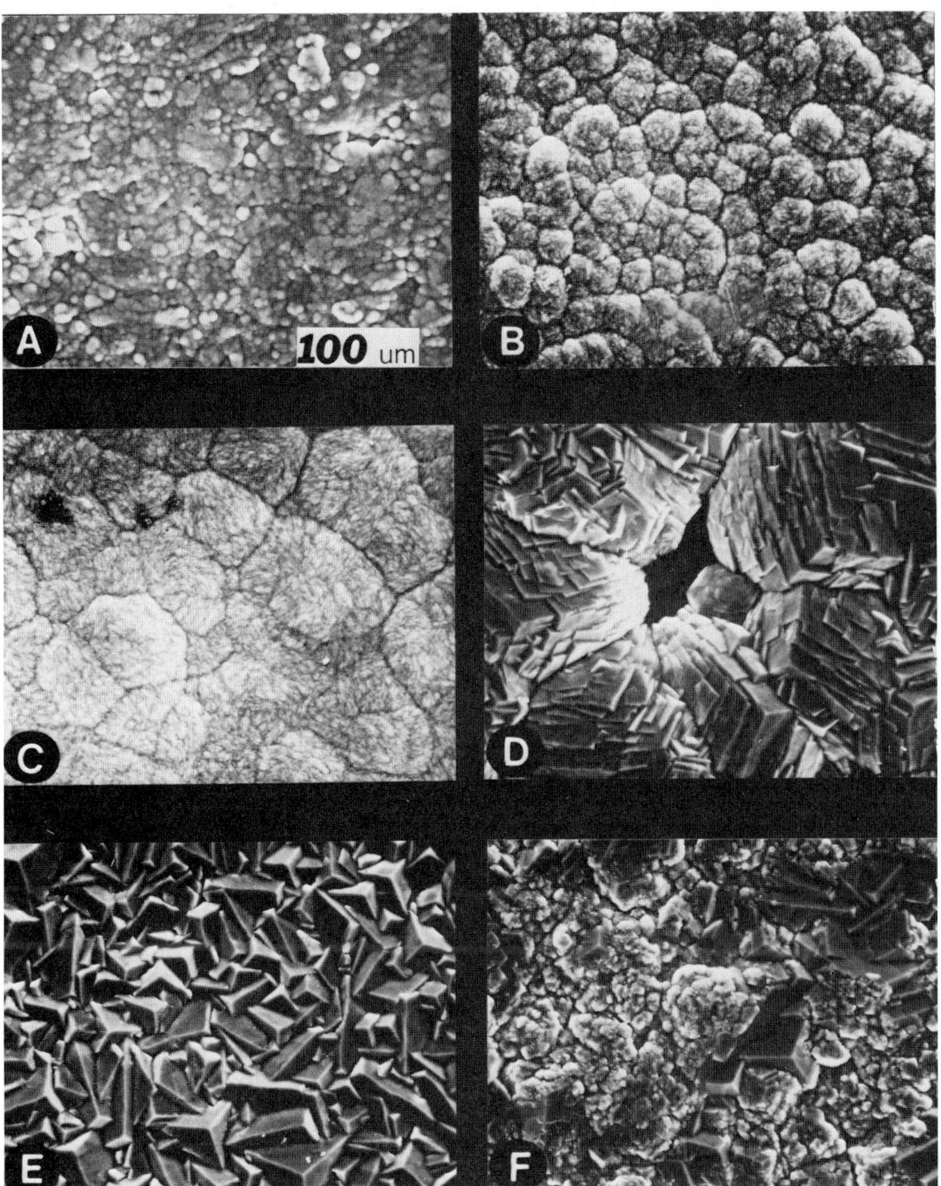

Fig. 3. Variation of deposit morphology with substrate temperature: (a) 1073K; (b) 1173K; (c) 1423K; (d) 1448K; (e) 1473K; (f) 1573K.

All of the deposits shown in Fig. 3 are crystalline by x-ray diffraction. X-ray diffraction was used to identify the presence of the hexaboride at substrate temperatures exceeding 1573K.[11b]

Effects of Total Pressure

The following conditions were held constant as total pressure was varied from 50 to 600 Torr;

Temperature = 1473K,
B:Si = 3:1,
H_2:(BCl_3 + SiH_4) = 25,
Total flow rate = 1040 sccm.

Figure 4 summarizes the observations of this study. All deposits were identified as SiB_4 by x-ray diffraction. The shape and orientation of the crystallites becomes more defined as the total pressure is increased from 50 to 100 Torr, and equiaxed clusters of these crystallites begin to appear. This "macrostructure" becomes more obvious in the form of a nodular structure as pressure is further increased.

Effects of the Reactant Gas B:Si Ratio

The following experimental conditions were held constant as the flow rates of BCl_3 and SiH_4 were adjusted to vary the B:Si ratio from 2:1 to 5:1;

Temperature = 1473K,
Total pressure = 50 Torr,
H_2:(BCl_3 + SiH_4) = 25,
Total flow rate = 1040 sccm.

The changes in morphology that were observed as the B:Si ratio was systematically changed are shown in Fig. 5. All deposits were identified as SiB_4 by x-ray diffraction. A variation of the nodular structure as shown in Figs. 3 and 4 is formed when the ratio is 2:1 (Fig. 5a). As the B:Si ratio is increased to 2.5:1, faceted crystallites appear which make up larger, equiaxed nodules. These crystallites grow approximately ten times in size as the ratio is increased to 4:1. At a B:Si ratio of 5:1, Fig. 5d shows that the edges of the crystallites begin to have a corroded appearance and become less well-defined. It is possible that this results because of a second phase (SiB_6) beginning to form, but in quantities too small to be identified by x-ray diffraction.

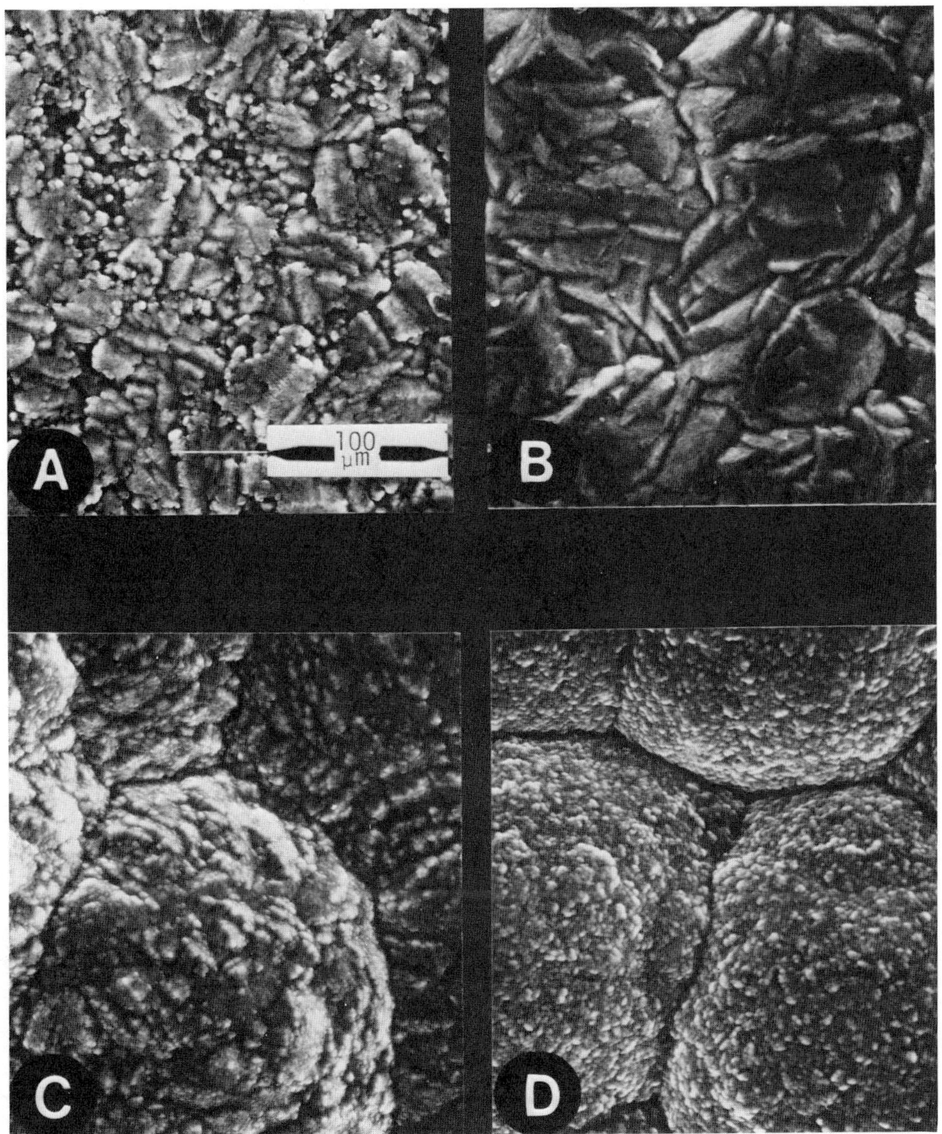

Fig. 4. Variation of deposit morphology with total pressure: (a) 50 Torr; (b) 100 Torr; (c) 300 Torr; (d) Torr.

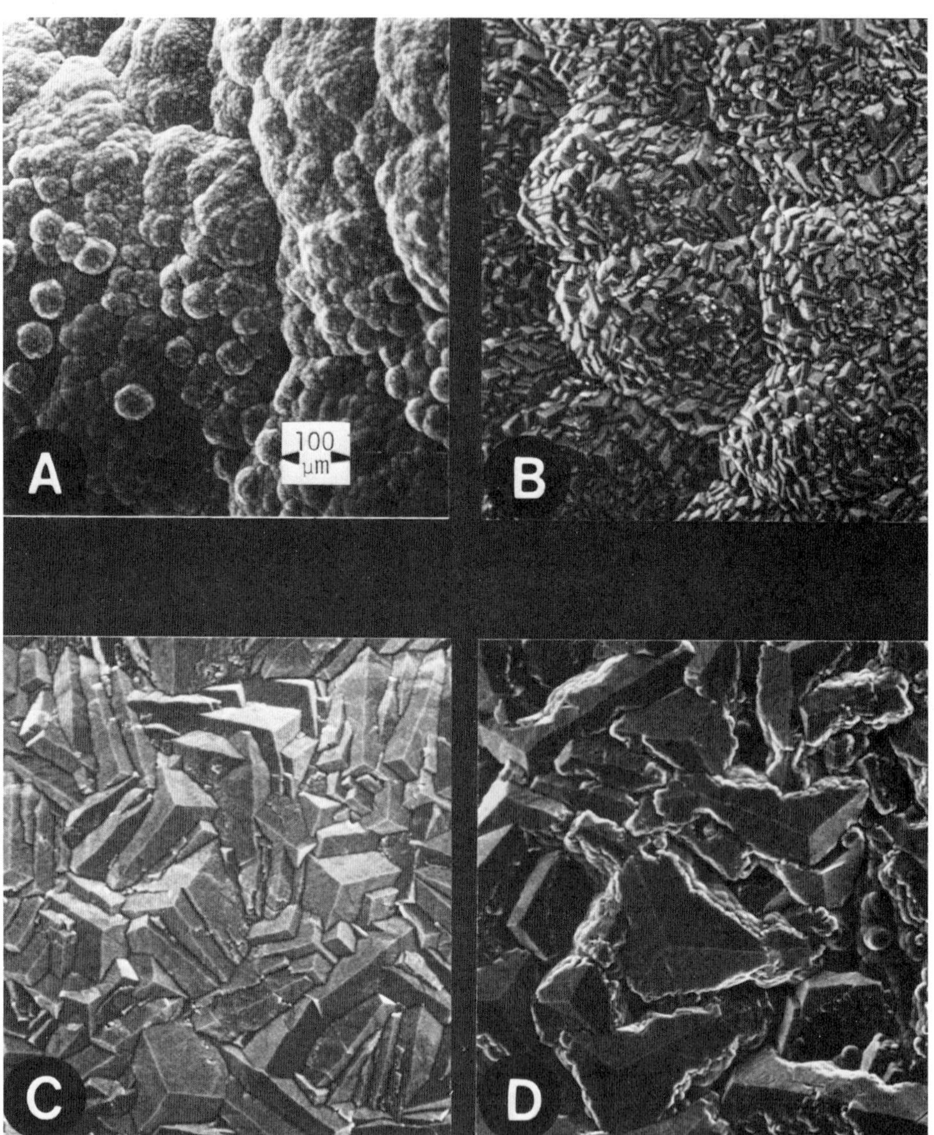

Fig. 5. Variation of deposit morphology with reactant gas B:Si ratio: (a) 2:1; (b) 2.5:1; (c) 4:1; (d) 5:1.

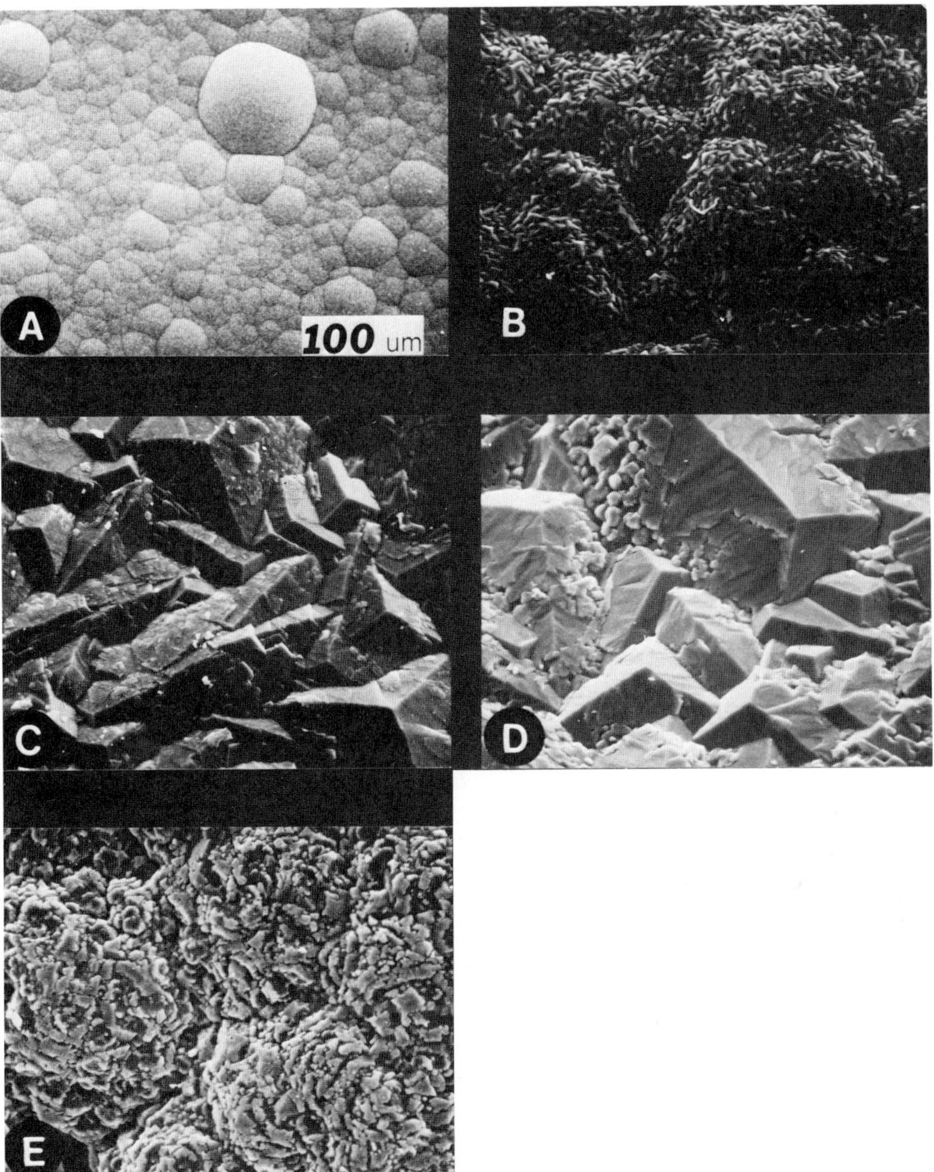

Fig. 6. Variation of deposition morphology with carrier gas composition:
(a) 1000 sccm H_2, 0 sccm Ar;
(b) 800 sccm H_2, 200 sccm Ar;
(c) 500 sccm H_2, 500 sccm Ar;
(d) 200 sccm H_2, 800 sccm Ar;
(e) 0 sccm H_2, 1000 sccm Ar.

Effects of Carrier Gas Composition

In the present CVD process being investigated, hydrogen serves the dual purpose of reducing the boron trichloride and also acting as a dilutent or carrier gas. In this set of experiments, hydrogen was systematically removed from the flow system and replaced by ultra high purity argon. The constant experimental parameters for this study were:

Temperature = 1473K,
Total pressure = 50 Torr,
B:Si = 3:1,
Carrier gas:(BCl_3 + SiH_4) = 25,
Flow of carrier gas (H_2 + Ar) = 1000 sccm,
Total flow rate = 1040 sccm.

The replacement of 200 sccm of the hydrogen with argon resulted in the formation of small, faceted crystallites of SiB_4, as shown in Fig. 6b. As more argon was added to the system at the expense of hydrogen, larger crystals were formed. Incomplete formation of the crystal edges began to occur when a flow rate of 800 sccm of argon was introduced to the system. Eventually, as shown in Fig. 6e, the regularly faceted crystals are no longer observed.

An identical set of experiments was conducted replacing the hydrogen with helium instead of argon. The flow behavior and thermal properties of helium closely resemble those of hydrogen. Therefore, any observed microstructural changes in these experiments could be attributed to the chemical behavior of hydrogen in the CVD reactor. However, no significant changes in microstructure were observed. This would suggest that the microstructural changes observed in Fig. 6 are a result of thermal and/or flow behavior changes caused by adding argon to the system, and are not due to chemical changes caused by the removal of hydrogen.

REFERENCES

1. A. R. Nicoll, U. W. Hildebrandt, and G. Wahl, Thin Solid Films, 64, 321-26 (1979).
2. G. Male and D. Salanoubat, Rev. Int. Hautes Temper. Refract., Fr., 18, 109-20 (1981).
3. L. C. McCandless, J. C. Withers, and C. R. Brummet, U.S. Patent 3,549,413, 1969.
4. S. Motozima, K. Sugiyama, and Y. Takahashi, Bull. Chem. Soc. Japan, 48 [5], 1463-66 (1975).
5. I. V. Petrushevich, L. A. Nisel'son, A. I. Belyaev, and M. A. Gurevich, Izv. Akad. Nauk SSR, Neorg. Mater., 3, 1389 (1967).
6. C. Powell and I. Campbell, Monatsh Chem., 88, 180 (1957).

7. B. Armas and C. Combescure, J. Less-Common Metals, 47, 135-40 (1976).
8. L. C. McCandless, J. C. Withers, and C. R. Brummet, U.S. Patent 3,607,367, 1971.
9. B. Magnusson and C. Brosset, Nature, 187, 54 (1960).
10. P. Ettmayer, H. C. Horn, and K. A. Schwetz, Mikrochim. Acta, Suppl. IV, 87 (1970).
11. Powder Diffraction File, Joint Committee on Powder Diffraction Standards, International Centre for Diffraction Data, Swarthmore, PA, 1982. (a) JCPDS card #13-210 from M. F. Rizzo and L. R. Bidwell, J. Am. Ceram. Soc., 43, 550-52 (1960). (b) JCPDS card #14-92.

DISCUSSION

W.-Y. Howng (Texas Instruments): Why are the microstructures shown in Figs. 4a and 6a different although they appear to have been prepared under identical conditions?

Author: This difference exists because the deposition process was started differently for these two cases. For the deposit shown in Fig. 4a, the substrate was heated to the desired temperature as the reactive gases passed over it. For the deposit shown in Fig. 6a, the substrate was heated to the desired temperature in an argon atmosphere and then the reactive gases were introduced into the reactor.

We are currently investigating the influence of the initial nucleation and growth processes on the final surface structure of the deposit.

A MORPHOLOGICAL STUDY OF SILICON CARBIDE PREPARED BY CHEMICAL VAPOR DEPOSITION[a]

P. Tsui and K. E. Spear

Materials Research Laboratory
The Pennsylvania State University
University Park, PA 16802

ABSTRACT

Surface morphologies of SiC deposits obtained via chemical vapor deposition (CVD) were studied as functions of substrate surface temperature and the concentration of the silicon and carbon source material, methyltrichlorosilane (MTS). Substrates of graphite and α-SiC crystals were used. Explanations of the observed morphologies on graphite substrates and their marked changes with temperature are given in terms of chemical kinetics and mass transport arguments. The results of thermodynamic calculations were used to help explain the observed morphologies of the deposits on α-SiC substrates.

INTRODUCTION

The technologically important mechanical and electrical properties of SiC warrant investigations into how these desired properties are achieved via specific processing techniques, such as chemical vapor deposition (CVD). The effect of the CVD parameters on the microstructure of the deposits are of interest because of the sensitive dependence of properties on microstructure. Weiss and Diefendorf[1] studied the deposit morphologies at low H_2/MTS ratios on graphite substrates and compiled a map of morphology vs. experimental parameters. Chin et al.[2] reported a similar morphology vs. experimental parameters map with temperature, total pressure and H_2/MTS ratios as variables. Christin et al.[3] evaluated such a deposition

---

[a]This research was supported by the NASA Lewis Research Center, Grant No. NAG3-177.

process through chemical equilibrium calculations and illustrated the effects of different parameters on the deposition process.

In this work, the temperature of the substrate and the concentration of the reactant gas, methyltrichlorosilane (MTS), were chosen as variables and the resulting microstructure of the deposits on two different substrates of graphite and α-SiC single crystals were studied and correlated.

EXPERIMENTAL

The reaction chamber used was a horizontal vycor tube with a wedge-shaped graphite susceptor which was heated externally by an rf coil. A simple schematic drawing of the reactor region is shown in Fig. 1. Methyltrichlorosilane (MTS) was used as both the carbon and silicon sources. The net chemical reaction producing silicon carbide is as follows:

$$CH_3SiCl_3(g) = SiC(s) + 3HCl(g) .$$

The MTS was injected into a manifold by a calibrated syringe pump, and was mixed with and carried by hydrogen into the reaction chamber. The total flowrate was kept constant at 1 litre/min (STP), and the pressure of the reactor was maintained at 70 Torr (approx. 0.09 atm) throughout this study. The substrate surface temperature was measured by an optical pyrometer, and the readings were corrected for absorption and scattering of the radiation by the reactor wall. The graphite substrates were preheated to the deposition temperature and maintained in a flow of hydrogen for a minimum of five minutes prior to each deposition run. The start-up and shut-down procedures were identical for all runs.

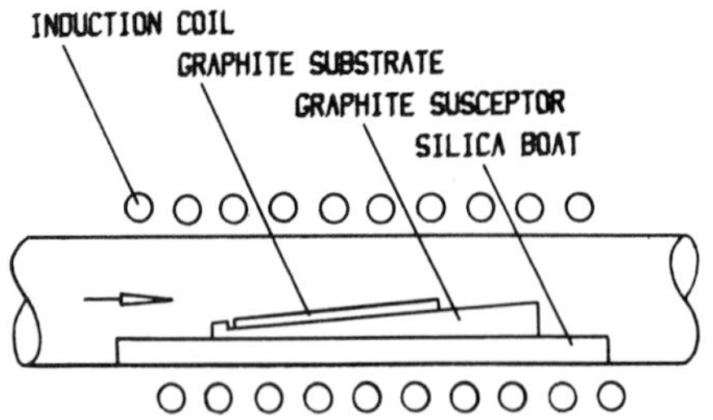

Fig. 1. A schematic of the reaction zone of the CVD reactor showing sample placement and support. The arrow indicates the direction of gas flow.

The process variables chosen were substrate temperature and the concentration of the MTS. Substrate surface temperatures ranged from 1423 to 1723K while the MTS concentration to the input gas varied from 0.1 to 0.6 mole %. A compilation of the experimental variables used is presented in Table 1.

Table 1. Experimental Conditions of This Study

Sample Number	Substrate Temp (K)	MTS (mole %)	Deposition Rate (mg/cm^2-min)	Equilibrium Deposition Rate
1	1723	0.1	0.2095	1.7143
2	1623	0.1	0.1394	1.6786
3	1523	0.1	0	1.6607
4	1723	0.3	0.5956	5.1696
5	1623	0.3	0.3576	5.1429
6	1523	0.3	0.2751	5.0893
7	1423	0.3	0.2362	5.0625
8	1723	0.4	0.7186	7.0000
9	1623	0.4	0.6512	6.9286
10	1523	0.4	0.4988	6.8571
11	1423	0.4	0.3301	6.7857
12	1723	0.6	0.9799	10.5000
13	1623	0.6	0.7629	10.4464
14	1523	0.6	0.5000	10.3929
15	1423	0.6	0.2125	10.2857
16*	1823	0.3	0.5662	5.3036
17*	1723	0.3	0.4320	5.1696
18*	1623	0.3	0.3355	5.1429

*α-SiC single crystal substrates; 1-15 are graphite substrates.

Depositions on α-SiC single crystals were carried out in the same apparatus. The crystal surface perpendicular to the c-axis was exposed by mounting it flush with the susceptor surface. The MTS concentration was chosen to be 0.3 mole % while the temperatures ranged from 1623 to 1823K. Total pressure was maintained at 70 Torr.

RESULTS AND DISCUSSION

The resulting deposits on graphite showed morphologies ranging from smooth to nodular to strongly faceted surfaces. Referring to the SEM photomicrographs in Fig. 2, it is observed that at lower temperatures (around 1423K), the deposits are fairly smooth and that the surfaces of the substrates are not completely covered. Coupling with the deposition rate data in Table 1, it is deduced that the deposition rates were significantly lower than the equilibrium limits obtained by thermochemical calculations. In the temperature range of 1523 to 1623K, surface coverage improved and the deposits assumed a nodular structure. The coarseness of this structure increased at the high end of this temperature range. Between 1623 and 1723K, a transition from the nodular structure to a faceted structure occurred. This transition in the morphologies can be attributed to a change in the rate-limiting mechanism of the deposition process, as is illustrated by the schematic Arrhenius plot shown in Fig. 3. This plot illustrates that low temperature deposition is governed by a high activation energy process, typically a surface reaction such as a step in the decomposition of MTS which subsequently yields SiC. Since this is a thermally activated process, the slope of the plot reflects the value of the activation energy.

At temperatures higher than where the bend of the plot in Fig. 3 occurs, the rate of SiC deposition is only weakly dependent on temperature and approaches that of the rate of mass transport of the active gaseous species through a compositional boundary layer to the surface. The surface reactions achieve equilibrium at these higher temperatures. This process has a low apparent activation energy and thus shows a smaller slope. As an active species is adsorbed on the surface, the surface mobility of that species plays an important role in determining the morphology of the deposit. The higher the mobility, the larger the crystallite size, which is manifested in the strongly faceted morphologies at temperatures above 1623K.

The change in concentration of MTS affected the surface morphologies to a lesser extent than temperature changes. Only the surface coverage seems to be improved at the higher concentrations. The resulting morphologies of this CVD study correlate quite well with the morphology-process relationships as described by Chin et al.[2] A three-dimensional plot of their process parameters vs. resulting morphologies of SiC deposits is presented in Fig. 4. The process parameters in this plot are the H_2/MTS ratio, total pressure, and temperature. The three regions marked in the plot to describe the deposit morphologies are faceted, angular/columnar, and smooth. The results of this work are represented by the shaded region of Fig. 4.

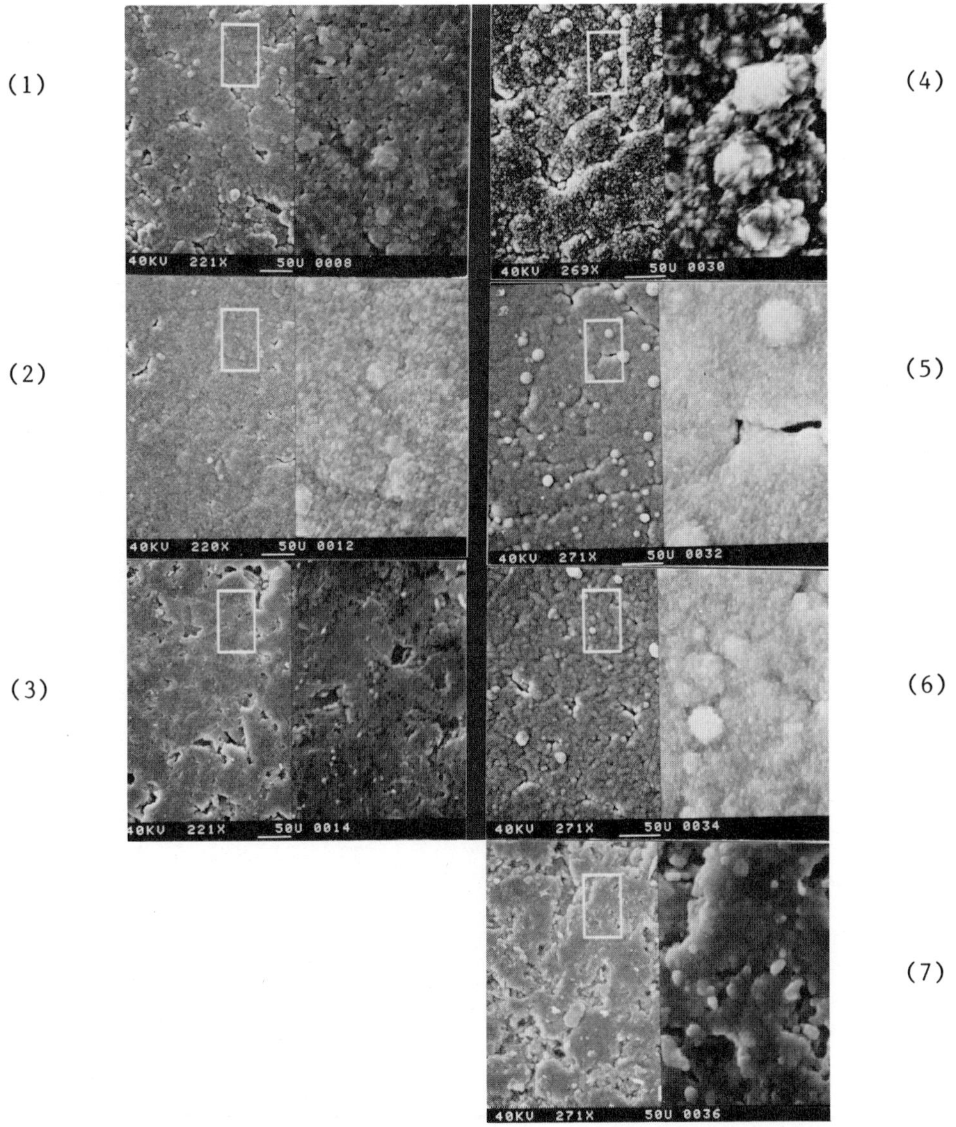

Fig. 2. Scanning electron micrographs of deposits surface morphologies on graphite substrates. The numbers alongside the micrographs are the sample numbers as indicated in Table 1.

(continued)

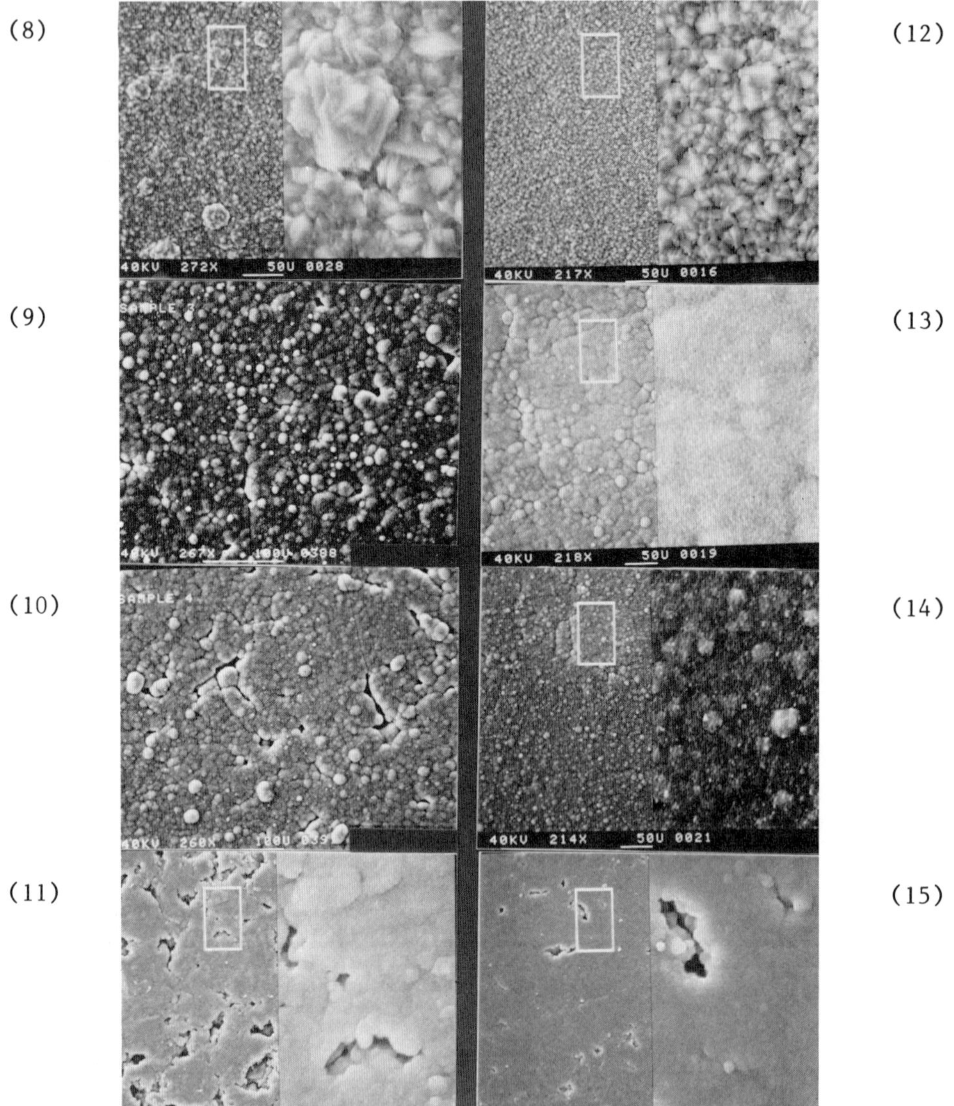

Fig. 2 continued.

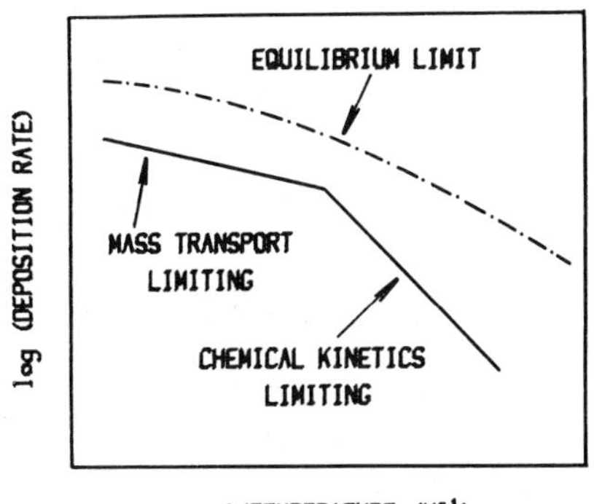

Fig. 3. An Arrhenius plot showing two different mechanisms operating in a deposition process as the temperature, in this case, varies. Deviation of this kind of process from equilibrium is indicated.

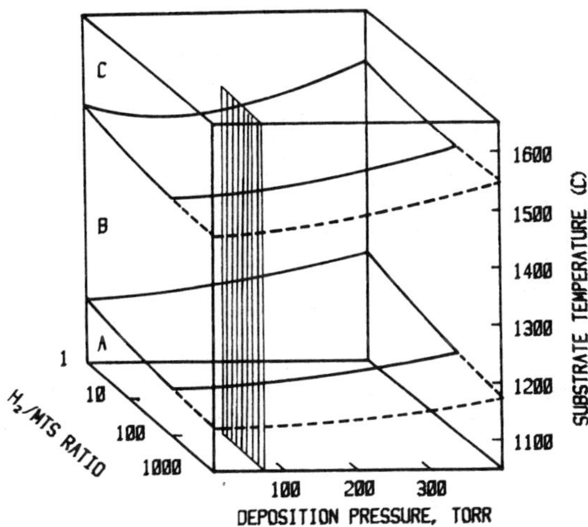

Fig. 4. The morphology-process relationships as described by Chin et al.,[2] dashed lines are extrapolations of Chin's data. Our results fall in the shaded region. Region A is the smooth region; B is the columnar/angular region and C is the faceted region.

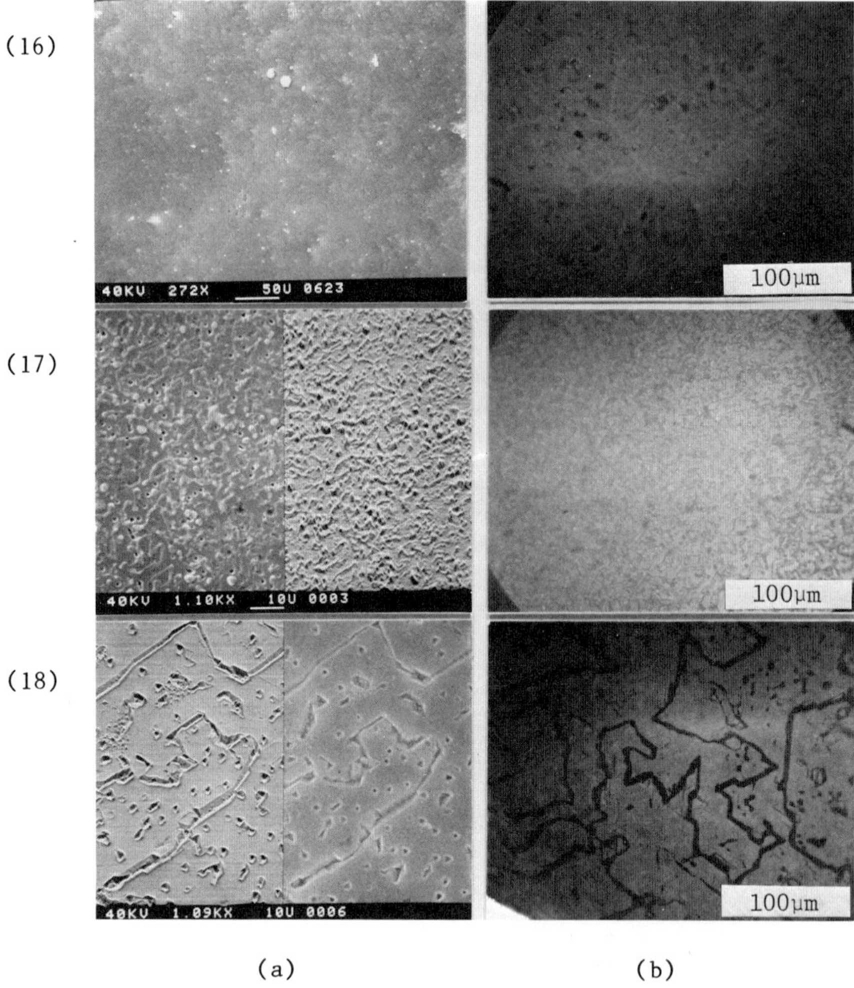

Fig. 5. (a) SEM and (b) optical micorgraphs of samples 16 to 18. Surface morphologies of deposits on single crystal α-SiC.

The deposit surface morphologies on single crystals of -SiC are shown in Fig. 5. Mosaic line patterns are observed in the 1623K sample. Jennings et al.[4] observed similar patterns in deposits on α-SiC at 1923K, a deposition temperature 300° higher than in our study. A y-modulation mode on the SEM revealed these mosaic lines were V-shaped grooves. Triangular pits on the surface showed that the deposited layer was cubic SiC, as was found by other investigators.[4,5] At 1723K, the deposited surface was rather smooth with isolated islands of materials on top. This may be explained by thermodynamic calculations carried out with the use of the SOLGASMIX-PV program initially developed by Eriksson.[6-8] As the vycor reactor tube was heated by radiation from the susceptor, etching of silica by the hydrogen may have occurred to a significant extent, and produced SiO along with water vapor according to the following equilibrium:

$$SiO_2(s) + H_2(g) = SiO(g) + H_2O(g) .$$

The SiO could further react with the substrate giving elemental silicon, which would be molten at the temperature of the substrate, and would produce the observed morphologies. This reaction can be written as follows:

$$SiC(s) + SiO(g) = 2Si(\ell) + CO(g) .$$

At 1823K, the deposit appeared relatively smooth and featureless. By utilizing underfocusing and overfocusing techniques on the optical microscope, it was found that certain "boundaries" existed that yielded grain-like regions of about 100 μm in diameter, but was otherwise structureless.

SUMMARY

The effects of two experimental variables on the morphologies of the deposits of silicon carbide via methyltrichlorosilane pyrolysis was studied using two different substrates of graphite and α-SiC crystals. The resulting deposit morphologies at high temperatures were explained in terms of the higher mobility of the active species, and thus the achievement of equilibrium at the surface and the resulting larger crystallites and faceted structure. At low temperatures, the low mobility and kinetic barriers of surface reactions led to finer structures and lower deposition rates. Results of the deposition on the single crystals was partially explained by the results of thermodynamic calculations. A more complete characterization of the deposited film should aid in developing better explanations of the observed morphologies.

REFERENCES

1. J. R. Weiss and R. J. Diefendorf, pp. 488-97 in Fourth International CVD Conference Proceedings, edited by G. F. Wakefield and J. M. Blocher, The Electrochemical Society, Princeton, NJ, 1973.
2. J. Chin, P. K. Gantzel, and R. G. Hudson, Thin Solid Films, 40, 57-72 (1977).

3. F. Christin, R. Naslain, and C. Bernard, pp. 499-514 in Seventh International CVD Conference Proceedings, edited by T. O. Sedgwick and H. Lydtin, The Electrochemical Society, Princeton, NJ, 1979.
4. V. J. Jennings, A. Sommer, and H. C. Chang, J. Electrochem. Soc., 113, 728-31 (1966).
5. W. V. Muench and I. Pfaffeneder, Thin Solid Films, 31, 39-51 (1976).
6. G. Eriksson, Acta Chim. Scand., 25, 2651-58 (1971).
7. G. Eriksson, Chemica Scripta, 4, 193-94 (1973).
8. G. Eriksson, Chemica Scripta, 8, 100-03 (1975).

LOW-TEMPERATURE PREPARATION OF PYROLYTIC CARBON

Richard W. Kidd, David A. Seifert and M. F. Browning

Battelle's Columbus Laboratories
505 King Avenue
Columbus, Ohio 43201

ABSTRACT

Previous studies have demonstrated that nuclear waste forms coated with chemical vapor deposited pyrolytic carbon (PyC) at ~ 1273 K can provide ground water leach protection. To minimize the release during coating of volatile material from the waste forms and permit the coating of waste forms with a low softening point, a study was initiated to develop parameters for the catalytic deposition of PyC at low temperatures (LT). The parameters surveyed in a fluidized-bed coater were deposition temperatures, carbon precursor, catalyst, diluent gas, concentration, and pressure.

INTRODUCTION

The program covered by this paper was carried out during the period October 1, 1979 to September 30, 1981 at Battelle's Columbus Laboratories (BCL) where it had been demonstrated earlier[1-3] that coatings of pyrolytic carbon (PyC) and alumina applied to simulated supercalcine nuclear waste pellets by chemical vapor deposition (CVD) provided protection against leaching and oxidation.

One of the objectives of the recent program is discussed in this paper. This objective was to reduce the coating temperature for carbon from that used previously (1173 -1373 K) to a level compatible with less refractory forms of nuclear waste materials such as glass marbles and more volatile components such as cesium.

The prospect of decreasing the carbon coating temperature resulted from the work of Neumann and Kofler[4] wherein the addition of readily dissociated nickel carbonyl [$Ni(CO)_4$] to the hydrocarbon coating atmosphere permitted the chemical vapor deposition of carbon at temperatures of 723 K and below. This result was consistent with the low-temperature deposition of carbon at nickel, iron, and cobalt surfaces, which has been studied earlier, primarily by Lobo and Trimm,[5-7] Lobo, Trimm and Figueiredo,[8,9] Lobo,[10] Bernardo and Lobo,[11,12] and Bernardo and Trimm.[13]

The deposition of carbon from acetylene is actually favored thermodynamically at temperatures in the range of interest. The fact that carbon does not deposit on non-catalytic surfaces until higher temperatures implies a kinetic barrier to nucleation and growth. However, the barrier for precipitation of carbon from solid solution in nickel is much lower than that on non-catalytic surface, and the barrier to the decomposition of acetylene to form carbon in solid solution in nickel is also apparently low. Accordingly, carbon from the decomposing acetylene can dissolve in the nickel to form a slightly supersaturated solution, be transported initially to a site where carbon has not deposited, and then precipitate. After the formation of a monolayer of carbon, the supersaturated solution would precipitate carbon on a carbon site. The rate of carbon deposition has been shown to be limited by the rate of diffusion through the nickel.

If this is indeed the mechanism, that is, if reaction of the hydrocarbon occurs where the nickel has access to the gas (i.e., is uncovered by carbon) and precipitation occurs at some random nucleation site on the carbon surface, the fundamental feasibility of forming an impervious coating by this method is brought into question. Despite this reservation, it was believed that in view of the potential results, efforts should be made to explore the technique fully, i.e., to cover the full range of parameters and to minimize other contributions to porosity, such as by occlusion of gas-phase precipitated material.[a] This material, if incorporated in the coating, would contribute to porosity and therefore its formation should be suppressed. Accordingly, consideration was given to means of minimizing gas-phase precipitation throughout the program.

---

[a]Carbon will not precipitate as such from acetylene at e.g., 723 K. However, gas-phase precipitation of nickel from the nickel carbonyl can occur at temperatures above about 473 K (depending upon concentration) [Oxley and Carlton[14]]. Once the nickel precipitates from the gas phase, it would be available to nucleate carbon.

EXPERIMENTAL APPROACH

Coating Substrates

While the ultimate goal of the program was to develop low-temperature (LT) PyC coating parameters and then use these parameters to coat simulated nuclear waste forms, this paper deals only with the development of LT-PyC coating parameters. In an effort to maximize coating quality it was decided to perform the parameter study using a fluidized-bed coater. A fluidized-bed coater is the most effective means of uniformly exposing a large surface area of particulate material to a reactant gas for CVD coating. Since the simulated nuclear waste forms are too large to be conveniently handled in a fluidized-bed coater, it was decided for the parameter study to use stand-in substrates of a size and density which could be conveniently handled in a fluidized-bed coater. Quartz sand (0.297 to 0.590 mm) and uranium dioxide (UO_2; 0.009 to 0.011 mm) were the principal stand-in substrates.

Coating Quality

Two characteristics of the coatings are critical for success in the intended application:

(1) permeability, an intrinsic property;

(2) integrity, i.e., uniformity and adherence.

The permeability is a function primarily of coating conditions: temperature, pressure, nature and concentration of reactants, and nature and concentration of catalyst. In the fluidized bed, the coating structure and hence the permeability is also influenced by the frequent collisions of particles with the surface on which the deposition of coating material is taking place.

Coating integrity is a complex function of coating/substrate geometry, thermal expansion differential between coating and substrate, thermal cycle to which the particle is subjected during coating and cooling at the conclusion of the coating step, and the intrinsic stress in the coating, all of which, together with the modulus of elasticity, determine the stresses tending toward coating failure. Whether or not coating failure or loss of integrity occurs under that stress is a function of coating strength and strength of adhesion to the substrate.

Coating quality was primarily assessed by visual and metallographic examination and from leach tests of the coated materials. The approach taken was to adjust the equipment design and the coating parameters in the ranges and in the direction which, on the basis of prior experience with CVD systems, would tend toward a dense deposit of carbon on the substrates of interest. It was recognized that a trade-off might be involved between coating density and rate of deposition as, in general, high CVD rates tend toward decreased density, particularly where gas-phase-nucleated material is incorporated in the deposit. However, it was hoped that the trade-off could be resolved in a practical range.

Apparatus

The fluidized-bed equipment used in the present work is pictured schematically in Fig. 1. In the initial work with nickel carbonyl [$Ni(CO)_4$] as a catalyst, the carbonyl and hydrocarbons (C_xH_y) were introduced together through the bottom inlet, but plugging of the inlet by premature reaction led to separation of the reactants either by introduction of the catalyst stream from the top of the bed through a water-cooled center tube or by introduction of the catalyst through the bottom of the reactor and C_xH_y through the top.

An important part of the equipment is the decomposer in the exhaust line, which is operated at 673 K to decompose the residual nickel carbonyl. In normal CVD coating from nickel carbonyl, the additional precaution of venting the exhaust through a gas burner would be taken to destroy traces of the residue. However, since the concentrations used in the present work were relatively low, it was deemed adequate to discharge the product of the decomposer directly to the laboratory hood in which the apparatus was housed.

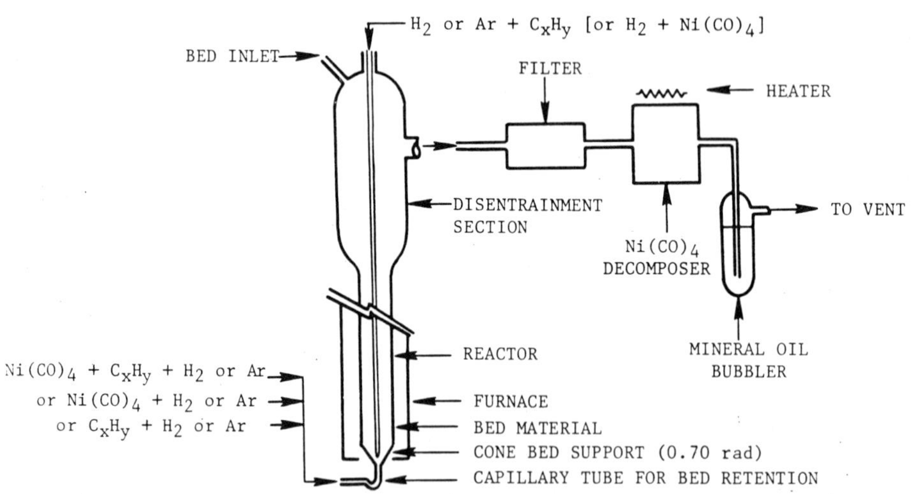

Fig. 1. Schematic diagram of the fluidized-bed coater and associated equipment showing alternatives for introducing reactants separately.

Cobalt acetylacetonate and iron pentacarbonyl were also investigated as catalysts. However, their use is not depicted in Fig. 1.

RESULTS AND DISCUSSION

Effect of Process Parameters on Coating Rate

With the general objective of optimizing coating rate and quality, a series of exploratory fluidized bed runs were made with a quartz sand (0.297 to 0.590 mm) substrate in which the following parameters were varied:

Bed temperature: ~ 598 to ~ 989 K
Nature of carbon precursor: acetylene (C_2H_2), methane (CH_4), propane (C_3H_8), heptane (C_7H_{16}), toluene (C_6H_5-CH_3)
Concentration of carbon precursor: ~ 1.5 to ~ 15 m/o
Nature of catalyst: nickel carbonyl [$Ni(CO)_4$], cobalt acetylacetonate [$Co(C_5H_7O_2)_2$], iron carbonyl [$Fe(CO)_5$]
Catalyst concentration: 0.0 to ~ 12 m/o
Diluent gas: argon, hydrogen

In addition, alternating injections of catalyst and carbon precursor were explored, as well as intermittent (in place of continuous) injections of catalyst. Obviously, not all permutations and combinations of these parameters were used. However, the range covered was believed sufficient to justify a number of useful conclusions.

Figure 2 shows data, taken with similar conditions except for temperature, for deposition from acetylene. A maximum in the deposition rate is indicated in the neighborhood of 723 to 773 K, similar to the maximum at ~ 648 K reported by Baird[15] for carbon deposits grown on nickel foils, where the character of the material changed from an open (C_o = 0.344 nm) columnar deposit to a tighter (C_o = 0.336 nm) structure at a higher temperature. At the higher temperature, the access of gaseous carbon precursor to the catalyst through the tighter structure may become limiting.

Figure 3 shows data for methane as a precursor, again showing a peak near 773 K. Based on past experience at BCL, it is somewhat difficult to believe that the actual carbon formation rate at 903 to 913 K would drop to 3 percent of its value at 803 K. Although Lobo and Trimm[5] show a drop to 10 percent between the 773 K peak and the 833 K valley for deposition from acetylene on a nickel foil, a decrease to 3 percent suggests the negative contribution of mechanical attrition to the net deposition rate.

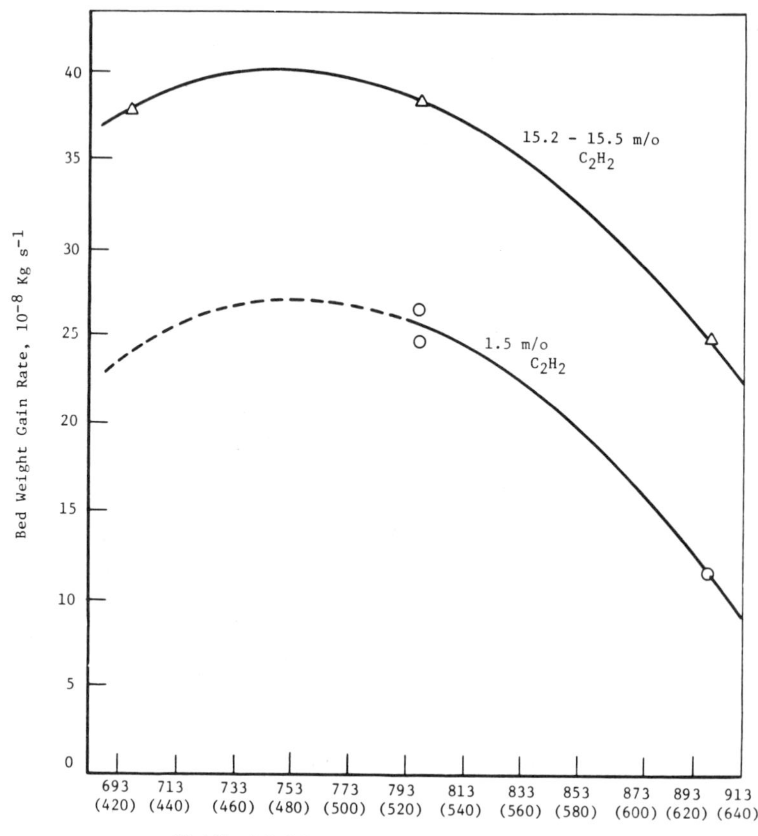

Fig. 2. Effect of temperature on deposition from acetylene. Bed = 0.037 Kg of quartz sand (0.297 to 0.590 mm) in 25-mm ID reactor. Total gas flow = 44.2×10^{-6} to 57.3×10^{-6} m^3 s^{-1} (2.65 to 2.44 ℓ min^{-1}). C_2H_2 as indicated plus 0.05 to 0.1 m/o $Ni(CO)_4$, balance Ar.

For purposes of orientation, it should be noted that the rate of bed weight gain [~ 58.5×10^{-6} Kg s^{-1} (~ 35 mg min^{-1})] at the peak in Figure 3 corresponds to the deposition of 2.8 nm s^{-1} of carbon of density = 1,200 Kg m^{-3} (1.2 g cm^{-3}) on 0.420 mm spheres.

Figure 4 shows the effect of substituting hydrogen for argon in the deposition of carbon from acetylene. In contrast with the observation of Koss, Kofler and Neumann,[16] who found a maximum in the deposition rate versus hydrogen content of the H_2/Ar diluent, no maximum was found in the present work. No similar series of experiments was carried out in the present program for the other hydrocarbons; however, Table 1 contrasts the available results for bed weight gain from methane and propane in argon versus hydrogen. In each case the deposition rate was suppressed by the hydrogen. It will be noted that because of the lower gas density, higher volumetric flows were needed with hydrogen for equivalent fluidization. The rate of deposition was lower for the hydrogen atmosphere in spite of the increased rate of supply of carbon precursor (increased by a factor of $6.66 \times 10^{-5}/4.28 \times 10^{-5} = 1.56$).

Figure 5 indicates that the rate of carbon deposition from acetylene in terms of bed weight gain at ~ 708 K increased linearly with concentration to 15 m/o. However, this conclusion must be tempered by the recognition that the "bed" collected at the end of the run contained an unknown quantity of 0.297 to 0.590 mm agglomerates of porous carbon, not deposited as coating material.

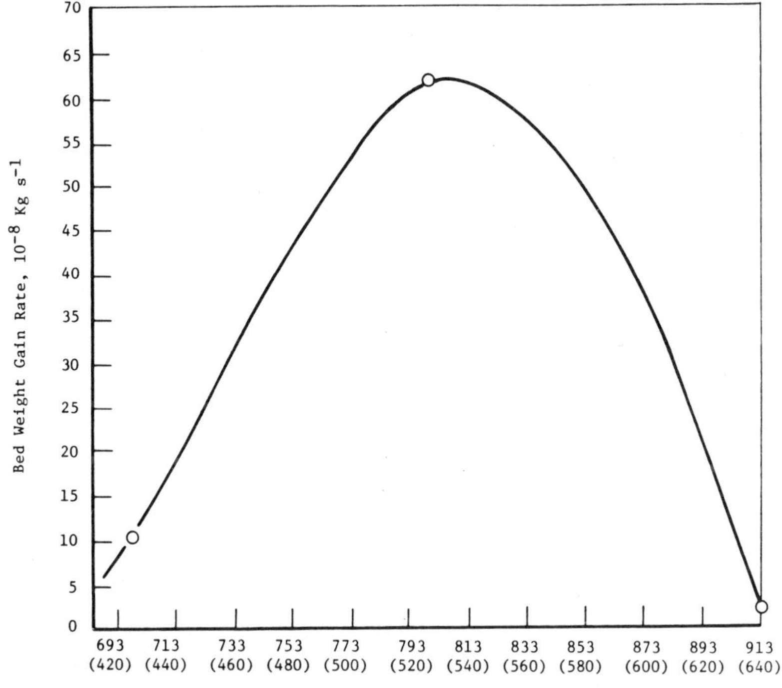

Fig. 3. Effect of temperature on deposition from methane. Bed = 0.032 Kg of quartz sand (0.297 to 0.590 mm) in 25-mm ID reactor. Total gas flow = 35.8×10^{-6} to 42.8×10^{-6} m^3 s^{-1} (2.15 to 2.57 ℓ min^{-1}), 15.0 m/o CH_4, 0.08 to 0.10 m/o $Ni(CO)_4$, balance Ar.

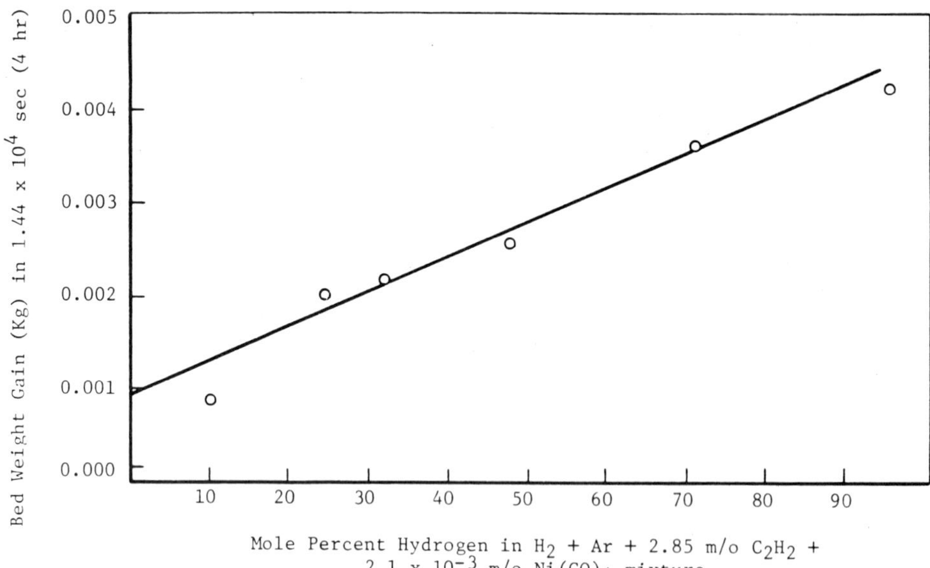

Fig. 4. Effect of hydrogen addition on the catalyzed deposition o carbon from acetylene. Solid line: Bed weight gain = 3.56×10^{-5} (m/o H_2) + 9.6×10^{-4}. Flow rate = 9.33×10^{-5} m^3 s^{-1} (5.6 ℓ min^{-1}). Quartz sand (0.297 to 0.590 mm) weight = 0.025 Kg.

Table 1. Effect of Hydrogen on the Catalyzed Deposition of Carbon from Methane and Propane, T = 768 K

Reactants, m/o			Total Flow, m^3 s^{-1}	Bed Weight Gain Rate, 10^{-8} Kg s^{-1}
Diluent	Precursor	Catalyst		
Ar, 84.9	CH_4, 15.0	$Ni(CO)_4$, 0.08	4.28×10^{-5}	59.7
H_2, 84.8	CH_4, 15.0	$Ni(CO)_4$, 0.20	6.66×10^{-5}	29.7
Ar, 84.9	C_3H_8, 15.0	$Ni(CO)_4$, 0.12	4.28×10^{-5}	458.3
H_2, 84.8	C_3H_8, 15.0	$Ni(CO)_4$, 0.20	6.66×10^{-5}	58.2

(Fluidized bed: 0.032 Kg of quartz sand (0.297 to 0.590 mm) in 25-mm ID reactor.)

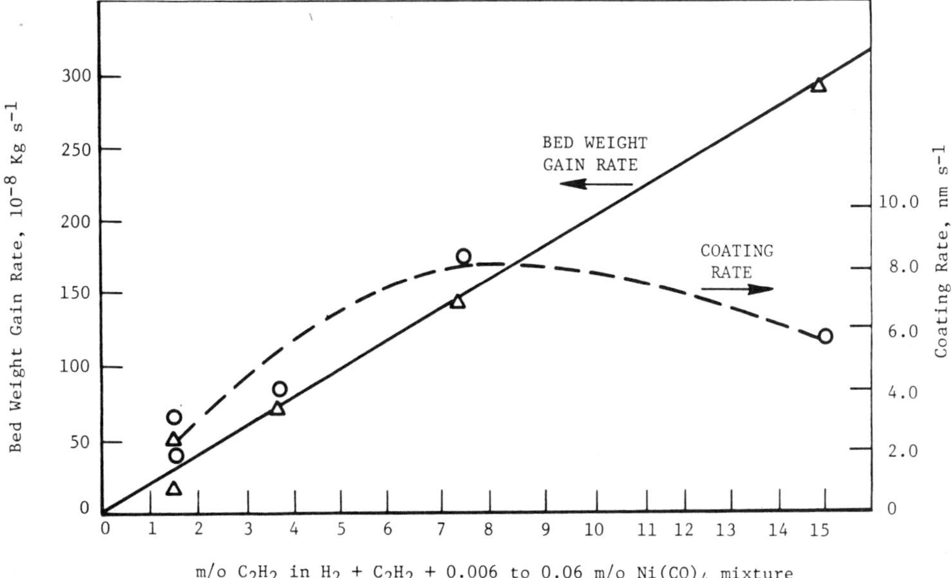

Fig. 5. Effect of C_2H_2 concentration on carbon deposition rate. T = 708 K. Bed: 0.032 Kg of quartz sand (0.297 to 0.590 mm) in 25-mm ID reactor.

Also plotted in Fig. 5 is the deposition rate in terms of rate of coating thickness increase as calculated from measured coating thickness at the end of the run. The drop-off at 15 m/o may reflect an attrition of the more porous carbon deposited under those conditions.

A leveling off of deposition rate (in terms of coating thickness) as a function of C_2H_2 concentration was also observed in a series of runs at higher temperatures (783 K) and at lower catalyst concentration [$Ni(CO)_4$ = ~ 2 x 10^{-3} m/o] as shown in Fig. 6. However, the leveling off occurred at lower C_2H_2 concentrations. The data shown in Fig. 6 suggest that the rate may become catalyst-limited at the higher hydrocarbon concentrations. In contrast with the data of Fig. 5, the bed weight gain also dropped off proportionately.

It is logical to assume that there exists some relationship between catalyst concentration and hydrocarbon concentration, such that the deposition rate could be either catalyst limited or hydrocarbon limited. This factor was never fully explored. However, the series of runs summarized in Table 2 suggest that the effectiveness of the catalyst can reach saturation at a very low level.

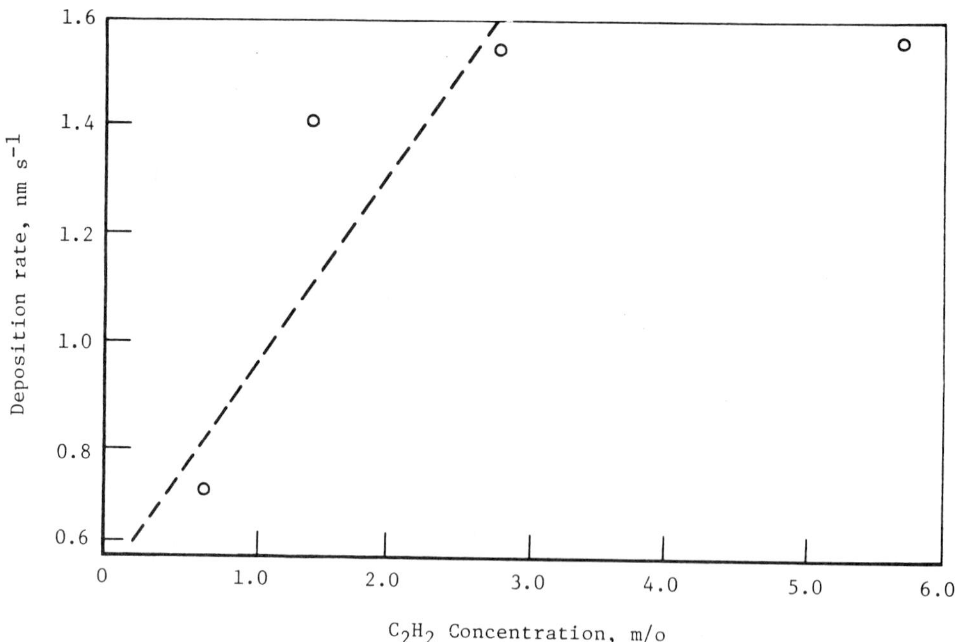

Fig. 6. Effect of C_2H_2 concentration on deposition rate at T = 783 K. $Ni(CO)_4$: $2.1 \pm 0.2 \times 10^{-3}$ m/o, 0.032 Kg of quartz sand (0.297 to 0.590 mm), + 0.010 Kg of UO_2 (0.009 to 0.011 mm) in 25-mm ID reactor, rate of deposition measured on UO_2.

Table 2. Effect of Catalyst Concentration on Deposition Rate at T = 798 K, C_2H_2 Concentration = 2.9 m/o

Run No.	$Ni(CO)_4$ Concentration, m/o	Deposition Rate on UO_2, nm s^{-1}
151	0.35×10^{-3}	0.61 (2.2μm hr^{-1})
150	0.74×10^{-3}	1.11 (4.0μm hr^{-1})
149	2.2×10^{-3}	1.11 (4.0μm hr^{-1})

(Bed = 0.032 Kg quartz sand (0.297 to 0.590 mm) + 0.010 Kg UO_2 (0.009 to 0.011 mm) in 25-mm ID reactor, total flow rate = 9.33×10^{-5} m^3 s^{-1} (5.6 ℓ min^{-1}), $Ni(CO)_4$ plus H_2 introduced through capillary at apex of bed-support cone, C_2H_2 introduced downward from center tube terminating in bed 0.9 cm above apex.)

The use of $Fe(CO)_5$ as a catalyst was explored briefly in a series of runs in the temperature range 693 to 753 K, at ~ 1.5 m/o C_2H_2 and from 0.001 to 0.08 m/o $Fe(C))_5$. Because of the fact that the coatings adhered poorly to the quartz sand (0.297 to 0.590 mm) particles (spalling of coating upon cooling), it was difficult to draw any firm conclusions other than that the $Fe(CO)_5$ appeared to be generally less effective as a catalyst in terms of conversion of C_2H_2 to carbon under comparable conditions, and that gas-phase precipitation appeared to be less pronounced, owing to the greater stability of $Fe(CO)_5$ relative to $Ni(CO)_4$.

Effect of Process Parameters on Coating Quality

It became quite evident early in the work that the highest rates of coating formation tended toward the least dense and most friable coatings. In fact, it was not until the C_2H_2 concentration in hydrogen was reduced to 1.5 m/o and the $Ni(CO)_4$ catalyst concentration to 0.03 m/o that the coating (Fig. 7) obtained could be classed as "hard" (i.e., dense and resistant to probing with a needle). That coating, pictured in Fig. 7, is represented by the lower point at 1.5 m/o in Fig. 5.

10μm

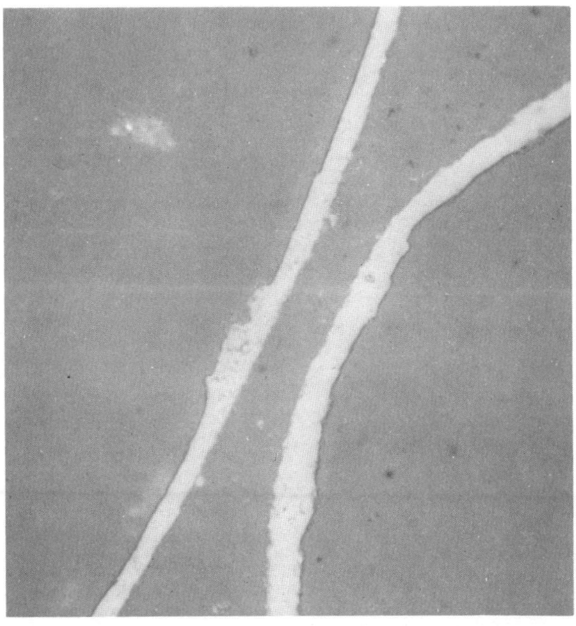

Fig. 7. Nominal 4- to 5-μm carbon coating on 0.297 to 0.590 mm quartz sand (lot 35445-88-54, 5093VD).

The data of Fig. 5 were taken in conjunction with a series of runs made to determine how high the C_2H_2 concentration (and deposition rate) could be raised without losing the integrity of the coating obtained at 1.5 m/o C_2H_2. However, evaluation of the coatings visually and by probing was not sufficiently quantitative to permit a correlation of coating quality with C_2H_2 concentrations. Therefore, the leach testing of UO_2 particles added to the fluidized bed was undertaken as a means of coating evaluation.

In earlier work at BCL the permeability of pyrolytic carbon coated UO_2 was assessed by oxidation in air at ~ 673 K, at which temperature UO_2 was readily converted to U_3O_8, but PyC was not oxidized. Leach testing of the coated particles was also of value, as low concentrations of uranium in the leach solution could be accurately measured (Browning et al.[17]).

A basic difference in the low-temperature pyrolytic carbon and the uncatalyzed high-temperature variety was demonstrated by the fact that the low-temperature coatings oxidized readily, indicating the presence of a large fraction of "dangling" bonds in the a-direction. Accordingly, leach testing in nitric acid was adopted. However, except for very low permeabilities, the leach test did not give a graded response. Once the LT-PyC coating was breached, the reaction was disruptive of the PyC coating. It has been speculated that the HNO_3 acid may have oxidized the low-temperature PyC coating. However, one coating lot (BCL Lot 35715-44-80) did show promising and reproducible results in the leach test. This lot was distinguished by having been deposited at the lowest deposition rate and at the lowest catalyst concentration. Unfortunately, the unique results of BCL Lot 35715-44-80 were difficult to reproduce, despite a large number of subsequent coating runs at comparable or presumably more favorable conditions. Part of the problem is due to the limited capability of the leach test.

Figure 8 is a cross-section of the BCL Lot 35715-44-80 product showing a reasonably uniform 15-17 μm coating with very little porosity observable at the 540X magnification.

It is probable that the permeability of the present coatings parallels their density. Thus techniques for measuring coating density were of interest. Measurement by the normal pycnometric technique would give a higher-than-actual result because of penetration of the open pores by the pycnometer fluid. Probing of the porosity with a mercury porosimeter has the disadvantage that a fragile coating can be crushed and densified during measurement by the higher pressures.

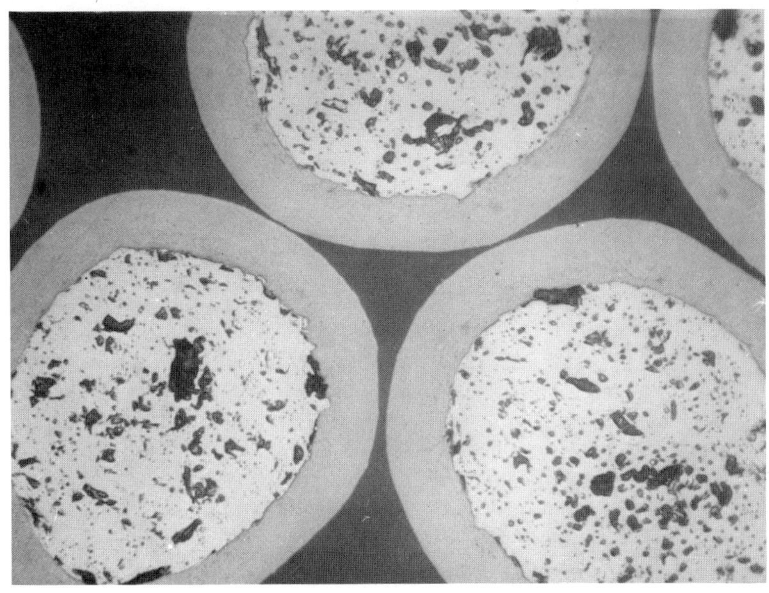

Fig. 8. Nominal 15- to 17-μm PyC coating on 0.009 to 0.011 mm uranium dioxide (Lot 35715-44-80, 5156VD).

The expedient of measuring the change in bulk density due to coating was adopted as a simple means of approximating the coating density. Such measurements were made for some of the material prepared with different concentrations of hydrogen, whose rates of deposition were analysed in Table 1 and Fig. 4. The data are given in Table 3.

Table 3. Coating Density Calculated from the Bulk Density of Coatings Corresponding to the Data of Table 1 and Fig. 4

Hydrogen Content of H_2 + Ar, m/o	W_i, Initial Bed Weight, Kg	V_{Bi}, Initial Bulk Volume, m^3	W_f, Final Bed Weight, Kg	V_{Bf}, Final Bulk Volume, m^3	Calculated Coating Density, Kg m^{-3} (a)
25	0.025	14.8×10^{-6}	0.02712	17.6×10^{-6}	1180
32	0.025	14.8×10^{-6}	0.02739	18.0×10^{-6}	1170
50	0.025	14.8×10^{-6}	0.02749	18.6×10^{-6}	1020
75	0.025	14.8×10^{-6}	0.02876	19.5×10^{-6}	1250

(a) 2650 Kg m^{-3} (2.65 g cm^3) taken as density of quartz sand.

Two conclusions can be reached from the data:

(1) There appears to be no trend of coating density with the hydrogen content of the fluidizing gas or with the resultant increase in deposition rate over the range studied, and

(2) When compared with the theoretical density of graphite, 2,260 Kg m^{-3} (2.26 g cm^{-3}) and with that of commercial grades of "dense" graphite (Union Carbide Corporation,[18] e.g., CS 1.710 Kg m^{-3} (1.71 g cm^{-3}) and ATJ 1,730 Kg m^{-3} (1.73 g cm^{-3}), the material depicted in Table 3 is concluded to be quite porous.

The density of the LT-PyC coatings of Table 3 is similar to that of porous "high-temperature" PyC prepared in an intermediate temperature range with high reactant concentrations where gas-phase precipitated material is occluded and is much lower than the 2,200 Kg m^{-3} (2.2 g cm^{-3}) and 2,100 Kg m^{-3} (2.1 g cm^{-3}) observed at higher and lower temperatures (Diefendorf[19]).

CONCLUSIONS

A primary goal of this work was to evaluate the potential of applying carbon at temperatures (e.g., ~ 773K) lower than those typically used (~ 1273 K). Due to its advanced stage of development, a fluidized-bed coater was selected to be used as the coater to evaluate the potential of applying LT-PyC. A parameters study using quartz sand (0.297) to 0.590 mm) confirmed that:

(1) Operation without a catalyst in the acetylene/argon or acetylene/hydrogen precursors yielded practically no carbon in the temperature range studied. LT-PyC could be deposited at readily detectable rates of temperatures above 620 K.

(2) Consistent with the observations of Lobo, Trimm and Figueiredo,[9] cobalt acetylacetonate was less effective (1 to 10 percent as effective) as a catalyst than nickel carbonyl.

(3) $Ni(CO)_4$ catalyst concentrations at the 1 m/o level resulted in relatively large quantities of gas-phase precipitated carbon.

(4) Coating rates as high as 11 to 14 nm s^{-1} (40 to 50 μm/hr^{-1}) could be obtained from propane, 15 m/o; $Ni(CO)_4$, 0.12 m/o; argon, 94.9 m/o at 803 K. However, the coatings were porous, friable and non-adherent.

(5) A carbon coating prepared on nickel-coated sand was very porous and friable, confirming the advantage of introducing the catalyst and precursor simultaneously or intermittently.

(6) No single hydrocarbon, of those explored, stood out as being markedly more effective than the others in terms of resulting in both a high deposition rate and a good quality coating.

(7) The addition of hydrogen (substitution for argon) to the acetylene plus $Ni(CO)_4$ system increased the deposition rate, consistent with the observations of Bernardo and Lobo.[11] However, the addition of hydrogen suppressed deposition from the saturated hydrocarbons, propane and methane.

(8) The rate of LT-PyC deposition appeared to exhibit a maximum and then decrease as the temperature was increased. This decrease in rate may be interpreted to be due to the catalyst and the carbon precursor reacting at sites other than on the substrate to produce gas-phase-nucleated carbon dust.

(9) Dust formation, as expected, was found to be detrimental to the coating quality.

Although a fairly wide range of conditions were explored for the catalyzed deposition of LT-pyrolytic carbon, no really satisfactory coatings were obtained. The system which showed the most promise was $H_2/C_2H_2/Ni(CO)_4$. It appeared possible to obtain coatings of varying degrees of porosity, density, and strength. However, those of greatest porosity and least strength were the ones obtained at the economically attractive high-rates of deposition. Whether a very low rate of deposition at a very low catalyst concentration would ever result in acceptable impermeability is questionable. The question may be purely academic in view of the sacrifice in deposition rate required and the correspondingly increased cost.

The porosity may be inherent if the only deposition mechanism that operates is the precipitation of carbon from solid solution in nickel which has become supersaturated at another location by the decomposition of the hydrocarbon in contact with the nickel. Only if the particle were surrounded by a retreating and expanding nickel outer shell could such a mechanism operate to yield a dense coating. Even in that case, the attainable density would appear to be limited by the high interlayer spacing, C_o, of the material obtained previously by such mechanisms.

REFERENCES

1. M. F. Browning and R. W. Kidd, unpublished annual report to Battelle's Pacific Northwest Laboratories, 1977.
2. M. F. Browning and R. W. Kidd, unpublished quarterly report to Battelle's Pacific Northwest Laboratories, 1978.
3. R. W. Kidd, M. F. Browning, and J. M. Rusin, Proceedings of the Seventh International Conference on Chemical Vapor Deposition, 1979, the Electrochemical Society, Inc., Princeton, New Jersey.
4. W. Neumann and D. Kofler, presentation at the American Ceramic Society's 81st Annual Meeting and Exposition, April 29-May 2, 1979.
5. L. S. Lobo and D. L. Trimm, Nature Phys. Sci., 234 (1971).
6. L. S. Lobo and D. L. Trimm, J. Catal., 29, 15-19 (1973).
7. L. S. Lobo and D. L. Trimm, Progr. Vac. Microbalance Tech., 2, 108 (1973).
8. L. S. Lobo, D. L. Trimm, and J. L. Figueiredo, Catal. Proc. Int. Congr. 5th, 1125-37 (1972).
9. L. S. Lobo, D. L. Trimm, and J. L. Figueiredo, Physical Org. Chem., 81, 3178 (1974).
10. L. S. Lobo, Acta Cient. Venez., Supl., 24 [2], 219-22 (1973).
11. C. A. Bernardo and L. S. Lobo, J. Catal., 37, 267-78 (1975).
12. C. A. Bernardo and L. S. Lobo, Carbon, 14 [5], 287-88 (1976).
13. C. A. Bernardo and D. L. Trimm, Carbon, 14 [4], 225-28 (1976).
14. J. H. Oxley and H. E. Carlton, AiChE, Jnl., 13 [1], 86-91 (1967).
15. T. Baird, Carbon, 15, 379-82 (1977).
16. P. Koss, O. Kofler, and W. Neumann, presentation at Third International Carbon Conference, June 30-July 4, 1980, Baden Baden, Germany.
17. M. F. Browning et al., USAEC Report CONF-117, 1963.
18. Union Carbide Corporation, Catalog Section S-5950, February 28, 1977.
19. R. J. Deifendorf, J. Chim. Phys., 57, 815 (1960).

DISCUSSION

A. S. Gates, Jr. (Fansteel VR Wesson): Would you comment on the volatility and thermal stability of cobalt acetylacetonate.

Author: To obtain appreciable concentrations in the gas phase a temperature of ~ 150 to ~ 200°C is required. Cobalt can be deposited from cobalt acetylacetonate in the temperature range of ~ 300 to ~ 400°C (H_2 atmosphere).

LASER CHEMICAL VAPOR DEPOSITION (LCVD)

Susan D. Allen

Center for Laser Studies
University of Southern California
Los Angeles, CA 90089-1112

INTRODUCTION

Laser chemical vapor deposition (LCVD) is one of several recently developed deposition techniques using laser sources. The two predominant characteristics of a laser light source--its directionality and its monochromaticity--can both be used to advantage in the deposition of materials. The directionality inherent in a laser source allows energy to be aimed very precisely at an area with dimensions on the order of the wavelength of the particular laser, causing localized deposition. The monochromaticity can be used to deposit energy directly into reacting molecules by exciting either electronic or vibrational energy levels in the reacting species. This precise control of energy flow in the system allows the deposition to occur at substrate temperatures much below those required for thermal equilibrium.

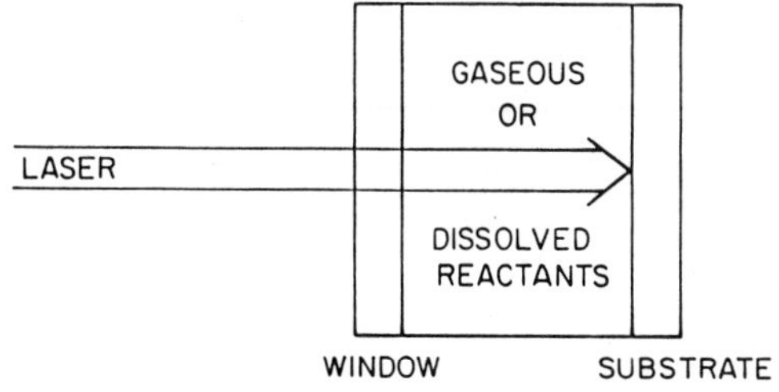

Fig. 1. Generalized laser deposition experimental arrangement.

The several kinds of laser deposition can be conveniently divided into two types: thermal and photochemical. A generalized laser deposition experimental arrangement is given in Fig. 1. For thermally driven laser deposition such as LCVD[1-8] and laser enhanced electroplating (LEEP)[9] the laser wavelength is chosen such that the reactants are transparent and the substrate absorbent. Focusing the laser on the substrate creates a localized hot spot at which the deposition takes place. For laser photochemical deposition (LPD)[10-15] the reverse conditions apply--the laser is absorbed by the reactants and the substrate is transparent. The focused laser creates a high concentration of reactive species near the substrate, causing localized deposition. In an alternative geometry, large areas can be coated using LPD by directing the laser parallel to the substrate. Many of these techniques can also be utilized in reverse, i.e., to cause localized or low temperature etching. Table 1 gives the range of experimental conditions and the materials deposited for three types of laser deposition. Other variations of both thermal and photochemical laser deposition are, of course, possible, e.g., photochemical deposition[17] and LCVD from solution on semiconductor

Table 1. Experimental Conditions for Laser Deposition

	LCVD	LEEP	LPD
Laser Source	IR, VIS, near UV	VIS	UV, Multiphoton IR
Intensity (W/cm^2)	10^2-10^6	10^4-10^5	10^{-3}-10
Reactant Concentration Torr	10-760	--	0.1-10
Deposition Rate (μm/sec)	0.1-100	$\leq$10	$\leq$0.01
Deposited	Ni[1], W[1,2] Cr[3], Mo[3], Al[3,4,5], Zn[4] Cd[4], Sn[3], Fe[2] Si[6], C[7], TiO_2[1] TiC[8], GaAs[8]	Ni[9], Cu[9] Au[9]	Cd[10], Zn[10]; Sn[10], Bi[10]; Al[10], W[10], Cr[10,1], Mo[1], Ga[10,13], As[13], GaAs[13] Si[10,14,15], Ge[10,15], SiO_2[12]

substrates.[18] The advantages of laser deposition are, in general, those associated with other laser processing techniques: (a) spatial resolution and control; (b) availability of rapid, nonequilibrium deposition conditions; (c) localization of heating in the thermal processing or low temperature processing in LPD; (d) increased purity of the deposits; and (e) ability to interface easily with other laser processing techniques such as laser annealing and contact formation in semiconductors and laser processing of metals and ceramics.

EXPERIMENTAL

In this laboratory we have investigated a wide range of LCVD reactions (Table 2) on metallic, semiconductor and insulator substrates using pulsed and cw infrared and visible lasers. Not all of the reactions given in Table 2 yielded good results under the conditions used in the preliminary survey. For example, the LCVD films from the metal alkyls were heavily oxidized due to the poor vacuum conditions employed in the early work. In order to concentrate on the characteristics unique to LCVD, only those systems which yielded films of good quality under a wide range of conditions such as the metal carbonyls and the H_2 reduction of WF_6 will be discussed in detail.

Table 2. Typical LCVD Reactions

$$Sn(CH_3)_4 \longrightarrow Sn + CH_4$$

$$TiCl_4 + H_2 + CO_2 \longrightarrow \underline{TiO_2} + HCl + CO$$

$$TiCl_4 + CH_4 \xrightarrow{(H_2)} \underline{TiC} + HCl$$

$$TiCl_4 + C_2H_2 + H_2 \longrightarrow TiC + HCl$$

$$Ga(CH_3)_3 + As(CH_3)_3 \xrightarrow{(H_2)} GaAs + CH_4$$

$$Ni(CO)_4 \longrightarrow \underline{Ni} + CO$$

$$WF_6 + H_2 \longrightarrow \underline{W} + HF$$

$$W(C))_6 \longrightarrow \underline{W} + CO$$

$$Fe(C))_5 \longrightarrow \underline{Fe} + CO$$

$$Al(CH_3)_3 \xrightarrow{(H_2)} Al + CH_4$$

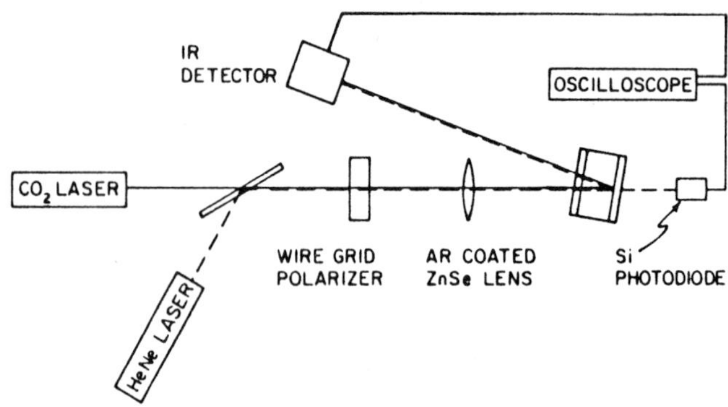

Fig. 2. Schematic diagram of the LCVD apparatus.

The cw CO_2 LCVD apparatus is shown in Fig. 2. Similar optical arrangements are used for LCVD with the pulsed CO_2 and cw Kr lasers. The laser is an electrically pulsed, line tunable cw CO_2 with maximum output power of 40 W. Attenuation of the beam is provided by current control and a wire grid polarizer. A removable power meter and or fluorescent viewing plates are used to check power stability and beam quality. A He-Ne beam is folded into the optical path with a ZnSe Brewster angle beam splitter to allow optical monitoring of the thickness of the LCVD film. Both beams are focused through a NaCl window onto the substrate with a 10 in. focal length 10.6 µm Au-coated ZnSe lens. The CO_2 beam profile at the substrate was measured by pinhole scans and is approximately gaussian with a D_{1/e^2} = 600 µm. The beam diameter of the He-Ne laser at the focus is much smaller. The temporal intensity profile of the CO_2 pulses of several ms or longer was essentially a step function with some initial overshoot as measured using a HgCdTe detector. Because the amount of reactant used during LCVD is a small fraction of the total concentration, the depositions were carried out with closed reaction cells. Fused quartz is an excellent substrate material for LCVD using a CO_2 laser as it has a high absorption coefficient at 10.6 µm and is transparent in the visible for ease of optical monitoring, both visually and with the He-Ne laser. In addition, quartz has a low thermal conductivity which tends to further localize the deposition and a high resistance to thermal stress.

LCVD Ni

In earlier work in the LCVD of Ni from $Ni(CO)_4$ on quartz substrates using the cw CO_2 laser,[1] it was found that the initial deposition rate was constant with irradiation time for a constant intensity laser pulse, as is shown in Fig. 3. This behavior is not what would be expected for deposition using an optical heat source. For the case of a reflecting (i.e., metal) film deposited on a highly absorbing substrate, the absorbed laser intensity and therefore the substrate temperature decreases as the film is deposited as shown in Fig. 4. As a result, the deposition rate should also decrease. The expected "optical self-limiting" of the film thickness has been observed in other systems, however, and will be discussed below.

Several other important points characteristic of LCVD are illustrated in Fig. 3. The initial deposition profile thickness is a truncated gaussian shape and reflects the temperature profile generated by the gaussian laser beam on the substrate. The diameter of the deposit can be much less than the corresponding beam diameter, however, because the deposition rate is a highly nonlinear, usually exponential, function of the substrate temperature. The result is a deposit thickness profile which has the shape of the central portion of a gaussian curve. This "resolution enhancement" effect is also observed in other laser deposition methods to some degree. LCVD spot diameters as small as 1/10 of the laser beam diameter have been observed.

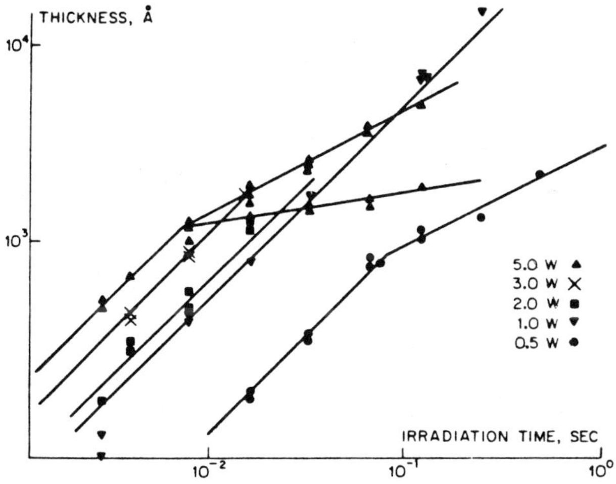

Fig. 3. Thickness of LCVD Ni films as a function of irradiation time. For the 5.0 W data, the upper curve is the maximum thickness and the lower curve is the thickness at the center of the spot. Similar branching is observed in the 3.0 and 2.0 W data but is not shown for reasons of clarity.

OPTICAL HEAT SOURCE

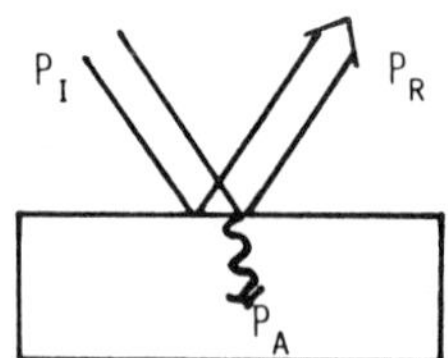

1. REFLECTOR/ABSORBER ------------ P_R INCREASES
   P_A DECREASES
2. ABSORBER/REFLECTOR ------------ P_R DECREASES
   P_A INCREASES
3. REFLECTOR/REFLECTOR
4. ABSORBER/ABSORBER ------------ P_R, P_A CONSTANT

Fig. 4. Absorptivity changes during LCVD for different film/substrate combinations.

At some point the curves of Fig. 3 change slope and split into two curves. The thickness profile for these "long" irradiation times changes smoothly from the truncated gaussian to a double humped "volcano" shape. The two curves therefore represent the maximum thickness and the center thickness of the volcano thickness profile. As the laser intensities used are not sufficient to melt either quartz or Ni and no evidence of melting is observed, the change in profile is ascribed to a depletion of reactants and build up of products at the center for the irradiated area and entrainment of reactants from the edges.

LCVD Fe

The thickness of LCVD Fe from $Fe(C))_5$[2] as a function of irradiation time, τ, for several laser powers is given in Fig. 5a. For irradiation times longer than the maximum given for each case, the deposition thickness profiles deviated significantly from the truncated gaussian shape observed for short irradiation times. In contrast to the similar data for Ni in Fig. 3, the thickness is not a linear function of the irradiation time and the lower intensity data (2.2 W) exhibits a steeper slope (thickness proportional to $\tau^{0.8}$) than the higher intensity points (5.5 W, thickness proportional to $\tau^{0.5}$). The same data is replotted in Fig. 5b as the apparent deposition rate, defined as the deposit thickness divided by the irradiation time. The actual deposition rate is a more complicated function of the heating time and the changing absorptivity of the surface during deposition, but this definition serves to illustrate qualitatively the optical self-limiting of the film thickness. For the shortest irradiation times, 2 ms, the higher incident laser intensity yields a faster deposition rate, in accord with the higher laser generated surface temperature. As the irradiation time increases, however, the deposited film reflects increasing amounts of the incident energy and the surface temperature decreases. The deposition rate should therefore approach a constant for long irradiation times. Before this point, however, other limiting mechanisms come into play. For example, as discussed above, diffusion processes may lead to a decreased reactant concentration at the center of the laser heated spot for relatively long irradiation times, resulting in decreased deposition at the spot center.

Another indication that optical self-limiting is occurring is given by experiments in which multiple irradiations of 3 ms each at 5.5 W incident power were made on a single site. After the initial deposition of approximately 500 Å, no additional deposition is observed. For the multiple depositions, the site is allowed to cool to ambient between each irradiation and the change in reflectivity from the uncoated quartz (R=0.10) to several hundred Å of Fe (R=0.94)[19] would decrease the laser generated surface temperature by almost an order of magnitude. For the continuously irradiated site, the decrease in surface absorptivity occurs on an already heated substrate and the surface temperature decreases but may remain above the deposition temperature.

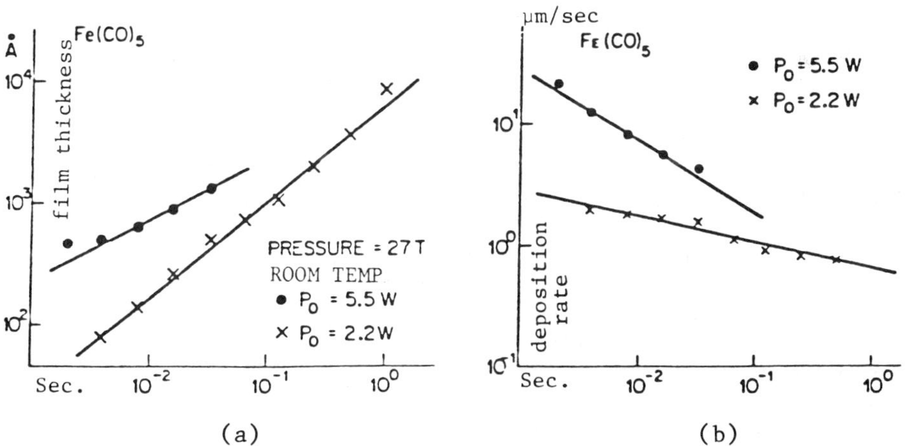

Fig. 5. (a) LCVD film thickness as a function of irradiation time (Fe from $Fe(CO)_5$). (b) Apparent deposition rate (film thickness irradiation time) as a function of irradiation time (Fe from $Fe(CO)_5$).

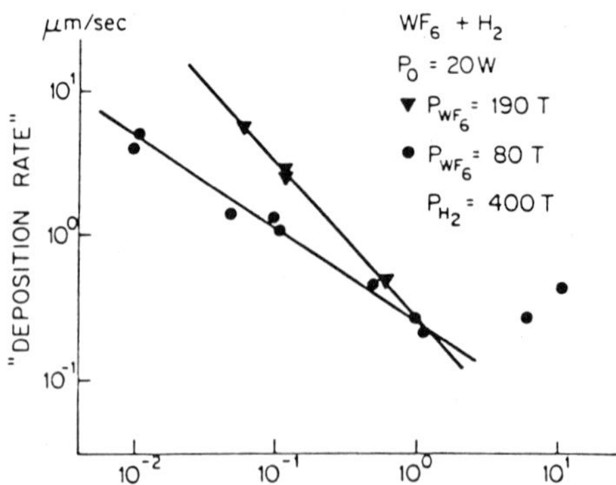

Fig. 6. Apparent deposition rate (film thickness/irradiation time as a function of irradiation time (W from WF_6 + H_2).

The LCVD Ni and Fe films were adherent, hard, shiny and metallic looking. Surface microstructure as determined by scanning electron microscopy (SEM) was essentially featureless to less than the resolution of the instrument (100 Å). The electrical resistivity of a 550 Å thick LCVD Ni film was measured as less than 4×10^{-5} ohm-cm but similar films of LCVD Fe showed high resistivities presumably due to oxidation of the relatively thin film.

LCVD W

LCVD of W from WF_6 + H_2 was carried out as a function of irradiation conditions for WF_6 pressures ranging from 40 to 200 T with a constant H_2 pressure of 400 T. The apparent deposition rate is plotted in Fig. 6 as a function of the irradiation time. In this case the deposition was carried out with the same laser intensity but with different concentrations of WF_6. The higher concentrations of WF_6 yielded greater initial deposition rates, as expected from kinetic considerations, which decreased faster with increasing irradiation time. For irradiation times on the order of 1 sec, the apparent deposition rate is the same for both reactant concentrations. (The lower pressure data for times greater than 1 sec is spuriously high because of buckling of the W films). This functional behavior is consistent with the optical self-limiting mechanism.

Lines were generated by translating the substrate under the laser beam at speeds of 1-10 cm sec. Cross sectional profiles were approximately trapezoidal with widths varying from 150 to 400 μm depending on scan speed and incident intensity. Resistivity measurements of lines 0.5 cm long and 600 Å thick, 400 μm wide and 1500 Å thick, 160 μm wide yielded 2×10^{-5} and 6×10^{-5} ohm-cm, respectively. These numbers compare favorably to the bulk resistivity of W ($\rho = 6 \times 10^{-5}$ ohm-cm[19]) particularly since the LCVD conditions have not been optimized. With the exception of irradiation times much greater than 1 sec, the LCVD films were quite adherent. Scanning electron microscope examination of the LCVD films revealed surface structure that varied from smooth (no observable features greater than 100 Å), to granular nodules approximately 2000 Å in diameter (Fig. 7a) to small crystallites approximately 4000 Å (Fig. 7b) in diameter. This change in surface structure appears to be related to both irradiation conditions and diffusion processes with larger crystallites observed for longer irradiation times, as would be expected. The optical self-limiting of the $WF_6 + H_2$ LCVD reaction is graphically illustrated in Fig. 7. The film shown in Fig. 7b is only about 40% thicker than that of Fig. 7a although the reactant concentration is higher and the irradiation time is almost two orders of magnitude longer.

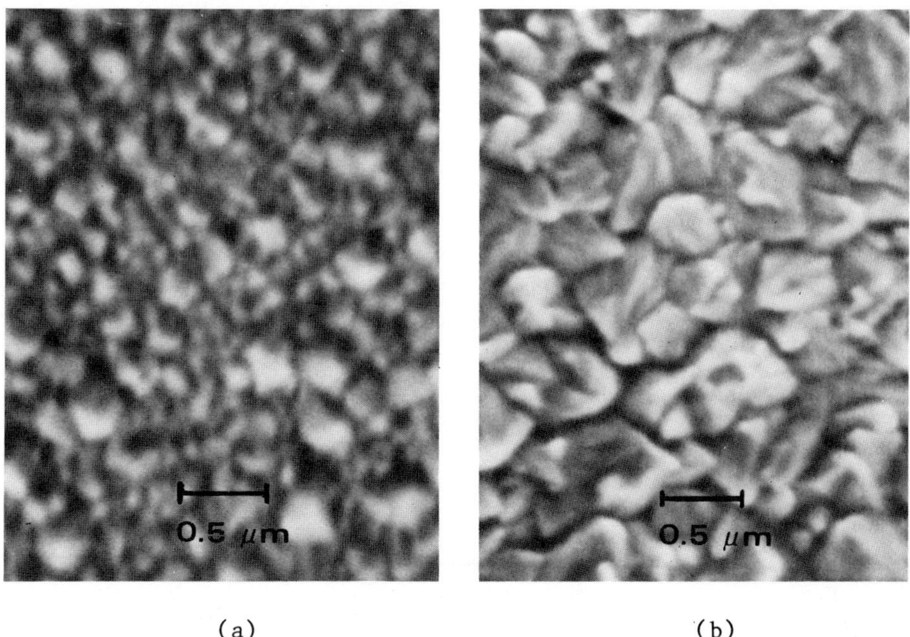

Fig. 7. (a) SEM photograph of LCVD W film (W from $WF_6 + H_2$, P = 20 W for 50 ms 5 irradiations). (b) SEM photograph of LCVD W film (W from $WF_6 + H_2$, P = 20 W for 2 sec.).

Experiments similar to those described above for LCVD W from WF_6 + H_2 have recently begun using Si substrates and a Kr ion laser. The Blue-green lines (468-530 nm) of the Kr laser are strongly absorbed by the Si and most gaseous reactants are transparent in the visible, satisfying the LCVD requirements. The laser is focused with a 6.5 cm focal length lens onto the substrate with a beam diameter of approximately 20 μm at the focus. Lines and spots were deposited at scan speeds of 0.2-2.0 mm/sec at laser powers of 3-4 W. A curious growth pattern was observed for some spot depositions. Single crystal W spikes could be deposited with dimensions on the order of 40-50 μm wide and 50-120 μm tall. These are similar in aspect ratio to the polycrystalline C and Si rods deposited by LCVD by Bauerle et al.,[6,7] but the appearance of one or, at most, a few large crystals has not previously been observed. The growth rates measured for these crystallites approach a mm/sec. The reason such structures have not been seen in the LCVD of W using the CO_2 laser is presumably due to the large difference in spot sizes--600 μm for CO_2 and 20 μm for the Kr laser. If the incident laser beam impinges on only a few growing crystallites, it is reasonable that one or more would begin to dominate in the growth. Additional experiments are planned to quantify the irradiation conditions which favor such single crystal deposition.

LCVD GaAs

GaAs was deposited using the reactions: $Ga(CH_3)_3 + As(CH_3)_3 + H_2 \rightarrow GaAs + CH_4$. Both $Ga(CH_3)_3$ and $As(CH_3)_3$ were distilled into the side-arm of the deposition cell and 150 T of H added. The side arm of the cell was maintained at 0°C and the vapor pressures of $Ga(CH_3)_3$ and $As(CH_3)_3$ were 65 and 98 T, respectively. The range of irradiation conditions over which deposition was observed was quite narrow: $D_1/e^2 \cong 0.6$ mm; $P_0 = 2\text{-}3$ W; $\tau = 2\text{-}12$ s. For higher incident powers, deposition occurred in a ring around a center devoid of measurable GaAs. The surface temperatures achieved under these irradiation conditions are approximately the same as used for conventional MO CVD[20] 600-900°C.

The LCVD films were not continuous and showed a strong effect of nucleation sites on the substrate. The scratches on the surface of the substrate are readily apparent in the growth of the film. This type of deposition has been observed in conventional CVD on similar substrates.[21] Deposition rates were much slower than observed for other LCVD materials:[1] 150-500 Å/s, but were comparable to rates observed in conventional MO CVD.[20] Film thickness could be built up by multiple irradiations of the same site, resulting in films several microns thick and better but not complete coverage. In one case, the laser irradiated a site where some of the liquid material had condensed. The deposition rate observed was several orders of magnitude greater than in a "clean" site. The composition of the LCVD GaAs films was analyzed using SEM/EDAX. The Ga/As ratio varied with irradiation conditions from 0.3 to 2.0, bracketing the value of pure GaAs, 0.73.

LCVD TiC and TiO_2

In order to deposit TiC on carbon steel substrates, several orders of magnitude more laser power is necessary due to the decreased absorptivity and increased thermal diffusivity of the steel relative to quartz. TiC was deposited from $TiCl_4$ at the room temperature vapor pressure of 12 T and CH_4 or C_2H_2 (200 T) with and without H_2 (200 T). The high power LCVD apparatus consisted of a stainless steel tank electroless Ni plated with x and y translation stages driven by an external stepping motor. Coating tracks were deposited by translating the substrate under the stationary CO_2 laser beam. Incident powers were 500-1200 W, D_1/e^2 = 1-2 mm and translation speeds of 3-6 mm/s.

A SEM of the fracture cross section of a typical LCVD TiC track is given in Fig. 8a along with the Ti x-ray map of the same area (Fig. 8b). The Ti is concentrated in the fine grained region shown in Fig. 8a which corresponds to the LCVD TiC coating. The fine grained material has a grain size of 500-1000 Å. These LCVD tracks show increased surface hardness but are not thick enough (typically 0.5-1.0 μ) for the hardness to be measured directly.

Fig. 8a. SEM photograph of fracture cross section of LCVD TiC film on carbon steel ($TiCl_4$ + CH_4).

Fig. 8b. Ti x-ray map of the same area as Fig. 8a.

TiO_2 was deposited from $TiCl_4 + CO_2 + H_2$. $TiCl_4$ was vacuum distilled into a side-arm tube of the deposition cell and 205 T each of CO_2 and H_2 added. The partial pressure of $TiCl_4$ was 12.4 T, the room temperature vapor pressure. The irradiation conditions were: laser beam diameter D_1/e^2 = 0.8-3.0 mm; laser power P = 5-10 W; and irradiation time τ = 0.1-1.0 s. The measured deposition rate was 300-3000 Å/s. The LCVD TiO_2 films were clear and adherent. Other properties are given in Ref. 1.

Optical Monitoring of LCVD

Using the apparatus shown in Fig. 2, the LCVD film growth was monitored in real time by monitoring both the reflected CO_2 beam using a Ge:Au or HgCdTe detector and the transmitted He-Ne beam using a Si photodiode.

Representative transmission and reflection curves are given in Fig. 9 for the W from $WF_6 + H_2$ system for different laser intensities and an irradiation time of nominally 100 ms. The upper curve is the He-Ne intensity and the lower curve the reflected CO_2 intensity as measured by a Ge:Au detector. At the lower incident power (Fig. 9a) a delay of approximately 67 msec is observed before significant deposition takes place as evidenced by a steep decrease in the He-Ne transmission and an increase in the CO_2 reflectance. (The initial decay in the transmission is due to substrate heating and subsequent defocusing of the He-Ne beam.) A comparison of the delay times for several different incident intensities showed that the data could be fit by a very simple model assuming no deposition until a "threshold temperature" was reached and one dimensional heat flow in the laser heated substrate. Under these conditions, the surface temperature is proportional to the incident intensity and the square root of the irradiation time if we further assume that the optical and thermal properties of the quartz substrate are constant. As shown in Table 3, the product of the incident power and the square root of the delay time before deposition takes place is a constant, indicating that there is some validity to the simple model. Similar calculations of the actual surface temperature for a gaussian beam do not yield reasonable numbers however.[22] For quartz it is necessary to take into account the temperature dependence of the thermal properties such as heat capacity and thermal diffusivity. Such nonlinear calculations yield surface temperatures for the experimental conditions of Table 3 of 800-950°C,[23] in good agreement with temperatures used in conventional CVD.[24]

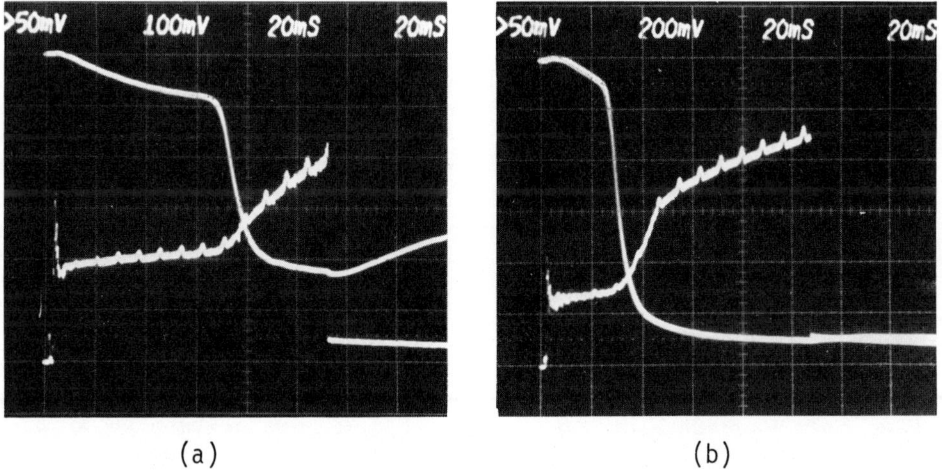

(a) (b)

Fig. 9. Transmission and reflection vs. time curves for cw LCVD of W from $WF_6 + H_2$ (rt T, 400 T). The upper curve is the He-Ne transmission intensity and the lower curve the CO_2 reflected intensity. Laser power: (a) 4.7 W; (b) 7.0 W.

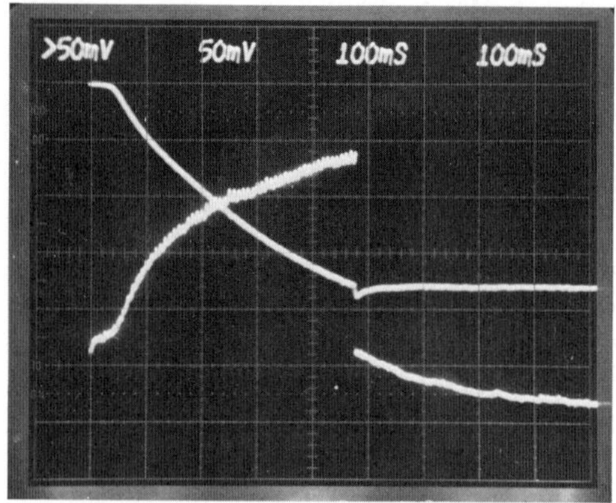

Fig. 10. Transmission and reflection vs. time curves for cw LCVD of Fe from $Fe(CO)_5$ (27 T). The upper curve is the He-Ne transmission intensity and the lower curve the CO_2 reflected intensity. Laser power is 2 W.

Table 3. Relative Surface Temperature, T_s, for Different Irradiation Conditions (W from WF_6 + H_2)

P (watts)	τ (msec)	$P\tau^{1/2} \propto T_s$
4.7	55.8	35.1
7.0	22.5	33.2
10.0	10.9	33.0
15.0	4.9	33.2

A representative transmission and reflection curve for LCVD Fe from $Fe(CO)_5$ is given in Fig. 10. In this case there is a similar delya time before deposition begins (approximately 35 msec) but the deposition is much slower and continues for several hundred milliseconds. The calculated surface temperature before deposition begins for the irradiation conditions of Fig. 10 is 500°C. For the lower laser power used in the Fe from $Fe(CO)_5$ LCVD experiment there is no observable thermal defocusing of the He-Ne beam.

CONCLUSIONS

From a comparison of both the thickness vs. irradiation time data and the real time optical monitoring of the deposition initiation and rate for the several deposition systems examined, some general conclusions can be drawn. For the low temperature reactions such as the Ni from $Ni(CO)_4$ deposition, the laser generated substrate temperature after the deposition of a reflecting metal film is still above the threshold temperature for the reaction. The initial LCVD rate in this case must be very rapid and the optical self-limiting must occur for irradiation times shorter than the 3 msec used.[1] The linear dependence of deposition rate with irradiation time can be attributed to the substrate temperature reaching an effective equilibrium. For the higher temperature reactions such as W from WF_6 + H_2, there is a delay (Table 2) before the substrate temperature reaches the threshold temperature of the reaction. The change in optical absorption with LCVD film growth of several hundred Å lowers the substrate temperature below the threshold temperature, and film growth ceases. The Fe from $Fe(CO)_5$ case lies between these two extremes. This simple picture ignores any kinetic or chemical effects which may be observed. The model does, however, give a general picture of the optical and thermal properties which must be taken into account in LCVD.

Applications for LCVD films will depend not only on the localization property of the technique but also on the ability to generate different film microstructures using the rapid heating and cooling rates available with a laser heat source. It is possible to deposit films with dimensions less than the diameter of the depositing laser focal spot due to the exponential dependence of the deposition rate on temperature. For properly chosen irradiation conditions, the LCVD film will only be deposited in the center of the laser heated hot spot on the substrate surface.[1] Spot sizes of 1/10 of the beam have been observed. In order to utilize this resolution enhancement, short irradiation times are necessary, where short is defined in terms of the thermal diffusivity of the substrate. LCVD experiments with a pulsed CO_2 TEA laser have shown that irradiation times as short as 1 μsec can be used to deposit useful film thicknesses for some reaction systems, thereby minimizing lateral thermal spreading.

The rapid deposition rates measured for LCVD are a consequence of several factors. High surface temperatures can be achieved because only a small area is heated, effectively suppressing gas phase reactions in favor of surface reactions. For the same reasons, high concentrations of reactants can be utilized. A third effect is the enhanced diffusion of reactants inherent in the spot geometry of the reaction site as opposed to the planar geometry of conventional CVD. The increase in diffusion of reactants should be a function of the irradiated spot size with smaller deposit geometries exhibiting faster deposition rates.

Initial results on the LCVD of compounds have not been as promising as those obtained for metal films, with deposition rates being an order of magnitude or more smaller. Additional experiments are planned for these more complex systems.

Possible applications for LCVD coatings include: IC mask and circuit repair; writing interconnects for IC circuits; one-step ohmic contacts; localized protective coatings; localized coatings for waveguide optics; and localized wear and corrosion resistant coatings.

ACKNOWLEDGMENTS

This work was supported in part by a grant from AFOSR under the technical cognizance of H. Schlossberg.

REFERENCES

1. S. D. Allen, J. Appl. Phys., 52, 6501 (1981) and references therein.
2. S. D. Allen and A. B. Trigubo, J. Appl. Phys. (Feb. 1983).
3. S. D. Allen and A. B. Trigubo, unpublished results.
4. Y. Rytz-Froidevaux, R. P. Salathe, and H. H. Gilgen, Phys. Lett., 84A, 216 (1981).
5. R. W. Bigelow, J. G. Black, C. B. Duke, W. R. Salaneck and H. R. Thomas, Thin Sol. Films, 94, 233 (1982).
6. C. P. Christensen and K. M. Lakin, Appl. Phys. Lett., 32, 254 (1978); V. Baranauskas, C. I. Z. Mammana, R. E. Klinger, and J. E. Greene, Appl. Phys. Lett., 36, 920 (1980); D. J. Ehrlich, R. M. Osgood, Jr., and T. F. Deutsch, Appl. Phys. Lett., 39, 957 (1981); D. Bauerle, P. Irsigler, G. Leyendecker, H. Noll, and D. Wagner, Appl. Phys. Lett., 40, 819 (1982).
7. G. Leyendecker, D. Bauerle, P. Geittner, and H. Lydtin, Appl. Phys. Lett., 39, 921 (1981).
8. S. D. Allen and A. B. Trigubo, J. Vac. Sci. Tech., 20, 469 (1982).
9. R. J. von Gutfeld, R. E. Acosta, and L. T. Romankiw, IBM J. of Res. and Dev., 26, 136 (1982); J. C. Puippe, R. E. Acosta, and R. J. von Gutfeld, J. Electrochem. Soc., 128, 2539 (1981); R. J. von Gutfeld, E. E. Tynan, R. L. Melcher, and S. E. Blum, Appl. Phys. Lett., 35, 651 (1979).
10. T. F. Deutsch, D. J. Ehrlich, and R. M. Osgood, Jr., Appl. Phys. Lett., 35, 175 (1979); D. J. Ehrlich, R. M. Osgood, Jr., and T. F. Deutsch, J. Vac. Sci. Tech., 21, 23 (1982); D. J. Ehrlich, R. M. Osgood, Jr., and T. F. Deutsch, IEEE J. Quant. Elec., QE-16, 1233 (1980) and references therein.

11. R. Solanki, P. K. Boyer, J. E. Mahan, and G. J. Collins, Appl. Phys. Lett., 38, 572 (1981); R. Solanki, P. K. Boyer, and G. J. Collins, Appl. Phys. Lett., 41, 1048 (1982).
12. P. K. Boyer, G. A. Roche, W. H. Ritchie, and G. J. Collins, Appl. Phys. Lett., 40, 716 (1982); P. K. Boyer, W. H. Ritchie, and G. J. Collins, J. Electrochem. Soc., 129, 2155 (1982).
13. J. G. Berg, P. Yeung, and S. D. Allen, TRW, El Segundo, CA, unpublished results.
14. M. Hanabusa, A. Namiki, and Keitaro Yoshihara, Appl. Phys. Lett., 35, 626 (1979).
15. R. W. Andreatta, C. C. Abele, J. F. Osmundsen, J. G. Eden, D. Lubben, and J. E. Greene, Appl. Phys. Lett., 40, 183 (1982).
16. See also Proceedings of the MRS Symposium on Laser Diagnostics and Photochemical Processing for Semiconductor Devices (11-82), North Holland, NY.
17. R. H. Micheels, A. D. Darrow, and R. D. Rauh, Appl. Phys. Lett., 39, 418 (1981).
18. R. F. Karlicek, V. M. Donnelly, and G. J. Collins, J. Appl. Phys., 53, 1084 (1982).
19. American Institute of Physics Handbook, edited by D. E. Gray, McGraw-Hill, NY, 1972.
20. P. D. Dapkus, H. M. Manasevit, and K. L. Hess, J. Cryst. Growth, 55, 10 (1981).
21. P. D. Dapkus, private communication.
22. J. F. Ready, Effects of High-Power Laser Radiation, Academic Press, NY, 1971.
23. S. D. Allen and J. Goldstone, to be published.
24. C. F. Powell, J. H. Oxley, and J. M. Blocher, Jr., Vapor Deposition, Wiley, NY, 1968.

PART V

ION BEAM DEPOSITION

ION BEAM TECHNIQUES FOR THE DEPOSITION OF CERAMIC THIN FILMS

James M. E. Harper

IBM Thomas J. Watson Research Center
Yorktown Heights, NY 10598

ABSTRACT

Broad-beam multiaperture ion sources are capable of generating collimated ion beams of narrow energy spread and high flux at low background pressure. These features allow the preparation of compound thin films under highly controlled conditions. The operation of broad-beam ion sources is summarized, followed by two examples of compound formation using ion beams: surface compound layer formation (e.g., ion beam oxidation of Ni); and reactive compound deposition (e.g., dual ion beam deposition of Si_3N_4*). In each case, the quantitative information inherent in the ion beam technique adds to the understanding and control of the film growth environment. Areas of ion source development are also discussed.*

INTORDUCTION

Broad-beam ion sources were invented by Kaufman for space propulsion in 1961.[1] Extensive development of ion source technology has yielded designs capable of generating large diameter (many cm) ion beams, operating in the range of tens of eV to several keV with ion flux values up to several mA/cm^2 (1 mA/cm^2 = 6.25 x 10^{15} ions/cm^2-sec).

The development of ion source technology and its application to thin film processing is described in recent review articles.[2,3] Ion beam techniques are being applied to the synthesis of ceramic materials because they allow the preparation of compound thin films under highly controlled conditions of material flux, energy and direction in a low pressure environment (typically 0.01 Pa background pressure). Examples of ceramic materials prepared by ion beam processing are: Al_2O_3, SiO_2, perovskites,[4] BN,[5,6] TiN,[7] AlN,[8] NbN,[9] Si_3N_4,[10] and hard amorphous carbon.[6,11]

BROAD-BEAM ION SOURCES

The design and operation of broad-beam ion sources is described in detail elsewhere[2,12] and will be only summarized here. As shown in Fig. 1, a broad-beam ion source consists of a discharge plasma region, an extraction grid region, and an ion beam region. The discharge plasma is maintained by electrons emitted from a hot cathode and accelerated by a discharge voltage of typically 50 V to an array of anodes. To enhance the ionization efficiency, a multipole magnetic field is configured above the anode surfaces to increase the path length of electrons in the discharge chamber.

Ions are extracted through a pair of grids with a large number of aligned apertures. The potential difference between the grids (determined by the anode potential V_{anode} and the grid potential V_{accel}) accelerates and directs the ions into a collimated beam. The ion energy is determined by the anode voltage, and is variable from tens of eV up to several keV, with a low energy spread of about 10 eV. The beam intensity (ion flux) is independently controllable by varying the heating current to the hot cathode. When directing the ion beam at an insulating target, a neutralizing filament is added to the beam, to compensate the positive charge of the beam ions. Ion beams of most gases can be generated including inert species (e.g., Ar, Kr, Xe), reactive species (e.g., O_2, N_2, H_2) and more complex molecular species (e.g., CH_4, CF_4, $B_3N_3H_6$). The latter undergo fragmentation in the discharge, giving a distribution of ion species in the beam.

As compared with radio frequency driven plasma processes, ion beam processing has the advantage of physically and electrically separating the region of ion generation from the point of use, thereby minimizing the exposure of samples to undesired energetic species. Broad ion beams provide the control needed to perform quantitative analysis of thin film growth processes.

DEPOSITION CONFIGURATIONS

The four major configurations used in ion beam deposition are shown in Fig. 2.

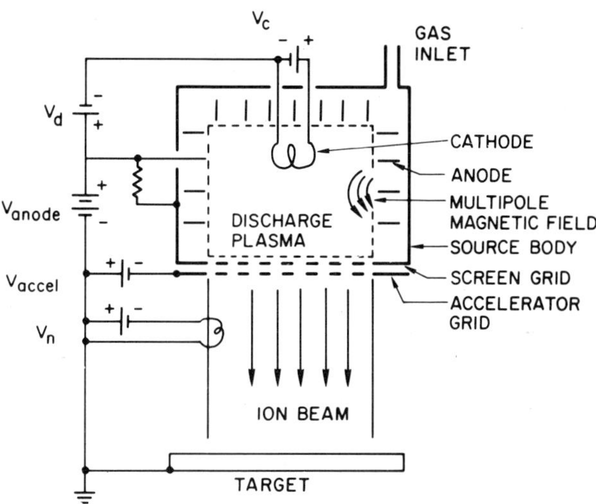

Fig. 1. Schematic diagram of a Kaufman-type multiaperture ion source.

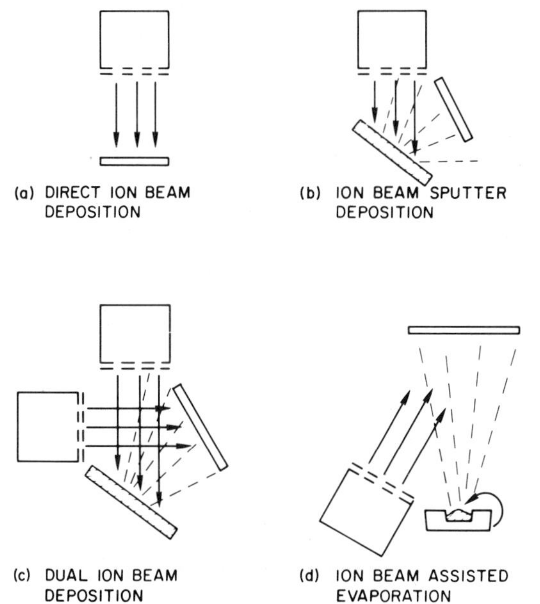

Fig. 2. Configurations used in ion beam deposition of thin films.

Direct ion beam deposition (Fig. 2a), in which the beam species is deposited directly on the substrate, has limited usefulness with metal species, due to severe limitations on beam energy and size.[13,14] However, with certain molecular gases, such as CH_4, hard insulating carbon films have been deposited at reasonable deposition rates.[10,15] Alternatively, a reactive ion beam directed at a sample surface can be used to grow surface compound layers.[16]

Ion beam sputter deposition (Fig. 2b), uses an inert or reactive ion beam to sputter a target material for deposition onto a substrate. Since almost any solid material can be sputtered, this process is as widely applicable as r.f. sputtering, and has been used in many thin film applications.[2] For compound deposition, however, this configuration is limited by the low energy of arriving species at the substrate, which may not provide full stoichiometry.

A more controllable technique for compound deposition is the dual ion beam configuration (Fig. 2c), in which an inert ion beam sputters a target material (e.g., Si) and a reactive ion beam (e.g., N_2^+) is directed at the substrate to continuously form a compound film (Si_3N_4). Introduced by Weissmantel,[8] this technique has also been applied successfully to the control of alloy film compositions[17] and to improvement of step coverage[18] in multilayer devices. The dual beam technique has the advantage of delivering the reactive species to the film with enough energy (tens to hundreds of eV) to activate the chemical reaction needed to obtain stoichiometry. The impact zone is characterized by extensive atom motion for a short duration, simulating a rapid quench from a high temperature state, while not significantly raising the bulk temperature.

The fourth configuration (Fig. 2d), ion beam assisted evaporation, combines the advantage of ion bombardment with the versatility and high rates of electron beam or resistance-heated evaporation. This configuration is also suited to reactive compound formation at the substrate. In addition, inert ion bombardment during thin film evaporation has been shown to dramatically alter the intrinsic stress of the film.[9,19]

EXAMPLES OF COMPOUND FORMATION

Surface Compound Layer Formation

Using the configuration of Fig. 2a, a beam of reactive gas ions such as O_2^+ or N_2^+, can be directed at a material surface to grow a surface compound layer, such as an oxide or nitride. This process has been recently reviewed, and compared with the related techniques of plasma oxidation, r.f. oxidation, and ion implantation of buried oxide layers.[20]

The growth rate of the surface compound layer depends on the fraction of incident ions incorporated in the film, the diffusion rate through the growing layer, and the sputter etching rate of the compound layer under ion bombardment. The possibilities are shown schematically in Fig. 3, which shows growth occurring for 100% incorporation, < 100% incorporation (if ion reflection and sputtering occur), and > 100% incorporation (if an adsorbed gas layer is also activated by the ion impact). Saturation of the compound layer thickness will occur when a balance is reached between the oxide growth rate and the sputter etching rate. By using a modified multi-aperture ion source capable of high ion flux (1 mA/cm^2) at low energy (< 100 eV), Harper et al.[16] demonstrated this self-limiting oxide growth on Ni surfaces, as shown in Fig. 4. This figure shows the resistance of Ni-oxide-Ni junctions formed by ion beam oxidation followed by sputter deposition of Ni, as a function of oxidation time. The oxide thickness is approximately given by the logarithm of resistance. At an ion energy of 80 eV, a self-limiting oxide thickness is achieved, whereas, at 45 eV, growth continues for the time shown, because the sputter etching rate is much lower at low ion energy. Varying the ion energy over the range of 30-200 eV allowed control of junction resistance over a range of five orders of magnitude.

Low energy ion beam growth of surface compound layers has been applied to superconducting tunneling devices,[20] metal-oxide-metal devices,[16] and to the growth of Si_3N_4 surface layers on Si for delineating small device structures.[21] This process may be viewed as the low energy extension of ion implantation modification of surface properties,[22] a process capable of chemically and physically altering the near-surface region without affecting bulk material properties.

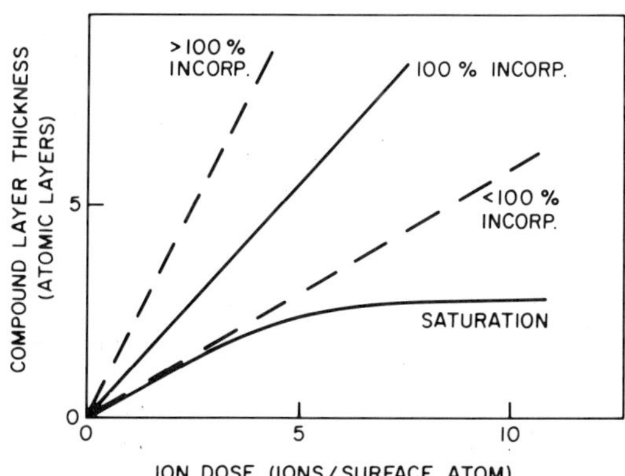

Fig. 3. Thickness of surface compound layer as a function of ion dose for low energy ion beam surface layer formation.

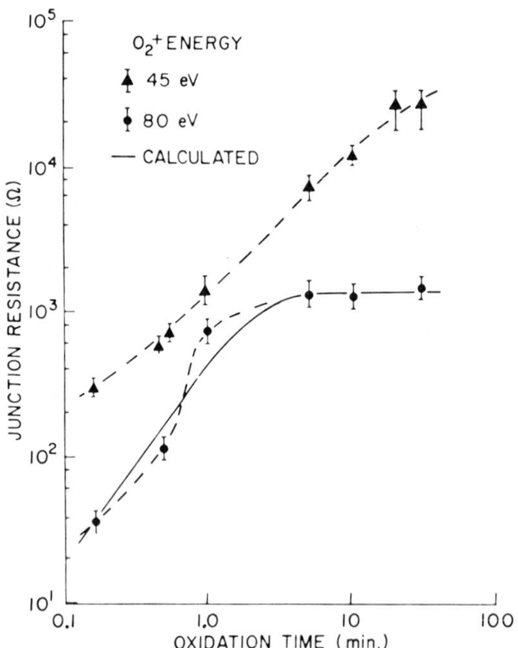

Fig. 4. Resistance of Ni-oxide-Ni junctions as a function of oxidation time, for low energy ion beam oxidation at 45 eV and 80 eV, and 0.7 mA/cm^2.

Reactive Compound Deposition

Using the configurations of Fig. 2c and 2d, thin films may be deposited under reactive ion bombardment to form stoichiometric compound phases during deposition. This process is essentially surface compound layer formation while supplying additional metal atoms to form a continuously fresh surface. This technique is of interest in the fabrication of ceramic materials because of the control of stoichiometry available, and the low substrate temperatures (generally near room temperature).

The general behavior is summarized in Fig. 5. For a fixed deposition rate of material A (e.g., a metal atom flux), as the flux of low energy reactive ions of species B is increased, the concentration of species B in the film will increase at a rate determined by the fraction incorporation. This will be related to the reflection coefficient and the sputtering yield. When the arrival rate ratio of B ions to A atoms reaches a value at which a stable compound phase (e.g., AB_2) is formed, a further increase in the flux of B ions will not necessarily cause further incorporation. The compound structure may reject excess atoms of B by diffusion from the film surface. Such a self-limiting composition has been observed by Erler et al.[10] in the dual ion beam deposition of Si_3N_4 as shown in Fig. 6. In this work, 630 eV N_2^+ ions were directed at a growing film of Si, sputtered from a Si target by an Ar^+ ion beam. The film composition is shown in Fig. 6 as a function of the ratio of nitrogen ion dose to Si atom arrival rate. Evaluation of the film composition as a function of ion/atom arrival rate ratio disclosed that only one-third of the flux of impinging N atoms were incorporated in the film. By decreasing the ion impact energy of the reactive beam, it is expected that this incorporation probability will increase. The dependence of film properties on the degree of excess nitrogen ion bombardment remains to be fully studied.

AREAS OF DEVELOPMENT

The application of broad ion beams to compound material processing is rapidly developing, as the control inherent in ion beam techniques yields information about materials parameters, and thereby leads to control of material properties. One area yet to be fully exploited is the directionality of ion beams, which enables surface structures to be selectively treated according to their orientation. The operation of broad beam ion sources with molecular gas species is also an area of development, and is expected to result in techniques for controlling beam composition in addition to energy and flux.

The development of compact ion source units[2] has made it possible to evaluate ion beam processing techniques in conjunction with existing vacuum processes, such as evaporation, without the necessity of completely reconfiguring the system. New materials for ion extraction grids, such as single crystal Si,[23] enable greater control of uniformity and intensity of the ion flux. The intrinsic current limitations of dual-grid extraction systems[12] are also being overcome with fine mesh single grid designs,[2,16] and with a gridless design based on the Hall effect,[2] in which a circulating electron current creates a field configuration performing the extraction function of a conventional grid.

The emphasis on ion source design is to increase the ion flux available at low ion energy (< 100 eV), and to improve the performance with reactive gas species. These developments will create a bridge between high temperature chemical processing of materials and high energy (> 1000 eV) ion beam processes. This merging of fields will be based on the chemical processing of materials by low energy ion beams.

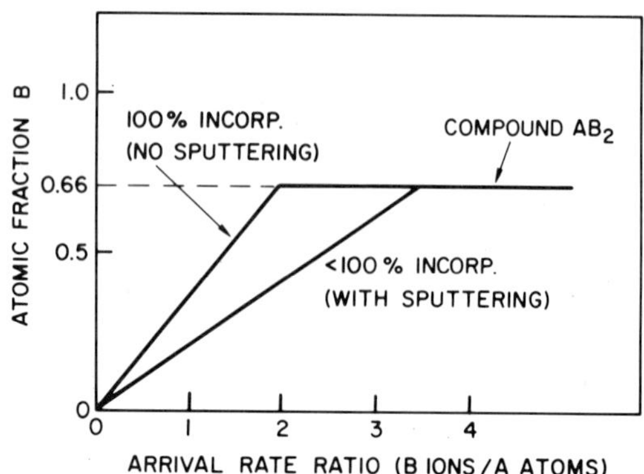

Fig. 5. Fraction of reactive species B incorporated in film of material A as a function of the arrival rate ratio of B ions to A atoms, in dual reactive ion beam deposition of compounds.

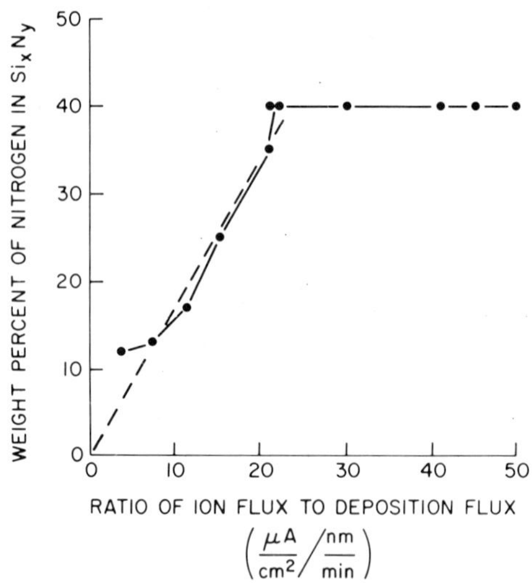

Fig. 6. Concentration of nitrogen in Si_xN_y films deposited by dual reactive ion beam deposition, using 630 eV nitrogen ion bombardment of the film, as a function of the ratio of ion dose to Si arrival rate. From reference 10.

ACKNOWLEDGMENTS

The author thanks J. J. Cuomo and H. R. Kaufman for collaboration in the development of ion beam applications to materials.

REFERENCES

1. H. R. Kaufman, NASA Tech. Note TND-585, 1961.
2. H. R. Kaufman, J. J. Cuomo, and J. M. E. Harper, J. Vac. Sci. Technol., 21, 725 (1982); J. M. E. Harper, J. J. Cuomo, and H. R. Kaufman, J. Vac. Sci. Technol., 21, 737 (1982).
3. J. M. E. Harper, J. J. Cuomo, and H. R. Kaufman, Ann. Rev. Materials Science (to be published 1983).
4. R. N. Castellano and L. G. Feinstein, J. Vac. Sci. Technol., 16, 184 (1979).
5. S. Shanfield and R. Wolfson, J. Fac. Sci. Technol. (to be published 1983).
6. C. Weissmantel, K. Bewilogua, K. Breuer, D. Dietrich, U. Ebersbach, H. J. Erler, B. Rau, and G. Reisse, Thin Solid Films, 96, 31 (1982).
7. M. Satou, S. Fukui, and F. Fujimoto, Proc. of Int'l Workshop on Ion-based Techniques for Film Formation, p. 349, Tokyo, 1981.
8. C. Weissmantel, Thin Solid Films, 32, 11 (1976); C. Weissmantel, G. Geisse, H. J. Erler, F. Henny, K. Bewilogua, U. Ebersbach, and C. Schurer, Thin Solid Films, 63, 315 (1981).
9. J. J. Cuomo, J. M. E. Harper, C. R. Guarnieri, D. S. Yee, L. J. Attanasio, J. Angilello, C. T. Wu, and R. H. Hammond, J. Vac. Sci. Technol., 20, 349 (1982).
10. H. J. Erler, G. Reisse, and C. Weissmantel, Thin Solid Films, 65, 233 (1980).
11. T. J. Moravec and T. W. Orent, J. Vac. Sci. Technol., 18, 226 (1981).
12. H. R. Kaufman, pp. 265-373 in Electronics and Electron Physics, Vol. 36, edited by L. Marton, Academic, NY, 1974; H. R. Kaufman, J. Vac. Sci. Technol., 15, 272 (1978).
13. J. M. E. Harper, p. 175 in Thin Film Processes, edited by J. L. Vossen and W. Kern, Academic, NY, 1978.
14. J. Amano and R. P. W. Lawson, J. Vac. Sci. Technol., 14, 831 (1977); R. B. Fair, J. Appl. Phys., 42, 3176 (1971).
15. P. Oelhafen, J. L. Freeouf, J. M. E. Harper and J. J. Cuomo, J. Appl. Phys. (1983).
16. J. M. E. Harper, M. Heiblum, J. L. Speidell, and J. J. Cuomo, J. Appl. Phys., 52, 4118 (1981).
17. J. M. E. Harper and R. J. Gambino, J. Vac. Sci. Technol., 16, 1901 (1979).
18. J. M. E. Harper, G. R. Proto, and P. D. Hoh, J. Vac. Sci. Technol., 18, 156 (1981).
19. D. W. Hoffman and M. R. Gaerttner, J. Vac. Sci. Technol., 17, 425 (1980).

20. A. W. Kleinsasser, J. M. E. Harper, J. J. Cuomo, and M. Heiblum, Thin Solid Films, 95, 333 (1982).
21. R. Hezel and N. Lieske, J. Electrochem. Soc., [Feb.], 379 (1982).
22. AVS Symposium on Ion Implantation--New Prospects for Materials Modification, edited by W. L. Brown, American Institute of Physics, NY, 1978.
23. J. L. Speidell, J. M. E. Harper, J. J. Cuomo, A. W. Kleinsasser, H. R. Kaufman, and A. H. Tuttle, J. Vac. Sci. Technol., 21, 824 (1982).

DISCUSSION

L. Toth (National Science Foundation): (1) Have you done any measurements to determine the crystalline state of the surface compounds? If so, which techniques were used? (2) What is the gas composition of the ion beam source used for reactive compound formation? (3) How is the reactive gas introduced and by what mechanism do compounds form?

Author: (1) Ion beam sputter deposition: Metals--generally polycrystalline; Si--amophous up to 450°C, polycrystalline ~ 450-750°C; epitaxial on (100) above 750°C. (2) Ar/O_2 mixtures or pure O_2 (or N_2). Compounds form on target and substrate with single beam, or only on substrate with dual beam.

IONIZED-CLUSTER BEAM DEPOSITION AND EPITAXY

Toshinori Takagi

Ion Beam Engineering Experimental Laboratory
Kyoto University
Sakyo, Kyoto 606, Japan

INTRODUCTION

The ionized cluster beam (ICB) deposition and epitaxial process is an ion-assisted technique by which high quality films of metals, dielectrics and active semiconductor materials can be formed at a low substrate temperature in a technical-grade vacuum system. In the ICB process, film material is vaporized from a confinement crucible under conditions which result in the formation of aggregate clusters of atoms held together by weak forces. Clusters can be ionized by electron impact and subsequently accelerated by high potentials. Through selection of available parameters, it is possible to control the average energy of depositing species over the range from thermal ejection to above 100 eV per atom. It is within this range that optimum conditions for film growth are generally achieved. In the ICB deposition, characteristics of the deposition are mainly caused by both the structural characteristic of the clusters and the effects of ionization and acceleration of the clusters.

In this paper, the fundamental effects of kinetic energy and charge content of the cluster beams on film formation are summarized. The formation and the properties of the cluster beam and the film formation mechanism are subsequently mentioned. Lastly, characteristics of the films deposited by ionized cluster beams are discussed.

ROLE OF IONS IN FILM FORMATION

In ion-assisted film formation, ions transfer energy and charge to a substrate and a depositing film surface.[1] Therefore, the role of ions becomes primarily important and may be described in terms of kinetic energy and ionic charge as shown in Table 1.

Table 1. Influence of Ions on Film Formation

(A) Kinetic energy is converted to:
1. Sputtering energy
2. Thermal energy
3. Implantation energy
4. Migration energy on substrate surface
5. Creating energy of activated centers for nuclear formation.

(B) Presence of ions has a great influence on:
1. Critical parameters in the condensation process of film formation such as nucleation, coalescence, etc.
2. Chemical reaction even without additional acceleration voltage, and even when only a few percent of ionized particles are included in the total flux.

The optimum value of the kinetic energy will be different for different combinations of deposits, substrate materials and desired film properties. The factors to determine the optimum conditions are shown in Table 2.

CLUSTER BEAM FORMATION

The dimensions and factors for the design and operation conditions of the cluster source are shown in Fig. 1.[2]

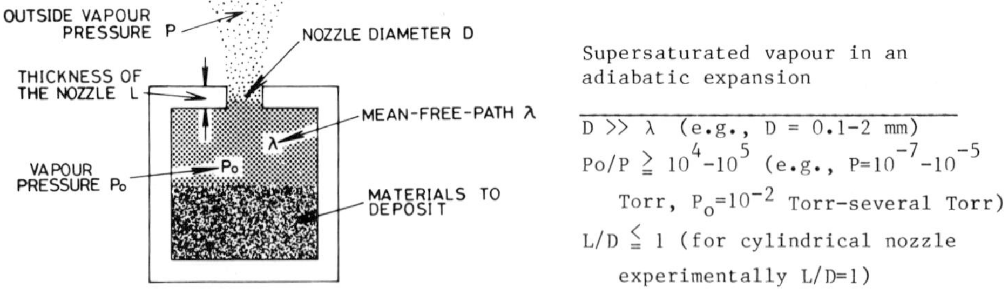

Fig. 1. Dimension and parameters for the design of the cluster source.

Table 2. Optimum Conditions for the Kinetic Energy of Ions Incident on the Substrate for Film Formation

Conditions	Required Incident Ion Energy	Result
Deposition	Less than the energy corresponding to the sputtering rate S(E) = 1. Larger than the energy at which the sticking probability becomes too low	Optimum value of kinetic energy: a few hundred electron volts
Surface Cleaning	Larger than the energy of adsorption on the substrate surface, i.e., 0.1– 0.5 eV for physically adsorbed gases and 1–10 eV for chemically adsorbed gases	
Good Quality Film Formation	In a range where enhanced adatom migration influences properties of the deposited film suitable ion bombardment affects the growth of nuclei a suitable amount of defects or atomic displacement near the substrate surface contributes to film formation during the initial stage	

The size of a cluster N (the number of atoms per cluster) was estimated by energy measurement using the equation $E = NmU^2/2$ where m is the atomic mass of a material and U is the velocity of the cluster. Figure 2 shows the energy distribution of the ionized clusters measured by an electrostatic energy analyzer.[3]

Two kinds of peaks are observed in the spectra; weak peaks close to zero correspond to monatomic species and strong peaks at 80–170 eV correspond to the clusters. The intense peak from the clusters appears in the pressure range where the formation conditions shown in Fig. 1 are satisfied. By combining the results with ejection velocity measurement by a rotating disk method, cluster size was calculated to be 500 - 2000 atoms/cluster. A retarding field[4,5] and time-of-flight[6] methods and electron microscopy[7] were also used to measure the cluster size.

Similar results were obtained not only for elemental materials (Ag, Au, Cu), but also compound materials (CdTe).

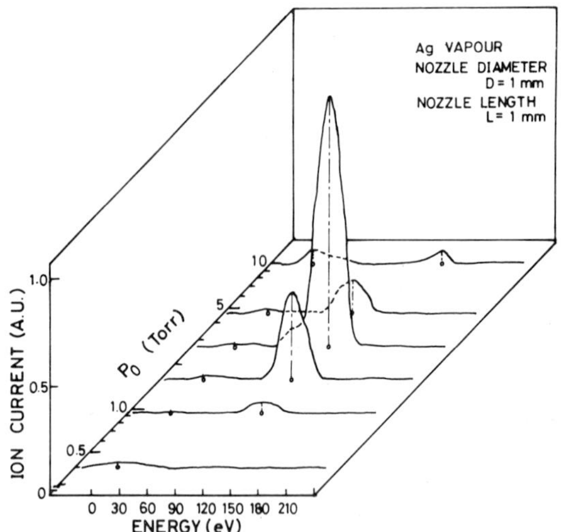

Fig. 2. Energy distribution of the cluster beam as a function of the vapour pressure in the crucible.

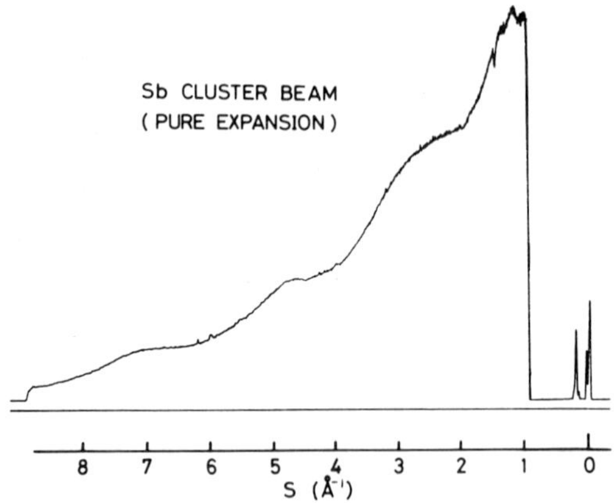

Fig. 3. Microdensitometer trace of the electron diffraction photograph.

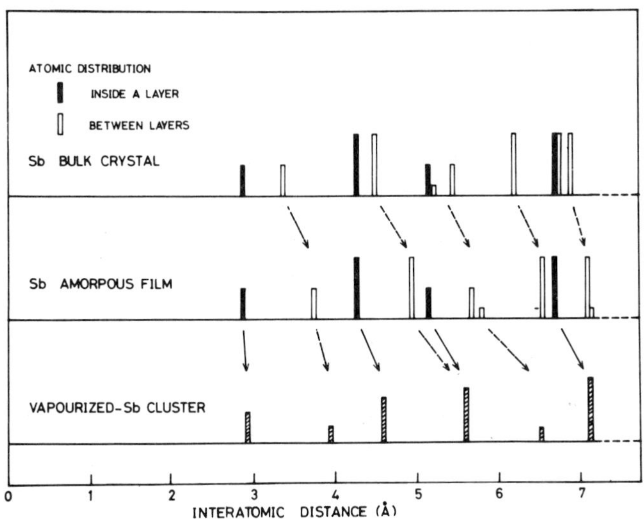

Fig. 4. Comparison of the interatomic distances in various kinds of states.

One of the most important properties of the cluster ion is that problems caused by the space charge effects can be eliminated because of the extremely small charge to mass ratio, as the clusters formed by ejection through the nozzle are ionized to be singly charged by electron bombardment. In addition, the cluster beam is well oriented at high intensity. These features allow the cluster beam to transport an equivalently large mass within a very low energy range, as shown in Table 2.

Figure 3 shows the microdensitometer trace of the electron diffraction photograph for an Sb cluster beam. The pattern shows that the formed cluster is amorphous. The patterns were analyzed by radial distribution functions to obtain the interatomic distance in the cluster.[8] Figure 4 shows a comparison of the interatomic distance in the Sb cluster with those of a bulk crystal and amorphous films.[9] The atomic distance in the amorphous film is longer than that in the crystal. Further spread of the interatomic distance can be seen in our clusters. This characteristic results in unique film formation kinetics such as enhanced adatom migration at low substrate temperature.

ICB DEPOSITION SYSTEM

In ionized cluster beam deposition, ions of macroaggregate atoms (clusters) are utilized instead of ions in the atomic state. The ratio of ionized clusters to total clusters can be adjusted by changing the electron current (Ie) for ionization. For example, the degree of ionization is 5 - 7% at Ie = 100 mA, 7 - 15% at Ie = 150 mA and 30 - 35% at Ie = 300 mA.[10] The ionized clusters are accelerated by an acceleration voltage and bombard a substrate surface together with the neutral clusters which have some kinetic energy corresponding to the ejection velocity.

In reactive ICB (RICB) deposition for hydride, oxide, nitride or carbide film formation, deposition is conducted in the presence of a reactive gas having a partial pressure in the range of 10^{-5} - 10^{-4} Torr. This prevents the production od a plasma in the chamber.[11] If a plasma occurs in the chamber, clusters are destroyed by collision with energetic particles, and the advantages of the ICB technique are lost.

A typical schematic diagram of the ICB apparatus is shown in Fig. 5. A view of commercial ICB experimental deposition system is shown in Fig. 6.[12] This system has four crucibles. Each crucible can be placed sequentially into the ion source via external controls. Deposition condition such as crucible temperature, deposition rate, acceleration voltage, electron ionization current, substrate temperature, film thickness, etc., can be controlled automatically by a computer system. A dual cluster beam system having two groups of of crucibles, has also been developed. In Fig. 7, a single crucible system is shown which contains multiple nozzles used to form a ribbon beam for producing a wide sheet of deposited material with high uniformity on large industrial substrates.[13]

FILM FORMATION MECHANISM BY ICB

The process of film formation by ICB is conceptionally shown in Fig. 8. The partially ionized clusters are accelerated in the direction of the substrate. Neutral clusters also drift to the substrate with ejection velocity. Upon bombarding the substrate surface, both ionized and neutral clusters are broken up into atoms and scattered over the surface as a result of their increased energy of surface diffusion.

Migration of the atoms on the substrate surface was investigated by electron microscopy of the films.[17] The results show that migration can be enhanced by increasing acceleration voltage.

The mass M deposited on the substrate as a function of substrate temperature is given by,

$$M = \dot{M}\, t - \frac{\dot{M}\, N_o}{I^*} \exp\,(-\, U/kT)$$

where $U = \Phi_{ad} - \Phi_d$, Φ_{ad} is the activation energy for desorption and Φ_d is the activation energy for surface diffusion, $\dot{M}$ is the mass impingement rate on the substrate, I^* is the rate of formation of critical nuclei, and N_o is the density of adsorption sites on the substrate surface. Since Φ_{ad} is generally larger than Φ_d, the curve of mass versus reciprocal temperature will have a positive slope. The solid line in Fig. 9 shows the mass deposited vs the reciprocal substrate temperature in silicon deposition by ICB.[14] The dotted line shows the transition temperature of the crystalline properties of the films deposited at different acceleration voltages. For unionized clusters (corresponding to aconventional deposition method), the mass deposited increases with decreasing substrate temperature.

The mass deposited at a given substrate temperature decreases with increasing acceleration voltage because of the sputtering effect. Moreover, the slope of the line which shows the mass deposited changes with increasing acceleration voltage. The change of the slope from positive to negative with increasing acceleration voltage is considered to be caused by the change in the values of N_o, I^* and U. This is one of the important features of the ICB deposition process. In conventional ion beam deposition which uses atomic neutral or ionized particles, the energy U does not change.[15]

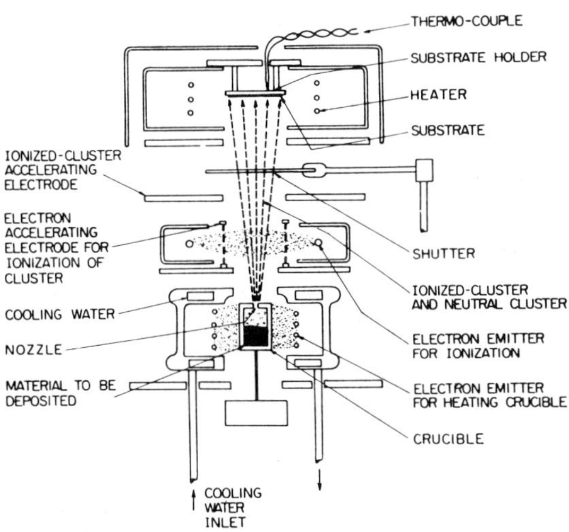

Fig. 5. Ionized cluster beam source with single crucible and single nozzle.

Fig. 6. ICB experimental deposition system.

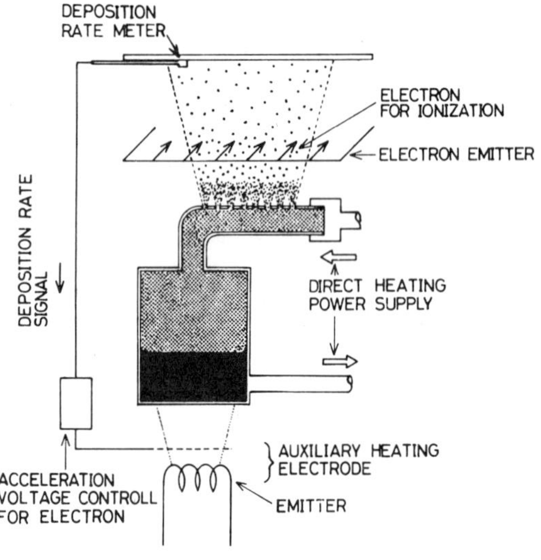

Fig. 7. Ionized cluster beam source having a single crucible with multiple nozzles.

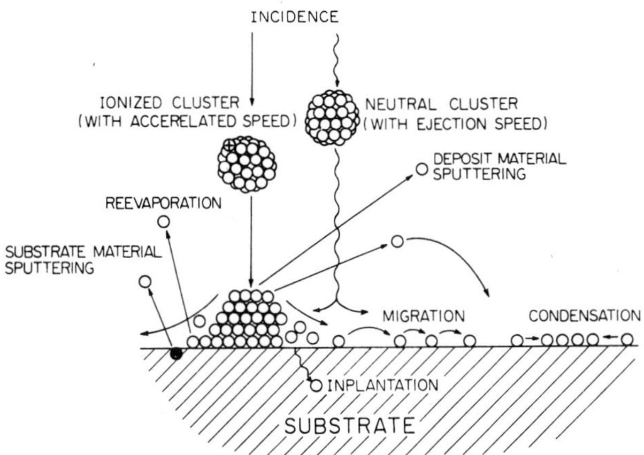

Fig. 8. A schematic diagram of film formation by the ICB deposition technique.

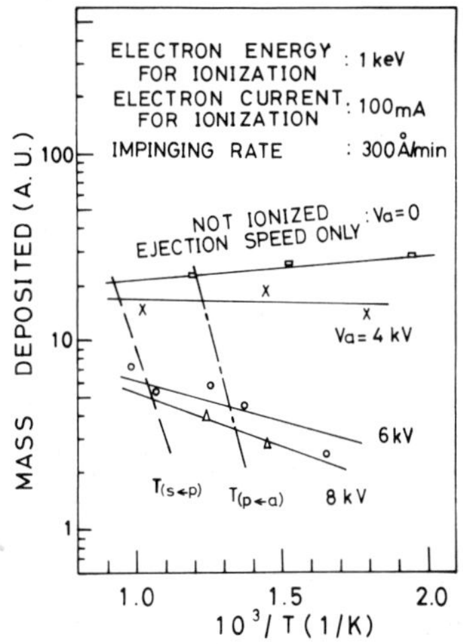

Fig. 9. Mass deposited vs. reciprocal substrate temperature, and the change of transition temperature for different acceleration voltages.

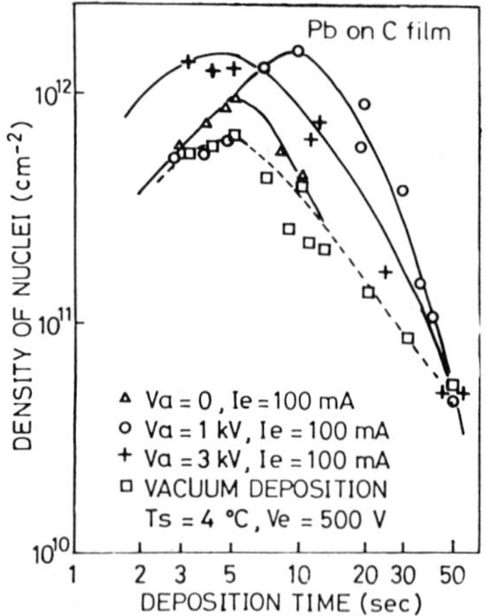

Fig. 10. Density of nuclei as a function of deposition time at different acceleration voltages.

Figure 10 shows the change in the density of nuclei as a function of deposition time at different acceleration voltages. Lead clusters in this experiment were deposited onto a copper grid covered with carbon films keeping the deposition rate constant by monitoring a quartz thickness meter.[16] The density of nuclei increased monotonically at first by increasing the acceleration voltage up to about 1 kV; however, it subsequently became saturated. But the rise-up time of saturation became shorter by increasing the acceleration voltage. This suggests that both the nuclei formation effect and the migration effect are enhanced by bombardment.

Silicon epitaxial films could be formed at a reasonably high deposition rate using 8 kV acceleration voltage and a 633-893K substrate temperature range in a chamber which was evacuated with an oil diffusion pump to a base pressure of 10^{-7} - 10^{-6} Torr, as shown in the lower part of Fig. 11. In this deposition, no special cleaning process was used before the deposition, except chemical cleaning. The diffraction pattern obtained with a 100 kV electron beam was used to evaluate the crystalline structure. The results show that an amorphous or polycrystalline structure was formed in a range of 0 - 4 kV acceleration voltage[17,18] and that the crystallinity could be improved using a higher acceleration voltage. In another case of deposition, epitaxial growth of Si on an atomically clean and well-ordered silicon surface in an ultra high vacuum chamber was obtained using only 200 V acceleration voltage and a substrate temperature of

773K.[16] The film formed by this method was examined by *in situ* electron diffraction at 5 kV. The results are shown in the upper part of Fig. 11. In the former case, the acceleration voltage necessary to obtain epitaxial deposition was about 6 - 8 kV at 10^{-7} - 10^{-6} Torr, because it was necessary to sputter the native oxides on the surface and the impinged residual gas atoms during the deposition. On the other hand, in the deposition on an atomically clean surface, a single crystalline pattern was observed at 200 V acceleration, because it was not necessary to sputter the oxides and residual gas atoms during the deposition. By increasing acceleration voltage, improvement of the crystalline quality could be observed. This might be caused by an enhanced adatom migration effect.

FILM PROPERTIES

Many experiments have been undertaken to prepare various kinds of films by ICB and reactive ICB (RICB) methods. The films deposited by these methods and their features are summarized in Table 3. Some of the recently obtained information on the films is shown here.

It has already been reported that the ICB technique makes Si epitaxy possible at a lower substrate temperature than that of conventional methods.[19] The crystallinity of the silicon films is dependent on the substrate temperature and the acceleration voltage. These experimental results suggest the possibility of controlling the range of the ordered atoms by adjusting the operating conditions. Applying this idea, hydrogenated amorphous silicon films with thermally stable characteristics have been obtained.[11] Silicon was deposited in relatively low gas pressure of hydrogen in a range of 10^{-5} - 10^{-4} Torr. The hydrogen in this pressure region does not collide with silicon beams in its path. Therefore the chemical reaction occurred on the depositing surface. The reaction increased with the acceleration voltage. For doped film formation, hydrogen gas mixed with phosphine or diborane on the order of 5000 ppm was available sequentially in the same chamber without any trouble as a result the residual reactive gas used in previous processes. P- or n-type films could be reproducibly formed at practical deposition rates. The structural analysis showed that the film mainly consisted of monohydrides. The density of monohydrides can be increased by depositing at a higher acceleration voltage. In contrast, the SiH_2 structure decreased at higher acceleration voltages. Figure 12 shows the dependencies of the acceleration voltage on photo- and dark-conductivities. Figure 13 shows the dependence of the absorption coefficient after cyclic heating. These results show that the films have good electrical properties and thermally stable characteristics. Because deposition is carried out at relatively low gas pressures, gas inclusion which is seen in the films deposited by conventional glow discharge methods is extremely uncommon.

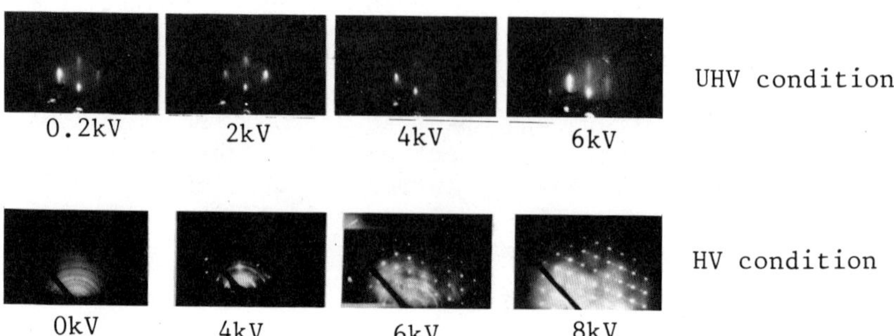

Fig. 11. Electron diffraction patterns of epitaxial silicon films produced by ICB deposition at different acceleration voltages.

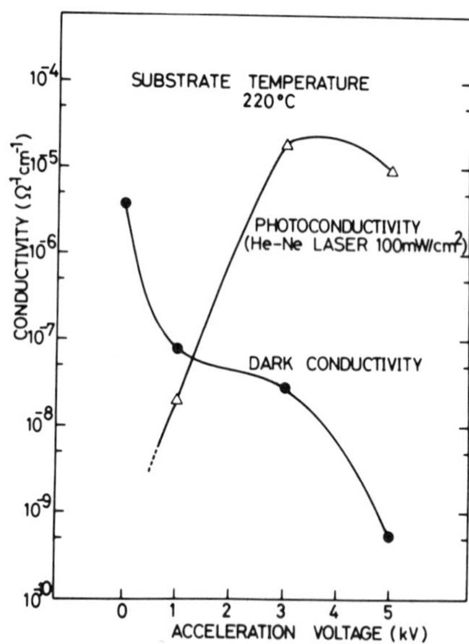

Fig. 12. Photo- and dark-conductivities as a function of acceleration voltages.

Table 3. Film Properties and Their Features

Film/ Substrate	Acc. Volt. (kV)	Subst. Temp. (K)	Advantage (Applications)	Refs.
Au, Cu/glass, capton	1-10	RT	Strong adhesion, high packing density, good electrical conduction in a very thin film (high resolution flexible circuits, optical coating).	6 22
Pb/glass	5	RT	Controllable crystal structure, strong adhesion, smooth surface, improved stability on thermal cycling (super conductive device).	3 7 16
Ag/Si(n-type)	5	RT	Ohmic contact without alloying.	20
Ag/Si(p-type)		673	Ohmic contact at low temperature (semiconductor metallization).	21 22
Ag-Sb/GaP	2	673	Ohmic contact, strong adhesion (semiconductor metallization).	23
Al/SiO_2, Si	0-5	RT-473	Controllable crystal structure strong adhesion, ohmic contact at low temperature, electromigration resistant (semiconductor metallization).	24
c-axis preferentially oriented MnBi/glass	0**	573**	Spatially uniform magnetic domain, high density optical memory (magnetooptical memory)	25
GdFe/glass	0	473	Thermally stable and uniform amorphous (magnetooptical memory).	26
c-axis preferentially oriented PbTe/glass	0-3	473	High Seebeck coefficient, low thermal conductivity (high efficiency thermoelectrical converter).	27

(continued)

Table 3 (continued)

Film/ Substrate	Acc. Volt. (kV)	Subst. Temp. (K)	Advantage (Applications)	Refs.
Preferentially oriented ZnSb/ glass	1	413	Thermally stable amorphous, high Seebeck coefficient (high efficiency thermo-electric converter).	28
ζ-$FeSi_2$	0-5	433	Thermally stable amorphous, p- or n-type controllable, high Seebeck coefficient (high efficiency thermo-electric converter).	29
Si/(111)Si Si/(100)Si Si/(1$\bar{1}$02) sapphire	6	893	Low temperature epitaxy in a high vacuum (10^{-7} - 10^{-5} Torr) shallow and sharp p-n junction (semiconductor devices).	18 19 20 30
Amorphous Si/ glass	2	473	Thermally stable (solar cell, thin film transistor).	10 17
GaAs/Cr-doped GaAs	6	823	Epitaxy in a high vacuum (10^{-7} - 10^{-5} Torr) (semi-conductor devices).	30
GaP/GaP GaP/Si	4	823 723	Low temperature epitaxy in a high vacuum (low cost LED).	31
ZnS/NaCl	1	473	Single crystal formation, (optical coating).	32
ZnS:Mn/glass	1	473	(ac-dc electroluminescent cell of the low impedance)	32
CdTe/glass	2	523	Improved monocrystalline domains (infrared detector).	5
Li-doped Zn/O (1$\bar{1}$02) sapphire	0.5-1	503	Single crystal formation (optical wave guide).	33

Film/ Substrate	Acc. Volt. (kV)	Subst. Temp. (K)	Advantage (Applications)	Refs.
c-axis preferentially oriented ZnO/glass	0	423	Controllable crystal structure controllable optical transmission (optical wave guide, SAW).	33
BeO/(0001) sapphire	0	673	Single crystal formation, transparent film (semiconductor device, heat sink, electrical insulator).	34
C-axis preferentially oriented BeO/glass	0	673	High electrical resistivity and high thermal conductivity coating, low temperature growing (semiconductor device, heat sink, electrical insulating and thermal conduction).	35
PbO/glass	3	RT	Low temperature growth, smooth surface, accurately controllable thickness, good adhesion (superconductor).	3
SnO_2/glass	0	673	Transparent low resistivity film, strong adhesion (NESA glass).	36
GaN/ZnO/glass	0	723	Low temperature growth of GaN film on amorphous substrate (low cost LED, photo cathodic electrode).	37

*Annealing temperature. **Ejection velocity only.

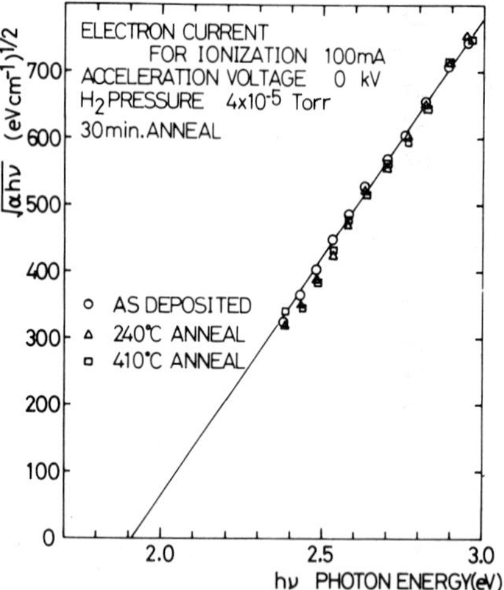

Fig. 13. Absorption characteristics as a function of acceleration voltages.

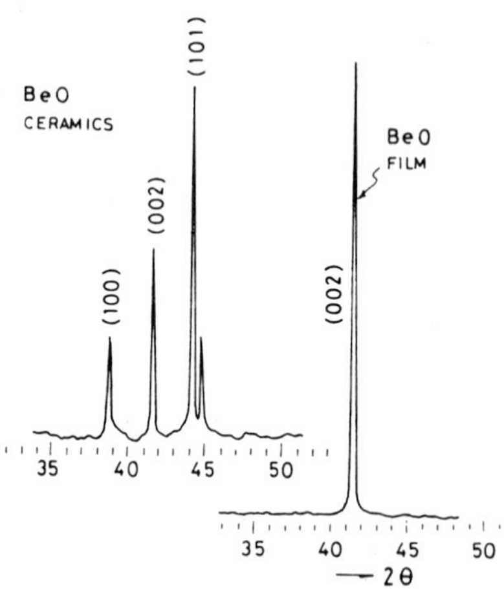

Fig. 14. X-ray diffraction patterns of BeO films and ceramics.

In the RICB deposition in oxygen, various kinds of oxides can be formed. In this case, a gas pressure of less than 10^{-5} - 10^{-4} Torr was also sufficient to induce chemical reaction without generating a plasma. A transparent BeO film with a good crystalline structure was obtained.[34] BeO has a high thermal conductivity of the same order as that of aluminum in spite of being an excellent insulating material with a band gap of 11.2 eV. Figure 14 compares an x-ray diffraction pattern of the grown films with that of thermally baked BeO ceramics. As seen in this figure, there is a remarkable difference between the results of the ceramic and the film. The films deposited on a glass substrate showed strong c-axis preferential orientation. Epitaxial films can also be prepared if the (0001) plane of a sapphire substrate is used. The anisotropic thermal conductivities parallel to and perpendicular to the c-axis were measured. These are shown in parallel to and perpendicular to the c-axis were measured. These are shown in Fig. 15.

Nitride films can also be formed in a nitrogen atmosphere in the pressure range of 10^{-5} - 10^{-4} Torr. GaN films with a hexagonal structure were formed by RICB. In this experiment, a ZnO film which was formed on a glass substrate by RICB was used as an intermediate layer to serve an adequate nucleation seed for growing a GaN film.[37] The film was formed at the substrate temperature of 723K under Ie = 100 - 300 mA without accelerating the cluster ions of Ga. The strong diffraction peak of the (002) plane was obtained.

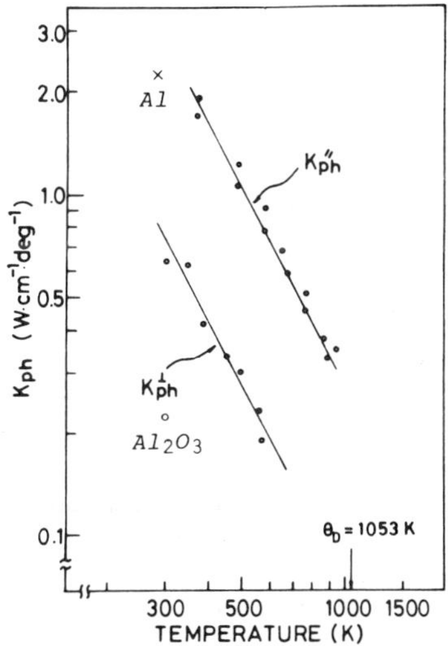

Fig. 15. Temperature dependence of anisotropic lattice thermal conductivities of BeO films.

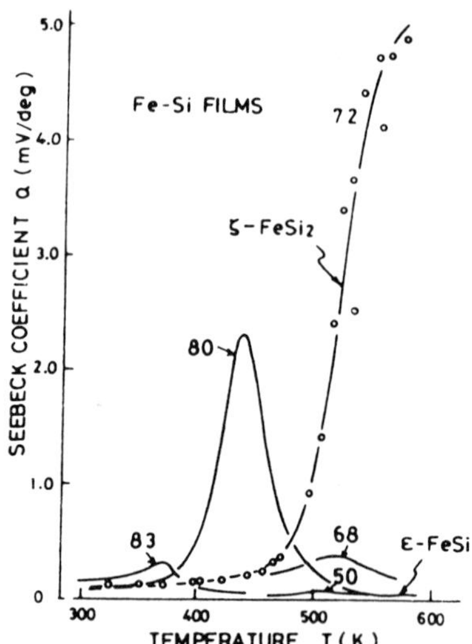

Fig. 16. Temperature dependence of the Seebeck coefficient for amorphous iron silicide films.

Amorphous iron disilicide films with high Seebeck coefficient (thermoelectric power) have also been formed.[29] Films were deposited by using dual cluster beams, wherein only the Fe beam was ionized but the Si beam only ejected. An acceleration voltage for the Fe beam was not applied. Figure 16 shows the temperature dependence of the Seebeck coefficients of the films containing 50 - 83 atomic percent Si.

The film containing 50 at.% Si showed the ζ-$FeSi_2$ structure, and the Seebeck coefficient was of the order of 40 μV/deg. In contrast, films containing Si at 72 at.% showed the ζ-$FeSi_2$ structure with the Seebeck coefficient being +5000 μV/deg at 550 K. Further increasing the Si content in the film decreased the Seebeck coefficient. These films showed p-type conduction.

A film with n-type conduction could be formed by introducing a small amount of oxygen gas in the chamber (8×10^{-5} Torr oxygen). Figure 17 shows the temperature dependence of the Seebeck coefficient of the films with 70 at.% Si. This film is slightly p-type to the temperatures around 400K, but beyond this temperature strong n-type conduction appeared. At 580K, an enormously high value of -20000 μV/deg was obtained. The films also showed thermally stable characteristics under repeated heat cycles. Several analyses have been done to explain such high Seebeck coefficients and their thermal stability. It may be caused by the interaction between current carriers and a long wavelength magnon, which has been observed in many magnetic semiconductors.

G-axis oriented MnBi films[25] and amorphous GdFe films[26] for magneto-optical memory applications have been formed. In the MnBi films the annealing temperature to initiate Faraday rotation using a He-Ne laser was found to be lowered by 1/3 of that of the films prepared by the conventional evaporation technique. The intensity of the Faraday rotation could be controlled by changing the electron current for ionization, that is, the content of ionized clusters in the total flux. For amorphous GdFe films, a calculated reduced-density function G(r) indicated that Gd-Fe pairs were dominant, even when 30% of the Fe-clusters were included. This is in contrast with the results obtained by sputtered films, in which the densities obtained by of direct pairing (Fe-Fe and Gd-Gd pairs) are fairly large.

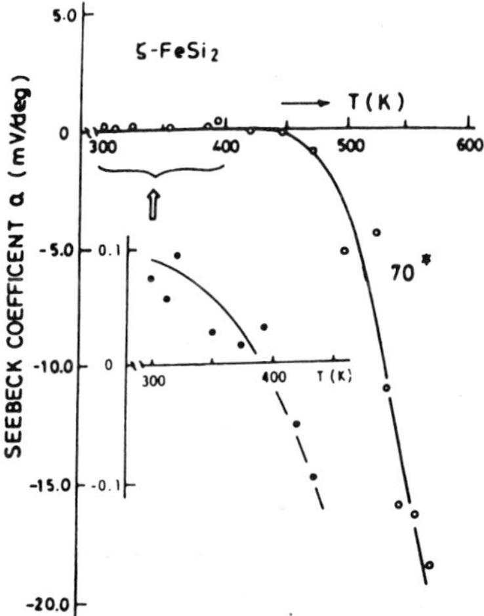

Fig. 17. Temperature dependence of the Seebeck coefficient for disilicide films containing small amount of oxygen.

CONCLUSIONS

The ICB technique allows growth of not only single element or compound material films, but also hydrides, oxide and nitride films. The formation mechanism of the cluster beam, the fundamental film formation mechanism and some examples of the film properties have been shown. The films can be deposited at a lower substrate temperature than those produced by other deposition methods. Vacuum in the chamber during the deposition of these films was in the range of 10^{-7} - 10^{-5} Torr which can be obtained by conventional vacuum equipment. In the case of RICB deposition, the residual gas pressure is approximately in the range of 10^{-5} - 10^{-4} Torr. The remarkable feature of the ICB versus the conventional techniques is due to the unique film formation mechanism which is accompanied by sputtering, implantation, heating, a suitable amount of defect formation and migration effects etc.; and the influence of the presence of ionized clusters each having an equivalent intense current and low kinetic energy which is suitable for deposition.

From these experimental results it is clear that the ICB and RICB methods have high potential for the preparation of high quality amorphous to monocrystalline coatings, and in many fields of application including electron devices.

REFERENCES

1. T. Takagi, Thin Solid Films, 92, 1 (1982).
2. T. Takagi, I. Yamada, M. Kunori, and S. Kobiyama, Proc. 2nd Int. Conf. on Ion Sources, 1972, Vienna, Osterreichiche Studiengesellshaft fur Atomenergie, Vienna, 790 (1972).
3. I. Yamada and T. Takagi, Thin Solid Films, 80, 105 (1981).
4. T. Takagi, I. Yamada, and A. Sasaki, Int. Phys. Conf. Ser., 38, 229 (1978).
5. J. B. Theeten, R. Madar, A. Mircea-Roussel, A. Rocher, and G. Laurence, J. Cryst. Growth, 37, 317 (1977).
6. T. Takagi, I. Yamada, and A. Sasaki, J. Vac. Sci. Technol., 12, 1128 (1975).
7. I. Yamada, H. Takaoka, H. Inokawa, H. Usui, S. C. Cheng, and T. Takagi, Thin Solid Films, 92, 137 (1982).
8. I. Yamada, G. D. Stein, H. Usui, and T. Takagi, Proc. 6th Symp. on Ion-Assisted Technology, 1982, Tokyo, IEEJ, 47 (1982).
9. H. Krebs and F. Schultze-Gebhardt, Die Naturewissenshaften, 20, 474 (1954).
10. T. Takagi, I. Yamada, and A. Sasaki, Inst. Phys. Conf. Ser., 38, 142 (1978).
11. I. Yamada, I. Nagai, S. Ishiyama, and T. Takagi, Proc. 4th Symp. on Ion Sources and Ion Application Technology, 1980, Tokyo, IEEJ, 117 (1980).

12. Technical Data, Eaton Corporation, 16 Tozer Road, Beverly, MA, USA.
13. Technical Data, Sumitomo Bakelite Co., Totsuka, Yokohama, Japan.
14. T. Takagi, I. Yamada, K. Matsubara, and H. Takaoka, J. Cryst. Growth, 45, 318 (1978).
15. V. O. Babaev, J. V. Bykov, and M. V. Guseva, Thin Solid Films, 38, 1 (1976).
16. T. Takagi, Preprint of Ion Assisted Surface Treatments, Techniques and Processes, 1982, Coventry, The Metal Society, London, 1.1 (1982).
17. T. Takagi, I. Yamada, and A. Sasaki, Thin Solid Films, 39, 207 (1976).
18. T. Takagi, I. Yamada, and A. Sasaki, Technical Digest 1976 Int. Electron Devices Meeting, 976, Washington, D.C., 605 (1976).
19. I. Yamada, F. W. Saris, T. Takagi, K. Matsubara, H. Takaoka, and S. Ishiyama, Japan J. Appl. Phys., 9, L181 (1980).
20. T. Takagi, I. Yamada, and A. Sasaki, Proc. Int. Conf. on Ion Plating and Allied Techniques, 1977, Edinburgh, CEP Consultants, Edinburgh, 50 (1977).
21. T. Takagi, I. Yamada, and A. Sasaki, Proc. 7th Int. Vacuum Congr. and 3rd Int. Conf. on Solid Surfaces, 977, Vienna, Berger, Vienna, 195 (1977).
22. I. Yamada, K. Matsubara, M. Kodama, M. Ozawa, and T. Takagi, J. Cryst. Growth, 45, 326 (1978).
23. T. Ishida, S. Wako, and S. Ushio, Thin Solid Films, 39, 227 (1978).
24. H. Inokawa, K. Fukushima, I. Yamada, and T. Takagi, Proc. 6th Symp. on Ion Sources and Ion-Assisted Technology, 1982, Tokyo, IEEJ, 355 (1982).
25. T. Takagi, K. Matsubara, N. Kondo, and H. Takaoka, Japan J. Appl. Phys., 9 [Suppl. 9-1], 507 (1980).
26. N. Kondo, K. Matsubara, and T. Takagi, J. Mag. Soc. of Japan, 5, 105 (1981).
27. K. Matsubara, H. Takaoka, K. Shigeno, Y. Kuriyama, and T. Takagi, Proc. 6th Symp. on Ion Sources and Ion-Assisted Technology, 1982, Tokyo, IEEJ, 399 (1982).
28. T. Koyanagi, K. Matsubara, H. Takaoka, and T. Takagi, Proc. 6th Symp. on Ion Sources and Ion-Assisted Technology, 1982, Tokyo, IEEJ, 409 (1982).
29. T. Takagi, K. Matsubara, M. Oura, and T. Koyanagi, Proc. 6th Symp. on Ion Sources and Ion-Assisted Technology, 1982, Tokyo, IEEJ, 39 (1982).
30. T. Takagi, I. Yamada, and A. Sasaki, Thin Solid Films, 45, 569 (1977).
31. K. Morimoto, H. Watanabe, and S. Itoh, J. Cryst. Growth, 45, 334 (1978).
32. T. Takagi, I. Yamada, and A. Sasaki, p. 275 in Ion Implantation in Semiconductors, edited by S. Namba, Plenum Press, NY (1974).

33. K. Matsubara, I. Yamada, N. Nagao, K. Tominaga, and T. Takagi, Surface Science, 86, 290 (1979).
34. T. Takagi, K. Matsubara, and H. Takaoka, J. Appl. Phys., 51, 5419 (1980).
35. K. Matsubara, I. Yamada, H. Takaoka, and T. Takagi, Japan J. Appl. Phys., 1 [Suppl. 1-], 403 (1981).
36. H. Takaoka, K. Matsubara, and T. Takagi, Proc. 4th Symp. on Ion Sources and Ion Application Technology, 1980, Tokyo, IEEJ, 143 (1980).
37. K. Matsubara, T. Horibe, H. Takaoka, and T. Takagi, Proc. 4th Symp. on Ion Sources and Ion Application Technology 1980, Tokyo, IEEJ, 137 (1980).

DISCUSSION

L. Toth (National Science Foundation): For semiconducting films such as GaAs prepared by your technique, how do parameters such as electron mobility or the concentration of defect states in the gap compare with similar films prepared by MBE techniques.

Author: Detailed evaluations of many materials have shown that films deposited by ICB have very good technical qualities. In the case of some materials such as gallium arsenide our evaluations have been preliminary and more investigation is still required. But we have examined crystallinity by Rheed and Rutherford backscattering and have measured electron mobility. Results are essentially equivalent to those on films by other methods.

ION BEAM DEPOSITION OF CERAMIC-LIKE COATINGS

C. Weissmantel, K. Bewilogua, K. Breuer, J. Erler, B. Rau, G. Reisse, and D. Roth

Sektion Physik/EB, Technische Hochschule Karl-Marx-Stadt, German Democratic Republic

INTRODUCTION

Over the last decade, the deposition of thin and thick films by condensing suprathermal, energetic particles has roused steadily increasing interest. The most convenient way to generate particle fluxes with excess energies in the range from a few eV to some keV is offered by using adequately accelerated ions that can be extracted from external sources or from a plasma sustained inside the deposition chamber. Although ion beam and plasma deposition processes have much in common, the term "ion beam deposition" should be applied only if the work pressure in the deposition chamber is below some 10^{-2} Pa. Then the mean free path of the particles is large compared with the usual dimensions, and gas phase interactions may often be neglected. A particular advantage of ion beam techniques is that the process parameters can be controlled rather independently within wide ranges of particle energies and flux densities.[1-3] On the other hand, it must be emphasized that systematic investigations by ion beam techniques can contribute to optimize high-rate deposition processes, which on a technological scale are performed by plasma ion plating or magnetron sputtering.

In ion beam deposition, in addition to the energetic film-forming species, thermal neutrals and accelerated inert gas ions also may be present. Furthermore, a flux of neutrals with energies in the order of 10 eV can be created by sputtering from a suitable target. If the films are grown by condensing these sputtered species that normally contain only a small fraction of ions, the technique might be characterized as "secondary ion beam deposition," following a proposition by Harper.[4] According to the basic mechanisms of ion-solid interaction, the most important effects caused by impinging energetic species during the film growth may be listed as follows:[3,5]

(a) sputter cleaning or "ion scrubbing" of the substrates and recoil implantation of impurities or film-forming species;

(b) condensation with high sticking probability accompanied by a strong local activation;

(c) defect creation, atomic mixing and secondary sputtering due to collisions and atomic displacements;

(d) the promotion of phase transitions and/or chemical reactions.

During the last year evidence has been accumulated that these ion actions are in the first place responsible for the good adherence, the dense coverage and the structural properties of films and surface coatings prepared by ion beam techniques. In this paper, after a brief discussion of the principal ion beam configurations, some results of our work on the deposition of cermet films and hard and/or protective coatings will be presented and further prospects of ion beam deposition in the field of ceramic-like coatings will be outlined.

EXPERIMENTAL CONFIGURATIONS

Figure 1 shows schematically the experimental configurations that have been used so far in ion beam deposition and allied work. Different types of ion sources are suitable for these techniques, in particular the hot cathode multiaperture sources developed by Kaufman[6] and in our laboratory by Fiedler et al.,[7] the duoplasmatron,[8] saddlefield sources,[9] postionization systems based on electron impacts[3,10] and microwave sources now under development. The sources should be capable of delivering uniform ion beams of several cm diameter at ion currents of more than 10 mA and ion energies ranging from some 10^2 to 10^4 eV. An important requirement is that contaminations by evaporated or sputtered source materials must be as low as possible. As some of the mentioned sources and even complete ion beam systems are now offered commercially, a widespread utilization of ion beam techniques has become feasible. It can be anticipated that applications on a technological scale will soon emerge.

Prior to almost any ion beam deposition, the substrates are cleaned by operating the source with argon or krypton. In the case of ceramic-like coatings, we have obtained optimum adhesion and growth conditions when the process was changed continuously from etching to deposition by gradual substitution of the work gas. Moreover, directional ion beam etching by means of a rare gas, oxygen or fluorocarbon ions, as is illustrated in Fig. 1(a), has become an important method for producing sharp microstructure patterns on solid surfaces.[11]

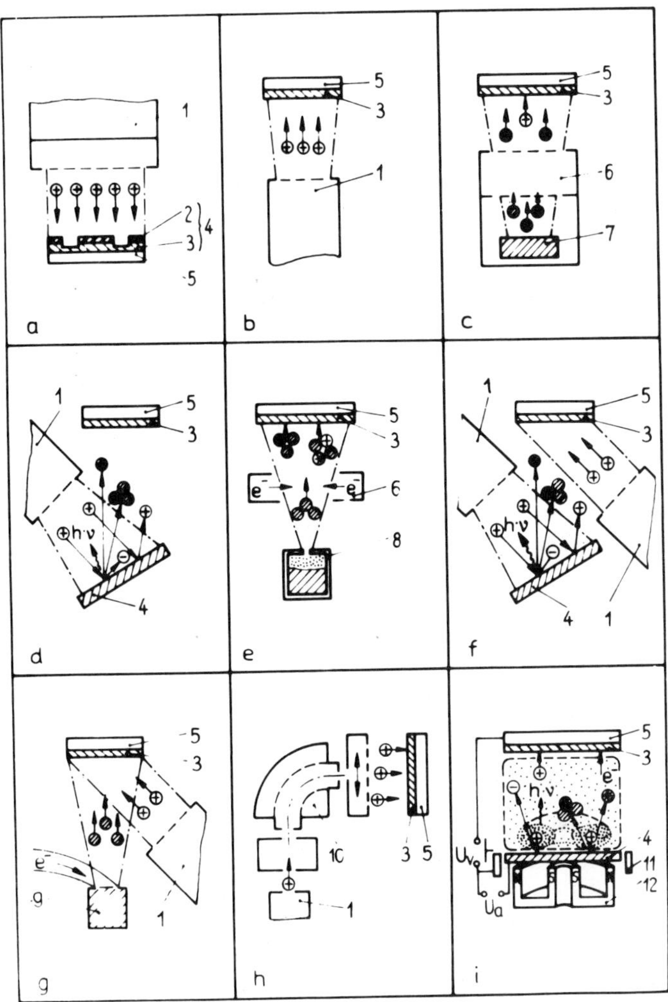

Fig. 1. Ion beam configurations. (a) ion beam etching; (b) ion deposition; (c) ion beam plating; (d) ion beam sputtering; (e) ionized-cluster deposition; (f) dual beam sputtering; (g) ion-induced evaporation; (h) ion implantation; (i) magnetron sputtering. (1 - ion source; 2 - mask; 3 - film; 4 - target; 5 - substrate; 6 - postionization system; 7 - evaporator; 8 - cluster ion source; 9 - electron beam evaporator; 10 - mass separator; 11 - ring anode; 12 - ring magnet.)

At the forefront of fundamental research, considerable efforts have been made by several groups[12-14] to study systematically the primary deposition of nearly monoenergetic ions as a function of the particle energy. This sophisticated technique, which for very low ion energies requires the addition of energy/mass filters and retardation devices, is indicated in Fig. 1(b). Most of this work has been done with metals or semiconductors, however, an extension to components of ceramic-like films would be desirable.

A much easier and more productive approach to primary ion deposition is offered by the concept of ion beam plating, as is illustrated in Fig. 1(c). The film-forming material is evaporated and a fraction of the particles becomes postionized by passing crossing electron beams.[15] When the ions are accelerated by a suitable substrate bias voltage, the beam splits into an energetic branch consisting of ions and a branch of thermal neutrals. By proper adjustment of the bias voltage and the ionized particle fraction, a wide range of the mean energy of the condensing species can thus be realized. In our laboratory, special ionization devices comprising a large circular cathode filament, a wide-mesh anode, a base reflector electrode and a narrow-mesh ion extraction electrode were developed.[3,15] These units enable the generation of broad and intense beam bundles having a high degree of ionization at low total pressures (10^{-3} - 10^{-2} Pa). By operating the system with benzene vapor or mixtures of benzene vapor and evaporated metals hard i-carbon coatings or metal-carbon composites were deposited efficiently on substrates of relatively large areas. The properties of these coatings will be discussed below.

Fig. 1(d) illustrates the principle of ion beam sputtering or secondary ion beam deposition. An intense ion beam, typically of Ar^+ at 5 - 10 keV energy, is used to sputter material from a target. The films grow by condensing the sputtered species, mainly consisting of neutral atoms with minor fractions of clusters and ions, on suitable substrates. In our early publications[7,16] it was demonstrated that ion beam sputtering due to the separation between the ion-generating plasma and the deposition region offers specific advantages compared with conventional plasma sputtering techniques. Numerous reports of applications, including deposition in ultrahigh vacuum, coupled with the use of in situ diagnostics[17] have been published. It should be noted that the mean energy of the sputtered particles is on the order of 10 eV. However, much more energetic particles originating from the high-energy tail of the spectrum will impinge upon the substrates. These energetic species together with backscattered, mostly neutralized primary ions may create defects in the growing films. For ion energies above 1 keV the probability of backscattering decreases sharply with increasing mass ratio M_1/M_2, where the suffix 1 denotes the primary ion and the suffix 2 the target atom. Recent applications of ion beam sputter deposition will be presented in later sections in conjunction with Ag/SiO_2 cermet films and BN coatings.

The ionized-cluster deposition shown in Fig. 1(e) has been developed by Takagi et al.[2,18] Multiatomic clusters are formed by adiabatic expansion of the vapor through nozzles and become ionized by passing an electron bombardment system. Films of different materials including ceramic-like coatings were obtained by condensing the accelerated cluster ions. An advantage of the method is that the excess energy of the particles is shared by many atoms and, hence, the likelihood of defect creation is small.

The dual beam sputter technique according to Fig. 1(f) was originally developed in our laboratory to study the kinetics of reactive film growth. Whilst ion beam sputtering of a silicon target by nitrogen ions yielded only deposits of poor quality, stoichiometric Si_3N_4 films were obtained by sputtering the target with Ar^+ ions of 8 keV and simultaneous bombardment of the growing film by reactive N_2^+ ions of 0.68 keV.[19] Later it was shown that inert gas ion bombardment of growing carbon films obtained by ion beam sputtering from a graphite target causes a transition to hard and nearly colorless deposits of i-C[1].

Effects of ion bombardment on the growth and structure of thin films have also been investigated by combining an evaporator with an ion gun, as illustrated in Fig. 1(g). A more powerful device of this kind is now being used in our laboratory to study the possibilities of preparing hard i-BN coatings.[20]

Although in this paper we will not deal with ion implantation, it should be emphasized that the surface regions of materials can be modified and even buried compound layers may be prepared by implanting ions, as indicated in Fig. 1(h).

Finally, Fig. 1(i) showing the principle of magnetron sputter deposition has been added to emphasize that compound and ceramic films can be deposited at high rates by using the magnetron sputter gun.[21] Since it is difficult to investigate the mechanism of film growth in magnetron sputtering, typical ion beam techniques may be used to elucidate important process parameters.

CERMET FILMS

Composite film structures consisting of ceramic or dielectric material and a metal exhibit interesting electrical and mechanical properties that imply various applications. So far, cermet films have been prepared mainly by continuous or flash evaporation, by chemical vapor deposition or by plasma sputtering. The intention of our investigations was to study the possibilities of depositing cermet structures by ion beam sputtering. The system Ag/SiO_2 was chosen since reliable data for cermets of this composition and for the comparable Au/SiO_2 system have been published.[22-24] In this context it should be mentioned that the specific resistivity of cermets on the base of silicon oxides, as was stated by Neugebauer,[25] is largely determined by the volume fraction of SiO_2. Gasperic et al.[26] have demonstrated the validity of a similar rule for cermets on the base of other dielectrics.

The deposition conditions were as follows: targets composed of high purity silver and silica glass segments were positioned relative to the broad ion beam (Ar^+ at 8 - 10 keV and a total ion current of 30 - 40 mA) in such a way that the sputtered particle flux contained the desired fraction of metal species. The accumulation of disturbing space charges on the insulating silica surfaces was avoided by flooding the target with electrons that were generated by a hot-filament gun attached to the target holder.[27] With an ion incidence angle of 45°, deposition rates of about 10 nm/min were measured. Normally substrates of glass, germanium or silicon were used, but for structure investigations by transmission electron microscopy the layers were deposited on NaCl. The substrate temperature was 20°C, 200°C or 400°C, respectively, and the layer thickness was varied from 20 to 500 nm.

Figure 2 shows transmission electron micrographs (TEM) of typical films deposited at a substrate temperature T_S = 20°C. For higher temperatures up to 400°C no significant difference was found. The film with the highest silver content is built up of coalescent islands; whereas, the layer with the lowest silver concentration exhibits a structure of nearly globular islands. The intermediate deposits of Fig. 2(b and d) represent developmental stages in the size and shape of the metallic islands, which are known as "capillary structures." The composition of the films was determined by X-ray microanalysis. Inspection by electron diffraction indicated the presence of some Ag_2O at the surface but not in the bulk of the films. On the other hand, a careful evaluation of the resistivities and the analytical data revealed an oxygen deficit in the SiO_2 matrix of up to 10%.

Figure 3 shows a plot of the measured thermal coefficients of resistivity (TCR) versus the electrical resistivity of the Al/SiO_2 cermet films as a function of the various silver contents for the numerous films deposited at a substrate temperature T_S = 20°C. In the transition region between metallic and tunnel conduction the TCR changes its sign, and a region of small temperature dependence of the resistivity is found.

STRUCTURE AND PROPERTIES OF i-CARBON

A spectacular result of the work on ion-enhanced film deposition is that extremely hard carbon layers can be obtained by condensing carbonaceous ionized species of sufficient energy or by ion bombardment of growing sputtered films.[3,15,20,28] Our proposition of the

the generic term i-carbon (i-C), where the "i" alludes to the ion action, has become widely accepted for these unusual deposits.

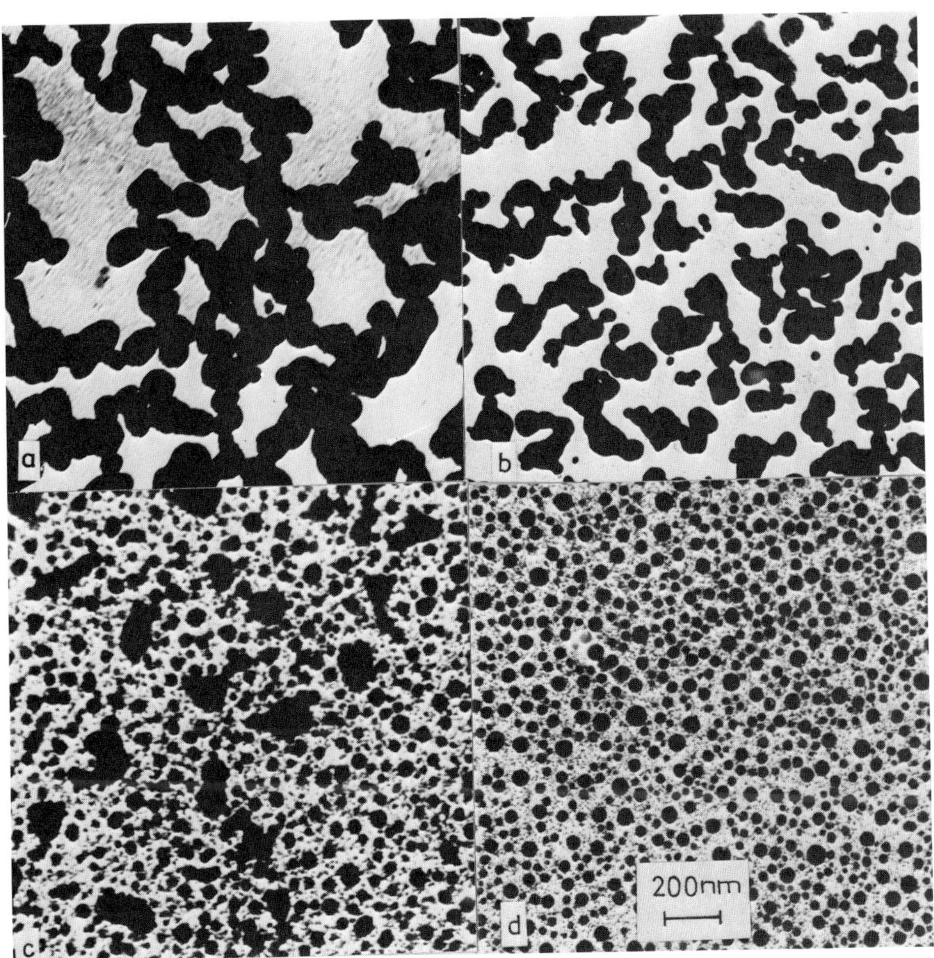

Fig. 2. TEM micrographs of Ag/SiO_2 cermet films of different silver contents, C_{Ag}, and various specific resistivities, ρ, deposited at T_S = 20°C.
(a) C_{Ag} = 93.4 mass-%, $\rho = 5 \cdot 10^{-7}$ Ohm·m;
(b) C_{Ag} = 91.8 mass-%, $\rho = 1 \cdot 10^{-6}$ Ohm·m;
(c) C_{Ag} = 78.0 mass-%, $\rho = 5 \cdot 10^{-4}$ Ohm·m;
(d) C_{Ag} = 64.7 mass-%, $\rho = 2 \cdot 10^{-1}$ Ohm·m.

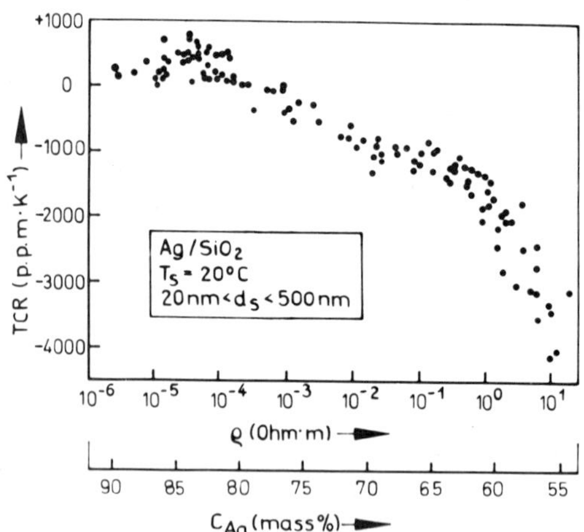

Fig. 3. Thermal coefficient of resistivity (TCR) of the Ag/SiO_2 cermet films versus the resistivity as a function of the silver contents, C_{Ag}.

Our efforts have been focused on two distinctly different processes to prepare i-C films: (1) ion beam sputter deposition from a graphite target under ion bombardment of the growing film, and (2) ion beam plating using ionized species obtained from hydrocarbons, preferably benzene. While films deposited by the first technique can be transparent and almost colorless, the layers obtained by the second method are yellowish or brownish. On the other hand, the ion plating process is much more productive; at work pressures of about 10^{-2} Pa deposition rates of 100 nm/min can be achieved, and an extension of the technique to large-area coating technology poses no principal problems.

Fig. 4(a) summarizes the nature of deposits prepared by the ion plating process as a function of the substrate bias voltage U. For energies above some hundred eV the energy spread can be neglected and the mean ion energy is eU_B. Films condensed at low bias voltages of some tens of volts were found to be polymer-like; they are soft and transparent, the electrical resistivity is very high ($\sim 10^{15}$ ohm·cm), and the infrared spectra resemble those of polystyrene. In the range of U_B from about 100 V to 800 V a drastic change of the physical properties was observed. The films become very hard and yellowish, the resistivity drops to about $5 \cdot 10^7$ ohm·cm, and the refractive index at a wave length of 546.1 nm increases to values of about 2.4. The fact that films belonging to this transition region are IR-transparent and that the refractive index can be adjusted within a certain range, suggests applications in infrared optics.

Coatings of maximum hardness are formed in the U_B range from 800 V to 1100 V. As these measurements were made using benzene, and the ionized species normally comprise six carbon atoms (only about 30% belong to fragmented aromatic rings), it follows that about 150 eV of kinetic energy per carbon atom is required to prepare i-carbon of maximum hardness by this technique. From yield measurements it could be deduced that under our conditions mainly the ions participate in the film growth.[29]

The hardness measurements were performed with layers of some μm thickness deposited on hard alloy (type "HG10") substrates by the following techniques: (1) Vickers microhardness testing, generally at a load of 0.5 N; (2) scratching tests using other hard materials; and, (3) dynamical recording of the indentation and the elastic recovery by the method of Frohlich.[30] The maximum values of the Vickers hardness were $VH_{0.5}$ = 60 GPa; from the scratching tests a Mohs hardness of 9.4 was derived. The dynamical tests indicated an unusually high elastic recovery.

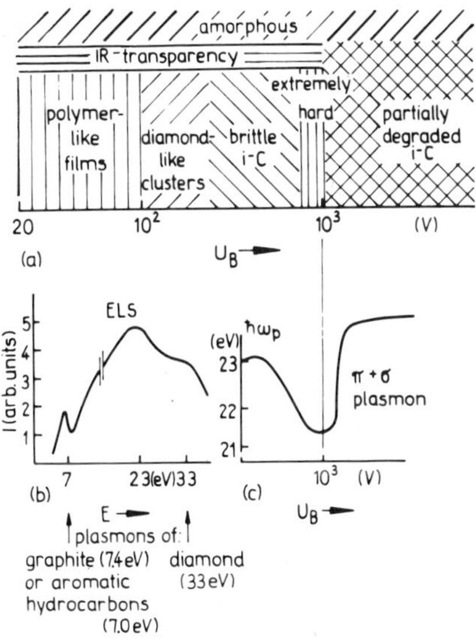

Fig. 4. Characteristic properties of i-C films deposited from benzene by the ion beam plating process at various values of the bias voltage U_B.
(a) General film properties;
(b) typical electron energy loss (ELS) spectrum corresponding to the plasmon energy region;
(c) position of the main plasmon peak versus U_B.

Films deposited at bias voltages above 1100 V were found to be less hard ($VH_{0.5} \simeq 30 - 35$ GPa), and they became increasingly dark due to radiation damage and graphite segregation. Finally, no layer growth was observed for $U_B > 3$ kV because then resputtering predominates.

Extensive structure investigations by means of transmission electron microscopy, electron diffraction techniques and electron energy loss spectroscopy (ELS) revealed that the i-carbon is basically amorphous. The ELS spectrum of a film deposited at $U_B = 500$ V is shown in Fig. 4(b). Besides the main plasmon peak a slight shoulder at higher energies and a small peak at about 7 eV can be discerned. The latter is interpreted as the π-plasmon peak originating from undissociated benzene rings. In accordance with this assumption it was found that the intensity of this peak decreases steadily with increasing bias voltage and attains zero for $U_B \geq 1$ keV. Hence the region of maximum film hardness corresponds to an almost complete dissociation of the aromatic molecules but a still limited radiation damage. A plot of the main plasmon peak position versus the bias voltage U_B, as shown in Fig. 4(c), reveals a characteristic minimum for the region of maximum film hardness. This minimum, which was reproduced many times, has been interpreted as a resonance shift due to diamond-like structural elements.[20]

What can be said at present about the nature of i-carbon? A first hypothesis, which is still maintained by many authors, was based on the assumption of a quasi-amorphous diamond structure, consisting of randomly arranged sp^3-bonded clusters. However, two recent results have led us to the conclusion that this conjecture is not true. First, measurements of the mass density by floating film particles in mixtures of benzene and bromoform yielded surprisingly low values of about 2.0 g/cm^3. Second, from intensity data obtained by scanning electron diffraction and using the measured mass density, the radial distribution function was calculated.[31] In conclusion, co-ordination numbers slightly smaller than 3 for the nearest neighbors and smaller than 6 for the second nearest atoms were derived. Therefore, despite its diamond-like macroscopic properties, i-carbon must be regarded as a rather loosely packed metastable structure that is formed through the action of the rapidly collapsing energy spikes created by the impinging energetic ions.[32] We have proposed a new structure model assuming puckered n-fold (n from 3 to 8) carbon rings, which are interlinked by strong cross-bonds.[31] It is further assumed that both fourfold co-ordinated diamondlike and twofold co-ordinated carbyne elements contribute to stabilize the structure and that broken bonds, at least in i-carbon prepared from hydrocarbons, are saturated by hydrogen. The results of recent measurements of the hydrogen concentration in i-C films prepared by ion beam plating from benzene are presented in Fig. 5. The analysis was made by using the nuclear resonant reaction $^{19}F(p,\alpha\gamma)^{16}O$ with 17 MeV N^+ projectiles. A similar method has been applied to determine hydrogen concentrations in amorphous silicon.[33]

The results show instructively that the hydrogen concentration decreases with increasing bias voltage U_B, but typical i-C films corresponding to the region of minimum hardness still contain more than 20 atom-% hydrogen.

The previously reported growth of small crystals of cubic diamond in i-C films annealed at 800 - 1000°C may be interpreted by the loss of hydrogen, which causes the structure to become unstable and to crystallize.[3,20]

As i-carbon can be deposited readily on nearly all kinds of substrates, it could find widespread applications for the surface-hardening of tools and mechanical parts and for the protection of metals or other materials against corrosion or erosion. The main hindrance consists in the high compressive stress that may cause failures due to cracking and lift-off of thick i-C films.

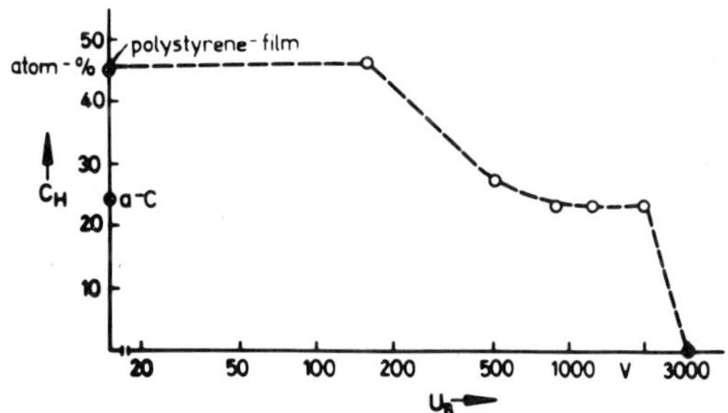

Fig. 5. Hydrogen contents of i-C films prepared by ion beam plating from benzene at different bias voltages U_B. (For comparison, results obtained for a film of polystyrene and an evaporated film of amorphous carbon, a-C, have been included.)

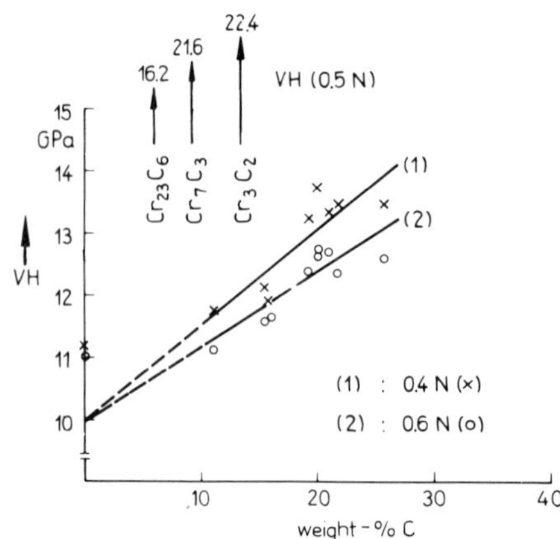

Fig. 6. Vickers hardness VH of Cr/C films. (The arrows indicate the hardness of equivalent carbides.)

METAL/CARBON COMPOSITE COATINGS

A promising way to overcome the stress problem associated with hard carbon coatings is offered by the concept of depositing alloy- or cermet-like films consisting of metals, for example Cr, Ti, Fe, or Al, and carbon. Tentative experiments in this direction confirmed that almost any mixture between metals and carbon can be deposited, yielding finely dispersed composite structures or amorphous alloys, by the ion beam plating technique. In the particular case of the system Al/C, sometimes decomposition was observed; however, under certain conditions homogeneous layers of golden color were formed. With respect to stoichiometric Al_4C_3, we have prepared films of the general composition Al_xC_{1-x}, with x ranging from 0.2 to 0.6, which were not attacked by water. The position of the main plasmon peak in the ELS spectra could be related linearily to the elemental composition.

The system Cr/C has been studied in some detail.[34] Films deposited at a bias voltage of U_B = 800 V on water-cooled hard alloy or silicon substrates proved to be smooth and amorphous up to relatively high carbon concentrations. Fig. 6 shows the measured Vickers hardness of such films. The electron diffraction data are consistent with the assumption that the carbon atoms are incorporated on interstitial sites, thus dilating the interatomic distances of the metal. The formation of microcrystals and eventually some carbide was observed only after annealing. It is remarkable that the hardness increases nearly linearly with the carbon concentration. The behavior at higher carbon concentrations is under investigation. It can also be expected that a continuous approach to i-carbon can be realized in this system.

In addition, attempts are being made to prepare similar structures on the basis of boron nitride instead of carbon. The feasibility of preparing hard i-BN has been demonstrated.

ACKNOWLEDGMENTS

The authors thank Dr. G. Holzhuter, WPU Rostock, for his contribution to the determination of the radial distribution functions and Dr. W. Rudolph, ZfK Rossendorf, for performing the hydrogen determination.

REFERENCES

1. G. Gautherin and C. Weissmantel, Thin Solid Films, 50, 135 (1978).
2. T. Takagi, Thin Solid Films, 92, 3 (1982).
3. C. Weissmantel, Thin Solid Films, 92, 55 (1982).
4. J. M. E. Harper, "Ion Beam Deposition," in Thin Film Processes, edited by J. L. Vossen and W. Kern, Academic Press, NY, 1978.
5. G. Carter and D. G. Armour, Thin Solid Films, 80, 13 (1981).
6. P. D. Reader and H. R. Kaufman, J. Vac. Sci. Technol., 12, 1344 (1975).
7. O. Fiedler, G. Reisse, B. Schoneich, and C. Weissmantel, Proc. 4th Intern. Vac. Congr., Manchester, 1968, 569.
8. G. Guatherin, P. Bouchier, C. Schwebel, and P. Vapaille, Le Vide, 182, 235 (1976).
9. J. Franks, Le Vide, Suppl. to 196, 53 (1979).
10. K. Bewilogua, D. Dietrich, L. Pagel, C. Schurer, and C. Weissmantel, Surf. Sci., 86, 308 (1979).
11. J. M. E. Harper, J. J. Cuomo, P. A. Leary, G. M. Summa, H. R. Kaufman, and F. J. Bresnock, J. Electrochem. Soc., 128, 1077 (1981).
12. J. H. Freeman, Nature, 275, 634 (1978).
13. K. Yagi, S. Tamura, and T. Tokuyama, Jap. J. Appl. Phys., 16, 245 (1977).
14. A. E. T. Kuiper, G. E. Thomas, and W. J. Schouten, J. Crystal Growth, 51, 17 (1981).
15. C. Weissmantel, G. Reisse, J. Erler, F. Henny, K. Bewilogua, and C. Schurer, Thin Solid Films, 63, 315 (1979).
16. C. Weissmantel, O. Fiedler, G. Hecht, and G. Reisse, Thin Solid Films, 13, 359 (1972).
17. G. Gautherin, C. Schwebel, and C. Weissmantel, Proc. 7th Intern. Vac. Congr., Berger, Vienna, 1977, 1579.
18. T. Takagi, I. Yamada, and A. Sasaki, J. Vac. Sci. Technol, 12, 1128 (1975).

19. H.-J. Erler, G. Reisse, and C. Weissmantel, Thin Solid Films, 65, 233 (1980).
20. C. Weissmantel, Thin Solid Films, 96 (1982), in print.
21. S. Schiller, U. Heisig, and K. Goedicke, Thin Solid Films, 40, 327 (1977).
22. J. E. Morris, Thin Solid Films, 11, 299 (1972).
23. N. C. Miller, B. Hardiman, and G. A. Shirn, J. Appl. Phys., 41, 1850 (1970).
24. E. B. Priestley, B. Abeles, and R. W. Cohen, Phys. Rev. B 12, 2121 (1975).
25. C. A. Neugebauer, Thin Solid Films, 6, 443 (1970).
26. J. Gasperic and B. Navinsek, Thin Solid Films, 36, 353 (1976).
27. P. Reinhardt, C. Reinhardt, G. Reisse, and C. Weissmantel, Thin Solid Films, 51, 99 (1978).
28. H. Vora and T. J. Moravec, J. Appl. Phys., 52, 6151 (1981).
29. D. Deitrich, Techn. Hochsch. Karl-Marx-Stadt, personal communication, 1982.
30. C. Weissmantel, C. Schurer, F. Frohlich, P. Grau, and H. Lehmann, Thin Solid Films, 61, L5 (1979).
31. K. Bewilogua, D. Dietrich, G. Holzhuter, and C. Weissmantel, phys. stat. sol. (a), 71, K57 (1982).
32. D. A. Thompson, Ratiation Effects, 56, 105 (1981).
33. M. H. Brodsky, Thin Solid Films, 50, 57 (1978).
34. K. Bewilogua, E. Bugiel, B. Rau, C. Schurer, and C. Weissmantel, Kristall und Technik, 15, 1205 (1980).

PART VI

LASER AND ION BEAM MODIFICATION OF SURFACES

LASER SURFACE MELTING OF METALS AND ALLOYS

David B. Snow

United Technologies Research Center
Silver Lane
East Hartford, Connecticut 06108

ABSTRACT

The use of laser irradiation to melt and rapidly self-quench metal surfaces has been extensively utilized as a research tool. The two principal objectives of this procedure have been to refine the scale of chemical inhomogeneity, and/or to change the composition of a surface layer by material addition. In both cases, sufficiently rapid solidification rates occur (usually $\geq 10^4$ °C/s) to produce metastable surface layers with desirable properties. The present status of this technology is reviewed, with emphasis on the physical metallurgy of the alloys to which it has been applied.

INTRODUCTION

It has been recognized for nearly two decades that the surface microstructure of metals and alloys can be substantially modified by laser-induced rapid melting and self-quenching.[1,2] The recent publication of several excellent review papers, which taken together treat this subject very comprehensively, demonstrate the intense research interest which it has engendered.[2-8] As mentioned by Breinan et al.[8] and Draper,[2] the wide range of laser beam power densities and irradiation dwell times which can be achieved with the various types of lasers presently available permit metal surface solidification rates of from 10^4-10^{10} °C/s. Since one investigator or research group tends to utilize one particular laser system to examine a relatively more limited processing regime than shown by the full range of parameters designated as "laser glazing" in Fig. 1, it is perhaps understandable that the differences in microstructure associated with such extremes of solidification rate are not always apparent upon

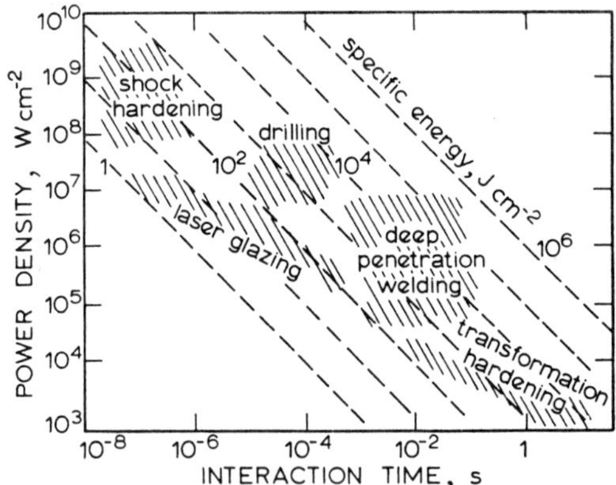

Fig. 1. Operational regimes for laser materials processing.

examination of the literature. By way of illustration, the virtually defect-free, 0.5 μm deep surfaces achieved in semi-conductor materials with nanosecond pulses from Q-switched ruby lasers[6] have little in common with the dendritic microstructures containing high dislocation densities which are characteristic of the ⩾ 25 μm surface layers created in nickel-base alloys with scanned, continuous CO_2 lasers and millisecond dwell times.[9] Generally, such microstructural characteristics as dislocation density, grain structure, degree of compositional refinement and/or the solid-state phase transformations in a laser-melted metal surface are strongly dependent upon both the energy input parameters and the material composition. Until recently, specific microstructural effects have often been more accurately determined by experimental observation than by predictive models of energy adsorption and heat flow. The following text will attempt to systematically describe some of the microstructural effects that have been derived from laser surface melting, with some commentary on property changes as well.

LASER CHARACTERISTICS

The more important operating characteristics of the laser systems which are commercially available for laser surface melting are described in Table 1. There are a number of trade-offs associated with the selection of any one of these. Since very rapid solidification rates are often of primary interest, and pulse duration is a reasonable approximation of the self-quenching solidification time, the Q-switched lasers offer the best opportunity to obtain very rapidly cooled microstructures. At the opposite end of the power range, continuous CO_2 lasers can rapidly process very large amounts

Table 1. Comparison of Laser Operating Characteristics[a]

Laser Source	Laser Wavelength, μm	Power Output, W	Dwell or Pulse Time, s
Continuous CO_2	10.6	500-20,000	10^{-1}-10^{-5}
Pulsed Nd:YAG	1.06	200 (average)	10^{-2}-10^{-3}
Pulsed Nd:glass	1.06	10^6 (peak)	10^{-2}-10^{-3}
Pulsed ruby	0.69	10^6 (peak)	10^{-2}-10^{-3}
Q-switched Nd:YAG	1.06,0.53	1-10 (average)	1.5×10^{-7}-3×10^{-7}
Q-switched ruby	0.69	10^8 (peak)	5×10^{-9}-5×10^{-8}

of surface area, particularly when focused above or below the irradiated surface, but are very expensive compared to pulsed lasers and cannot provide the same degree of day-to-day beam energy reproducibility. Surface reflectivity increases with beam wavelength, to the extent that most metals have a reflectivity of > 90% for wavelengths > 5 μm. This effect is offset, however, by the greater power of 10.6 μm CO_2 lasers, and becomes a problem only for very efficient reflecting surfaces such as those of gold, aluminum or copper. On the whole, pulsed ruby and Nd:YAG lasers offer a good balance of parameters for general surface melting, and can achieve melt depths of from 10-700 μm and solidification rates of approximately 10^5-10^7 °C/s by variation of beam focus and pulse energy.

BEAM-SURFACE ENERGY TRANSFER

Metals exhibit a more complex behavior than non-metals in the way incident optical energy is absorbed, and this is particularly true of the infared wavelength of CO_2 lasers (10.6 μm). Nevertheless, several useful heat flow models have been published,[10-13] which can be used to predict the efficiency and geometry of the energy transfer from the laser beam into the metal surface. For a laser beam of any wavelength, the initial temperature increase occurs within the electromagnetic skin depth of the metal (< 100 μm) as the optical energy is converted into heat. Using a one-dimensional approximation,[2] heat diffuses from this region to a distance of $(2Dt_p)^{1/2}$, the thermal diffusion distance, where t_p is the pulse or dwell time and D (the thermal diffusivity) = $K/C\delta$; and K, C and δ are the thermal conductivity, specific heat, and density, respectively. A square wave laser pulse or traversed laser beam will give an

[a]After Draper[2].

average temperature rise within this thermal diffusion layer of

$$\Delta T = (1-R_o)It_p/C\delta(2Dt_p)^{1/2} \tag{1}$$

where R_o is the reflectance and I is the incident laser intensity.[2] Other, more detailed heat flow models have used an analytical solution to the thermal diffusion equation using Green's functions,[10] or finite difference analyses incorporating the effect of the latent heat of melting. Examples of the predictive data from one such analytical solution, which assumes a semi-infinite plate and neglects the latent heat of melting and property changes due to the solid-liquid transformation,[9,12] are shown by Figs. 2 and 3, where the anticipated variation of temperature with time and of cooling rate with melt depth within a laser-melted zone are given. While models of this sort are quite useful for initial estimation of melt depths and cooling rates, the condition of the metal surface finish must be uniform if reproducible melting is to be obtained. Changes in surface roughness or even random scratches can significantly change the effective value of R_o and thus, the efficiency of energy absorption and the melt depth. This effect becomes more important at large values of intrinsic reflectance, and can be a major source of poor correlation between predicted and observed melted zone geometries.[13]

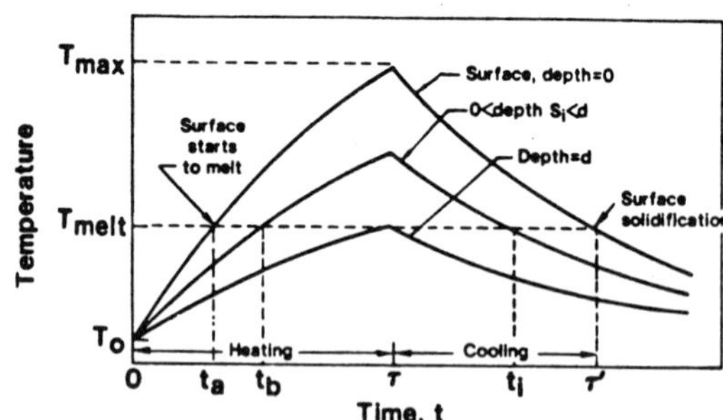

Fig. 2. Variation of temperature with time at selected melt depths.

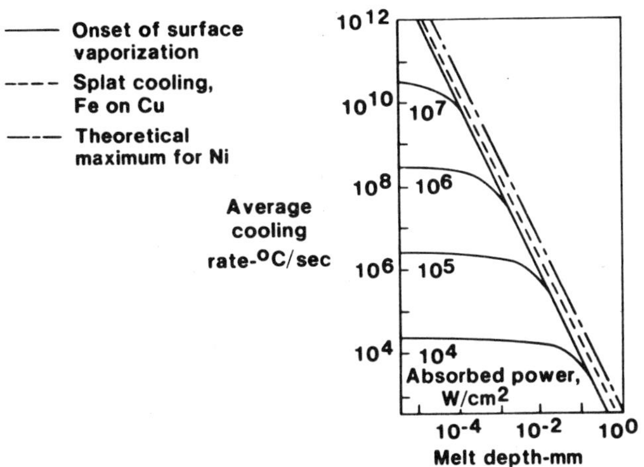

Fig. 3. Effect of power density and melt depth on cooling rate.

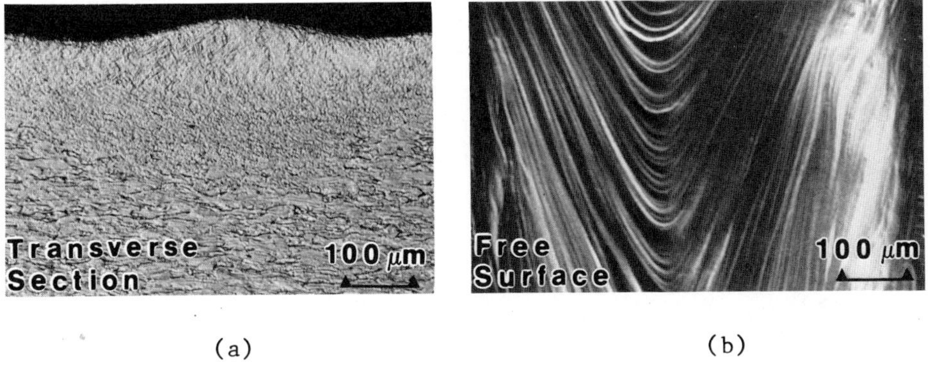

(a) (b)

Fig. 4. Topography of laser melted surface of Zircaloy-4. Continuous CO_2 laser; power density, 2.3 MW/cm^2; dwell time, 1 ms; scan rate, 50.3 cm/s.

INTRINSIC SURFACE EFFECTS

Surface Topography

The relatively short interaction times (~ $15\text{-}60\times10^{-9}$s) and consequent shallow melt depths (≤ 0.5 μm) associated with Q-switched laser surface melting are reported to smooth surface topography in some cases.[6] However, as interaction times increase to the millisecond range and melt depths exceed 10 μm, fluid flow effects usually create differences in height between the first and last surfaces to freeze, as well as fine scale surface ripples across the entire melted surface (Fig. 4). This surface topography results from both surface tension effects and fluid flow around the beam impingement point,[14,15] and becomes a problem only if fine-scale surface roughness is undesirable.

Thermal Stresses

The rapid solid state cooling of laser melted surfaces causes them to experience thermally-induced stresses considerably in excess of the yield point; these in turn impose high strain rates. Consequently, melted zone cracking is a much more common occurrence than is immediately apparent from a review of the literature, and in the case of melted depths exceeding ~10 μm which form brittle crystalline microstructures upon solidification, it is virtually unavoidable. However, there appear to be no cracking problems, or even measurable dislocation populations, in silicon melted to $\leqslant 0.5$ μm,[6] which implies that neither the yield nor the fracture stress is exceeded in such thin layers. On the other hand, a simple stress analysis in the case of deeper melt depths[16] shows that high ($> 10^{11}$ cm^{-2}) dislocation densities, and residual tensile stresses on the order of the yield stress, can be routinely expected in crack-free laser melted surfaces. Pure metals and most solid solutions usually deform plastically upon self-quenching, but in many alloys which experience rapid high temperature grain boundary segregation of impurities or second phase precipitation, cracking can occur with dramatic results.

Microstructure

Laser surface melting without the addition of new material is generally used to reduce the scale of, or to completely eliminate, undesirable microstructural features in metals and alloys. For the $< 10^{-7}$s pulse duration and < 1 μm melt depths generated by Q-switched lasers, the near-surface damage caused by ion-bombardment of semiconductor materials can be virtually eliminated and short-range redistribution of the injected species can be achieved.[6] However, much of the laser surface melting of other metal systems has utilized continuous CO_2 lasers with dwell times of $> 10^{-4}$s and melt depths of $\geqslant 25$ μm, as exemplified by the surface melting parameters of the LASERGLAZE™ process.[17] The melt areas thus generated all involve some degree of microstructural refinement, and are often intended to improve the mechanical or chemical properties of the surfaces. The most dramatic of these microstructural transformations is the creation of an amorphous layer on the surface of alloys of suitable glass-forming composition[3,18] such as $Ni_{60}Nb_{40}$. While this holds considerable potential for creating corrosion-resistant surfaces, it is not yet clear that devitrification at the interface between successively-melted regions and/or substrate cracking due to thermal stresses can be consistently avoided.

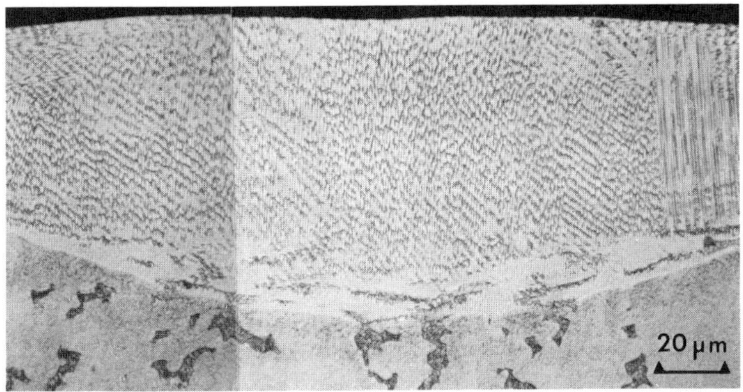

Fig. 5. Refinement of dendritic structure in laser surface melted nickel base superalloy. Chill cast substrate. Continuous CO_2 laser; 1.4 MW/cm^2; 0.6 ms.

A more frequently-documented result of the laser surface melting of many alloys is the creation of a refined dendritic structure, and in some cases of a refined grain structure as well. Examples of the former case have been reported quite frequently and in considerable detail.[3,7,18] The solidification modes close to the unmelted substrate often change rapidly from a narrow band of planar front to cellar to dendritic, indicative of the large gradient in cooling rate close in this region.[7,15] By examining such microstructures (Fig. 5) to determine the secondary arm spacing of the dendrites,[18] the variation in solidification rate throughout the melted zone can be determined experimentally.

Observations of grain refinement in various alloys have been reported, for example in steel,[13] but this effect has been infrequently documented; whereas, coarse (> 100 µm), columnar grain structures have also been reported in which heterogeneous grain nucleation at the substrate-melt interface is dominant.[9] However, the large amounts of plastic strain experienced by laser melted layers with thicknesses greater than several µm provides the potential for grain refinement by subsequent recrystallization, as has been demonstrated in the case of laser-processed nickel base superalloys.[19]

In a variety of alloy systems, the creation of refinement of second phase dispersion can be achieved, even with relatively low solidification rates of $\geqslant 10^4$ °C/s. One such example is the improved carbide dispersion created by dissolution and reprecipitation in laser-melted high speed steel surfaces.[7] Another is the extremely fine second phase dispersion of rare earth compounds created in titanium base alloys by laser processing,[20] without which these compounds are heavily segregated and detrimental to mechanical properties. Depending on the phase transformation kinetics, these dispersions can precipitate spontaneously from solid solution in alloys which experience cooling rates of ~ 10^6 °C/s or less; with more rapid cooling, supersaturation is often preserved.

SURFACE ALLOYING

Laser surface melting with material addition can be used to alloy metal surfaces so as to create a homogeneous layer of different composition and controlled thickness. This process has many attractive applications, and has become an area of steadily increasing research activity,[4,5,7,21] and the subject of a recent review paper.[2] This interest is due primarily to the potential benefits to be derived not only from a rapidly-solidified surface structure, but also the opportunity to utilize much smaller quantities of expensive alloying elements than might be required to form the same bulk material.

Most laser surface alloying is accomplished by prepositioning the material to be added in close contact with the substrate surface, then mixing the two to the desired extent by laser irradiation. The various techniques of material pre-placement have been described in detail elsewhere,[2,4] and include vapor deposition, ion bombardment, powder layers mixed with a binder and overlaid metal sheet or foil. Two of the more novel procedures involve the injection of abrasive particles into the laser-melted liquid prior to self-quenching[22] to create wear-resistant surfaces; and the continuous introduction of a stream of powder feedstock into the beam impingement site to create ~ 100 μm self-quenched layers, repeatedly overlaid to any desired depth or thickness (the LAYERGLAZE™ process).[21]

Two of the most important requirements for successful laser surface alloying are that the desired alloy composition be achieved with the desired degree of chemical homogenity. Extensive studies of the variation of composition with depth in specimens of gold-coated nickel illustrate the variation in alloying effects which result from different dwell times and melt depths (Fig. 6). Longer dwell times and lower surface reflectivities tend to create more dilute alloys through initial evaporative loss and a greater volume of substrate mixed in. In addition, if the alloyed surface is to be both very chemically homogeneous and highly alloyed, consideration must also be given to whether the mixing effect of liquid state convective turbulence may have been prematurely interrupted (Fig. 7).[23]

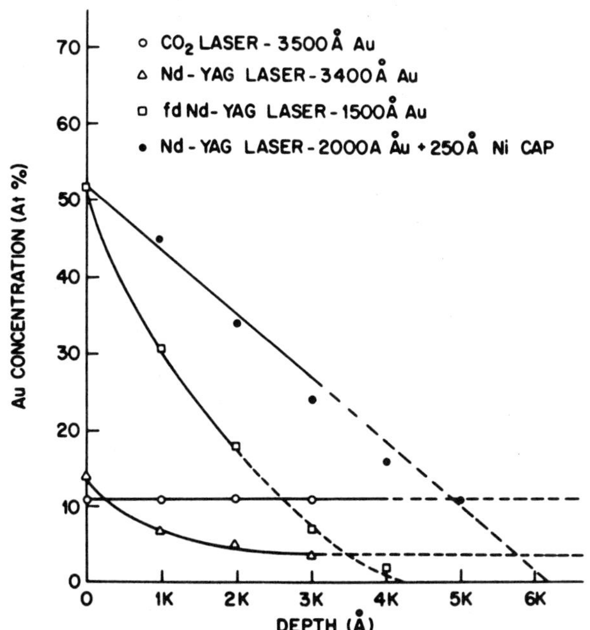

Fig. 6. Calculated and observed depth profiles for laser surface alloyed Au-Ni (from Draper, Ref. 2).

CONCLUDING REMARKS

The melting and rapid self-quenching of metal surfaces by laser irradiation can be utilized to achieve rapidly-solidified surface structures of unique composition and properties while leaving the underlying substrate unaffected. The forgoing text has reviewed the various aspects of this technique quite briefly, and inevitably contains many omissions of specific examples discussed by previous reviewers.[2-7] Neither the application nor the effects of laser surface melting are trivial, but thanks to the unusual and useful microstructures and properties which it can produce, it remains one of the more interesting and active areas of research in physical metallurgy.

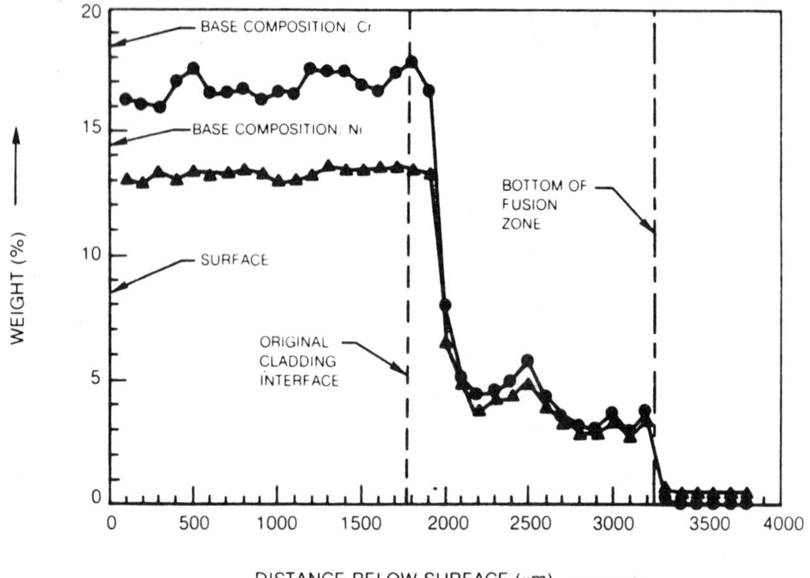

Fig. 7. Cr and Ni profiles determined by electron microprobe analysis; laser melted low carbon steel, clad with 317L stainless. Continuous CO_2 laser; 3.1 MW/cm^2; 10 ms dwell time.

REFERENCES

1. F. E. Cunningham, (MS Thesis), The Use of Lasers for the Production of Surface Alloys, Massachusetts Institute of Technology, 1964.
2. C. W. Draper, in Lasers in Metallurgy, edited by K. Mukherjee and J. Mazumder, The Metallurgical Society-AIME, Warrendale, PA, 1981.
3. E. M. Breinan, B. H. Kear, and C. M. Banas, Physics Today, 29 [11], 44 (1976).
4. D. S. Gnanamuthu, in Applications of Lasers in Materials Processing, American Society for Metals, Metals Park, Ohio, 1979.
5. K. Stanford, Metallurgia 3, 109 (1980).
6. J. Narayan, J. Metals, 32 [6], 15 (1980).
7. B. H. Kear, J. W. Mayer, J. M. Poate, and P. R. Strutt, in Metallurgical Treatises, edited by J. K. Tien and J. F. Elliot, The Metallurgical Society-AIME, Warrendale, PA, 1981.
8. E. M. Breinan and B. H. Kear, in Laser Materials Processing, edited by E. M. Bass, North-Holland, New York (1983) (in press).

9. D. B. Snow, E. M. Breinan, and B. H. Kear, in Superalloys 1980, edited by J. K. Tien et al., Am. Soc. for Metals, Metals Park, Ohio, 1980.
10. H. E. Cline and T. R. Anthony, J. Appl. Phys., 48, 895 (1977).
11. S. C. Hsu, S. Chakravarty, and R. Mehrabian, Met. Trans. 9B, 221 (1978).
12. L. E. Greenwald, E. M. Breinan, and B. H. Kear, in Laser-Solid Interactions and Laser Processing, edited by S. D. Ferris, H. J. Leamy, and J. M. Poate, Am. Inst. of Physics, New York, 1979.
13. T. R. Tucker, A. H. Clauer, C. T. Walters, and S. L. Ream, in Lasers in Metallurgy, edited by K. Mukherjee and J. Mazumder, The Metallurgical Society-AIME, Warrendale, PA, 1981.
14. H. E. Cline and T. R. Anthony, J. Appl. Phys. 48, 895 (1977).
15. S. M. Copley, D. Beck, O. Esquivel, and M. Bass, in Laser-Solid Interactions and Laser Processing-1978, edited by S. D. Ferris, H. J. Leamy, and J. M. Poate, American Institute of Physics, Schenectady, NY, 1979.
16. T. R. Anthony and H. E. Cline, J. Appl. Phys., 49, 1248 (1978).
17. C. M. Banas, E. M. Breinan, B. H. Kear, and A. F. Giamei, United States Patent 4,122,240, October 24, 1978.
18. H. W. Bergmann and B. L. Mordike, in Rapidly Solidified Amorphous and Crystalline Alloys, edited by B. H. Kear and B. C. Giessen, North-Holland, NY, 1982.
19. E. M. Breinan, D. B. Snow, and C. O. Brown, United Technologies Research Center Final Report R81-914346-8, DARPA-Sponsored ONR Contract N00014-78-C-0387, January 1981.
20. T. C. Peng, S. M. L. Sastry, and J. E. O'Neal, in Lasers in Metallurgy, edited by K. Mukherjee and J. Mazumder, The Metallurgical Society-AIME, Warrendale, PA, 1981.
21. E. M. Breinan, D. B. Snow, C. O. Brown, and B. H. Kear, in Rapid Solidification Processing, Principles and Technologies, II, edited by R. Mehrabian, B. H. Kear, and M. Cohen, Claitor's Publishing Div., Baton Rouge, 1980.
22. J. D. Ayres, T. R. Tucker, and R. J. Schaefer, Ibid.
23. D. B. Snow, L. E. Greenwald, and E. M. Breinan, United Technologies Research Center Annual Technical Report R-82-914763-6, ONR Contract N00014-79-C-0649, January 1982.

LASER PROCESSING OF CERAMICS

J. R. Spann, R. W. Rice, W. S. Coblenz, and
W. J. McDonough

Ceramic Branch
US Naval Research Laboratory

ABSTRACT

Results of experiments exploring the feasibility of using a laser heat source for sealing plasma sprayed coatings, toughening ceramic bars, and making ceramic alloy powders are reported. Chemical modification of the coated surfaces in conjunction with surface melting showed some promise of controlling craze type cracking. Precipitates which should be beneficial to the toughness of zirconia coatings were developed as a result of laser treatment. Fine, complex microstructures were developed in the surface of ceramic alloy bars and in spherical particles. Strengthening of some bars was indicated, but it was limited by the shape and character of the underlying material. Molten droplets of selected ceramic compositions typically solidified into spheres of the order of 40-200 μm for the range of rotational speeds used in the study. Spheres typically contained porosity which enhances their crushability to give powders for possible future body fabrications.

INTRODUCTION

Continued improvement of the performance of bulk ceramics and ceramic coatings to better achieve the significant potential that these materials have for a technological society requires that a variety of treatment and processing methods be investigated. Lasers are a convenient and sometimes unique tool for investigating some of the methods of treatment or processing ceramics that may significantly improve their properties. This paper summarizes partially completed work on evaluation of surface treatment of ceramic coatings and of bulk ceramic "alloys," as well as the making of ceramic "alloy" powders using a laser.[a] The specific concepts, and goals behind each of these investigations will be addressed in the introduction to the section on that technique.

EXPERIMENTAL PROCEDURE

The laser used throughout these experiments was a CO_2, 10.6 micrometer, continuous wave laser with a maximum power output of 10 to 15 kw depending upon the mode of operation. Figure 1 schematically illustrates one of the laser setups and is representative of the others used. The calorimeter was used to check beam power, before every run. The power density on the sample was controlled by changing the laser power and/or the position of the sample relative to the focal point of the final focussing mirror. The laser was operated in one of two modes: first; operating as a stable oscillator, whose rotating the spinning mirror gives a "top hat" beam, a cross section of whose energy profile is nearly a step function (Fig. 1). The other mode was as an unstable oscillator, which gives roughly a gaussian distribution with superimposed noise. Defocussing of the unstable oscillator beam gives a "doughnut" beam, i.e., with an annular distribution of energy (see Fig. 16). Specific procedures for each type of process are discussed with that process.

EXPERIMENTAL MOTIVATIONS, CONCEPTS, RESULTS, AND DISCUSSION

Laser Treatment of Ceramic Coating Surfaces[b]

The motivations for laser treatment of ceramic coatings was first to seal their surfaces, i.e., make them impervious, and second to enhance their physical survivability. Ceramic coatings have important potential for a variety of new applications. However, these require that a coating not be permeated by the environments they are exposed to. This is a particularly notable requirement in the use of ceramic coatings in heat engines with dirty fuels. Fuel contaminants can enter the pores and cracks that commonly exist in many ceramic coatings, resulting in degradation or destruction of the coating. Improving the mechanical capabilities of the coating in terms of thermal shock resistance, errosion, abrasion, etc., is also of broad interest.

---

[a]Note that other studies address the issue of possible use of lasers in machining of ceramics.[1,2]

[b]Further details on treatment of coatings with static beams can be found in Ref. 3.

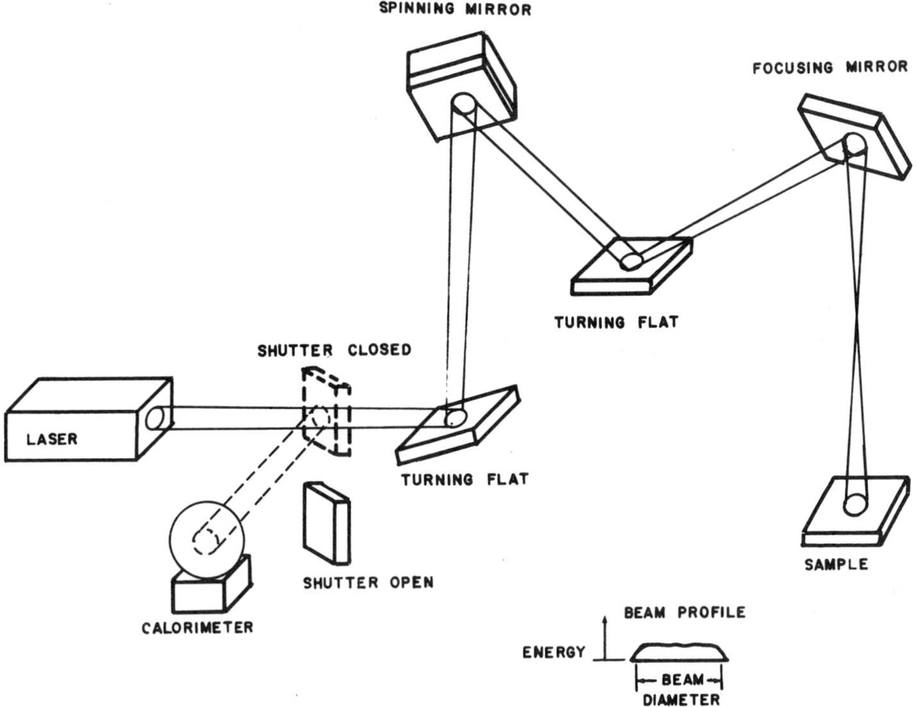

Fig. 1. Schematic of laser operation with the top hat beam (note inserted sketch of beam energy profile).

The basic concept employed for sealing such coatings was to melt each of their surfaces with the laser. In the case of partially stabilized ZrO_2 (PSZ) coatings one of the techniques for improving coating performance was to toughen them by developing precipitates within the individual grains (as will be shown, as-received plasma sprayed coatings typically contain none of the precipitate structures that have been associated with toughening in bulk PSZ). It was recognized that melting of the surface could lead to significant cracking problems due to the resultant thermal stresses and to the shrinkage of the material due to reduction of porosity. It was therefore proposed that chemical modification of the surface in conjunction with the laser treatment might adequately control these problems as well as contribute to the overall mechanical performance of the coating. The surface of the coating was coated and/or infiltrated with a sol, salt solution or slurry that would produce the desired constituents to be alloyed into the surface.

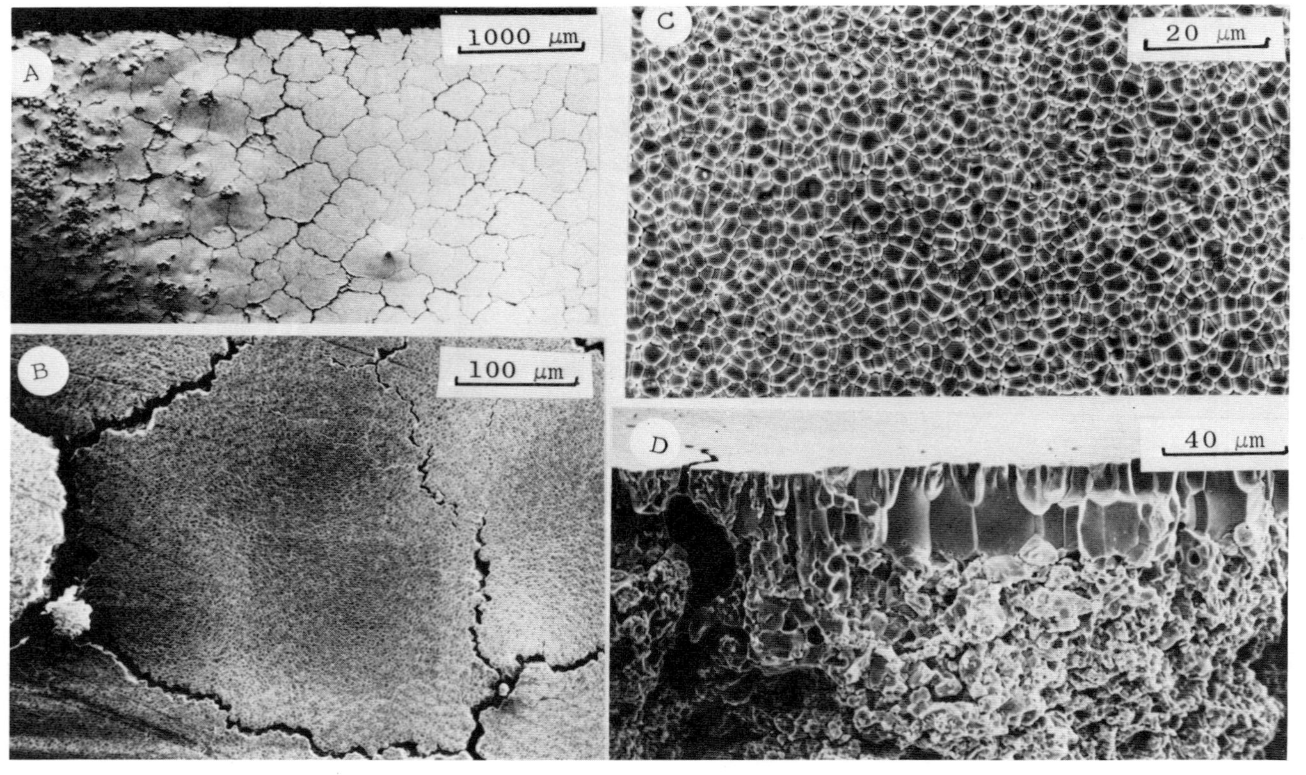

Fig. 2. Microstructure of laser treated ZrO_2 coatings. (A), (B), and (C) show the surface of a ZrO_2 +8 wt% ZrO_2 coating after melting with a static beam at 1.4 kw/cm^2 (140 J/cm_2). Note the craze cracking (A), (B) and fine microstructure (C). (D) shows a fracture cross section of a ZRO_2 + 6 wt% coating irradiated at 2 kw/cm^2 (200 J/cm_2) and the resultant columnar structue and intergranular fracture in the solidified surface.

Another part of the concept was that addition of any solids content to pores or cracks in the coating would reduce the amount of shrinkage and hence reduce cracking on resolidification. More generally, the intent was to produce lower melting surface compounds which would undergo less differential contraction relative to the lesser heated internal portions of the coating on resolidification and to produce compositions with smaller solidification contractions and/or a lower thermal expansion. The latter could potentially put the coating surface in compression not only suppressing cracking during cooling, but also possibly enhancing the mechanical reliability of the coating. It was recognized that some of the constituents could be chosen to actually toughen the coating surface. Examples of these approaches are: the addition of SiO_2 to ZrO_2 or Al_2O_3 coatings to produce $ZrSiO_4$ or $3Al_2O_3{\cdot}2SiO_2$ in the surface. Both of these compounds have substantially lower melting points than pure ZrO_2 or Al_2O_3 and have about half their thermal expansions. Other possibilities, for toughening, would be the addition of Al_2O_3 to the surfaces of zirconia coatings or ZrO_2 to the surface of Al_2O_3 coatings which would significantly lower the melting point due to the formation of the ZrO_2-Al_2O_3 eutectic and result in a tough Al_2O_3-ZrO_2 composite or "alloy" surface.

Table 1 lists the coatings used in this investigation and Table 2 lists the source of the additives. Consider first, the results on ZrO_2 plasma spray coatings which have been most extensively investigated, primarily using the stationary top hat beam. Of all the ZrO_2 coatings investigated, only the 2YSZ coating exhibited any significant spalling or separation from the substrate during or after laser melting. This commonly occurred by the coating slowly peeling up from either end of the substrate after exposure of the complete coating length to the beam. This peeling typically occurred at a slow rate within a few minutes after the coating had been exposed to the laser. The peeling and related spalling were eliminated when the coating and substrate were heated to approximately 300°C prior to, and during laser exposure and then slowly cooled for a few minutes after laser treatment. Besides surface melting, the other result common to all ZrO_2 coatings exposed to the laser was the occurrence of craze type cracking (Fig. 2). Use of a heater to raise the substrate-coating temperature to several hundred °C prior to laser irradiation did not appear to have any effect on the nature or extent of this craze cracking. The other macroscopically visible change in the coatings due to irradiation with a static top hat beam was that the irradiated area was often somewhat whiter and there was often significant enhancement of "rust spots" which are believed to be due to melting of iron rich areas.

Microstructural examination of the irradiated ZrO_2 coatings showed that the melted layer solidified with a columnar grain structure, e.g., see Fig. 2d. The length of these grains was consistent with the calculated depth of melting. Craze cracking which was

Table 1. Plasma Sprayed Coatings Used for Laser Experiments.

Coating					Substrate		Bond Coat	
Comp.	Desig.	Source	Thick. (mm)	X-ray Char.	Material	Thick. (mm)	Type	Thick
ZrO_2	2YSZ	US/UK[(1)]	0.43	Cubic	Steel	2.6		
ZrO_2-8Y_2O_3	8YSZ	NASA-Lewis[(2)]	0.36	Cubic	Ni-5.7AL-33Mo	3.1	Ni-17.1Cr-6Al-.31Y	0.12
ZrO_2-6Y_2O_3	6YSZ	UTC-P&W[(3)]	0.25	Cubic	(Ni-bass alloy) AMS5544 Waspaloy	8.0	NiCoCrAlY Ni-22-Co-18Cr-12.5 Al-0.5Y	0.12
ZrO_2-3CaO	3CSZ	SUNY[(4)]	0.81		mild steel	2.3	none	
Al_2O_3	A	SUNY[(4)]	0.20		mild steel	3.1	none	
TiO_2	T	SUNY[(4)]	0.33		mild steel	3.0	none	
Al_2O_3	13TSA	SUNY[(4)]	0.52		mild steel	2.4	none	

Samples courtesy of:
(1) R. N. Katz, AMMRC, Watertown, MA 02172.
(2) S. R. Levine, NASA-Lewis, Clev. OH 44135.
(3) G. W. Goward and D. S. Duvall, P & W, UTC, East Hartford, CN 06108.
(4) H. Herman, SUNY, Stony Brook, NY 11794.

Table 2. Chemical Modifiers Used on Coatings.

Modifier	Manufacturers Designation	Inorganic Constituent	Source
Polymerized Sol	Si5-1 MH	17 w/o SiO_2	Am Sol Co. Pgh., PA 15235
Cabosil	70A°	395 $m^2/gmSiO_2$	Cabot N. Bed., MA
Colloidal Zirconia	Prod. 54824	~5 w/o ZrO_2	TAM Div. NL Ind. Inc. Niagara Fls, NY 14305
Polymerized Sol	A'-N6-1	4 w/o Al_2O_3 (5% Na_2 0.95% Al_2O_3)	Am Sol Co. Pgh., PA 15235
Liquid Oxide	T4-EE2	5 w/o TiO_2	Oxitane Co. Akron, OH 44313

intergranular along these columnar grains often appeared to extend deeper than these columnar grains. Transmission electron microscopy showed that the fine ~ 0.1 μm grains in the as-received coating were eliminated by laser melting. The grain size of the treated coating was about 2 μm. More importantly, this TEM study showed no tetragonal precipitates in the as-received coatings, but a definite precipitate structure in laser treated coatings (e.g., Fig. 3).

Initial experiments moving the specimen relative to the beam have beeen conducted on some of the ZrO_2 coatings (e.g., Figs. 4,5). Here a much finer beam and higher power densities (e.g., 54 kw/cm^2) have been used to compensate for the motion of the sample to achieve similar overall average energy densities (50-400 J/cm^2). These initial experiments do not show cracking between adjacent passes, thus indicating that if the problem of craze type cracking can be solved, complete specimen surfaces can be treated by moving the specimen relative to the beam or by the translation of the beam over the surface. One difference observed between the moving and stationary beam experiments was that the material was commonly darkened in the moving beam experiments with the darkening generally increasing at higher power densities. It was also observed that deeper melting resulted in some laminar cracks beneath areas of deeper melting (Fig. 5), apparently, this was due to inhibited shrinkage by material on either side of the melted area toward the substrate.

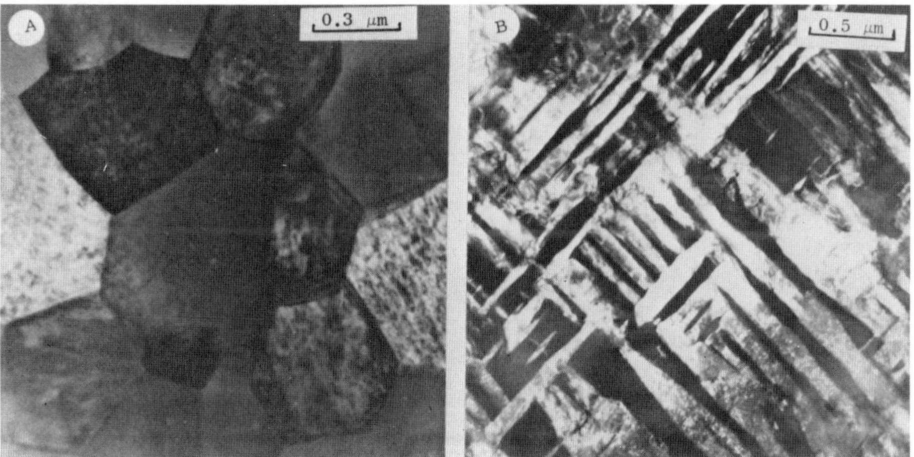

Fig. 3. Transmission electron micrographs of ZrO_2 (+ 8 wt% Y_2O_3) coatings. (A) as-received coating showing no precipitates; (B) laser melted area showing distinct tetragonal precipitates (elongated white areas). Photos courtesy of Mr. Barry Bender.

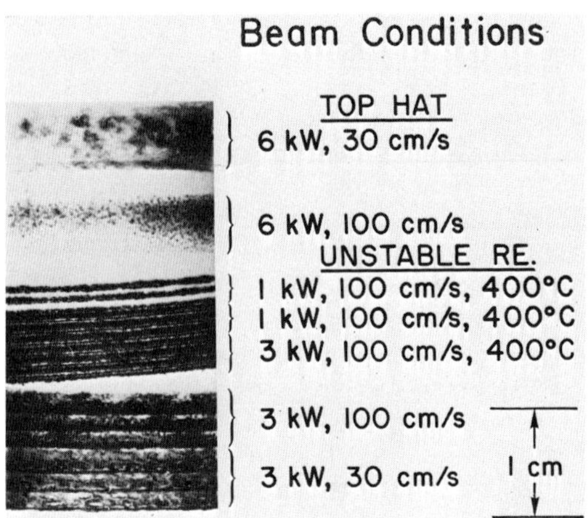

Fig. 4. Example of a ZrO_2 + 8 wt% Y_2O_3 coating with multiple beam passes.

Consider now experiments on the surfaces of ZrO_2 coatings which were chemically modified in conjunction with the laser treatment. Small amounts (a few %) of SiO_2 have been added to the surface as fine particles on the coating[c] (e.g., Fig. 6). However, even with this small amount of SiO_2 (which could not be detected by x-ray analysis) had a significant effect on the surface microstructure as shown in Fig. 6b and showed some promise of reducing crazing. When melted while moving under the laser beam, an initial experiment with ~ 5% SiO_2 added to a ZrO_2 coating surface showed evidence of breaking up the columnar grain structure (Fig. 7).

Next consider preliminary laser treatment of other coatings. Laser melted Al_2O_3 coatings also showed craze cracking, but these were very narrow (Fig. 8). Laser treated Al_2O_3 surfaces also were highly faceted (Fig. 8). Cracks were primarily intergranular along the distinct columnar grains (Fig. 9). The addition of very small amounts of TiO_2 to Al_2O_3 coating surfaces before laser treatment had a pronounced effect on resultant solidified microstructures, but little other apparent effect. TiO_2 coatings showed no craze cracking and substantial variation in surface microstructure attributed to progressively higher vaporization in higher power density regions (Fig. 8). The laser melted regions of the TiO_2 coatings had columnar grains and associated voids which are attributed to vaporization which was generally not observed in any of the other coatings (Fig. 9).

---

[c]The weight % additive was calculated by measuring the weight gain from the addition, assuming a 90% theoretical density for the coating, and measuring the dimensions of the coating.

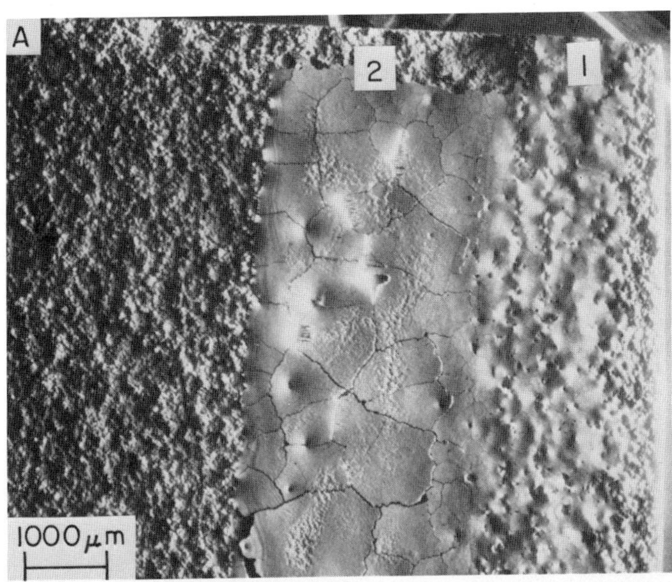

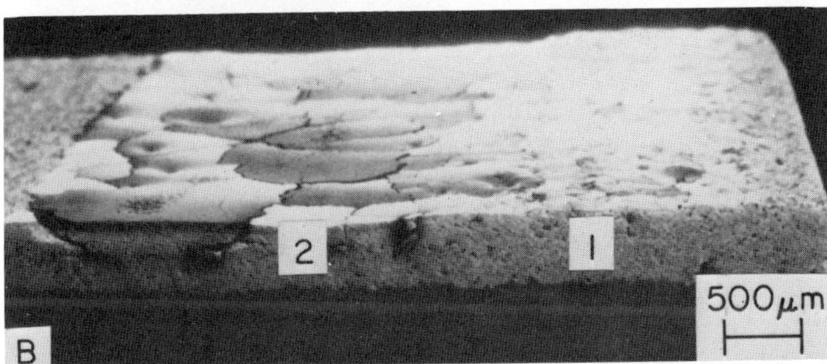

Fig. 5. Multiple beam passes on a ZrO_2 (+ 8 wt% Y_2O_3) coating. Note the normal craze cracking in 2A (lower magnification) and 2B (higher magnification). Also, note the laminar craze crack below the heavily melted area (2B, left) and the absence of specific cracking between passes. The heavier (higher energy) pass (2) and the lighter pass (1) were at 400 J/cm^2 and 50 J/cm^2, respectively.

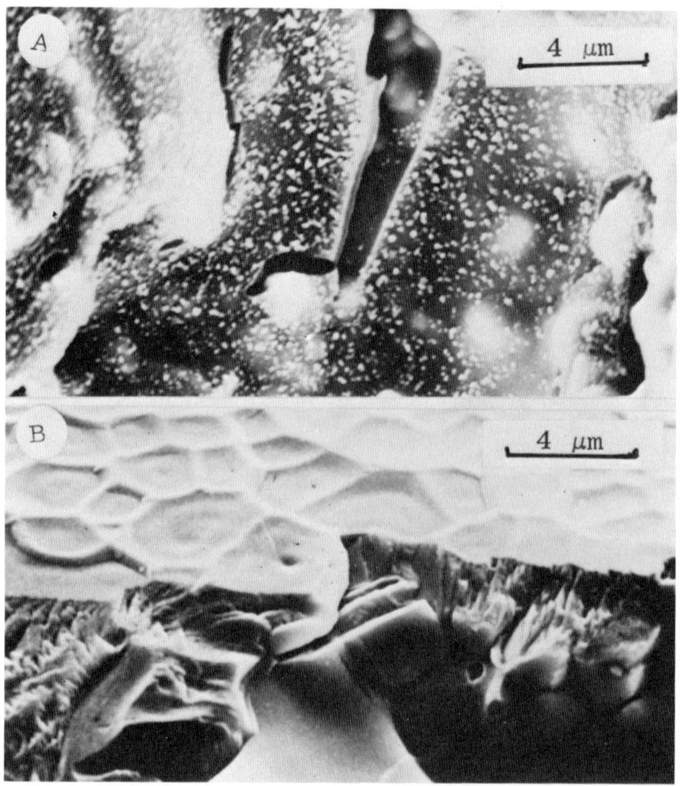

Fig. 6. Lower (~ 1%) SiO_2 addition to the surface of a Zr02 (+ 8 wt% Y_2O_3) coating. (a) shows the surface coated with SiO_2 (light spots) prior to laser treatments, (b) depicts it after laser treatment with 160 J/cm^2 (static top hat beam).

The results to date thus confirm that craze cracking in the melted layer (possibly extending down into unmelted material) is the major problem in laser surface treatment of these coatings. Some laminar cracking between the melted and unmelted material may also be a problem. However, even very small additions of material to the surface make significant changes in the surface microstructure and show signs of reducing cracking. Thus, if further development of the surface chemical modification in conjunction with the melting can successfully control cracking, laser surface treatment of coatings may be a viable method of sealing them. This would appear to bring other benefits with it as indicated by the presence of precipitates formed within grains in surface melted regions of partially stabilized ZrO_2 coatings, while maintaining a reasonably small grain size.

Fig. 7. Higher (~ 5%) SiO_2 addition to the surface of a ZrO_2 (+ 8 wt% Y_2O_3) coating. The columnar grain structure (revealed again by intergranular fracture) is clearly partially broken up. Specimen treated with a moving beam at 140 J/cm^2.

Laser Surface Treatment of Ceramic Alloy Bodies

The motivation for laser treatment of ceramic alloys was to improve the toughness in the surface region by obtaining finer, more homogeneous two-phase microstructures in the surface region that should lead to enhanced toughening. For example, sintered or hot-pressed Al_2O_3-ZrO_2 bodies have most of the ZrO_2 occurring as larger particles along grain boundaries which dominate their mechanical behavior. While such systems provide toughening, the lack of a significant volume fraction and size of ZrO_2 particles within the grains means that the material is not as effectively toughened as it might be. On the other hand, fused Al_2O_3-ZrO_2 materials make extremely tough abrasives, presumably due to their fine, laminar eutectic structure. However, factors such as thermal stresses apparently prevent large bulk bodies from being solidified to benefit from this apparently high toughness eutectic structure. Thus the concept was to melt the surfaces of Al_2O_3-ZrO_2 (or its close relative Al_2O_3-HfO_2) bodies to achieve tougher surfaces through the resultant refinement and homogenization of the material. In order to prevent thermal shock of the specimens, test bars (2 cm or more in length and

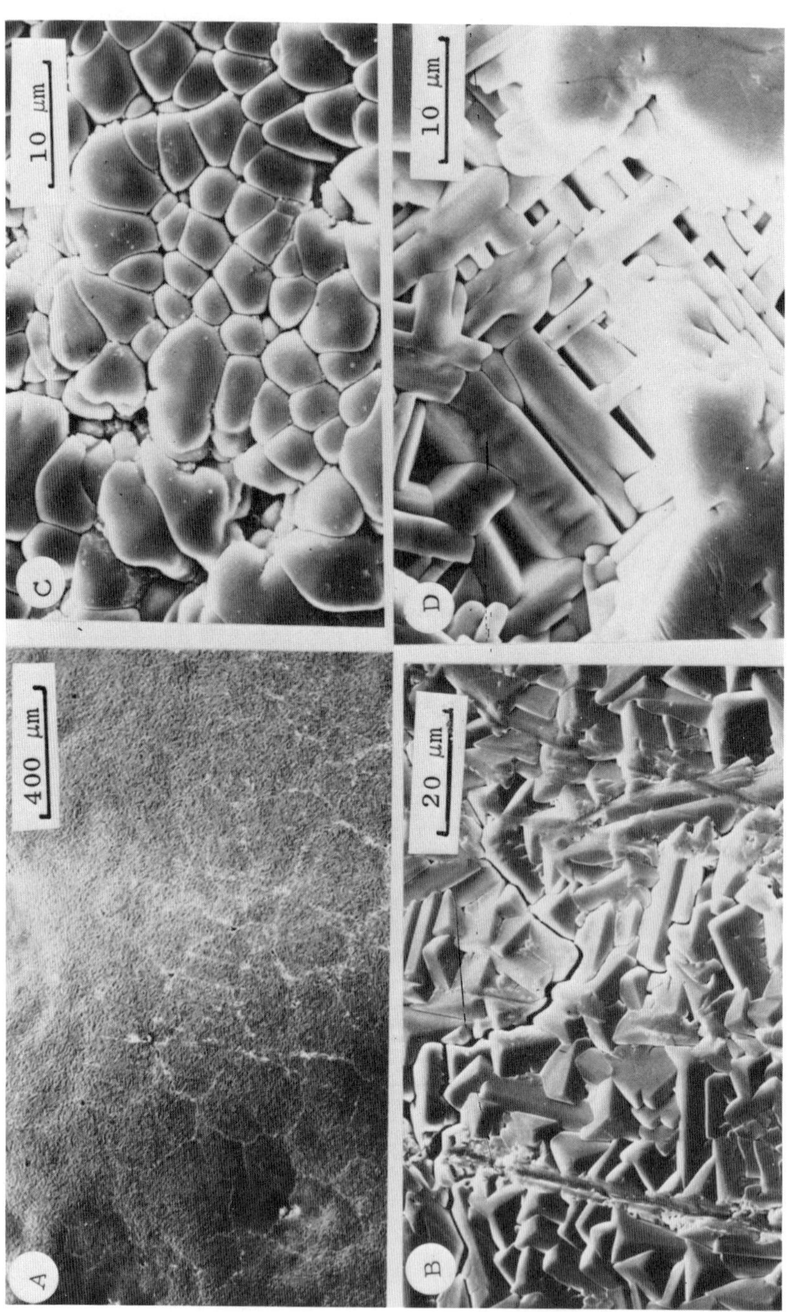

Fig. 8. Surfaces of static (top hat beam 180 J/cm^2) laser treated surfaces of Al_2O_3 and TiO_2 coatings. (A) shows the very fine craze cracks and (B) the faceted nature of the surface. (C) and (D) show respectively normal and faceted (vapor produced?) microstructure on laser treated TiO_2 surfaces.

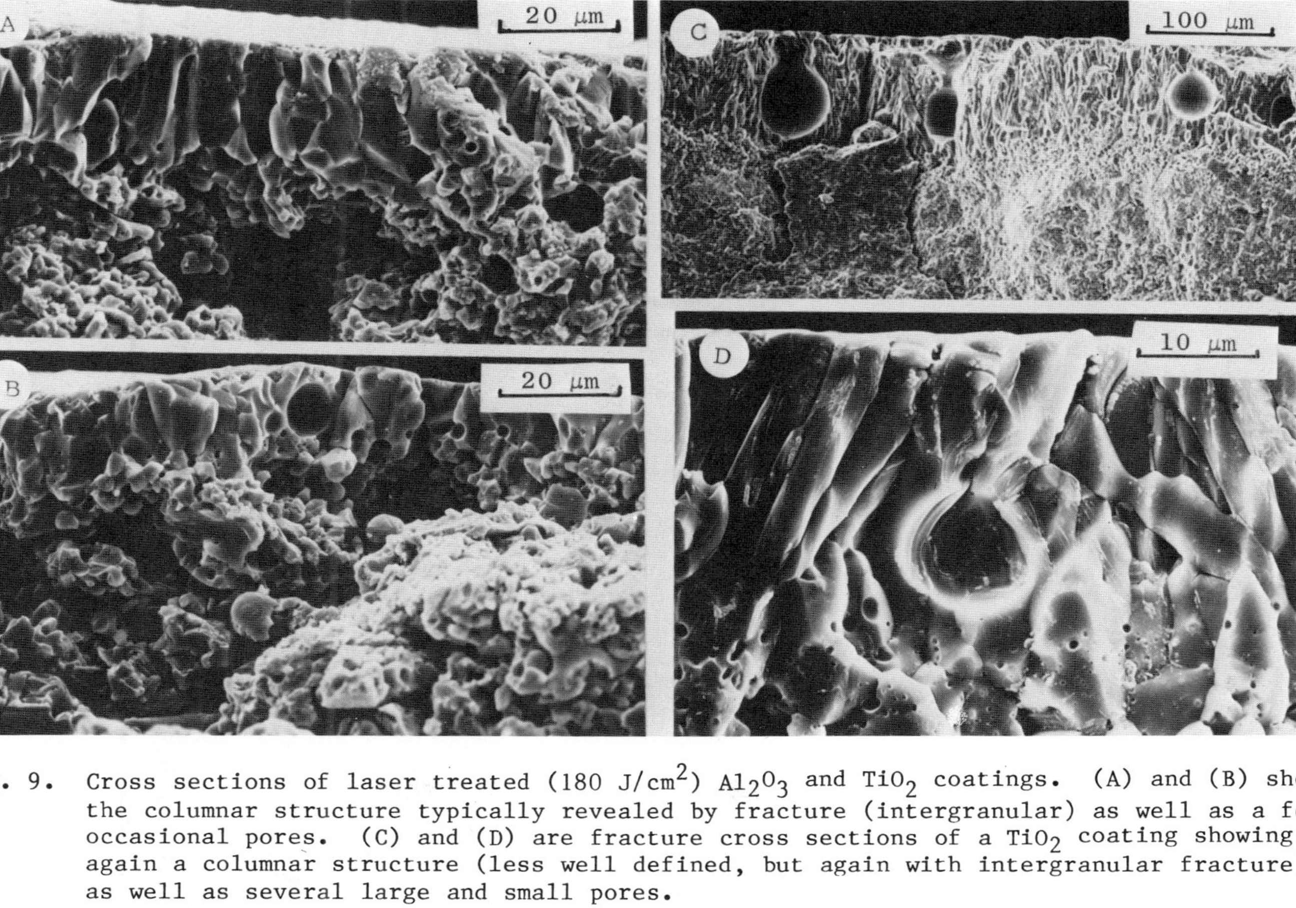

Fig. 9. Cross sections of laser treated (180 J/cm^2) Al_2O_3 and TiO_2 coatings. (A) and (B) show the columnar structure typically revealed by fracture (intergranular) as well as a few occasional pores. (C) and (D) are fracture cross sections of a TiO_2 coating showing again a columnar structure (less well defined, but again with intergranular fracture) as well as several large and small pores.

approximately 3 mm wide and 2 mm thick) were placed in a $MoSi_2$ resistance furnace which had a slot in the top of the furnace to allow entrance of the laser beam. The furnace (usually heated to ~ 1300°C) was mounted on a table which could be traversed (at ~ 1 cm/s) under the laser beam. Typically the laser was operated in the unstable oscillator mode with a beam diameter slightly greater than the specimen width, and with average power densities of the order of 0.5-2.0 kw/ cm^2.

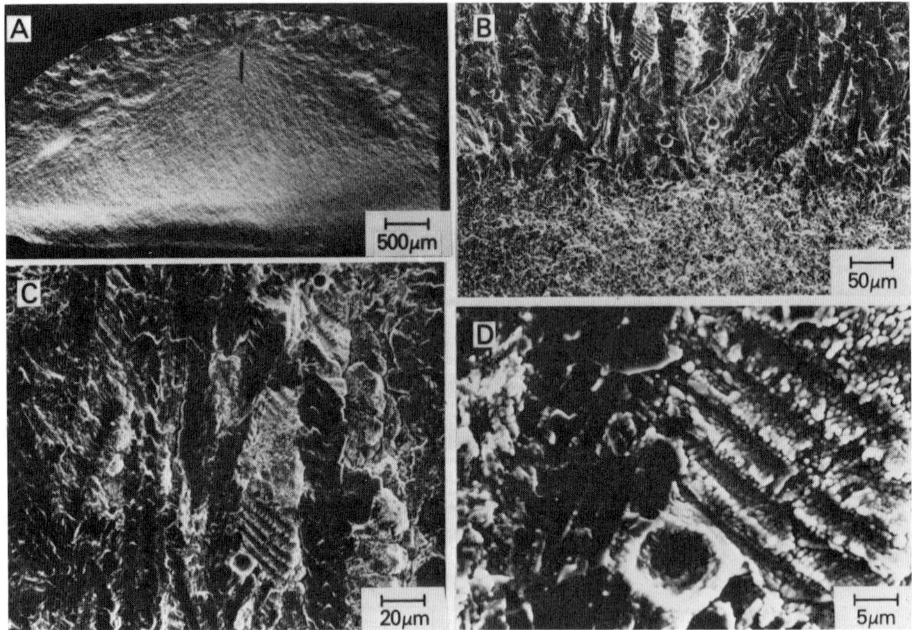

Fig. 10. Cross section of an Al_2O_3 - 45 wt% HfO_2 laser treated bar. Note doming of melt and fracture origin (vertical line) from center area of unmelted material in (A). (B)-(D) show details of the solidified microstructure.

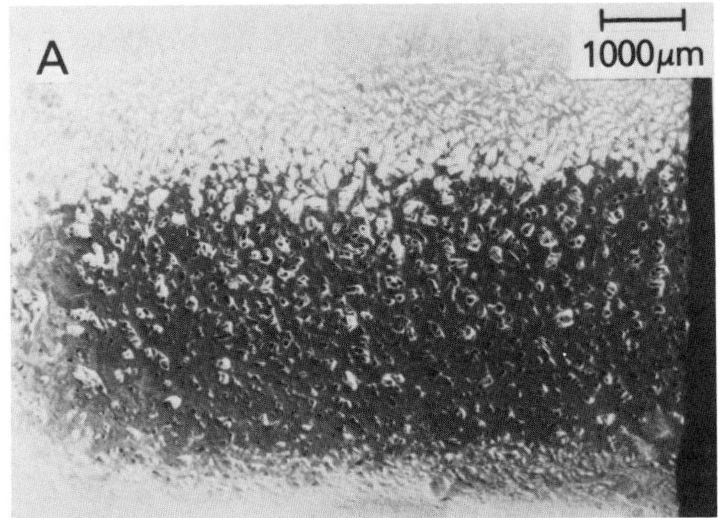

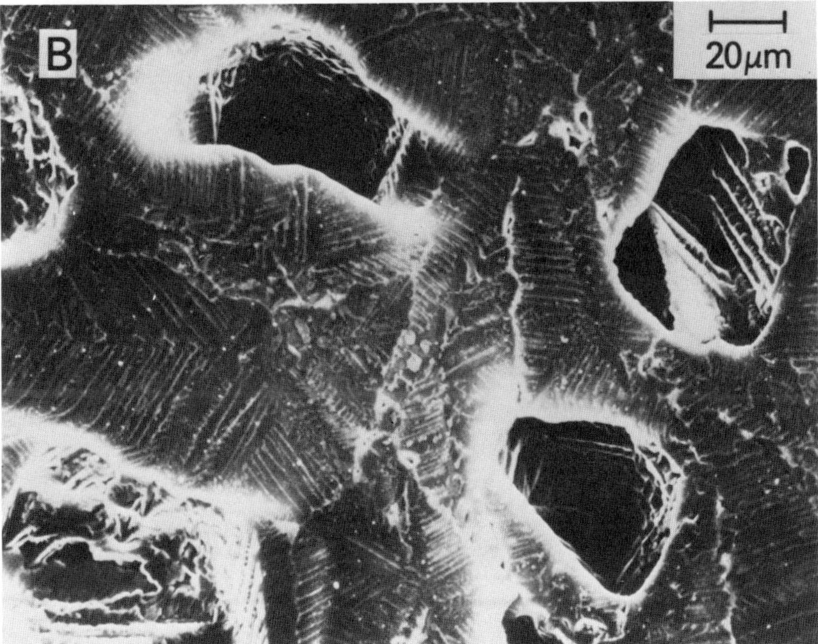

Fig. 11. Surface of an Al_2O_3 - 45 wt% HfO_2 bar. Note the pores opening onto the surface.

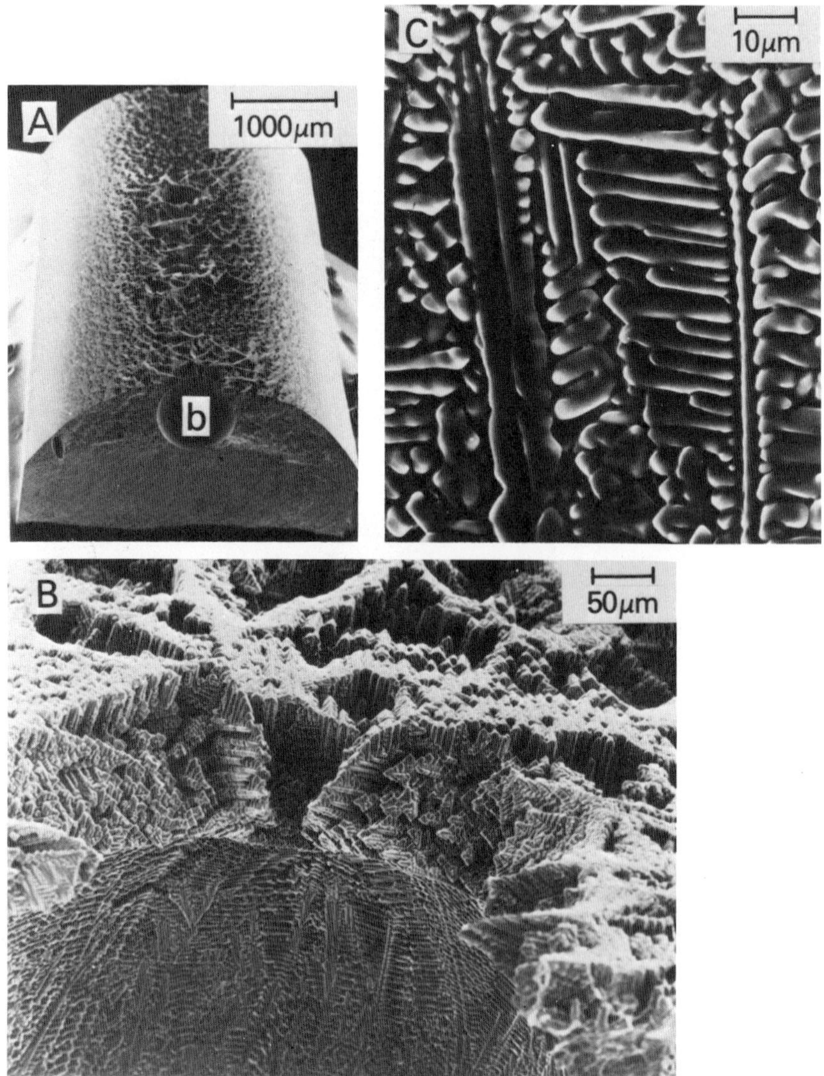

Fig. 12. Al_2O_3 - 10 volume % ZrO_2 laser treated bar. Note the large bubble (b in A) and the fine microstructure in B and C.

Figure 10, a fracture of an Al_2O_3-HfO_2 test bar, shows most of the promise and some of the problems of this technique. The low magnification photo shows doming of the sample due to surface tension of the melt. The resultant fracture initiation is near the interface between the melted and unmelted material and near the highest point of the unmelted material. The doming itself as well as its asymmetry presented difficulties in calculating accurate strengths. As an approximation, the average thickness was used in calculating the strength (3-point bend test), giving a strength for a sample of about 600 MPa (~ 90,000 psi) which is nearly twice the strength of machined bars that had not been laser treated. The fact that the fracture origin does not occur in the laser melted material is encouraging and is not surprising in view of the fairly fine microstructure achieved. However, the fracture initiation within the original material indicates that this remains a limitation, one which would be expected to diminish in an inverse fashion with increasing thickness of the melted material. However, the impact of the underlying material is accentuated by the doming characteristics. The laser melted surface of a similar specimen (Fig. 11) shows pores in the surface structure (between the colonies) left from solidification. These in combination with the long columnar grain structure indicate that solidification is occurring from the base material outward as expected. Such pores thus illustrate another limitation of the technique since they represent defects from which mechanical failure could initiate on a melted surface.

Figure 12 illustrates some similar effects and additional problems in an Al_2O_3-ZrO_2 alloy. Again a significant doming and a fine microstructure were observed along with some surface pores as well as a large bubble which served as the fracture origin for this sample (at a strength of the order of 25-125 MPa). More bubbles in the melted region as well as some delamination of the solidified melt from the bulk material were observed in some Al_2O_3-ZrO_2 bars. The bubbles are attributed to volatile materials in the body which presumably could be adequately removed by use of proper raw materials, firing and heat treatment conditions. Preliminary trials with Al_2O_3-TiO_2 bodies showed similar results, namely extensive doming combined with very fine complex microstructures (Fig. 13) and pores at the surface between the colonies (e.g., very similar to Fig. 12). Whether the low strengths (of the order of 25 MPa) were due to the surface roughness and porosity or other factors, cannot yet be determined.

Laser melting of partially stabilized ZrO_2 single crystals was used to study the precipitate structure in the resolidified surface as well as its effect on the development of precipitates within the material in subsequent heat treatments. Although cracking within the samples could not be totally eliminated, even with furnace temperatures of the order of 1400°C, one could study precipitate development. A curious result (Fig. 14) was consistently observed for samples containing 3 wt% Y_2O_3. These bars characteristically slowly curled up from the ends over a period of the order of a minute or more after having been cooled in the furnace and removed for one to several minutes. The slow curling was accompanied by very audible cracking and pinging sounds. This occurred only in crystals with 3 wt% Y_2O_3 (the lowest Y_2O_3 level investigated in single crystals). Since these crystals contained substantial monoclinic phase, it was postulated that the curling was due to differences in the tetragonal monoclinic phase content of melted and unmelted surfaces. Since the monoclinic phase has a greater volume than the tetragonal phase, the bottom would expand, curling the bar upward. A lack of a reverse curling when the top cools may be due to a reduced amount of tetragonal to monoclinic conversion in the surface due to the quenching.

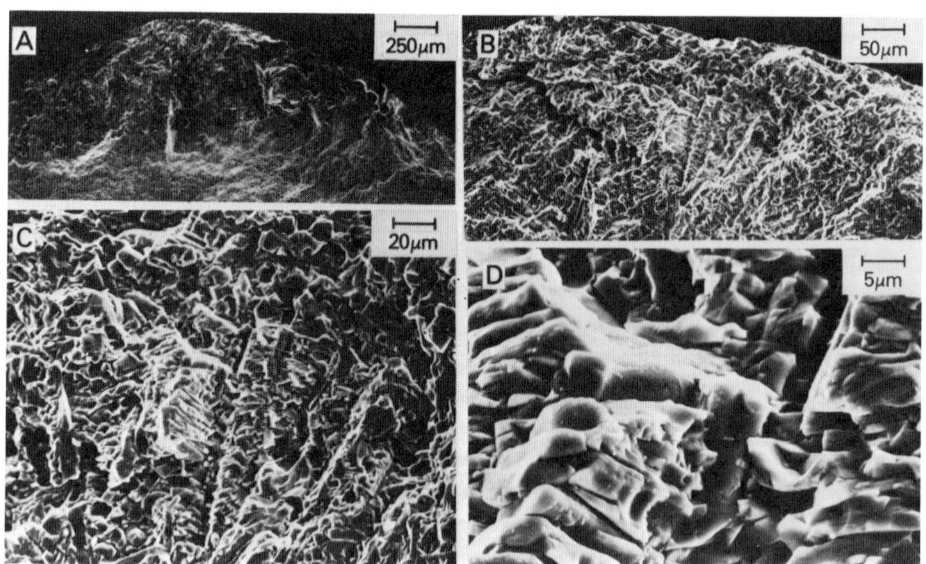

Fig. 13. Cross section of laser treated Al_2O_3 - 15 wt% TiO_2. Note the doming and some pores in (A) and fine microstructures in (B)-(D).

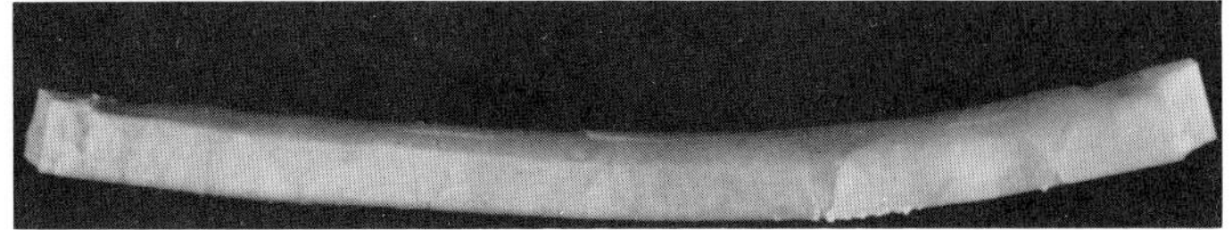

Fig. 14. Single crystal bar of ZrO_2 + 3 wt% Y_2O_3 after laser treatment at ~ 1300°C. Note the upward curling which occurred after removal from the furnace; i.e., the laser melted surface is on the concave side.

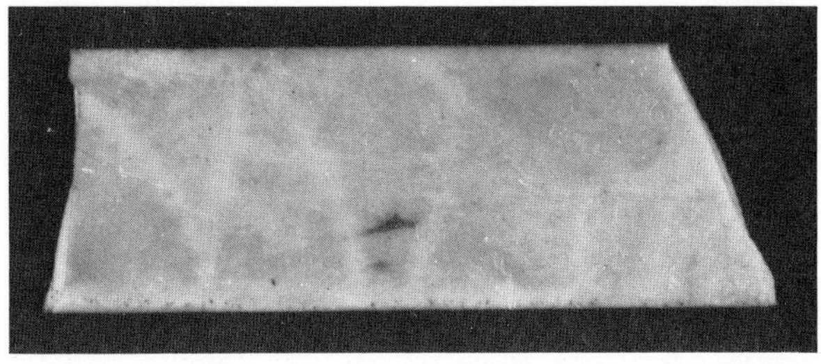

Fig. 15. Laser treated surface of Al_2O_3 - 25 volume % HfO_2 after machining. Note faint triangular shaped pattern.

A few laser treated polycrystalline test bars were machined after laser treatment to investigate the effect of machining, to reduce or eliminate the shrinkage voids and resultant roughness on the solidified surface, and to minimize some of the geometry related testing problems associated with the doming. The resultant strengths were about the same as the bars before the laser treatment. This is tentatively attributed to the effect of flaws in the base material. However, on some of the specimens on which the melted material was entirely machined away there was a curious triangular pattern which is as yet unexplained, e.g., see Fig. 15.

Ceramic Powders Derived Via Laser Melting

The motivation for making powders was again to obtain tougher materials. One of the primary concepts was again to obtain fine homogeneous microstructures of two phase systems in which toughening might be expected, again emphasizing Al_2O_3 with ZrO_2 or HfO_2. The second motivation was to obtain partially stabilized ZrO_2 bodies with toughening precipitate structures within finer grains. Currently toughening in polycrystalline PSZ bodies due to precipitation is achieved only after high temperature solution treatment of sintered or hot pressed bodies with substantial resultant grain growth, e.g., typically 50 to 100 μm. The concept here was to obtain particles by laser melting that were either in a state of solid solution, or that had precipitates. These particles could be fabricated into polycrystalline bodies with substantially reduced grain sizes over those obtained by solution and precipitation treatment after densification. A requirement would of course be that sufficiently fine powders be obtained so that densification could be achieved at temperatures below that of over aging, which means that one could conceivably densify materials in the ZrO_2-Y_2O_3 system at temperatures to approximately 1800°C.

The laser provided a convenient, concentrated energy source with the important attribute that it could be used in a completely oxidizing environment, thus avoiding or minimizing problems of reduction of materials such as ZrO_2 and HfO_2. Two types of specimens to be melted were investigated. The first were rods typically about 8 mm diameter. The laser beam was brought in on the axis of the rod, thus melting one end, while the other end was held in a chuck of a variable high speed motor. The other type of specimens were disks ~ 4 cm in diameter and 2 to 5 mm thick. They were mounted on a shaft mounted in a chuck of the same variable high speed motor with the laser beam striking the rotating disk nearly radially. The latter concept was an attempt to obtain finer size drops by having a substantially larger diameter to produce greater centrifugal force and to produce more material with much less movement of the point of melting from the focal point. It was recognized that a rotating disk presented two possible problems: (1) the centrifugal stresses, combined with possibly a higher thermal stresses associated with a much larger radius, could lead to thermal shock, and (2) the much higher surface area seen by the beam per unit time might significantly reduce or even eliminate melting. Supplemental heating by torches was therefore investigated to both spread out the thermal gradients and to reduce the amount of energy needed for melting. However, this approach was not successful in our trials due to shattering of the disks when the beam was turned on, regardless of which beam mode was used.

Two types of beam profiles were used on the rods, the first and most successful was the defocussed doughnut shaped beam, the other was a focussed, approximately gaussian shaped beam resulting in melt regions on the rods as shown in Fig. 16. The focussed beam often presented problems of shattering of the rods while the doughnut beam gave by far the most practical operation. This can be explained, qualitatively since the hot material ahead of the highly focussed beam is in effect a conical wedge being driven into the rod attempting to shatter it, whereas the doughnut beam is heating an annulus which is putting the remainder of the rod in compression. The conical cavity produced by the focused beam could have the advantage of producing a much more uniform size particle since the melt would move to the lip of the crater which would thus produce molten droplets at an approximate constant radius in contrast to the highly variable radius at which molten droplets can come off of the rod heated with the doughnut beam.

Aluminum cylinders of various sizes were placed around the vertically mounted rods to stop the particles and allow them to fall onto an aluminum tray. Aluminum was selected since any contamination of the particles by it would only lead to a limited amount of Al_2O_3 in the resultant ceramic body which for essentially all compositions considered should be of little or no consequence. Cylinders having inside diameters from 5 cm to nearly 15 cm were tried. The intent was to use small cylinders to allow molten droplets to splat against the walls before solidification, to increase cooling rates and especially to get a flake type material which would be much more crushable to the finer particle sizes desired for good densification. However, the molten droplets splatting on the wall typically adhered to it and despite recession of the rods with melting there was sufficient variability of the angle of the particles such that many landed on top of one another. Thus this approach requires a more sophisticated system to translate the collecting cylinder and assure adequate release of individual splatted particles from it. Since such development was not feasible in the early phases of this preliminary investigation the focus was on the use of larger cylinders where the particles had typically solidified enough that they simply bounced off the cylinder to be collected on a tray beneath it (Fig. 17).

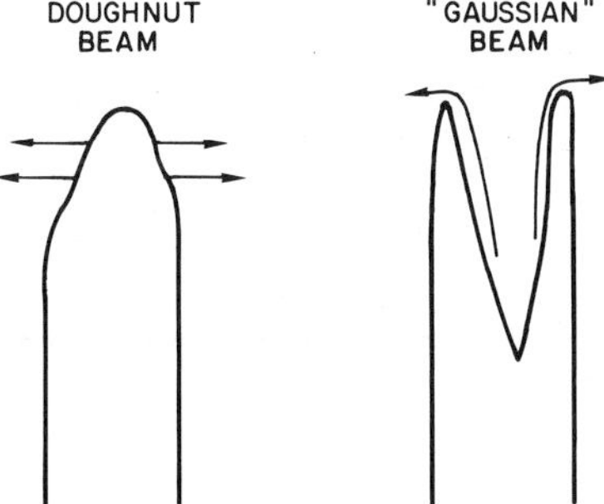

Fig. 16. Schematic of profiles left with ceramic rods after melting with the two types of laser beams used.

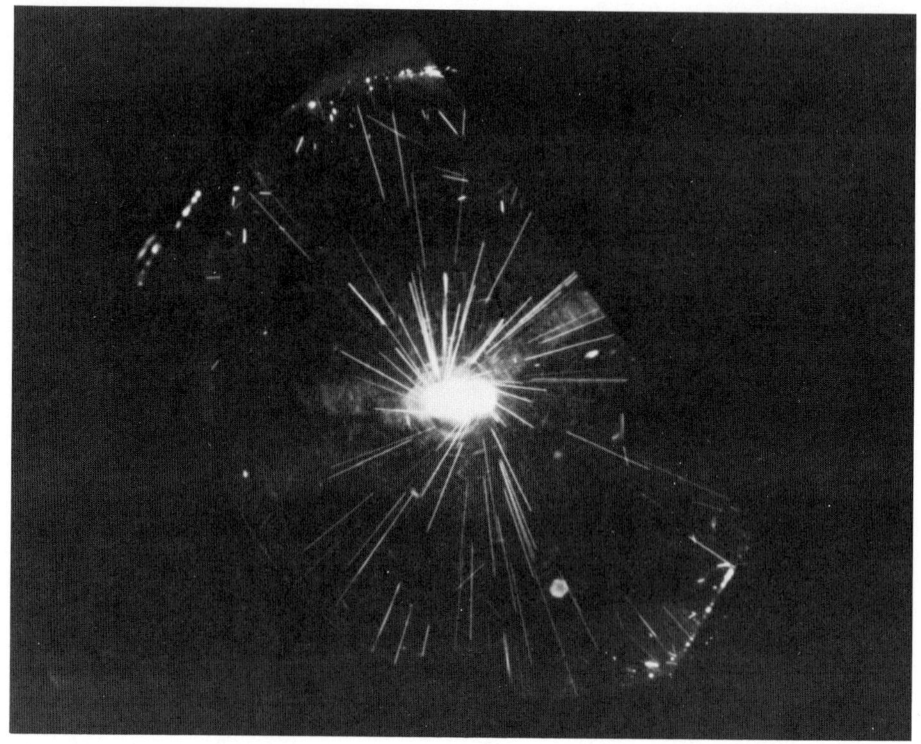

Fig. 17. Stop action photo showing rotating rod, molten particles in flight, and most of the surrounding Al cylinders.

Rods cut from a commercial ZrO_2 body partially stabilized with 4 weight % MgO (Zircoa 1027), ZrO_2 single crystals partially stabilized with Y_2O_3 (typically 5 or 6 wt%) and from Al_2O_3-25 volume % ZrO_2 hot pressed or sintered bodies caused little problem of thermal shock failure; the possibility of melting some of them with the highly focussed beam was also evident. Al_2O_3 10-25 volume % HfO_2 rods were readily shattered by the beam (apparently due to poor microstructural development in the original rods), as were Al_2O_3 15 wt% TiO_2 rods and rods of a few other compositions investigated.

The following results are for rods irradiated with average power densities of 2-4 kw/cm^2 which typically melted these rods at rates of approximately 5 mm/s. Figure 18 shows some examples of typical particles produced. They were approximately spherical, but with variations in size and shape. Surface wrinkling of some particles indicated that they had hit the walls with the exterior solid but the interior still fluid so that they underwent some plastic deformation. Some elongated particles were observed which are attributed to either some degree of splatting or to extreme cases of particles being joined before they fully solidified. Figure 19 shows that despite substantial scatter there is indeed a definite trend toward finer average particle size with increasing rotational speed. As noted earlier it is postulated that a much narrower range of particle sizes would be achieved with beam profiles producing a melt crater in the rods so that particles would form primarily at the radius of the rod.

Crushing of the particles reveals that they typically contain internal porosity, e.g., Fig. 20. While some of the porosity maybe due to volatilization, the fact that it also occurred in droplets from single crystals (i.e., previously melted) shows that most, or all, of the porosity is due to solidification from the outside inward with resultant internal porosity due to the shrinkage. However, note that a wide variety of distribution of this internal porosity is observed. Such porosity should make the spheres much easier to crush. A key question is whether or not the resultant fractures will always pass through voids so that no particles produced from crushing of the spheres will contain significant porosity trapped to be carried through into the final product.

While the microstructures have not been fully analyzed, Figs. 21 and 22 show that grain structures are observed in the material and that these are of the order of 2-40 μm. Relatively fine microstructures are observed in the Al_2O_3-ZrO_2 and Al_2O_3-TiO_2 systems which is consistent with results of surface melting of bulk bodies of these compositions (section on Laser Surface Treatment). Transmission electron microscopy has not yet been completed on the partially stabilized ZrO_2 samples to determine whether or not they contain solid solution or precipitate structures. Experiments to determine how fine the materials can be reasonably ground are now underway prior to investigating their densification.

Fig. 18. Example of particles from laser melting. These ZrO_2 + 4 wt% MgO particles are representative of the size and shape variations observed. Flattened particles with surface ridges or wrinkles are attributed to the particle hitting the outer cylinder before the interior solidified.

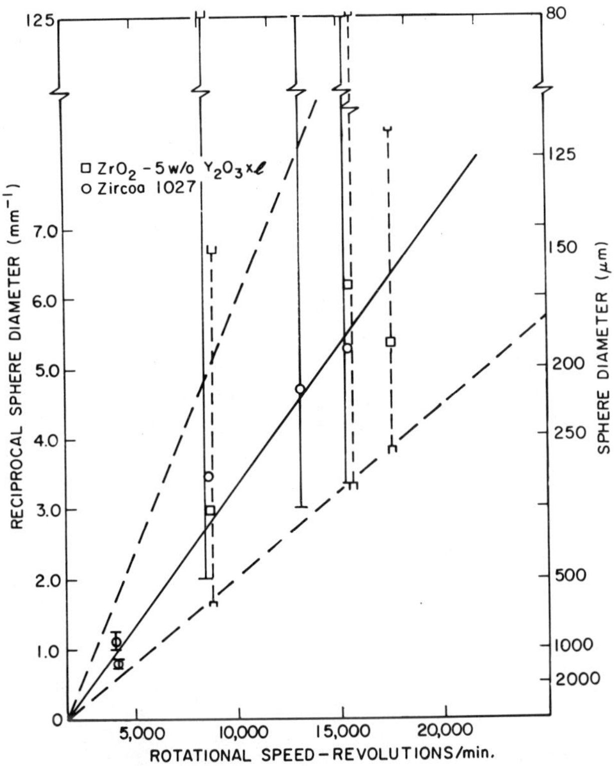

Fig. 19. Particle size versus rotational speed for ZrO_2 particles from laser melting.

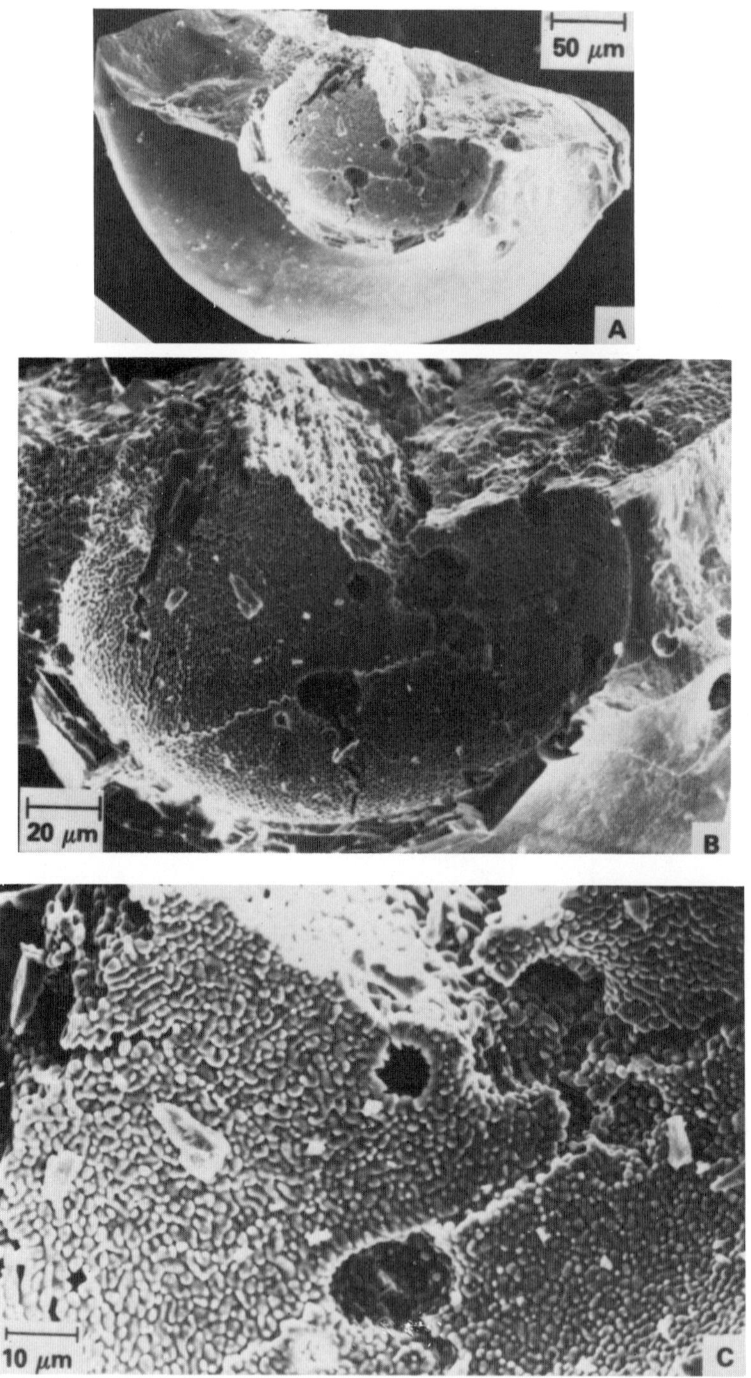

Fig. 20. Crushed sphere from laser melting of ZrO_2 + 4 wt% MgO. Note large central and smaller voids, and fine grain structures.

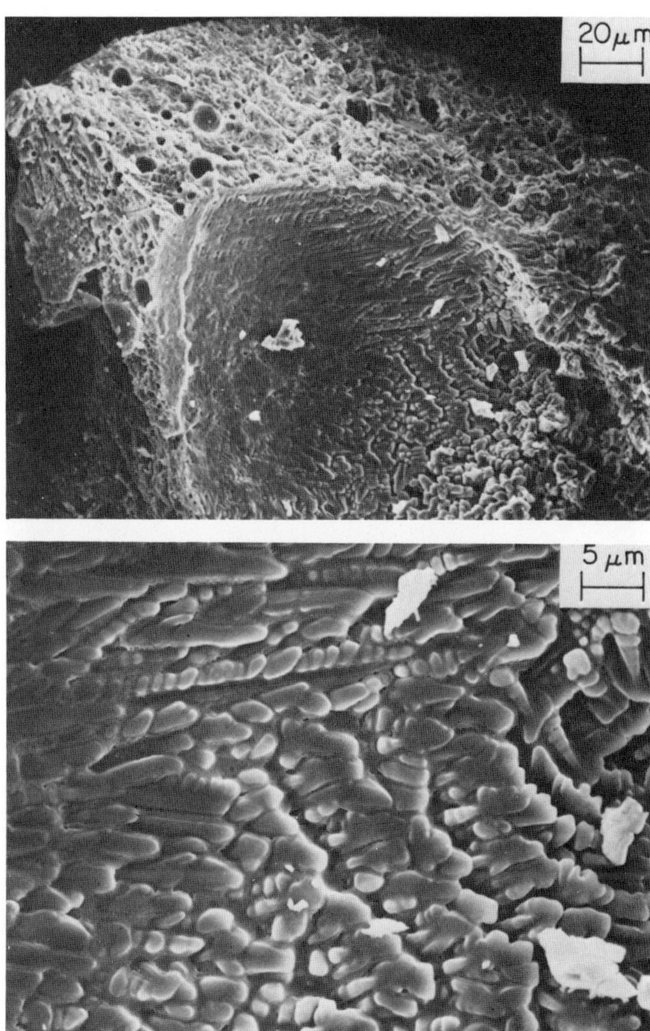

Fig. 21. Crushed sphere from laser melting of ZrO_2 - 5 wt% Y_2O_3 single crystal. Note the large central void, the smaller random voids and the fine grain structure.

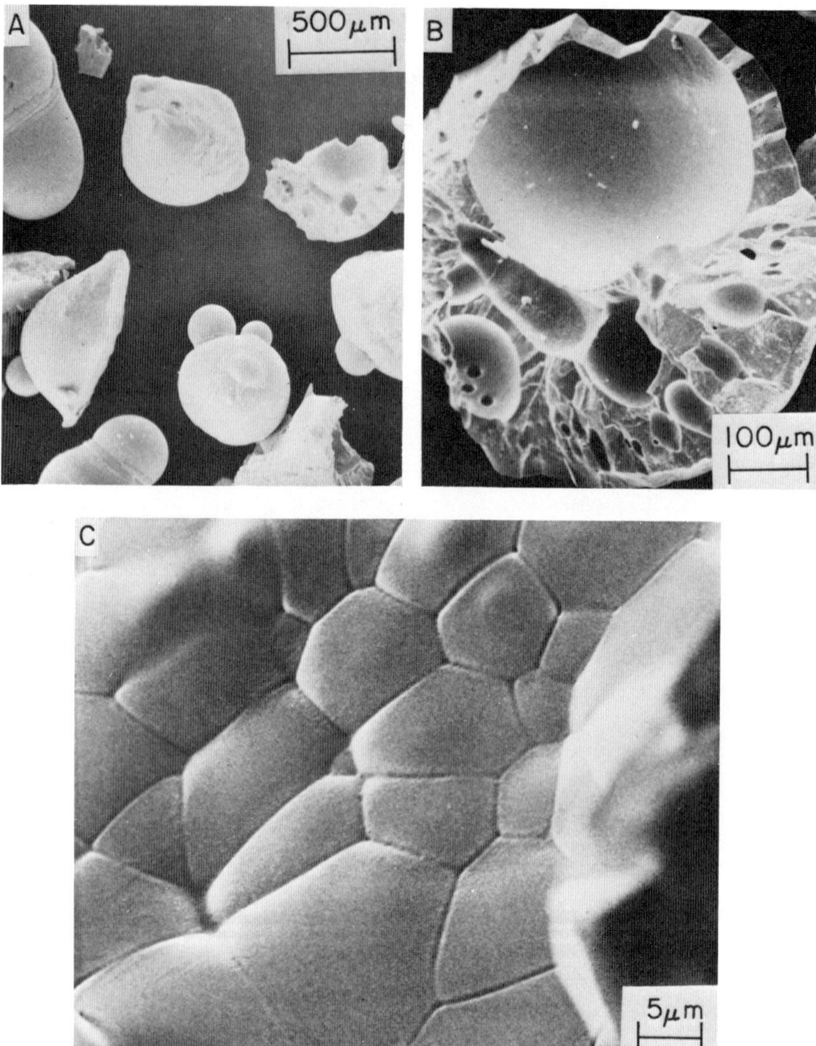

Fig. 22. Crushed sphere from laser melting of Al_2O_3 - 25 wt% ZrO_2. Note large central void, smaller random voids and fine grain structure.

GENERAL DISCUSSION, SUMMARY AND CONCLUSIONS

While craze type cracking in resolidified laser treated surfaces apparently did not occur in TiO_2 coatings, it was a common occurrence in other coating materials. Some darkening was observed in the surface and some evidence of delamination cracking was seen between resolidified and unmelted material in ZrO_2 coatings which were subjected to higher intensity moving beams, but there was no obvious problem of cracking between adjacent beam passes. Thus, the primary requirement to successfully seal ceramic coatings with laser treatment is to eliminate craze type cracking. Use of additives coated on, or infiltrated into, the surface of the coating prior to melting showed some promise of suppressing this cracking, but it remains to be demonstrated that it can be fully eliminated. Such cracking was not observed to be a problem in laser treatment of alloy bars, apparently due to the use of a furnace to heat them to temperatures of ~ 1300°C before and during laser treatment. (The metal substrate under the ceramic coatings greatly restricts the temperatures used to limit the crazing problem.)

Laser treatment of ceramic coatings as well as bulk ceramic bodies resulted in columnar grain structures indicating that solidification proceeded from the unmelted material outward as expected. This is also consistent with laser treated surfaces showing no entrapped porosity except near the final solidified surface in two phase materials. The observation of porosity in other materials, e.g., TiO_2 coatings and Al_2O_3-ZrO_2 bodies is attributed to vaporization of the material itself, e.g., in the case of TiO_2 or quite possibly of impurities in the case of Al_2O_3-ZrO_2 bodies. Laser treatment of two phase bodies such as Al_2O_3-HfO_2 clearly results in fine homogeneous microstructures with which toughening might well be associated. Some tests though substantially complicated by the doming of the melt material, do in fact suggest some strengthening. However, any such strengthening is limited by the high point of unmelted material under the melt dome.

There are still several possible ways that laser surface treatment of bulk materials can be improved. First the approach of adding chemical modifiers to the surface, as with coatings, could offer interesting opportunities and has yet to be explored. Second and more generally, multiple passes might significantly reduce the amount of doming; however, such improvements are clearly limited by the thinness of the melted layer. Increasing the thickness of the melted layer to reduce the maximum stress on the unmelted material below it runs the risk of coarser microstructures and especially of trapping solidification shrinkage pores within the melted layer due to solidification beginning both from the bulk of the body as well as the liquid surface.

Experiments used to produce ceramic powders by rapid solidification from the melt also produced interesting microstructures. A variety of practical problems have thus far limited the material made to roughly spherical shaped particles in the range of approximately 40 to 200 μm, but a variety of modifications of the techniques, including higher speed rotation, or a variety of other ways of generating liquid droplets should allow somewhat finer droplets to be made. Present droplets contain substantial amounts of porosity, most or all of which is due to solidification shrinkage. It remains to be seen whether or not this porosity allows the particles to be crushed into small enough fragments to allow adequate densification without retaining a significant amount of trapped porosity from the original spheres.

ACKNOWLEDGMENT

The authors are grateful for funding by Dr. J. Fairbanks of DOE for the coatings studies and by Dr. R. Pohanka of ONR for the toughening studies. We also wish to thank Dr. S. R. Levine of NASA-Lewis, Dr. G. W. Goward of P & W, Dr. R. N. Katz of AMMRC, and Dr. H. Herman of SUNY, Stony Brook, for supplying us with plasma sprayed coatings. Here at NRL we wish to thank R. Stegman, R. Beigle, and J. McDermott for operation of the laser, L. Green for the SEM pictures, and TID personnel for their invaluable assistance in preparation of this paper.

REFERENCES

1. S. M. Copley, N. Bass, and E. Garmire, R & D Status Report #6, ARPA Contract No. N00014-77-C-078 Order No. 3421, Nov. 1979.
2. J. P. Spann and R. W. Rice, "Laser Machining of Ceramics," to be published.
3. J. R. Spann and R. W. Rice, NRL memo report, in press.

DISCUSSION

R. L. Pober (Massachusetts Institute of Technology): (1) Please comment on compositional distribution across the particles which are some tens of microns in diameter. (2) How were the coatings produced and what were the substrates?

Author: (1) Studies of composition distribution have not been completed, in part due to operational problems with our SEM-probe facilities. (2) As detailed in the text, the coatings were plasma sprayed (from the acknowledged sources) onto superalloys.

E. A. Barringer (Massachusetts Institute of Technology): What do you hope to gain with this technique in subsequent processing and properties over methods which produce fine powders (e.g., oxidation of $ZrCl_4$ or decomposition of zirconium alkoxides)? That is, unless these larger spheres can be crushed to a fine, uniform size, then the properties of sintered materials will be dominated by structural defects, as is common for agglomerated powders.

Author: For systems involving precipitates for toughening, e.g., partially stabilized ZrO_2 (PSZ), the motivation was to get particles that were already in a solid solution or precipitated state in order to get a finer grain size precipitation-toughened PSZ than achievable by conventional sinter-solution-precipitate methods. For eutectic forming systems such as Al_2O_3-ZrO_2 and Al_2O_3-HfO_2, the motivation was to get fine laminar eutectic structures to give a more homogeneous and effective distribution of the toughening ZrO_2 or HfO_2, e.g., as indicated by the high toughness of Al_2O_3-ZrO_2 fused abrasive grain.

M. F. Bernard (Iowa State University, Ames Lab): What is the approximate smear density of these hollow particles? Does the interior void cause problems in sintering a compact of these particles to a high density?

Author: We have not yet measured the densities of the spheres, in part because of the known porosity in them. It is not our intent to sinter bodies from the spheres because of their size, but instead to crush them, as discussed in the text. As I noted, the key issue may be whether the particles fracture through all significant voids.

J. V. Portugal (Alfred University): You mentioned that you abandoned the idea of using plasma as your heating source due to the reducing nature of the atmosphere. Have you considered utilizing an oxygen plasma to help control the stoichiometry?

Author: We decided not to use conventional plasma torches because of their typical reducing conditions and the sensitivity of the ZrO_2 and HfO_2 constituents to such reduction. An oxygen plasma would be of interest if good oxygen stoichiometry can be maintained.

S. P. Mukherjee (Battelle Columbus Laboratory): Do you have any idea about the quenching rates of laser melted particles?

Author: The quenching rates have not yet been measured. A reasonable estimate of these rates may be feasible by calibration of the photographic record of the flight of the droplets.

MICROSTRUCTURAL ANALYSIS OF RAPIDLY SOLIDIFIED ALUMINA

John P. Pollinger and Gary L. Messing

The Pennsylvania State University
Department of Materials Science and Engineering
University Park, PA 16802

INTRODUCTION

Rapid solidification processing (RSP) of metals and semiconductors has become an established field over the last 10 years. The major emphasis of RSP has been the development of new and unique materials as a result of the fast cooling rates (> 10^4 °C/sec) of the molten materials. Generally, it has been shown that improved homogeneity, structural refinement, finer grain size, and non-equilibrium phases can be obtained for a wide variety of materials. These findings have been coupled with significant improvements in physical, electrical, and chemical properties of the metallic and semiconductor materials.[1-3]

One of the tools that has received considerable attention for RSP is the laser. Laser heating provides a method for concentrating sufficient heat into a localized region to affect surface melting. By limiting the volume of the melt zone it is possible to realize rapid solidification via thermal conduction to the underlying solid.

Although a vast literature is developing for RSP of metals and semiconductors via laser processing,[4-6] there is little mention of its application to crystalline ceramic materials.[7-9] The primary reason for this appears to be the inherent lack of thermal shock resistance of ceramic materials coupled with their high melting temperatures and low thermal conductivity. Because most ceramic materials are highly absorbing at the CO_2 laser emission wavelength (10.6 µm), laser processing can be an efficient method for introducing energy to a ceramic surface. It is suggested that despite the potential difficulties inherent with melt processing of crystalline ceramic materials, laser processing is a unique method for investigating the potential of RSP of ceramics.

In this study a CO_2 laser was used to investigate RSP of alumina. Alumina was chosen for this preliminary study as it is a well known system and will facilitate the development of a basic understanding of the process.

EXPERIMENTAL PROCEDURE

Alumina substrates of 1.5 mm thickness were prepared by tape casting a commercial alumina.[a] The substrates were fired to 70% of theoretical density. Low density substrates were prepared to avoid thermal shock during irradiation.

A 50 watt CO_2 laser[b] was used to obtain power densities of 50 to 10^4 watts/cm^2 by using either a focussed or unfocussed beam. Power densities of 10^5 to 10^6 watts/cm^2 were obtained with a focussed 525 watt CO_2 laser.[c] The lasers were operated in the TEM_{00} mode (i.e., a Gaussian intensity distribution) which was verified by measurements with a UV illuminated fluorescent plate for the 50 watt laser and by irradiating blocks of plexiglas for the 525 watt laser.

Total beam power was measured with two thermocouple type radiation detectors.[d] The beam power intensity distribution was measured by incrementally passing a knife edge through the beam. The fraction of the beam not blocked was measured with one of the thermocouple detectors. Knowing the total power of the beam, the maximum power intensity (I) can be calculated from a relation given by Cohen[10] for the intensity distribution of a Gaussian laser beam:

$$I = \frac{P}{\pi a^2} \qquad (1)$$

where:

I = maximum power density (watts/cm^2)
P = total power of beam (watts)
a = beam radius at which power falls to 1/e of maximum observed (cm).

Alumina substrates were rotated on a variable speed table under the laser beam (Fig. 1). Scan rates were 0.2, 2.0, and 20 cm/sec

---

[a]Reynolds RC-122-BM.
[b]Coherent Model 42.
[c]Coherent Model 525.

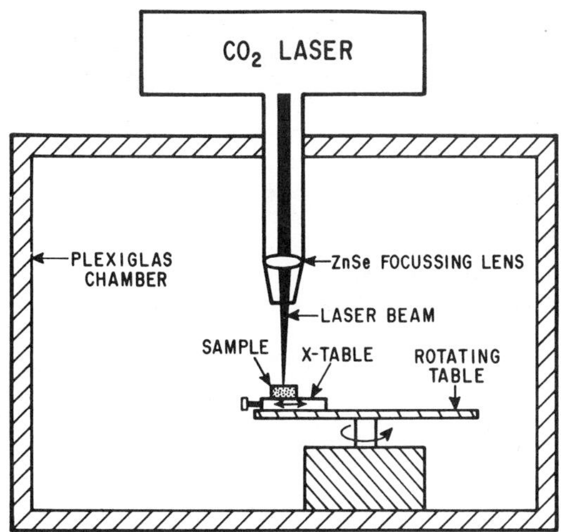

Fig. 1. The experimental apparatus for laser scanning of Al_2O_3 substrates.

with power densities of 50 to 10^6 watts/cm^2. An x-table was positioned on the rotating table to produce multiple scans of 1 mm spacing for each sample.

The surfaces of the irradiated samples and cross-sections of the melt zones were examined with SEM. The effected regions were also removed from the substrates, ground, and examined by x-ray diffraction.

RESULTS AND DISCUSSION

The surface morphology of the as-sintered Al_2O_3 substrates is shown in Fig. 2. The grain size is seen to vary from 2 to 5 μm.

The specific beam parameters used for these studies are given in Table 1. To obtain a range of power densities it was necessary to use either an unfocussed beam or a beam focussed with a Ge lens for the 50 watt laser or a ZnSe lens for the 525 watt laser. The laser was operated in the TEM_{00} mode giving a Gaussian beam intensity profile.

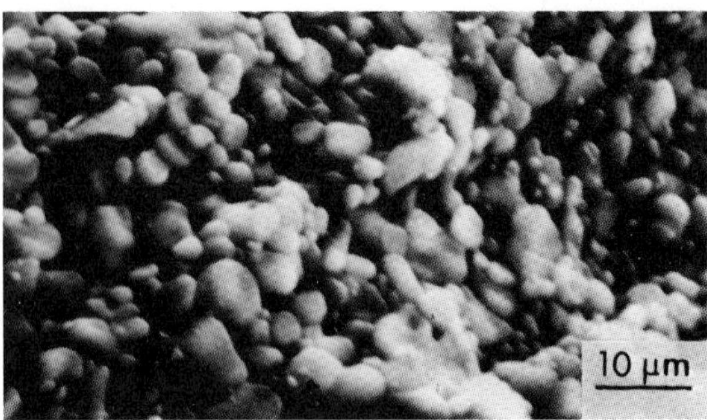

Fig. 2. The surface of the sintered substrate before irradiation.

Table 1. Laser Beam Parameters used in this Study

Laser Used	Beam Focussed	Total Beam Power (watts)	Beam Diameter of 1/e of Maximum Power (mm)	Power Density (watts/ cm^2
50 watt	No	3.7	3.0	$5.1x10^1$
	No	7.1	3.0	$1.0x10^2$
	Yes	2.0	0.5	$1.0x10^3$
	Yes	20.0	0.5	$1.0x10^4$
525 watt	Yes	200.0	0.5	$1.0x10^5$
	Yes	530.0	0.3	$7.5x10^5$

Macroscopic Observations

For the conditions studied, melting, thermal shock, or both were observed to occur to some of the substrates. For power densities of 50 to 10^3 watts/cm^2 and all three scanning speeds no macroscopic modifications were observed. The minimum condition at which melting occurred was 10^4 watts/cm^2 at a scan rate of 0.2 cm^2/sec. Faster scan rates at this power density did not produce melting because insufficient energy was transferred to the substrate to produce melting. At the highest power density (10^6 watts/cm^2) and slowest scanning speed (0.2 cm/sec) melting was accompanied by thermal shock and subsequent fracture of the substrate. Between these two extremes it was possible with a combination of sufficient power (10^4 - 10^6 watts/cm^2) and scan rates (0.2 -10 cm/sec) to affect melting without thermal shock of the substrate. This is obviously a condition where it

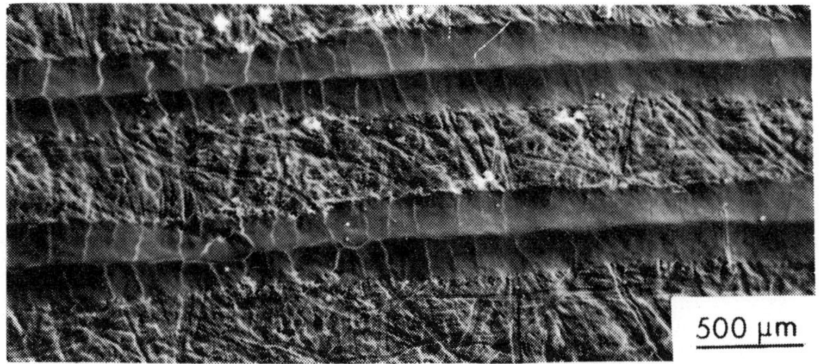

Fig. 3. Laser melted regions produced on the substrate surface under conditions of 10^5 W/cm^2 and 20 cm/sec.

is possible to deliver sufficient energy for melting in a short enough time so that the condition for thermal shock of the substrate is not attained.

Although thermal shock of the substrate was avoided for some conditions, the solidified alumina always shows extensive cracking parallel and perpendicular to the scanning direction (Fig. 3). While thermal shock cannot be totally ruled out as the cause of these cracks, the more likely reason is the large volumetric change (≈ 30%) that the alumina goes through on melting and solidification. Consequently, at the interface between the solidified region and the substrate a strain develops which is relieved by cracking of the weak crystallized region.

In most cases the solidified material could be easily lifted from the substrate. Interestingly, at 10^5 watts/cm^2 and scan rates of 0.2 and 20 cm/sec the solidified region was tightly bound to the substrate. These two conditions yielded extremes in the volume of the material melted. Microstructural examination reveals that for the fastest solidification condition that no cracks are present between the substrate and solidified zone. For the largest melt zone there is evidence of some microcracking at the interface but also areas of tight bonding. It appears that the liquid has permeated the substrate for both cases giving a region with a density gradient which is sufficiently strong to withstand any stresses developed during solidification.

Examination of the geometry of the solidified zones reveals two types of profiles perpendicular to the scanning direction. For the most rapid scanning rates it is observed that the center of the solidified zone is raised. Anthony and Cline[11] discuss the existence of surface tension gradients from the center of the molten zone to

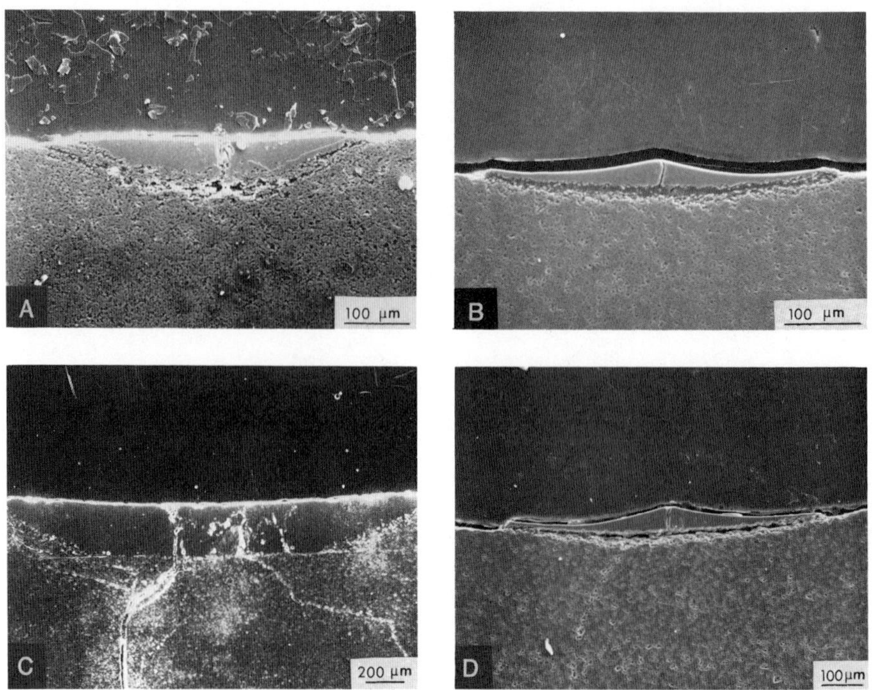

Fig. 4. Cross-section of melt region produced by irradiation at: (a) 10^4 W/cm^2 and 0.2 cm/sec; (b) 10^5 W/cm^2 and 20 cm/sec; (c) 10^5 W/cm^2 and 0.2 cm/sec; (d) 10^6 W/cm^2 and 20 cm/sec.

the edges that arises because of the temperature gradient produced by the laser. The surface tension differential results in liquid flow from the center to the edge during irradiation and flow back after the beam has passed. Thus, depending on the time of solidification, either a central peak or depression may be frozen in. If sufficient time is available the liquid level can attain equilibrium and the surface will be macroscopically flat. As seen in Fig. 4, all three conditions were observed in this study.

The dimensions of the melt zones can be correlated to the beam intensity profile. In Fig. 5 the profile of the melt zone cross-section is placed on a graph of the beam intensity profile for laser power densities of 10^4, 10^5, and 10^6 watts/cm^2. The thinnest surface zone produced was obtained at 10^5 watts/cm^2 scanned at 20 cm/sec and the largest melt zone corresponds to 10^5 watts/cm^2 scanned at 0.2 cm/sec. The shapes of these profiles are a function of not only the processing conditions but of the thermal transport characteristics of the substrate which will not be discussed here.

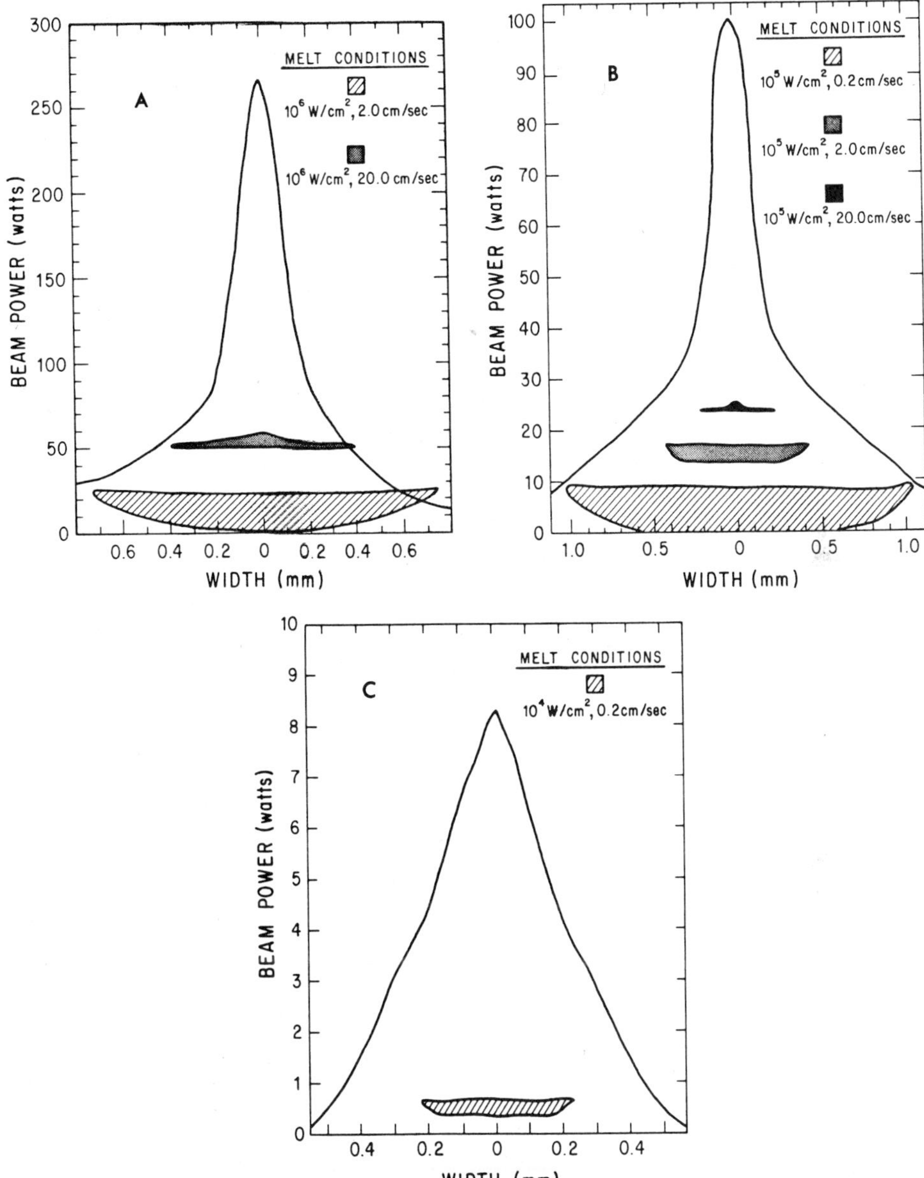

Fig. 5. Comparison of beam profile and the resulting melt region cross-sections produced at: (a) 10^6 W/cm^2, 2.0, and 20 cm/sec; (b) 10^5 W/cm^2, 0.2, 2.0 and 20 cm/sec; (c) 10^4 W/cm^2 and 0.2 cm/sec.

Cooling Rate Calculations

It has been well demonstrated in the RSP literature that the microstructure of a solidified region is a function of its cooling rate. Consequently, approximate cooling rates were calculated to determine whether the observed microstructures correlate with the predicted cooling rates.

A simple model for calculating the cooling rate of a finite layer of a liquid in contact with a heat sink is given by Sarjeant and Roy.[12] If it is assumed that the underlying Al_2O_3 substrate acts as the heat sink, the cooling rate (dT/dt) can be calculated as follows:

$$\frac{dT}{dt} = \frac{(T_m - T_s)h}{\rho Cpd} \tag{2}$$

where:

T_m = melting temperature (2045°C)
T_s = substrate temperature (25°C)
h = heat transfer coefficient at T_m (0.13 cal/cm^2 sec °C)
ρ = density of liquid at T_m (2.97 g/cc)
Cp = heat capacity of liquid at T_m (0.32 cal/g °C)
d = thickness of liquid layer (cm).

Unfortunately, the heat transfer coefficient between the liquid and the substrate is not accurately known. However, using a reasonable approximation proposed by Sarjeant and Roy that h is an order of magnitude greater than the thermal conductivity of the melt at T_m, an order of magnitude approximation for the cooling rate can be determined. The cooling rates were calculated for the maximum melt thickness for each condition studied (Table 2).

Table 2. Calculated Cooling Rates (°C/sec)

Scanning Rate (cm/sec)	Power Density (W/cm^2)		
	10^4	10^5	10^6
20	--	9.5×10^4	4.7×10^4
2.0	--	2.3×10^4	1.4×10^4
0.2	4.8×10^4	1.1×10^4	--

Microstructure Analysis

There exists a variety of different microstructures in each of the solidified melt regions as a result of the Gaussian distribution of the incident beam. However, for each change in cooling rate there is a different microstructure observed. The resultant microstructure is a function of the balance between solidification rate and the magnitude of the temperature gradient in the liquid just ahead of the liquid-solid interface. The product of solidification rate and temperature gradient is the cooling rate,[13] which can be varied by changing beam residence time and power density.

At the lowest cooling rate predicted, a dendritic structure is obtained exhibiting many chains of faceted crystals as shown in Fig. 6. (In all of the micrographs presented the scanning direction is left to right.) The chains are approximately 100 to 600 μm wide. At slightly higher cooling rates a crystalline faceted structure is produced in the center of the melt region while elongated oriented grains of length 60 to 180 μm by 0.8 to 1.6 μm wide are observed at the edges (Figs. 7 a and b). This difference in microstructure on the same sample is due to the Gaussian power intensity profile of the laser beam which produces a temperature gradient perpendicular to the scanning direction. As the predicted cooling rate increases, high aspect ratio oriented grains are seen sweeping back from the center of the melt region to the edges away from the direction of the laser beam (Fig. 8). For samples scanned at 10^6 W/cm^2 for 20 cm/sec and 10^4 W/cm^2 for 0.2 cm/sec, the predicted cooling rate is 4.7×10^4 °C/sec. Although considerably different processing conditions are used, the two microstructures are very similar and show small oriented groups of grains where the groups are randomly oriented with respect to each other (Figs. 9 and 10). The grains are approximately 10 to 25 μm long with aspect ratios of approximately 30. At the highest cooling rate attained in these studies, a microcrystalline structure of randomly oriented grains is produced (Fig. 11). The grains are 0.8 to 2 μm long and 0.2 to 0.4 μm wide. It is interesting that this is a much finer grain size than in the original substrate. Although cooling rates of 10^4 °C/sec can result in metastable and non-equilibrium phases for some systems, only alpha alumina was detected in these samples.

SUMMARY

It has been demonstrated that a CO_2 laser is a useful technique for the study of RSP of alumina. For the laser beam intensities and residence times studied, the cooling rate ranged from 1.1 to 9.5×10^4 °C/sec. Changes in the microstructure correlated well with the calculated cooling rates and ranged from a coarse dendritic structure for the slowest cooling rate to a microcrystalline structure for the fastest cooling rate.

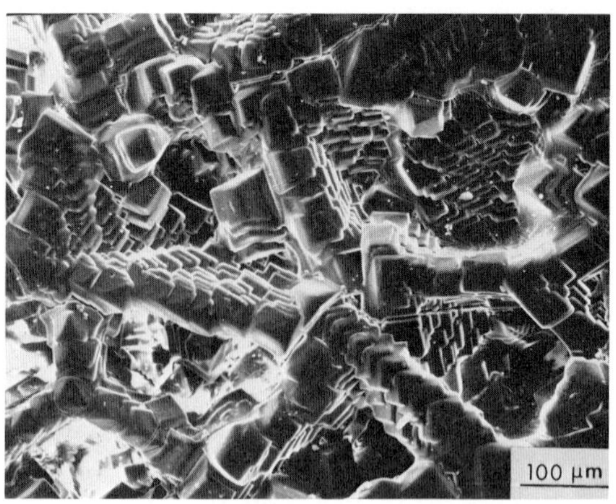

Fig. 6. Sample irradiated at 10^5 W/cm^2 and scanned at 0.2 cm/sec (cooling rate is 1.1×10^4 °C/sec).

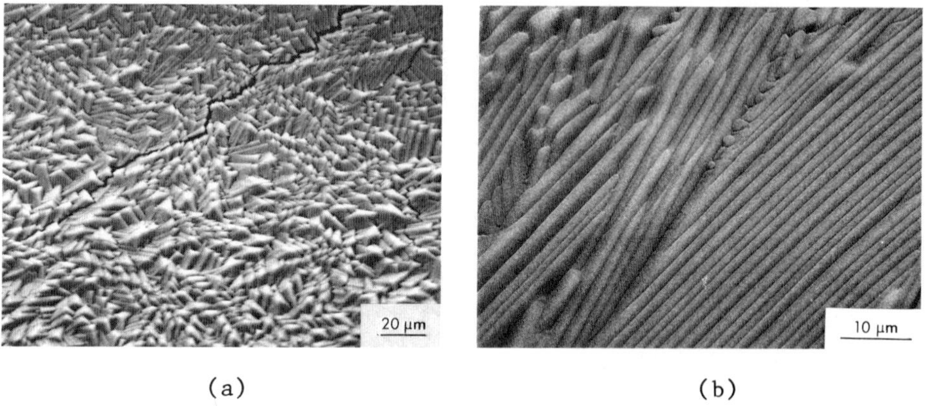

(a) (b)

Fig. 7. (a) Center of melt region irradiated at 10^6 W/cm^2 and scanned at 2.0 cm/sec (cooling rate is 1.4×10^4 °C/sec). (b) Edge of melt region irradiated at 10^6 W/cm^2 and scanned at 2.0 cm/sec (cooling rate is 1.4×10^4 °C/sec).

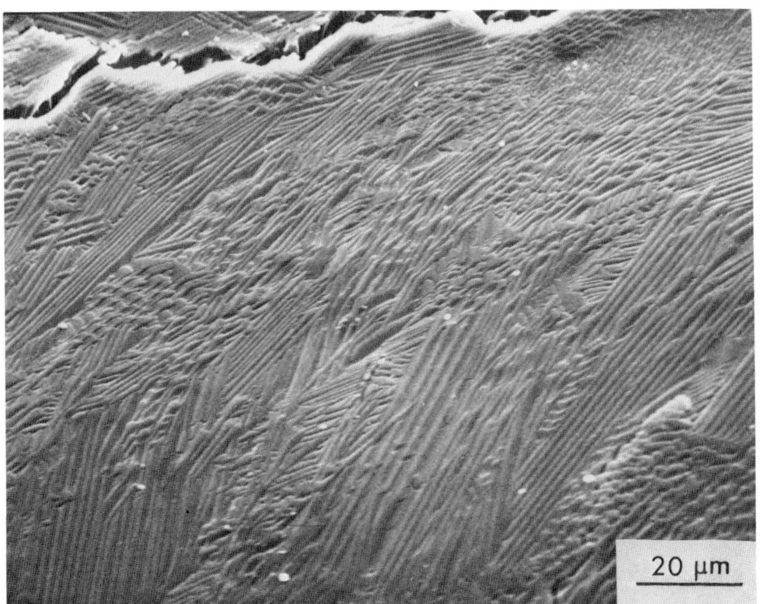

Fig. 8. Sample irradiated at 10^5 W/cm^2 and scanned at 2.0 cm/sec (cooling rate is 2.3×10^4 °C/sec).

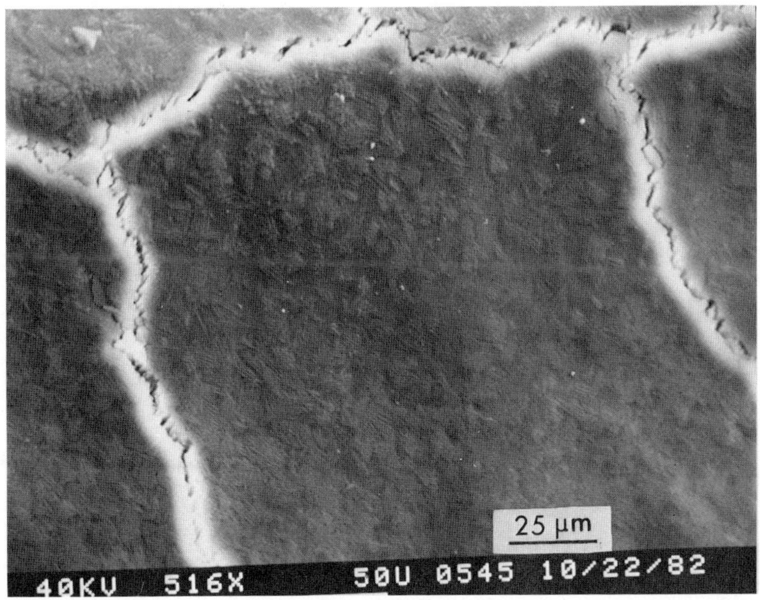

Fig. 9. Sample irradiated at 10^6 W/cm^2 and scanned at 20 cm/sec (cooling rate is 4.7×10^4 °C/sec).

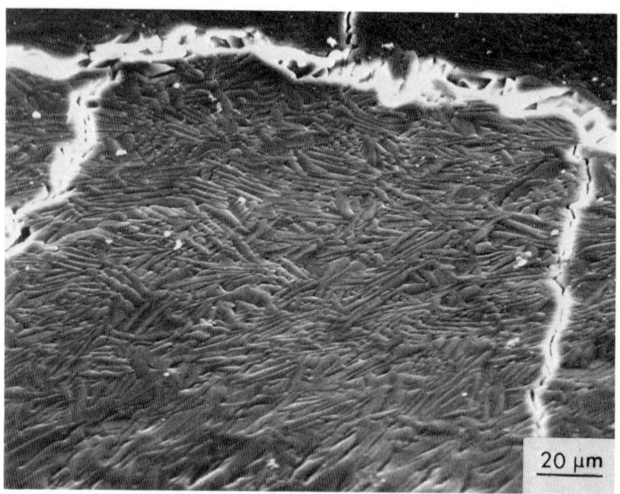

Fig. 10. Sample irradiated at 10^4 W/cm^2 and scanned at 0.2 cm/sec (cooling rate is 4.8×10^4 °C/sec).

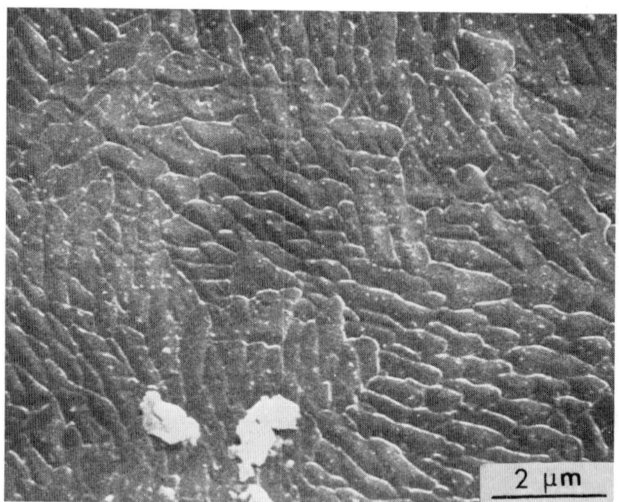

Fig. 11. Sample irradiated at 10^5 W/cm^2 and scanned at 20 cm/sec (cooling rate is 9.5×10^4 °C/sec).

ACKNOWLEDGMENTS

The authors acknowledge the financial support of the Basic Energy Sciences Division of the Department of Energy and the technical assistance of Mr. Richard A. Shellman.

REFERENCES

1. R. Mehrabian, B. H. Kear, and M. Cohen, ed., Rapid Solidification Processing, Principles and Technologies, II, Claitors Publishing Division, Baton Rouge, 1980.
2. R. Mehrabian, Inter. Metals Rev., 27, 185-208 (1982).
3. R. W. Cahn, Ann. Rev. Mat. Sci., 12, 51-63 (1982).
4. C. W. White and P. S. Peercy, eds., Laser and Electron Beam Processing of Materials, Academic Press, NY, 1980.
5. S. D. Ferris, H. J. Leamy, and J. M. Poate, eds., Laser-Solid Interactions and Laser Processing - 1978, American Institute of Physics, NY (1979).
6. J. Narayan, J. Metals, 32, 15-21 (1980).
7. L. E. Topol, D. H. Hengstenberg, M. Blander, R. A. Happe, N. L. Richardson, and L. S. Nelson, J. Noncryst. Solids, 12, 377-90 (1973).
8. L. S. Nelson, M. Blander, S. R. Skaggs, and K. Keil, Earth Planet Sci. Letters, 14, 338-44 (1972).
9. L. S. Nelson, S. R. Skaggs, and N. L. Richardson, J. Am. Ceram. Soc., 53, 115-16 (1970).
10. M. I. Cohen, 1577-1647 in Laser Handbook Vol. 2, edited by F. T. Areechi and E. O. Schulz-Dubois, North Holland Publishing Co., Amsterdam, 1972.
11. T. R. Anthony and H. E. Cline, J. Appl. Phys., 48, 3888-94 (1977).
12. P. T. Sarjeant and R. Roy, Mat. Res. Bull., 3, 265-79 (1968).
13. M. Cohen, B. H. Kear, and R. Mehrabian, 1-23 in Ref. 1.

STRUCTURE OF CERAMIC SURFACES MODIFIED BY ION BEAM TECHNIQUES[a]

C. J. McHargue, H. Naramoto,[b] C. W. White,
J. M. Williams, B. R. Appleton, P. S. Sklad, and
P. Angelini

Oak Ridge National Laboratory
Oak Ridge, Tennessee 37830

INTRODUCTION

Modification of the near-surface region of materials by use of energetic ion beams has been investigated extensively in recent years. The nature of the process allows one to introduce any element into the near-surface region of solids in a controlled and reproducible manner that is independent of most equilibrium constraints. Since the process is nonequilibrium in nature, compositions and structures unattainable by conventional methods may be produced.

In ion implantation, the dopant or alloying element is the ion beam which after an acceleration of tens to hundreds of kiloelectron volts impinges upon the surface of the target (host material). The energetic ion comes to rest by displacing atoms from their normal lattice sites by atomic collisions; thus producing a large number of point defects. After injection, the implanted ions have an approximate Gaussian distribution which is peaked at a fraction of a micrometer beneath the free surface.

Since ceramic materials are particularly sensitive to surface conditions, we have initiated a program to determine the structure

---

[a]Research sponsored by the Division of Materials Sciences, U.S. Department of Energy, under contract W-7405-eng-26 with the Union Carbide Corporation.

[b]On assignment from Physics Division, JAERI, Tokai-mura, Japan.

and properties of several classes of ceramics whose near-surface region have been modified by a variety of ion beam methods.

Previous studies of ion implantation into Al_2O_3 have dealt largely with changes in optical properties,[1-3] characterization of the disorder produced by the implantation process,[4-8] or the volume changes produced by implantation of gas atoms.[1,9,10]

The optical absorption band produced in Al_2O_3 by particle irradiation has been ascribed to aluminum vacancies[1] or to anion (oxygen) vacancies.[2] The lattice location experiments of Carnera et al.[6,7] showed implanted lead atoms to occupy octahedral interstitial sites that were significantly displaced along the <0001> axis.

Naguib and Kelly[11] include Al_2O_3 in their list of materials that become amorphous during ion bombardment. This conclusion was based on results from gas-release studies,[12] Rutherford backscattering (RBS),[4] and reflection electron diffraction patterns.[13] A transmission electron microscopy (TEM) study by Rechtin, however, indicated the structure of Al_2O_3 implanted with helium, oxygen, neon, or carbon to remain crystalline with a defect structure largely characteristic of that produced by electron or neutron irradiation.[14]

Investigations of the structure and electrical properties of ion implanted silicon carbide have been reported.[15-24] The amount of disorder, as measured by RBS, saturates at the value for a random sample for both nitrogen and antimony.[15-20] Once the amount of lattice disorder, as measured by ion scattering/channeling techniques, equals that from a randomly oriented specimen, it is considered that the material has become amorphous, although Raman spectroscopy indicates some characteristic crystalline bonding may remain.[22-23]

Krypton implantation into sputtered TiB_2 films has been reported to cause blistering and increased adherence to the substrate.[25] The latter effect may have been caused by ion beam mixing with the stainless steel substrate.

EXPERIMENTAL PROCEDURE

Single crystals of α-Al_2O_3 were obtained from Union Carbide Corporation (Crystal Products Division) and Crystal Systems, Inc. Crystals from both suppliers were of high purity (<100 ppm total cation impurities) and contained a low dislocation density ($\sim 10^3$ cm^{-2}). Specimens oriented within 2° of (0001) were polished and then annealed for five days in air at 1200°C to remove any residual mechanical damage. Single crystal [0001] samples of α-SiC were obtained from the Carborundum Company as platelets produced in an Acheson furnace.

The polycrystalline TiB_2 was prepared at ORNL from Starck powder which initially contained ~ 1% oxygen as the major impurity. Specimens having 98.4% theoretical density were prepared by vacuum hot pressing at 2050°C under 25 MPa uniaxially applied pressure for 4 h. The grain size was in the range of 75 to 100 μm.

An Extrion 200 kV ion implantation accelerator was used to implant 10^{16} to 10^{17} ions•cm^{-2} of ^{52}Cr (particle energy of 280 or 300 keV), ^{48}Ti (150 keV), and ^{90}Zr (150 keV) into Al_2O_3 at an orientation 7° off axis at nominally room temperature. The concentration of implanted ions corresponded to 1.6 to 10.0% of the cations initially present in the crystal. A part of each specimen was shielded from the ion beam by a metal mask to preserve a virgin region as a reference state. The same equipment was used for the SiC implants. Fluences from 2.7 x 10^{13} to 8.1 x 10^{16} ions•cm^{-2} of ^{14}N (62 keV) and 2.9 x 10^{14} to 3.1 x 10^{16} ions•cm^{-2} of ^{52}Cr (280 keV) were used. The ^{58}Ni implantations into TiB_2 were carried out on the ORNL 5 MV Van de Graaff facility to give a fluence of 1 x 10^{17} ions•cm^{-2} (corresponding to a peak concentration of 16.7% cations).

Annealing of the Al_2O_3 was conducted in air for periods of 1 h at temperatures between 600 and 1600°C. The TiB_2 was annealed for 2 h in vacuum at 1450°C.

The single crystal specimens were examined using Rutherford backscattering-ion channeling techniques (RBS-C) with 2.0 MeV $^4He^+$ to determine the depth profile of the implanted species, the depth distribution of damage in the host lattice, and the lattice location of the impurity. Transmission electron microscopy specimens, prepared by ion milling, with the plane of observation both parallel and perpendicular to the implantation beam were employed to determine the structural characteristics of the implanted zone. Surface profilometry[c] gave data on the volume changes introduced by implantation. Raman spectroscopy gave important information regarding the structure of implanted SiC.

DISCUSSION OF RESULTS

Figure 1 shows typical spectra of 2.0 MeV He^+ scattered from an Al_2O_3 single crystal in both a random and channeling orientation. The random reference spectrum was obtained from the as-implanted (1 x 10^{17} ^{52}Cr•cm^{-2}, 300 keV) crystal by continuously rotating it to average over many crystallographic orientations. Ion channeling spectra were obtained with the ion beam incident parallel to <0001> of both implanted and virgin Al_2O_3. The random spectrum is characteristic of the scattering yield from a completely disordered Al_2O_3

---

[c]DEKTAK, Sloan Technology Corporation, Santa Barbara, CA.

sample while the <0001> virgin shows the yield reduction caused by the channeling effect in a perfect Al_2O_3 crystal. Utilizing the mass specificity of the Rutherford scattering process and the known energy loss of He^+, the scattered energies of Fig. 1 have been converted to separate depth scales for Cr, Al, and O in Al_2O_3. The spectra in Fig. 1(a) show that the Al_2O_3 crystal lattice has been heavily damaged by the chromium implantation. Note, however, that the channeled yield from the as-implanted sample never reaches the random value, indicating that the near-surface region was not totally disordered. This is true for all the implanted species studied (Cr, Zr, Ti, Ni, Fe, and Ge).

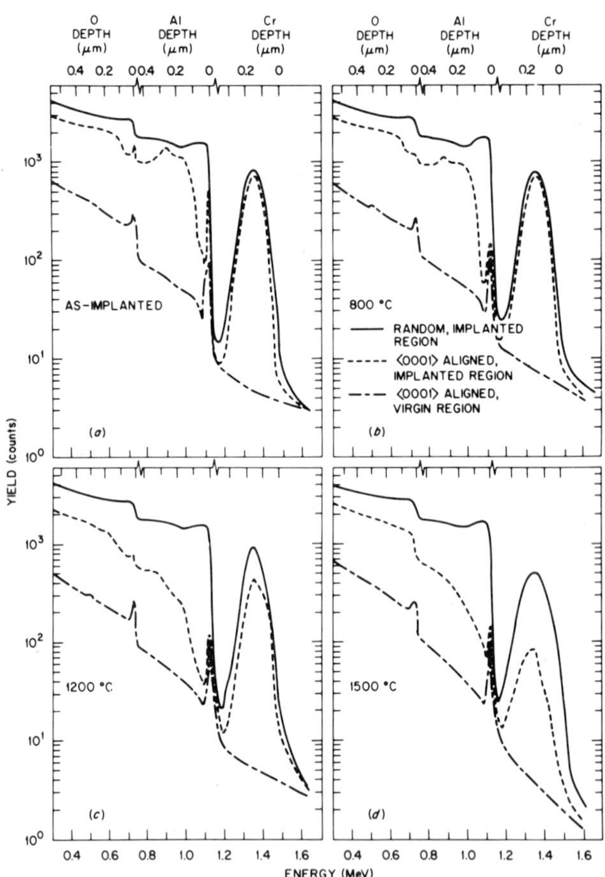

Fig. 1. Rutherford backscattering-channeling spectra for 2 MeV He^+ from chromium-implanted Al_2O_3 (300 keV, 1×10^{17} cm^{-2}). (a) As-implanted; annealed at (b) 800°C, (c) 1200°C, and (d) 1500°C.

The relative damage to the aluminum and oxygen sublattices was seen to saturate for implantation doses from 5×10^{15} to 1×10^{17} $Cr \cdot cm^{-2}$. Utilizing the ion-channeling geometry to detect preferential lattice locations of the implanted species,[28] it was found that chromium atoms showed only a slight bias toward substitutionality in the samples implanted with 1×10^{17} $Cr \cdot cm^{-2}$.

An α-Al_2O_3 specimen implanted to 2×10^{16} $Cr \cdot cm^{-2}$ was examined at 1 MV accelerating potential in the ORNL Hitachi 1000 microscope. The diffraction pattern showed the implanted region to be crystalline despite the large amount of damage. The TEM images contained a high density of "black spots," suggestive of point defect clusters. Attempts to determine the character of these clusters were unsuccessful due to large residual stresses in the thinned foils.

Figures 2(a) and 3(a) contain the RBS-C spectra for Al_2O_3 implanted with 2×10^{16} $Zr \cdot cm^{-2}$ and 3×10^{16} $Ti \cdot cm^{-2}$, respectively. The description given above for Fig. 1(a) applies to these curves also. There is substantial damage to both sublattices but the material is not amorphous. Both figures indicate random (interstitial and substitutional) solid solutions for concentrations of zirconium and titanium which exceed the equilibrium solid solubility.

An analysis of Figs. 1(b-d), 2(b-d), and 3(b-c) reveals the effects of thermal annealing on the lattice damage and the redistribution and lattice sites of the implanted species in α-Al_2O_3. Damage recovery began selectively in the aluminum sublattice at a temperature of ~ 800°C for the chromium-implanted material [Fig. 1(b)]. There was little or no change at this temperature in either the damage distribution in the oxygen sublattice or the degree of chromium substitutionality. At 1000°C, recovery of the oxygen sublattice began but again there was no significant change in the chromium substitutionality [Fig. 1(c)]. After a 1200°C anneal, there was significant incorporation of chromium into substitutional (aluminum) lattice sites and ~ 80% of the damage to the aluminum sublattice had recovered.

The effects of higher temperature annealing treatments are summarized in Fig. 4. Figure 4(a) shows that the concentration of chromium as a function of depth is the same after annealing to 1500°C as in the implanted condition. However, ion channeling showed that annealing to 1500°C caused 98% of the chromium to acquire substitutional sites.[28] Detailed angular scans about $\langle 0001 \rangle$, $\langle 1\bar{2}10 \rangle$, $\{10\bar{1}0\}$, $\{0001\}$, $\{1\bar{2}10\}$, and $\{10\bar{1}0\}$ showed conclusively that >98% of the chromium was substitutional in the aluminum sublattice after the 1500°C anneal.[28] However, after the 1300°C anneal some chromium and oxygen remained in interstitial positions. Measurements utilizing EPR showed that the substitutional chromium was largely, if not totally, in the Cr^{+3} state.[28] Further annealing to 1600°C [Fig. 4(a)] resulted in significant redistribution in depth of the chromium but with no loss in substitutionality.

From a detailed examination of the chromium spectra in Fig. 1(a-d), one can deduce that during annealing below 1200°C damage recovery in the aluminum sublattice competed with chromium incorporation. Chromium incorporation at higher annealing temperatures appears to be accomplished by oxygen diffusion in from the surface.

The annealing behavior of the titanium-implanted Al_2O_3 differed significantly from the Cr-Al_2O_3 specimens. Again, recovery in the aluminum sublattice began at ~ 800°C and in the oxygen sublattice at ~ 1000°C. After the 1500°C anneal, recovery was essentially complete [Fig. 2(c)]. However, the titanium spectrum of Fig. 2(b) (1300°C anneal) exhibited a double peak, showing that some of the titanium had diffused toward the surface. This specimen had about 60% of the titanium in substitutional sites. Optical microscopy and TEM revealed that two titanium-rich phases had precipitated. The TEM photograph of Fig. 5 shows one precipitate to be acicular and oriented parallel to $\langle 10\bar{1}0 \rangle$. From the work of Philips et al.[29] one would expect this phase to be TiO_2. There is a second precipitate which is much smaller (10-90 nm) and disk shaped. Diffraction patterns have not yet been identified but EDS shows these particles to contain a large amount of titanium. A third feature of these photographs is a large number of small (~ 15 nm) voids or bubbles.

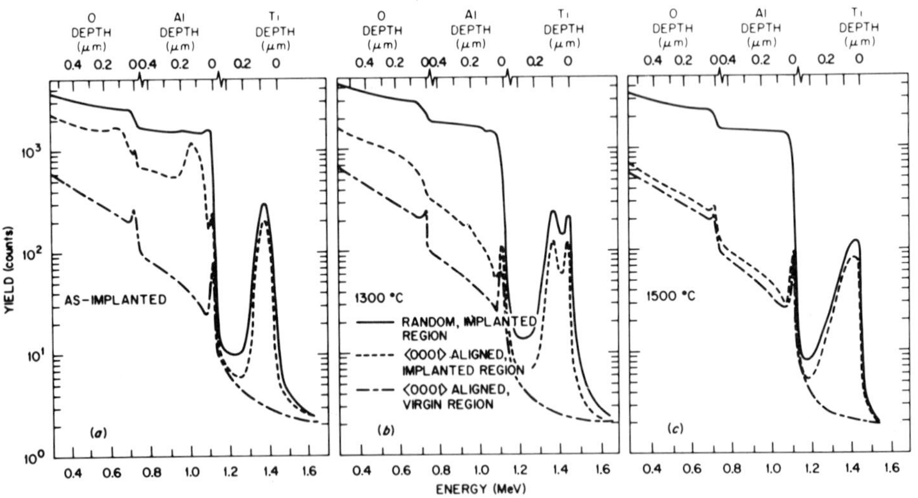

Fig. 2. RBS-C spectra from Ti-implanted Al_2O_3 (150 keV, 3×10^{16} cm^{-2}). (a) As-implanted; (b) 1300°C anneal; and (c) 1500°C anneal.

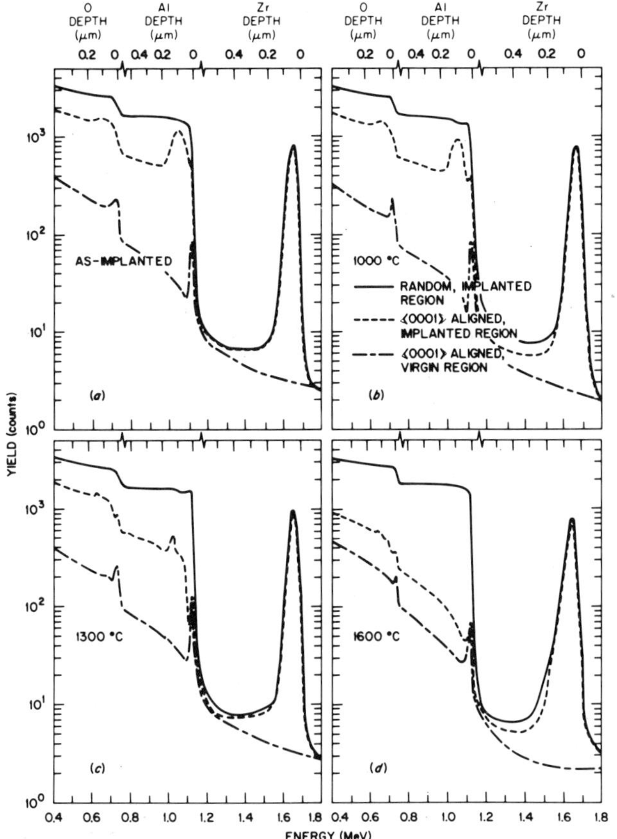

Fig. 3. RBS-C spectra from Zr-implanted Al_2O_3 (150 keV, 2 x 10^{16} cm^{-2}. (a) As-implanted; annealed at (b) 1000°C, (c) 1300°C, and (d) 1600°C.

Further annealing to 1500°C enhanced the damage recovery [Fig. 2(c)] and caused the bulk of the titanium to diffuse toward the surface [Fig. 4(b)]. Specimens implanted and examined along the <$1\bar{2}10$> direction did not exhibit this redistribution; hence, it can be concluded that diffusion along the c-axis is much faster than along the α-axis. The diffusion occurs in a temperature range where the titanium was largely distributed randomly suggesting, perhaps, an interstitial diffusion mechanism.

In contrast to the above results, annealing to 1500°C caused no change in the RBS-C spectra for zirconium and the onset of damage recovery was shifted to higher temperatures [Fig. 3(b-d)]. Damage recovery in the oxygen sublattice started at ~ 1300°C and, although recovery in the aluminum sublattice began at ~ 800°C, large amounts of disorder remained in both sublattices even after annealing at 1600°C. The zirconium exhibited no substitutionality, and no redistribution occurred from the implanted zone. Such observations would be consistent with the precipitation of the zirconium at an early stage of annealing.

The TEM photograph of Fig. 6 confirms that a zirconium-rich second phase was present after the 1300°C anneal. The precipitates were 5 to 20 nm in size but their structure or composition has not been determined. The channeling results suggest that the second phase formed at the lowest annealing temperature (600°C). The precipitates in the 1500°C annealed specimen had about the same size distribution as at 1300°C indicating that, whereas the phase nucleates very early, its growth is slow.

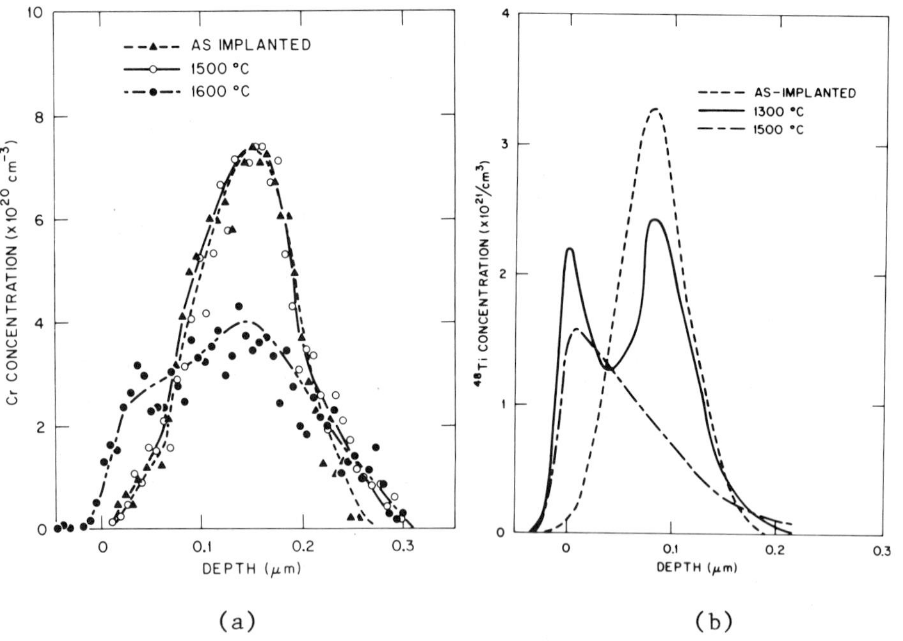

Fig. 4. Concentration profiles for (a) Cr (300 keV, 1 x 10^{16} cm^{-2}), (b) Ti (150 keV, 3 x 10^{16} cm^{-2}) in Al_2O_3 for as-implanted condition and after annealing.

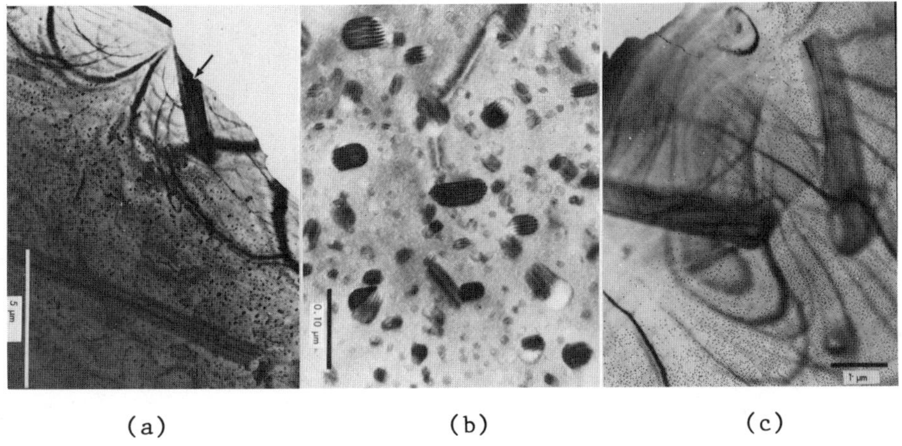

Fig. 5. TEM of back-thinned specimen Al_2O_3 implanted with 3 x 10^{16} Ti-cm^{-2} and annealed at 1100°C. (a) Large precipitates, (b) fine precipitates, and (c) voids or gas bubbles.

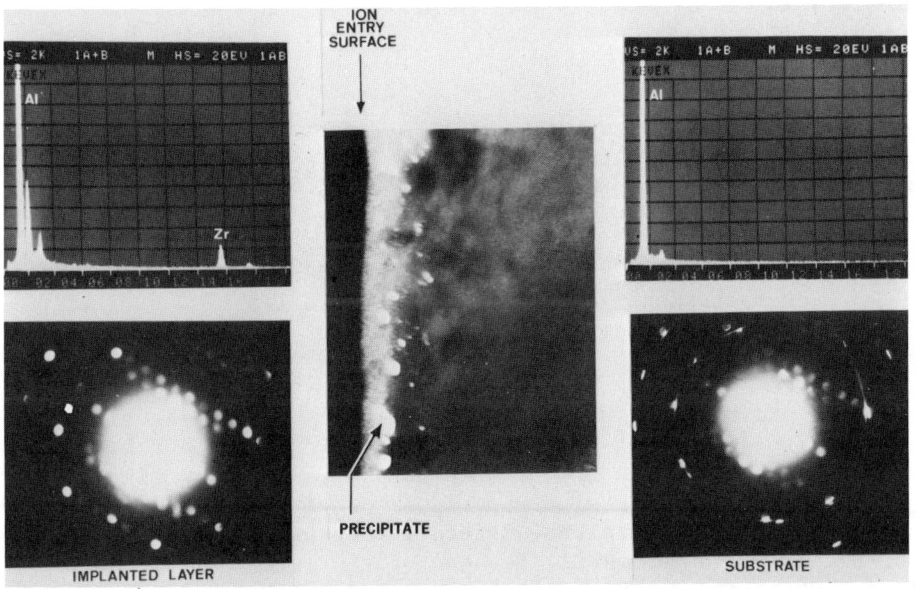

Fig. 6. Dark field TEM, electron diffraction patterns and EDS analysis of Zr-implanted Al_2O_3 (2 x 10^{16} cm^{-2}) after annealing at 1300°C.

Ion-channeling analyses of single crystals of SiC implanted with nitrogen or chromium to various fluences showed that the channeled ion-scattering yields reached the random yield when implanted doses corresponded to about 0.2 displacements per atom.[30] Figure 7 contrasts a [0001] channeling spectrum for a crystal implanted with 2.9 x 10^{14} $Cr \cdot cm^{-2}$ with the [0001] channeling and rotating random reference spectra taken from an unimplanted (virgin) SiC single crystal. For this fluence, the damage induced by the implanted chromium ions has "randomized" the crystal in a region 0.02 to 0.2 nm from the surface. This is the region where the damage energy was a maximum and brackets the range where it exceeded the critical value of 0.2 dpa. At higher fluences, the random region spreads in both directions. The dose dependence of randomization for each ion is given in ref. 30.

It is generally assumed that an overlapping of the aligned spectrum with the random spectrum indicates an amorphous structure. In order to determine if this was the case for the present study, TEM and Raman spectroscopy were used to examine the Cr-implanted SiC specimens. The TEM micrograph showed halos in the diffraction patterns characteristic of amorphous material to a depth of 0.25 μm and crystalline patterns at greater depths.[27] The range of 280 keV chromium ions in SiC is ~ 0.25 μm. The Raman spectra for the virgin region contained peaks at 768.13, 784.51, 796.15, and 959.95 cm^{-1} which are characteristic of crystalline SiC.[23,24] After implantation to 2 x 10^{15} $Cr \cdot cm^{-2}$ these crystalline modes were absent, confirming the ion channeling and TEM observations on the amorphous nature of the implanted region.

Because only polycrystalline TiB_2 was available, the techniques used to characterize its implanted structure were limited. Figure 8 shows the TEM results for TiB_2 implanted with 1 x 10^{17} $Ni \cdot cm^{-2}$ (1 MeV). The diffraction patterns show that the surface region remained crystalline. The TEM micrograph shows damage extending to ~ 0.8 μm from the surface which consists of a near-surface region with a "coarse damage" structure and an interior region of "fine damage." Studies to determine the nature of the damage are in progress.

The existence of damage to ~ 0.8 μm was surprising since the penetration depth of 1 MeV nickel ions in TiB_2 is about 0.4 μm. However, our calculations show that a boron ion which receives the maximum energy transfer in a primary knock-on with a 1-MeV Ni ion would have this range. The titanium recoil range would be about the same as the nickel range. Thus, because of the large difference in the masses of Ti and B, two damage regions are formed: the one nearer the surface due primarily to the displacement of the heavier ions and the other due to the displacement of the lighter ions.

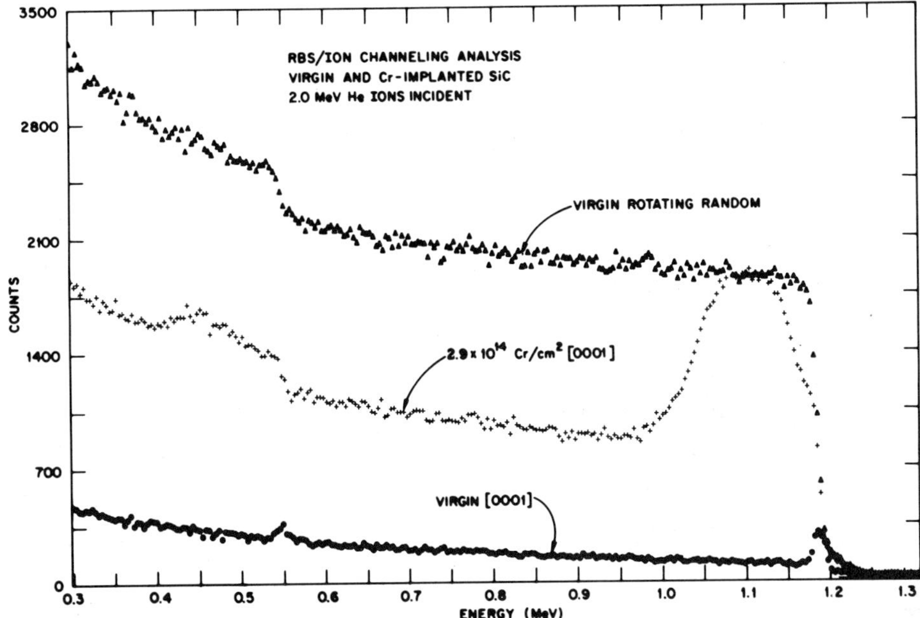

Fig. 7. RBS spectra from α-SiC implanted with Cr.

SUMMARY

A wide variety of structures are produced by ion implantation in ceramics. Random (substitutional and interstitial site occupancy) solid solutions with concentrations of solute that exceed the solubility limit can be produced in Al_2O_3. The changes that occur during annealing are complex and sometimes unpredictable.

Silicon carbide becomes amorphous in a manner analogous to Si for ion fluences that produce more than 0.2 dpa damage. Light (N) and heavy (Cr) ions produce similar results if the fluence is scaled to damage energy deposited.

Because of mass differences in the ions, two damage regions are developed in TiB_2. The structure remains crystalline to very high damage levels.

These structural alterations cause changes in the surface mechanical properties. Since virtually any chemical species can be implanted, one can independently control structural damage and chemical effects. When coupled with selective annealing, this technique has the potential for producing a wide range of surface structures and properties.

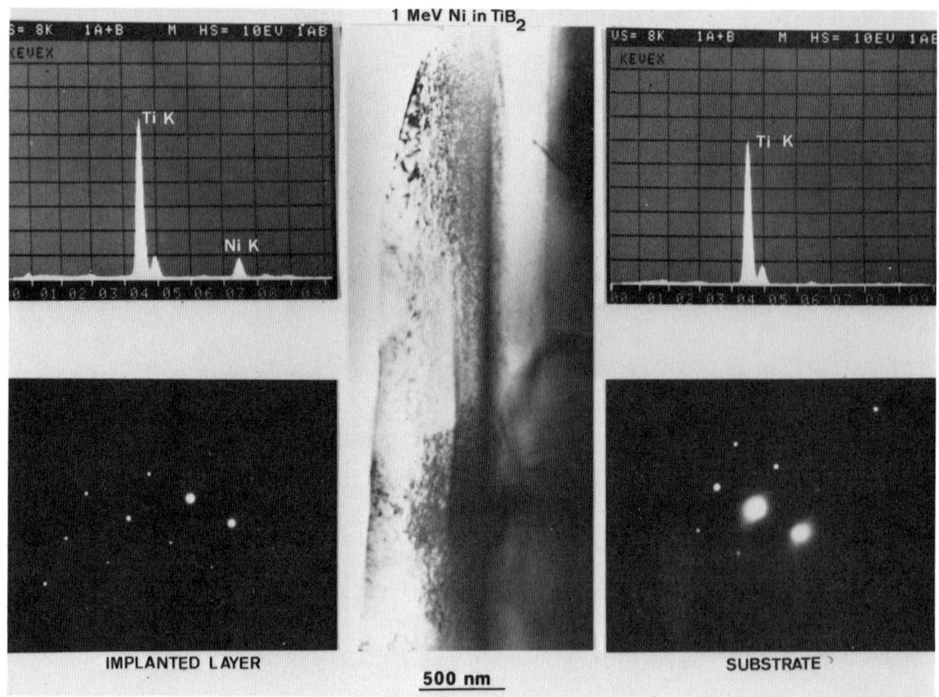

Fig. 8. TEM, electron diffraction and EDS results for TiB_2 implanted with 1×10^{17} $Ni \cdot cm^{-2}$ (1 MeV).

REFERENCES

1. G. W. Arnold, G. B. Krefft, and C. B. Norris, Appl. Phys. Lett., 25, 540-42 (1974).
2. B. D. Evans, H. D. Hendricks, F. D. Bazzarre, and J. M. Bunch, pp. 265-74 in Ion Implantation in Semiconductors-1976, edited by F. C. Chernow, J. A. Borders, and D. K. Brice, Plenum Press, NY, 1976.
3. T. F. Leura, J. A. Borders, and G. W. Arnold, Ibid., pp. 285-94.
4. H. M. Naguib, J. F. Singleton, W. A. Grant, and G. Carter, J. Mater. Sci., 8, 1633-40 (1973).
5. A. V. Drigo, S. Lo Russo, P. Mazzoldi, P. D. Goude, and N. E. W. Hartley, Radiat. Eff., 33, 161-71 (1977).
6. A. Carnera, A. Drigo, and P. Mazzoldi, Radiat. Eff., 49, 29-32 (1980).
7. A. Carnera, G. Della Mea, A. V. Drigo, S. Lo Russo, P. Mazzoldi, and N. E. W. Hartley, Radiat. Eff., 35, 201-8 (1978).
8. A. Turos, H. Matzke, and P. Rabette, Phys. Stat. Sol. (a), 64, 565-75 (1981.

9. G. B. Krefft, W. Beezhold, and E. P. EerNisse, IEEE Trans. Nucl. Sci., NS-22, 2247-49 (1975).
10. G. Krefft and E. EerNisse, J. Appl. Phys., 49, 2725-30 (1978).
11. H. M. Naguib and R. Kelly, Radiat Eff., 25, 1-12 (1975).
12. C. Jech and R. Kelly, J. Phys. Chem. Sol., 30, 465-74 (1969); 31, 41-48 (1970).
13. H. Matzke and J. L. Whitton, Canad. J. Phys., 44, 995-1010 (1966).
14. M. D. Rechtin, Radiat. Eff., 42, 129-44 (1979).
15. O. J. Marsh and H. L. Dunlap, pp. 285-95 in Ion Implantation, edited by F. Eisen and L. T. Chadderton, Gordon and Breach, NY, 1970.
16. R. Hart, H. Dunlap, and O. Marsh, Radiat. Eff., 9, 261-66 (1971).
17. O. J. Marsh, pp. 471-85 in Silicon Carbide-1973, edited by R. C. Marshall, J. W. Faust, Jr., and C. E. Ryan, University of South Carolina Press, 1973.
18. A. B. Campbell, J. B. Mitchell, J. Shewchun, D. Thompson, and J. A. Davies, Ibid., pp. 486-92.
19. W. J. Choyke, L. Patrick, and P. J. Dean, Phys. Rev. B, 10, 2554-65 (1974).
20. A. B. Campbell, J. Shewchun, D. A. Thompson, J. A. Davies, and J. B. Mitchell, pp. 291-98 in Ion Implantation in Semiconductors, edited by S. Namba, Plenum Press, NY, 1975.
21. D. A. Thompson, M. C. Chan, and A. B. Campbell, Canad. J. Phys., 54, 626-32 (1976).
22. R. B. Wright, R. Varma, and D. M. Gruen, J. Nucl. Mater., 63, 415-21 (1976).
23. R. B. Wright and D. M. Gruen, Radiat. Eff., 33, 133-40 (1977).
24. V. V. Makarov, T. Tuomi, and K. Naukkarinen, Appl. Phys. Lett., 35, 922-24 (1979).
25. K. Padmanabhan and G. Sorensen, Thin Solid Films, 81, 13-19 (1981).
26. C. J. McHargue, H. Naramoto, B. R. Appleton, C. W. White, and J. M. Williams, pp. 147-53 in Metastable Materials by Ion Implantation, S. T. Picraux and W. J. Choyke, North Holland, NY, 1982.
27. C. J. McHargue and J. M. Williams, Ibid., pp. 303-9.
28. H. Naramoto, C. W. White, J. M. Williams, C. J. McHargue, O. W. Holland, M. M. Abraham, and B. R. Appleton, submitted to J. Appl. Phys.
29. D. S. Philips, A. H. Heuer, and T. E. Mitchell, Phil. Mag. A, 42, 385-432 (1980).
30. J. M. Williams, C. J. McHargue, and B. R. Appleton, in Proceedings of Ion Beam Modification of Materials-1982, to be published in J. Nuc. Instr. Meth.

MICROSTRUCTURE AND MECHANICAL PROPERTIES OF ION-IMPLANTED CERAMICS

C. S. Yust and C. J. McHargue

Metals and Ceramics Division
Oak Ridge National Laboratory
Oak Ridge, Tennessee 37830

ABSTRACT

The effects of the direct implantation of cations into the lattice of alumina, silicon carbide, and titanium diboride have been studied. Alumina implanted with chromium, titanium, and zirconium ions remains crystalline and shows an increase in hardness and fracture toughness. Titanium diboride experiences similar trends. In contrast, silicon carbide becomes amorphous when implanted with chromium ions, becomes less hard, but shows an increased fracture toughness. The response to scratching and annealing is also discussed.

INTRODUCTION

Ion implantation is a technique for the direct introduction of ions into the surface layers of solids. In this manner, the surface characteristics of solids can be varied in a controlled way through the generation of nonequilibrium compositions and microstructures. The ability to produce solid surfaces having particular characteristics suggests the possibility for improvements in materials and is analogous to the use of special coatings or adhered layers.

The process of ion implantation is generally accompanied by the introduction of lattice defects arising from the interaction of the injected ions and the lattice.[1,2] The interstitial atoms, vacancies, voids, and dislocation loops which may be formed contribute to change in the properties of the implanted surface layer with respect to those of the unimplanted bulk lattice. Additional property changes may result from the formation of nonequilibrium phases or structures.

This paper presents evidence of such damage effects in three ceramic materials, the consequences of surface modification for mechanical contact, and the effect of annealing on the implanted body.

MATERIALS AND EXPERIMENT

Specimens of alumina, silicon carbide, and titanium diboride were utilized in this work. The alumina was obtained as single crystal discs from Crystal Systems, Incorporated. The disc surfaces were within 2° of the basal plane orientation. Prior to implantation, the specimens were annealed in air at 1200°C for 5 d to remove all traces of surface damage. Total impurities in the as-received specimens was reported to be less than 100 ppm.

Silicon carbide was obtained in both the single crystal and polycrystalline form. The single crystals, which were predominantly the 6H polytype, were from the Acheson furnace process. The polycrystalline body was sintered α-silicon carbide. Both were supplied by the Carborundum Company.

The titanium diboride samples were prepared in this laboratory by hot pressing titanium diboride powder at 1800°C for 2 h without additives. The powder was obtained from Hermann C. Starck, Berlin, and had an initial average particle size of 8 μm. The as-pressed density of the material was 99% of the theoretical value.

The specimens were ion implanted by use of a Varian-Extrion 200 kV ion implantation accelerator or the ORNL 5 MV Van de Graaf facility. The alumina single crystal samples were implanted with ^{52}Cr (particle energy of 300 keV), ^{90}Zr (150 keV), and ^{48}Ti (150 keV). The silicon carbide, both single crystal and polycrystalline forms, was implanted with ^{52}Cr (280 keV), while the titanium diboride specimen was implanted with ^{58}Ni (1 MeV). The Al_2O_3 and SiC samples were implanted using the Varian accelerator and were exposed to fluences of 2.9×10^{14} to 1.0×10^{17} ions cm^{-2}. The TiB_2 was irradiated in the Van de Graaf facility to a fluence of 1×10^{17} ions cm^{-2}.

The hardness of the samples was measured by means of the Knoop microhardness technique. With a 15 g load (0.147 N), this procedure creates indentations which are approximately 2500 to 3000Å deep. The implanted ion concentration peaks in these specimens range in depth from 1000 to 3000 Å; thus the hardness values obtained represent a composite response of the implanted layer and the underlying unmodified lattice. The hardness values are reported only as relative values, expressed as the ratio of the hardness of the implanted area to that of an unimplanted region on the same crystal, indicating magnitudes of hardness changes rather than absolute hardness values.

Unimplanted areas are retained on all implanted specimens by selective masking to allow the implanted/unimplanted hardness ratio to be determined on crystal volumes having identical histories.

Indentation fracture toughness values were determined from Vickers hardness indentations. Vickers indents produced by 0.49 and 0.98 N loads generate cracks at the corners of the indentor impressions. The crack lengths and indentation diagonals were measured and used to calculate an apparent fracture toughness value by the methods of Evans[3] and of Marion.[4] The fracture toughness values determined from the two methods differ somewhat, but the ratios of the values for implanted/unimplanted material were comparable.

An evaluation of the response of the surface to simulated mechanical abrasion was made by means of a scratch test. In this test, a stylus is slowly translated across the surface under normal loads of 0.098 to 0.49 N while the tangential force on the stylus is continuously measured. The scratch origin was positioned to cause the stylus to cross the unimplanted/implanted interface during the traverse of the surface. The stylus is a Vickers diamond indentor, positioned to move over the sample surface with one of the pyramid faces normal to the direction of motion. The stylus moved at a velocity of 28 μm/s.

RESULTS

Hardness

The hardness of single crystal alumina is significantly increased by ion implantation. Relative hardness values as a function of the ratio of implanted cation to aluminum ion are presented in Fig. 1. The lower line for Cr_2O_3-Al_2O_3 shows the hardness of homogeneous chromia-alumina solid solutions produced by conventional chemical techniques.[5,6] For chromium/alumina atomic ratios of up to 0.25, it is seen that a 10% increase in relative hardness is produced by alloying Al_2O_3 with Cr_2O_3.

In contrast, implantation of chromium, zirconium, or titanium ions yields greater hardness increases, the increase being as great as 50% in the case of implanted chromium. McHargue et al. have shown by Rutherford backscattering (RBS) measurements that chromium implantation introduces disorder into both the oxygen and aluminum sublattices, but does not totally disrupt the crystallographic lattice.[7] In the as-implanted state, the chromium ions are randomly distributed among interstitial and substitutional lattice sites. The increased lattice disorder and the presence of interstitial cations contribute to the observed hardening. The same effects are observed for the implantation of zirconium and titanium ions.

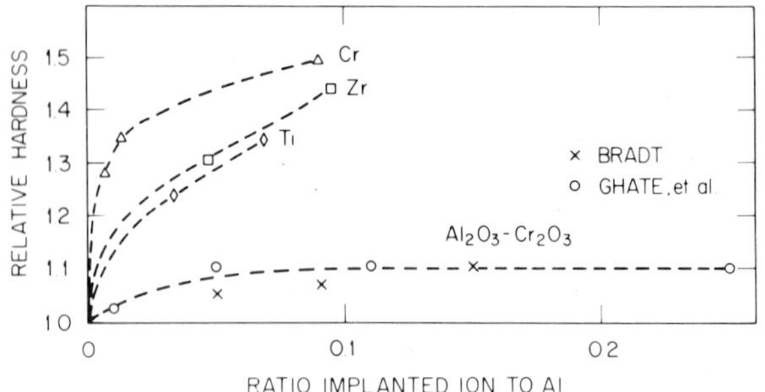

Fig. 1. Relative hardness ratios for alumina implanted with chromium, zirconium, and titanium as a function of implanted ion concentration. The hardness of alumina-chromia solid solutions is shown to be significantly lower than that of implanted surfaces.

In the implantation of ions into SiC, the RBS measurements show that the surface layer becomes amorphous during implantation, and the amorphous layer is found to increase in depth as the dose of implanted ions increases. The measurement of relative hardness of SiC as a function of irradiation dose reveals that the relative hardness decreases from 1.0 to 0.7 at the highest implantation level. This is consistent with the fact that the hardness measurement in the implanted layers is a composite measure of the layer and the underlying bulk; as the layer thickness increases, the influences of the layer hardness is proportionately greater. It is clear that the amorphous SiC is less hard than the crystalline form.

The implanted layer thickness in TiB_2 implanted with nickel is found to be greater than in the case of alumina or silicon carbide. Transmission electron microscopy shows the implanted region to be crystalline with a large dislocation content.[8] The resultant hardness ratio in the implanted material is found to be in the range 1.7 to 2.0. This is a significant increase in hardness for this material which even prior to implantation is a very hard material. At twice the unimplanted hardness, the hardness value approaches that of cubic boron nitride, one of the hardest materials known. Because the sample is polycrystalline and the hexagonal TiB_2 lattice is anisotropic with respect to hardness, it was necessary to make the hardness indents within individual grains which spanned the unimplanted/implanted interface. The hardness results for TiB_2 implanted with nickel are presented in Fig. 2.

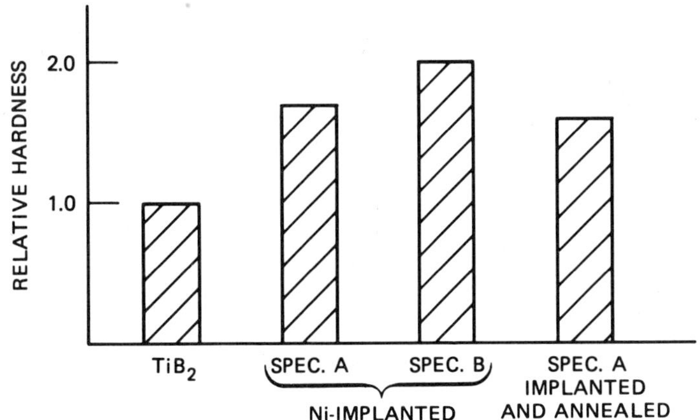

Fig. 2. The relative hardness of TiB_2 implanted with 1 MeV nickel.

On annealing for one hour periods, the relative hardness of ion implanted alumina partially recovers--the extent of recovery being determined by the annealing temperature and the specific implanted species. The recovered hardness as a function of annealing temperature is shown in Fig. 3. For each of the implanted ions, some hardness recovery is evident after the 800°C anneal, and for chromium and titanium implantations further softening is observed at 1000°C. The zirconium-implanted specimen recovers only modestly with annealing to temperatures as high as 1500°C, and a slight additional recovery takes place at 1600°C. Above 1000°C, the titanium-bearing sample shows increased hardness, while the chromium-bearing sample retains the same hardness developed at the 1000°C anneal until reannealed at 1600°C.

The titanium diboride specimens were annealed for 2 h at 1450°C and showed a hardness recovery from a relative hardness of 1.7 to 1.45. No annealing experiments were performed on the SiC specimen.

Fracture Toughness

Fracture toughness values were determined for each of the ion-implanted ceramics. For ion-implanted alumina, fracture toughness increases of 15 to 20% with respect to the unimplanted samples were recorded. The greatest increase in toughness was attained with titanium implantation (~20%), the least (~15%) with zirconium. The chromium-bearing samples fell between these values. This increase was independent of the implantation dose up to an implanted ion/aluminum ion ratio of about 0.1.

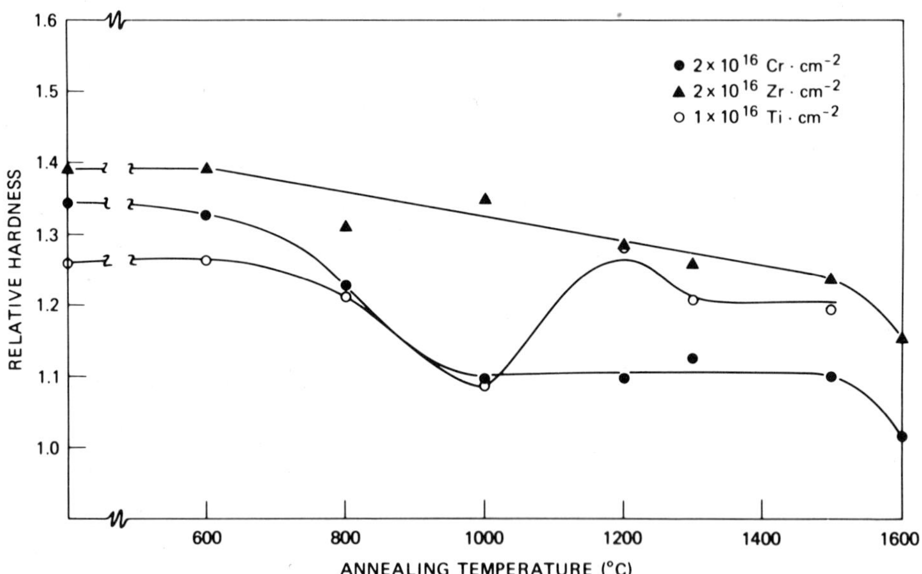

Fig. 3. Recovery of hardness as a function of annealing temperature for ion-implanted alumina.

The fracture toughness of TiB_2 implanted with nickel increased from 40 to 80%, while that of single crystal SiC implanted with chromium increased in apparent K_{IC} from 10 to 20%.

Scratch Tests

The traverse of both the implanted and the adjacent unimplanted surfaces by a diamond stylus gives an indication of the response of the surface to mechanical abrasion. In all instances, a well defined groove was formed by the moving indentor, and variations in the tangential force during scratching were observed. The scratches formed in chromium implanted alumina are shown in Fig. 4. Two scratches are shown in the region of the implantation interface, the implanted area lying to the right of the interface. The stylus motion was from left to right; the normal force applied to the stylus was 0.29 N for the upper scratch and 0.49 N for the lower. At the interface, marked by arrows in the figure, the degree of fracture is seen to diminish as the groove moves into the implanted region. The assessment of diminished fracture is based on the comparison of the size and number of fractures on each side of the interface. The measured tangential force is found to typically increase by 20 to 30% in going from the unimplanted to the implanted region.

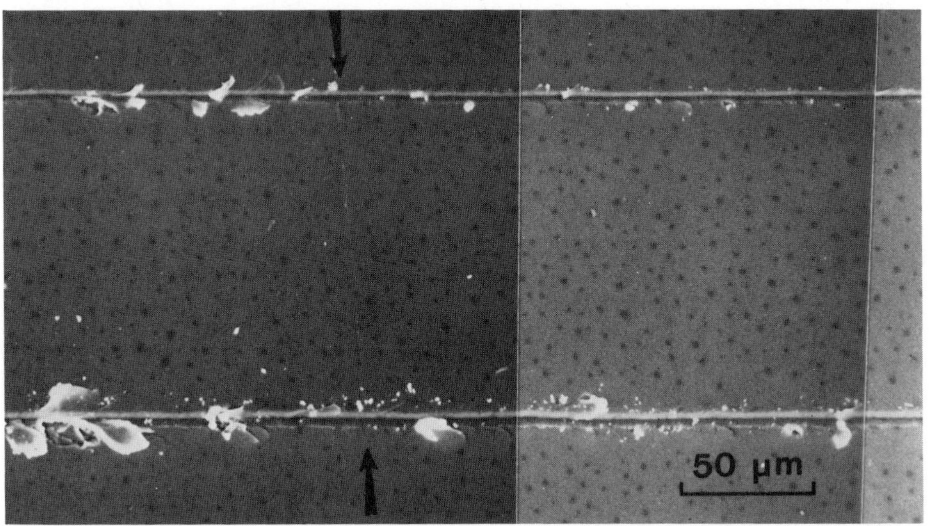

Fig. 4. Diamond scratches on an alumina single crystal. The implantation interface is marked by arrows. (Implanted region to the right of the interface.) The scratches show a lessened tendency to fracture in the implanted region. Normal force applied to the stylus in the upper scratch was 0.29 N; lower scratch, 0.49 N.

The scratch in the region of the implantation interface in polycrystalline silicon carbide is shown in Fig. 5. In this figure, the implanted region is at the right of the interface, indicated by arrows. A significant diminution in cracking is evident as the groove is traced from the unimplanted to the implanted surface. In addition, the groove widens in the implanted surface, and the groove edge shows distinct evidence of increased inelastic deformation. Tangential force increases of the order of 20 to 40% are experienced in traversing from the unimplanted to the implanted region on this surface. Figure 6 shows the nature of the groove formed in the implanted and unimplanted portions of a silicon carbide single crystal. In the unimplanted region, the groove edge contains fractures, as does the base of the groove. In the implanted region (to the right of the interface) fracturing in and adjacent to the groove rapidly disappears. At the end of the groove, the ledge which was formed at the upper side of the groove, probably from a slight misalignment of the diamond indentor, has become a small lip and the debris adjoining the groove has the appearance of ductile chips (Fig. 7). The tangential force variation on crossing from the unimplanted to the implanted part of the single crystal is a decrease of the order of 20 to 30%. This decrease is in contrast to the increase in tangential force observed in the polycrystalline specimen. The tangential force readings for both single and polycrystalline silicon carbide are recorded in Table 1.

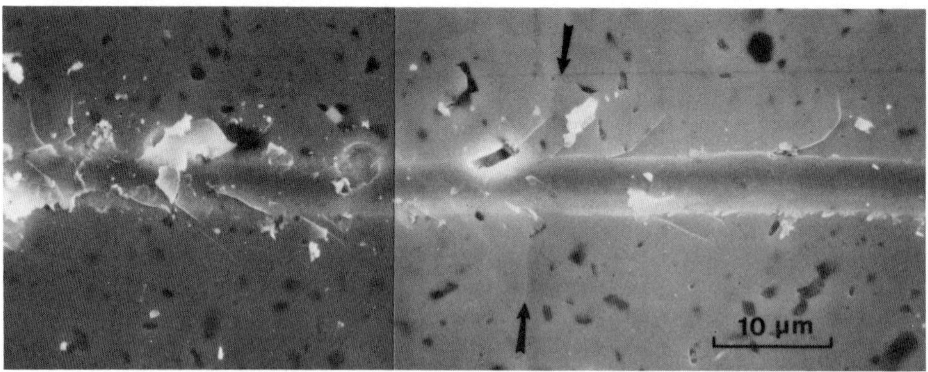

Fig. 5. Scratches in polycrystalline SiC; the implantation interface is indicated by arrows. The normal force on the stylus was 0.49 N. Fracture diminishes rapidly after crossing the interface and soon ceases. The groove also widens in the implanted region (right side) and shows evidence of inelastic deformation.

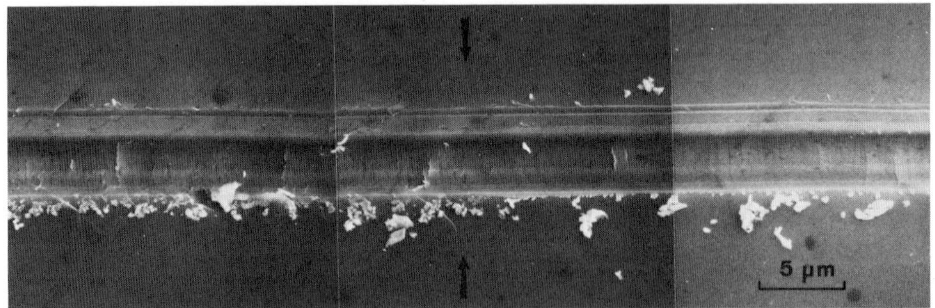

Fig. 6. Scratch formed in single crystal SiC. The implantation interface is located at the arrows; the region to the right of the interface is implanted with nickel. The force applied to the stylus in forming this scratch was 0.49 N.

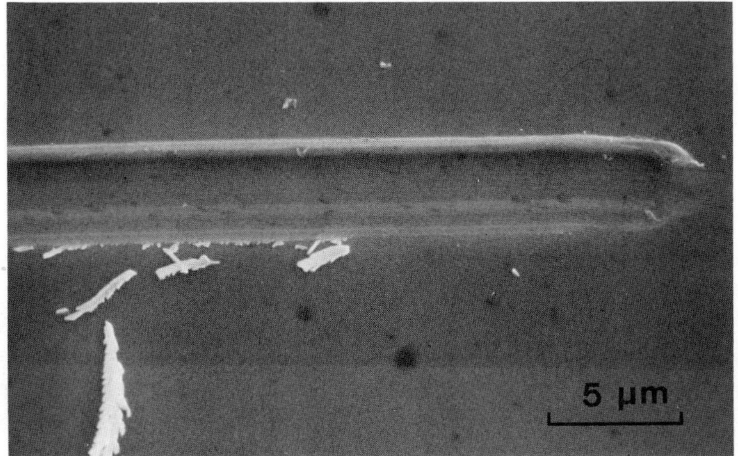

Fig. 7. End point of the scratch shown in Fig. 6, illustrating the deformed appearance of the groove edge and the formation of chip-like debris.

Table 1. Silicon Carbide Tangential Force Values

Normal force F_n $(N \times 10^2)$	Tangential force F_t $(N \times 10^2)$				Tangential force variation (%)	
	Unimplanted		Implanted			
	SC[a]	PC[b]	SC	PC	SC	PC
9.8	4.9	2.5	3.1	3.4	-37	+36
19.6	7.8	5.4	6.7	7.4	-14	+37
29.4	11.3	8.8	8.8	11.3	-31	+28
49.0	19.1	15.2	15.2	18.1	-20	+19

[a]Single crystal. [b]Polycrystal.

For titanium diboride implanted with nickel ions, the change in groove appearance on crossing the implantation interface is far less dramatic than that noted in silicon carbide. A careful examination of the groove on either side of the interface (Fig. 8), however, suggests that somewhat lessened fracture damage is incurred in the ion-implanted region. The tangential force reading did not change on traversing from the unimplanted to the implanted surface, but profilometer measurements of the grooves showed a decrease in the cross section.

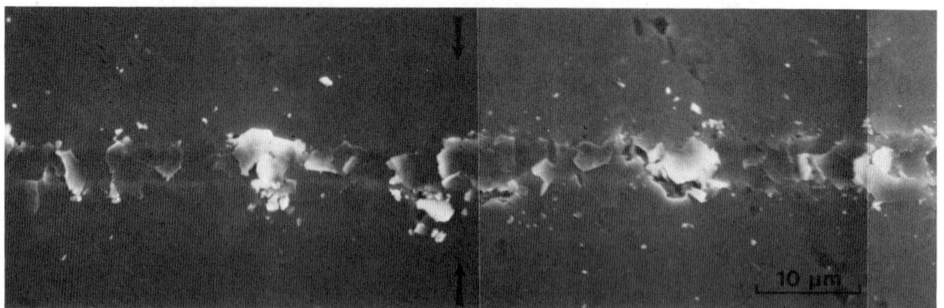

Fig. 8. Diamond scratch on titanium diboride surface (implantation interface at arrows; implanted region at left of interface). A lessened degree of fracture in the implanted region is suggested by this photo.

DISCUSSION OF RESULTS

The recovery of lattice damage with annealing, as evidenced by hardness variations (Fig. 3), can be related to recovery effects detected by RBS results. McHargue et al. have discussed the lattice recovery process indicated by RBS curves recorded after successive annealing treatments.[8] The RBS curves indicate the onset of recovery of the aluminum sublattice damage at 800°C, corresponding to the first indication of reduction in hardness for implanted alumina. At 1000°C, where further reductions in hardness are observed, the RBS curves indicate that lattice damage in the oxygen sublattice is recovering. For chromium-implanted alumina, further reductions in hardness are not encountered until the annealing temperature reaches 1500°C, at which point the RBS results indicate that chromium atoms have become substitutional rather than interstitial and are beginning to diffuse from the implantation zone.

In the case of titanium implanted into alumina, the initial softening with annealing temperature up to 1000°C is observed, again a consequence of the aluminum and oxygen sublattice recovery. Above 1000°C, however, titanium migrates toward the specimen surface and forms a precipitate, the increase in hardness corresponding to the formation of a precipitate hardened surface. Annealing of the zirconium-implanted alumina results in a gradual decrease in hardness, while the RBS curve remains unchanged up to annealing treatments at 1500°C. This result is attributed to an early precipitation of a stable, zirconium-rich precipitate which maintains the hardness of the implanted layer.

The scratch response of the alumina specimens tested by this method can also be related to the lattice condition as indicated by the several tests applied to the material. In crossing the implantation boundary, the stylus moves from a relatively defect-free annealed region into the implanted zone containing lattice defects and a concentration of implanted ions. The implanted chromium ions are found as a partially interstitial layer distributed from about 1000 to 2000 Å beneath the surface [see Fig. 4(a), ref. 6]. The hardness of the implanted region was found to increase to 50% greater than the annealed lattice. The fracture toughness was increased slightly. The stylus moving from the unimplanted lattice into the implanted, damaged material encounters a lattice in which plastic deformation becomes more difficult and requires higher shear stresses. The lattice remains crystalline, and the slip system arrangement with respect to the stress field accompanying the moving stylus remains the same, but the operation of the slip systems is hampered by the defected lattice. Accordingly, the observed tangential force required to form the groove increases. The fracture toughness increase of 15 to 20% in the implanted region is consistent with the observed slight decrease in fracture along the scratch.

Silicon carbide was implanted and scratched in both the single and polycrystalline form. The RBS analysis showed that the implanted regions became amorphous, and the hardness was found to decrease. The scratch response was dramatic in the case of these specimens, especially the observed scratch appearance. The fracture pattern commonly associated with scratch formation in the ceramic was supplanted by a crack-free, inelastically deformed groove in the implanted region. The tangential force values, collected in Table 1, were also observed to decrease at the implantation boundary in the single crystal, but to increase in the polycrystalline case. The tabulated tangential values indicate that groove formation in the implanted silicon carbide layer at low normal forces requires about the same tangential force in both the polycrystal and the single crystal. As the normal force on the indentor is increased, however, and more of the underlying bulk material is included in the groove formation process, the tangential force values diverge. The increasing influence of the underlying bulk is also seen in the polycrystalline case in the diminishing magnitude of the tangential force variation at the implantation boundary. A comparable diminution in tangential force variation is observed for the single crystal case, although the data are less regular.

Another instructive comparison is that of the tangential force values for the unimplanted regions of both specimen types. At all indentor loads, the tangential force is greater in the single crystal tests than in the polycrystalline tests. The tangential force changes for the single and polycrystalline forms, negative in the former and positive in the latter case, may therefore be related to the fact that with single crystal specimens only one lattice orientation is presented to the moving indentor while in the polycrystalline body many lattice orientations are available at any instant to respond to the applied stresses. Although specific mechanisms are not clear, the data suggest that grain size and lattice orientation (i.e., polycrystal versus single crystal) may be factors in determining the magnitude of the implantation boundary tangential force change.

For titanium diboride specimens, the lattice hardness was found to increase markedly (up to 2x), but the implanted region remains crystalline. The recorded tangential force is unchanged in moving from unimplanted to implanted surface, and the scratch appearance does not change significantly, although at the applied force level of 0.29 N, there is an indication of a decrease in fracture in the implanted region. At the 0.49 N level, there is significant fracture associated with the scratch in both implanted and unimplanted regions. Profilometer measurements of the scratch cross sections consistently show the implanted region grooves to have smaller cross sections. Consequently, if a constant groove cross section were maintained on crossing the implantation boundary in TiB_2, the tangential force would be expected to increase. In TiB_2, plastic deformation has been observed to be associated with the motion of a blunt stylus on the surface. The motion of the sharp diamond indentor as a stylus will likely also result in some subsurface plastic deformation as well as fracture. In the irradiated TiB_2, deformation and fracture are resisted in the implanted layer, but for those loads which impose stress systems significantly greater in extent than the implantation layer thickness, subsurface deformation and fracture may be initiated in regions beneath the implantation zone, the deformation and/or fracture subsequently being transmitted into the surface layers. Up to a normal force of 0.29 N the hardened surface layer results in a decrease in surface damage with scratching, whereas at the 0.49 N applied force the damage apparently initiates below the hardened surface, and a relative increase in surface disruption occurs.

SUMMARY

Ion implanted alumina and titanium diboride retain a crystallographic structure in the treated surface and become harder. Silicon carbide, in contrast, becomes amorphous in the implanted region and becomes less hard. Hardening of the crystallographic regions is associated with lattice damage created by the impinging ions and the presence of cations in interstitial lattice sites. Annealing at successively higher temperatures results in damage recovery and softening, except where precipitation reactions retain hardness, as in the case of zirconium and titanium implanted into alumina. The implantations modify the mechanical response of the surface to mechanical contact, as demonstrated by scratch tests.

ACKNOWLEDGMENTS

The authors acknowledge the support of the Division of Materials Sciences, U.S. Department of Energy (contract W-7405-eng-26). Appreciation is also expressed to B. C. Leslie for metallographic support and M. B. Lewis and J. M. Williams for ion implantations.

REFERENCES

1. D. A. Thompson, Rad. Effects, 56, 105-150 (1981).
2. S. M. Myers, Nuc. Inst. and Methods, 168, 26-274 (1980).
3. A. G. Evans, Fracture Mechanics Applied to Brittle Materials, edited by S. W. Frieman, American Society for Testing and Materials, Philadelphia, PA, 1979.
4. R. H. Marion, Fracture Mechanics Applied to Brittle Materials, edited by S. W. Frieman, American Society for Testing and Materials, Philadelphia, PA, 1979.
5. R. C. Bradt, J. Amer. Ceram. Soc., 50, 54 (1967).
6. B. B. Ghate, W. C. Smith, C. H. Kim, D. P. H. Hasselman, and G. E. Kane, Ceram. Bull., 54, 210 (1975).
7. C. J. McHargue, H. Naromoto, B. R. Appleton, C. W. White, and J. M. Williams, Metastable Materials Formation by Ion Implantation, edited by S. T. Picraux and W. J. Choyke, Elsevier Science Publishing Company, New York, 1982.
8. C. J. McHargue, H. Naramoto, C. W. White, J. M. Williams, and B. R. Appleton, Proceedings of the Nineteenth University Conference on Ceramic Science, November 8-10, 1982.

DISCUSSION

G. E. Gazza (AMMRC): (a) What indenter load was used for the fracture toughness measurements? (b) Was the same indenter load used for the microhardness comparison between conventional Al_2O_3-Cr_2O_3 material and the implanted material?

Author: (a) Two loads were used. One was 50 g and the other was 100 g. (b) Yes. Either 25 or 50 g loads were used.

H. Palmour III (NCSU): (a) Have loading rate sensitivities been observed in indentation or scratch tests? (b) If so, might they help in delineating shifts in mechanisms of deformation/cracking due to the ion implantations?

Author: (a) No. We have limited our experiments to date to a constant loading rate. (b) Perhaps so, but we plan to examine other variables as well, including the orientation of indentations and scratches with respect to the lattice orientation as a probe of deformation and fracture mechanisms.

C. Weissmantel (T. H. Karl-Marx-St.): The interpretation should consider high density cascade effects, as recently described by Thompson, because such effects are likely to occur in covalent solids.

Author: Yes.

E. K. Beauchamp (Sandia National Labs): I would expect the ion implantation process to produce a planar compressive stress. This would strongly influence the response in hardness, scratch, and indentation K_{Ic} measurements. Did you take this stress into account in your measurements.

Author: Yust said that they had not measured the stress, McHargue said that distortion of thin sections showed that stress existed.

R. Rice (NRL): Have you investigated Si_3N_4. I recall hearing of some data on this from Oak Ridge that showed little improvement in hardness upon implantation. This is consistent with initial experiments at NRL implanting hot pressed Si_3N_4 (NC-132) with no definite increase in hardness.

Author: Yes. We have implanted Si_3N_4 at levels of 2 to 3 x 10^{16} ions cm^{-2} of chromium and zirconium. The optical appearance, optical metallography, and the form of the scratches is similar to that of implanted SiC. This suggests that Si_3N_4 may also be amorphous. TEM is in progress to verify the nature of the surface. The knoop microhardness showed a decrease of about 10 per cent, which is also consistent with the SiC result.

MICROHARDNESS OF N-IMPLANTED YTTRIA STABILIZED ZrO_2

J. K. Cochran, K. O. Legg, and G. R. Baldau

Georgia Institute of Technology
Atlanta, Georgia 30332

ABSTRACT

Cubic ZrO_2 stabilized with 9.8 m/o yttria was implanted with 50 keV nitrogen ions at dose levels 10^{15} to 3×10^{17} cm^{-2}. Microhardness measurements indicated hardness increases up to 3×10^{16} N^+ cm^{-2} at low indenter loads for unannealed crystals but no hardness change occurred for annealed samples. At $f=3 \times 10^{16}$ N^+ cm^{-2}, a critical dose was exceeded resulting in softening of both annealed and unannealed films. Exceeding this critical dose resulted in shearing of the implanted film due to compressive strain, as evidenced by blistering and chipping at indentation edges. The sheared film thickness was 1300 Å. Sputter Auger profiles indicated constant N concentration from 150 to 1300 Å rather than the expected Gaussian distribution. Stress induced diffusion was thought responsible for the square profile.

INTRODUCTION

Ion implantation of the surface of ceramic materials is being investigated at an accelerating rate to determine the effects on mechanical, physical, chemical, and electrical properties. The possibility that ion implantation might significantly enhance the wear resistance of ceramic materials comes at a time when Al_2O_3 linings for abrasive materials transfer systems is developing into a major industry. Also, the emergence of partially stabilized zirconia as a viable material for internal combustion engines could be augmented if even low levels of improvement could be produced by implantation. Dearnaley[1] observed improvements in wear and friction characteristics of WC/Co composites and steels implanted with 4×10^{17} N_2^+ cm^{-2}.

Considering observed implantation improvements and the current level of interest in zirconia-based ceramics, this work was initiated to characterize the effects of N_2^+ implantation in yttria stabilized zirconia single crystals. The fully stabilized cubic structure was chosen since it is the host lattice for high strength partially stabilized zirconia, is the electrolyte for oxygen ion sensors, and is being used for heating elements, as well as serving in a variety of tool, die, and bearing applications.

Hardness by microindentation has been reported for a variety of implanted brittle materials. Microhardness of silicon carbide and silicon single crystals did not change until a critical dose of $\sim 4 \times 10^{17}$ N_2^+ cm^{-2} was exceeded.[2] Above this critical dose level, softening of the implanted layer occurred, resulting in an approximate 25% hardness reduction. Alumina implanted with 10^{16} -10^{17} ions cm^{-2} of Cr and Zr exhibited $\sim$ 40% increases in hardness and fracture toughness increased 10-15%.[3] When Al_2O_3 was implanted with Fe^+, microhardness increased 15%.[4]

Ion implantation depths are of the order of 1000 to 5000 Å and as the implanted ion comes to rest by a series of collisions with the host lattice, the implanted film may or may not exhibit crystalline structure modification. Naguib and Kelly[5] predicted implantation-induced structural changes for a large number of non-metallic solids. Zirconia (ZrO_2) was predicted to remain crystalline under irradiation but the cubic form was indicated rather than monoclinic. When amorphous ZrO_2 was irradiated, crystallization to a cubic structure did occur[6] but it was later shown that the crystallization resulted from thermal heating rather than from individual ion collisions.[7]

EXPERIMENTAL PROCEDURE

Single crystals of skull-grown, cubic zirconia fully stabilized with 9.8 m/o yttria were obtained from Ceres Corp. and sliced approximately perpendicular to the <110> axis. Surfaces were ground on a series of finer SiC grits and final polished with 1 µm diamond. Samples were irradiated as polished, "unannealed," and also after the crystals were annealed in air at 1200°C for 6 hours and cooled to room temperature at 50°C/hour.

Implants were made with a 100 keV mass analyzed beam of N_2^+, scanned to assure uniformity. Average beam current was maintained below 20 µA cm^{-2} for the higher doses and below 5 µA cm^{-2} for the lowest doses. Ambient pressure was below 2×10^{-6} Torr during implantation which prevented carbon build-up.

Hardness measurements using a Reichart microhardness tester equipped with a Knoop profile indenter were made over a load range of 10-100 grams. This produced indentation lengths ranging from 9 to

36 μm corresponding to penetration depths at the indenter tip of 0.3 to 1.2 μm. Each hardness data point represented three measurements on annealed samples and two for unannealed. Error was reported as mean deviation. Selected indentations were observed in a field emission SEM on surfaces coated with 300 Å of AuPd to prevent charging.

Sputter Auger profiles were taken with a 3 keV Ar^+ beam and a 1.5 keV electron beam to eliminate charging. Approximate concentrations of the atomic species were measured from the peak heights using sensitivity factors derived from standard data.[8] Multiple beam interferometry was used to determine the depth of the sputter pit which provided absolute depth calibration.

RESULTS

Microhardness

Microindentation has been used to measure implant hardness effects because at low loads the implanted layer is a significant fraction of the indented volume. As indenter depth increases with load, hardness approaching bulk values is expected. Unannealed zirconia exhibited decreased hardness at the lowest N^+ dose and increased thereafter up to a dose of 3×10^{16} N^+ cm^{-2}, Fig. 1. In this range, hardness at low loads was greater than at high loads indicating that the implanted film was harder than bulk zirconia. Above a "critical" dose of 3×10^{16} N^+ cm^{-2}, the hardness decreased to 10.8 $\pm$ 1 GPa regardless of load. Annealing resulted in a general softening of unimplanted and implanted samples, Fig. 2. Up to dose levels of 10^{16} N^+ cm^{-2}, neither implantation nor load appeared to have hardness effects with all values being 12.5 $\pm$ 1.3 GPa except for two 10 g load values. Between 10^{16} and 3×10^{16} $N^{\mp}$ cm^{-2}, critical dose was exceeded and again softening occurred irrespective of load. Here values were 9.8 $\pm$ 1.3 GPa or a decrease of ~ 22%.

SEM observation of indentations showed well formed, crisp geometries for all samples below the critical dose, Fig. 3. However, for all samples exceeding critical dose, the implanted surface "blistered" and the indentation edges were chipped revealing a partially separated uniform film, Fig. 4. Measurements from micrographs indicated a film thickness between 1200 and 1400 Å, Fig. 5. Once blistering occurred, the number of blisters cm^{-2} increased with dose but the blister diameter remained relatively constant. For the unannealed samples, most blisters cracked at 3×10^{17} N^+ cm^{-2}, Fig. 6, but no blister cracking was observed on annealed samples. This indicates the unannealed implanted film was more brittle than the annealed film. Further evidence for a more ductile annealed film is provided by comparing the deformed indentation edges for unannealed, Fig. 6, and annealed, Fig. 7, samples. In the unannealed state, edges chipped more and blistered close to the edge fractured. The annealed film appeared to be extruded by the indenter, possibly sliding at the failed interface between the bulk material and indentation edges appeared more plastically deformed. Finally, it should be noted that blisters in Fig. 7 are smaller and less numerous indicating that the critical dose has just been exceeded at 3×10^{16} N^+ cm^{-2}. For the unannealed samples, critical dose had not been reached at 3×10^{16}. Thus, critical dose, i.e., the implantation level needed to compressively fracture the film, must be $\sim 3 \times 10^{16}$ N^+ cm^{-2}.

N Depth Profile

From the typical Auger spectrum shown in Fig. 8, no major contaminants were present. (The argon and carbon signals both derived from the sputtering ambient.) Auger analysis indicated the oxygen to zirconium ratio was unchanged in the implanted region, showing that this region did not become anion deficient. The N depth profile of Fig. 9 showed an unusual implant distribution. The LSS theory would predict a Gaussian N distribution as shown in Fig. 9. In reality, sputtering would be expected to make the peak lower, broader and closer to the surface with a lower retained dose. However, the profile observed was flat from 150 Å to 1300 Å with a nitrogen concentration of 13.0 a/o (10^{22} N atoms cm^{-2}). The total retained dose (measured as the integral beneath this curve) was $\sim 1.6 \times 10^{17}$ N atom cm^{-2}, which was low but a reasonable proportion of the incident dose.

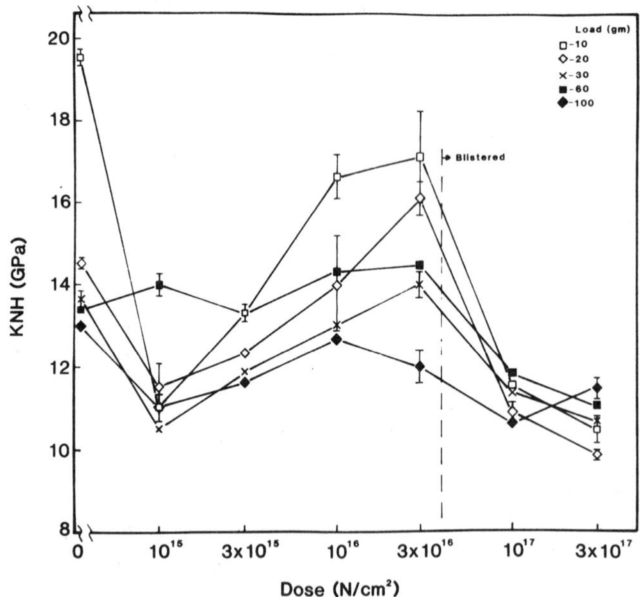

Fig. 1. Effect of 50 keV N_2^+ implantation on Knoop's microhardness of diamond polished yttria stabilized zirconia.

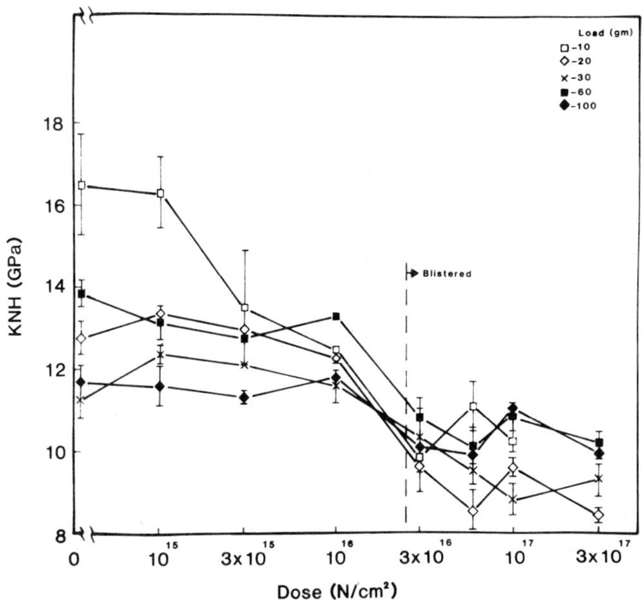

Fig. 2. Effect of 50 keV N_2^+ implantation on Knoop's microhardness of polished and annealed yttria stabilized zirconia.

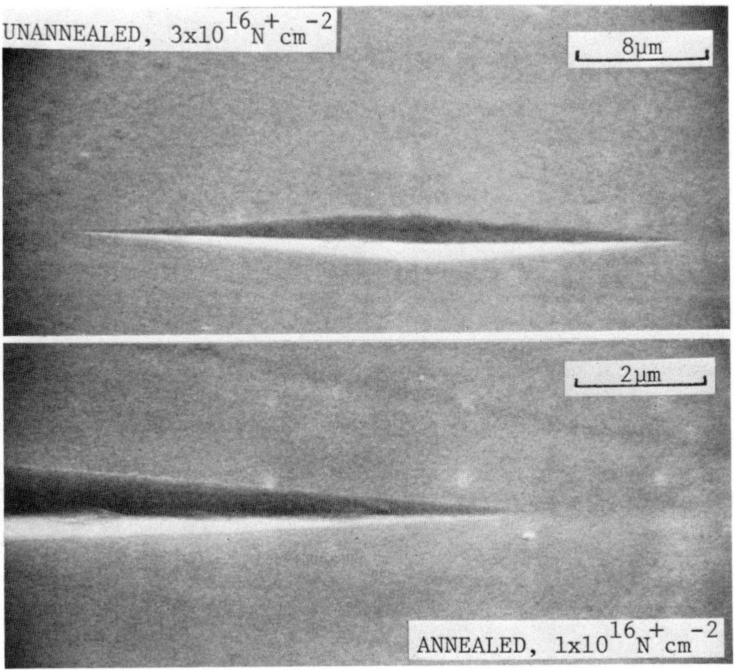

Fig. 3. Indentations on N implanted cubic ZrO_2 from 100 g loads.

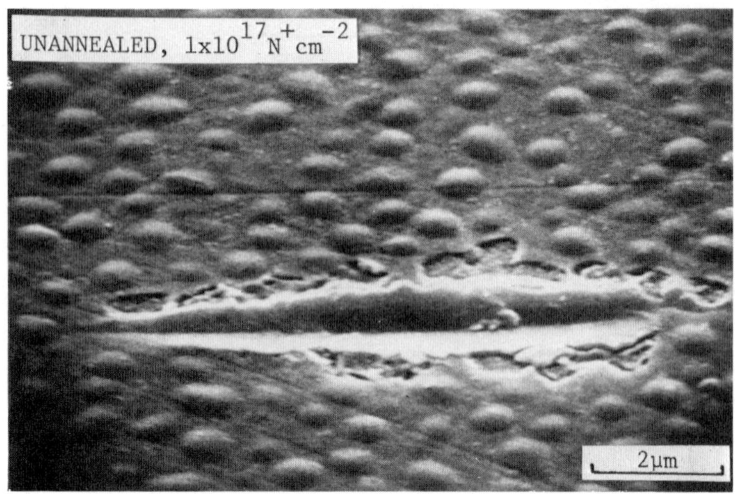

Fig. 4. Partially sheared N implanted film above critical dose showing blisters and fracture at indentation edges.

Fig. 5. Fractured blister showing N implanted film thickness.

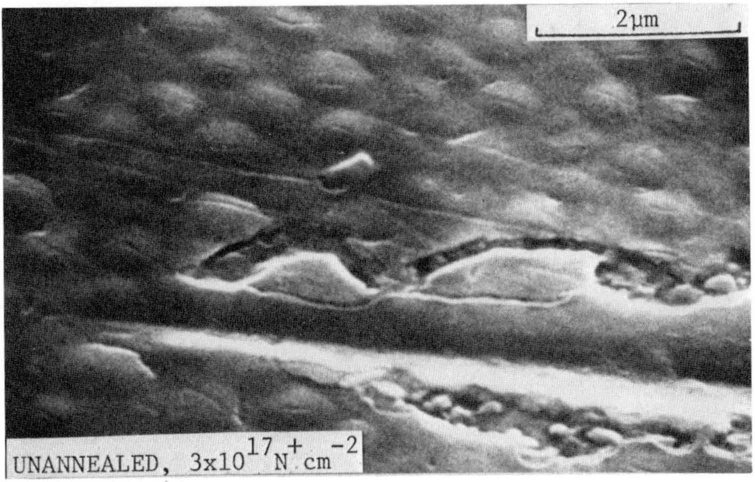

Fig. 6. Unannealed N implanted film showing blister cracking and brittle failure at indentation edge.

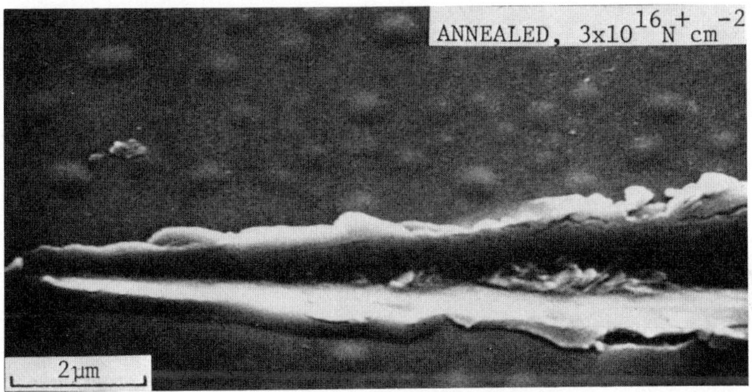

Fig. 7. Annealed N implanted surface showing onset of blistering and plastic failure at indentation edge.

Since the measurement technique is not accurate to better than about 50%, the retained dose may well be somewhat higher. Furthermore, substantial diffusion of the implant into the bulk may have occurred, reducing the surface concentration. The square profile was indicative of considerable redistribution of the implant, perhaps by stress-enhanced diffusion. Assuming a covalent radius[9] of 0.7 Å, the volume strain from this measured N concentration was crudely estimated as 1.3%.

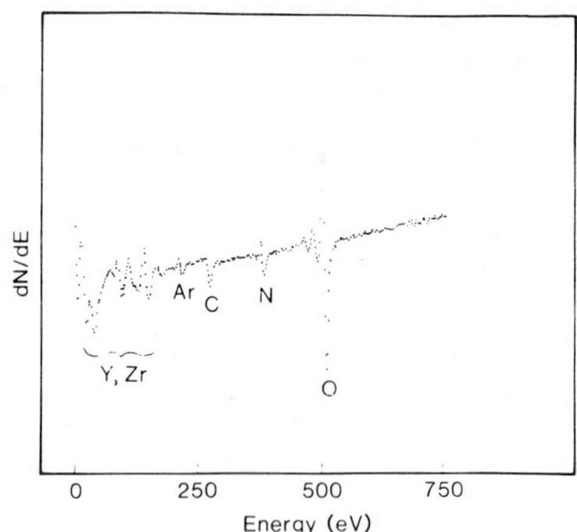

Fig. 8. Auger spectrum for yttria stabilized zirconia implanted with 3 x 10^{17} N^+ cm^{-2}.

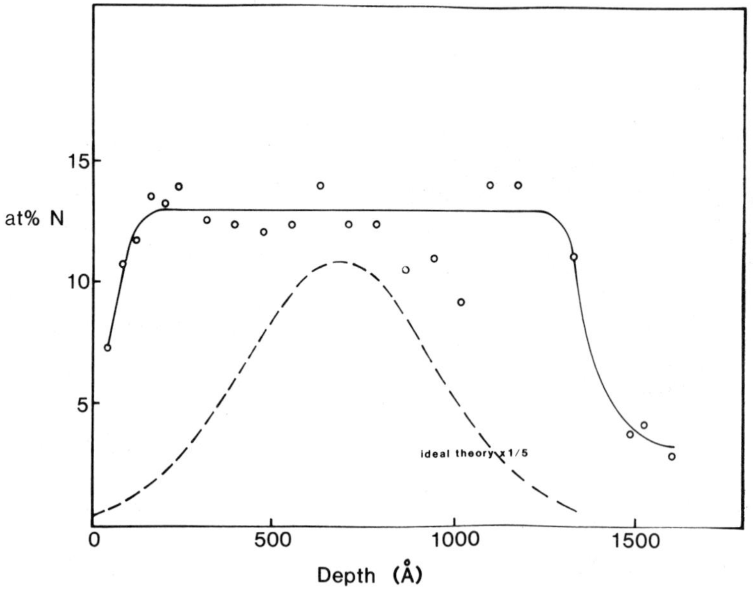

Fig. 9. Sputter auger profile for yttria stabilized zirconia implanted with 3 x 10^{17} N^+ cm^{-2}.

Multiple beam interferograms of unannealed samples showed a clear upward expansion of the surface in the blistered regions of 430 Å and 740 Å in the areas containing 10^{17} and 3 x 10^{17} N atoms cm^{-2}, respectively. Neither swelling nor sputtering could be detected in any other region. This swelling probably reflects the raising of the average surface position resulting from blistering.

DISCUSSION

From the micrographs of the blisters, the implanted films above critical dose appeared to have failed due to excessive compressive strain. It is surprising that the annealed samples did not show increased hardness if compression levels were sufficient to cause fracture. However, the unannealed films did show hardness increases at lower indentation penetrations as would be consistent with compressive stress. If all the N were in the 1300 Å film at the critical dose level of 3×10^{16} N^+ cm^{-2} and the N were covalent, isotropic volume strain would be ~ 0.41% or a linear strain of 0.14%. Considering the assumptions made, implant-induced strain levels sufficient for fracture were reasonable. Large strain levels were also consistent with the square N depth profile which could be produced by stress-induced diffusion.

As a final note, it was noticed that the higher implant dose films were darkened, i.e., the visible spectrum absorption had increased. Also, it was evident during SEM observation that the 3×10^{17} N^+ cm^{-2} films charged less, and thus were more electrically conductive than other areas. Consequently, the resistance of the unannealed 3×10^{17} N^+ cm^{-2} film was measured by four point probe. Assuming all the conduction was in the 1300 Å film, the implanted region had a resistivity of 60 Ωcm.

REFERENCES

1. G. Dearnaley, Mater. Eng. Applic., 1, 28 (1978).
2. S. G. Roberts and T. F. Page, in Ion Implantation into Metals, edited by V. Ashworth, W. A. Grant, and R. P. M. Proctor, Pergamon Press, 1982.
3. C. J. McHargue, H. Naramotor, B. R. Appleton, C. W. White, and J. M. Williams, in Metastable Materials Formation by Ion Implantation, edited by S. T. Picraux and W. J. Choyke, Elsevier Science Publishing Co., Inc., 1982.
4. M. B. Lewis and C. J. McHargue, in Metastable Materials Formation by Ion Implantation, edited by S. T. Picraux and W. J. Choyke, Elsevier Science Publishing Co., Inc., 1982.
5. H. M. Naguib and R. Kelly, Radiation Effects, 25, 1 (1975).
6. H. M. Naguib and R. Kelly, J. Nuclear Mater., 35, 293 (1970).
7. J. Leteurtre and J. Soullard, Radiation Effects, 20, 175 (1973).
8. L. E. Davis, N. C. MacDonald, P. W. Palmberg, G. E. Riach, R. E. Weber, Handbook of Auger Electron Spectroscopy, Physical Electronics Industries, Eden Prairie, MN, 1976.
9. L. Pauling, The Nature of the Chemical Bond, Cornell University Press, 1972.

PART VII

HOT ISOSTATIC PRESSING

HOT ISOSTATIC PRESSING OF CERAMIC MATERIALS

R. R. Wills, M. C. Brockway, and L. G. McCoy

Battelle, Columbus Laboratories
505 King Avenue
Columbus, Ohio 43201

INTRODUCTION

In 1955 the concept of hot isostatic pressing (HIP) was invented at Battelle's Columbus Laboratories by Saller, Dayton, Paprocki and Hodge[1] as a means of diffusion bonding nuclear fuel elements. The principal reason for using a gas as the pressure transmitting medium was to effect bonding in three dimensions. Since many of the experimental fuel materials were powder products most of the early studies were performed with metallic matrix dispersion fuels, cermets and ceramics.[2]

The HIP process is now a relatively well known production process in the metallurgical industry but the ceramics industry is just beginning to become interested in the process. In this review paper the main elements of HIP equipment and the methods of HIP processing of ceramics are briefly described. The potential for improved material properties is also discussed with reference to the available technical data. It should, however, be noted that very little technical data exists in the open literature. Applications and likely future developments are also described.

Before discussing the equipment and methods of HIP processing it is pertinent to define the pressure range used in HIP in view of the number of fabrication processes employing pressure, for example gas pressure sintering. High pressure is defined here as above the typical 20.7-34.5 MPa (3,000-5,000 psi) range used in uniaxial hot pressing and below approximately 1035 MPa (150,000 psi). HIPing is usually performed in the range 69-207 MPa (10,000-30,000 psi). The general incentives for HIPing are given in Table 1.

Table 1. Incentives for HIP

• Superior Materials Properties
• Near Net Shape Capability
• Lower Cost
• Densification of Otherwise Unsinterable Materials
• Better Uniformity Within Parts and Part to Part

HIP EQUIPMENT

A HIP system is basically a pressure vessel containing an internal heater, as shown schematically in Fig. 1. Gas is pumped from storage through compressors into the pressure vessel. Power is supplied to the furnace through SCR-type devices and controlled on the basis of the inputs from the furnace and work-load thermocouples.

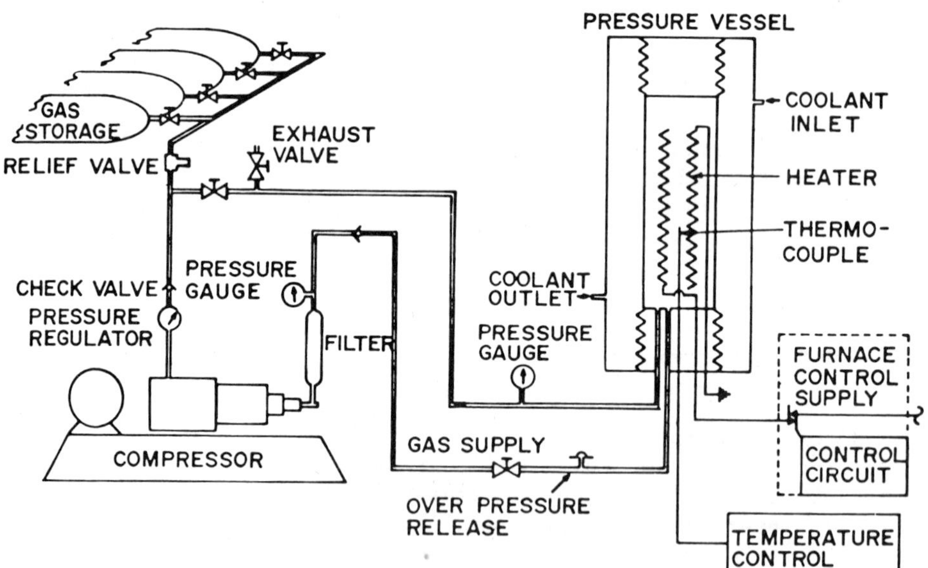

Fig. 1. Simplified schematic diagram for a hot-isostatic pressure system.

HIP systems today are of the cold-wall design where the internal resistance heater (furnace) is insulated from the wall of the vessel in some manner. The pressure vessel is penetrated, normally through the end closures, to provide power, instrumentation, and auxilliary services to the furnace. The pressure vessel is normally cooled to maintain an acceptable operating temperature. Temperature is measured by thermocouples placed throughout the volume of the furnace.

METHODS OF HIP PROCESSING CERAMIC MATERIALS

Selection of the appropriate method of HIPing is dependent primarily upon desired material properties, component geometry and acceptable manufacturing costs. There are five methods of HIP processing ceramics:

1. elevated temperature impregnation of porous ceramics;
2. diffusion bonding;
3. sinter + HIP;
4. HIP powder consolidation; and
5. hot isostatic forging (HIF).

ELEVATED TEMPERATURE IMPREGNATION OF POROUS CERAMICS

The equipment used for this mode of HIP processing is slightly different than that used in conventional HIPing. Basically the equipment (see Fig. 2) consists of a furnace with an isolation chamber enabling pressure to be applied while evolved gases are exhausted. This process has been applied to pitch impregnated graphite[3] preforms using temperatures of about 700°C and pressures of about 103.5 MPa (15,000 psi). During heatup the pitch becomes liquid and is forced into the small pores. At 400°C when pyrolysis occurs, hydrogen is liberated, and diffuses through the wall of the steel container holding the graphite preform. There is no indication that this type of HIP processing has as yet been applied to classical ceramics.

DIFFUSION BONDING

HIP was originally invented at Battelle's Columbus Laboratories as a diffusion bonding process. Metal-ceramic bonding using HIP has been achieved in numerous systems (for example, Cr-Al_2O_3, W-ThO_2, UO-Zircalloy 2, Rh-BeO, Nb-Al_2O_3, Nb-ZrO_2 and Ni-SiO_2)[4] but the application of HIP for ceramic-ceramic bonding has been limited to alumina[5] nuclear waste isolation containers in which HIP processing is used to bond the alumina lid to the large alumina container. The strength of the joint is reported to be similar to other parts of the wall.[5]

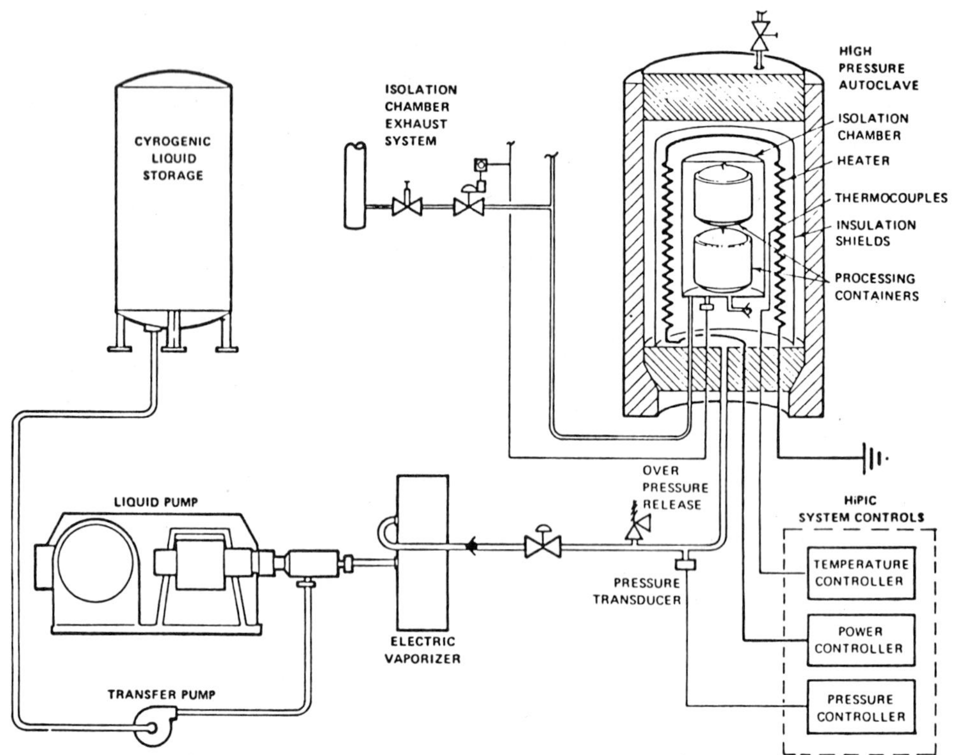

Fig. 2. Simplified schematic of a HIPIC system.

SINTER PLUS HIP

The prime objective of this mode of HIP processing is to remove the residual porosity in sintered ceramics. The microstructure of the ceramics must contain only closed porosity to ensure densification occurs during HIP. The presence of microcrack networks in parts of the samples will prevent complete densification if these are interconnected. Otherwise healing of individual microcracks of flaws is anticipated. This form of HIP processing is the most widely used method today for ceramics.

Sintered parts are easily handled and do not require canning. In contrast to the HIP powder consolidation process the packing density of components in the autoclave is significantly higher because (1) the "can" or encapsulant surrounding each powder compact occupies part of the available space, and (2) the green pressed component occupies a larger volume than the equivalent sintered part. There is however less control over the final microstructure and properties using this mode of HIP processing. For example surface defects will not heal during HIP.

In Fig. 3 the effects of HIP on selected properties of sintered ceramics are predicted. HIP is expected to be beneficial to density, strength, Weibull modulus and optical transmission. Although the technical data are not extensive these predictions are confirmed by the experimental results. For example, Engel and Hubner[6] increased the average bend strength of a cobalt bonded tungsten cermet from 1240 MNm^{-2} to 2330 MNm^{-2} (see Table 2) and increased its Weibull modulus from 8.4 to 14.8. These changes were due to removal of small pores and movement of the tungsten carbide grains during HIPing. It should be noted that this cermet fractures in a purely brittle manner at room temperature, and thus the data shows the improvements in strength and Weibull modulus that would be expected by HIPing a classical sintered ceramic. Improvements in the strengths of sintered silicon nitride,[7] silicon carbide,[7] and alumina-zirconia[8] composites have also been reported.

PROPERTY	EFFECT OF HIP
Density	Increased
Strength	Increased
Weibull Modulus	Increased
Thermal Expansion/Conductibity	Insignificant Effect
Hardness	Insignificant Effect
Fracture Toughness	Insignificant Effect
Creep Resistance	Insignificant Effect
Resistance to Slow Crack Growth	Insignificant Effect
Modulus of Elasticity	Insignificant Effect
Optical Transmission	Improved

Fig. 3. Predicted effects on HIP on selected properties of sintered ceramics.

HIP POWDER CONSOLIDATION

The elements of the HIP powder consolidation process are shown in Fig. 4. Ceramic powders are generally submicron in size and possess a low bulk density. Consequently to minimize the possibility of shape distortion during HIP the powder is first green formed by pressing, injection molding or another green forming method. The green formed component is then encapsulated in an encapsulant or "can" which acts as a pressure transmitting membrane during HIP. Metal or glass "cans" are generally used but the trend in ceramics has been to use glass "cans"primarily because of cost considerations. After HIPing the "can" must be removed either by mechanical or chemical means without damaging the component.

Table 2. Effects of HIP Processing on the Properties[6] of Sintered Cobalt Bonded Tungsten Carbide

Material Treatment	Bend Strength, MN/m^2	Weibull Modulus	Fracture Toughness, $MN/m^{3/2}$
As Sintered	1240	8.4	8.96
HIP 1260°C, 103 MPa, 1 hr	2540	8.8	8.99
HIP 1360°C, 103 MPa, 1 hr	2330	14.8	8.68

In the ceramics industry uniaxial hot pressing is a standard fabrication process. One must therefore ask the question "Why HIP?". There are three main reasons for considering HIP instead of uniaxial hot pressing: (1) better temperature control; (2) the application of pressure isostatically enables more complex shapes to be readily fabricated; and, (3) the higher pressures employed in HIP when used in conjunction with a knowledge of a microstructure-property relationships in ceramics facilitates the fabrication of materials with superior properties.

In contrast to the sinter + HIP fabrication method disadvantages are inherent in the lower packing density of components in the autoclave and the need to encapsulate all specimens. Preforms also require greater care in handling than sintered specimens. However the potential for achieving improved properties is substantially greater. For example, grain growth which readily occurs during sintering can be avoided. The use of an encapsulant also enables volatile species to be readily handled so that material properties and furnace components are not affected in an adverse manner.

HIP powder consolidation has received only moderate attention to date probably because of the lack of a commercial application and the need to form a reliable encapsulant. Early work by Hodge[2] described the fabrication of transparent ceramics and the avoidance of grain growth in HIPed alumina and beryllia. Carmichael[9] was the first to measure the properties of a HIPed ceramic. Barium titanate fabricated by the HIP powder consolidation technique was found to have a higher dielectric constant than if it was processed either by sintering or uniaxially hot pressing.

The most extensive data exist for silicon nitride. Wills et al.[10,11] have shown that it is possible to produce a silicon nitride ceramic with unusual microstructural features and properties superior to any other form of additive based silicon nitride. These properties include high temperature, strength, creep resistance, and subcritical crack growth rate.

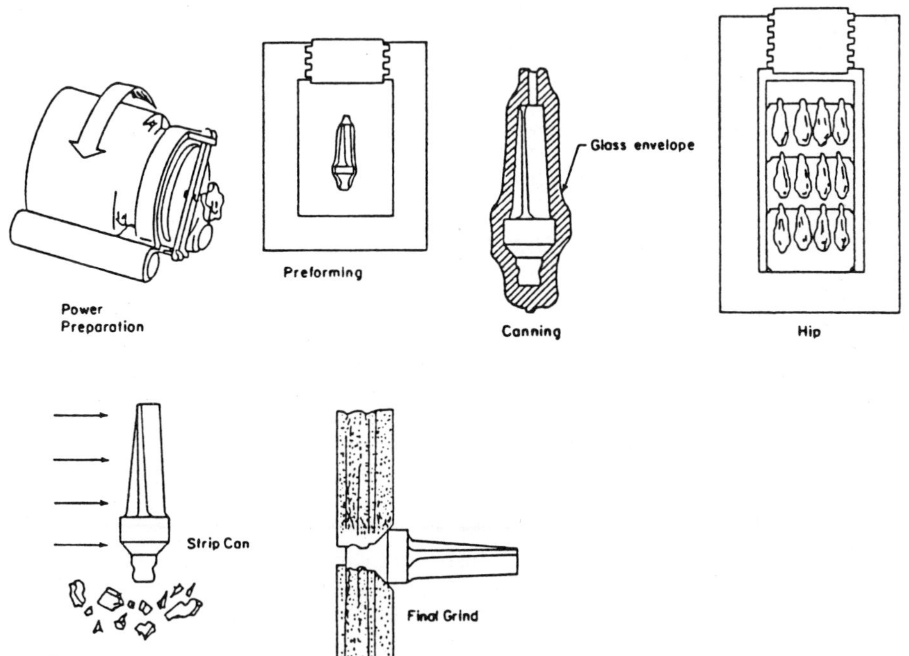

Fig. 4. HIP powder consolidation process.

Metal "cans" have traditionally been used as encapsulants for HIPing powder preforms but in recent years considerable attention has been directed towards glass "cans" primarily because of cost considerations. Much of the effort derives from the interest in HIPing structural ceramics. The candidate metals, tantalum and molybdenum, are relatively expensive but the use of glass "cans" imposes certain constraints on the HIP equipment since during the initial part of the cycle the equipment must operate at low pressures or under vacuum if a glass frit canning approach[12,13] is employed. Not all industrial equipment is capable of operating under these conditions.

The fabrication of complex ceramic shapes by HIPing powder preforms has received increasing attention in recent years because of the potential demand for nuclear waste containers and heat engine components. Silicon nitride ball bearings,[14] turbine blades,[11] stators,[14] and rotors[15] have been fabricated together with alumina containers[5] 2.5 meters long with an outside diameter of 0.5 metres. It appears that good dimensional control can be achieved during HIPing.

HOT ISOSTATIC FORGING (HIF)

Hot isostatic forging was first conceived as the HIP related process at Battelle's Columbus Laboratories by John Mueller in the late 1970s as a means of producing crack-free forged halide crystals. In essence HIF (see Fig. 5) is a forging operation performed in an autoclave so that during forging a gas pressure is applied to the surfaces outside as the platens so the crystal is effectively isostatically forged.

APPLICATIONS AND FUTURE DEVELOPMENTS

Figure 6 lists numerous existing and potential applications for HIPing ceramics. The list is probably incomplete, but it does indicate the tremendous potential for the application of HIP processing to ceramics. Existing commercial applications (cutting tools, drill bit inserts, multilayer capacitors and magnetic tape heads) employ only the sinter plus HIP processing mode. This is the simplest form of HIP processing and an excellent method of improving the quality of a ceramic. With the current concerns over quality and productivity all modes of HIP processing should find increasing usage. Large potential applications in heat engine ceramics and nuclear reactor technology exist, but the timing of these opportunities is difficult to quantify. Many more mundane applications are likely to emerge as the technical community gains a better understanding of the various modes of HIP processing and the cost-performance tradeoffs inherent in process selection.

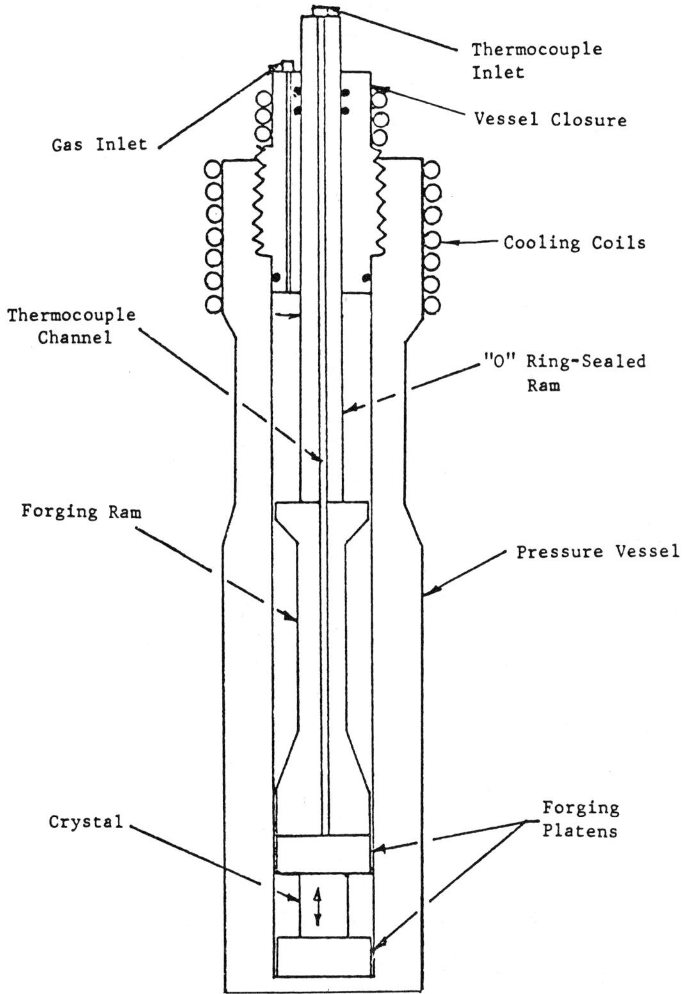

Fig. 5. Design of experimental hot isostatic forging vessel.

There is little doubt that interest in HIP of ceramics has increased sharply in the last 6 years. Much of this has been due to the strong interest in structural ceramics but progress in high temperature HIP equipment, "canning" technology and near-net shape technology have advanced it to an "off the shelf-technology" for many ceramics. In the next few years, developments (see Fig. 7) are likely to focus on wider application of the five modes of HIP processing to more ceramics, refinements in equipment and canning technology, and the definition of more applications. A considerable amount of development and engineering work still needs to be done, but the basic research is likely to have been largely completed within the next five years.

- Letdown Valves for Coal Liquefaction
- Turbine Disks
- Turbine Blades
- Turbine Vanes
- Sterling Heater Heads
- Cutting Tools
- Orthopedic Implants & Dental Ceramics
- Magnetic Tape Heads
- Transducers
- Laser Windows
- Sputtering Targets
- Drill-Bit Inserts
- Nuclear Waste Consolidation
- Nuclear Waste Containers
- Bearings
- Radomes & infrared Domes
- β-Alumina Tubes
- Nuclear Reactor Core Supports
- Fusion Reactor Insulators
- Special Dies
- Multilayer Capacitors
- Semiconductor Dopant Sources

Fig. 6. Existing and potential applications.

- Refinement of Equipment and Canning Technology
- Wider Application to Other Ceramics, Glasses and Composites
- Better Understanding of Dimensional Control Achievable in HIP Power Consolidation
- Greater Exploitation of Hot Isostatic Forging
- Enhanced Use of HIP for Joining
- Generation of Predictive Models for Processing, Properties and Dimensional Control

Fig. 7. Likely future developments.

REFERENCES

1. H. A. Saller, S. J. Paprocki, R. W. Dayton, and E. S. Hodge, U.S. Patent No. 687,842 (Pending) and Canadian Patent No. 680,160, February 18, 1964.
2. E. S. Hodge, Powder Metall., I [4], 168-201 (1964).
3. W. Chard, M. Conaway, and D. E. Niesz, in Petroleum Derived Carbons, edited by M. L. Deviney and T. M. O'Grady, American Chemical Society, Symp. 21, 1976.
4. Battelle Columbus Laboratories, unpublished data.
5. M. T. Larker, ASEA J., 54 [4], 85-90 (1981), ASAE Co., Vesters, Sweden.
6. U. Engel and H. Hubner, J. Mater. Sci., 13, 2003-12 (1978).
7. M. Moritoki, paper presented at the HIP Conference, Goreham International, Danvers, Mass., March 30-April 1, 1982.
8. F. F. Lange, paper presented at the 84th Annual Meeting of the American Ceramic Society, Cincinnati, OH (1982).
9. D. C. Carmichael, P. D. Ownby, and C. B. Boyer, report to U.S. Army Electronics Research and Development Laboratory, 1964.
10. R. R. Wills, L. G. McCoy, M. C. Brockway, and D. E. Niesz, Ceramic Engineering and Science Proceedings, I (7-8), (3), 534-39 (1980).
11. R. R. Wills and M. C. Brockway, Proceedings of British Ceramic Society, Special Ceramics, 7 [31], 233-47 (1981).
12. J. Adlerborn and H. Larker, U.S. Patent 4,112,143, 1978.
13. W. Hintler, K. Schweitzer, and A. Rossman, U.S. Patent 4,242,294, 1980.
14. R. R. Wills, paper presented at the HIP Conference, Goreham International Meeting, Danvers, Mass., March 30-April 1, 1982.
15. M. Larker, J. Adlerborn, and H. Bohman, paper presented at the Society of Automotive Engineers Meeting, Detroit, 1977.

DISCUSSION

R. Rice (NRL): Will you discuss the issue of outgassing. While many metals and non-oxide ceramics have solid products from reaction with atmosphere species some powders such as oxides and halides often have substantial outgassing problems. MgO and CaF_2 (or BaF_2?) are specific examples of this. There one has the limitation that finer powders for easier consolidation, finer grain size, etc. have more outgassing problems. On the other hand, one may be able to overcome some of these problems by sintering--then HIPing?

Author: As I have discussed, there are several methods of HIP processing. Selection of the appropriate method is dependent upon the desired microstructure. Sinter plus HIP is one way of avoiding outgassing problems, but this method will generally result in a microstructure exhibiting grain growth. If this is not acceptable grain growth inhibitors must be added to the powder or the HIP powder consolidation technique used. Outgassing powders is not a general problem, but for certain powders outgassing requires several hours.

R. Arons (Celanese Research Co.): (Comment pertaining to R. Rice question about in-can outgassing problems.) In early work at Argonne National Laboratory on HIPing of nuclear waste ceramics such as SYNROC, we observed voids in the finished products which were caused by volatilization of cesium oxide. Subsequent microprobe work showed the voids to be lined with a layer of condensed cesium oxide. The message is, of course, to be cautious with the use of any potentially volatile component in HIP systems.

Author: The two main variables in HIP powder consolidation are temperature and pressure. These must be optimized to obtain any given density and microstructure. In the SYNROC example the material may have been HIPed at too high a temperature.

DENSE CERAMIC PARTS HOT PRESSED TO SHAPE BY HIP

Hans T. Larker

High Pressure Laboratory
ASEA AB
S 915 00 Robertsfors, Sweden

ABSTRACT

An alternative way to carry out encapsulation and HIPing can produce fully dense and accurately shaped parts. Only areas of particularly narrow tolerances have to be machined after HIP. The process can therefore be developed for economically competitive mass production. The method uses a green powder body to determine a final shape of the part. This is in contrast to the prevailing technique to fill a powder into a glass or metal container and allow the interior shape of the latter to determine the shape of the HIPed body. In the new method injection molding is often found to be suitable for the fabrication of the green body. A temporary containment of glass taking the shape of the green body is used to prevent the gaseous pressure medium from penetrating the porous body during the HIP process.

INTRODUCTION

Hot isostatic pressing (HIP) has many features which should be valuable in the production of components of advanced ceramic materials. As in uniaxial hot pressing a virtually pore free product can be produced at a relatively low temperature. As the pressure level used in the HIP process typically is 100-300 MPa to compare with 30-50 MPa in uniaxial hot pressing and because the isostatic mode of application of pressure is generally more efficient than the uniaxial one, this effect is even more pronounced.

High purity ceramic powders which are impossible to densify by normal sintering or even by hot pressing can be well densified by HIP.

The amount of sintering aids can be reduced to a minimum if desired. As the temperature generally can be kept at a lower level also the amount of grain growth inhibitants can be reduced. The use of a gas tight container or surface layer around the ceramic body during the HIP processing offers very efficient means to prevent dissociation or decomposition of the ceramic material. As the outside gas pressure on this container is very high even materials with high partial pressures of the species formed upon decomposition can be successfully processed.

The total cost of a finished part fulfilling the requirements in a given application will however always determine the commercial success of a certain fabrication process. HIP has in addition to the ability of producing high performance materials also the inherent feature to make parts to net or near net shape.

THE GAS IMPERVIOUS SURFACE BARRIER

A gas tight container or a gas tight surface layer has to completely surround the porous body during the hot isostatic pressing (unless only closed porosity is present in the body). Such a surface barrier will, e.g., in the case of silicon nitride, prevent dissociation and weight loss upon sintering. Silicon nitride has a nitrogen partial pressure of one bar at about 1870°C. As the nitrogen gas cannot escape through the surface barrier during HIPing an automatic adjustment of the nitrogen partial pressure inside the powder body to the equilibrium pressure will take place and no further dissociation occurs.

It is, however, important to realize that processing methods, which require that gaseous species leave the body after the surface barrier has been sealed, will not be successful. For example, processes like deoxidation of SiC by excess C forming volatile CO must be carried out before the body is isolated from the environment by the gas tight surface barrier.

A very clear demonstration of the isolation of the powder body from its environment during HIPing was experienced during our work on large canisters of a-Al_2O_3. The water content of the alumina powder in the mild steel container had to be reduced to below 200 ppm to avoid exaggreated grain growth during the HIP process at 1350°C.[1] With a content of around 1000 ppm confined water an abnormal grain growth up to a grain growth rate of 100-200 μm/h was fond. A bend strength reduction from 500 MPa to 75 MPa was experienced. This example demonstrates that a thorough knowledge of the process often is needed for the best results from HIP processing. However, generally the good control of the process offered by the presence of the gas impermeable barrier is of great advantage.

GENERIC HIP METHOD

The development work on HIP of silicon nitride began at ASEA about 10 years ago and the first basic process patent was filed in 1974.[2] Already at this stage it was realized that the method to allow the powder body to determine the final shape of the part had a great potential for ceramic components. The realization of the principle which was published in 1977[3] demonstrates the method in a way which may be easy to understand. The green powder body was in this case formed by cold isostatic pressing followed by machining in the green state. It was placed in a container of high silica glass. The only requirement of the shape of the container is that its internal size must be large enough to receive the green body. The glass container with its content was outgassed and sealed by melting. It was then heated to such a temperature that the glass softened. Pressure was gradually applied with the result that the glass container folded over the green body and took its shape. The outside gas pressure and the temperature were then increased to the processing parameters, typically 1750°C and 200 MPa. After a desired dwell time, usually 0.5-2 hours, cooling and depressurization occurred. Remaining glass after cooling could be removed by sand blasting.

ENCAPSULATION USING GLASS POWDER

The process using glass ampoules can be convenient for material development and to make a limited number of parts of simple shape. It is, however, hardly suitable for further development to high volume production. Furthermore, the ability to make complex parts is limited. The high viscosity of the glass when it folds over the body will put too much stress on weak protruding parts of a green body with a complex shape. Another realization of the principle which gives a very good solution to this problem has however been developed. It is based on the use of glass powder covering all surfaces of the green body. The glass is sintered to an impermeable state before the high external pressure is applied. The basic technique was documented already in 1971[4] but it has of course been much further developed.

FACTORS INFLUENCING PRODUCT PRECISION

There are many factors which determine the final shape of a body densified from powder. Consider a powder body and designate the initial density relative to the theoretical density with D_0 and the relative final density D_f. The powder mass is considered to be homogeneous and isotropic. Furthermore, the powder mass is not in contact with any foreign material which can interfere with its shrinkage upon densification. If finally the temperature throughout the powder body is constant during the densification, the linear shrinkage can simply be calculated by the equation

$$f = \left(\frac{D_o}{D_f}\right)^{1/3}$$

where f is the relation between the final and the initial linear dimension of the part in any direction.

A glass encapsulation can be made in such a way that its influence on the shrinkage of the powder body can be neglected. Temperature gradients during densification have the effect that the areas at higher temperature densify before parts of lower temperature. In some favorable cases the effect of a temperature gradient during densification cannot be detected on the final part if the time for complete densification also of the parts initially of lower density is sufficient. This may be the case for parts of simple geometry but generally there is a remaining effect on the shape. A remedy is of course to keep the temperature gradient in the part sufficiently low when the high external gas pressure is applied.

Variations in the final density D_f, will of course directly affect the shrinkage and consequently the final dimensions. As D_f in a HIP process generally is very reproducible and close to one, this factor does not affect the dimensions after HIPing. This is favorable compared to other sintering techniques for which usually a variation of D_f occurs, which gives rise to additional variation of the product tolerances. The uniformity of the initial density of the powder body and its isotropy is of course important. One prefers, of course, a uniform density without any anisotropy, because the calculation of the suitable dimensions of the powder body to give a product with a certain final dimension is easiest. It is, however, in the HIP process acceptable if the density distribution and a possible anisotropy is the same from part to part in a production series. This can be compensated for by altering the shape of the green body.

One further factor which is very important in the making of precision parts of complicated shape is the distortion by gravity effects (sagging) during the densification process. The usual remedy is of course to support all parts of the green powder body very carefully in the sintering furnace. This is, however, not possible with a part like a turbine wheel consisting of a conical hub with protruding blades. In the HIP process the surface of the powder body is sealed and a high pressure difference between the outside and the interior of the porous powder body is prevailing. This gives the same effect as is observed in powders like ground coffee contained in an evacuated plastic bag. In the HIP process, the pressure difference is much higher and the relative effect of gravity forces on such a part as a horizontally protruding turbine blade during HIPing will be negligible.

GREEN BODY FORMING METHODS

Dry Methods or Low Binder Content

A process which is well known to give green powder bodies of uniform density is cold isostatic pressing (CIP). The CIP process can be used without or with very low binder contents. It has a correcting effect on, e.g., density variations in the filled powder but cannot fully eliminate such variations. A typical silicon nitride powder (H. C. Starck grade H 1) can without binder be densified to 66% of theoretical density (TD) at 1 GPa pressure. Also lower pressures can be used but it gives a weaker green powder body. A CIPed powder body can be machined by turning, milling or grinding with silicon carbide or alumina wheels. As the powder is very hard and abrasive the wear of the tool tips will, however, be rather high. This technique was used during our initial studies to find out if also fairly complex parts could be HIPed with the use of the glass powder encapsulation process (Fig. 1).

Fig. 1. As-HIPed parts made by the glass encapsulation process. Green powder bodies made by CIP without binder and subsequently turned or milled in the green state.

The study demonstrated that even relatively weak protruding sections and sharp corners could be made by the glass encapsulation process. CIP is hardly optimal for series production of parts of the type shown in Fig. 1. It could, however, be suitable for parts of relatively simple shape or which have an interior shape suitable for pressing on a mandrel. Dry bag presses followed by automated exterior grinding could be used (cf the machines used for spark plug insulators).

Dry pressing in rigid tools could be suitable for other relatively simple shapes, e.g., plates.

Wet Methods or High Binder Contents

Such methods are particularly suitable for parts of complicated or irregular shapes. Injection molding with a binder system and slip casting are important examples of this technique. Any method for forming a powder body which is highly reproducible from part to part can in principle be used. The requirement on the green strength is minimal. Any part that can support its own weight appears sufficiently strong to be used. The requirements seem to be very close to what is needed for normal sintering. Injection molding is particularly well suited for high volume production. The tooling is, however (particularly for complicated products), very expensive so a long series production may be needed to amortize it. Slip casting may be of interest for medium size production.

In the injection molding process as high percentage as possible for the ceramic material is desired. With a suitable binder system it has been found possible to injection mold masses containing up to 70 volume percent silicon nitride material. This can be compared with 66% of TD obtained during CIP at 1 GPa of unlubricated powder. It is, however, generally found suitable to reduce the amount of ceramic material to 64 to 66%. This gives a wider margin in the practical operation. It also allows a final adjustment of the size of the product simply by varying the amount of binder to ceramic material ratio.

With a suitable binder system a very uniform and homogeneous powder distribution appears to be obtained by injection molding throughout even a complex part with both very thin and very thick sections. Figure 2 shows a turbine wheel made in our cooperation with United Turbine, Malmo, Sweden, a company in the Volvo group. The trailing edges of the aerodynamically shaped and twisted blades are as thin as about 0.25 mm. The thickest section is up to 30 mm in the green body.

Fig. 2. Integrated turbine wheel with twisted blades made by the ASEA method for United Turbine, Sweden. Left, injection molded silicon nitride preform; right, as-HIPed to full density and cleaned by sand blasting.

The temporary binder must be removed without distortion of the green powder body in any way. A shrinkage of less than 1% linearly is obtained during this operation which is a low temperature treatment at or below atmospheric pressure. It is important that the binder is completely removed, not leaving carbonaceous remains. Otherwise the mechanical properties of the final product will be reduced. The removal of binders is completed below 400°C but a higher temperature under vacuum may be needed for complete removal of adsorbed gaseous compounds from the powder.

Over the last few years several integrated axial turbine wheels with gradually more advanced shape have been produced in cooperation with United Turbine. Figure 3 shows three generations of such wheels. The wheel to the left shown in a publication in 1980[5] had 24 blades with a constant cross section. It was made in a simple low cost injection molding tool which was manually assembled and disassembled but allowed a couple of injection molded wheels to be made in a day. The efficiency of the wheel was, of course, not high but in a gas generator it could drive its own compressor. The wheel in the middle has 33 straight blades, which are slightly thicker toward the root. The tool was still relatively cheap and manually assembled.

In spite of the relatively simple shape this wheel was used as the first turbine stage in a three shaft KTT engine which was used for driving a Volvo car on the streets of Malmo in March 1982. The only machining after HIP and sand blasting was grinding of the stub shaft to fit the axle attachment and of the outer periphery to reduce gas losses. Balancing was made by selecting the position of the shaft machining. The wheel to the right has 37 blades and is made in a very advanced injection molding tool, which requires simultaneous drawing of all the 37 tool segments. The blades have a thermodynamically correct shape with only slight modifications to allow drawing of the tool. This type of wheel is now undergoing testing at United Turbine.

Slip casting may have an advantage for parts of very thick cross sections because the temporarily added materials are easier to remove than from injection molded parts. Results from HIP indicate, however, that the uniformity in the density distribution in the green body may be more difficult to control than in injection molding. However, components with a cross-section of the green body of up to 90 mm have, as a matter of fact, been injection molded. The binder has been removed from the parts and they have been HIPed to full density. The time needed for binder removal from such components is, however, very long.

Radial turbine wheel shapes do not appear to raise any additional difficulties in the process. Figure 4 shows a radial turbine rotor for a turbo charger. The injection molding has been made in a modified tool originally made for the wax pattern for investment casting of such a wheel.

UTILIZATION FOR WEAR PARTS

Components like integrated turbine wheels have been very important for the development of HIP using the glass powder encapsulation and injection molding techniques. The methods can, of course, be used for many other parts, particularly for solving wear or corrosion problems. A selection of such components is shown in Fig. 5.

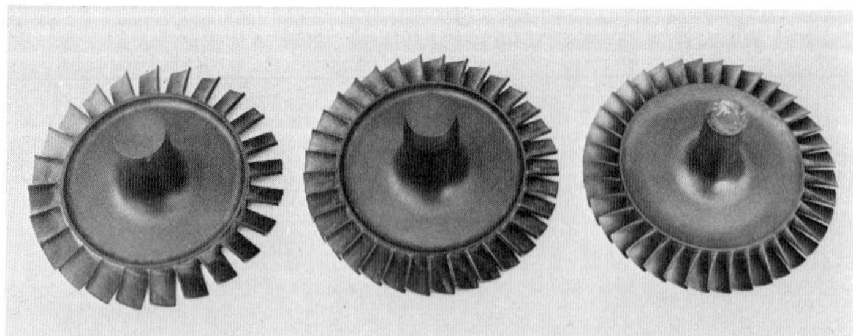

Fig. 3. Three generations of integrated silicon nitride turbine wheels made by the ASEA-method. Courtesy of United Turbine.

Fig. 4. Turbo charger impeller: right, injection molded with binder; left, after HIP to full density. (Injection molding tool courtesy Sterling Metals.)

Fig. 5. Different parts for wear and corrosion applications. As-HIPed and sand blasted.

CERAMIC MATERIALS OTHER THAN SILICON NITRIDE

Exploratory studies made using ceramic materials other than silicon nitride, e.g., hexagonal boron nitride, boron carbide and partially stabilized zirconia, show that the method can, with advantage, be used and parts HIPed to shape with modifications of the glass powder encapsulation process. The density obtained is in many cases higher or the temperature needed for full densification lower than what is reported to be needed in uniaxial hot pressing of similar materials. Hexagonal boron nitride of high purity was HIPed to a density of 2.22 g/cm^3 at 1900°C and 200 MPa for 1 h. Boron carbide, B_4C, with approximately 0.1% impurities and 0.25% O, 0.15% N was HIPed to a density of 2.48 g/cm^3 (99% TD) at 2000°C, 200 MPa for 1 h. The hardness H_v (0.5 kg) was 4300 kp/mm^2 and the 3-point bending strength 590 MN/m^2. Partially stabilized zirconia (PSZ) with 2.5% Y_2O_3 was densified at 1300°C, 200 MPa, 1 h to 6.1 g/cm^3.

MANUFACTURING ABILITIES

The practically obtainable tolerances on parts like integrated turbine wheels produced by injection molding and HIP with the ASEA method using glass encapsulation seem to be similar to what is reported for investment casting.[6] A surface finish of 1 μm CLA has been obtained on the HIPed part starting with green bodies with very good surfaces. Up to now the maximum weight of a single part has been about 1 kg for silicon nitride products, but it is believed to be possible to make considerably bigger parts if needed. The basic processing steps in injection molding and glass particle encapsulation can be highly automated in mass production. The HIP cycle has to be batch wise but the throughput for a single unit could be high. As an example, a single hot isostatic press of the size now available at ASEA Robertsfors could, if modified for series production of axial turbine wheels of the size previously discussed, produce 50,000 to 100,000 such wheels a year. Production cost estimates indicate that an acceptable cost level can be achieved for such components for passenger car gas turbines.

CONCLUDING REMARKS

HIP using the green powder body as the shape determining factor and glass particle encapsulation dring HIP has been demonstrated to be feasible. It is capable of producing advanced components like integrated turbine wheels with thick hub and very thin trailing edges of the blades. This versatile process could be used for many other ceramic materials to make parts of complicated shape and high material performance. It is believed that the process will soon begin to find a growing range of industrial applications.

REFERENCES

1. R. Tegman, pp. 803-6 in High Pressure in Research and Industry, 8th AIRAPT and 19th EHPRG Conf., edited by C. M. Backman, T. Johannisson, and L. Tegner, Uppsala, Sweden, 1982.
2. J. Adlerborn and H. Larker, "Method of manufacturing bodies of silicon nitride," priority date 11 Nov. 1974, G.B. Patent No. 1522705.
3. H. Larker, J. Adlerborn, and H. Bohman, SAE Paper 770335 (1977).
4. S-E. Isaksson and H. Larker, "Method of manufacturing a sintered powder body," priority date 15 March 1971, U.S. Patent No. 4339271.
5. H. T. Larker, AGARD CP-276, 18-1ff (March 1980). Available from NTIS, Va. 22161, USA.
6. E. Green-Spikesley, Materials in Engineering Applications, Vol. 1, 328-34 (1979).

DISCUSSION

S. Dutta (NASA-Lewis): Did you do any HIPing of silicon carbide utilizing the glass powder canning technique?

Author: Yes, but the experience is very limited.

W. S. Coblenz (Naval Research Lab): Is shrinkage common during the glass encapsulation stage of green pressed bodies?

Author: No.

S. Dutta (NASA Lewis): You mentioned that the HIP process heals flaws. Our experience indicates it is very difficult to remove any gross flaws (volume or surface) by HIPing. Could you comment on what kind of flaws you are talking about?

Author: Any flaw with a clean surface can be healed.

J. Haggerty (MIT): Please comment on the degradation resulting from the glass encapsulant interacting with the parts and/or its post HIPing removal. Also, how much does this surface damage degrade strength?

Author: We have found the as-HIPed surface to be at least comparable with the bulk as measured on diamond-machined surfaces. This is found both for bend strength and erosion/corrosion resistance of silicon nitride.

Both an as-HIPed and "properly" sand blasted surface and a diamond-machined surface can, however, be considerably degraded in strength by "improper" sandblasting.

M. Koizumi (I.S.I.R., Osaka Univ.): Can you tell me the dimension for the rotor which you showed in your slide fabricated by using HIP technique?

Author: The diameter is approximately 110 mm.

G. E. Gazza (AMMRC): Have you tried to HIP reaction bonded silicon nitride containing a sintering aid?

Author: It has been done and does not present any particular problems. Results have been published by J. Heinrich et al., AGARD CP-276 (1980). Available from NTIS, VA 22161, USA.)

FABRICATION OF Si_3N_4 CERAMICS WITH ADDITIVES OF METAL NITRIDES BY HIGH PRESSURE HOT-PRESSING AND HIPING

Masahiko Shimada, Norimasa Uchida, and Mitsue Koizumi

Institute of Scientific and Industrial Research
Osaka University, Suita, Osaka 565, Japan

ABSTRACT

High dense Si_3N_4 ceramic bodies with additives of nitrides such as AlN, TiN, YN, VN, NbN and ZrN were fabricated by high pressure hot-pressing and HIPing under the conditions of 0.15-3.0 GPa and 1650°C -1800°C for 1 hour. Temperature dependences of Vickers micro-hardness (Hv) and fracture toughness (K_{IC}) were measured from room temperature to 1200°C. Only little change in Hv and K_{IC} was observed in this temperature range.

INTRODUCTION

In recent years, efforts have been made to develop Si_3N_4 ceramics for engineering use, especially for the use as gas turbine components. However, the consolidation of Si_3N_4 into usable, dense, high strength ceramic bodies pose difficult problems because of the high degree of covalent bonding in Si_3N_4 and the small self-diffusion coefficients of constituent elements.

A major method to manufacture dense Si_3N_4 ceramics is conventional hot-pressing in the presence of additives such as MgO, Al_2O_3 and Y_2O_3.[1-3] These additives play an important role in promoting densification, but cause the precipitation of glassy phases that have a deleterious effect on mechanical properties at high temperature.

In order to evaluate the intrinsic thermal and mechanical properties of Si_3N_4 ceramics, attempts to consolidate Si_3N_4 into highly dense self-bonded ceramic bodies without additives have been made using high pressure hot-pressing and hot isostatic pressing techniques. Prochazka and Rocco performed high pressure hot-pressing of Si_3N_4 and obtained fully dense self-bonded Si_3N_4 ceramics.[4] Recently, we have studied the relations between densification and phase transformation under high pressure using different kinds of Si_3N_4 powders as starting materials,[5] and examined the thermal and mechanical properties of Si_3N_4 ceramics without additives.[6-8]

It is expected that the addition of a non-oxide to Si_3N_4 can prevent the formation of an oxide glassy phases at the grain boundary which occurs when oxides are used as additives. In the nitride system AlN-Si_3N_4-Be_3N_2, the existence of a solid solution was reported between AlN and $BeSiN_2$.[9] Recently, Greskovich[10] prepared highly dense Si_3N_4 ceramics with additives of $BeSiN_2$ and SiO_2 using a gas pressure sintering technique. Although many studies on fabrication and characterization of Si_3N_4 ceramics with oxide additives were conducted, only few reports of Si_3N_4 ceramics with nitride additives have been published.

The present paper deals with the fabrication of Si_3N_4 ceramics having some nitride additives and the temperature dependencies of Vickers microhardness and fracture toughness from room temperature to 1200°C.

EXPERIMENTAL PROCEDURE

In the high pressure hot-pressing and the hot isostatic pressing of Si_3N_4 with 5 mole% nitride additives, Si_3N_4 (sample H1, H. C. Starck) and AlN, TiN, YN, VN, NbN or ZrN were weighed, and then mixed by ball-milling with Al_2O_3 balls for 16 h. After the milling, a slight increase in weight (0.3 wt%) was observed. Starting powders were compressed to form pellets 5 mm by 2.5 mm and 7 mm by 20 mm at 150 MPa at room temperature and prefired in an N_2 atmosphere at 1200°C for 1 h for high pressure hot-pressing and for HIPing, respectively.

High pressure hot pressing was conducted using a cubic anvil type apparatus. The powder compacts were put into a BN capsule and charged in the high pressure cell assembly.[11] Then they were heated in the range of 1500 -1800°C under 3.0 GPa for 1 h. The temperature of the sample was monitored with a Pt20RH/PT40Rh thermocouple inserted in the cell. After being maintained under the desired high pressure hot-pressing conditions, the samples were quenched to room temperature and the pressure was released.

For HIPing, the green compact was put into a BN capsule, which was placed in a container made of high silica glass. The silica container was evacuated to 0.1 Pa and sealed. The glass-sealed specimen was placed into the hot zone of a high pressure vessel, and HIPing experiments were performed using Ar gas as the pressure transmitting medium. The specimen was heated to the softening temperature of silica glass under a pressure of 2 MPa, and then pressure and temperature were simultaneously raised to the desired conditions. The experiments were conducted at 1650°C or 1800°C under 150 MPa for 1 h. After the run was over, the container was stripped and the specimen was cleaned by polishing the surface.

The bulk density of the hot-pressed bodies was measured by means of a water displacement method. The density of porous specimens was measured after coating the surface with nitro-cellulose lacquer. The phases present in the hot-pressed bodies were determined by X-ray powder diffractometry using Ni-filtered CuKα radiation.

The microstructure of the fracture surface of the hot-pressed bodies was observed by scanning electron microscope (SEM). Vickers microhardness was measured using a diamond Vickers indenter from room temperature to 1200°C in a 4.05 Pa vacuum. The load and loading time were 200 g and five seconds, respectively. Before the measurement, the specimen surface was polished with several grades of diamond paste (8-0.3 μm). The critical stress-intensity factor, K_{Ic}, was measured using the indentation method of Evans et al.[12] A high-temperature microhardness tester (Model QM, Nikon) and a Vickers diamond pyramid indenter were used from room temperature to 1200°C in a 4.05 Pa vacuum. An indenter load of 1000 g was applied to the specimen for 5 seconds. The Young's modulus of Si_3N_4 with nitride additives was determined to be 307.5 GN/m^2 for Si_3N_4 (5 mol% YN), 298.5 GN/m^2 for Si_3N_4 (5 mol% AlN), 310 GN/m^2 for Si_3N_4 (5 mol% ZrN), respectively.

EXPERIMENTAL RESULTS AND DISCUSSION

A summary of some of the sintering results is given in Table 1. The sintering results of specimen numbers 1 and 2 show that Si_3N_4 without additive can be densified to 92% of theoretical density by HIPing and 100% by high pressure hot-pressing at the temperature condition of 1850°C. In our previous investigation,[5] a remarkable effect of pressure on the relative density of Si_3N_4 hot-pressed bodies without additives was pointed out and, in case of HIPing at pressures lower than 200 MPa (0.2 GPa), a small amount of additives must be mixed with the Si_3N_4 powder to produce fully dense Si_3N_4 ceramics. As seen in Table 1, it is expected that the additives of nitrides such as AlN, TiN, YN, VN, NbN and ZrN are quite effective to consolidate Si_3N_4 into a fully dense body. In high pressure hot-pressing, in fact, dense Si_3N_4 bodies with nitride additives were obtained.

Scanning electron micrographs of the fracture surface of some hot-pressed bodies are shown in Fig. 1. As for Si_3N_4 ceramics with additives such as AlN, TiN, VN and ZrN, the grains had hexagonal shape and were similar to those of Si_3N_4 bodies with oxide additives

Table 1. Density and Content of β Phase for Si_3N_4 Sintered Body

Specimen No.	Additive	Sintering Conditions (GPa	°C	h)	Density (R.D.)* (g/cm^3)	Phase (wt%)
1	none	0.15	1850	1	2.93(0.92)	100
2	none	3.0	1850	1	3.19(1.00)	100
3	5 mol% AlN	0.15	1850	1	3.06(0.96)	100
4	"	3.0	1650	1	3.13(0.98)	15
5	"	3.0	1850	1	3.19(1.00)	100
6	5 mol% TiN	0.15	1850	1	3.04(0.95)	50
7	"	3.0	1650	1	3.03(0.94)	20
8	"	3.0	1850	1	3.19(0.99)	100
9	5 mol% VN	0.15	1850	1	3.10(0.96)	100
10	"	3.0	1650	1	3.07(0.95	42
11	"	3.0	1850	1	3.23(1.00)	100
12	5 mol% YN	3.0	1650	1	3.24(1.00)	20
13	"	3.0	1850	1	3.24(1.00)	100
14	5 mol% ZrN	3.0	1650	1	3.19(0.98)	43
15	"	3.0	1850	1	3.26(1.00)	100
16	5 mol% NbN	3.0	1650	1	3.06(0.94)	14
17	"	3.0	1850	1	3.25(1.00)	100

*R.D.: relative density.

such as MgO and Al_2O_3.[6] The Si_3N_4 body without additives consisted of elongated grains. These grains were more irregular in shape and were smaller in size compared with those for the bodies with additives. The difference in microstructure was probably caused by the mechanism of $\alpha \rightarrow \beta$ phase transformation.

The results of high temperature Vickers microhardness measurements are shown in Fig. 2. The microhardness of a Si_3N_4 sintered body without additives (specimen No. 2) decreased gradually with increasing temperature and no drastic decrease was found. The temperature dependence of microhardness of Si_3N_4 bodies with nitride additives, as seen in Fig. 2, showed no drastic decrease, but the tendency of degradation of high temperature microhardness was greater than that of Si_3N_4 without additives. Tsukuma et al.[6] reported that the drastic decrease in microhardness occurred at ≃ 750°C for Si_3N_4 + 4 wt% MgO, and at 950°C for Si_3N_4 + 4 wt% Y_2O_3 ceramics, and that the rapid degradation was due to the softening of glassy phase at grain boundaries. The fact that no drastic decrease was found for Si_3N_4 bodies with nitride additives indicates that the dense Si_3N_4 ceramics with nitride additives are suitable for high temperature use.

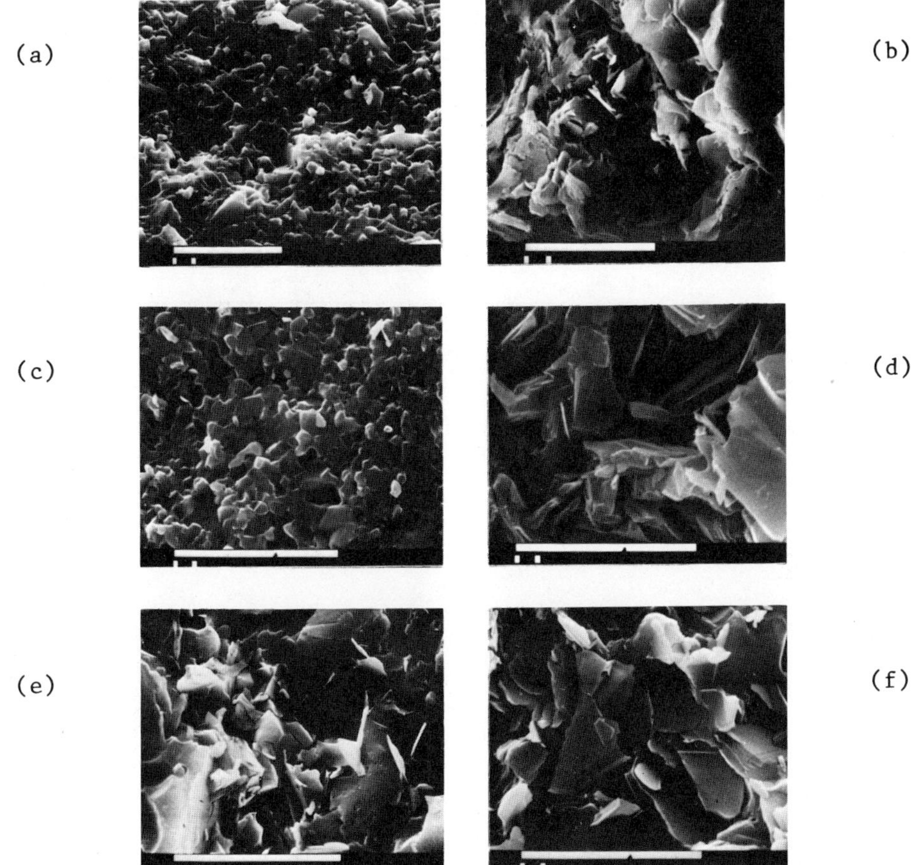

Fig. 1. Scanning electron micrographs of fracture surface of Si_3N_4 (bar = 10 μm).

Variation of the critical stress-intensity factor, K_{Ic}, as a function of temperature for fully dense Si_3N_4 without and with nitride additives is shown in Fig. 3. As seen in Fig. 3, K_{Ic} for Si_3N_4 ceramics under present investigation was nearly temperature independent. The results of temperature dependence of K_{Ic} for Si_3N_4 with MgO indicated that K_{Ic} increased sharply at 1100°C.[13] This was thought to be resulted from the softening of the secondary glassy phases existed in the grain boundaries.

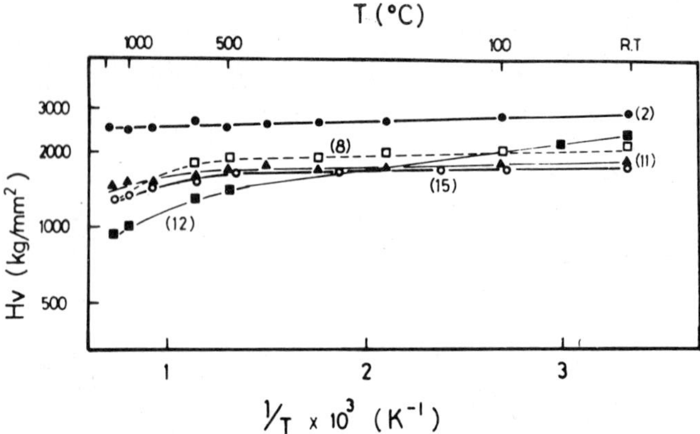

Fig. 2. Temperature dependence of microhardness. Number in parentheses indicates the specimen number listed in Table 1.

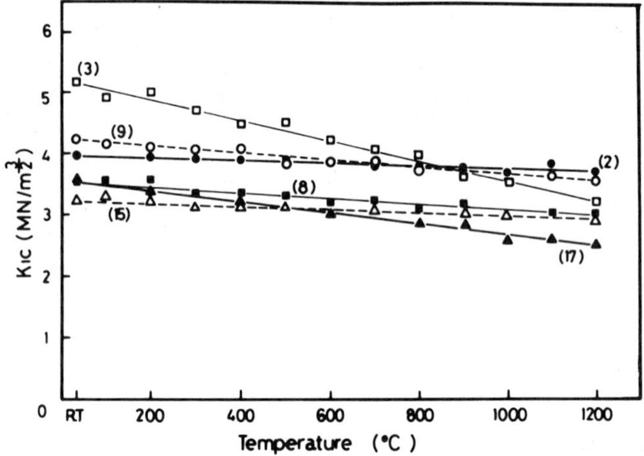

Fig. 3. Temperature dependence of K_{Ic}. Number indicates in parentheses the specimen number listed in Table 1.

From the present results on densification, high temperature microhardness and temperature dependence of K_{Ic}, it is expected that dense Si_3N_4 ceramics with nitride additives are more suitable for the application of high-temperature structural materials.

REFERENCES

1. G. G. Deely, J. M. Herbert and N. C. Moore, Powder Metall., 8, 1451-151 (1961).
2. Y. Oyama and O. Kamigaito, Jpn. J. Appl. Phys., 10, 1637 (1971).
3. G. E. Gazza, J. Am. Ceram. Soc., 56, 662 (1973).
4. S. Prochazka and W. A. Rocco, High Temp. High Pressures, 10, 87-95 (1978).
5. T. Yamada, M. Shimada, and M. Koizumi, Am. Ceram. Soc. Bull., 60, 1281-83 (1981).
6. K. Tsukuma, M. Shimada, and M. Koizumi, Am. Ceram. Soc. Bull., 60, 910-12 (1981).
7. T. Yamada, M. Shimada, and M. Koizumi, Am. Ceram. Soc. Bull., 60, 1225-28 (1981).
8. M. Shimada, M. Koizumi, A. Tanaka, and T. Yamada, Comm. Am. Ceram. Soc., C-48 (1982).
9. I. C. Husebt, H. L. Lukas, and G. Petzow, J. Am. Ceram. Soc., 58, 377-80 (1975).
10. C. Greskovich, J. Am. Ceram. Soc., 64, 725-30 (1981).
11. M. Shimada, N. Ogawa, M. Koizumi, F. Dachille, and R. Roy, Am. Ceram. Soc. Bull., 58, 519-21 (1979).
12. A. G. Evans and E. A. Charles, J. Am. Ceram. Soc., 59, 371-73 (1976).
13. R. K. Govila, J. Am. Ceram. Soc., 63, 319-26 (1980).

DIFFUSION BONDING OF Al_2O_3 AND Si_3N_4 CERAMICS BY HIPING

Masahiko Shimada, Kimiaki Tanihata, Takahiro Kaba and Mitsue Koizumi

Institute of Scientific and Industrial Research
Osaka University, Suita, Osaka 565, Japan

ABSTRACT

Diffusion bonding of Al_2O_3 and Si_3N_4 was attempted using high pressure technology. The high pressure and temperature conditions were 100 MPa and 1650°C for HIPing and 3.0 GPa and 1650°C for high pressure experiments. The results of SEM observations and Vickers microhardness measurements for the interface of the bonded specimens suggested that the strong bonding between these sintered ceramic bodies was completed and that the high pressure technique is one of the useful methods for diffusion bonding of ceramics.

INTRODUCTION

There has been much effort in recent years in the application of ceramic materials to specific engineering problems in which the intrinsic brittleness of these materials had to be conceptually accounted for in the original design. Silicon nitride (Si_3N_4), silicon carbide (SiC), zirconia (ZrO_2), alumina (Al_2O_3) have emerged as candidate materials with outstanding potential for engineering applications. Much research and development has been performed on preparation of powdered samples as starting materials and on fabrication of ceramic bodies. In the area of fabrication, especially, many kinds of sintering techniques such as reaction bonding, hot-pressing, HIPing, high pressure hot-pressing, pressureless sintering and gas pressure sintering have been developed.

The technique employed to bond ceramic parts to each other is important in the fabrication of high temperature structural ceramic components with a complicated shape. There are many kinds of bonding methods such as liquid phase diffusion bonding, eutectic bonding, thermocompression bonding and explosive bonding for metal matrix components. Among these methods, hot isostatic pressing (HIP) has been developed in recent years as a useful way to bond metal and ceramic components.[1] HIP was initially originated as an isostatic bonding process for nuclear fuel elements. The even application of pressure to all surfaces allows bonding of highly complex structures which can incorporate brittle materials with little damage. Hanes et al.[2] has reported the details of the HIP bonding process and the many experimental results on metals such as stainless steel, Ti and superalloys. Larker has reported the application of the diffusion bonding technique to ceramics for the fabrication by HIPing of an Al_2O_3 container to preserve nuclear waste.[3]

The present paper presents the experimental results of the diffusion and reaction bonding of Al_2O_3 and Si_3N_4 under the application of pressure.

EXPERIMENTAL PROCEDURE

The starting sintered and additive-free bodies of Al_2O_3 and Si_3N_4 were fabricated by high-pressure hot-pressing to theoretical density.[4,5] After polishing the surface of the hot-pressed bodies with several grades of diamond paste (8–0.3 μm), diffusion bonding experiments were conducted by the application of high pressure techniques such as high pressure hot-bonding and HIP bonding. High pressure hot-bonding was carried out using cubic anvil type equipment. The sintered bodies were put into a BN capsule and charged in the high pressure cell assembly.[6] After being maintained at 1650°C and 3.0 GPa for 1 h, the pressure was released, and the temperature was decreased to room temperature at a rate of 30°C/min. For HIP bonding experiments, the sintered bodies were put into a BN capsule, which was put into a container made of high silica glass. The silica container was evacuated to 0.1 Pa and sealed. The glass-sealed specimen was placed into the hot zone of a high pressure vessel, and HIP bonding experiments were performed using Ar gas as a pressure transmitting medium. The specimen was heated to the softening point of the silica glass under a pressure of 2 MPa. Pressure and temperature were then simultaneously raised to 100 MPa and 1650°C, respectively, and maintained under these conditions for 1 h. The container was then stripped and the specimen cleaned by polishing the surface. A representative experimental procedure is shown in Fig. 1.

After cutting the sample perpendicular to the bonding interface using a diamond cutter, the microstructure of the specimen was observed by scanning electron microscopy.

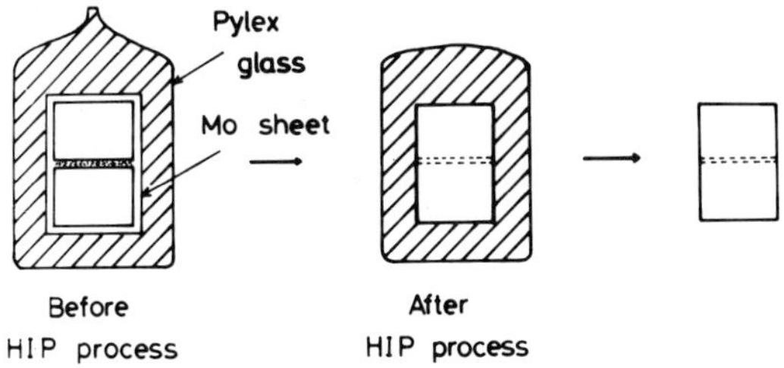

Fig. 1. Representative experimental procedure for HIP bonding.

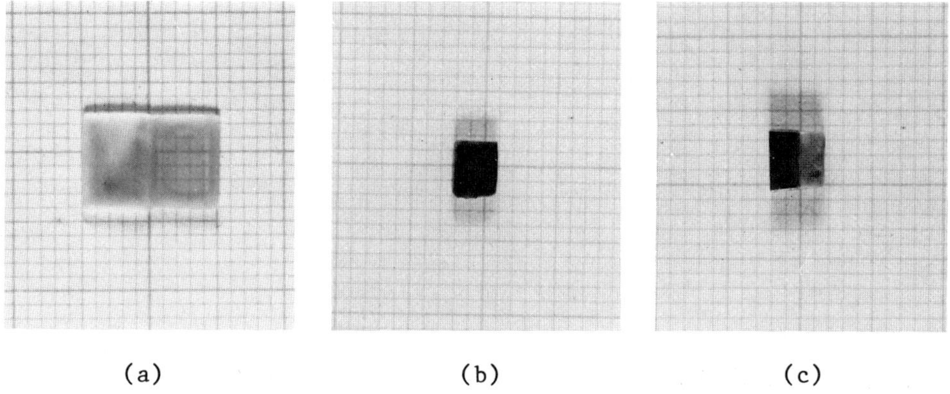

(a) (b) (c)

Fig. 2. Representative examples of diffusion bonded ceramics. (a) Al_2O_3-Al_2O_3 (100 MPa, 1650°C 2 h); (b) Si_3N_4-Si_3N_4 (3.0 GPA, 1800°C 1 h); (c) Si_3N_4(left)-Al_2O_3(right) (3.0 GPa, 1800°C 1h)

The specimens were polished with several grades of diamond paste (8-0.3 μm). This was followed by room temperature microhardness measurements using a Vickers diamond indenter with a 200 g load.

EXPERIMENTAL RESULTS AND DISCUSSION

Representative examples of diffusion bonded Al_2O_3 and Si_3N_4 ceramics are shown in Fig. 2. Present experimental results are tabulated in Table 1. As seen in this table, when a small amount of powder was placed at the interface between the sintered bodies, the tightly and strongly bonded samples of Al_2O_3 and Si_3N_4 ceramics were fabricated at the completion of diffusional reaction bonding under application of pressure.

Table 1. Present Experimental Results

Experimental Number	Materials	Experimental Conditions			Bonding Feature
		pressure (GPa)	temperature (°C)	time (h)	
1	Al_2O_3-Al_2O_3	0.1	1650°C	1	weak
2	Al_2O_3-Al_2O_3	0.1	1650°C	2	good
3	Al_2O_3-Al_2O_3	3.0	1650°C	1	good
4	Si_3N_4-Si_3N_4	3.0	1800°C	1	good
5	Si_3N_4-Si_3N_4	3.0	1600°C	1	weak
6	Si_3N_4-$Si_3N_4(Y_2O_3$	3.0	1800°C	1	good
7	Si_3N_4-Si_3N_4(MgO)	3.0	1800°C	1	good
8	Si_3N_4-Al_2O_3	3.0	1800°C	1	good

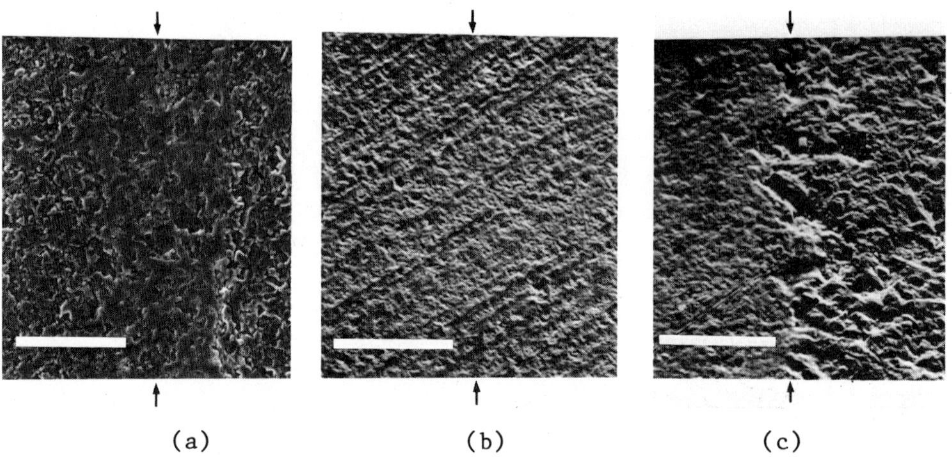

(a) (b) (c)

Fig. 3. Scanning electron micrographs of the interface of bonded samples: (a) Al_2O_3-Al_2O_3; (b) Si_3N_4-Si_3N_4; and (c) Si_3N_4(left)-Al_2O_3(right). The pair of arrows indicates the interface position (bar = 100 μm).

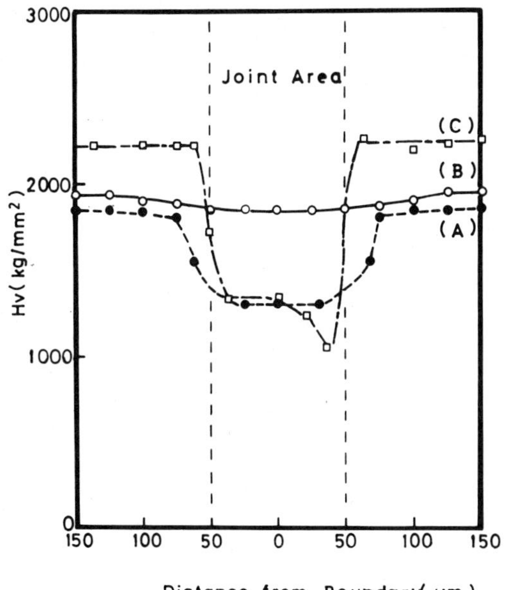

Fig. 4. Microhardness profile of a joint area at room temperature: (a) Al_2O_3-Al_2O_3; (b) Si_3N_4-Si_3N_4; and (c) Si_3N_4-Al_2O_3.

Scanning electron micrographs of the interface of bonded samples are shown in Fig. 3. It is generally said that conventional fusion welding often results in a significant microstructural change at the interface zone. In the case of the bonding of similar materials, it is almost impossible to discern a well-bonded interface as seen in (A) and (B) of Fig. 3. Since the area of the interface was identical in terms of microstructure with the parent material, it is expected that basic ceramic properties are normally achieved. The interface microstructure between alumina and silicon nitride shown in (C) of Fig. 3 indicated that it was possible to join these highly incompatible materials using a diffusion bonding technique.

The results of Vickers microhardness measurements are shown in Fig. 4. As seen in this figure, the values of microhardness of the joint area were slightly smaller than those of the parent sintered bodies. In the case of Al_2O_3-Al_2O_3 bonded ceramics, the values of Vickers microhardness were 1850 kg/mm^2 for parent bodies and 1300 kg/mm^2 within the joint area. The value of Vickers microhardness for the parent bodies near the interface zone was 1550 kg/mm^2 and the distance dependence of Vickers microhardness abruptly changed at the boundary between the interface zone and the parent bodies. On the other hand, as seen in Fig. 4(B), the values of Vickers microhardness were 1940 kg/mm^2 for the parent bodies and 1850 kg/mm^2 within the joint area in the case of Si_3N_4-Si_3N_4 bonded ceramics. The Vickers microhardness gradually changed from the parent body to the joint area.

From the aforementioned results, it may be seen that pressure bonding is one of the useful methods for diffusion bonding of ceramics.

REFERENCES

1. H. D. Hanes, High-Pressure Science and Technology, 2, 633-50 (1979).
2. H. D. Hanes, D. A. Seifert, and C. R. Watts, MCIC Report, 77-34 (1977).
3. H. T. Larker, High-Pressure Science and Technology, 1, 329-37 (1981).
4. Y. Ishitobi, M. Shimada, and M. Koizumi, Am. Ceram. Soc. Bull., 59, 1208-11 (1980).
5. Y. Ishitobi, M. Shimada, and M. Koizumi, Am. Ceram. Soc. Bull., 60, 1281-83 (1981).
6. M. Shimada, N. Ogawa, M. Koizumi, F. Dachille, and R. Roy, Am. Ceram. Soc. Bull., 58, 519-21 (1979).

RELATIONSHIP BETWEEN DENSIFICATION AND HIGH TEMPERATURE MECHANICAL PROPERTIES OF HIPed SILICON NITRIDE

R. R. Wills, M. C. Brockway, and G. K. Bansal

Battelle, Columbus Laboratories
505 King Avenue
Columbus, Ohio 43201

ABSTRACT

The higher pressures available in the HIP process permit the fabrication of yttria-based silicon nitride that cannot be prepared by conventional ceramic processing. Since both the densification and high temperature properties of silicon nitride are controlled by a similar mechanism the slow rate of densification in HIPed silicon nitride corresponds to lower creep rates and subcritical crack growth rates than in other additive-based silicon nitrides.

INTRODUCTION

Silicon nitride is a ceramic that does not readily sinter unless the powder contains a critical oxygen content and a sintering aid. The oxygen content of the powder and the additive type and concentration are major factors controlling the sintering kinetics because the sintering mechanism is a liquid phase one in which silicon nitride grains first dissolve in the silicate liquid formed at the sintering temperature by both the oxygen on the surface of each silicon nitride grain and the additive. The sintering kinetics are dependent upon: (1) the solubility of silicon nitride in the melt; (2) the rate of reaction at the silicon nitride-liquid interface; (3) the diffusion rate through the melt; (4) the volume of liquid present during sintering; and, (5) the temperature and pressure.

The high temperature strength and creep properties of silicon nitride are dependent on the nature of the grain boundary glass phase since the viscosity of this phase is a critical parameter determining grain boundary sliding, one of the prime mechanisms responsible for deformation and crack growth in additive-containing silicon nitrides at elevated temperatures.

In order to fabricate a silicon nitride ceramic with improved high temperature properties the remnant grain boundary glass phase must possess a high softening point and the volume of this phase must be kept to an absolute minimum. This presents a processing challenge since the transport mechanisms operative during creep are the very same ones responsible for sintering. In this paper we discuss the use of hot isostatic pressing to form a high purity yttria-based silicon nitride together with its high temperature strength and creep properties.

Experimental

The high purity grade of silicon nitride powder manufactured by Kawecki-Berylco Industries was chosen for use because of its low oxygen content (see Table 1). Silicon nitride powder was mixed with

Table 1. Silicon Nitride Powder Characteristics*

Oxygen	0.7 wt percent
Magnesium	10 ppm
Aluminum	200 ppm
Calcium	$\leq$ 30 ppm
Iron	400 ppm
Free Silicon	0.5 wt percent
Carbon	0.22 wt percent
BET Surface Area	1.74 m^2/g
Fisher Size	0.96 μm
-325 Mesh	99.1 percent
Alpha Si_3N_4	79 percent

*High purity grade Kawecki-Berylco, Inc., Pennsylvania, U.S.A.

1-5 weight percent yttria[a] by ball milling in "nanograde" hexane[b] in polyethylene containers for 4 hours. Sintered silicon nitride milling media[c] were used to avoid powder contamination. The powder was vacuum dried at 180°C and sieved through a minus 40 mesh sieve. Rod samples 15.9 mm diameter were cold isostatically pressed at 201 MPa (30,000 psi), "canned" in tantalum, and subsequently HIPed at 1725°C for 1 hour at 201 MPa (30,000 psi). For comparative purposes, the powder mixture was also uniaxially hot pressed at the same temperature for 1 hour using an applied pressure of 40 MPa (6,000 psi).

Four-point bend strengths were determined using rectangular 2.5 x 5.1 x 39 mm test bars using an Instron testing machine at a crosshead speed of 0.127 mm/min. All specimens were ground parallel to the span direction using a 320-grit diamond wheel, the edges of the tensile face being rounded and polished to eliminate edge flaws. Cylindrical compression creep specimens 8.64 mm long and 4.32 mm diameter were used to determine the steady state creep rate at 1350°C to 1500°C in air.

Subcritical crack growth is described by the relation $V=AK^n$, where V is the crack velocity, A is a material constant and K the stress intensity at the crack tip. The value of the exponent n is a measure of the subcritical crack growth, large values indicating very little slow crack growth and time invariant strength. Differential strain rate tests were performed on four point bend test specimens at crosshead speeds of 0.00508 mm per min., 0.12 mm per min. and at 0.508 mm per min. in order to generate the data necessary to calculate the value of n. All measurements were made in air at 1400°C.

Results

Five weight percent yttria additive level was needed to achieve high density silicon nitride. The percentage of the theoretical density was calculated to be 99% based upon the presence of the oxygen rich fluoroapatite phase $Y_5(SiO_4)_3N$ detected in the microstructure of this silicon nitride.[1] The uniaxially hot pressed silicon nitride containing 5 wt. per yttria exhibited much less sintering. Sample density was 2.59 g/cm^3 (79.2 percent theoretical). Thus it can be readily seen that HIP processing has enhanced the rate of sintering, enabling high densification to be achieved in a ceramic that does not readily sinter even when uniaxially hot pressed.

In initial creep testing for 146 hours under 67 MPa (10,000 psi) stress no deformation was observed. The data in Fig. 1 confirms the excellent creep resistance of this HIPed silicon nitride in

[a]Research Chemicals, Inc.
[b]American Hospital Supply Company.
[c]GTE Sylvania, Inc.

comparison with other additive based silicon nitrides. The steady state strain rate can be expressed in the form:

$$\varepsilon \propto e^{-E/RT} \sigma^n$$

where ε = steady state strain rate,
E = activation energy,
σ = applied stress,
n = a constant.

From determinations of steady state creep rates at 134 MPa (10,000 psi) to 536 MPa (80,000 psi) at 1325°C to 1500°C the stress exponent n was calculated to be 2.3 and the activation energy E 87.9 Kcals/mole. Comparable data[2] for uniaxially hot pressed magnesia and yttria-based silicon nitride designated NC132 and NCX-34 by Norton Company is given in Table 2. The stress exponent values for these three materials are quite different but the activation energies are similar.

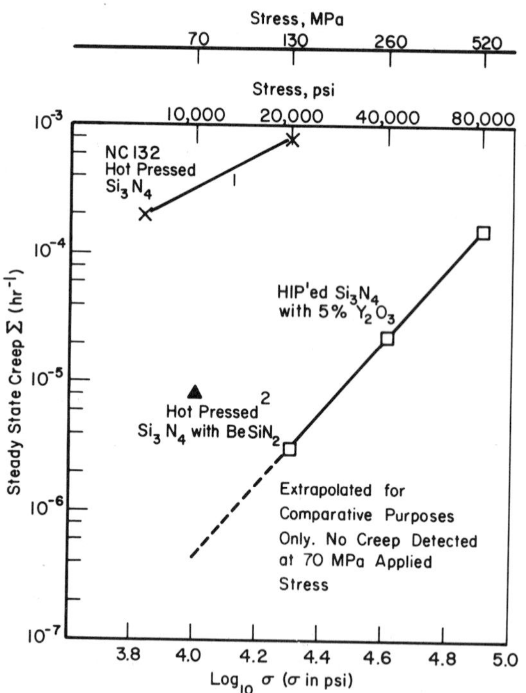

Fig. 1. Comparison of creep rates of HIPed and hot-pressed silicon nitrides. (Ref. 1, 2)

Table 2. Comparison of High Temperature Creep Properties of HIPed and Hot Pressed Silicon Nitrides

Material	Temperature	Stress Exponent	Activation Energy Kcals/Mole
HIP Si_3N_4	1400°C	2.3	88
*Norton NC-132	1350-1500°C	1.6	130
*Norton NCX-34	1350-1500°C	0.8	91

*Norton Co., Worcester, Massachusetts.

In Fig. 2 the high temperature behavior of HIPed silicon nitride is compared with that of these other two additives containing hot pressed silicon nitrides. The HIPed silicon nitride possesses both higher strength and greater resistance to subcritical crack growth.

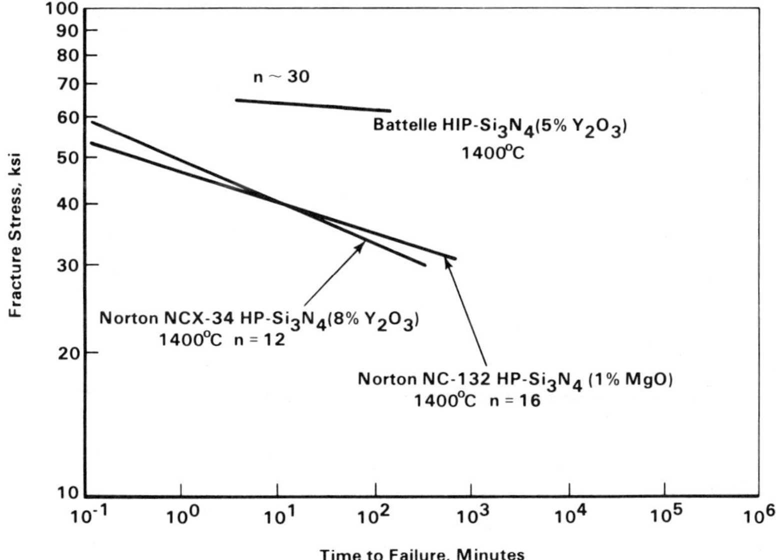

Fig. 2. Strength degradation by subcritical crack growth.

DISCUSSION

The higher pressures used in HIPing during densification are likely to generate higher stresses in individual particles, principally at their contact points. Evidence for this is seen in approximately 15% of the silicon nitride grains which exhibit plastic deformation in the form of a very high dislocation density.[1] Although we have not examined this feature of HIPed silicon nitride in detail it appears that the dislocation density is higher than that observed in uniaxially hot pressed silicon nitride. Evans and Davidge[3] have estimated the approximate yield stress at room temperature to be ≥ 7 GM/m^{-2} and at temperatures of about 1800°C the stress required for gross plastic flow to be greater than 450 MN m^{-2}. The local plastic deformation observed in approximately 15% of the grains would appear to be consistent with this since during HIPing at 1750°C and 202 MN m^{-2} stresses at some of the particle interfaces are likely to be higher than the applied stress. Plastic deformation however can be ruled out as the principal densification mechanism since densification is a function of additive content.[1]

In the early stages of liquid phase sintering densification occurs as a result of grain rearrangement. Wills et al.[4] observed unstable liquid formation at 1600°C in the silica-rich section of the Si_3N_4-Y_2O_3-SiO_2 ternary system in the region of composition 13 mo ℓ% Si_3N_4 + 19 mo ℓ% Y_2O_3 + 68 mo ℓ% SiO_2. Melting was also observed at 1650°C at the yttria-rich end of the phase diagram. The silica-rich liquid is probably responsible for enhancing densification during HIP since the second crystalline phase observed in the microstructure of HIPed silicon nitride is $Y_5(SiO_4)_3N$, the oxygen rich ternary compound found in this system. This phase could only have formed by precipitation from a silica-rich solution. Further evidence for this argument is given by analysis and estimation of the composition of the glassy grain boundary phase which is discussed later. The composition of this phase is approximately 6 $SiO_2 \cdot N_{0.37} \cdot Y_2O_3$. In view of the low oxygen content of the silicon nitride powder (see Table 1) this observation is unexpected. In contrast, uniaxially hot pressed silicon nitride usually contains the compound $Si_3N_4 \cdot Y_2O_3$ as the second crystalline phase because densification is caused by the yttria-rich liquid of approximate composition 25 mo ℓ% Si_3N_4 + 75 Y_2O3.

Wills et al.[4] also reported that this high silica content, silicon yttria oxynitride liquid is unstable at atmospheric pressure. It is not expected to be a good vehicle for liquid phase sintering since it is not likely to readily wet silicon nitride. The high HIP pressure probably plays a dual role in retarding the decomposition rate of this liquid and accelerating movement of the liquid between silicon nitride grains thus facilitating rearrangement. Since the silicon nitride powder compacts contains only 3 percent liquid the rearrangement mechanism alone cannot account for complete densification, although the extent of densification may be higher than that predicted by Kingery's model.[5]

Previous work on the sintering and hot pressing kinetics of silicon nitride has generally assumed a liquid phase mechanism involving solution and reprecipitation. For example, Mitomo[6] studied the sintering of silicon nitride containing 5 wt% magnesia at 1450°C to 1900°C, and interpreted his data in terms of Kingery's theory, concluding that the contribution of rearrangement to densification was about 10%; the remainder of densification being due to solution precipitation. Bowen et al.[7,8] studied kinetics of hot pressing of silicon nitride containing several different additives, and concluded that the rate controlling mechanism during this stage of densification is diffusion through the liquid phase.

Cannon and Chowdhry[9] reviewed the deformation behavior and mass transport in silicon nitride containing oxide additives in an effort to construct a consistent mechanistic view of the role of the viscous grain boundary phase. They concluded that atom transport may be limited by the solution precipitation step rather than by diffusion within the boundary phase. In more recent work Raj et al.[10-13] derived similar conclusions from experiments on the dissolution of $\beta\text{-}Si_3N_4$ in Mg-Si-O-N glass and from creep experiments on several materials. Differentiation between these mechanisms can be achieved by monitoring the densification kinetics as a function of grain size since the rate of densification in diffusion-controlled sintering is inversely proportional to the cube of the grain size whereas, if the reaction rate is controlled by an interface reaction, the rate of densification is inversely proportional to the grain size. The basic relations developed by Coble[14] are:

$$\frac{dp}{\rho dt} = \frac{14\pi\ \sigma\ D\ell\ Co\Omega}{d^3\ KT}\left(\frac{P}{\rho} + \frac{\gamma\ell v}{r}\right)$$

$$\frac{dp}{\rho dt} = \frac{4K\ Co\ \Omega}{e\ d\ KT}\left(\frac{P}{\rho} + \frac{\gamma\ell v}{r}\right)$$

where:

σ = thickness of the boundary phase,
K = Boltzmann's constant,
T = temperature,
t = time,
$D\ell$ = controlling diffusivity in the liquid,
Co = solubility of Si_3N_4 in the liquid,
D = grain size at zero stress,
$\gamma\ell v$ = liquid surface energy,
p = applied pressure,

ρ = fractional density,
r = effective pore radius,
Ω = molecular volume.

The equations recently derived by Raj[11] for interface and diffusion controlled liquid phase sintering show the same dependency on grain size.

Creep of silicon nitride is thought to occur by grain boundary sliding with accommodation by cavitation.[15] Dislocation motion can be disregarded as shown in the calculations by Evans and Davidge[3] and by Kossowsky.[16] These show that the contribution to creep strain is several orders of magnitude below the observed strain.

A stress exponent of 2 has generally been associated with a grain boundary sliding mechanism[17] but as shown in Table 2 the exponent varies quite considerably for the three silicon nitrides. Similar variation has been found by Larsen and Adams[2] in their evaluation of twelve hot pressed and sintered silicon nitrides. This may be due to different amounts of cavitation in these materials. In his analysis of the compressive creep of silicon nitride magnesia alloys Lange[18] has shown that the contribution of cavitational creep increases with $Vi^2 \sigma t/\eta$, i.e., larger volume fractions of the liquid, higher stresses, longer times and lower viscosity. Changes in the volume fraction of the liquid will have the greatest effect. Detailed microstructural analysis of these three materials needs to be done in conjunction with appropriate experiments to determine the contribution of cavitation to deformation.

The data in Table 2 suggest that the mechanism of creep in HIPed silicon nitride is similar to that in hot pressed silicon nitride. The lower creep rate exhibited by the HIPed silicon nitride is probably due to the presence of a very refractory grain boundary phase. At 1400°C this phase is presumed to possess a higher viscosity than the comparable amorphous phases present in the two uniaxially hot pressed silicon nitrides. Evidence for this is found in the composition of this amorphous phase.

Analysis of the glassy grain-boundary phase in HIPed silicon nitride by EDAX gave the following composition $Al_{4.27}Si_{65.31}CR_{2.31}$ $Fe_{3.26}Ni_{1.12}Y_{23.73}$. The approximate composition of the glass phase is 6 $SiO_2 \cdot Y_2O_3$ if the impurities are ignored. This is an extremely refractory oxide composition that partially melts at 1660°C. It does not become totally liquid until 2200°C. This phase probably also contains nitrogen since the solution precipitation liquid phase sintering process involves nitrogen dissolution in the liquid. Silicon metal oxynitride glasses typically contain 5 to 10 weight percent nitrogen. Assuming the nitrogen content of this glass is 5 weight percent, its composition would be 6 $SiO_2N_{0.37} \cdot Y_2O_3$. Addition of nitrogen to silicate glasses increases their viscosity,[19] and consequently, this glass would be expected to be more refractory than the 6 $SiO_2 \cdot Y_2O_3$ glass. Although the exact composition of the amorphous grain boundary phase is not known, the analysis indicates it will be a high softening-point glass.

The rate of creep is controlled by the applied stress and by the rate of mass transport through the viscous boundary phase. The latter process consists of two sequential steps: transfer of atoms across the fluid/grain interface and diffusion of silicon nitride through the fluid glass phase. Creep equations for both the interface reaction-controlled and diffusion-controlled processes have been reported.[11] The main difference between them is in the grain size (d) dependence of the creep rate (ε); it is $\varepsilon \propto d^{-1}$ for interface control and $\varepsilon \propto d^{3}$ for diffusion control. Identical relationships with respect to grain size describe interface controlled and diffusion controlled pressure enhanced densification, as discussed earlier. Tsai and Raj[10] have calculated the creep rate in uniaxially hot pressed silicon nitride, and shown that good agreement exists between their theoretical values and the measurements made by Lange.[18] These calculations were made assuming that the creep reaction is interface controlled. It thus appears likely that the creep rate of HIPed silicon is interface controlled. On this basis a similar mechanism controls the rate of densification during HIPing.

The subcritical crack growth rate in HIPed silicon nitride is lower than that exhibited by the two uniaxially hot pressed silicon nitrides (see Fig. 2). Larsen and Adams[2] confirmed this data by examining the fracture features of these three silicon nitrides. Slow crack growth features extended over 20 percent, 10 percent and 3 percent respectively of the sample cross section for NC-132, NCX-34 and the HIPed silicon nitride.

Slow crack growth in ceramics is largely independent of environment and is consequently an intrinsic property of the material. The mechanism of slow crack growth in additive-containing silicon nitrides is probably associated with plastic processes occurring in the vicinity of the crack tip and it is generally believed that this plasticity is related to the viscosity of the amorphous grain boundary phase present in the microstructure. Thus differences in viscosity of the glass phase at 1400°C in these silicon nitrides probably accounts for their slow crack growth behavior. Evans and Weiderhorn[20] attributed this time-dependent cracking to the formation of secondary cracks ahead of the main crack by a process of grain boundary sliding. Tsai and Raj[21] determined the time constant for grain boundary sliding and compared the experimental slow crack growth rate data with estimates of crack growth rates controlled by grain boundary sliding in an effort to specify the rate-controlling mechanism in high temperature fracture. They concluded that sliding, although necessary, is not the rate-limiting step in microcrack nucleation.

ACKNOWLEDGMENTS

The data on silicon nitride was obtained on a program sponsored jointly by the Defense Advanced Projects Research Agency under ARPA Order 3562 and by the Air Force Materials Laboratory under Contract Number F33615-78-C-5118. The authors wish to thank Mr. J. E. Reichlderfer, Mr. R. Shaw, Mr. C. B. Boyer and Mr. R. Palmer for the preparation of HIPed silicon nitride samples and the IIT Research Institute for determination of the subcritical crack growth rate exponent. Helpful discussions with Dr. R. Raj are also acknowledged.

REFERENCES

1. R. R. Wills and M. C. Brockway, Special Ceramics 7, Proc. Brit. Ceram. Soc., 31, 233-47 (1981).
2. D. C. Larsen and J. W. Adams, IIT Research Institute, November 1981, Interim Technical Report No. 11, Contract F-33615-79-C-5100, U.S. Air Force Wright Aeronautical Laboratories.
3. A. G. Evans and R. W. Davidge, J. Mater. Sci., 5 , 314-25 (1970).
4. R. R. Wills, S. Holmquist, J. M. Wimmer, and J. A. Cunningham, J. Mater. Sci., 11, 1305-09 (1976).
5. W. D. Kingery, J. Appl. Phys., 30, 301-06 (1959).
6. M. Mitomo, J. Mater. Sci., 4, 1103-07 (1976).
7. L. J. Bowen, T. G. Carruthers, and R. J. Brook, J. Amer. Ceram. Soc., 61 [7-8], 335-39 (1978).
8. L. J. Bowen, R. J. Weston, T. G. Carruthers, and R. J. Brook, J. Mater. Sci., 13, 341-50 (1978).
9. R. M. Cannon and H. Chowdhry, in Nitrogen Ceramics, edited by F. L. Riley, Noordoff International, Reading, Massachusetts, 1977.
10. R. L. Tsai and R. Raj, Communications of J. Amer. Ceram. Soc., June 1982.
11. R. Raj and C. K. Chyung, Acta. Metall., 29 [1], 159-96 (1981).
12. R. Raj and P. E. D. Morgan, J. Amer. Ceram. Soc., 64 [10], C143-45 (1981).
13. R. L. Tsai and R. Raj, J. Amer. Ceram. Soc., 65 [5], 270, 274 (1982).
14. R. L. Coble, J. Appl. Phys., 41 4798 (1970).
15. R. Kossowsky, D. G. Miller, and E. S. Diaz, J. Mater. Sci., 10, 983 (1975).
16. R. Kossowsky, in Ceramics for High Performance Applications, Second Army Materials Technology Conference, Brook Hill Publishing Company, 1974.
17. J. Weertman, Trans. ASM, 61, 680-93 (1968).
18. F. F. Lange, B. I. Davis, and D. R. Clarke, J. Mater. Sci., 15 [3], 601-10 (1980).
19. R. E. Loehman, J. Amer. Ceram. Soc., 62 [9-10], 491-94 (1979).

20. A. G. Evans and S. M. Weiderhorn, J. Mater. Sci., 9 [2], 270-78 (1974).
21. R. L. Tsai and R. Raj, J. Amer. Ceram. Soc., 61 [9-10], 513-17 (1980).

DISCUSSION

S. Dutta (NASA-Lewis): Do you think that HIPing of silicon carbide is a viable process in view of the fact that it requires very high temp (> 2000°C) for densification and there exists a "canning" problem at that temperature because of reaction with the canning materials?

Author: Yes. We have HIPed silicon carbide above 2000°C in "cans." With appropriate effort the microstructure of HIPed silicon carbide will be optimised and reliable production HIP facilities for ⩾ 2000°C developed. Work is already going on in this area.

R. Rice (NRL): Have you seen any suggestion of an inverse relation between room temperature $K_{1}c$ and high temperature creep resistance? Our work on $K_{1}c$, although having considerable scatter, suggests that there is such an inverse trend.

MICROSTRUCTURAL CHANGES DURING HOT ISOSTATIC PRESSING OF SINTERED LEAD ZIRCONATE TITANATE

K. G. Ewsuk and G. L. Messing

Department of Materials Science and Engineering

The Pennsylvania State University

INTRODUCTION

The production of dense, fine-grained ceramics is dependent on the control of both powder properties and fabrication. Although considerable effort is expended to obtain these controls, it is still difficult to consistently produce ceramics with dense, homogeneous microstructures. Efforts to gain greater control over microstructure have centered around the production of highly reactive powders with consistent properties, the utilization of densification aids during sintering, and hot pressing. An alternative approach is to utilize hot isostatic pressing as a post sintering processing technique for the removal of residual porosity in sintered ceramics. It should be understood, of course, that it is necessary for the residual porosity to exist as closed pores to be affected by pressure transmission during HIP. HIPing of sintered ceramics has been shown to yield higher densities in some ceramic systems;[1-4] however, there has been little effort to determine what happens to the microstructure during this process.

To determine the effect of HIPing on microstructure of sintered ceramics, macropores of 100 μm were placed in the microstructure of PZT and changes in this porosity were monitored as a function of HIP time and pressure. Pores of such large size were used to facilitate their examination and to determine whether defects of this magnitude could be eliminated by HIPing.

EXPERIMENTAL

A commercial lead zirconate titanate powder (UPI 401[a]) of composition $(Pb_{0.94}Sr_{0.06})(Zr_{0.53}Ti_{0.47})O_3$ was used exclusively throughout this study. To enhance sintering reactivity, the greater than 6 micrometer fraction of the particle size distribution of this powder was removed with an Acucut air classifer.[b]

The powder was prepared for pressing by adding 2 volume percent of an acrylic wax emulsion (Rhoplex B-60A[c]). Macropores of a distinct size and geometry were introduced into the microstructure by adding 100±5 micrometer polymethyl methacrylate spheres[d] to the green powder. Each sample was prepared by hand mixing 0.0025 grams of these spheres (approximately 5,000 in number) into 3 grams of the powder-binder mixture. To reduce preferential pressing defects, pellets were first formed by uniaxially pressing at 35 MPa in a 1.27 cm die, followed by isostatic pressing at 172 MPa. The organic spheres and binder were burned out by heating at a rate of 10°C/minute to 500°C and holding for 1 hour. To determine the effects of HIP on samples without gross porosity, pellets were also formed without the macropore addition.

Prior to HIPing, samples were sintered for either 1 hour at 1320°C or for 15 minutes in a gradient furnace. The former technique was used to produce relatively high density samples while the latter was used to produce samples having a range of densities. In both cases, the pellets were sintered in an oxygen atmosphere so that the gas entrapped in closed pores could readily diffuse through the sample during HIPing. Volatilization of lead oxide during sintering was reduced by packing the pellets in a combination of green and sintered PZT powders and enclosing in a platinum-lined alumina crucible.

For HIPing the sintered samples were packed in a platinum-lined alumina crucible with a lead oxide source and HIPed with argon at 1300°C. Low pressure HIP experiments were conducted at 20.7 MPa for times of 7.5 to 60 minutes in a non-commercial HIP vessel (Fig. 1). High pressure experiments were conducted at 138 MPa for 1 hour in a laboratory HIP unit.[e] A typical HIP cycle involved first heating the samples to 1300°C, applying pressure for the requisite time, then rapidly reducing both temperature and pressure (Fig. 2).

---

[a]Ultrasonic Powders, Inc., South Plainfield, NJ 07080.
[b]Donaldson Company, Inc., Minneapolis, MN 55440.
[c]Rohm and Haas, Philadelphia, PA 19105.
[d]Polysciences, Inc., Warrington, PA 18976.
[e]IsoHipper, Autoclave Engineers, Inc., Erie, PA 16512.

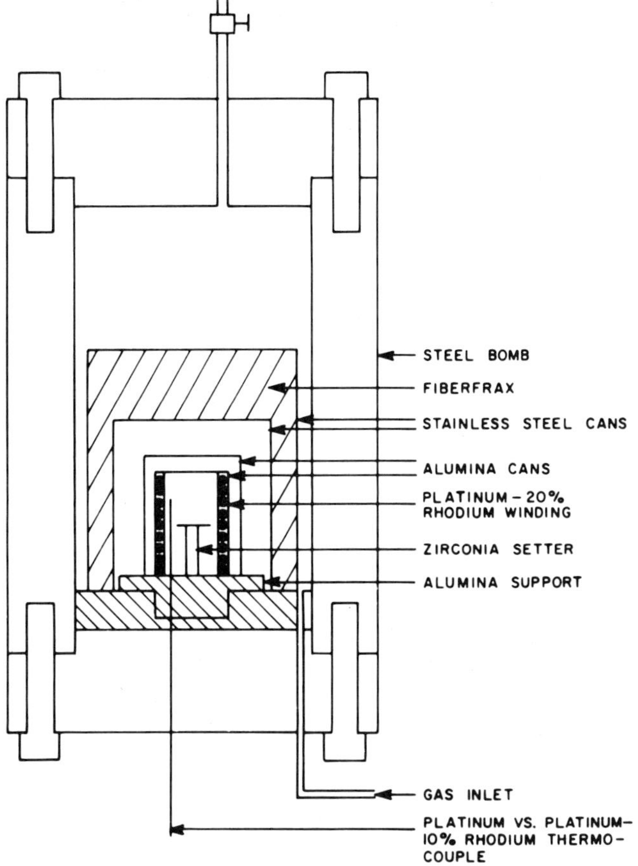

Fig. 1. The hot isostatic press used for low pressure experiments.

Densities of the sintered and HIPed specimens were determined by Archimedes' method. Samples were prepared for microstructural analysis by cutting pellets in half, polishing with 0.25 μm diamond paste, and etching with a solution of 70 v/o H_2O -29.5 v/o HNO_3 -0.5 v/o HF at 80°C for 40 seconds. Macropore size was determined by the average linear intercept method.[5]

RESULTS AND DISCUSSION

Sintering

An average density of 97.3% of theoretical was obtained for samples sintered for 1 hour at 1320°C. Samples with and without macropores sintered to the same relative density, indicating that the addition of macropore formers does not significantly affect the sintering of PZT. The average diameter of macropores was 123 micrometers, which is substantially larger than the 100 micrometer diameter spheres used to form them. Green samples were examined after organic removal and it was substantiated that all of the pore growth occurred during sintering.

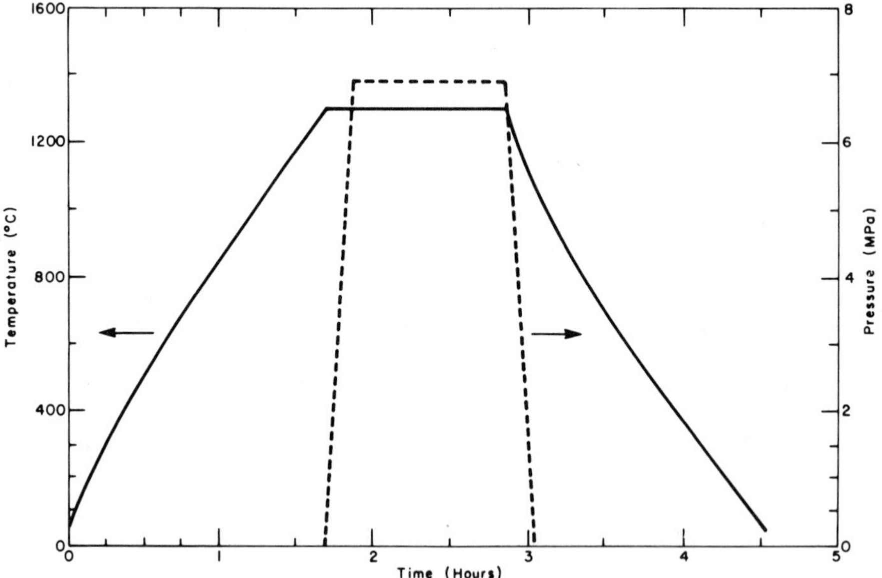

Fig. 2. A typical pressure-temperature cycle used in the HIP studies.

A characteristic microstructure of the sintered PZT is shown in Fig. 3. The distinct spherical geometry of the macropores, and the surrounding dense microstructure are clearly shown in Fig. 3a. At higher magnifications, the grains at the pore surface are seen to be equiaxed and smooth (Fig. 3b).

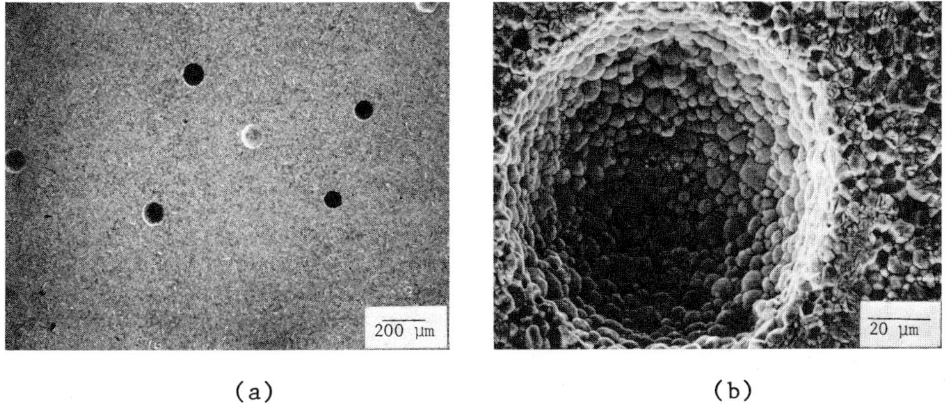

(a) (b)

Fig. 3. Photomicrograph of macroporosity in sintered PZT at (a) 45x and (b) 650x.

Low Pressure HIPing

The densities of the samples after sintering and HIPing as a function of time are shown in Tables 1 and 2. HIPing is seen to improve the density of sintered PZT for all times studied, with the majority of densification occurring in less than 15 minutes. Little or no additional densification is observed at the longer HIP times. This suggests that densification occurs by a two-stage process, with a rapid process controlling the initial stage and a slower process controlling final densification. The results also indicate that 98% of theoretical density is an end point density for these HIP conditions.

Table 1. Density of PZT Without Macropores, After Sintering for 1 Hour at 1320°C, and After HIPing at 1300°C and 20.7 MPa.

HIP time (min)	Sintered Density (%)	HIPed Density (%)
0.0	97.32	---
7.5	97.10	97.67
15.0	97.80	97.87
30.0	97.70	97.82
60.0	96.10	97.80

Table 2. Density of PZT With Macropores, After Sintering for 1 Hour at 1320°C, and After HIPing at 1300°C and 20.7 MPa.

HIP time (min)	Sintered Density (%)	HIPed Density (%)
0.0	97.32	---
7.5	97.63	98.05
15.0	97.07	97.66
30.0	96.90	97.63
60.0	97.80	97.92

A characteristic microstructure of HIPed PZT is shown in Fig. 4. Macropores in HIPed PZT do not show the spherical perimeter characteristic of those in the sintered samples. Instead, macropore surfaces have a rough appearance, indicating that structural rearrangement occurred (Fig. 4a). This may be due to particle rearrangement during HIPing, a process that would also explain the rapid initial densification. Localized microstructural variations are also apparent in HIPed PZT, including what appears to have been a liquid phase during HIPing. Microprobe analysis has shown this localized concentration and the crystals on the pore surfaces to be $PbTiO_3$-rich PZT (Fig. 4b). Previous studies[6,7] have reported the presence of a liquid phase during the sintering of PZT, but only at grain boundaries, and not in such large amounts. The crystals are observed only in HIPed PZT. Therefore, it is concluded that pressure is responsible for these localized variations in microstructure. The presence of a liquid phase gives additional evidence to support grain rearrangement during HIPing as this is commonly observed in liquid phase sintering.[8,9] Solution-precipitation also frequently occurs when sintering in the presence of a liquid phase.[8,9] Thus, it is likely that a combination of both grain rearrangement and solution-precipitation are responsible for the rapid initial densification observed on HIPing.

The change in macropore size as a function of HIP time is given in Table 3. It is seen to continually decrease with HIP time, shrinking from 123 to 85 micrometers in diameter in 1 hour. This amounts to a 67 volume percent reduction of the macroporosity. The majority of this reduction occurs in a relatively short period of time, the macroporosity shrinking to half its initial volume within the first 15 minutes of processing. Additional shrinkage is observed at longer times, but the rate of shrinkage is reduced considerably. These trends are clearly demonstrated in the plot of volume percent reduction of macroporosity versus HIP time shown in Fig. 5. As was the case for the change in density as a function of HIP time, the kinetics again indicate a two-stage process. The rapid initial shrinkage of the macroporosity supports the belief that grain rearrangement and solution-precipitation control the initial stage.

Table 3. Change in macroporosity as a function of HIP time at 1300°C and 20.7 MPa.

HIP Time (min)	Average Macropore Diameter (μm)	Volume (%) Reduction of Macroporosity
0.0	123	0.0
7.5	100	46.3
15.0	98	49.3
30.0	90	60.8
60.0	85	67.0

High Pressure HIPing

To determine the effects of HIPing with high pressure as a function of starting density, sintered samples with densities ranging from 90 to 98% of theoretical were HIPed for 1 hour at 138 MPa. As was evident in the low pressure experiments, the densities of the sintered samples improve with HIPing (Fig. 6).

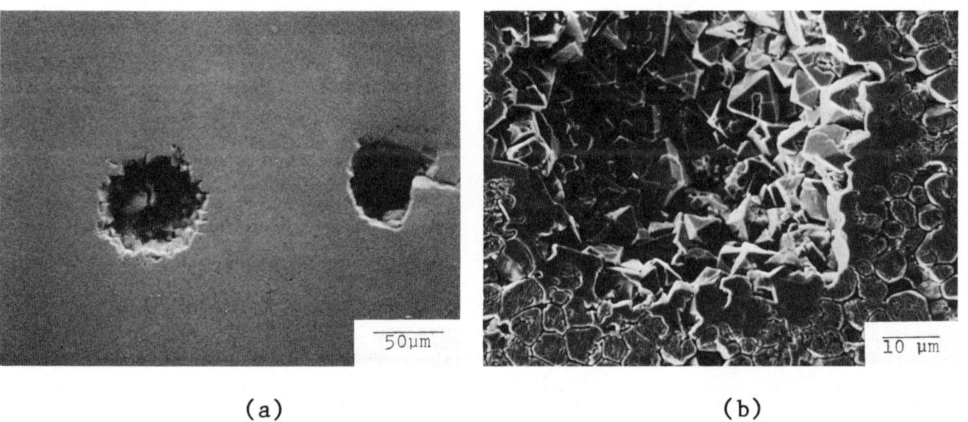

(a) (b)

Fig. 4. Photomicrograph of PZT HIPed for 7.5 minutes at 1300°C and 20.7 MPa: (a) indicating grain rearrangement at macropore surfaces; and (b) showing the crystals and liquid phase present at macropore sites.

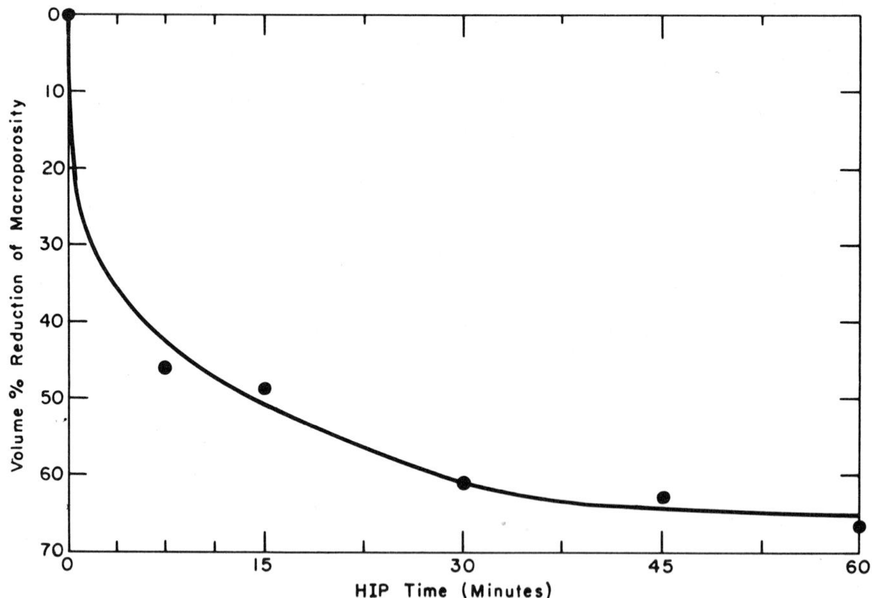

Fig. 5. The volume percent reduction of macroporosity as a function of HIP time at 1300°C and 20.7 MPa.

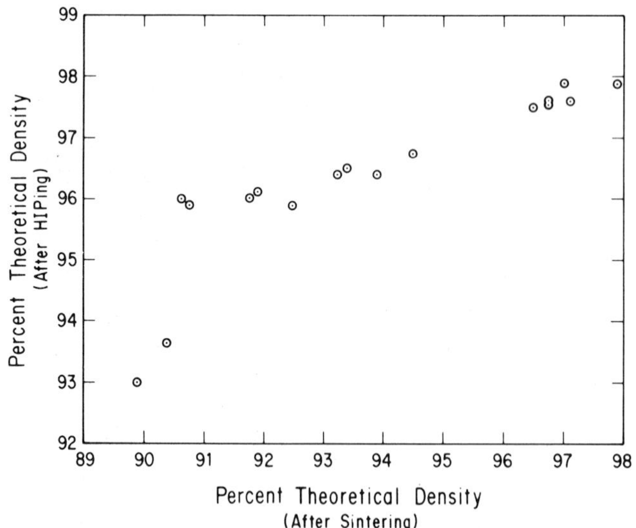

Fig. 6. HIPed density as a function of the sintered density of PZT after HIPing at 138 MPa for 1 hour.

As expected, the largest improvements in density were seen in the lower density sintered samples. The lowest density samples improved only a few percent during HIPing. This increase is attributed to additional sintering and not to HIPing as these samples have open porosity at the beginning of the HIP cycle. There is an appreciable increase in HIPed density as a function of sintered density at approximately 90.5% of theoretical. This sintered density closely corresponds to the value for the closed porosity stage as predicted by Budworth.[10] Interestingly, there is not a plateau in HIPed density above this critical value as has been reported in the literature.[11] Instead, there is a small slope in the curve with the highest density being obtained in the sample with the highest sintered density. This suggests that 1 hour may be insufficient time for pores to shrink completely to obtain theoretical density in the samples, even with 138 MPa pressure. Microstructural examination of these samples revealed macropores of similar size to those placed in the samples HIPed at the lower pressure. Thus, the end point density of 98% can be partially attributed to the inability to remove these large pores during HIPing.

CONCLUSIONS

It is concluded from these studies that macropores in sintered PZT shrink appreciably by a two-stage process during hot isostatic pressing. In the first stage, a rapid size reduction occurs via a combination of rearrangement and solution-precipitation due to the high concentration of a liquid phase at pore surfaces. The results of high pressure HIP experiments indicate that processing related macropores are responsible for end point densities that are less than theoretical. Thus, it is concluded that HIP cannot be used to completely remove large processing related voids. However, with the proper control of powder processing and sintering to avoid these large defects, HIPing can be a viable means of removing residual porosity in sintered ceramics.

ACKNOWLEDGMENTS

The authors acknowledge the financial support of the Applied Research Laboratory under contract with the U.S. Naval Sea Systems Command at Pennsylvania State University.

REFERENCES

1. K. H. Hardtle, Phillips Tech. Rev., 35 [2/3], 65-72 (1975).
2. K. H. Hardtle, Am. Ceram. Soc. Bull., 54 [2], 201-207 (1975).
3. U. Engel and H. Hubner, J. Mater. Sci., 13, 2003-12 (1978).

4. L. J. Bowen, W. A. Schulze, and J. V. Biggers, Powder Met. Int., 12 [2], 92-95 (1980).
5. R. L. Fullman, Trans. AIME, 197 [3], 447-52 (1953).
6. A. I. Kingon, Studies in the Preparation and Characterization of Selected Ferroelectric Materials (Ph.D. Thesis), University of S. Africa (1981).
7. E. K. W. Goo, R. K. Mishra, and G. Thomas, J. Ceram. Soc., 64 [9], 517-19 (1981).
8. W. D. Kingery, J. Appl. Phys., 30 [3], 301-10 (1958).
9. F. Thummler and W. Thomma, Metal. Rev., 12 [115], 69-108 (1967).
10. D. W. Budworth, Trans. Br. Ceram. Soc., 69 [1], 29-31 (1970).
11. H. Fischmeister, Powder Met. Int., 10 [3], 119-22 (1978).

DISCUSSION

R. Rice (NRL): Is there an increase in local porosity around the larger pores? It appeared to me that there may have been an increase in smaller pores around the larger pores after HIPing, e.g., suggestive of grain boundary sliding as a factor in the attempted closure of the pores. Also, what was your evidence for a liquid phase around the pores?

Author: The smaller pores observed in the microstructure after HIP are due to grain pullout during sample preparation. Pores that would occur concurrently with grain boundary sliding would be too small to be observable by SEM. The evidence for a liquid phase is the dense region around the pore. This material is observed only in HIPed samples and cannot be etched to reveal any grain boundaries. EDXA in the SEM indicates this region has a different chemical composition than the PZT grains.

R. Raj (Cornell Univ.): (a) The beautiful, round micropores you have shown are suggestive of a vapor bubble. If so, this may explain why full density is not obtained in spite of applied isostatic pressure. (b) You show how starting densities also lead to low finish densities after HIPing. Normally the HIPed density should be insensitive to the starting density as long as open porosity is not present. Do you have a microstructural basis for interpretation of your results?

Author: (a) Invalid question; these were the macropores intentionally introduced for modelling purposes. (b) No, we do not have a microstructural basis for the sharp increase in the HIP density versus the sintered density. However, the density at which a pronounced effect of sintered density is observed is consistent with theoretical prediction and other investigators' density values for the closed porosity stage.

W. S. Coblenz (NRL): Have you observed reduction or lead loss during HIPing of PZT?

Author: No. Precautions were taken to limit weight loss during both sintering and HIPing. Small weight losses were observed only during sintering (~ 0.3 wt%).

R. J. Gottschall (U.S. Department of Energy): As a complementary effort, you might wish to consider characterizing the dynamics of the porosity structure (pore size, distribution, number of pores) by means of small angle x-ray or neutron scattering at the Center for Small Angle Scattering at Oak Ridge National Laboratory.

D. R. Biswas (IT & T, Roanoke, VA): (a) Do you see any differences in electrical and mechanical properties between the porous sintered and HIPed PZTs? (b) What are the limitations of achieving more than 99% theoretical density in both cases?

Author: (a) Property measurements were made on samples sintered for 1 hour and those HIPed as a function of time at 3000 psi. As would be expected from the small difference in density between the two (~ 1%), HIPing had no effect on the unclamped dielectric constant, dissipation factor, piezoelectric coefficient d_{33}, radial coupling coefficient, radial frequency constant, and Young's modulus. However, by reducing the size of large voids in the samples, some improvement in dielectric breakdown strength was observed. (b) The major limitation to achieving greater than 99% of theoretical density seems to be the macroscopic porosity that is introduced as a function of processing.

PART VIII

DYNAMIC COMPACTION

DYNAMIC COMPACTION OF POWDERS

Rolf Prümmer

Fraunhofer-Institut fur Werkstoffmechanik
Rosastr. 9
7800 Freiburg, West-Germany

INTRODUCTION

The destructive application of explosives unfortunately is well known, although the peaceful application for instance in mining, quarrying and excavating is several centuries old. The constructive application of explosives has been developed in the last decades. Nowadays explosive forming, welding and cutting are in common use and the most unique method is explosive compaction. This method profits from the direct application of high detonation pressures developed by explosives and ranging up to about 300 kbars. High densities of the compact up to 100% of the theoretical density with considerable green strength are attainable.

The first device was developed by La Rocca and Pearson in 1958 using an explosively driven piston being accelerated into a porous material or powder.[1] A modified gun barrel for compaction of powders was built by Breycha and McGee in 1962.[2] In most cases, projectiles accelerated by a gun to high velocities impacting with a powdered material contained in a die are used to study the dynamic behavior of porous substances.[3]

The direct method is rather simple. The arrangement consists of a mild steel tube with end plugs containing the powder and surrounded by a uniform layer of explosive.

Such arrangements have only been used on a commercial basis. It is the purpose of this paper to describe the procedure, its fundamental aspects and the properties of the compacts.

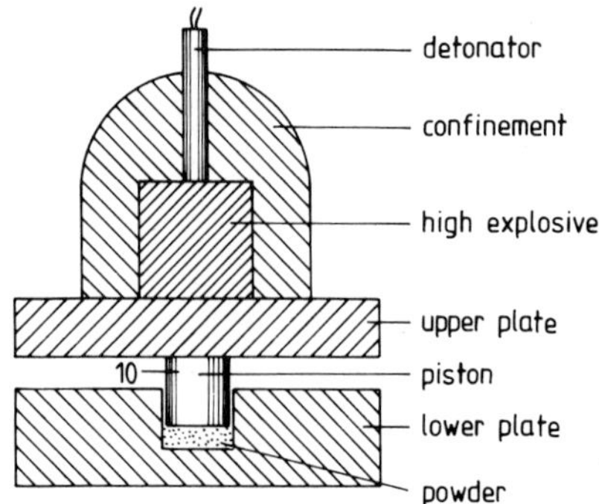

Fig. 1. Single piston arrangement for explosive compaction of powders.

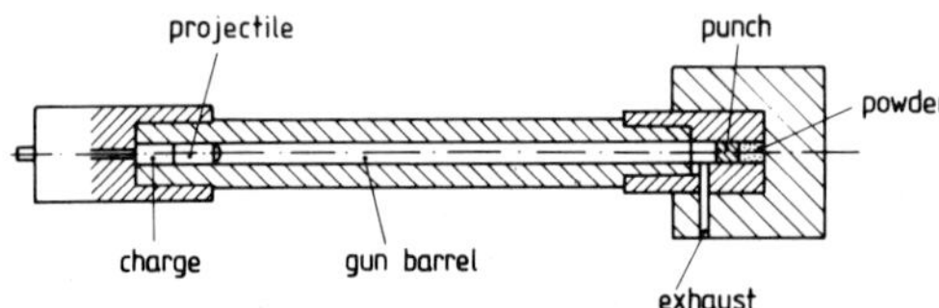

Fig. 2. Modified gun barrel used for explosive powder compaction.

SHAPE OF THE SHOCK WAVE FRONT

The arrangement of the direct method of explosive compaction is shown in Fig. 3. The detonation is initiated at the upper end and a detonation wave proceeds in the axial direction. The detonation velocity depends on the type and amount of explosive and the higher it is the greater is the pressure acting on the container wall and the powder. The amount of the explosive is a measure for the duration of the pressure.

Both the pressure and its duration determine the starting intensity of the shock wave acting on the container wall. As it is proceeding toward the center of the cylindrical sample it is meeting powder in front of it and leaving compacted material behind it. There are two opposing effects.

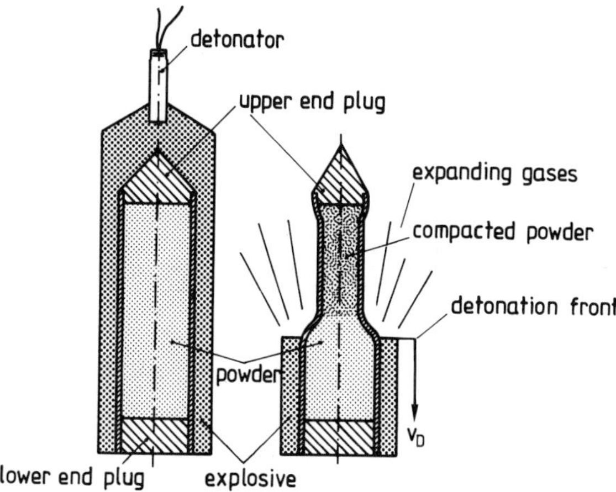

Fig. 3. Assembly for direct explosive compaction of powders to cylindrical rods.

(1) Due to the cylindrical arrangement the intensity of the shock wave is increasing toward the center, becoming infinite in the center.

(2) As compaction of the powder occurs, work is performed, leading to an attenuation of the shock wave.

Both effects oppose and in the ideal case there is balance between the two.

If the absorption during compaction is too severe and a decreasing intensity of the shock wave is the result of it, uncompacted material can be left in the center. On the other hand, if absorption is low the convergive nature can overbalance. A reflected wave at the center lead to fracture of the already compacted sample. In severe cases, when the converging shock wave is forming a "Mach-stem" in the center, molten material is thrown out leaving a hole in the center of the compacted cylindrical sample. Several types of the shape of the shock wave are shown in Fig. 4. As the shock wave velocity is increasing with pressure three principal types are possible. These were studied by means of x-ray flash patterns,[4-6] electric pins or by computer modeling.[7] An empirical rule is able to qualitatively describe the different configurations, and to optimize experimental results iteratively.[8] Figure 5 shows a sample of compacted IN-100-powder with compacted material in the outside and center region and insufficiently compacted powder in between. In this case it is required to increase the pressure to some extent in order to obtain a uniform compaction. This can be achieved by the same amount but different type of explosive with a higher detonation velocity or by increasing the tap density of the same explosive. The ideal case is the one with uniform pressure over the cross section of the sample, the invariant shock wave velocity leading to a conical shape of the shock wave front. In this case it is a single shock which does not reach high values at the center of the specimen, so that no substantial reflection of shock waves upon themselves can occur.

DYNAMIC COMPRESSIBILITY OF POWDERS

During the passage of the shock wave high loading rates occur. The associated pressures greatly exceed the shear strength of the material to be compacted. Also high transient temperature rises occur. Therefore, the state of the material under the high pressure shock loading can be considered as hydrostatic loading. The pressure-volume relationship for powder materials, of course, differs quantitatively from the corresponding curve for solid materials. Qualitatively the shock wave behavior of a porous material can be described as follows (Fig. 6). Starting from the specific volume V_0, powder compaction takes place along the Hugoniot curve A up to the state $(V_1 P_1)$. The release adiabat, curve B, describes the unloading up to the end volume V_E at ambient pressure. The end volume V_E is a measure for the achieved compaction.

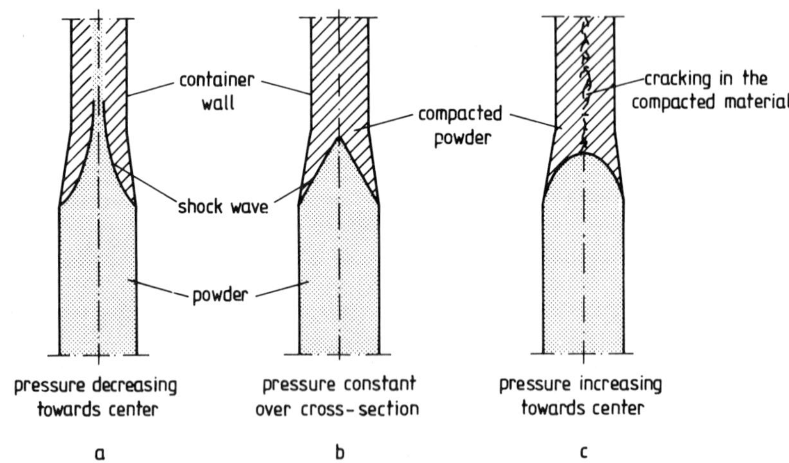

Fig. 4. Typical shapes of shock wave fronts in explosive compaction: (a) underimpacted; (b) correct; (c) overcompacted.

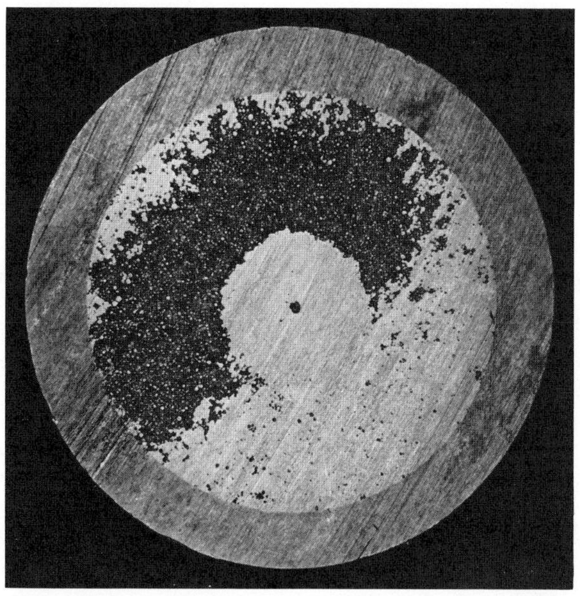

Fig. 5. IN-100-powder, explosively compacted at slightly too low pressure, 20 mm diameter.

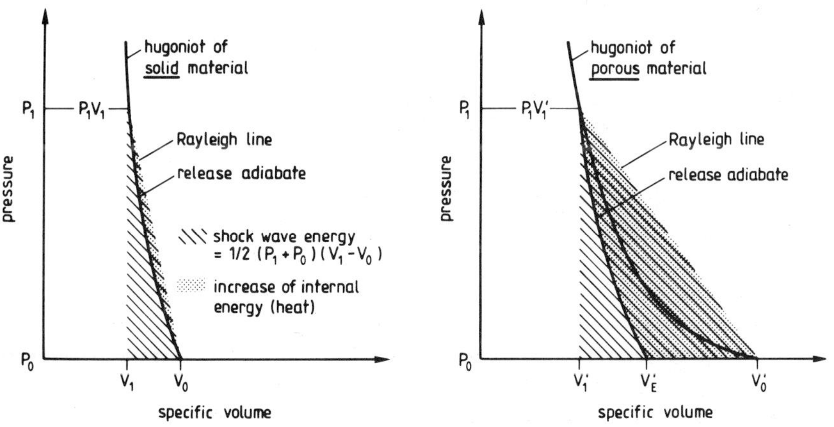

Fig. 6. Pressure-volume relationship for explosive compression of solid and porous matter.

In the case of a solid material the Hugoniot and the release adiabatic curves are approximately the same, providing the pressures are not too high. The total work necessary for compaction of the powder is represented by the triangle V_0-P_1-V_1. A part of it is released during unloading, area V_E-P_1-V_1. The difference (dotted area) of both represents the increase in internal energy due to one cycle (loading and unloading). As can be seen from the schematic diagram the increase of internal energy is much higher in powdered substances than in solid materials. Its increase is higher the greater the starting volume $V_{0,\ porous}$. Most of the internal irreversible energy is dissipated as heat.

The heat is not generated uniformly over the whole volume of the sample during explosive compaction. It is more or less limited to the surface of the individual particles. As the shock wave passes, there exists a lot of relative movement of the particles over one another. Therefore friction causes rapid temperature rises. It is also possible that individual particles collide under such conditions that explosive welding and jetting is likely to occur. The effect of friction is enhanced even more when mixtures of powders of different materials, exhibiting different values of shock impedance, are explosively compacted. It is indeed observed that mixtures of powders seem to have a larger "window" for explosive compaction than the corresponding pure substances do.

It is also easy to understand that not only the type of the powder but also its grain size, grain size distribution and grain shape is of great importance to the performance of explosive compaction.

THE PARAMETERS FOR EXPLOSIVE COMPACTION

In the first pioneering attempts of explosive compaction[1,9] high explosives were used. High explosives develop high detonation pressures and it was the wish to achieve high densities in the green compacts.

However, it was observed that ceramic substances did not cohere, and in order to cohere, some ductility of the powder was required.

It was shown later[10] that the necessary amount of explosive applied during direct explosive compaction is related to the compressive yield strength of the powdered material. Later a procedure[8] of optimizing the parameters for explosive compaction was described emphasizing the importance of the type of explosive, as its detonation velocity is a measure for the pressure acting on the powder. In an E/M versus the square of detonation velocity plot (upper part of Fig. 7) a curve is established experimentally which gives the parameters for correct compaction with uniform density over the cross-section of the sample with no cracking. It separates the region of undercompacted and overcompacted samples. The higher the detonation velocity v_D (the higher the pressure) the lower is the amount of explosive E/M necessary for compaction. The detonation pressure is related to the detonation velocity and the specific weight by the following equation

$$p = \left(1 - \frac{\rho_o}{\rho_s}\right) \cdot \rho_o \cdot v_D^2 \approx \frac{1}{4} \cdot \rho_o \cdot v_D^2$$

where ρ_o is the density of the crystalline explosive and ρ_s is the density of the explosive in the reaction zone, which is approximately 4/3 of the density of the crystalline explosive. Therefore the abscissa in Fig. 7 is linearly related to the detonation pressure. The lower part of Fig. 7 shows the achieved density (relative density) versus the square of the detonation velocity. The most important feature of this curve consists of a maximum of density at a certain value of the square of the detonation velocity (pressure). If higher detonation velocities are chosen, the compact is overcompacted: due to a strong release wave the already achieved bonding between individual particles as described above is disrupted. Voids created this way lead to smaller densities of the compact even though the applied pressure was higher. This effect is more pronounced the higher the (excess) detonation pressure. In order to achieve optimum results an explosive has to be chosen with a detonation velocity where maximum density is observed (point D). The appropriate E/M value can be taken from the upper graph in Fig. 7, following the dashed line up to point F. The optimal pressure value, where maximum density is observed at point D, is dependent on the type of powder material to be compacted. Aluminum powder, for instance, only requires a low pressure (explosive with a low detonation velocity) whereas, for instance, a highly alloyed steel powder, needs a high pressure (high explosive) in order to achieve a compact with optimum density. Different explosive compaction curves for a variety of materials and mixtures (metal and ceramic powders) were established.[8,11,12] It is self-evident, that the observed values of E/M and detonation velocity are only valid as long as the container wall of the container is thin compared to its diameter. In case of thick-walled tubes, additional energy is required for plastic deformation of the container material with the result that higher E/M-ratios have to be applied. It also has to be taken into account that the convergence of the shock wave already increases the shock wave intensity in the container wall itself, leading to higher pressures.

The fact that excessive pressures can create temperature spots in the compact is dramatically shown during compaction of metallic glasses. Metallic glasses are alloys maintaining the amorphous non-crystalline state after rapid cooling in the order of 10^6°K/s. They can be produced as thin strips (~ 40 um thick) or powders.

There is no other way to consolidate such powders but explosive compaction. There is rapid heating of the surface of the particles only whereas the interior remains at lower transient temperatures during passage of the shock wave. In this way melting at the interface of individual grains can occur and subsequent rapid cooling leads to an amorphous state of the interphase. Amorphous phases observed in AlN[13] after explosive compaction are due to rapid quenching from high temperatures, too. If the pressure is chosen too high, excessive heat is generated, leading to recrystallized spots in the joint areas between the individual particles of the compact.

Figure 8 shows micrographs of a metallic glass powder, consisting of 89% Ni and 11% P, explosively compacted at different pressures. When compacted at 37 kbars high porosity is observed. The porosity disappears after increasing the pressure to 44 kbars, the compact remaining amorphous. At a pressure of 51 kbars, however, locally recrystallized areas in the micrograph can be observed (dentritic structure) and a starting recrystallization is observed by means of x-ray analysis.

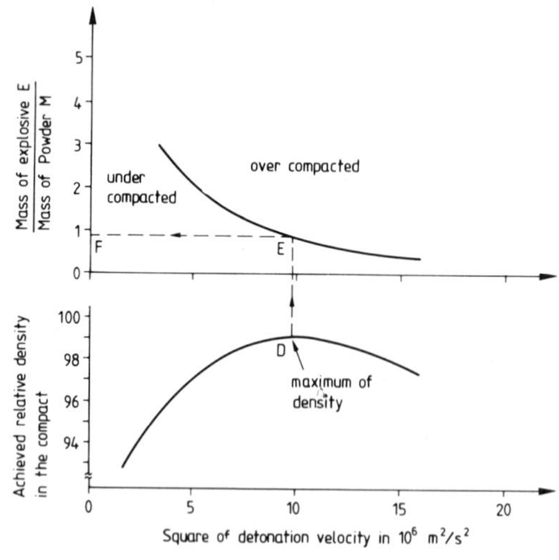

Fig. 7. Determination of the parameters for explosive compaction.

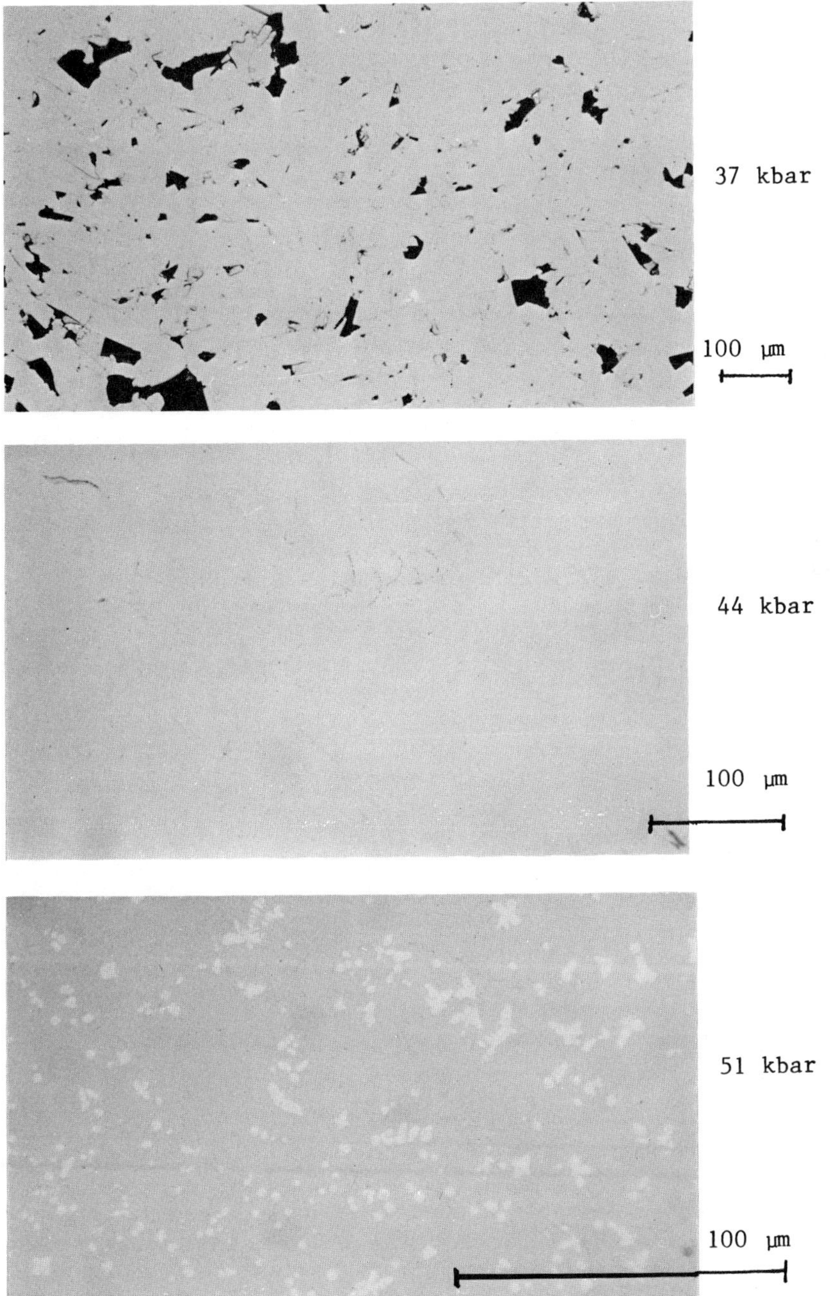

Fig. 8. Metallic glass Ni 89 P11 explosively compacted at different pressures.

For a given material, a shift towards smaller pressures in the upper part of the diagram in Fig. 7 material is possible by means of heating the powder during compaction. It is performed this way so that a heated sample is rapidly conveyed into a hollow explosive and the detonation is initiated then by means of an electric contact. As the strength of the material decreases with temperature, the use of lower pressures (explosives with lower detonation velocity) and lower quantities of explosive, compared to a "cold" compaction, leads to good compaction results.

For instance, the pressure required to obtain a maximum density of 99% of theoretical in cold explosive compaction of pure iron is 6.5 kbars at an E/M ratio of 1.3 whereas in hot compaction (800°C) the pressure is less than 3.6 kbars at an E/M-ratio of 0.8 and the obtained density is 100% of theoretical. Hot explosive compaction especially can be applied to high strength materials such as highly alloyed steel powder or chromium- or nickel-based superalloys and refractory alloys.

A set of compaction curves of the type shown in Fig. 7 is shown in Fig. 9 for different materials. At the lower part, the maximum density achieved uniformly over the cross-section of cylindrical samples of 25 mm starting diameter is shown.

With metallic materials the achieved densities in the compacts are very close to 100% of theoretical density. However, with ceramic materials maximum obtainable density of the compact with uniform cross sectional density is around 95% of theoretical or less. The reason, therefore, is the fact that in explosive compaction the process of densification starts outside and is progressing towards the center with the shock wave front (Fig. 10). The outside material has to undergo plastic deformation in order for the process of consolidation to proceed toward the center of the sample. With metallic materials the plastic deformation is no problem, however, ceramic materials are essentially brittle. If the density of the compacted outer portion of the cylindrical sample of ceramic material is too high, it has to be crushed during the compaction and this happens in a type, which is often observed, it is shear banding. Figure 11 shows such a shear banding in explosively compacted boron carbide B_4C and Fig. 12 in Al_2O_3. Spirally coiled Nb foils of 0.1 mm thickness put into the Al_2O_3-powder before compaction are an indicator of the amount of shearing.

The effect of adiabatic shearing is less pronounced at lower densities of the sample and can be avoided by limiting the desired density of the cylindrical compact. The maximum density achievable in alumina without shear bands in the compact is dependent on the grain size of the alumina powder.

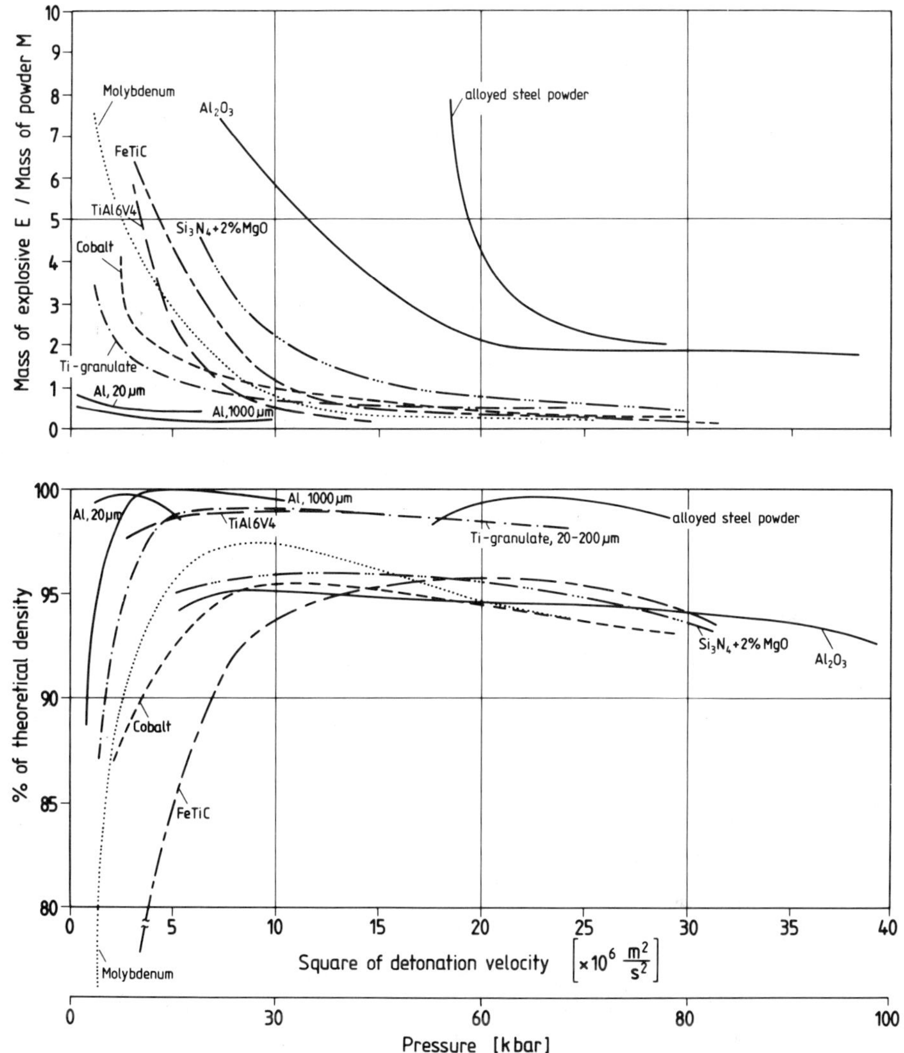

Fig. 9. Determination of parameters for explosive compaction of different metal and ceramic powders.

STRUCTURAL CHANGES IN CERAMICS AFTER EXPLOSIVE COMPACTION

It was shown by means of x-ray analysis that after explosive compaction of powders of ceramic materials severe lattice distortions are introduced, as in Al_2O_3, ZrO_2, SiC and B_4C,[14] MgO,[15,16] TiN[16] and several refractory compounds.[17]

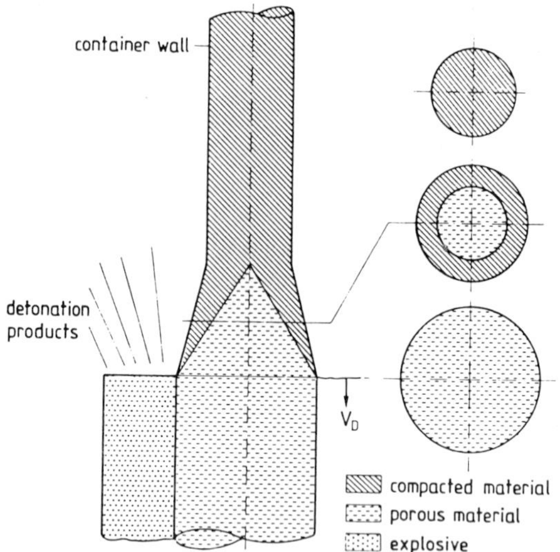

Fig. 10. Dynamic compaction and cross section of compacted area growing from outside to the center.

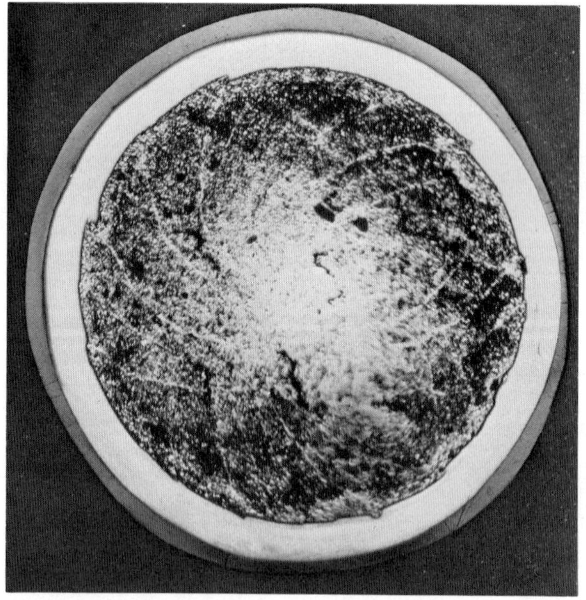

Fig. 11. Cross section of sample of B_4C after compaction with shear-band formation, ~ 20 mm ϕ.

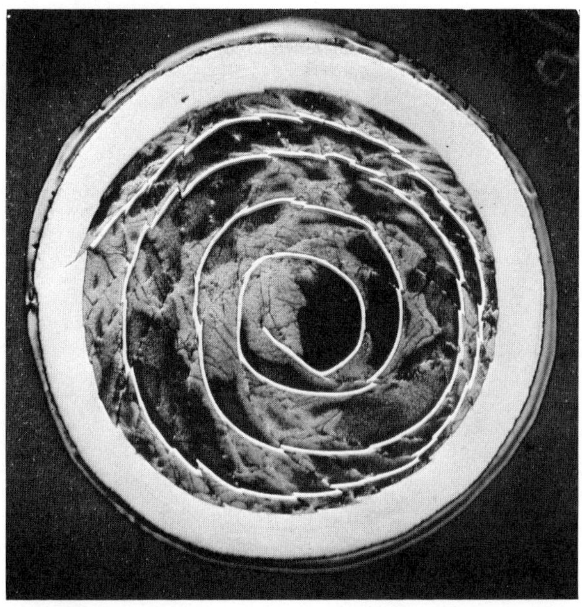

Fig. 12. Cross section of sample of Al_2O_3 after compaction; with Nb-foil spiral, indicating the shearing, occurring at too high a chosen density.

Investigations taking into account the pressure during explosive compaction were performed with alumina powders of different grain sizes.[18] It is shown that the explosive compressibility is very much different for a fine and a coarse grained alumina powder; e.g., note that maximum density is obtained at different pressures. As a function of pressure three ranges are found for the occurrence of lattice defects, the variation of density, particle size and crystallite size for both powders. Similar ranges are also found for metal powders in explosive compaction.[19] Figure 13 gives the lattice distortions introduced into the cubic lattice of Al_2O_3 after explosive compaction as a function of the square of detonation velocity of the explosive, which is linearly related to the pressure. The saturation in the lattice distortion in the coarse grained alumina sets in earlier. The ultimate value of the lattice distortions is greater in the fine grained alumina and amounts to 0.33%. For comparison, the numbers are given for a severely milled powder (157 hrs.) and for a shocked sintered body of alumina. The activation of the powder is highest after explosive shock treatment. From these x-ray data the distortion energy in the fine grained alumina can be estimated to be 0.77 cal/g. This is a high value and comparable to such values obtained in metals after severe plastic deformation. It was also shown in

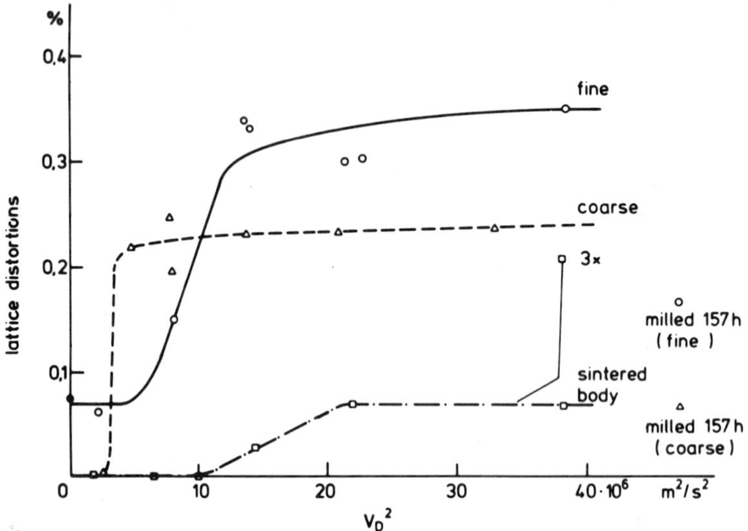

Fig. 13. X-ray investigation of lattice distortion in explosively compacted alumina of 5 μm and 300 μm grain size as a function of square of detonation velocity.

annealing experiments, that the explosively activated alumina shows a much faster growth of crystallite size with temperature (Fig. 14). The data are in qualitative agreement with sinter-investigations performed on explosively shocked alumina at NCSU.[20] These encouraging features of explosive consolidation of ceramic materials are considered to provide motivation for future investigations.

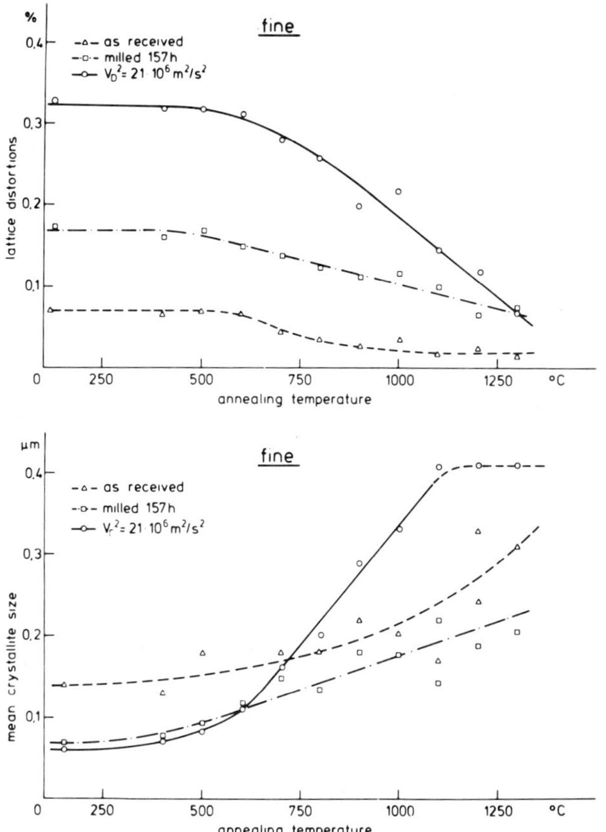

Fig. 14. Variation of lattice distortion and mean crystallite size of explosively compacted alumina powder of 5 µm and 300 µm grain size with annealing temperature.

REFERENCES

1. J. Pearson and E. W. La Rocca, Rev. Sci. Instr., 29, 848-50 (1958).
2. R. J. Brejcha and S. W. McGee, Am. Machinist, 106, 63-65 (1962).
3. R. R. Boade, J. Appl. Phys., 40 [9], 3781-90 (1969).
4. R. W. Leonard, V. Linse, and D. Laber, Proc. 2nd Int. Conf. of the Center for High Energy Forming, Estes Park, CO, 1969, 8.3.1-8.3.23.

5. A. A. Deribas and A. M. Staver, Fizika Gorenija i Vzryva, 10 [4], 568-78 (1974).
6. R. Prümmer, pp. 814-20 in High Pressure Science and Technology, edited by K. D. Timmerhaus and M. S. Burber, Plenum Press, 1979.
7. M. Wilkins and C. Cline, this conference.
8. R. Prümmer, Proc. 4th Int. Conf. Center for High Energy Forming, Vail, CO, 1973; see also: Ber. DKG, 50, 73-81 (1973).
9. J. Pearson, ASTM Creative Mfg. Seminar, Paper SP 60-158, 1960-1.
10. A. M. Staver, Paper 50 in Shock Waves and High Strain Rate Phenomena in Metals, Plenum Press, NY, 1981.
11. R. Prümmer and G. Ziegler, Proc. 6th Intl. Conf. on High Energy Fabrication, edited by R. Prümmer, Essen, Germany, 1977.
12. R. Prümmer, p. 814 in Proc. 6th AIRAPT Conf. High Pressure Science and Technology, edited by K. D. Timmerhaus and M. S. Barber, Plenum Press, NY, 1979.
13. C. L. Hoenig and C. S. Yust, Ceram. Bull., 60 [11], 1175-1224 (1981).
14. O. R. Bergman and J. Barrington, J. Am. Ceram. Soc., 49, 502 (1966).
15. R. W. Heckel and J. L. Youngblood, J. Am. Ceram. Soc., 51, 398 (1968).
16. A. Sawaoka, K. Kondo, and A. Tamotsu, Rept. Tokyo Inst. Tech., 4, 109-15 (1979).
17. G. A. Adadurov, O. N. Brensov, A. N. Dremin, and V. F. Tatsil, Poroshkovaja Metallurgija, 107 [11], 7-9 (1971).
18. R. Prümmer and G. Ziegler, Powder Met. Intl., 1 (1977).
19. R. Prümmer, Chapter 10 in Explosive Forming, Welding and Compaction, edited by T. Blazynski, 1983.
20. K. Y. Kim, A. D. Batchelor, and H. Palmour III, this conference.

DISCUSSION

W. H. Gourdin (LLNL): The explosive pressure increases as v_D^2, one expects that the stress delivered to the powder increases accordingly. From the Hugeniot curve the found density should then continue to increase. Can you explain why you find a maximum?

Author: It is often observed that the density of the compact decreases after passing a maximum value (Fig. 7 and Fig. 9), a condition called by some authors an anomalous compressibility. It is mainly due to a reflected wave from the center of the cylindrical specimen. At pressures up to the maximum of the density-pressure relationship, however, dynamic compaction curves seem to correspond fairly well[19] with isostatic compaction curves.

A. Sawaoka (Tokyo Inst. of Tech.): Could you explain the possibility of non-crack powder compaction with near theoretical density, using cylindrical technique?

Author: The great number of slides of compacted samples containing cracks perhaps gave the impression that crack-free samples cannot be obtained in explosive compaction. This is not true. Using the proper parameters, i.e., detonation velocity (= pressure) and E/M-ratio, cylindrical samples and samples with rotational symmetry[19] can be produced without revealing cracks. The parameters are not only dependent on the type of powder material but also on its grain size, grain shape and their distributions. With metallic materials densities very close to theoretical density are obtainable. With ceramic materials this density limit is lower.

DYNAMIC COMPACTION OF CERAMIC POWDERS

J. H. Adair, R. R. Wills, V. D. Linse

Battelle's Columbus Laboratories
505 King Avenue
Columbus, Ohio 43201

INTRODUCTION

Dynamic compaction is a powder compaction process that involves a combination of high pressures and high strain rates to densify powder materials. Emphasized in this review paper on dynamic compaction of ceramic powders are pressures obtained by explosive detonations which generate strain rates typically from 10^3 to 10^4 s^{-1} and reported pressures from 0.1 GPa to 100 GPa.

There are several potential advantages of dynamically compacting ceramic powders. For example, this process may provide an alternate powder consolidation route to conventional techniques. Many ceramic powders, most notably nonoxide ceramics, are difficult to consolidate using conventional processes or require sintering aids that are detrimental to final properties of the ceramics. Since dynamic compaction may be performed at relatively low temperatures, unique microstructures may be retained that are ordinarily lost through homogenization or recrystallization at the higher temperatures used in conventional consolidation techniques.

The major disadvantage of dynamic compaction is the lack of understanding of the fundamental material and processing variables required for consolidation of a particular material. Furthermore, despite the large amount of theoretical and practical research performed on the response of solid materials to shock waves,[1-4] there is little fundamental work on the response of powders to intense dynamic loading.[5,6] As Clyens and Johnson[6] point out, there has been "little attempt to relate macroscopic dynamic compaction behavior of a powder to the microscopic behavior of its constituent powder particles."

Another problem is the microcracks and/or macrocracks typically present in dynamically compacted bodies. Unless this basic technological problem is overcome, dynamic compaction will remain a scientific curiosity, not a feasible technological process. In this paper we review the research performed to date on the dynamic compaction of ceramics.

THEORETICAL ASPECTS

To obtain insight into the dynamic compaction process for ceramic powders it is worthwhile to start with a brief discussion of the response of solid materials at different pressure levels. There are a number of reviews on this topic.[1,2,10-13] The response of solids to different pressure regimes is summarized in Table 1. Chou[10] has given an excellent, but perhaps dated, discussion of these effects. For pressures below the yield point of a solid, the material behaves in an elastic manner. Materials subjected to pressures much greater than the yield point respond as compressible fluids. Equations of state may be used to describe solid response in a hydrodynamic manner to very high pressures. However, at intermediate pressures, the situation becomes much more complex. In moderate pressure regimes above the material yield point, the constitutive equations must take into account rate and hysteresis effects due to plastic deformation. Quite a few constitutive equations are available for solids, a few for foams, but virtually none for powders.[1-4,13-16]

Table 1. Responses of Solid Materials at Different Pressure Regimes (from Chou[10])

Pressure Regime	Material Response	Constitutive Equation	Governing Equation
High	Compressible Fluid (hydrodynamic)	Equation of State	Nonlinear
Above Yield	Finite-strain plastic	Complicated (rate effects, hysteresis, etc.)	Nonlinear
Below Yield	Linear Elastic	Hooke's Law	Linear

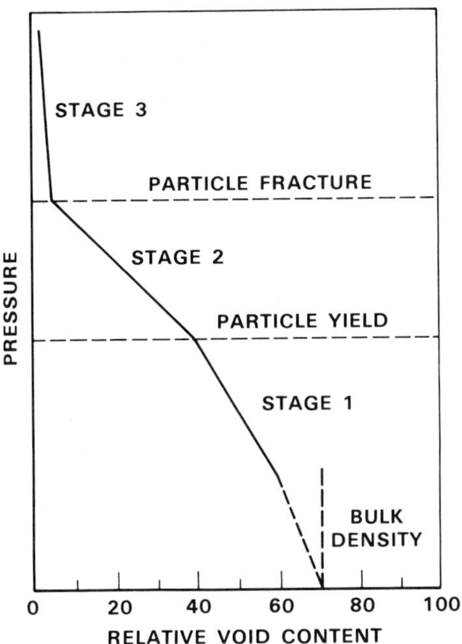

Fig. 1. Schematic compaction curve showing densification mechanism stages (from Donachie and Burr[17]).

The powder compaction work of Donachie and Burr[17] on metal powders has provided insight into the mechanistic steps in powder compaction. To resolve the mechanisms involved in powder compaction it is useful to show pressure as a function of relative void content. A schematic of a typical compaction curve plotted in this manner is given in Fig. 1. Three mechanisms of powder compact densification are indicated in the figure.

Void elimination in stage 1 is due to particle rearrangement and restacking. Practically, this mechanism may reduce relative void content to about 40% before particles lock together into a rigid three dimensional matrix. Densification and void elimination in stage 2 are due to plastic deformation of the particles when the yield point of the particle matrix is exceeded. If the fracture stress of the particle matrix is not exceeded, stage 2 plastic deformation should lead to total void elimination. However, for ceramic materials this is rarely the case. Stage 3 of powder compaction is due to particle fracture and cold working. This mechanism gives only a slight reduction in relative void content most of which is due to particle rearrangement after fracture.

Studies of strain rate effects on powder compaction behavior in the literature are scant.[5,6,18,19] Furthermore, there are few studies available for strain rate effects in the dynamic compaction strain rate regime from 10^3 to 10^4 s^{-1}. Roman et al.[20] have briefly discussed the relevance of mass velocity to the time required to plastically deform particles in a shock wave. Qualitatively, these studies have shown that strain rate effects are material-, particle shape-, and particle size-dependent. Generally, from studies reported for small strain rates, increasing strain rate reduces relative densification at a given pressure. At higher strain rates, apparently particle rearrangement does not have time to occur prior to plastic deformation and particle breakage.[5,6,17,19]

Reconciling the responses of solids with the responses of powder to high pressures is qualitative at best. The major difference between solids and powders is the particle breakage encountered in the latter stages of powder compaction. If strain rates are too high, void entrapment may occur. When the particle strength is exceeded, fracture may occur from the inside walls of voids. This effect has been demonstrated experimentally at low strain rates with hollow closed cylinders composed of brittle materials subjected to large hydrostatic pressures.[21] Spalling of material from cylinder inside surfaces occurred with eventual filling of cylinder centers by fracture particulates.

The discussion above indicates that for maximum densification of a powder the pressure should exceed the yield stress but be smaller than the fracture stress. This may be analogous to the weldability window found in explosive welding of metals.[21,22] Some insight into this requirement might be obtained using the concepts of deformation and fracture mechanism maps for specific materials developed by Ashby and coworkers.[7,8,9] Examples of a deformation map and fracture mechanism map for magnesia are given in Figs. 2 and 3, respectively. Stresses are normalized with respect to shear modulus for deformation maps and Young's modulus for fracture mechanism maps. Stresses are given as a function of homologous temperature (temperature divided by melting point).

Among the advantages of this approach is that strain rate and grain size are taken into account in the calculations. Comparisons of deformation stresses to fracture stresses may indicate criteria for dynamic compaction. However, a weakness of this method of representation is that maps are prepared using tensile stresses, not the compressive stresses generated in dynamic compaction. Furthermore, they are calculated using hydrostatic stresses, not the uniaxial compressive stresses usually resulting from shock waves. Therefore, some modifications to the equations used to generate deformation and fracture mechanism maps are probably required.

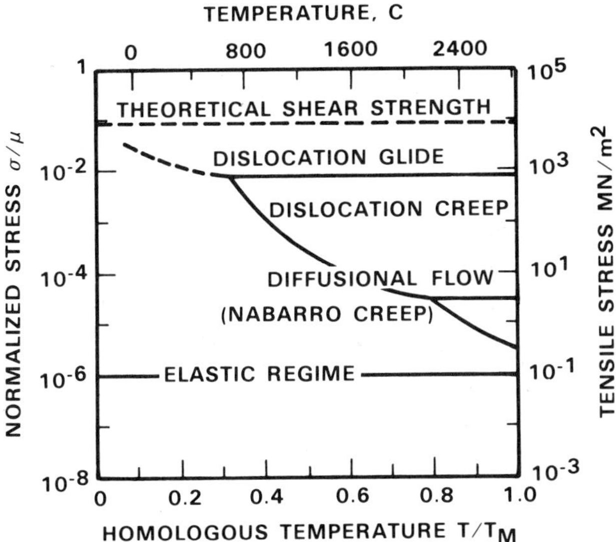

Fig. 2. Deformation map for magnesium oxide. grain size and strain rate used in calculations were 32 μm and 10^{-8} s^{-1}, respectively (from Ashby[7]).

The primary processing objectives for successful dynamic compaction of powders are void elimination (densification) and interparticle bonding (final body integrity). As described above, to achieve maximum void elimination, some plastic deformation of the particles must occur. Although some interparticle bonding may be achieved without total void elimination, consolidation to full density requires both void elimination and maximum interparticle bonding. Therefore, a prerequisite for successful dynamic compaction of powders is to maximize mass transport processes such as plastic deformation during the process.

Microstructural investigations of dynamically compacted aluminum nitride[24-26] and alumina and boron,[25,26] indicate the relative flow of particles past one another during compaction is important. Particle surface temperatures generated by this relative movement are high enough to melt or recrystallize surface zones on the particles which results in grain boundary formation. However, melt zones on particle surfaces may not be necessary for some ceramic materials to consolidate to dense bodies from dynamic powder compaction. Microstructure of a dynamically compacted aluminum nitride powder[24] indicates consolidation may be achieved without localized melting. Whether optimum interparticle bonding is attained through localized melting or some other mechanism must be resolved.

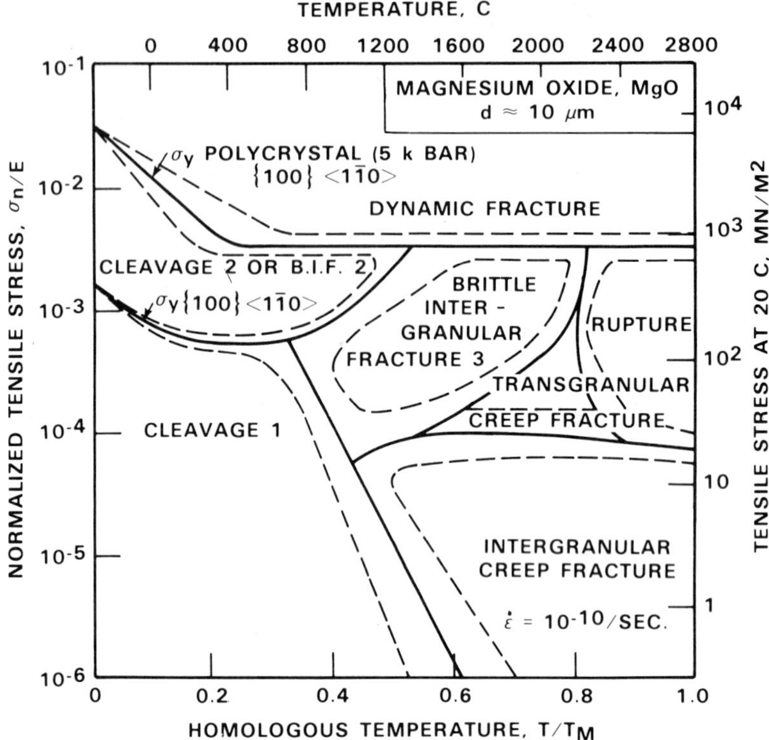

Fig. 3. Fracture mechanism map for magnesium oxide (from C. Gandhi and M. F. Ashby[9]).

REVIEW OF EXPERIMENTAL RESULTS

In this section, dynamic compaction results for various ceramic powders are reviewed. Unfortunately, many results reported in the literature do not give all the compaction conditions or powder characteristics necessary to clearly define the process. Therefore, experimental results correlated with corresponding compaction conditions will be presented only for selected ceramic powders.

A complete discussion of dynamic compaction experimental techniques and conditions is beyond the scope of this paper. The interested reader is referred to various reviews of dynamic compaction techniques.[2,6,27-29]

Ceramic Powders Dynamically Compacted

Many of the ceramic powders dynamically compacted and reported in the open literature are shown in Table 2. As indicated in the table, there has been an obvious emphasis on dynamic compaction of non-oxide ceramic powders which are usually difficult to sinter.

Table 2. Ceramic Powders Dynamically Compacted

Ceramic Powder	References	Ceramic Powder	References
AlN	24-26,30,31	Si_3N_4	30
Al_2O_3	25,26,30-35	Si_3N_4/2% Mgo[a]	33
B	25,26,30	TiB_2	31,39
B_4C	33	TiC	31,38,39
BeO	36,37	TiS_2	39
FeTiC	33	UO_2	41
MgO	36,38	WC	39
$MoSi_2$	36,37,39	W_2B_5	39
$MoSi_2$/20% ZrB_2	37	ZrB_2	39
NbC	39,40	ZrB_2/20% $MoSi_2$[a]	36,37
SiAlON	41	ZrC	39,40
SiC	38	ZrO_2	33
		$ZrSi_2$	39

[a]% = weight percent.

The Roles of Powder to Explosive Mass Ratios, Detonation Velocity, and Particle Size

Several different investigators[32-34] have established that the relative mass of powder (M) to explosive mass (E) and explosive detonation velocity (V_D) play a significant role in dynamic compaction of ceramic powders. The results of Leonard, Laber and Linse[32] are summarized in Fig. 4 showing M/E as a function of compressive yield strength for five different metal and ceramic powders. The M/E for a given powder was that required to obtain 97 percent theoretical density or greater in the final body. The relationship between M/E and compressive yield stress is linear. Peak pressures in the steel container adjacent to the powder were 3.5 to 4.0 GPa, only slightly above the compressive yield strength of alumina (~2.9 GPa). The pressures indicate that compaction occurred in the finite-strain plastic pressure regime at least for the alumina powder.

Pruemmer[33,34] has extended the work of Leonard, Laber and Linse on M/E to include the effects of detonation velocity and particle size. Figure 5 summarizes some of the results obtained on two different particle size alumina powders, 3.5 μm particles and 300 μm particles, compacted as cylinders.

E/M as a function of V_D^2 has the shape of a hyperbola in Fig. 5 (top). E/M is a measure of the time duration of the acting pressure while V_D^2 is a measure of the acting pressure[33,34] shown in the ordinate of the figure. For a given V_D, therefore, the hyperbolic relationship indicates a specific impulse for each type of powder material compacted to a particular density. For the 300 μm alumina powders, E/M values shown are less at a given V_D than the 3.5 μm alumina powders. The general shape of both curves are the same.

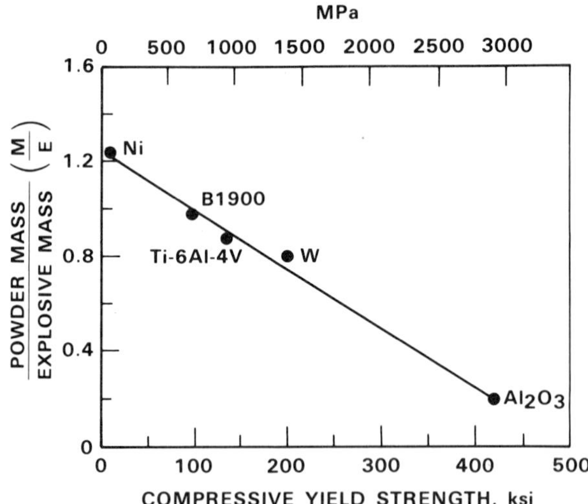

Fig. 4. Ratio of powder mass to explosive mass as a function of compressive yield stress for various metal and alumina powders (from Leonard, Laber, Linse[32]).

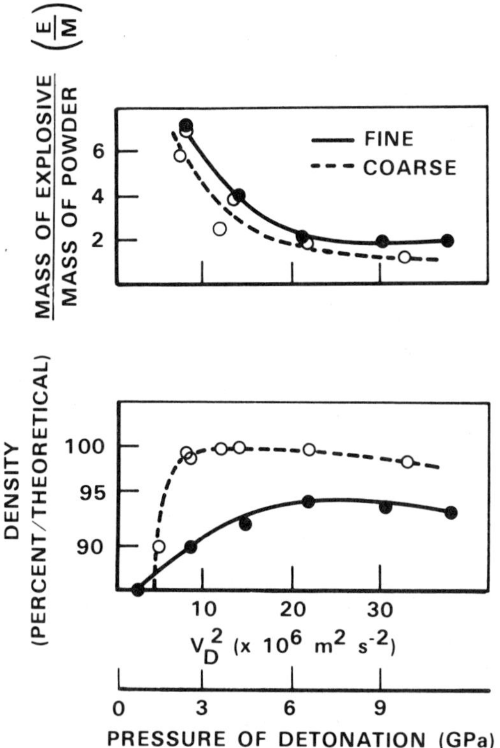

Fig. 5. Ratio of explosive mass to powder mass (top) and compaction density (bottom) as a function of the square of detonation velocity for fine (3.5 µm) and coarse (300 µm) alumina powders. Calculated detonation pressure is also given in ordinate (from Pruemmer[34]).

Coarse powders compact more efficiently than fine powders. Figure 5 (bottom) summarizes percent theoretical density as a function of V_D^2 for the different particle size alumina powders. The shapes of the resulting curves for both powders are roughly parabolic. Achieved densities for the 300 µm powders were substantially larger than densities obtained from compaction of the 3.5 µm powders, i.e., close to 100 percent theoretical compared to about 95 percent theoretical at maximum density values, for a given pressure. Roman et al.[20] have found similar results with the dynamic compaction of coarse and fine tungsten powders.

The parabolic shape of final body densities as a function of V_D^2 can be explained in terms of compaction curve behavior. If the pressure is too low, maximum densification is not achieved. On the other hand, if the pressure is too high, particle fracture results from overcompaction and density is lost. Therefore, there is an optimum pressure at intermediate levels which results in maximum densities.

Densities and Mechanical Properties

Densities and mechanical properties achieved for selected dynamically compacted ceramic powders are summarized in Table 3. Care must be taken in comparing data since different sample geometries and energy delivery systems were used in these studies. Furthermore, powder characteristics such as particle size, shape, and even composition may be different for a given material used by different investigators. A few general observations will be made but these reservations must be kept in mind.

The highest reported percent theoretical density (98 percent) was achieved with the dynamic compaction of aluminum nitride.[25,26,30] Lowest densities reported in the table are with alumina[31] (82 percent) and titanium carbide[40] (83 percent). A number of the investigations report final densities greater than 90 percent theoretical.

Correlations between the processing variables, E/M, V_D, and pressure (P) are difficult because of their interdependency. Work on titanium carbide[40] indicates little effect of pressure or V_D on relative densification for powders with the same initial density. However, a lower initial compact density results in a lower final density for titanium carbide powders compacted with otherwise the same process conditions.[40] Temperature effects are also important. One study [31] gives a higher final density (89 percent theoretical) for an alumina powder compacted at lower E/M, V_D, and P, but higher starting temperature (1423 K) than a sample (final density 82 percent theoretical) prepared using the same geometry and energy delivery system but compacted at 293 K.

Pressure required to obtain final densities greater than 90 percent theoretical vary considerably based on geometry. A pressure of 23-24 GPa calculated at the container wall-powder interface[40] of a hollow cylindrical sample was required for a final density of 91 percent theoretical. A pressure of only 3.5 GPa[25,26,30] was needed to achieve a final density of 91 percent theoretical in a cylindrical sample. In these examples, however, both the energy delivery systems and initial temperatures were different (see Table 3 for specifics).

Mechanical properties of dynamically compacted powders are also given in Table 3. Some hardness values[31] for aluminum nitride and alumina have the same range of values as hardness values found in the

Table 3. Selected Dynamically Compacted Ceramic Powders with Process Variables and Properties.

Ceramic Powder	Sample Geometry[a]	E/M	V_D ($km\ s^{-1}$)	P^b (GPa)	Initial Density (% TD)	Final Density (% TD)	H (GPa)	K_{IC} ($MPa\ m^{1/2}$)	Refs.
AlN[c]	R-DL	5.7	5.7	7.4	28	98	20.1±0.8[d]	-	25,26,30
AlN	FD-DL	0.96	4.6	4.2	55	89	8.6[f]	4.6[g]	31
Al_2O_3[e]	HC-DT	-	4.6	∿24	57	91	18.1[f]	3.5[g]	31
Al_2O_3[e]	FD-DL	0.94	4.6	4.2	60	89	12.6[f]	2.8[g]	31
Al_2O_3	FD-DL	1.83	4.8	4.9	58	82	6.4[f]	3.2[g]	31
Al_2O_3[c]	R-DL	5.7	4.2	3.6	18	92	20.3±2.4[d]	-	25,26,30
Al_2O_3[c]	R-DL	3.8	5.5	7.7	24	91	22.4±0.1[d]	-	25,26,30
Al_2O_3	R-DL	4.98	3.0	∿2.8	29	>97	-	-	32
B[c]	R-DL	13.8	5.8	7.5	38	97	31.1±1.2[d]	-	25,26,30
$NbC_{0.99}$	R-DT	-	5.2	6.7	70	96	-	-	40
Si_3N_4[c]	R-DL	8.0	4.7	4.3	28	90	6.8±0.7[d]	-	25,26,30
$TiC_{0.98}$	R-DT	-	5.2	6.7	70	93	-	-	40
$TiC_{0.98}$	R-DT	-	6.2	12.0	70	94	-	-	40
$TiC_{0.98}$	R-DT	-	6.2	12.0	30	83	-	-	40
$TiC_{0.98}$	R-DT	-	6.8	19.0	70	93	-	-	40
$ZrC_{0.96}$	R-DT	-	5.2	6.7	70	94	-	-	40

[a] R = rod or cylindrical sample; FD = flat disc shaped sample; HC = hollow cylinder sample configuration; DL = energy delivered with explosive directly loaded on container; DT = energy delivered with an impacting driver tube or plate.

[b] Pressures are reported based on different calculation methods. For specific calculation methods, see specific references.

[c] Sample was not evacuated prior to compaction. Alumina samples were not evacuated prior to compaction and were annealed in air at 1375 C for 36 hours prior to container remover after compaction. Silicon nitride powder is amorphous.

[d] Hardness value reported as average Knoop microhardness with 100 gram load taken at central region of cylindrical samples.

[e] Compaction took place with sample preheated to 1423 K.

[f] Hardness value reported is average diamond pyramid hardness from a range of loads.

[g] Fracture toughness determined using the microindentation technique (see text for specifics).

literature[42-45] for materials consolidated by conventional techniques or single crystals. Hardness values also show similar variations with respect to porosity found in the literature. However, hardness values determined at the center of cylindrical samples in another study[25,26,30] for aluminum nitride, alumina, and boron samples are significantly higher than would be expected for the reported final densities of these materials. The higher hardness values for these specimens were attributed to decreases in porosity and strain increases due to cold working at the central core regions because of converging shock waves at the center of cylindrical samples. Hardness values taken near the edge of samples in this study (not shown in Table 3) were generally lower than hardness values determined at central cores.

Few studies report fracture toughness values for dynamically compacted powders. Fracture toughness values for alumina in Table 3 are consistent with values found in the literature. The fracture toughness of aluminum nitride equal to 4.6 MPa $m^{1/2}$ in the table is similar to that for dense material[47] (4.2 MPa $m^{1/2}$). The microindentation technique [46] was used to determine fracture toughness values in this study.[31]

Microcracks and Macrocracks in Dynamically Compacted Materials

One of the most challenging problems to overcome in the dynamic compaction processing of powders is the elimination of cracks in final bodies. Most microstructures of dynamically compacted ceramic powders reported in the open literature show cracks on either a fine or coarse scale.[24-26,35,40,41]

Figure 6 shows the microstructure of an alpha silicon carbide powder explosively compacted.[a] Pressure used was equal to 27.5 GPa, V_D equal to 7 km s^{-1}, and E/M equal to about 4.5. The structure shows extensive microcracking throughout the particles.

In Fig. 7, cracks on a coarse scale are displayed in an aluminum nitride sample explosively compacted.[31] Details of the compaction process variables are given in Table 3. So long as cracked microstructures similar to that shown in Fig. 7 are obtained from dynamic powder compaction, it is not a viable consolidation technology for ceramics. However, few systematic studies have been performed and reported at this writing to eliminate or minimize cracking.

---

[a]Unpublished data from Battelle's Columbus Laboratories, Columbus, Ohio.

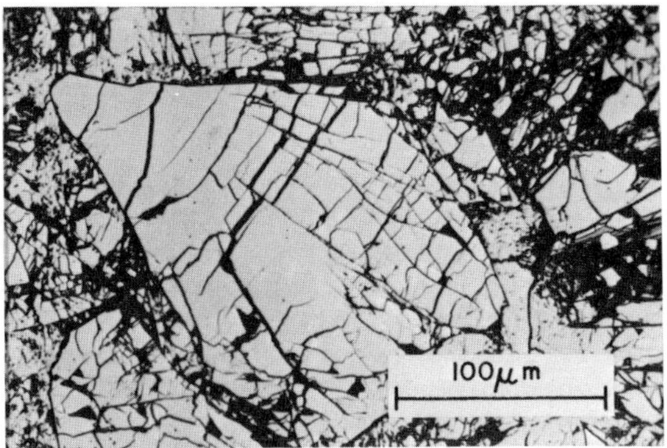

Fig. 6. Microstructure of explosively compacted α-SiC[a].

Fig. 7. Macrostructure of explosively compacted aluminum nitride.[31]

---

[a]Unpublished data from Battelle Columbus Laboratories, Columbus, Ohio.

CONCLUSIONS AND SUGGESTED RESEARCH AREAS

Dynamic compaction has obvious potential as a consolidation technology, particularly for ceramic powders, such as nonoxide materials that exhibit poor sinterability. Although conceptually simple, successful dynamic compaction depends in a complicated way on processing variables, powder features, and particle material characteristics. Among the areas where there is scant theoretical and experimental work is the combined effect of strain rate and pressure on powder compaction in the dynamic compaction regimes. However, from experimental data, a prerequisite for successful dynamic compaction is plastic deformation of the particles. For dynamic compaction to move beyond the conceptual stage and into technological production will require at least a qualitative theoretical understanding of the process.

Experimental evaluations of the dynamic compaction of ceramic powders indicate the potential of the process. However, the best densities reported fall short of theoretical. Some studies indicate enhanced mechanical properties may be achieved in dynamically compacted powders. Other studies give mechanical properties of dynamically compacted powders as about the same as those achieved with more standard consolidation technologies. Unfortunately, comparisons of results from different studies may be misleading because of different starting materials and dynamic compaction techniques.

Included in future research on dynamic compaction should be the following:

(1) Development of theoretical guidelines for dynamic powder compaction in terms of the process variables and the mechanical and physical characteristics of the powders which are process limiting.
(2) Determination of the effects of strain rate and pressure duration on powder compaction mechanisms in the dynamic compaction regimes.
(3) Determination of the relative compactabilities of powders composed of different materials with different material properties.
(4) Systematic studies to determine the influence of processing variables on final body properties.
(5) Comprehensive and systematic studies of various sample geometries, energy delivery systems, and sample container materials to eliminate or minimize cracking.

Dynamic compaction studies on ceramic powders should at the very least report the process and material variables given in Table 3. In addition, powder properties, sample geometry, sample container material, and energy delivery systems should be described. Chemical, phase, and microstructural analyses should be evaluated so that properties can be related to processing conditions and microstructural features. Particularly for certain materials, X-ray diffraction analyses should be performed and line broadening and any phase transformations reported.

REFERENCES

1. L. Davison and R. A. Graham, Phys. Reports (Review Section of Physics Letters), 55 [4], 255-379 (1979).
2. W. J. Murri, D. R. Curran, C. F. Petersen, and R. C. Crewdson, Advances in High Pressure Research, 4, 1-163 (1964).
3. D. E. Grady, J. Geophys. Res., 85 [B2], 913-924 (1980).
4. D. J. Steinberg, Lawrence Livermore Laboratory, Univ. of California, Livermore, CA., Report No. UCID-16946, November 7, 1975.
5. J. D. Campbell, Mat. Sci. Eng., 12, 3-21 (1973).
6. C. Clyens and W. Johnson, Mat. Sci. Eng., 30:121-139 (1977).
7. M. F. Ashby, Acta Met., 20, 887-897 (1972).
8. H. J. Frost and M. F. Ashby, Division of Engineering and Applied Physics, Harvard University, Cambridge, Mass., Report No. NR-031-732, Office of Naval Research Contract N00014-67-A-0298-0020, August 1973.
9. C. Gandhi and M. F. Ashby, Acta Met., 27, 1565-1602 (1979).
10. P. C. Chou, pp. 1-6 in Dynamic Response of Materials to Intense Impulsive Loading, edited by P. C. Chou and A. K. Hopkins, Air Force Materials Lab., Wright Patterson AFB, Ohio (1972).
11. B. D. Trott and J. G. Dunleavy, DCIC Cer. Awareness Bull., 25, 13-17 (1970).
12. J. G. Dunleavy, DCIC Cer. Awareness Bull., 21, 18-22 (1970).
13. T. J. Ahrens, W. H. Gust, and E. B. Royce, J. Appl. Phys., 39 [10], 4610-4616 (1968).
14. C. L. Mader, Los Alamos Scientific Laboratory, Univ. of California, Los Alamos, New Mexico, Rept. No. LA-4381, May 1970.
15. J. R. Rempel and D. N. Schmidt, pp. B111-B119 in Proceedings of the Fourth Symposium (International) on Detonation, ACR-126, Office of Naval Research-Dept. of the Navy, Washington, D.C. (1965).
16. O. M. Roman and I. M. Pickus, Sixth International Conference on High Energy Rate Fabrication, Haus der Tecnik, Essen, West Germany, 5.6.1-5.6.9, 1977.
17. M. J. Donachie and M. F. Burr, J. Met., 15, 849-855 (1963).
18. E. E. Walker, Trans. Faraday Soc., 19 [1], 614-622 (1923).
19. D. B. Leiser and O. J. Whittemore, Am. Ceram. Soc. Bull., 49 [8], 714-717 (1970).
20. O. V. Roman, V. F. Nesterenko and I. M. Pikus, Fizika goreniya i Vzryva, 15 [5], 102-107 (1979).
21. P. W. Bridgman, Studies in Large Plastic Flow and Fracture, Harvard University Press, Cambridge, Mass., 164-173 (1964).

22. A. A. Deribas and I. D. Zakharenko, Fizika Goreniya i Vzryva, 10 [3], 409-421 (1974).
23. R. H. Wittmen, pp. 153-167 in Papers Presented to the Second International Symposium on the Use of Explosive Energy in Manufacturing Metallic Materials of New Properties, Vol. 1, Marianski Lazne, Czechoslovakia (1973).
24. W. H. Gourdin, C. J. Echer, C. F. Cline, and L. E. Tanner, Lawrence Livermore National Laboratory, Univ. of California, Livermore, CA, Rept. No. UCRL-85274 (1981).
25. C. Hoening, A. Holt, M. Finger, W. Kuhl, and L. Dengler, Sixth International Conference on High Energy Rate Fabrication Haus der Tecnk, Essen, West Germany 6.3.1-6.3.25 (1977).
26. C. Hoening, A. Holt, M. Finger, and W. Kuhl, Lawrence Livermore Laboratory, Univ. of California, Livermore, CA, UCRL-79345, September 1977.
27. W. A. Bassett and T. Takahashi, Adv. in High Pressure Res., 4, 165-247 (1974).
28. G. R. Fowles, pp. 405-480 in Dynamic Response of Materials to Intense Impulsive Loading, edited by P. C. Chou and A. K. Hopkins, Air Force Materials Lab., Wright Patterson Air Force Base, Ohio (1972).
29. A. N. Staver, pp. 865-880 in Shock Waves and High-Strain-Rate Phenomena in Metals, edited by M. A. Meyer and L. E. Murr, Plenum Press, New York (1981).
30. C. L. Hoenig and C. S. Yust, Am. Ceram. Soc. Bull., 60 [11], 1175 (1981).
31. Work Performed under U.S. Dept. of Energy Contract No. W-7405-eng-92 at Battelle's Columbus Laboratories, Columbus, Ohio.
32. R. W. Leonard, D. Laber, and V. D. Linse, pp. 8.3.1-8.3.23 in Proceedings of the Second International Conference on the Center for High Energy Forming, sponsored by U.S. Army Materials and Mechanics Research Center and Advanced Research Projects Agency, Estes Park (1969).
33. R. A. Pruemmer, pp. 9.2.1-9.2.27 in Proceedings of the Fourth International Conference of the Center for High Energy Forming, sponsored by the University of Denver, Vail, CO (1973).
34. R. A. Pruemmer, Powder Met. Int., 9 [1], 11-14 (1977).
35. C. S. Yust and L. A. Harris, pp. 881-894 in Shock Waves and High-Strain-Rate Phenomena in Metals, edited by M. A. Meyers and L. E. Murr, Plenum Press, New York (1981).
36. R. J. Carlson, S. W. Porembka, and C. C. Simons, Am. Ceram. Soc. Bull., 45 [3] (1966).
37. S. J. Paprocki, C. C. Simons, and R. J. Carlson, American Society of Tool and Manufacturing Engineers SP62-29, 1962.
38. V. V. Skorokhod, G. I. Savvakin, S. M. Solonin, and L. L. Kolomiets, Sov. Powder Metall. and Metal Cer., 13 [8], 667-669 (1974).
39. G. A. Adadurov, O. N. Breusov, A. N. Dremin, and V. F. Tatsil, Sov. Powder Metall. and Metal Cer., 10 [11], 859-861 (1971).

40. G. V. Samsonov, V. P. Alekseevskii, S. A. Bozhko, and V. V. Yarosh, Powder Metall. and Metal Cer., 11 [1], 74-79 (1972).
41. R. W. Rice, Naval Research Lab., Washington, D.C., Report No. NRL Memorandum Report 4637, 1981.
42. Engineering Property Data on Selected Ceramics, Vol. I-III, MCIC Pub. No. MCIC-HB-07, Metals and Ceramics Info. Center, Battelle's Columbus Laboratories, Columbus, Ohio (1976-81).
43. K. Niihara and T. Hirai, J. Mater. Sci., 12 [16], 1233-1242 (1977).
44. K. M. Taylor and C. Lenie, J. Electrochem. Soc., 107 [4], 308-314 (1960).
45. H. Palmour III, W. W. Kriegel, and J. J. DePlessis, pp. 313-331 in Mechanical Properties of Engineering Ceramics, edited by W. W. Kriegel and H. Palmour III, Wiley-Interscience, New York (1961).
46. A. G. Evans, pp. 112-135 in Fracture Mechanics Applied to Brittle Materials, ASTM STP 678, edited by S. W. Freiman, American Society for Testing and Materials, Philadelphia, PA (1979).
47. W. F. Adler, Effects Technology, Inc., Santa Barbara, CA, Final Report No. CR-82-1075, Office of Naval Research Contract N00014-76-C-0744, May 1982.

DISCUSSION

W. H. Gourdin (LLNC): What pressures do you refer to? The stress in the clad, the pressure in the explosive, or the average stress in the powder?

Author: Battelle stresses are those in the clad.

W. H. Gourdin (LLNC): Considering the number of possible variables--explosive type, density, energy powder type and density material properties--can you rationalize the use of a single parameter, E/W?

Author: No.

EXPLOSIVE CONSOLIDATION OF ALUMINUM NITRIDE CERAMIC POWDER: A CASE HISTORY

W. H. Gourdin, S. L. Weinland, C. J. Echer and
S. L. Huffsmith

Lawrence Livermore National Laboratory
P. O. Box 808 L-369
Livermore, CA 94550

INTRODUCTION

Prummer[1] and Hoenig and Yust[2] have demonstrated that high densities can be achieved in explosively formed compacts made from various ceramic powders. Of those materials which have been tried, aluminum nitride (AlN) seems to be particularly suited to dynamic compaction. Cline and Heard[3] have demonstrated that AlN becomes plastic under combined hydrostatic and uniaxial stress, suggesting that substantial flow may occur under shock loading. This is borne out by the observations of Hoenig and Yust[2] as well as Gourdin et al.[4] who report considerable deformation of individual grains in explosively consolidated material. These authors also report the presence of an amorphous phase at grain boundaries, indicating that local energy deposition occurs during the passage of the shock wave. This amorphous phase may be responsible for the cohesiveness of dynamically consolidated AlN much like the interparticle melting observed in metal compacts.[5,6]

In this paper, a case study of the explosive consolidation of AlN is presented. Measured and calculated stress histories show a two-wave structure during the initial compaction, consisting of an ingoing pulse and an outgoing pulse reflected from the cylinder axis. The macrostructural features of the consolidated specimen are rationalized and discussed in terms of this observed stress history. It is suggested that the second shock determines, at least in part, the final density of the compact. Examination of heat treated specimens in the transmission electron microscope (TEM) shows that recovery of the heavily dislocated microstructure is rapid at 1500°C, and, further, that the amorphous intergranular phase[4] disappears. A dispersion of fine precipitates appears at grain boundaries during heat treatment as well. The fracture toughness of explosively consolidated specimens, 3.0 $MPa \cdot m^{1/2}$, is the same as that of hot-pressed material, indicating that explosive consolidation is a viable means of producing dense, well bonded AlN.

COMPACTION EXPERIMENTS AND MACROSTRUCTURE

Aluminum nitride powder[a] was compacted in the conventional cylindrical apparatus shown schematically in Fig. 1. A layer of the explosive PETN of thickness δ_{HE} was packed around a mild steel tube of 1.27 cm OD by 0.16 cm wall thickness which contained the AlN powder tapped to an initial density of 1.76 gm/cc (0.54 of theoretical). The explosive was detonated so as to produce an axisymmetric detonation wave which propagates parallel to the cylinder axis. The lateral pressure produced by the expanding product gases generates an oblique shock in the powder. A carbon film piezo-resistive pressure sensor[b] was placed at the powder-tube interface, as indicated in Fig. 1, to record the stress-time history experienced by the powder. A second sensing element was placed between the explosive and the tube so that the gauge response could be calibrated against the known properties of the explosive.

The gauge record for experiment 061181-1 is shown in Fig. 2. A two-wave structure is observed, characteristic of compaction in the configuration shown in Fig. 1. The first pulse is an ingoing wave determined by the properties of the explosive and the steel cladding. Calculations predict a pressure in the steel cladding of 4 GPa, which is not inconsistent with an initial stress in the AlN powder of 0.7-1.0 GPa, as indicated in Fig. 2. The second pulse is clearly much stronger than the first. When the ingoing wave reaches the axis, it reflects "off of itself," behaving as if the interaction occurred with an infinitely stiff material. The stress amplification upon reflection can be particularly large for a porous material because the particle velocity induced by the ingoing wave is quite high relative to the stress. A reflected stress of 10 GPa or more is not unreasonable. As this pulse propagates outward, its amplitude is decreased somewhat by the divergence imposed by the cylindrical geometry. More significant, however, is the interaction with the release waves from the explosive. As these propagate inward they decrease the stress rapidly. In the case of 061181-1, the layer of explosive was quite thin, δ_{HE} = 0.64 cm, so that a relatively rapid release is expected, an observation borne out by the sensor record at the outside of the capsule. A return pulse of 3.0 to 3.5 GPa is consistent with this description.

---

[a]Materials Research Corporation, Orangeburg, NY.
[b]Dynasen, Inc., Goleta, CA.

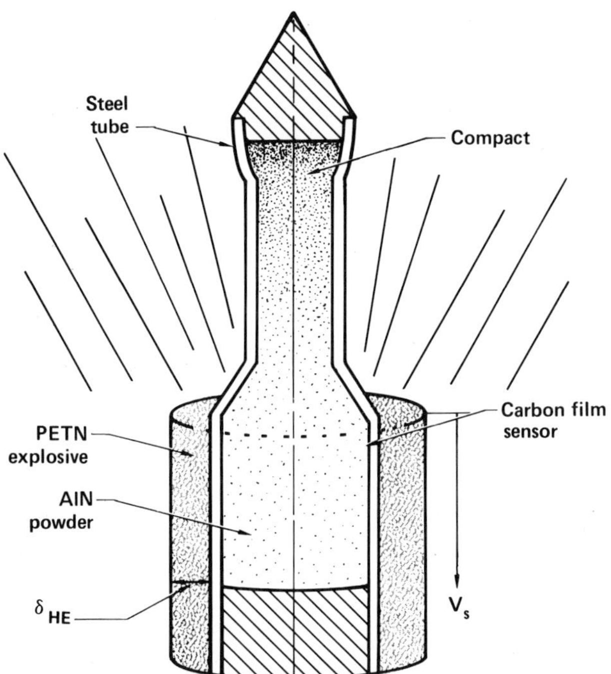

Fig. 1. Schematic of explosive consolidation in a cylindrical geometry. The detonation velocity of the explosive, v_D, is parallel to the cylinder axis.

Also shown in Fig. 2 is a pressure-time history calculated using the hydrocode HEMP, into which a P-α model[7] for the powder has been incorporated. The equation-of-state of AlN was constructed for mechanical and thermomechanical data, but the crush-up (P-α) relationship was simply a best guess based on the shell theory developed by Carroll and Holt.[8] The calculated curve shows the two-wave structure clearly and is in quantitative agreement with the measured curve both in the overall duration of the two pressure pulses and in the magnitude of the second pulse. These observations suggest that the powder response may not be a sensitive function of the crush-up relationship for the powder. In light of the crudeness of the model, however, the agreement may be fortuitous. It is significant that the apparent stress level of the ingoing wave is calculated to be substantially higher (~1.8 GPa) than the measured value (~1.0 GPa). Uncertainties in sensor response and calibration and the complexity of the two-dimensional flow make the resolution of this discrepancy difficult.

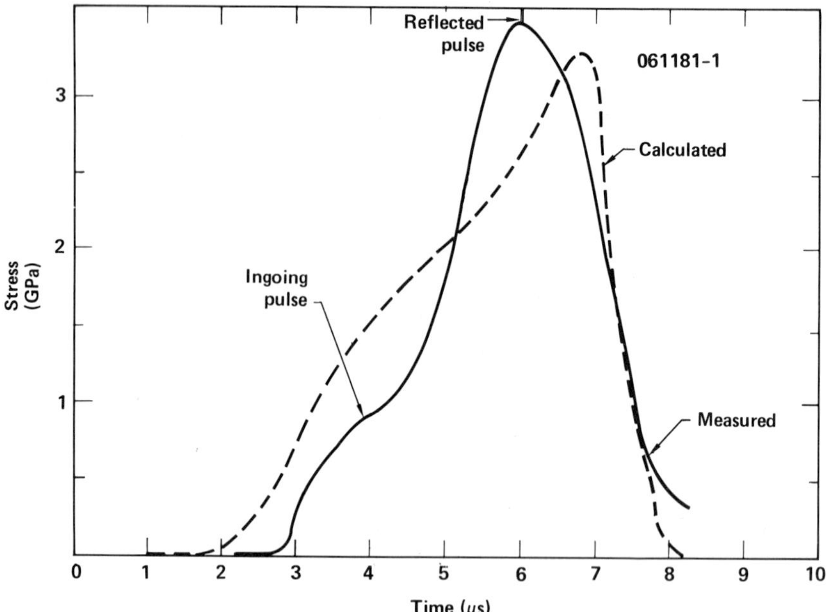

Fig. 2. Measured and calculated stress-time history at the powder-capsule interface for experiment 061181-1. The explosive was PETN at a density of 0.97 gm/cc, $P_{CJ} = \sim 8.0$ GPa, $\delta_{HE} = 0.64$ cm.

One effect of the two-wave structure on the macrostructure of the specimen is illustrated in Fig. 3. As the initial compressive pulse passes through the powder, partial densification occurs. No cracking occurs because all stresses, including the circumferential (hoop) stress, are compressive. Upon reflection at the axis, the radial stress is magnified substantially, but remains compressive. The hoop stress for the outgoing wave, however, becomes tensile, so that, depending upon the amplitude of the stress, the state of release, and the strength of the consolidated material, radial cracking can occur. Radial cracks on the axis are a common feature in specimens compacted to uniform density in a cylindrical geometry (Figs. 3 and 4). Prummer[1] demonstrated, however, that by varying the pressure and time of application of the explosive, material apparently free of cracking could be obtained. If an analogy is drawn with the compaction of tungsten,[9] the parameter "window" for such consolidation is narrow, and small variations during consolidation may have a substantial effect on the integrity of the final piece.

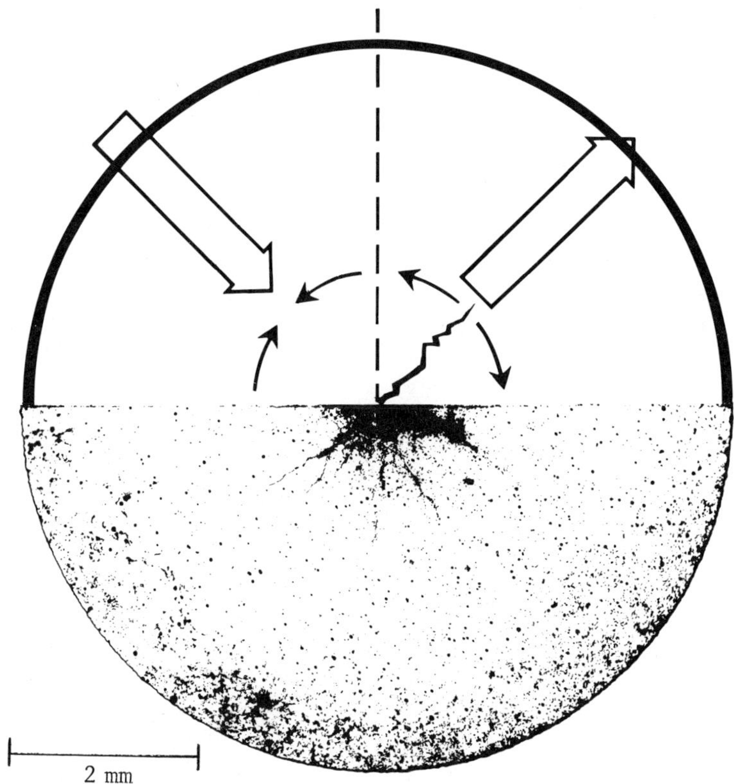

Fig. 3. The radial cracking pattern observed in consolidated specimens is the result of tensile hoop stresses which develop upon reflection of the ingoing wave at the cylinder axis.

The stress amplification which occurs at the axis can produce unique effects. In metals, material at the center can melt and be expelled, leaving a uniform hole along the axis. Hoenig and Yust[2] found equiaxed recrystallized grains of AlN near the center of their specimens, indicating temperatures high enough to effect substantial recovery. The disruptive effect of the high stress at the axis is illustrated in the radiographic density profile shown in Fig. 5. Although the average density of this specimen, 080780-3, is 3.1 gm/cc (0.95 TD) a sharp decrease is noted near the axis over a region corresponding roughly to the extent of cracking shown in Fig. 4.

Fig. 4. Macrograph of sample 080780-3, compacted with the explosive PETN at a density of 0.82 gm/cc, producing a pressure (P_{CJ}) of 5.0 GPa. In this case δ_{HE} = 1.9 cm. The final density is 95% of theoretical (3.26 gm/cc).

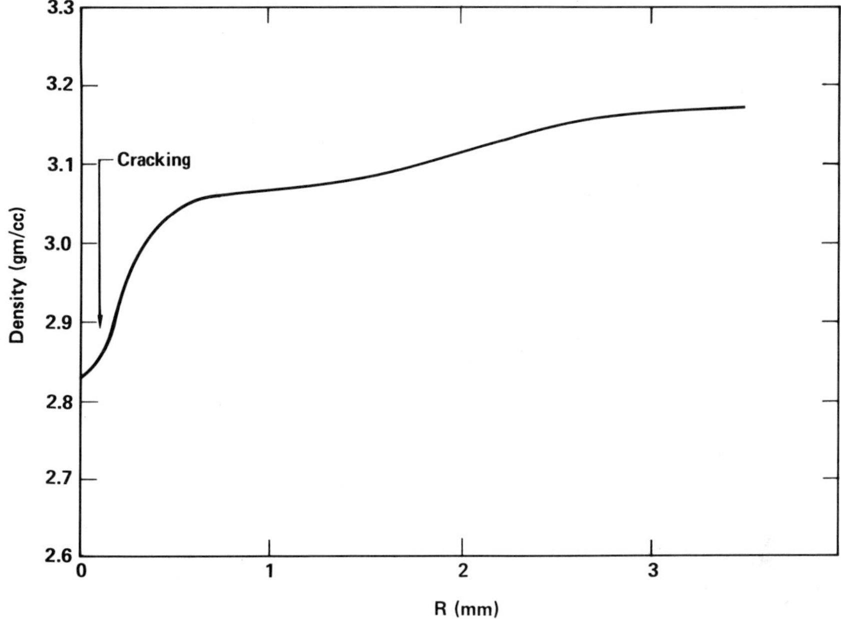

Fig. 5. Radiographic density profile of 080780-3. The density change at about 2 mm may be related to changes in the stress amplitude as the initial pulse travels into the powder.

It is interesting to note that the average relative density of sample 080780-3 is 95% while that of 061181-1 is only 85-90% despite the fact that the explosive "pressure" (P_{CJ}) was less for 080780-3 (5.0 GPa) than for 061181-1 (8.0 GPa). Experiments with carbon sensors show that, all other conditions being the same, a thicker layer of explosive produces a larger second pulse, without substantially changing the amplitude of the first "knee" in the response curve. A larger amount of explosive produces a slower release and there is less attenuation of the returning pulse. The final density is apparently determined in part by the second and, to a lesser degree, subsequent shocks in the powder. Control of the relative amount of explosive is thus important.[1,9]

MICROSTRUCTURE

Specimens were cut from 080780-3 at approximately the mid-radius position and were ion-thinned for transmission electron microscopy. The as-compacted microstructure, shown in Fig. 6, has been described in detail elsewhere.[2,4] The individual grains are heavily dislocated and a mottled phase is apparent between many grains. Analysis in dark field[4] has shown that this material is amorphous, although its composition has not been established. Specimens were subjected to isothermal heat treatments in flowing nitrogen at 1500°C for increasing times, and the microstructural changes were observed. After 5 minutes (Fig. 7a) noticeable recovery has occurred and the amount of amorphous material has decreased, although some regions are still apparent. At 16 minutes (Fig. 7b) grains are well defined and equiaxed, although still dislocated. Note that small precipitates have appeared at grain boundaries and the amorphous phase has disappeared. After 31 minutes there is some evidence of recrystallization (Fig. 7c) and precipitates can be clearly seen. A heat treatment of 8 hours yields the well recovered structure shown in Fig. 7d, although a few grains still show some strain. Comparison with the hot-pressed microstructure shown in Fig. 7e shows that the explosively consolidated material has a finer grain and precipitate size.

The rapid disappearance of the amorphous intergranular phase indicates that it is not an equilibrium phase at the heat treatment temperature. Without observations of individual powder particles, it is impossible to determine whether amorphous layers are present prior to explosive consolidation. However, the observations presented here are consistent with the suggestion of Gourdin et al.[4] that the intergranular phase is metastable, formed as a result of the very rapid surface heating and cooling which occurs during compaction.

Although precipitate growth occurs during heat treatment, it is not clear whether precipitates are present in the as-consolidated specimens. The presence of unidentified "insoluble residues" in the starting powder[10] would seem to suggest this, but the dark contrast induced by the lattice strains in Fig. 6 makes the identification of precipitates ambiguous at best. The precipitates in heat treated and hot-pressed specimens are electron opaque and hence cannot be structurally analyzed with selected area diffraction. X-ray energy dispersive spectra (XEDS), however, show significant quantities of tungsten, iron, cobalt and chromium. One of the strong lines of tungsten masks that of silicon, reported in a preliminary analysis of precipitates in hot-pressed AlN.[10]

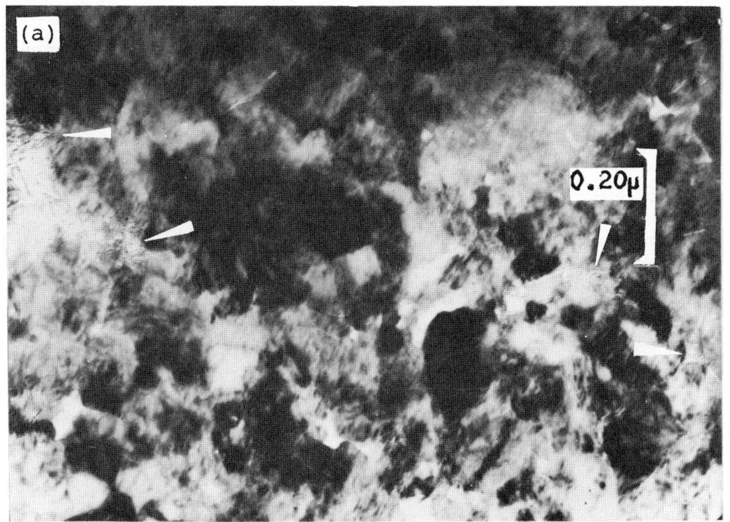

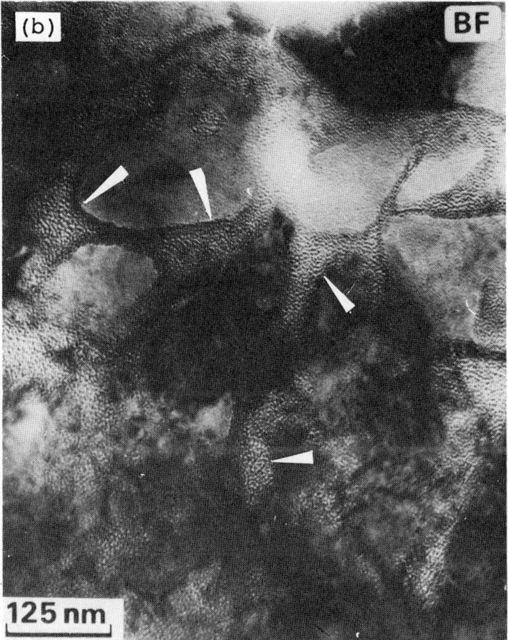

Fig. 6. As-compacted microstructures. Arrows indicate amorphous regions between grains.

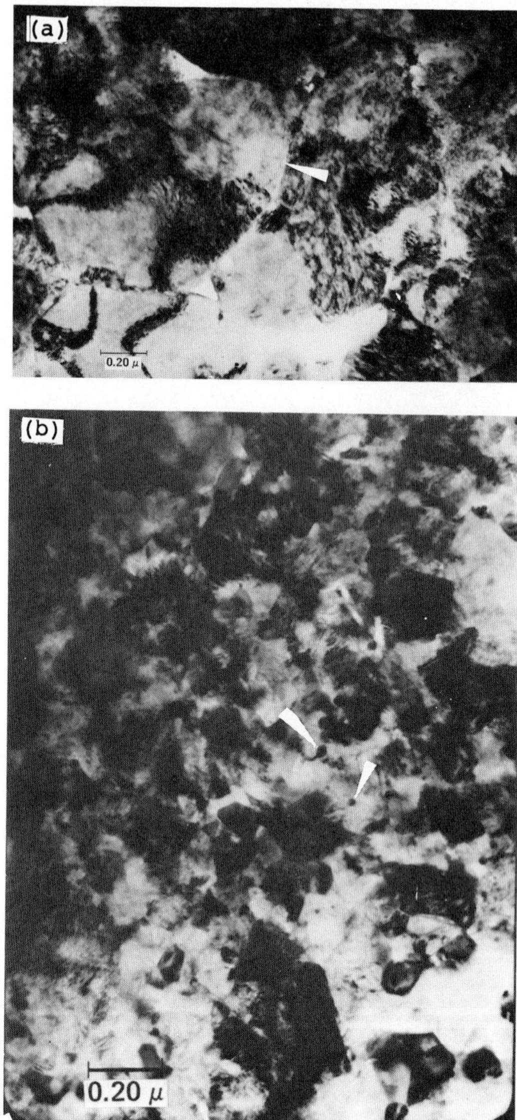

Fig. 7. Microstructures following isothermal heat treatments at 1500°C in flowing nitrogen: (a) 5 minutes, arrow indicates mottling which may be residual amorphous material; (b) 16 minutes, note the appearance of fine precipitates at grain boundaries.

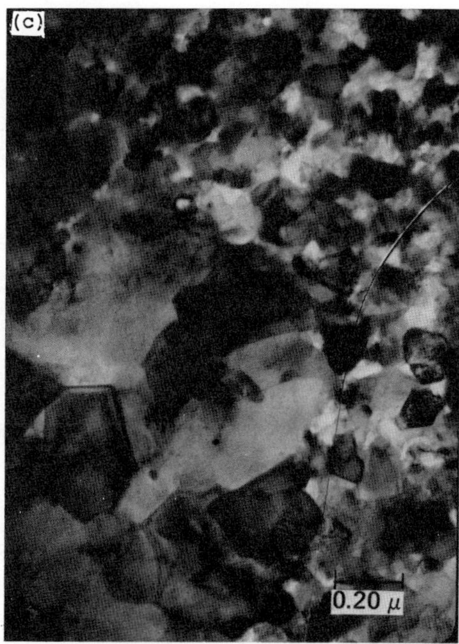

Fig. 7 (continued). Microstructures following isothermal heat treatments at 1500°C in flowing nitrogen: (c) 31 minutes; (d) 8 hours.

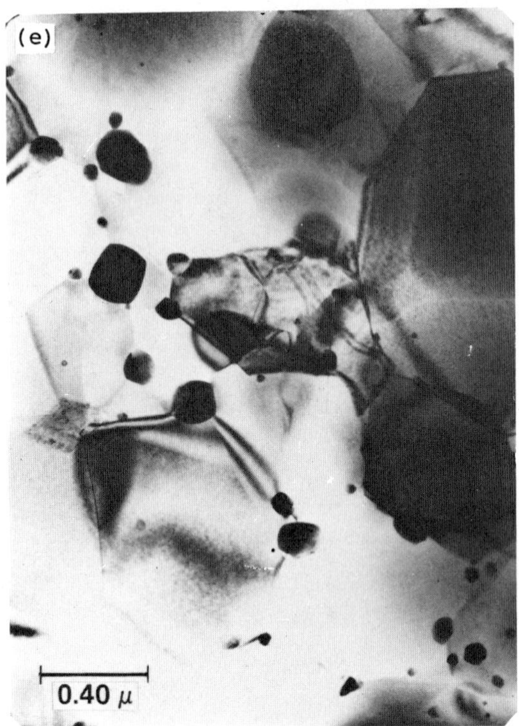

Fig. 7 (continued). (e) Microstructure of material hot-pressed in flowing nitrogen at 1700°C and 27 MPa for 1 hour.

FRACTURE TOUGHNESS

As-compacted material has considerable strength, despite the cracks evident in Fig. 4, and it can be handled as would any well bonded ceramic. However, because of the prevalent cracking and the small size of the available specimens, conventional specimens for fracture toughness measurements could not be fabricated. As a result, the Vickers indentation technique[11-15] was used to evaluate the toughness of the as-compacted AlN relative to hot-pressed material. The critical stress intensity factor, K_{IC}, is given by[15]

$$K_{IC} = 0.036\, E^{0.4} P^{0.6} a^{-0.7} (c/a)^{-1.5} \quad \text{(SI units)} \tag{1}$$

where E is Young's modulus, a the half diagonal of the Vickers hardness indentation, P the load, and c the half span of the radial cracks produced at the hardness indentation. The fracture toughnesses of hot-pressed specimens 95% dense were measured in a commercial device[c] for comparison with indentation measurements. The results of these experiments are summarized in Table 1. A comparison of the KIC value obtained from bulk samples of the hot-pressed material and that determined from hardness data shows that the expression (1) apparently provides an adequate description of material 95% dense for 1000 gm loadings, although loading to 2000 gm produces consistently higher values. The considerable scatter in the indentation data results in part from inconsistent cracking in some specimens. Nevertheless it is clear that the explosively consolidated and the hot-pressed material have the same toughness, about 3.0 $MPa \cdot m^{1/2}$.

Table 1. Fracture Toughness

Sample	Description	Technique	K_{IC}	σ^*	Comments
HP95	Hot-pressed, 95% dense	Terra Tek	3.0	---	Bulk specimen
		Indentation	3.1	0.4	1000 gm load
		Indentation	3.4	0.3	2000 gm load
080780-3	Explosively consolidated, 95% dense	Indentation	3.1	0.5	1000 gm load
		Indentation	3.7	0.3	2000 gm load

*Standard deviation.

DISCUSSION

Explosive consolidation produces a fine-grained, well bonded AlN comparable to hot-pressed material. However, it is not clear that the considerable plasticity of AlN is typical of other ceramic powders under shock loading. Experience with other oxide and nonoxide ceramics (Al_2O_3, ZrO_2, Si_3N_4) indicates that their behavior during dynamic compaction is considerably different, depending not only on the mechanical and thermal properties of the material itself, but on the properties of the powder as well. Initial density, surface area, surface chemistry, and morphology of the powder may be important parameters. To make a quantitative link between the stress history during consolidation, such as outlined above, and the structure and properties of the final piece, dynamic responses must be determined for well characterized powders. Likewise, the careful design, or "tuning" of experiments to consistently produce crack-free compacts

[c]Terra Tek, Salt Lake City, Utah.

of high quality requires that the dynamic properties of the powder be well understood. Experiments to determine the Hugoniot curve for the AlN powder used here are planned.

The intergranular amorphous phase[2,4] is a prominent feature in the as-compacted microstructure of explosively compacted samples (Fig. 6), and might be expected to affect the fracture toughness. Heat treatment, during which the phase disappears, appears to embrittle the specimens. Indentation measurements on heat treated material are thus of interest, and have been undertaken.

SUMMARY/CONCLUSIONS

1. Explosive consolidation can produce dense, fine grained, and well bonded aluminum nitride ceramic having a fracture toughness comparable to that of hot-pressed material, $\sim 3.0\ \mathrm{MPa \cdot m^{1/2}}$.

2. Recovery of the highly dislocated as-compacted structure occurs rapidly during heat treatment at moderate temperature (1500°C), and is accompanied by the disappearance of the amorphous intergranular phase.[2,4] The grain size following an 8 hour heat treatment is $\sim$ 0.25-0.50 μm.

3. A fine dispersion of precipitates forms during heat treatment. XEDS analysis shows these precipitates to contain W, Fe, Co and Cr. Similar precipitates are observed in hot pressed material.[10]

4. The two-wave structure observed for the configuration shown in Fig. 1 is a composite of an ingoing pulse and an outgoing pulse reflected from the cylinder axis. The amplitude of the outgoing pulse is considerably enhanced upon reflection, causing radial cracking near the axis and determining in part the final compact density.

REFERENCES

1. R. A. Prummer, Ber. Dt. Keram, Ges., 50, 75 (1973).
2. C. L. Hoenig and C. S. Yust, Amer. Cer. Soc. Bull., 60, 1175 (1981).
3. C. F. Cline and H. C. Heard, J. Mat. Sci., 15, 1889 (1980).
4. W. H. Gourdin, C. J. Echer, C. F. Cline, and L. E. Tanner, p. 233ff in Proc. 7th International Conf. on High Energy Rate Fabrication, University of Leeds, 1981.
5. D. Raybould, D. G. Morris, and G. A. Cooper, J. Mat. Sci., 14, 2523 (1979).
6. D. G. Morris, Metal Sci., 116ff (March 1981).
7. W. Herrmann, J. Appl. Phys., 40, 2490 (1969).
8. M. M. Carroll and A. C. Holt, J. Appl. Phys., 43, 1626 (1972).

9. H. P. Balzerowiak, Fr. Bock-Nussbaum, R. Prummer, and D. Schmidt, High Temp.-High Press., 3, 517 (1971).
10. W. H. Gourdin and L. H. Tanner, J. Mat. Sci., to be published (also Lawrence Livermore National Laboratory UCRL-87304).
11. G. R. Anstis, P. Chantikul, B. R. Lawn, and D. B. Marshall, J. Amer. Cer. Soc., 64, 533 (1981).
12. P. Chantikul, G. R. Anstis, B. R. Lawn, and D. B. Marshall, J. Amer. Cer. Soc., 64, 539 (1981).
13. A. G. Evans and E. A. Charles, J. Amer. Cer. Soc., 59, 371 (1976).
14. K. Niihara, R. Morena, and D. P. H. Hasselman, J. Mat. Sci. Lett., 1, 13 (1982).
15. D. B. Marshall and A. G. Evans, Comm. Amer. Cer. Soc., C-182 (December 1981).

COMPUTER SIMULATION OF DYNAMIC COMPACTION*

Mark L. Wilkins and Carl F. Cline

University of California, Lawrence Livermore Natl. Lab.
P. O. Box 808
Livermore, California 94550

ABSTRACT

The shock compression of powder materials by high explosives using cylindrical geometry has been simulated with a two-dimensional computer program. The flow fields in the powders are analyzed for different initial conditions. An objective is to determine the design of the high explosive assembly to yield the desired pressure-time history in the powders. The calculations provided the means to translate results from one experimental geometry to another. An example is given for the compaction of a metallic glass powder.

INTRODUCTION

Dynamic compaction is becoming an increasingly important technique to synthesize new material systems. With this technique high pressures are applied over a short period of time. Powder guns and high explosives are convenient ways to drive a material system to the desired pressure and temperature. The successful consolidation of a material starting from a powder involves a number of parameters associated with the material itself, such as the thermal and mechanical properties of the grains as well as the initial grain size. The hydrodynamic flow, which is governed by the driving system as well as the material properties of the sample material, results in temperature and pressure distribution within the volume of material being

---

*Work performed under the auspices of the U.S. Department of Energy by Lawrence Livermore National Laboratory under contract #W-7405-Eng-48.

compressed. A minimum requirement for the successful consolidation of a material is that the release of high pressures does not permit tensile stresses to be set up which would destroy the material.

Some of the technology developed for explosive welding can be used to design the dynamic high pressure system for consolidation of powders. The conditions for a successful explosive welding are well documented in the literature, for example the work of Cowan and Holtzman[1] in the U.S. and the work of Deribas[2] in the Soviet Union. Most of the research to identify geometrical arrangements for application and release of high pressures while avoiding tensile stresses has been done experimentally. In working with very high pressures obtained by flying plates, there are very critical parameters connecting the collision angle, collision velocity, as well as the original state of the material to be consolidated. Recovered samples from high explosive compaction, for example, often do not have uniform material properties through a cross section.

Computer simulations can provide a means to design experiments and to relate the distribution of material properties in the sample to flow history during consolidation. With this information the original experimental configuration can be redesigned. Another objective of using computer simulation programs is to determine mechanical properties of materials by correlating calculations with experiments. In this application the displacement of an observable position in the experiment is related to a material property by a constitutive model.

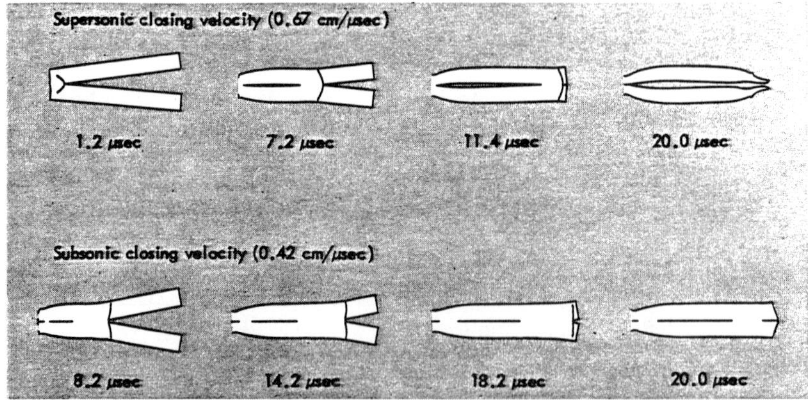

Fig. 1. HEMP calculations of collisions of aluminum plates. Time is measured from the first contact of the plates. The Lagrange calculational grid has been omitted for clarity. The plates are 1 cm thick and 8 cm long; velocity: 0.7 km/s.

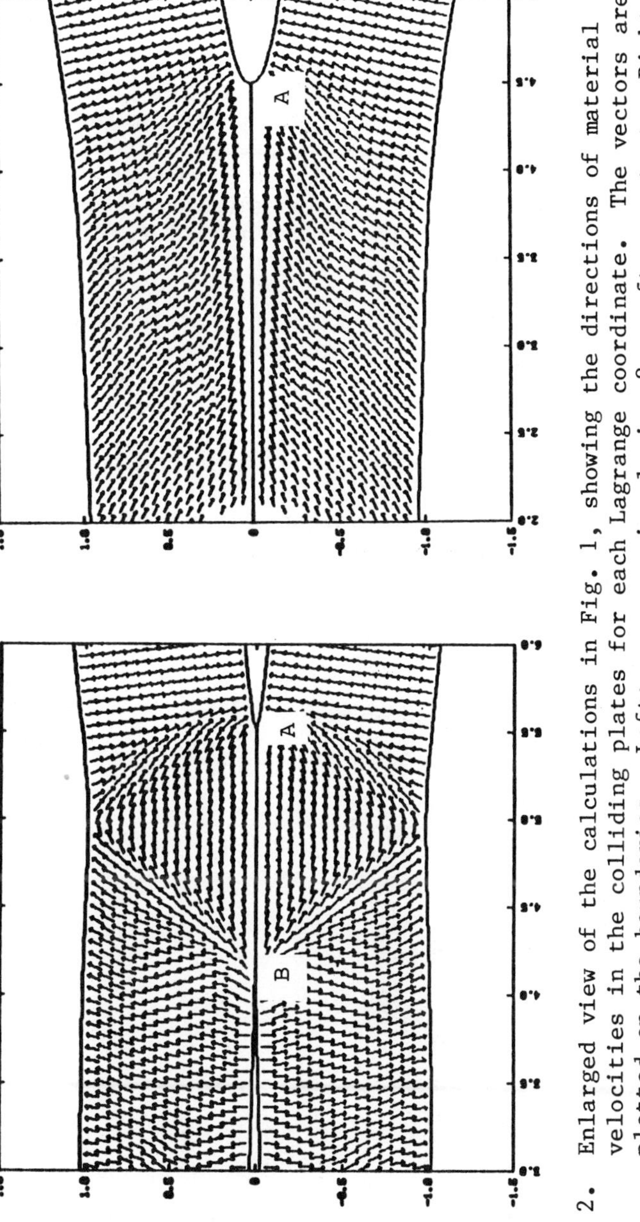

Fig. 2. Enlarged view of the calculations in Fig. 1, showing the directions of material velocities in the colliding plates for each Lagrange coordinate. The vectors are not plotted on the boundaries. Left: supersonic closing 8 µs after contact. Right: subsonic closing 10 µs after contact. A = collapse point. B = beginning of rebound.

COLLIDING PLATES

The analysis of the impact of colliding plates, as occurs in explosive welding, provides a means of understanding how high pressures can be applied and subsequently released while still avoiding large tensile stresses. Figure 1 shows a calculation simulating the collision of aluminum plates. The velocity normal to each plate is 0.7 km/s. At the top of Fig. 1 the original angle between the two plates has been chosen such that the point of contact moves at a velocity supersonic with respect to the sound speed in the aluminum. At the bottom of Fig. 1 the angle has been chosen so that the contact point is moving subsonic with respect to the aluminum sound speed. It can be seen that for the supersonic closure the plates rebound while for the subsonic case the plates remain in contact after closing.

Figure 2 shows the material velocity fields just behind the point of contact corresponding to the calculations in Fig. 1. In the case of the supersonic closing, Fig. 2, a shock wave turns the flow through a large angle and a large pressure field exists behind the shock wave. The pressure field interaction with the lateral boundaries subsequently turns the flow in a direction normal to the collision plane and the plates rebound. For the case of a subsonic closing, Fig. 2, rarefactions from the lateral boundaries can reach the point of contact. The flow field is turned through a smaller angle and the pressure field is lower than the corresponding supersonic case.

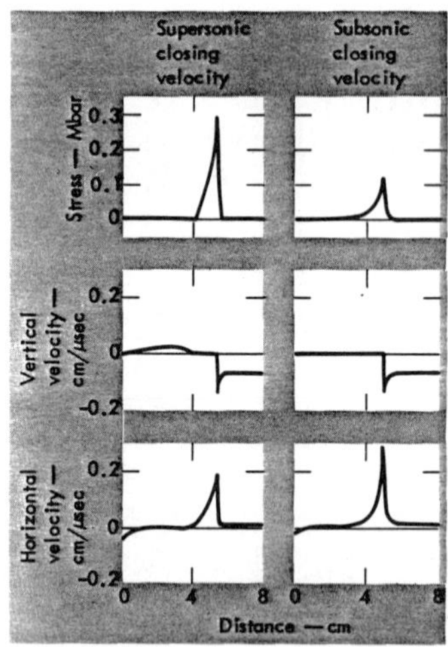

Fig. 3. HEMP calculations of conditions at the collision plane. The two cases correspond to those in Figs. 1 and 2. The data at the left are for 8 μs after first contact; the data on the right are for 11.4 μs.

Figure 3 shows conditions calculated at the interface between the colliding plates. Figure 4 compares the subsonic closing calculation of Figs. 1 and 2 with a calculation where the plates have been divided into two parts in the ratio 4:15. A frictionless interface separates the split plates. This geometry is similar to that given given in references 3, 4, and 5. In these works experiments were done with colliding plates where a thin wire had been imbedded in the plates through the thickness. Recovery experiments were performed so that the position of the wires could be measured after collision. The viscosity of the metals was determined from the relative displacement of the imbedded wires. The calculations presented here use Lagrange coordinates; hence a constant Lagrange coordinate can be used to monitor the relative motion of the material. Figure 4 shows the Lagrange grid near the collision point. A Lagrange coordinate has been darkened to simulate a wire marker. In the calculations shown here an elastic-plastic constitutive equation was used to describe the aluminum. The flow stress was 4.5 kbars. It was found that the position of the constant Lagrange coordinate depended on the magnitude of the flow stress. A thermal dependence on the flow stress showed an even greater displacement of the Lagrange coordinate along the axis due to the reduction in flow stress from thermal softening and melting.[6] These results were similar to those obtained in references.[3-5] Strength of materials can be thought as a limiting case of viscosity. Figure 5 shows an application of the simulation program to design the original setup geometry of an experiment so as to obtain the collision of a metal plate upon a second state without a rebound.

DYNAMIC COMPACTION OF POWDERS

A convenient geometry for studying the shock compression of powders is to place the powder in a metal cylinder which is surrounded by a high explosive.[7] The metal cylinder is imploded onto the powder by detonating the cylindrical charge of explosive from one end. We studied this geometry with the simulation program using PETN as the explosive to implode a steel cylinder onto copper powder. The calculation assumed the inside and outside radii of the steel cylinder were 0.630 and 0.795 cm, respectively. The PETN was surrounded by a plastic cylinder, inside radius 1.9 cm and outside radius 2.3 cm. The equation of state of the PETN detonation products was taken from Table 8-7, reference 8. The Chapman-Jouguet parameters were: density, 0.88 gm/cc; C=J pressure, 62 kb; and detonation velocity, 0.517 cm/ μs.

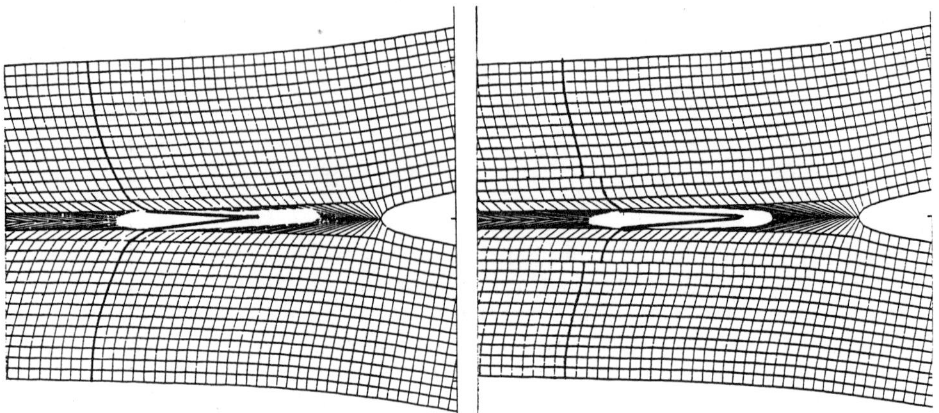

Fig. 4. Left: Lagrange grid of the subsonic closing velocity calculations given in Figs. 1 and 2 at time 10 μs. A constant Lagrange coordinate has been darkened. Right: The same calculations with the plate split into two pieces.

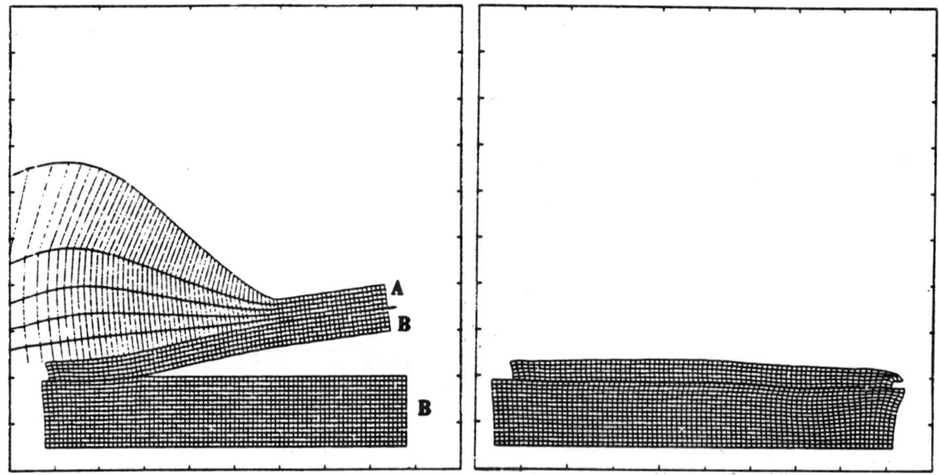

Fig. 5. Simulation of an explosive welding experiment. A = 0.5 cm PETN high explosive; B = metal plates. Left: 10 μs after detonation. Right: 50 μs after detonation. The grid of the high explosive has been deleted.

An elastic-plastic model described the steel cylinder. The porous model of Carroll and Holt[9] fitted to the experimental data of Boade[10] was used to describe the copper powder. After pore collapse an equation of state for solid copper was used. Several calculations were performed to assess the sensitivity of results to the assumption of the material models. The calculated hydrodynamics flow behind the collapse front of the powder did not seem to be sensitive to the parameters used to describe the porous response of the copper. Very similar results were obtained using a Gruneisen equation of state for the porous region. However, the hydrodynamic parameters were very sensitive to the flow stress, Y^0, of the consolidated powder.

Figure 6 shows two calculations where the starting density of the copper was 0.62 theoretical density. The collapse front, shown outlined in Fig. 6, as well as the originally parallel Lagrange coordinate j = 100, have different shapes. Figure 7 shows the calculated axial velocities vs time for different radial positions on the constant Lagrange coordinate j = 100 of Fig. 6. The stress in the axial direction has the same shape as the axial velocity, Fig. 7. The calculated peak stresses at $R/R_0 = 0$ were 450 kb and 260 kb for calculations with $Y^0 = 1$ kb and $Y^0 = 40$ kb, respectively. The large velocities along the center line and rapid drop for positions away from the center are consistent with results of Deribas and Staver.[7] The axial velocities are seen to reverse directions for the calculation with $Y^0 = 40$ kb, Fig. 7. Figure 8 shows the central channel with large axial velocities for $Y^0 = 1$ kb and the velocity reversal for $Y^0 = 40$ kb.

Two experiments were performed with the steel cylinder and PETN dimensions corresponding to the calculations (Figs. 6, 7, 8). The center portion of a steel cylinder contained a column of metallic glass powder and at either end columns of copper shot, each tapped to uniform density. The loaded steel cylinders were sealed at the base plug with epoxy and evacuated to 25-50 microns. The cylinders were mounted concentrically in lengths of PVC tubing 3.8 cm inside diameter which served as the explosive containers. The explosive was hand packed to a density of 0.95 g/cc. The detonation velocity was determined by placing coaxial self-shorting pins in the wall of the PVC tubing. A detonation velocity of 0.535 cm/μs was recorded for both experiments. The initial data for the two experiments are given in Table 1. The steel cylinders containing the powders are shown after compaction in Fig. 9. Sectioning the container revealed a small central channel in both the copper and the metallic glasses that contained melted material. The copper in both experiments was consolidated to 99.5% theoretical density. The Metglas™ 7025 and 3065 materials were consolidated to 97.4% and 98.6% respectively. The Vickers hardness values for the original ribbon materials obtained by rapid solidification technology RST were 108 kb and 116 kb for Metglas™ 7025 and 3065, respectively. The Vickers hardness for the compacted metallic glass samples were 100 kb and 115 kb for

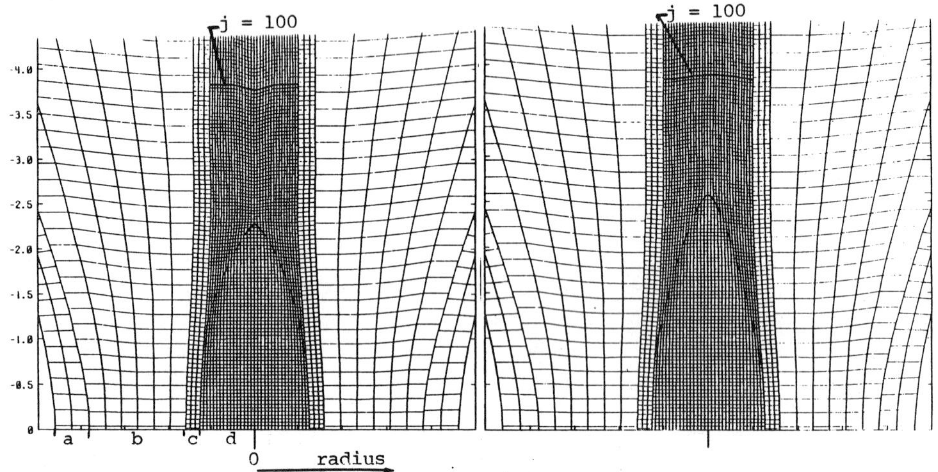

Fig. 6. Lagrange grid during compaction. Detonation is in the downward direction. The wave front has been outlined and Lagrange coordinate j = 100 behind the front darkened. Left: Y^0 = 1 kb; Right: Y^0 = 40 kb.

Metglas™ 7025 and 3065, respectively. This result is consistent with other metallic glass alloys[11] where the ribbon bulk properties, i.e., compressive strength and elastic constants, remain unchanged after consolidation of powder made from the ribbon.

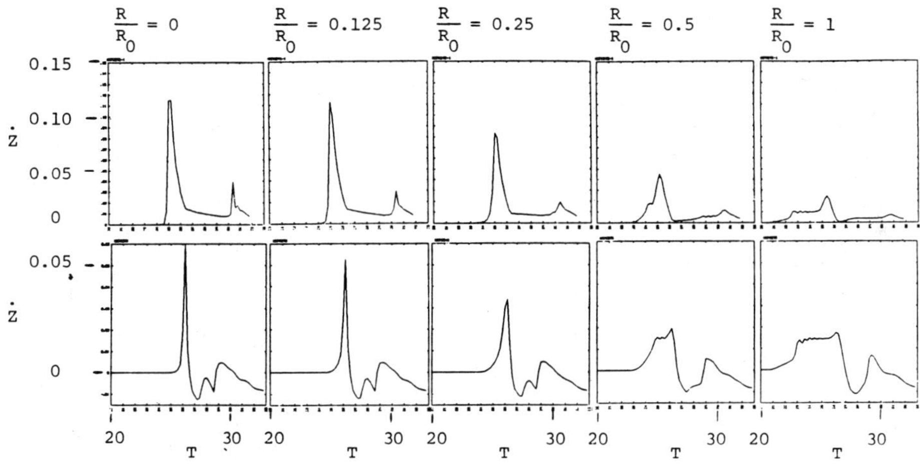

$\dot{Z}$ - Axial velocity cm/μs
T - Time μsec
R - Radial position in powder
R_0- Outside radius of powder

Fig. 7. Axial velocity vs time for positions along the constant Lagrange coordinate of Fig. 6. Top Y^0 = 1 kb; bottom Y^0 = 40 kb.

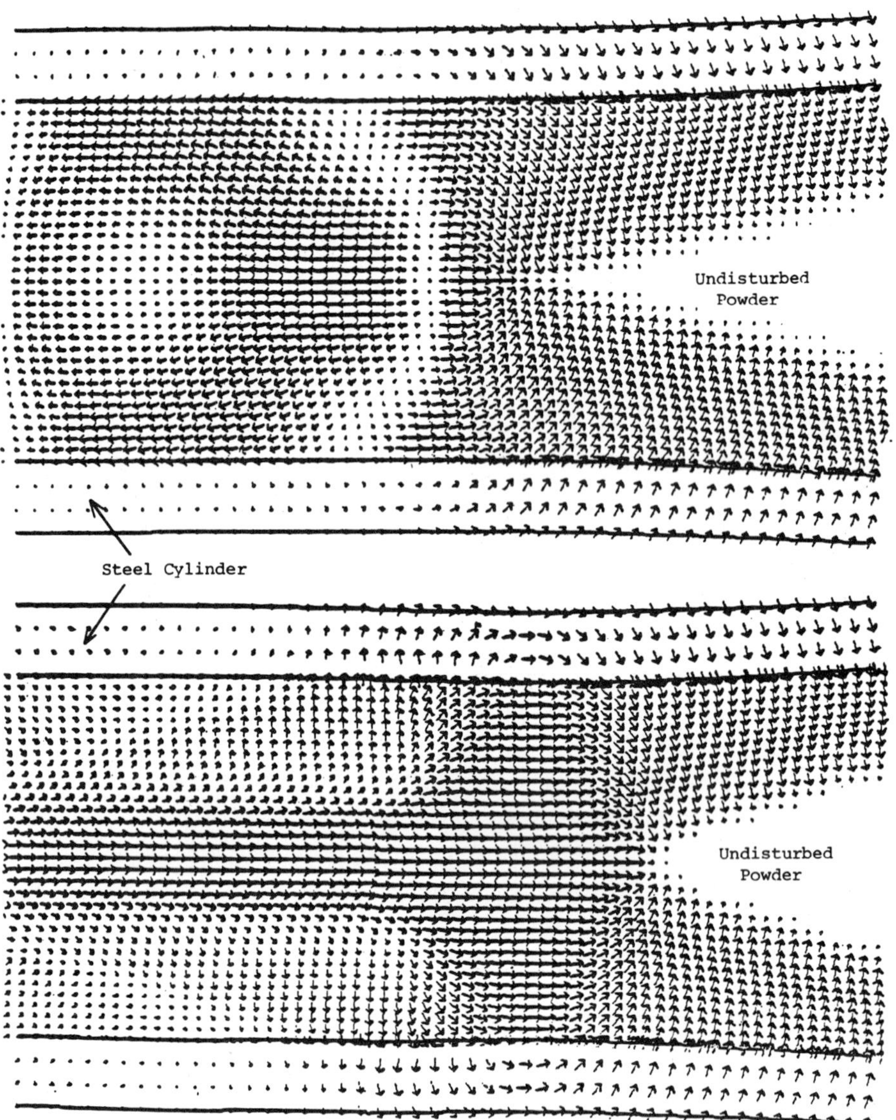

Fig. 8. Direction of material velocities at the same time as Fig. 6. Top Y^0 = 40 kb; bottom Y^0 = 1 kb.

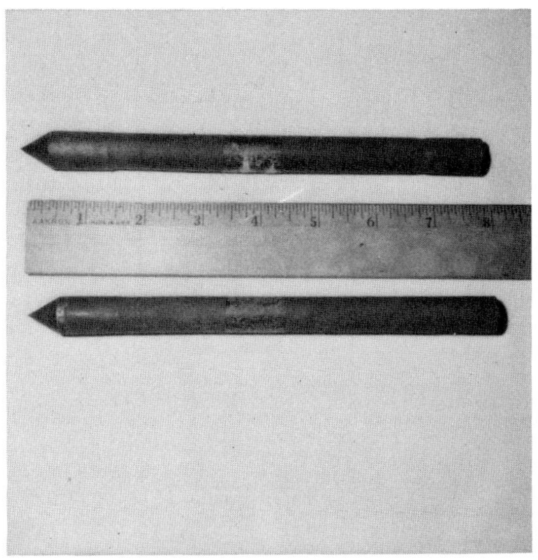

Fig. 9. Steel containers after explosive compaction. Detonation was from left to right.

Table 1. Materials Compacted

Metglas™ 3065		
Theoretical Density		9.23 g/cc
Bulk Density, Tapped		5.08 = 0.55 T.D.
Composition:	Mo	30.0 atom%
	B	10.5
	Ni	59.5
Metglas™ 7025		
Theoretical Density		9.05 g/cc
Bulk Density, Tapped		4.80 = 0.530 T.D.
Composition:	Mo	23.5 atom%
	Fe	9.0
	B	10.5
	Ni	57.0
Copper Shot		
Theoretical Density		8.95 g/cc
Bulk Density, Tapped		5.58 = 0.623 T.D.

™ Trademark Allied Corporation.

THE USE OF NITROMETHANE AND POWDERED HMX FOR COMPACTION EXPERIMENTS

Quantitative stress time conditions in powders could be calculated if accurate equations of state existed for the materials. It is not practical to determine the equation of state of the various powders to be studied and it is not necessary, provided that the implosion geometry can be correctly calculated. This latter condition places a requirement for knowledge of the equation of state for the detonation products of the high explosive. The previous experiments used PETN as explosive and an equation of state is available for certain packing densities.[7] However, in practice it is not always convenient to load the explosive to a specific density. Two explosives that can provide consistent loading densities are nitromethane and HMX powder. An equation of state for the detonation products of nitromethane is given in ref. 7: detonation velocity = 0.628 cm/μs, Chapman-Jouguet pressure = 125 kbars, density = 1.128 g/cc. The equation of state was checked by calculating the expansion of a copper cylinder and compared with streak camera results for the displacement of the outside of the cylinder, see Fig. 10.

Stock stainless steel pipe was selected for the container of powders to be studied. Pipe dimensions: O.D. 5.08 cm, I.D. 3.81 cm. The implosion history of the pipe was checked using nitromethane and HMX powder. Figure 11 shows results of the implosions obtained by flash x-ray and a calculation corresponding to the experiment with nitromethane. The agreement of the calculated 8° collapse angle with the experiment verified that the implosion could be accurately calculated with the known equation of state for nitromethane. An equation of state for the detonation products of powdered HMX is not available. However, since the detonation velocity of the powdered HMX for the density used, ρ_0 = 1.15 g/cc, is about the same as the detonation velocity of the nitromethane, it would be possible to obtain an approximate equation of state by normalizing the nitromethane equation of state to obtain the same collapse angle as the HMX.

COMPACTION USING A CENTRAL MANDREL

In the experiments described earlier a central hole was drilled through the consolidated sample due to the high pressures developed along the axis of symmetry. The calculations of the experiments showed that the high pressures, due to the stagnation of the powder on the axis of symmetry, resulted in large axial velocities. Figure 12 shows the calculated axial velocity, $\dot{Z}$, as a function of the radial and axial position, R,Z. The peak velocity on the axis is of the order of 1 km/s. The placement of a mandrel on the axis of cylindrical symmetry prevents the large collapse velocities of the powders to occur which produce the large pressures on the axis. Figure 13 compares the peak axial material velocity profile with and

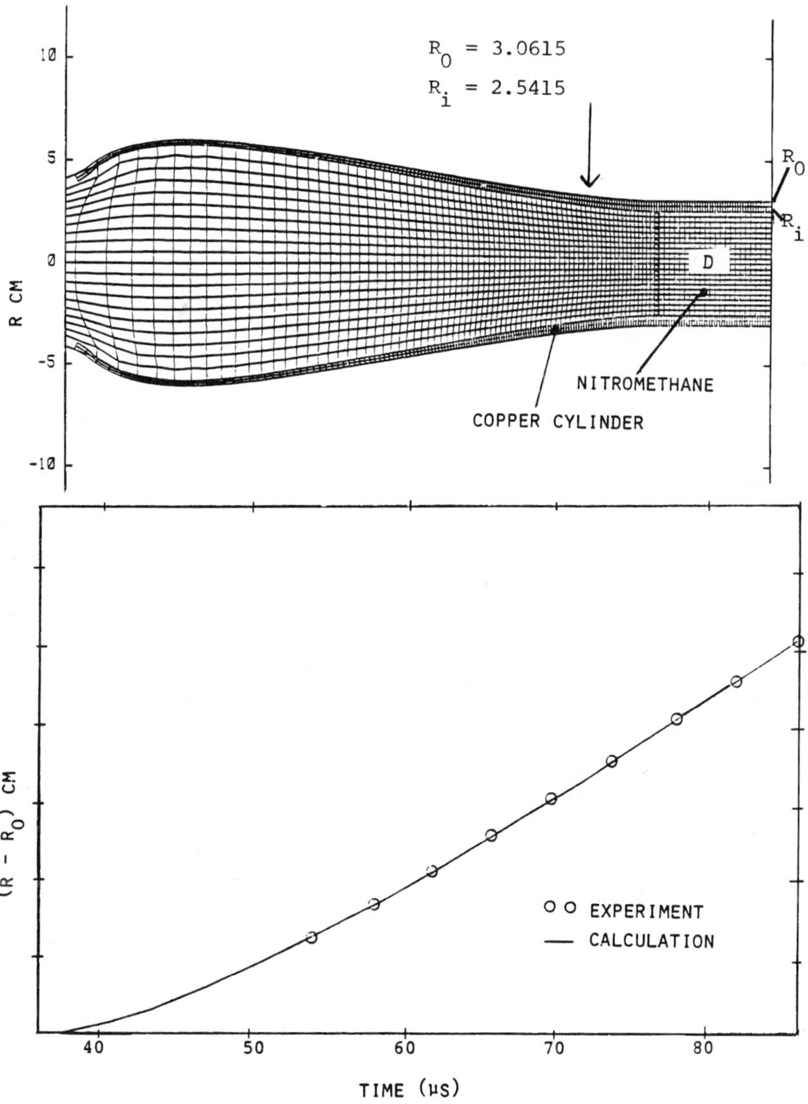

Fig. 10. Cylinder test of nitromethane. Top: Calculation of 45 μs after detonation from left end. Bottom: Radial displacement at axial position of arrow above.

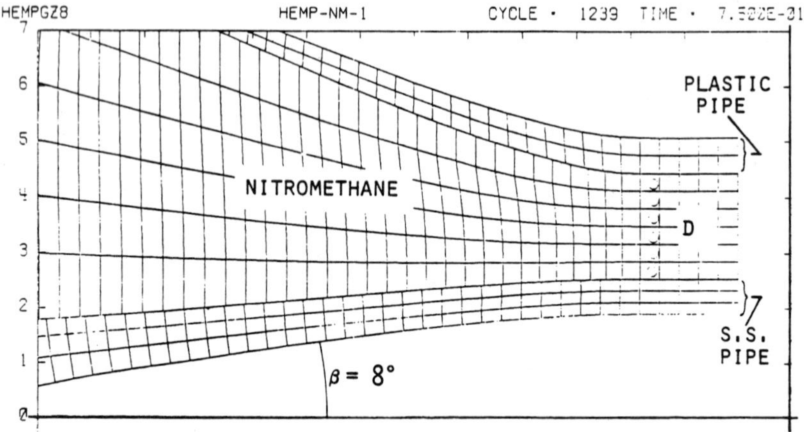

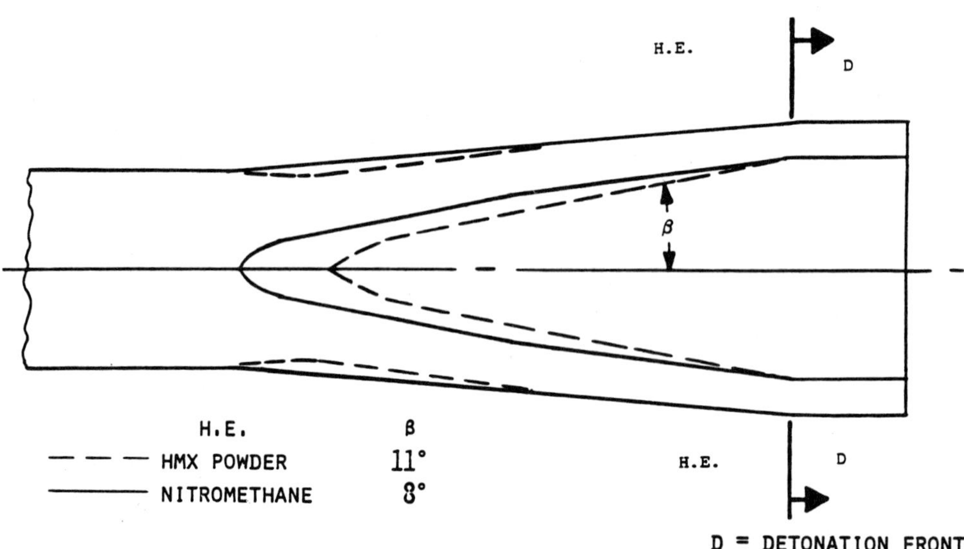

Fig. 11. Top: Calculation of the collapse of a stainless steel, S.S., pipe by nitromethane. Bottom: Experimental results from flash x-rays.

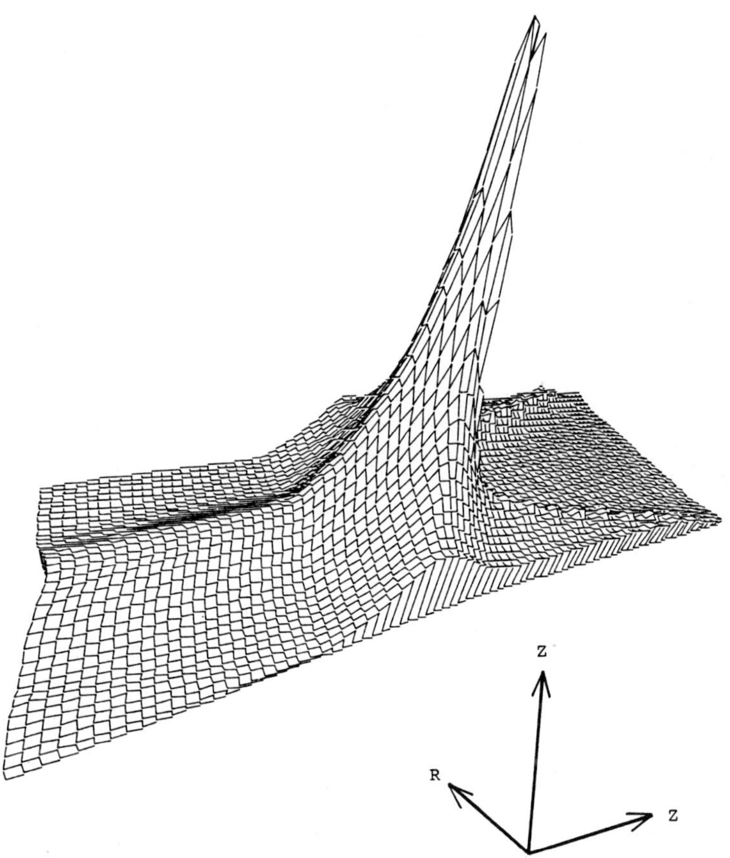

Fig. 12. Axial velocity $\dot{Z}$ in powder compacted without a central mandrel.

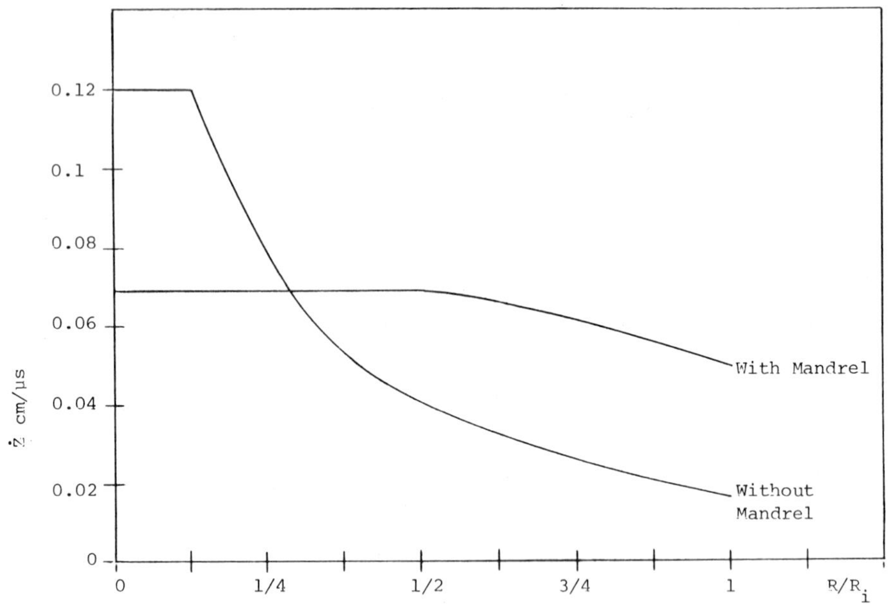

Fig. 13. Calculated maximum axial material velocity $\dot{Z}$, across a section of the sample container. R_i = inside radius of stainless steel container.

without a mandrel. In the calculation a mandrel was introduced with material properties appropriate to mild steel. The radius was approximately 1/2 the inside radius of the stainless steel tube used to contain the powder. It is seen that the large localized shear that would result from a sharp velocity gradient is not present when a mandrel is introduced. Figure 14 shows the cross sectional dimensions for experimental geometries selected for testing. The cylinders were 50 cm long and detonated from one end as previously described. The working volume was divided into three sections with the central section containing metallic glass and powdered copper on either side.

The calculated radial stress vs time in the metallic glass is shown in Fig. 15. It is seen that for both high explosive systems the powder is compressed and then undergoes tension before the radial stress returns to 0. The peak stress for the small and large HE systems are 150 kbars and 300 kbars, respectively. Figure 16 shows the corresponding relative volumes. The relative volume is defined here as the density of the solid divided by the actual density. After the peak compression, the relative volume overshoots and dialates to approximately 10% before settling to a value of 1.01 for the small HE system and 1.02 for the large system. In the calculations the flow stress of the metallic glass, Y, is assumed to be equal to Y = 1.2 p, where p is the pressure. A maximum Y of 40 kbars is imposed. It is further assumed that the powders forged together from the compression so that the flow stress, Y, remains at its largest value if the pressure should drop. Figure 17 shows the calculated flow field during compaction; it is seen that the metallic glass has separated from the mandrel after compaction. Point a, Fig. 17, marks the origin of the rarefaction waves. When the rarefaction waves reflect from the axis of symmetry tension occurs in the mandrel. The calculation predicts the mandrel will spall. Figure 18 shows a section perpendicular to the axis of the cylinder after compaction for the experimental geometry with HE radius 4.45 cm (Fig. 14). Figure 19 shows cross sections containing metallic glass and copper for the large HE system radius 6.67 cm. The metallic glass was consolidated to 99% of theoretical density for the large HE system and to 97% for the small HE system. The copper reached 97% compaction for the small HE system and 100% for the large HE system. The initial powdered copper size for these experiments was much greater than the size used in the experiments described earlier.

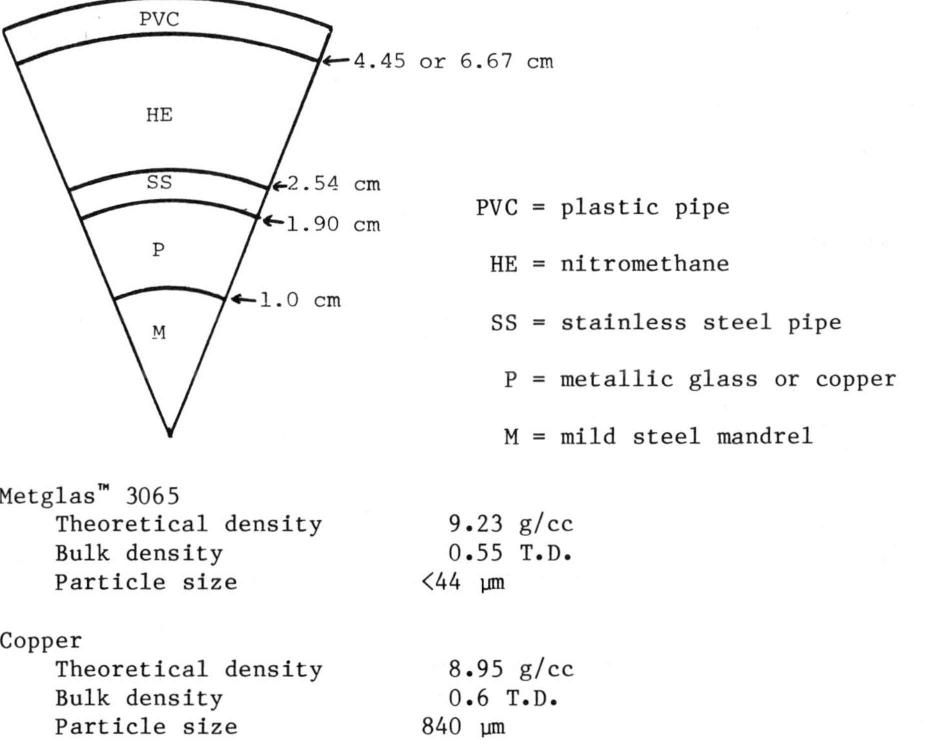

Fig. 14. Cross section of experimental geometry.

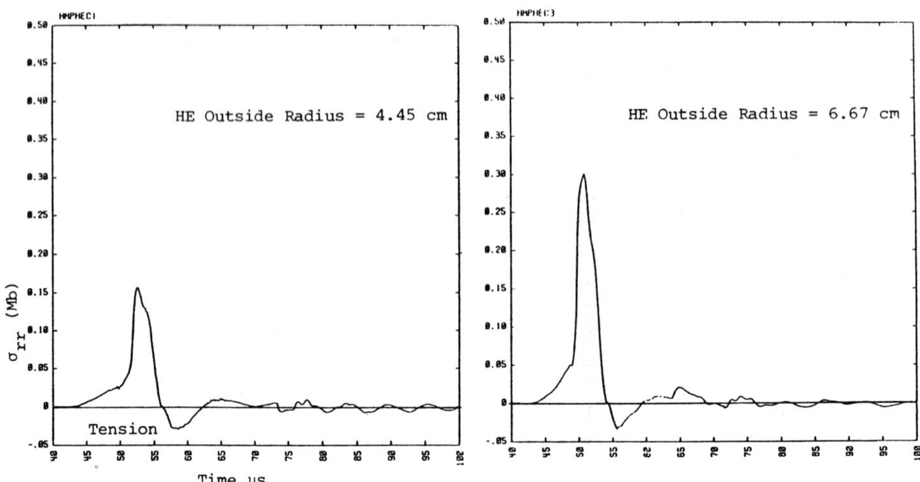

Fig. 15. Radial stress vs time in powder.

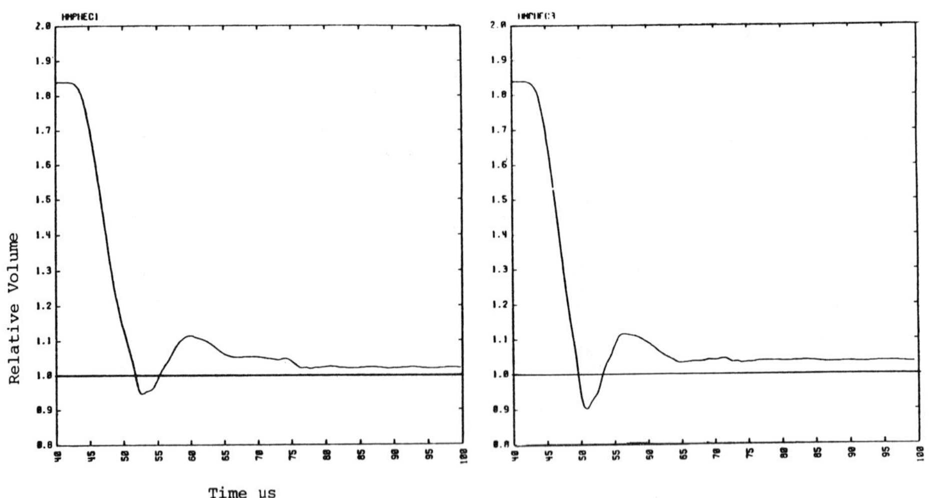

Fig. 16. Relative volume corresponding to Fig. 15.

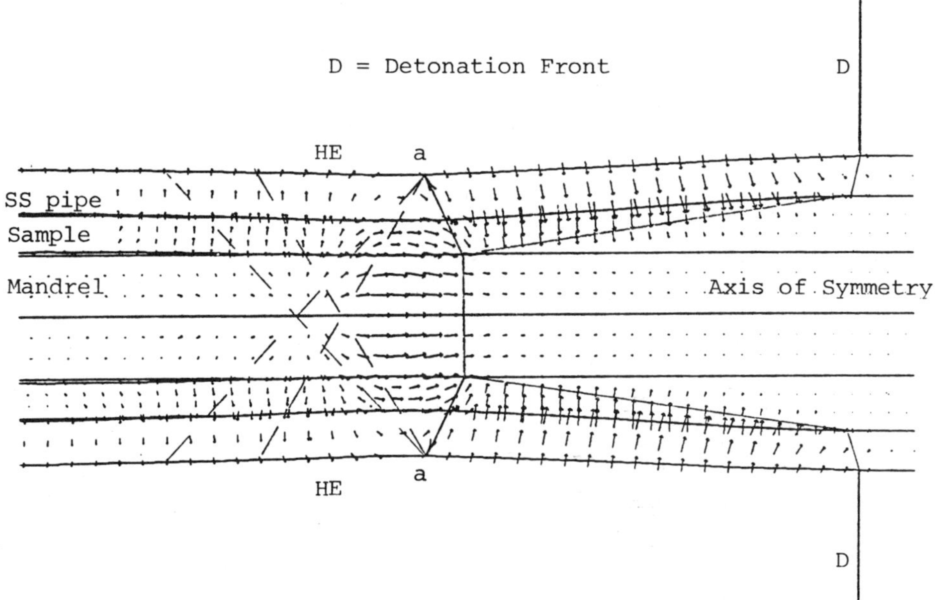

Fig. 17. Computed material velocity vectors during cylindrical compaction. Wave pattern is sketched in. Solid lines: shock front. Broken lines: rarefaction fan. Mandrel radius 1 cm; HE outside radius 4.45 cm.

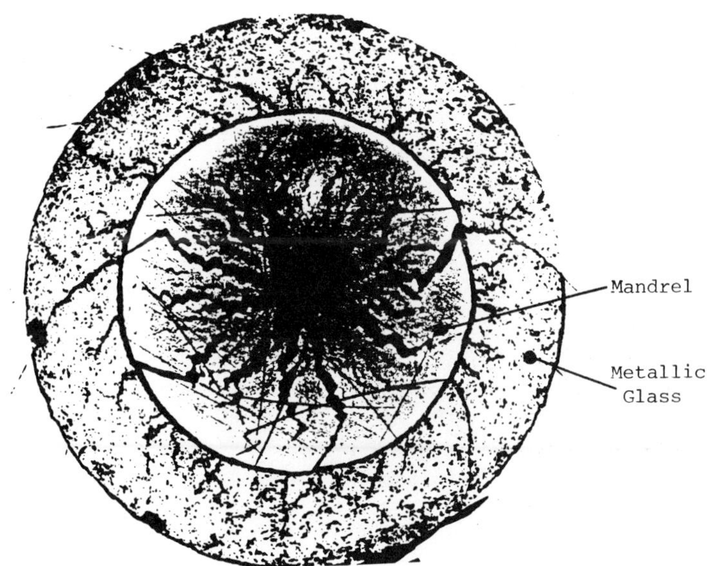

Fig. 18. Section perpendicular to the axis of cylindrical symmetry after compaction. Mandrel radius, 1 cm; nitromethane radius, 4.45 cm.

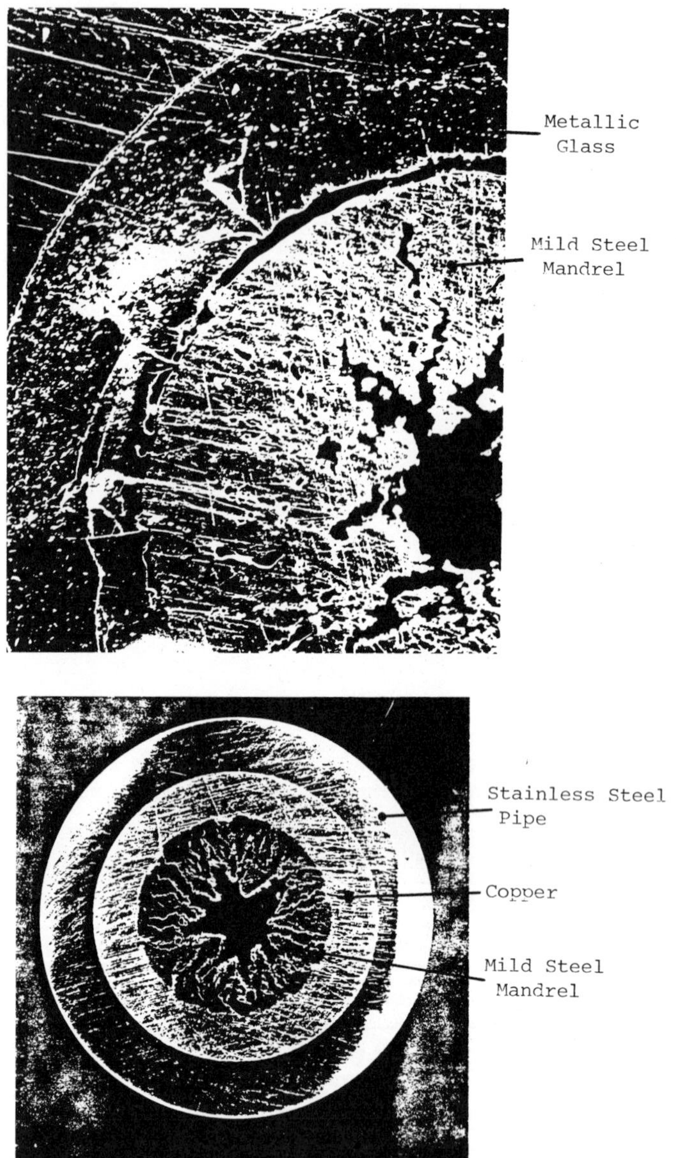

Fig. 19. Cross sections for experiment with HE system radius 6.67 cm.

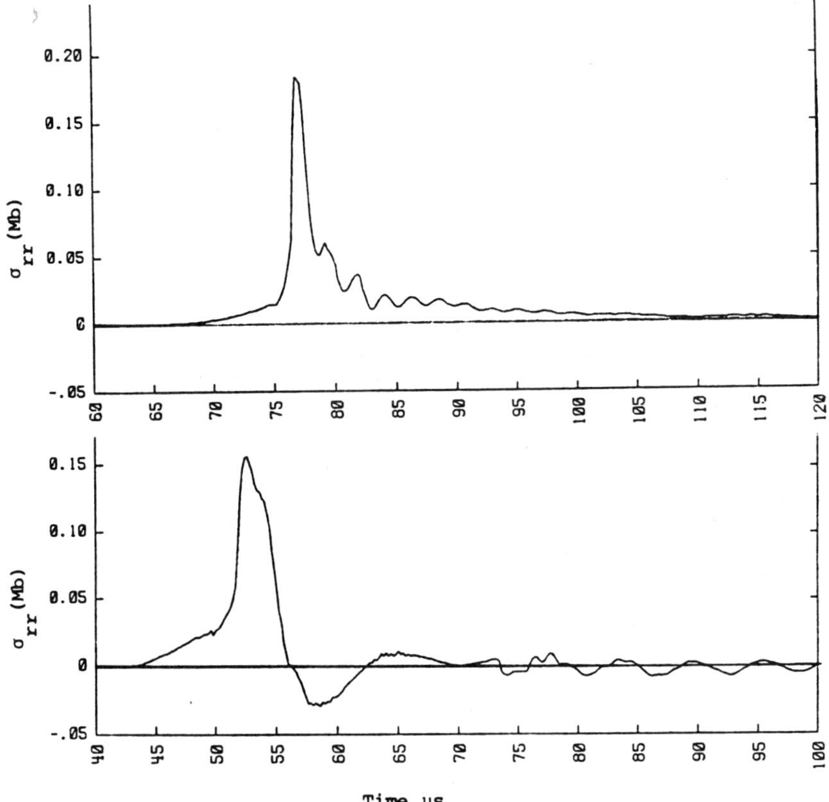

Fig. 20. Radial stress-vs-time in powder. Top: HE with detonation velocity 0.4 cm/μs. Bottom: Nitromethane, detonation velocity 0.638 cm/μs.

Referring to Fig. 17, the magnitude of the release waves would be reduced if the reflected shock encounters a higher pressure from the explosive, position a. This could be achieved, in principle, by employing a lower detonation velocity explosive. In this case the reflected compression wave would reach the explosive/steel container interface at a position nearer to the detonation front where the pressure would be higher. To check the understanding a calculation was done with an assumed equation of state for a high explosive with a low detonation velocity. Specifically, the detonation velocity was $D = 0.4$ cm/μs, density $\rho_0 = 1.02$ g/cc. A γ-law equation of state was used with $\gamma = 3$. With these parameters the Chapman-Jouguet pressure, P_{CJ}, and the chemical energy E_0 are:

$$P_{CJ} = \frac{\rho_0}{\gamma+1} D^2$$

$$E_0 = \frac{P_{CJ}}{2(\gamma-1)}$$

Figure 20 compares a calculation using the above equation of state with the calculation that simulated the experiment with nitromethane. The thickness of the low detonation velocity explosive was increased to provide the same peak pressure in the powder as was obtained in the nitromethane calculation. It can be seen that the pressure is released with no tension and correspondingly no dilation for the calculation with the low detonation velocity explosive.

ACKNOWLEDGMENT

We would like to acknowledge the work of Frank Helm, S. L. Weinland and C. R. Henry for the experiments and M. Kamegai for the porous model to describe powdered materials. Estella Salinas performed the HEMP calculations.

REFERENCES

1. George A. Cowan and Arnold H. Holtzman, J. Appl. Phys. 34 [4], (Part 1), 928 (April 1963).
2. A. A. Deribas, "Explosive Welding," report from Institute of Hydrodynamics, Novosibirsk (1967).
3. S. K. Godunuv, A. A. Deribas, I. D. Zakharenko, and V. I. Mali, Fizika Goreniya i Vzryva 1, 135-141 (January-March 1971).
4. N. S. Kozin, V. I. Mali, and M. V. Rubtsov, Fizika Goreniya i Vzryva 13 [4], 619-625 (July-Aug. 1977).
5. I. D. Zakharenko, V. I. Mali, "Combustion and Detonation Material," presented at the 3rd All-Union Symposium on Combustion and Detonation, July 5-10, 1971, 575-578 (Nauko, Moscow, 1972).
6. D. J. Steinberg, S. G. Cochran, and M. W. Guinan, J. Appl. Phys. 51 [3], 1498 (March 1980).
7. A. A. Deribas and A. M. Staver, Fizika Goreniyz i Vzryva 4, 568-578 (July-Aug. 1974).
8. B. M. Dobratz, LLNL Explosives Handbook, UCRL-52997 (March 16, 1981).
9. M. M. Carroll and A. C. Holt, J. Appl. Phys. 43, 759 (1972).
10. R. R. Boade, J. Appl. Phys. 41, 4542 (1970).
11. Carl F. Cline, Fourth International Conference on Rapidly Quenched Metals, Sendai, Japan, August 24-28, 1981.

INVESTIGATION OF A METHOD TO CONSOLIDATE HARD MATERIALS IN A TOUGH MATRIX

Jim D. Mote and Joan J. Fitzpatrick

Chemical and Materials Sciences Division
Denver Research Institute, University of Denver
Denver, CO 80208

INTRODUCTION

The production of the so-called cemented carbides, a class of very hard, wear resisting materials manufactured by powder metallurgy processes, is well known. The uses to which these materials are put range from cutting tools to drawing dies for tungsten wire. The processes used to produce such products are combinations of cold pressing, hot pressing, regular sintering and liquid phase sintering. The resulting material is usually a hard substance imbedded in a tough matrix.

It was the objective of this investigation to examine an alternative method of producing compacts of hard materials in a ductile matrix, namely using the explosive powder compaction method at elevated temperatures.

BACKGROUND

Conventional powder metallurgy methods, such as compressing a metal powder into a die to produce a green compact, and then sintering the compacts at elevated temperature to form some sort of particle bonding is also used with powder mixtures of hard materials and ductile materials.

The consolidation of powdered materials requires some mass transport mechanism and those customarily distinguished are: (1) evaporation and condensation; (2) viscous flow; (3) diffusional flow; (4) plastic flow. Diffusional flow is the most important mechanism in conventional sintering of metallic powders. However, consolidation can be enhanced, especially in ductile materials, by the application of high static pressures at elevated temperatures.

It is also known that dynamic pressure can be used to consolidate certain metallic powders and it is observed that there is considerable plastic deformation associated with the consolidation. The simplest means of imposing dynamic pressure on powder materials is through the detonation of an explosive in contact with a cylindrical metal container that encapsulates the powder. This method was described by Leonard et al.[1] and later elaborated by Pruemmer[2] and dubbed the direct method. Leonard et al. argued that there are two competing mechanisms operating during the direct consolidation process: (1) the pressure tends to increase because of the cylindrically converging wave front; (2) the pressure tends to decrease because of energy dissipation caused by heat generation, plastic deformation, comminution, etc. In order to produce a uniform and completely densified cylindrical compact, these two competing mechanisms must be properly balanced. If this balance is not achieved, undercompaction or overcompaction may occur.

Where undercompaction occurs the pressure wave is convex to the uncompacted powder. In the case of overcompaction the pressure wave is concave to the uncompacted powder and interior cracks may be observed. Also under some conditions of overcompaction a so-called "mach stem" is formed leaving behind a hole along the cylinder axis.

Attempts have been made to correlate powder densification with the detonation pressure of the explosive and the ratio of the powder mass to the explosive mass. By implication and sometimes by express statement[2] these investigators assign to the powder container the role of merely transmitting the shock wave with no consideration of the effect of the container mass on the compaction process.

Pruemmer[2] has used the procedure described below to investigate the explosive compaction of powder materials. He shows that the maximum density obtained in a compact is a function of the detonation velocity of the explosive used. We note that the detonation pressure is proportional to the square of the detonation velocity, i.e.,

$$\underline{P} \approx \frac{1}{4} \rho_o V^2$$

where $\underline{P}$ is the detonation pressure, ρ_o is the density of the explosive and V is the detonation velocity. Pruemmer chose as his other parameter the ratio of the explosive mass to the powder mass (E/M). Figure 1a shows the typical shape of an experimentally determined curve of E/M vs. detonation velocity squared. A point on the curve represents a sample that has uniform compaction across the cross section.

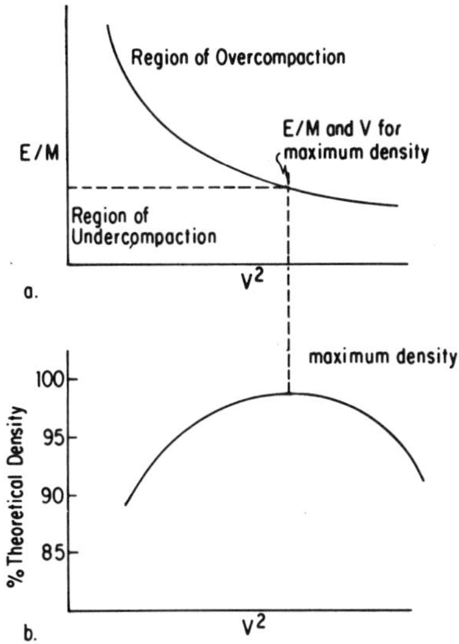

Fig. 1. Procedure for experimentally determining proper E/M and V^2 (after Pruemmer).

A point on the curve is determined by first selecting a detonation velocity. Subsequently, a series of experiments are conducted at various E/M with the chosen detonation velocity. The cross section of each of these samples is examined and the one that exhibits uniform compaction is selected. This provides the coordinates of a point on the curve representing uniform compaction across the cross section of the sample. All the other samples are discarded and the density of the uniformly compacted sample is measured. The relative density (measured density as a percentage of the theoretical density of the material) is plotted vs the detonation velocity squared as shown in Fig. 1b.

Pruemmer found that there is a maximum in the relative density vs detonation velocity curve. Thus the detonation velocity of the explosive and the E/M that will produce the maximum density in a given powder material can be determined experimentally. It should be pointed out, however, that the detonation pressure is not the pressure transmitted through the container wall. This pressure is a function of the explosive, the container material and the orientation of the detonation front.[3-5]

Bhalla and Williams[6] argue that a better correlation is obtained when the results of direct explosive compaction are interpreted in terms of the kinetic energy imparted to the container tubes. This approach accounts for the tube mass.

They determined an empirical expression relating the compacted powder density to the container tube impact energy. Later Lennon et al.[7] revised the expression to one wherein the compacted density was related to the specific energy absorbed by the powder. The form of the relationship was found to be

$$D_c = D_T - \Delta D \exp^{-\beta E^{\gamma}}$$

where Dc = the compact density
D_T = the theoretical density of the powder
$\Delta D = D_T - D_i$, D_i being the initial packing density
E = the corrected energy of compaction per unit volume of powder
β and γ are characteristic constants for the individual powders.

Unfortunately the characteristic constants β and γ must be determined experimentally.

Hoenig et al.[8] attempted hydrodynamic modeling of right circular cylinder compaction of ceramic powders with detonation at one end using a two-dimensional computer code. However, complications led them to do experiments using plane shock waves which were subsequently analyzed by two-dimensional hydrodynamic codes.

Hoenig et al. noted that the primary bonding mechanism in their experiments was due to surface melting of the powder particles. However, they did experience significant sample cracking which they attributed to unloading stresses.

Several authors[1,7] have pointed out that during the cylindrical compaction of a powder by a conical wave, as shown in Fig. 2, the outside of the cylinder is compacted first. This fact requires that the outer portion of the cylinder either deforms plastically or fractures during the compaction of the innermost regions. Leonard et al.[1] suggest that the spiral cracks found in cylindrical samples of brittle materials compacted by the direct method result from the aforementioned phenomena.

A variant to the approaches reviewed above is to preheat the powder prior to dynamic compaction. The rationale for this approach is based on the premise that materials usually show a decrease in flow stress and an increase in ductility with increasing temperature. It should be noted that the surface melting observed by Hoenig et al. may not provide the required temperature conditions for bulk plastic flow in the powder particles since the bulk temperatures may not reach the required level. Further, the consolidation pressure probably attenuates before bulk heating is achieved.

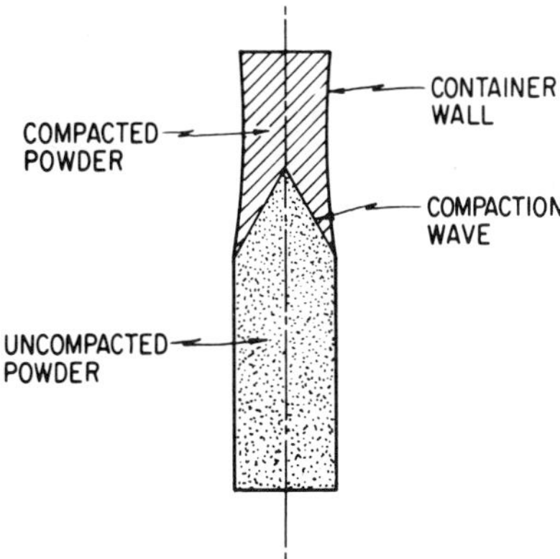

Fig. 2. Conical compaction wave showing uniform compaction.

By creating conditions for plastic flow of the particles, pore collapse can be accomplished without comminution and there is less chance for fracture by relief waves. The relief waves can be further modified by changes in the container material and its wall thickness. Also, it has been found that when the powder is preheated the E/M ratio and the tube mass can be increased significantly and still obtain uniform compaction over the cross section of the specimen.

Some survey experiments are described in the following sections. The materials compacted consist of powders of tungsten and mixtures of tungsten and tungsten carbide and tungsten and titanium carbide. The explosive detonation velocity was not varied but was selected to be in the range known to give a relatively high density in tungsten.[9]

EXPERIMENTAL PROCEDURE

As yet, no analytical model has been derived that adequately characterizes the explosive powder compaction process. Even with relatively simple geometries there are a large number of variables to be considered. Some experimental programs have provided useful guidelines such as (1) the compacting pressure should exceed the yield strength of the powder being compacted; (2) consolidation improves with increasing time of pressure application; and (3) elevated temperature induces ductility at lower pressure and, therefore, enhances the consolidation of hard materials. Pruemmer's empirical approach shows how to obtain the maximum density while maintaining the container material and wall thickness fixed as well as holding the pre-shock temperature constant.

In this investigation we limited out attention to the effect of elevated temperature, E/M ratio and container wall thickness using only the cylindrical geometry. The explosive used had a detonation velocity in the range reported to give the maximum density in tungsten powder compacted at room temperature. A summary of the experimental parameters is given in Table 1. The explosive to tube weight ratio (E/tm) is about twice that used by Bhalla and Williams[6] to compact nickel powders to 97.7% of theoretical density and the tube wall thickness is about 1.5 times the thickness they used. The E/M ratio ranges from 4.7 to 8.7, or 6 to 10 times the value used by Pruemmer[10] to consolidate pure tungsten powder at room temperature.

The details of the elevated temperature compaction method have been described elsewhere.[11] Only an outline of the procedure is given below.

The powder is encapsulated in a stainless steel tube after precompaction to about 55-65% of theoretical density. The sample is evacuated while preheating to the chosen temperature. When the selected temperature is reached, the capsule is sealed, placed in an insulated container, transported to the explosive assembly, dropped through a tube and the explosive auto detonates, thus compacting the powder.

Table 1. Explosive Compaction Parameters

Shot No.	Composition % Volume		E/M	E/tm	Preheat Temp., °C	Crystal Density gm/cc
1bWTiC1	TiC W	68.8 31.2	8.7	4.2	1320	9.41
1	TIC W	50 50	7.2	4.2	1320	12.12
2	TIC W	25 75	7.3	4.2	1320	15.7
8bW7	W	100	4.7	4.2	1320	19.3
4	WC W	29 71	5.7	4.2	1320	18.28
1bWCW1	WC W	30 70	4.5	3.8	1320	18.24
3	WC W	55 45	5.9	4.2	1320	17.35

STARTING MATERIALS

Three commercial powders of tungsten (W), tungsten carbide (WC) and titanium carbide (TiC) were acquired and characterized for the purposes of this work. Morphological and physical characteristics of the three starting materials are shown in Figs. 3 through 6.

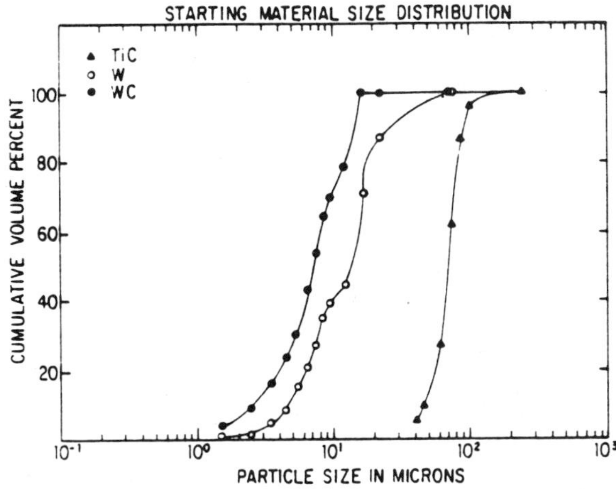

Fig. 3. Size distribution of starting powders.

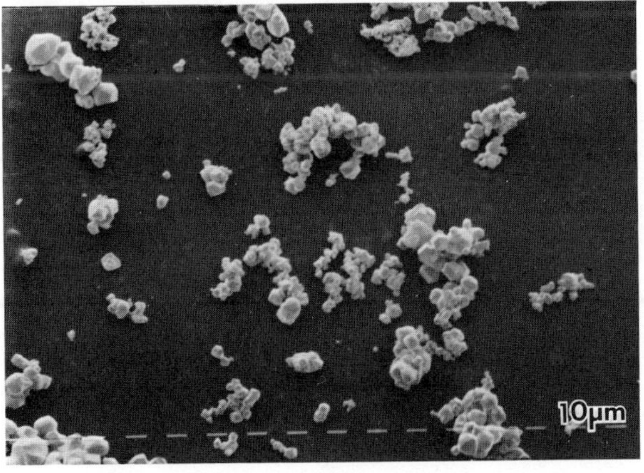

Fig. 4. Tungsten powder, as-received.

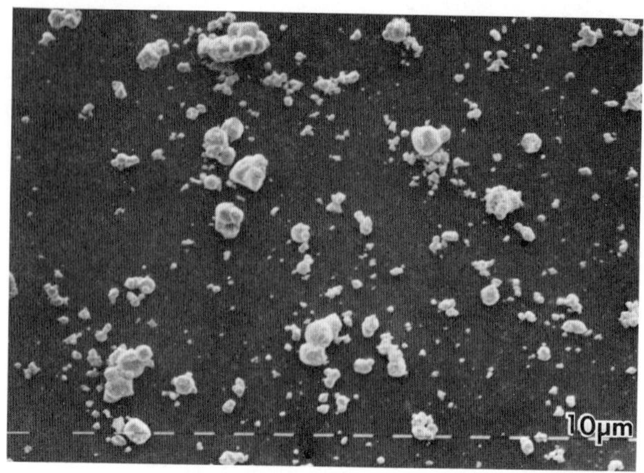

Fig. 5. Tungsten carbide powder, as-received.

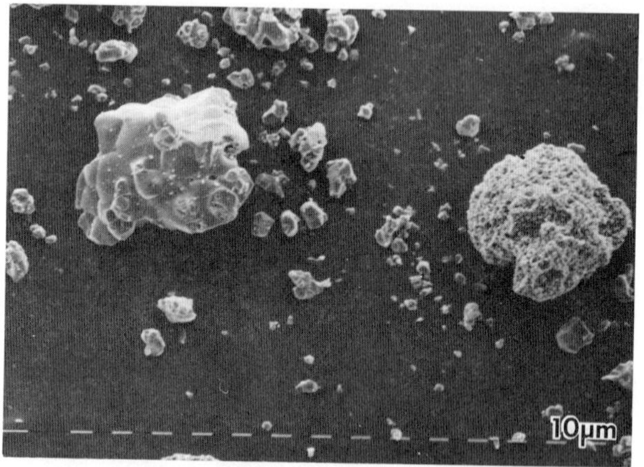

Fig. 6. Titanium carbide, as-received.

The tungsten and tungsten carbide powders were shown to be free from chemical contaminants above the limits of detection for the methods of analysis. The titanium carbide showed chemical impurities totalling 0.11%. In addition, the reported amount of total carbon in the titanium carbide falls 1.51% short of that required by stoichiometry. It is therefore likely that some titanium is present in forms other than the carbide.

Both the tungsten and tungsten carbide powders show bimodal particle size distributions (Fig. 3). An examination of the powders by scanning electron microscopy (SEM) shows that the apparent bimodality of the tungsten powder is due to a population of particle aggregates, not single grains, in the 10 to 25 micron size range (Fig. 4). The bimodality of the tungsten carbide, however, appears substantiated by the appearance of a population of well-sutured composite grains in the 10 to 15 micron size range (Fig. 5). Examination of the titanium carbide showed it to contain a wide range of particle sizes and types (Fig. 6). The morphological character of the TiC powder is dominated by the presence of large aggregate grains which vary in character from well-sutured composite grains to large, spongy, weakly bound aggregates of very small individual TiC particles.

RESULTS

Analysis of the explosively compacted materials consisted of scanning electron microscope characterization of polished sections to determine the mechanism of consolidation, hardness and density determination.

A total of six mixed powder combinations of tungsten/tungsten carbide and tungsten/titanium carbide and a single powder charge of tungsten were explosively compacted after pre-heating to 1320°C in an induction furnace by the method described in the previous section. Compositions and shot conditions for the six samples are given in Table 2. Density and hardness results on all samples are given in Table 3.

Table 2. Compositions and Shot Conditions for Compacted Powders

Shot No.	Composition % Volume		Vacuum (μm)	Time to Shot (seconds)
1bWTiC1	W	31%	< 1	24
	TiC	69%		
1	W	50%	45	18.5
	TiC	50%		
2	W	72%	40	18.5
	TiC	25%		
3	W	45%	40	18.5
	WC	55%		
1bWCW1	W	70%	< 1	19.6
	WC	30%		
4	W	71%	500	18.8
	WC	29%		
8bW	W	100%	<1 μm	28.9

Table 3. Physical Measurements on Test Shots

Shot No.	Composition % Volume		$\rho_{meas.}$	% $\rho_{crystal}$	Average D.P.H.
1bWTiC1	W TiC	31% 69%	8.64	91.7	819
1	W TiC	50% 50%	11.69	96.5	510
2	W TiC	75% 25%	11.3	71.9	588
3	W WC	45% 55%	16.2	93.1	676
1bWCW1	W WC	70% 30%	16.1	88.3	649
4	W WC	71% 29%	13.9	75.8	1255
8bW7	W	100%	18.8	97.3	542

The consolidation experiment on the pure tungsten powder yielded a sample at 97.5% theoretical density. SEM examination of a polished section from this shot shows uniform compaction with low porosity (Fig. 7). Plastic flow was evident in the interstices of a radial crack. With the successful consolidation of the tungsten powder, combination shots of TiC/W and WC/W were carried out with the intention of investigating the possibility of utilizing the tungsten as a binder to produce a hard and tough compact.

Tungsten/Tungsten Carbide Shots

Three combinations of tungsten and tungsten carbide powders were explosively compacted. SEM examination of these samples showed good consolidation in all three cases. Samples containing 44.97 and 69.9 percent tungsten by volume consolidated by identical mechanisms and hence, have similar appearances (Figs. 8 and 9). Small, dark regions of WC can be distinguished in a continuous matrix of W. The porosity of the 69.9% W shot is somewhat lower than that of the 44.97% W shot although the densities and hardnesses are very similar. The dominant mechanism of consolidation appears identical to that of the pure tungsten sample, namely plastic flow. The 71.03% W, however, has consolidated by a very different mechanism. X-ray analysis of areas of the polished section for this sample in the SEM shows that the consolidating matrix is a tungsten-steel alloy with a high iron content which has formed during compaction by jetting and reaction of the steel casing which holds the sample. Individual grains of tungsten and tungsten carbide rather than floating individually in the matrix have become segregated into patterned areas with locally high concentrations of starting materials (Fig. 10). The result is an extremely hard, low density compact (Table 3) with low comparative porosity.

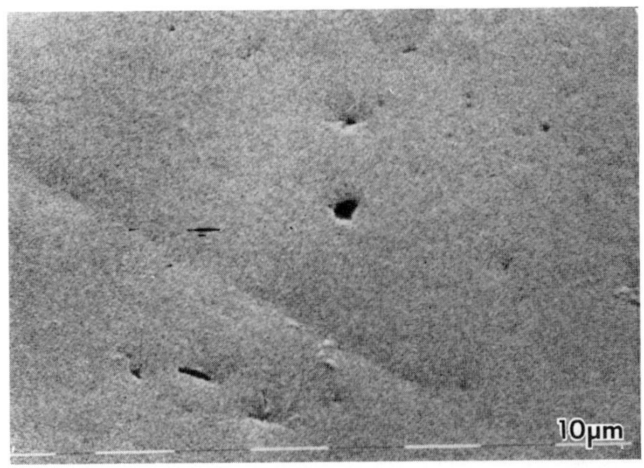

Fig. 7. Polished section, Sample 8bW7.

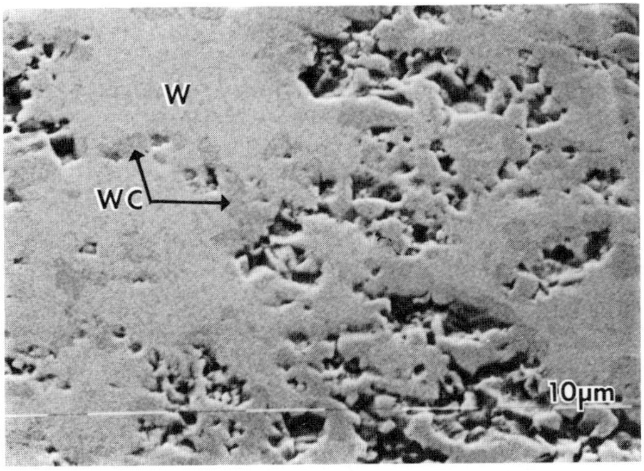

Fig. 8. Polished section, Sample 3.

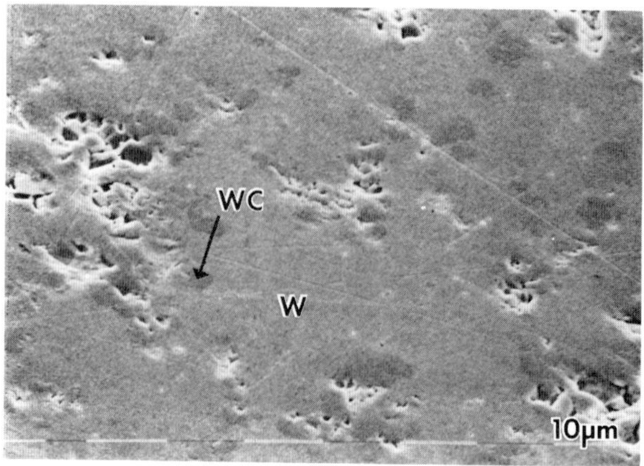

Fig. 9. Polished section, Sample 1bWCW1.

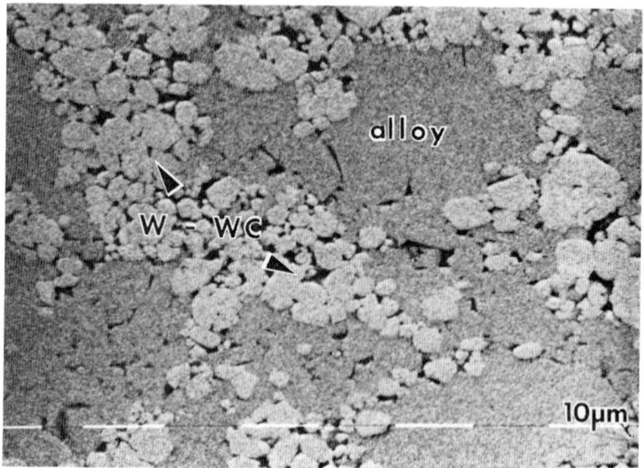

Fig. 10. Polished section, Sample 4.

Tungsten/Titanium Carbide Shots

Three combinations of tungsten and titanium carbide powders were explosively compacted at tungsten concentrations of 31.33%, 50%, and 75% tungsten by volume. All three shots again consolidated well although the porosity of these shots was, in general, higher than the WCW shots. SEM examination of polished sections from the three shots shows that the method of consolidation is gross plastic flow of tungsten and shock sintering of individual TiC grains (Figs. 11-14).

Unlike the WCW shots, production of a contiguous matrix of tungsten was not achieved in these samples although this condition was approached in the 75% W shot. It is of some interest to note that the large, spongy aggregates shown to be present in the starting TiC powder are completely absent in the shock compacted material. Also, in the case of the 31.23% W shot, several other modes of consolidation were seen. All of these involved reaction of the compact with the steel casing around the periphery of the sample, (Figs. 11 and 12).

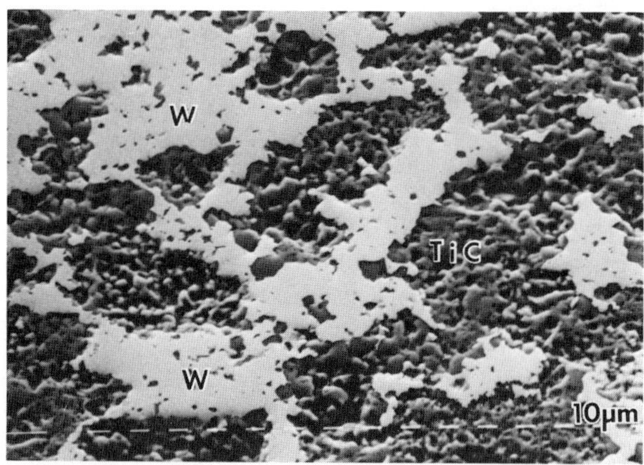

Fig. 11. Polished section, Sample 1bWTiC1, center.

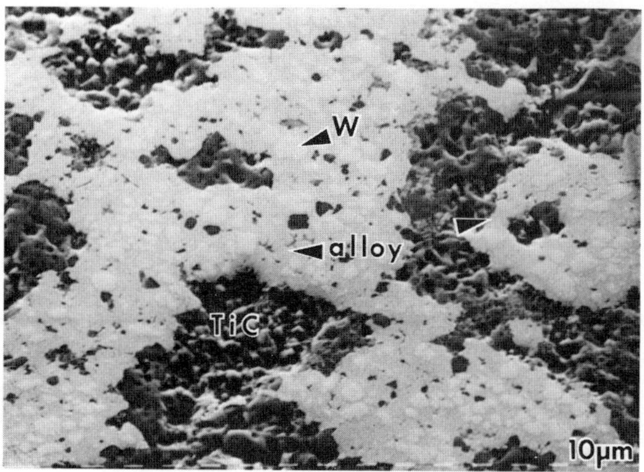

Fig. 12. Polished section, Sample 1bWTiC1, edge.

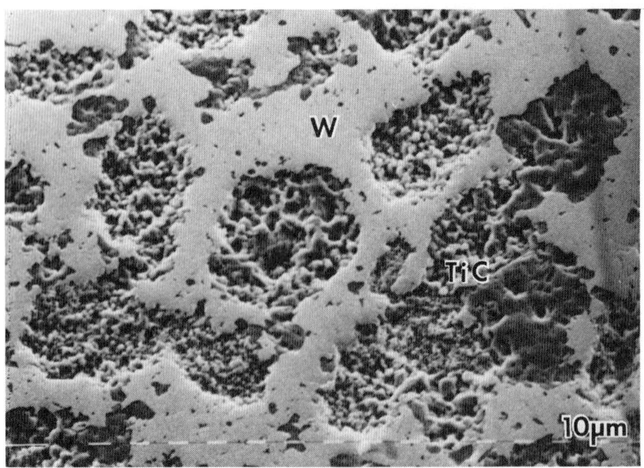

Fig. 13. Polished section, Sample 1.

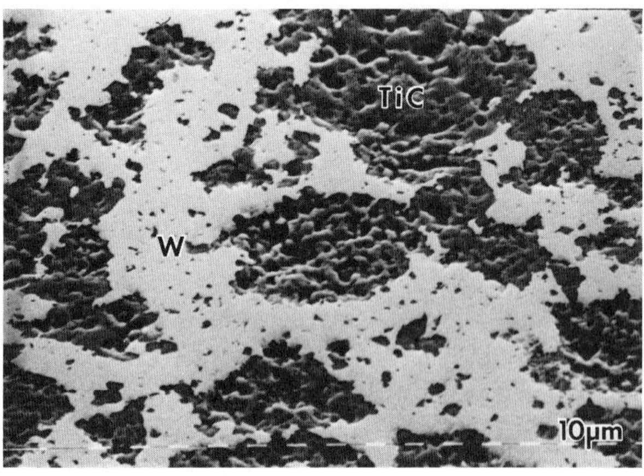

Fig. 14. Polished section, Sample 2.

OBSERVATIONS AND CONCLUSIONS

It has been demonstrated that mixed powders of titanium carbide/tungsten and tungsten carbide/tungsten can be successfully consolidated in cylindrical stainless steel containers at elevated temperatures by detonating a uniform layer of explosive in contact with the outer surface of the container.

In particular:

- Pure tungsten powder was consolidated to near theoretical density. The mechanism of consolidation was plastic flow at elevated temperature.
- Combinations of tungsten carbide/tungsten consolidated by plastic flow of the tungsten around the tungsten carbide particles.
- Combinations of titanium carbide/tungsten did not consolidate as well as the tungsten carbide/tungsten combinations. This may be due to the presence of large, spongy particles of titanium carbide which may have acted as energy sinks during the shock consolidation process. The mechanism of consolidation in these powder mixtures included both plastic flow of tungsten and apparent shock sintering of individual titanium carbide particles.
- In several cases, the steel casing holding the sample became involved in the consolidation process by partial melting and reaction with the powders. These compacts showed low densities and high hardness.

Further work in this area will concentrate on utilizing iron powder as a matrix material in order to alleviate the necessity of utilizing pre-heat temperatures which approach the melting point of the steel container. Using Pruemmer's empirical approach, limited experiments will be conducted to verify the maximum density at room temperature. With this value of E/M and V_{OD} the effect of temperature and container wall thickness on the density will be investigated.

REFERENCES

1. R. W. Leonard, D. Laber, and V. Linse, Second Int. Conf. Center for High Energy Forming, Estes Park, CO, 1969.
2. R. A. Pruemmer, _Proc. of the 4th Int. Conf. of the Center for High Energy Forming_, 1973.
3. G. E. Dieter, Am. Society for Metals, Reinhold Publishing Co., 1962.
4. G. E. Duvall, in _Response of Metals to High Velocity Deformation_, Metallurgical Society Conferences, Vol. 9, Interscience, 1961.
5. R. N. Orava and R. H. Wittman, _Proc. of the 5th Inter. Conf. of the Center for High Energy Forming_, 1975.
6. A. K. Bhalla and J. D. Williams, _Proc. of the 5th Inter. Conf. of the Center for High Energy Forming_, 1975.
7. A. R. C. Lennon, A. K. Bhalla, and J. D. Williams, _Proc. of the 6th Inter. Conf. of the Center for High Energy Forming_, 1977.
8. C. Hoenig, A. Holt, M. Finger, and W. Kuhl, _Proc. of the 6th Inter Conf. of the Center for High Energy Forming_, 1977.

9. R. Pruemmer, Proc. 6th AIRAPT Conf. High Pressure Technology, edited by K. D. Timmerhaus and M. S. Barber, Plenum Press, NY, 1979.
10. R. Pruemmer, Proc. of the 6th AIRAPT Inter. High Pressure Conf., 1979.
11. J. D. Mote, J. J. Fitzpatrick, and D. Chandra, Annual Report DARPA contract 682301.

PART IX

SHOCK SYNTHESIS: SHOCK CONDITIONING AND SUBSEQUENT DENSIFICATION

MODERN USES OF EXPLOSIVE PRESSURE--FROM ROCK BLASTING TO SYNTHETIC DIAMOND

O. R. Bergmann

E. I. Du Pont de Nemours & Co.
Petrochemicals Department, Explosives Products Division
Wilmington, Delaware, USA

ABSTRACT

Conventional and unconventional commercial uses of high pressures generated by detonation of chemical explosives are described. Recent basic research on factors affecting the performance of explosives in rock blasting are discussed.

The mechanisms and commercial applications of the explosion bonding process for metals are described. In particular, the jetting phenomenon and the processes responsible for bond zone formation are explained.

Experimental results from pioneering research on the effect of explosive shock waves on ceramic powders are presented. Such powders are unusually responsive to sintering and explosive shocking may be a unique way of "cold-working" ceramics.

One of the latest "constructive" commercial uses of explosive pressure, the manufacture of polycrystalline synthetic Diamond, is described. This unique material has established itself firmly as a commercial specialty super abrasive.

INTRODUCTION

For many centuries explosives have been used almost exclusively for fracturing and fragmentation of structures and materials, primarily in military applications and in mining and quarrying. Truly "constructive uses" of the extremely high pressures generated by the detonation of chemical explosives have only been developed during the last few decades.

It is the purpose of this paper to: Illustrate how the age-old art of rock blasting with explosives has recently been put on a much better scientific foundation as a result of basic experimental research and to describe the most recent developments for "constructive" uses of explosives, namely, the processes for bonding metals with explosives, for explosive shock treatment of ceramic powders and synthesis of polycrystalline diamond using explosive shock techniques.

ROCK BLASTING

The most important industrial use for chemical explosives at the present time is for rock blasting operations in quarries and mines. Until recently, effective use of explosives in rock blasting was mainly an art developed over many decades by practical experience. There are good reasons why quantitative scientific understanding of the interaction of key variables in the blasting process was meager or nonexistent for a long time.

Although rock blasting with explosives is a process that requires only a few, relatively simple operations (drilling of holes, loading of holes with explosives and detonating via blasting cap from remote location), the detailed physical processes that lead to the observable end result are anything but simple and occur in such a short time that observation becomes difficult. Particularly the factors that cause most of the rock breakage, namely the transient stresses and their distribution set up in the rock by the detonation of the explosive, are complex and are influenced by many variables.

For the above stated reasons, the important practical problem of relating explosives properties to blasting results, particularly fragmentation, remained relatively obscure until recently. To provide answers to these important questions Du Pont's Explosives Research group conducted an in-depth experimental study over a period of several years.[1,2] The study involved instrumented model blasting experiments in large, crack-free blocks of rock such as granite, limestone and sandstone. By normal laboratory standards the size of the blocks of rock was quite large. Typically, each block weighed about 15 tons, but this size was deemed necessary to retain realistic scaling ratios.

Results from these experiments showed that rock fragmentation is not controlled by a single explosive property but by a combination of several properties. Empirical fragmentation equations were developed for granite, limestone and sandstone. From these relationships an expression was developed that may be used to rate fragmentation performance of different explosives. The heart of this expression is an empirical term which allows comparison of the relative fragmentation performance of equal volumes of different explosives at the same burden distance and borehole diameter. This term is defined as the EPT:

$$EPT = (0.36 + \rho_e) \cdot \frac{D_e^{2}}{\left(1 + \frac{D_e^{2}}{V_r^{2}} - \frac{D_e}{V_r}^{1.33}\right)} \cdot R_v^{-1} E_{Wk/g} \cdot \rho_e \qquad (1)$$

where ρ_e = density of explosive, g/cm^3
D_e = detonation velocity of explosive, km/sec
V_r = sonic velocity of rock to be blasted, km/sec
R_v = volume decoupling ratio = shot-hole volume:explosive volume
$E_{Wk/g}$ = calculated maximum expansion work of explosive in Kcal/g.

R_v in the above equation becomes unity when the explosive fills the borehole completely. If the numerical value of the EPT for a given explosive is higher than that of the standard, a better fragmentation performance, i.e., smaller average fragment size, is inferred. Likewise, a smaller number than that of the standard indicates inferior performance.

Examination of the EPT of Eq. (1) tells us a great deal about the relative importance of different explosives properties in influencing fragmentation in rock blasting operations. The following main points are brought out:

Detonation velocity, up to a certain point, is of great importance in producing smaller fragment size. This is shown by the fact that the detonation velocity exponent is twice that of the energy (expansion work) exponent. However, this is only true when D_e is below the sonic velocity of the rock. When D_e exceeds the sonic velocity, the value of the term

$$\frac{D_e^{2.00}}{\left(1 + \frac{D_e^{2}}{V^{2}} - \frac{D_e}{V}^{1.33}\right)}$$

does not change much and actually begins to decline for D_e/V_r ratios above 1.3. This means that detonation velocities greatly in excess of the sonic velocity will not lead to any further improvement in fragmentation.

Charge energy, in terms of maximum expansion work, is very important. Equation (1) shows clearly that increased charge energy (expansion work) per unit length of shot-hole will produce smaller fragment size under otherwise comparable conditions.

Density of the explosive is also an important factor. High density explosives will give smaller fragment size provided the increased density does not result in lower expansion work (energy) or lower detonation velocity values.

Coupling and loading density also play significant roles. Equation (1) shows clearly that good coupling between explosives and shot-hole wall will produce better fragmentation (decoupling ratio R_v as small as possible, or loading density as high as possible).

The overall conclusion from this work is that fragmentation in blasting is not controlled by a single explosives property such as energy but by a combination of explosives energy, detonation velocity, density, degree of coupling between explosive and borehole wall, by the sonic velocity of the rock by the powder factor and by the shot geometry.

The EPT (Eq. (1)) can be used as the basis of a rating system for blasting explosives which accounts realistically for the interplay of several explosive parameters and rock properties in contrast to the traditional systems which are generally based on explosive energy alone. This improved insight into the effect of different explosive properties on blasting action also constitutes a highly useful tool in guiding development work on new explosives formulations which are specifically tailored for blasting of different types of rock.

A second subject that was clarified considerably by the results of a related and similar research program was the question of what are the quantitative effects of delays and of shot-hole patterns on rock fragmentation in typical bench blasting operations.[2] This is a question of great practical importance.

In order to provide quantitative experimental information on this subject instrumented model rock blasting experiments were carried out on crack-free granite blocks. The shots were arranged as multi-hole bench blasting shots and delay times between holes were controlled precisely by using a single cap to simultaneously initiate parallel lines of different lengths of "low energy" detonating cord leading to each individual shot hole.

The results showed that best fragmentation results were obtained when delays between holes were 1 to 2 milliseconds per ft. of burden and when a rectangular shot-hole pattern was used with spacings larger than the burden. Effective fragmentation in multiple hole blasts seemed to depend primarily on the full development of the crack network around each hole before the charge in the next hole was detonated. The test results showed that simultaneous or almost simultaneous initiation of shot holes in bench blasting resulted in poor fragmentation. Optimum delay times, as derived from the results of this work, are consistent with present field practice and these model blasting experiments also confirmed the beneficial effect of rectangular shot-hole patterns compared to square patterns.

In summary of this section one key point emerges clearly and is sufficiently illustrated by the above described examples of recent research on rock blasting with explosives: Namely, that great strides have been made during the last two decades by the leading organizations in this field to put this age-old art on a sound scientific basis.

METAL BONDING WITH EXPLOSIVES

The most important process for metal working with explosives is that of bonding, or cladding, of metals with explosives. Different variations in technique have been described in the literature but the basic technique that is practiced commercially throughout the world by a number of manufacturers is the "parallel" technique. This technique was developed and patented by the Du Pont Company[3,4] and has been highly successful on the fully commercial scale for almost twenty years now. The process has the unique capability to bond a wide variety of different metals via a direct metallurgical bond that is characterized by the virtual absence of diffusion between the metals at the bond zone. This permits the bonding of otherwise incompatible combinations such as titanium/steel and aluminum/steel.

The mechanism of bond formation between the metal plates involves formation of a forward "jet" as the metals collide[5] and, simultaneously, severe plastic deformation and flow of the two metals in the solid state under the influence of extremely high transient stress gradients. This severe plastic flow of the metals, which under certain conditions results in the preferred "wavy bond" configuration is analogous to fluid flow phenomena observed in viscous fluids.[6]

A unique feature of this bonding process is the fact that bonding occurs in an extremely short time, in the order of microseconds, and that the bulk of the metals to be clad is never subjected to high temperatures. Compared with many conventional cladding processes, such as hot rolling or brazing, in which the metals are exposed to high temperatures for long times, the explosion cladding process is a "cold" process, therefore no substantial diffusion occurs across the bond zone between clad metal and backer metal. Explosion clads have ductility, formability, and high bond strength in systems that are considered incompatible in conventional hot roll bonding techniques where the high processing temperatures lead to extensive diffusion and formation of brittle intermetallic compounds between the metals. For these reasons, explosion cladding can be applied to a much wider variety of similar and dissimilar metal combinations.

SHOCK-CONDITIONED CERAMIC POWDERS

During the early 1960s Du Pont conducted a basic research program on the effects of explosive shock waves on ceramic powders.[7] Changes in the physical properties of the powders resulting from the shock treatment were studied. Very fine particle size SiC and B_4C could be produced without introducing impurities. Strong line broadening in the x-ray diffraction pattern of shocked powders was observed and correlated with lattice strain and crystallite size reduction. Shocked $CaCO_3$ turned blue when exposed to x-rays. Shocked powders gave higher tapped densities, higher green densities after pressing, and were unusually responsive to sintering. Explosive shocking may be a unique way of "cold-working" ceramics and may permit subsequent control of their microstructure through primary recrystallization.

Hayne Palmour III at North Carolina State University subsequently conducted some further experimental and theoretical studies on explosively shocked ceramic powders and he was able to develop a more quantitative model describing the higher sintering "activity" of explosively shocked ceramic powders.

POLYCRYSTALLINE SYNTHETIC DIAMOND

One of the latest and most fascinating new uses of explosive pressure and energy is the synthesis of polycrystalline diamond from graphite. The process and product, which resulted from the Du Pont Company's research on new uses for explosive energy and commercialization, was achieved during the late 1960s. Using the controlled detonation of chemical explosives to produce transient shock pressures from two to seven million psi (150 to 500 kilobars) graphite is converted to diamond almost instantly in the order of microseconds.[8-11] The technique currently used by Du Pont is based upon a system of concentric cylindrical pipes that are surrounded by the explosive charge. The detonation dynamically collapses the outer tube onto the inner tube, thus generating an intense, transverse shock wave across diameter of the inner tube. This shock wave travels axially along the inner tube and causes the graphite to convert into diamond. After the shocking operation, the assembly is subjected to a series of mechanical and chemical operations to extract the diamond in the form of a powder. The diamond powder is then cleaned, shaped, graded, inspected, packaged and shipped. The commercial product is available in grades ranging from 1/8 micron to 60 microns in particle size.

Shock synthesized diamond is unique because of its polycrystalline characteristics, even in the smallest particles. In microstructure and appearance, Du Pont diamond resembles carbonado, a rare natural polycrystalline diamond. Both are black because they contain trace elements and are polycrystalline. The microcrystals that make up each particle of Du Pont diamond are in the order of 100 Å in size. As a result, this type of diamond consists of particles which contain from several hundred to many thousands of microcrystallites each, depending on particle size. These microcrystallites are bonded together in a random polycrystalline structure without significant cleavage planes. Therefore, the particles are equally strong in all directions.

During the past ten years this particular type of synthetic diamond has established itself firmly as a specialty super abrasive of outstanding performance in a growing number of industrial precision polishing and lapping applications. This unique utility is related to its polycrystalline structure. Each particle is blocky and has thousands of microcutting points. During use, the fracture and removal of clusters of microcrystals generate new microcutting edges. The resultant subparticles are blocky like their parent particles, but lacking a "hard" and a "soft" directionality, they will abrade uniformly on all surfaces and cannot create a sliver or flake which can embed in a work surface and cause random scratches. For certain sapphire, ceramic and ferrite polishing applications, this behavior is particularly useful, and shock synthesized polycrystalline diamond powder has shown vastly superior performance in these applications.[12]

FUTURE OUTLOOK

The examples of modern uses of explosive pressure described in this paper illustrate that remarkably swift progress has been made during the last few decades in harnessing the violent forces generated by detonating explosives for a variety of diverse industrial uses. The author is confident that the future will bring many more developments in this fascinating area of technology.

REFERENCES

1. O. R. Bergmann, J. W. Riggle, and F. C. Wu, Int. J. Rock Mech. Min. Sci. & Geomech. Abstr., 10, 585-612 (1973).
2. O. R. Bergmann, F. C. Wu, and J. W. Edl, in Engineering and Mining Journal, McGraw-Hill, Inc., NY, 1974.

3. U.S. Patents 3,137,937 and 3,233,312 assigned to E. I. Du Pont de Nemours & Co.
4. U.S. Patents 3,397,444 and 3,493,353 assigned to E. I. Du Pont de Nemours & Co.
5. O. R. Bergmann, G. R. Cowan, and A. H. Holtzman, Trans. Metall. Soc. AIME, 236, 646-53 (1966).
6. G. R. Cowan, O. R. Bergmann, and A. H. Holtzman, Metall. Trans. AIME, 2, 3145-55 (1971).
7. O. R. Bergmann and J. Barrington, J. Am. Ceram. Soc., 49 [9], 502-07 (1966).
8. U.S. Patent 3,667,911.
9. U.S. Patent 3,608,014.
10. U.S. Patent 3,568,248.
11. U.S. Patent 3,401,019.
12. O. R. Bergmann, N. F. Bailey, and H. B. Coverly, Metallography, 15, 121-39 (1982).

SHOCK-INDUCED MODIFICATION OF INORGANIC POWDERS[a]

R. A. Graham, B. Morosin, E. L. Venturini,
E. K. Beauchamp, and W. F. Hammetter

Sandia National Laboratories
Albuquerque, New Mexico 87185

INTRODUCTION

Early exploratory work in Japan,[1,2] the Soviet Union[3,4] and the United States[5,6] demonstrated that high pressure shock-wave loading such as produced by the detonation of high explosives or high speed projectile impact can substantially alter solid state reactivity. Such effects are manifest in catalytic activity,[2,7,8] compound synthesis,[1-3,9] enhanced sinterability,[6,9-11] reduction in reaction start temperature,[12,13] rapid growth of dense phase crystallites[5,14] and strong bonding in dynamic compaction and explosive welding.[15] That such effects can be utilized in industrial operations is well demonstrated in synthetic diamond production,[16-17] polycrystalline cubic boron nitride production[17] and explosive welding.[18] Scientists in the Soviet Union have maintained a continuing effort in shock-induced chemistry and shock modification[13] but mechanisms to account for and control shock-induced solid state reactivity are poorly understood. Presently, significant efforts in this area are developing in Japan and the United States.

It is the purpose of this paper to report results of studies undertaken to provide quantitative characterization of powders of TiO_2, ZrO_2 and Si_3N_4 which have been subjected to controlled explosive loading and preserved for post-shock analysis. Titanium dioxide (Johnson-Matthey Puratronic grade) was chosen as a high purity research grade material (metallic impurities < 20 ppm) whose properties have been studied under shock compression. No structural transitions

[a]This work performed at Sandia National Laboratiries supported by the U.S. Department of Energy under contract #DE-AC04-76DP00789.

are expected. Zirconia (Lot Sp 97310A Teledyne Wah Chang) was chosen as an important technological material expected to transform under pressure to either the tetragonal or orthohombic form.[19] Silicon nitride (GTE Sylvania SN 502) is also an important technological material for high temperature applications. The studies are directed toward identifying the mechanisms responsible for shock-induced enhancement of solid state reactivity.

It is likely that enhancements of solid state reactivity in shock-loaded samples result from: particle comminution, crystallite size reduction, point defects, line defects, higher-order defects, particle morphology changes, structural transitions or alteration of surface chemistry. In the present work these effects are studied with x-ray diffraction, electron spin reasonance, particle size and particle morphology measurements and specific surface measurements as indicated in Table 1. Quantitative measurements show that shock

Table 1. Schedule of Experiments†

Experiment(a)	XRD(b)	ESR(c)	Particle Size(d)	Specific Surface(e)	Remarks(f)
TiO_2					
7G800(B)	x	x	-	x	mixed TiO_2
18G800(C)	x	x	-	x	#1, $\rho_o = 2.1$
5G810(B)	x	x	-	x	#1, $\rho_o = 2.1$
17G810(B)	x	x	-	-	#2, $\rho_o = 2.3$
1M820(B)	x	x	x	x	#2, $\rho_o = 2.3$
2M820(C)	x	x	x	x	#2, $\rho_o = 2.3$
Silicon Nitride					
13G810(B)	x	x	x	x	$\rho_o = 1.75$
14G810(C)	x	x	x	x	$\rho_o = 1.75$
Zirconia					
16G810(B)	x	-	x	x	$\rho_o = 3.41$
8G810(C)	x	-	x	x	$\rho_o = 3.41$

†All shock loading carried out in Baby Bear fixtures. x indicates measurement; - indicates no measurement.
(a) (B) indicates baratol explosive, while (C) indicates composition B explosive. (b) x-ray diffraction. (c) electron spin resonance. (d) particle size distribution and morphology with Quantimet. (e) BET surface measurement by F. Williams, University of New Mexico. (f) ρ_o is the starting powder compact density in Mg/m^3.

modification varies strongly with shock-loading conditions and varies among the various materials.

SHOCK LOADING AND RECOVERY FIXTURES

Studies of shock-induced modifications of powders require that the shock loading systems produce reproducible, quantifiable conditions that can be systematically varied over pressure ranges of interest. Furthermore, the sample must be preserved, in place, in a fixture after shock loading. Because of the two-dimensional geometrical influences, the pressure and temperature within samples are expected to be different for various fixtures subjected to the same nominal loading conditions.

In the work reported here shock waves are produced with 55 mm diameter plane wave generators and 12.5 mm thick explosive pads of baratol or composition B. The wave is shaped by the presence of a steel disk which undergoes the 13 GPa transition. The shock wave from the steel then impinges upon the copper recovery fixture and then into the powder sample. One dimensional calculations with the computer code CHART D[19] indicate peak pressures in the copper fixture of 13 and 17 GPa, respectively. Calculations for pressure in the powder are based on extrapolations from existing shock compression data on rutile and indicate peak pressures of 11 and 17 GPa, respectively.

In the present work the recovery fixtures utilized are code named "Baby Bear" and are the same utilized in previous work.[10,12,19] Two dimensional calculations now in progress with the computer code CSQ indicate peak pressures in the center of this fixture approach 40 GPa due to the radial compression controlled by the powder cavity configuration. Similar behavior is predicted for the "Momma Bear A" fixture used for the work of Beauchamp et al.[11]

The calculations of pressure and accompanying shock-induced increase in temperature show that starting powder compact density has a strong influence. The increase in temperature is dominated by the powder "crush up" phenomenon accompanying the first wave passage through the powder which compresses the powder to densities close to theoretical. A bulk temperature can be calculated which corresponds to that of the pressure-volume work. Although this calculation does not address issues as to localization of deformation and temperature it does provide a first-order estimate of the temperature changes encountered as powder compact density is varied. The mean temperature is certainly an important factor in the recovery capsule after the shock loading when annealing can occur for a relatively long time. If the capsule does not seal well and vents the compressed gases, one would expect a reduced post shock annealing.

X-RAY DIFFRACTION STUDIES

Diffraction studies yield information (1) on structural phase transitions and chemical alterations of the starting materials, (2) on the retained lattice strain (principally via broadened high 2Θ diffraction lines), and (3) on coherent crystallite size (via broadened low 2Θ lines) provided such sizes are below several thousand angstroms. Contributions to the broadening of the diffraction profiles attributed to the sample involve more than crystallite size and lattice strain, the quantities of main interest in the present study. They also include defects arising from misplaced atoms, anisotropy in shape, crystallographic faults and dislocation distributions. Diffraction profiles on a particular shock-loaded powder probably contain a mixture of such contributions arising from the catastrophic nature of the shock process, and to date these do not appear to have been rigorously defined.

Limitations of determination of crystallite size and strain broadening are imposed from both theoretical and practical considerations. Crystallite size broadening of x-ray diffraction lines in powders is limited to sizes in the interval 2000 Å to 100 Å, though under ideal conditions both limits might be extended by a factor of two. Residual strain values are limited between 10^{-4} to 10^{-2} under ideal conditions. The type of materials under study, and symmetry and crystal structure complexities, are important factors. In shock-loaded materials, combinations of high values of apparent residual strain and small crystallite sizes also restrict quantification of values.

For the studies reported here, a simplified method, rather than the more complex Fourier transform techniques, is employed to obtain values of strain and size. The half-maximum widths, that is, the overall width of the line profile at half-maximum intensity measured above the background, for low, mid, and high two-theta lines are determined and fit upon curves of line breadth as a function of two-theta calculated as a sum of the Scherrer relationship ($\delta(2\Theta)_c = k\lambda/L \sin \Theta$). In this equation L is the "average" crystallite size and the strain is based on the small incremental change in the average d value ($\delta(2\Theta)_s = -2\varepsilon \tan \Theta$).

The crystal structure of ZrO_2 at ambient condition is monoclinic, $P2_1/c$, identical to that occurring naturally as the mineral baddeleyite and is often referred to by that structure type. Several forms of ZrO_2 have been described. The most familiar is the cubic fluorite form, normally stable above about 1400°C but which can readily be preserved by quenching to room temperature, provided large amounts of di- and trivalent impurities are present. That technologically important material is known as "stabilized" zirconia. A tetragonal and a closely related orthorhombic form have been reported in temperature-pressure phase studies.[20] Impurities appear to stabilize the tetragonal form so that it can exist at ambient conditions.

Significant differences between shocked and unshocked powders are readily observed. The most dominant feature of the x-ray patterns is the "washed out" or diffuse nature in both the 11 and 17 GPa shock-loaded samples due to the large lattice strain component which affects the high 2Θ lines (particularly for d values below 1.4Å). This is evidence of very large crystallite strain in these shock-loaded ZrO_2 powders, consistent with values of $4x10^{-3}$ and $8x10^{-3}$ for the low and high pressure loading, respectively. Crystallite size reduction is also present with average values of 1000Å and 1500Å for the low and high pressure loading, respectively. This apparent reversal in the variation of crystallite size with shock loading may be an artifact of the difficulty of determining values of size in samples with such high residual strain.

Significant amounts of the tetragonal phase have been found in shocked ZrO_2 samples. Because of the amounts and line broadening effects it is difficult to determine if this phase is truly tetragonal or orthorhombic; however, consideration of the observed breadths of lines and expected positions, favors the tetragonal form. As mentioned previously, such a phase generally exists at ambient conditions only if impurities are present. No detectable impurities were identified; however, the unusually large lattice strains caused by defects and possible local crystallite size reduction introduced by shock loading probably stabilize the tetragonal lattice. The amounts transformed to this tetragonal form are 5-10% for the low pressure shocked ZrO_2 and approximately 20% for the high pressure shocked sample.

The GTE Sylvania SN 502 Si_3N_4 employed in our studies consists of material 40% amorphous with about 8% of the crystalline phase in the β-form and the remaining in the α-form. Both of these forms are hexagonal, with silicon atoms tetrahedrally surrounded by nitrogen atoms; however, the α-form consists of a larger, more complex cell. The conversion from α to β at high temperature is observed at atmospheric pressure only by a vapor phase or solution precipitation reaction.

The x-ray line broadening studies show significant crystallite size reduction for the α-form, yielding values near 600Å. No significant differences were noted in crystallite size between the 11 and 17 GPa pressure samples. The lowest pressure shock loading resulted in residual strain of $2x10^{-3}$ while the higher pressures resulted in values near $4x10^{-3}$. The β-form also shows line broadening, though its lower concentration levels do not permit a quantitative measurement. The broadening gives the impression that the concentration was reduced; however, upon heating near 1800°C the pattern reverts to that observed for the initial concentration of β-form as all of the lines sharpen.

Two pure TiO_2 materials were studied, one consisting of a mixture of two phases, 80% rutile/15% anatase, and the other pure rutile phase. Though such shock-loaded materials showed large color changes, due to production of large Ti^{+3} defect concentration levels, no changes in structural or chemical phases were found by our diffraction studies. Further, the shock-loaded materials, whether subjected to low or high pressure or those of low or denser powder packing in the capsule, showed only statistical variation in crystallite size (~ 2000Å) and residual strain (~ 1×10^{-3}) among the various samples. X-ray diffraction patterns on annealed, shock-loaded powders retained strain-broadened lines to 1100 K, well above the temperature (600-700 K) at which the darker colors are bleached and ESR defect concentration levels drastically decrease.

PARAMAGNETIC POINT DEFECTS

The damage produced by shock-loading a solid includes numerous point defects, some of which are paramagnetic. Such paramagnetic defects can often be detected and concentrations quantitatively measured by electron spin resonance (ESR) techniques.

The ESR spectra were recorded at 9.8 GHz and 2 K, the low temperature providing an improved signal/noise ratio. The samples were sealed in quartz capillary tubes of 0.2 cm diameter, each tube containing 25 to 50 mg of powder. The quartz tubes provide a convenient method to handle the samples, and allow ESR measurements on the same material following annealing. The digitized spectra were numerically analyzed to determine the absorption intensity, g-factor, and linewidth. The intensity was converted to a defect concentration assuming random distribution throughout the powder and an effective electron spin of 1/2. The concentration uncertainty is estimated to be a factor of 2.

No ESR signals were detected in the TiO_2 starting powder. No information on impurity content is available on the Si_3N_4 powder. The as-received Si_3N_4 powder has a single ESR absorption with an isotropic g-factor of 2.0038(5), a linewidth of 8(1) Oe, and a concentration of $4 \times 10^{16}/cm^3$. This resonance is discussed below in connection with shocked Si_3N_4. The shock-modified ZrO_2 samples were found to be electrically conductive, and no ESR signals were detected.

At liquid helium temperatures two ESR signals are observed in shock-loaded TiO_2, a dominant absorption with axial symmetry and a weaker resonance with an isotropic g-factor. Only a relatively weak isotropic signal is detected at room temperature, and this does not agree with prior room temperature ESR data on shocked TiO_2.[21,22]

These authors do not provide details on lineshapes or linewidths, and show no spectra for shocked TiO_2, making comparisons with our results difficult. In our low temperature spectra for shock-loaded TiO_2 the axial signal has g-values of 1.969(3) perpendicular to the applied magnetic field and 1.935(5) parallel to the field, a line-width of 15 to 25 Oe in different powders, and concentrations from $2x10^{18}$ to $3x10^{19}/cm^3$ depending on shock conditions (see Table II). This signal is due to Ti^{+3} ions in the substoichiometric TiO_{2-x} powder produced by the shock-loading, similar to the Ti^{+3} ESR resonance reported in reduced rutile by various authors.[23-25] This identification is strengthened by the dull gray color of the post-shock powder in contrast to its original white color. The weaker isotropic signal in shocked TiO_2 has a g-factor of 2.0028(3), a linewidth of 2 to 4 Oe, and concentrations $1x10^{15}$ to $5x10^{16}/cm^3$. It is tentatively attributed to an electron trapped at an oxygen vacancy at or near the surface since this signal is not reported in reduced TiO_2 crystals,[24-25] and is strongly influenced by exposure to various gases.[26]

Two ESR signals are found in shock-loaded Si_3N_4 powders, both with isotropic g-factors. The dominant absorption has a g-factor of 2.0038(5), a linewidth of 9.3(5) Oe, and a concentration of $9x10^{17}/cm^3$ in the powder loaded to 17 GPa. As this powder is annealed up to 1000°C for one hour, this resonance shifts to a g-factor of 2.0046(5) with a linewidth of 7.5(5) Oe. A similar resonance has been reported in plasma-deposited SiN,[27] and is observed in amorphous Si as well. In both cases, it was attributed to silicon "dangling bonds," and the same assumption is made here. A weaker isotropic ESR signal is detected in shock-loaded Si_3N_4 with a g-factor of 2.0028(3), a line-width of 4.2(5) Oe, and a concentration of $1x10^{17}/cm^3$ in the powder loaded to 17 GPa. The defect responsible for this narrower resonance has not been identified.

In Fig. 1 the annealing behavior of the dominant defects in shock-loaded TiO_2 and Si_3N_4 powders are compared. In both cases the annealing study employed the powder with the highest post-shock defect concentration. For TiO_2 this corresponds to a shock pressure of 11 GPa, and a copper recovery capsule that vented to the atmosphere during shock-loading, leaving $2.8x10^{19}$ axial defects/cm^3. For Si_3N_4 the maximum defect concentration was $9.4x10^{17}/cm^3$ in a powder loaded to 17 GPa contained in a copper capsule that remained sealed during shock. Each data point in Fig. 1 represents the defects remaining after a one hour anneal at a successively higher temperature (all ESR spectra were recorded at liquid He temperatures). The important point is the monotonic decrease in defect concentration in TiO_2 at much lower temperatures compared with Si_3N_4. An hour at 775 K reduced the TiO_2 defect level to roughly 3% of its post-shock value, while an hour at 1275 K for Si_3N_4 leaves 25% of the initial defects present. Note that there is essentially no change in the Si_3N_4 defect level for anneals up to 1200K.

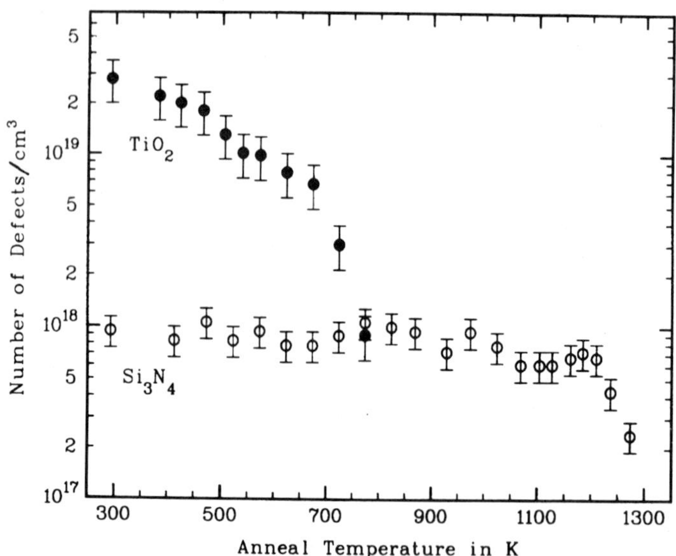

Fig. 1. Annealing of the dominant defects in TiO_2 and Si_3N_4 shows substantially different thermal stability for the defects.

Table 2 contains a summary of the ESR results for shock-loaded TiO_2 and Si_3N_4. The quoted defect concentrations refer to the dominant defect which is a Ti^{+3} ion in shock-loaded TiO_2 and most likely a silicon "dangling bond" in Si_3N_4 (discussed above). The absolute accuracy is estimated to be a factor of 2. It should be noted that for the first two experiments with TiO_2 powders, the recovery fixture vented during loading, resulting in considerably lower post-shock temperatures than calculated for a sealed capsule.

Two important points can be made concerning the data in Table 2. First, the defect levels in shock-loaded TiO_2 are consistently higher than in Si_3N_4, despite the fact that during loading the TiO_2 powder reaches temperatures where significant defect annealing takes place (Fig. 1). Second, the Si_3N_4 defect concentrations increase with increasing pressure and temperature during loading, suggesting that even higher peak pressures would yield more point defects. At lower TiO_2 powder densities an increase in shock pressure decreases the residual defect concentration. When higher powder densities are used the tendency is reversed; an increase in pressure causing a higher defect concentration.

Table 2. Results - Electron Spin Resonance for TiO_2 and Si_3N_4

Experiment	Peak Pressure GPa	Mean Bulk Temperature	Defect Concentration (cm^{-3})	Remarks
TiO_2				
7G800	11	570 K	2.8×10^{19}	capsule not sealed
18G800	17	800 K	5.7×10^{18}	"
5G810	11	570 K	2.4×10^{18}	well sealed
17G810	11	475 K	5.4×10^{18}	"
1M820	11	475 K	3.6×10^{18}	"
2M820	17	700 K	1.1×10^{19}	"
Silicon Nitride				
13G810	11	--	3.2×10^{17}	"
14G810	17	--	9.4×10^{17}	"

POWDER MODIFICATION

In every instance the shocked sample was recovered as a severely fractured disc which could be removed from the recovery fixture only as fragments. For several of the materials, e.g., Si_3N_4, fragments often adhered strongly to the surfaces of the recovery fixture and were difficult to break loose. Usually the smaller fragments had considerable mechanical integrity. This was particularly evident in Si_3N_4, which was, after shocking, lightly ground with a mortar and pestle to prepare it for sintering studies. Some of the fragments were extremely difficult to crush.

In general, shock-treated powders were free flowing with better packing characteristics than the starting material. The most striking example was Si_3N_4. The starting material is a very fluffy powder with a tap density of 6%. Shock treatment at 11 GPa raised the tap density to 53%; 17 GPa raised it to 50%.

An appreciation of the degree of aggregation of the shocked powders can be gained from particle size analysis, which shows particle size distribution (obtained by SEM image analysis) for Si_3N_4. The modal value for the starting material was 8.2 μm. Shock treatment, plus light milling, gave a final distribution with a modal value of ~ 60 μm independent of shock pressure. There is no significant difference in particle size distribution between the results of 11 GPa and those for 17 GPa pressure.

The ZrO_2 powder initially had a modal value for size distribution (by volume) of 97 μm. The shocked powder, as recovered, had a particle size of 230 μm. The TiO_2 powder, before shocking, had a modal value of 54 μm. Shock treatment produced essentially no change.

The shock treatment changed the Si_3N_4 powder morphology dramatically. As the SEM micrographs in Fig. 2 show, the starting material consisted primarily of loose aggregates of crystalline whiskers. The explosive shock wave fragmented most of these whiskers into roughly equiaxed grains and compacted the fragments into dense aggregates. Other powders which consisted of denser agglomerates of small grains were changed only slightly.

(a) (b)

Fig. 2. SEM micrographs of unshocked (a) and high pressure shocked powder (b) indicates a significant comminution and shock bonding process. The bars in both cases indicate a 5 μm size.

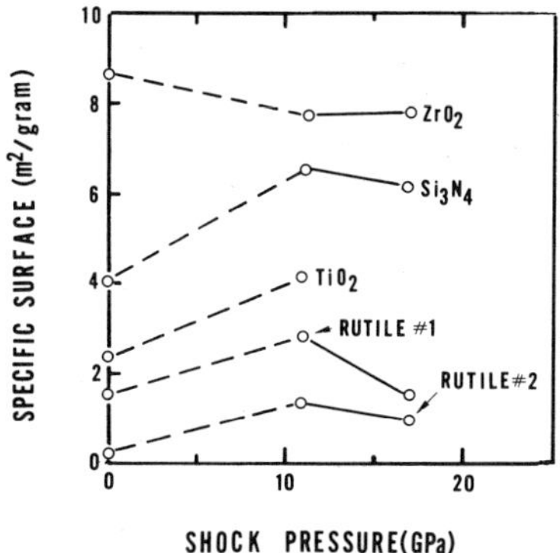

Fig. 3. The specific surface as indicated by B.E.T. measurements shows a significant but not substantial change in shock-modified powders. The tendency to peak at the lower pressure is representative of many powders.

The fracture surfaces on the shocked powders showed major changes in the fracture mechanism as a result of shocking. Powders which were initially loosely bound aggregates of fine particles and which fractured at grain boundaries were converted into dense aggregates with enough integrity to lead to transgranular failure. As an example, the SEM micrograph of Si_3N_4 in Fig. 2b shows a high level of transgranular failure after a pressure of 17 GPa. SEM examination of the 11 GPA shocked Si_3N_4 showed somewhat less transgranular failure. The implication of this increasing trend to transgranular failure as shock pressure increases is that bonding between grains is developed in the shock treatment and the degree of that bonding increases with pressure.

As Fig. 3 shows, the surface areas of the Si_3N_4 and rutile powders initially increase with shock pressure and then decrease. (Surface area measurements were made using the B.E.T. method.) ZrO_2, on the other hand, decreases slightly. The observation of an initial increase and then decrease in Si_3N_4 is consistent with the SEM observations that revealed that the shock comminuted the starting materials and then consolidated it into an aggregate. The implication from the surface area measurement is that although the 11 GPa shock produced an aggregate from the particles after comminution occurred, that aggregate remained porous enough to allow the adsorbant gas to reach the new surfaces generated. However, the 17 GPa shock, which almost certainly produced at least as much comminution as the 11 GPa shock also compacted the particles well enough so that not all of the surface generated was accessible to the gas. The high values of surface area for the starting powder confirm what was apparent in the

SEM micrographs of the powder. All the powders, except rutile #2, are porous aggregates of submicron grains.

From the surface area measurements and the observations of the morphology of the shock-compacted powders, we conclude that the shock wave not only comminutes the powders, but also compacts the powders into aggregates with enough intergranular bonding to force some transgranular fracture when the compact fails mechanically. The development of strength at grain boundaries is evidence for enhanced reactivity and substantial diffusion that is likely due to localized high temperature and the presence of shock-induced defects to activate the necessary deformation processes.

SUMMARY

The present work, in which shock-induced modifications of three inorganic powders are studied, quantifies the existence of extensive modifications, and at the same time demonstrates that modifications are strongly pressure and temperature dependent and vary strongly from material to material. Indeed, with the present limited data it is premature to develop a general model for material response to shock loading and its dependence on material structural characteristics.

It has been shown that in all cases the shock-modified particles are larger than the beginning particles because the comminution observed early in the shock loading process is followed by particle bonding. Enhanced reactivity[8,10,11] is not the result of classic comminution with enhanced specific surface since specific surfaces of the shocked powders are not greatly altered. Nevertheless, the combined effects of crystallite size reduction and formation of newly bonded surfaces may be significant.

Residual strain due to defects as indicated by x-ray line broadening is found to be pressure dependent in ZrO_2 but not in TiO_2. Nevertheless, the unusual nature of shock deformation is indicated by values of residual strain from 10^{-3} to 10^{-2}, which show that extensive plastic deformation occurs readily in materials otherwise considered to be brittle.

The strong influence of starting powder compact density on residual point defection concentrations indicates that post shock annealing in the elevated temperature state is significant. Indeed, the shock modification process produces a number of effects whose pressure and temperature influences may be quite different.

Although the causes for enhanced reactivity cannot yet be confidently identified, it appears that defects play a major role. Further study of shock-induced defects and their influence on solid state reactivity appears to be essential for understanding the effect of shock modification on material processing.

The authors are pleased to acknowledge the particle size measurements by Ken Varga and the explosive handling by C. J. Daniel.

REFERENCES

1. Y. Kimura, Japan. J. Appl. Phys., 2, 312 (1963).
2. Y. Horiguchi and Y. Nomura, Carbon, 2, 436-37 (1965).
3. Yu. M. Ryabinin, Sov. Phys. Tech. Phys., 1, 2575 (1956).
4. S. S. Batsanov, A. A. Deribas, E. V. Dulepov, M. G. Ermokov, and V. M. Kudinov, Combustion, Explosion and Shock Waves, 1, 47-49 (1965).
5. P. S. DeCarli and J. C. Jamieson, Science, 133, 1821-22 (1961).
6. O. R. Bergmann and J. Barrington, J. Amer. Ceramic Soc., 49, 502-07 (1966).
7. S. S. Batsanov, G. K. Boreskov, G. U. Gridasova, N. P. Keier, L. M. Kefeli, V. M. Kudinov, V. I. Mali, and I. S. Sazonova, Kinetics and Catalysts, 8, 1140-46 (1967).
8. J. Golden, F. Williams, B. Morosin, E. L. Venturini, and R. A. Graham, pp. 72-76 in Shock Waves in Condensed Matter - 1981 (Menlo Park), AIP Conference Proceedings No. 78, edited by W. J. Nellis, L. Seaman, and R. A. Graham, American Institute of Physics, 1982.
9. See B. Morosin and R. A. Graham, in Shock Waves in Condensed Matter, loc cit, pp. 4-13.
10. J. D. Keck, D. L. Hankey, R. A. Graham, and B. Morosin, in Shock Waves in Condensed Matter, loc cit, pp. 82-86.
11. E. K. Beauchamp, R. E. Loehman, B. Morosin, E. Venturini, and R. A. Graham, this Proceedings.
12. D. L. Hankey, R. A. Graham, W. F. Hammetter, and B. Morosin, J. Mat. Science Lett., accepted.
13. G. A. Adadurov and V. I. Bol'danskii, Russian Chemical Reviews, 50, 948-57 (1981).
14. G. E. Duvall and R. A. Graham, Rev. Mod. Phys., 49, 523-79 (1977).
15. See e.g., M. A. Meyers and L. E. Murr, eds., Shock Waves and High-Strain Rate Phenomena in Metals, Plenum, NY, 1981.
16. Dupont Diamond, Characteristics, Performance and Product Grade, Dupont Company Bulletin E-37692.
17. See e.g., Report of the Research Laboratory of Engineering Materials, Tokyo Institute of Technology, No. 6, 1981, Nagatsuta, Yokohama, Japan.
18. Dupont's Guide to Standard Products of Detaclad Explosive-Bonded Clad Metals Detacouple Welding Transition Joints, Dupont Company Bulletin E-15953.
19. L. Davison, D. M. Webb, and R. A. Graham, in Shock Waves in Condensed Matter, loc cit, pp. 67-71.

20. L. M. Lityagina, S. S. Kabilkina, T. A. Pashkina, and A. I. Khozyainov, Sov. Phys. Solid State, 20, 2009-10 (1978).
21. R. K. Linde and P. S. DeCarli, J. Chem. Phys., 50, 319-25 (1969).
22. M. C. R. Symons, J. Chem. Soc. A, 1648-52 (1971).
23. R. R. Hasiguti, pp. 69-92 in Annual Review of Material Science, Vol. 2, edited by R. A. Huggins, R. H. Bube, and R. W. Roberts, Annual Reviews, Palo Alto, CA, 1972, and references therein.
24. L. N. Shen, O. W. Johnson, W. D. Ohlsen, and J. W. DeFord, Phys. Rev., B10, 1823-25 (1974).
25. G. V. Chandrashekhar and R. S. Title, J. Electrochem. Soc., 123, 392-95 (1976).
26. E. Servicka, R. N. Schindler, and R. Schumacher, Ber. Bungsenges. Phys. Chem., 85, 192-95 (1981); E. Servicka, M. W. Schlierkamp, and R. N. Schindler, Z. Naturforschung, 36a, 226-32 (1981).
27. S. Yokoyama, M. Hirose, and Y. Osaka, Japan. J. Appl. Phys., 20, L35-L37 (1981).
28. R. S. Title, M. H. Brodsky, and J. J. Cuomo, p. 424 in Amorphous and Liquid Semiconductors, edited by W. E. Spear, CICL, Univ. of Edinburgh, 1977.

DISCUSSION

J. J. Petrovic (Los Alamos National Laboratory): We at Los Alamos National Laboratory have shock loaded H. C. Starck LC-12 alpha-Si_3N_4 in the range of 100-300 kbar and observe internal strain levels of the order of 0.3-0.4%, similar to the strain levels observed by Sandia in SN 502 Si_3N_4. Can hot microhardness testing, which is a combination of both elevated temperature environment as well as high compressive stresses, be employed as a possible approximation to particle contact conditions under the shock loading of powder arrays.

Author: Yes, hot microhardness testing might provide some insights into defect formation.

W. S. Coblenz (Naval Research Laboratory): Your results on shock activated powders is reminiscent of work done on milled powders correlating line broadening measurements with "activated sintering." These results have been explained in two ways: first, activated sintering due to defect formation; or secondly, due to reduction of agglomerates size and strength. Have you investigated changes in agglomerate structure in shock activated powders?

Author: I don't think point defects are persistent enough, or strain large enough to effect densification. Densification of agglomerates noted.

D. P. Dandekar (Army Mechanics and Materials Research Center): Was the stress hydrostatic to 30 GPa in your experiment? Could the scatter in the data (P-V) be due to the existence of hydrostatic and nonhydrostatic stress below and above ~ 10 GPa, respectively?

Author: The pressure transmitting medium used in these measurements was a mixture of 4:1 methanol:ethanol which remained hydrostatic up to 10.4 GPa. The medium rapidly became nonhydrostatic at pressures above 10.4 GPa, and that condition contributed significantly to the scatter in the data.

DENSIFICATION KINETICS OF SHOCK-ACTIVATED NITRIDES[a]

E. K. Beauchamp, R. E. Loehman, R. A. Graham, B. Morosin and E. L. Venturini

Sandia National Laboratories[b]
Albuquerque, New Mexico

INTRODUCTION

In 1966, Bergmann and Barrington[1] reported that, for a variety of oxide and carbide powders, improvement in the sintering behavior was obtained when the powders were subjected to explosive shock treatment. They concluded that, for the carbides, the improvement resulted primarily from the reduction in particle size, but for the oxides, the change was associated with a substantial increase in the internal defect population. Subsequent transmission electron microscopy showed high concentrations of dislocations in the shock-modified powders.[2] Later work has shown that very high concentrations of point defects are generated in shock-loaded rutile powders.[3]

Explosive shock treatment of nitrides has been carried out primarily by workers in the Soviet Union. Adadurov et al.[4] explosively shocked niobium, zirconium, silicon, chromium and gallium nitride powders and obtained strong compacts. More recently, Pruemmer[5] shock-compacted cylindrical samples of Si_3N_4 to 96% of theoretical density, and Hoenig and Yust[6] compacted AlN to 98% of theoretical density.

Although many of the samples in those early experiments were recovered as monolithic pellets, the pellets were always microcracked, presumably because of the tensile stresses developed by the release

---

[a]This work was sponsored jointly by the Department of Energy and the Defense Advanced Research Projects Agency.

[b]A U.S. DOE facility.

wave in the shock treatment. Anan'in et al.[8] examined the subsequent sintering behavior of those compacts and found that AlN pellets, which had a density of 86% after shocking, could be sintered at 1800°C to 97% of theoretical density. Kawada and Onodera[7] have reported sintering shock modified AlN and Si_3N_4 to 100% density at high pressure (5 GPa) and 1200°C.

The objectives of the present study were to determine whether AlN and Si_3N_4 powders densify more readily after shock-loading, and to identify the shock-induced modifications responsible for the improvement. For both AlN and Si_3N_4, bulk diffusion processes are generally too slow to contribute substantially to the grain deformation required for densification. It is generally believed that sintering of these materials occurs only by mass transport through a grain boundary phase.[9] For silicon nitride, oxide additives such as MgO, Y_2O_3, or Al_2O_3 combine with SiO_2 present as a surface oxide layer to form a grain boundary phase that is liquid at the processing temperature. In both sintering and hot-pressing, the solution-precipitation process that leads to densification is always associated with a crystallographic change in the Si_3N_4 from an amorphous or α-phase to the β-structure. The β-form of Si_3N_4 cannot be sintered or hot pressed to full density and the crystallographic change appears essential to the formation of dense Si_3N_4 ceramics.

For AlN, surface oxidation of the powder has been identified as the agent responsible for grain boundary transport.[10] In hot pressing, however, AlN shows a change from linear dependence of densification rate on applied pressure to a process with a stress exponent of about 10. This change has been associated with a transition from deformation controlled by grain boundary diffusion to dislocation creep.[10]

EXPERIMENTAL

Shock Treatment

To obtain a well-defined shock pressure with a high degree of uniformity through the sample, powders for the current work were shocked in disc form with the shock wave moving axially through the disc. For the present studies, the powders were shock-loaded in a new recovery fixture with a sample size of 5 cm.[3] A high explosive plane wave generator was used to obtain a planar shock front of uniform pressure.[11] The lower pressure shock loading used baratol explosive while the higher pressure loading was achieved with composition B explosive.

The nominal shock conditions for the AlN and Si_3N_4 powders are shown in Table 1. The powders were Starck Grade E AlN and GTE Sylvania SN502 Si_3N_4. Starting densities were obtained from sample

Table 1.

Starting Density*	Explosive	Nominal Shock Pressure+ Copper	Powder	Mean Bulk Temperature Increase
55	Baratol	13	11	475 K
55	Comp B	20	17	875 K

*Percent of crystal density.
+Calculated one-dimensional pressures in copper fixture and powder using rutile shock compression data.

mass and recovery fixture cavity dimensions and are shown as the fraction of crystal density. The shock pressures indicated are obtained from one-dimensional calculations based on available shock compression data on rutile powder. The "copper" values are those in the copper front plate at the copper-powder interface. The "powder" values are those calculated for TiO_2 powder under the same conditions. Equation-of-state data for AlN and Si_3N_4 were unavailable, but pressures are expected to be similar to those for TiO_2. The temperature increase is generated by the work of compacting the powder and is presented as a mean bulk value. Local temperatures at contacts between particles and at imperfections within grains could be much higher.[12] Although powder characteristics for AlN and Si_3N_4 are reported for both 11 and 17 GPa shock treatments, only 11 GPa shocked AlN was used in hot pressing and only 17 GPa Si_3N_4 was used in the transformation study. It is to be expected that two-dimensional effects, characteristically different in various fixtures, will be significant.

Hot Pressing Procedures

Hot pressing of the powders was conducted in a tungsten-mesh heated hot press equipped with a programmable controller. Hot press punches and dies were of graphite, lightly sprayed with boron nitride to reduce reaction between the graphite and the powders. Most runs were made with 0.3 g samples in a 0.64 cm diameter die. (The small quantity of powder produced in individual shock treatments imposed a limit on the size of the hot-pressing samples.) Ram travel was continuously monitored with an LVDT and recorded on an x-y recorder.

For the AlN samples, temperature was ramped to 1000°C at 50°C/min under vacuum with a 3 MPa load. At the end of a 20 minute hold at 1000°C, the chamber was backfilled with N_2 to 110 kPa. The temperature was then raised at 50°C/min to the final temperature. After a brief hold (2-3 min) in which thermal equilibrium was established, the ram pressure was increased to the final value and was held constant for the 60 minute run.

Phase Transformation Experiments

Because densification of Si_3N_4 is always accompanied by a crystallographic transformation to the β form, we decided to investigate the relative rates of the α to β transformation in shocked and unshocked Si_3N_4 as a surrogate for densification. The α to β conversion can be studied easily by x-ray diffraction techniques, which has the advantage that the early stages of the reaction can be detected readily.

Two types of experiments were performed. In the first set, cold-pressed cylinders of shocked and unshocked Si_3N_4 powder were heated under N_2 in a graphite-resistance furnace to temperatures between 1500 and 1800°C and for times from 30-300 min. The resulting specimens were ground in a TiB_2 mortar and pestle and subjected to x-ray diffraction analysis for phase content. The second set of experiments was similar to the first, except a tungsten mesh resistance furnace was used, and the Si_3N_4 powders were blended with 5 wt% MgO before heating.

The heating experiments on undoped Si_3N_4 were designed to test whether the shock-induced defects were sufficient to activate the α to β transition. The transition generally does not occur in Si_3N_4 except through the mediation of a liquid phase or by a vaporization-condensation process, both of which were ruled out by the experimental conditions. The experiments on the 0.95 Si_3N_4-0.05 MgO mixtures tested whether shock activation enhances an alternate conversion mechanism--dissolution in a liquid phase followed by precipitation. MgO was chosen as the additive because the MgO-SiO_2 liquid phase it forms dissolves less Si_3N_4 than other possible reactive liquids such as Y_2O_3-Al_2O_3-SiO_2 or MgO-Al_2O_3-SiO_2. The 5 wt% level was selected to provide a volume fraction of liquid phase near the minimum at which densification reactions proceed at any appreciable rate. Thus, if the defects produced by shock activation increase the solubility of Si_3N_4 in silicate liquids, there should be a discernible difference in the rates of the α to β transformation between the shocked and unshocked Si_3N_4 powders.

Fig. 1. SEM micrographs of SN502 Si_3N_4 (a) as received, and (b) after shocking at 17 GPa.

RESULTS

Powder Characteristics

As received, the SN502 Si_3N_4 is a very fluffy powder composed of porous aggregates of submicron grains and some large aspect-ratio whiskers. Figures 1a and 1b are SEM micrographs of the powder as received and after shock-loading. The shock treatment fragmented the whiskers and compacted the powder into a disc of greater than 90% theoretical density, which then fractured during release of shock pressure into irregular fragments approximately a millimeter in size. Light milling was used to reduce the particle size to 50-60 µm. In the scanning electron microscope, fracture of the powder shocked at 11 GPa appeared to be partially intergranular and partially transgranular. Fracture of Si_3N_4 shocked at 17 GPa appeared to be primarily transgranular, suggesting that significant intergranular bonding had been developed in the shock treatment.

The AlN powder as-received consisted of approximately 200 µm diameter agglomerates of 0.1 to 1 µm particles. As with the Si_3N_4 powder, the AlN was recovered from the shock treatment as a fragmented disc which was subsequently lightly milled in an alumina mortar and pestle to 200 µm particle size. Also as with Si_3N_4, the fracture of the shocked powder was mixed transgranular and intergranular for the low shock pressure and primarily transgranular for the higher pressure.

Surface area (BET) measurements on the Si_3N_4 powders showed a large initial increase with shock pressure (from a starting value of 4 m^2/g up to 7 m^2/g with 11 GPa) and a subsequent decrease (to 6 m^2/g with 17 GPa). For AlN, the surface area decreased from an initial value of 1.7 m^2/g to 1.4 m^2/g after an 11 GPa shock and to 1.0 m^2/g after a 17 GPa shock.

X-ray diffraction analyses (XRD) of the powders before and after shock treatment showed significant line broadening for high 2Θ lines, indicating, for Si_3N_4, shock-induced strain of 10^{-3}, corresponding to the introduction of 10^{15} dislocations/m^2. Broadening of the low 2Θ lines showed a reduction of the size of the coherent scattering zone from 200 nm to 120-160 nm. In AlN, XRD showed that strain of 2×10^{-3} was introduced at both high and low shock pressures while the coherent scattering region was reduced to 70 nm at 11 GPa shock pressure and to 100 nm at 17 GPa.

Electron spin resonance (ESR) measurements on the AlN powder show a single absorption with a g-factor of 2.003. The concentration of the defects responsible for this absorption increases from $10^{22}/m^3$ to $10^{23}/m^3$ as a result of shocking.

The Si_3N_4 also showed an ESR absorption at a g-factor of 2.003. The increase in intensity of this absorption corresponded to an increase in defect density from $6 \times 10^{22}/m^3$ in the starting material to $3 \times 10^{24}/m^3$ in low pressure shocked material and to $6 \times 10^{24}/m^3$ by the high pressure shock. The free electron-like g-factors suggest that, for both Si_3N_4 and AlN, the defect is probably an electron trapped at an anion vacancy.

Hot Pressing of Aluminum Nitride

A sampling of the AlN hot press densification curves is shown in Fig. 2. For a ram pressure of 47 MPa, the density after 60 minutes at 1730°C was 97.5% of theoretical (3.26 Mg/m^3) for the shocked powder compared with 91.5% for the unshocked powder. A similar difference is apparent for the 26 MPa data.

The hot pressing data are summarized in Fig. 3, which contains plots of the densification rate at 80 percent of theoretical density as a function of the ram pressure. The curve for unshocked powder at 1730°C shows two regimes of behavior, with a shallow slope up to 47 MPa and rapid increase of densification rate above that pressure. The curve for the shocked powder (at 1730°C) roughly parallels the upper portion of the unshocked curve but is displaced to substantially lower pressures. Plots of densification rate as a function of ram

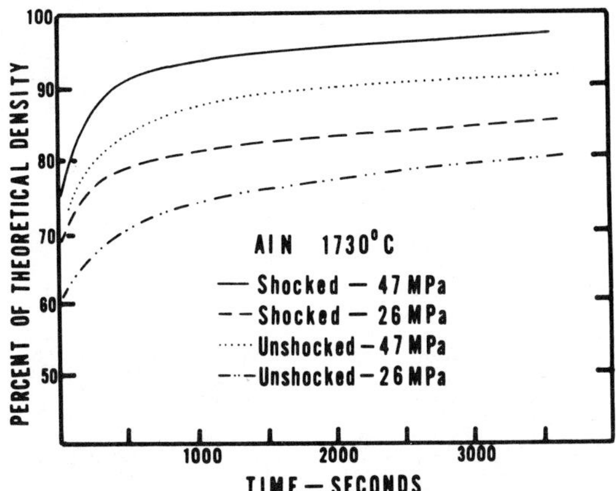

Fig. 2. Hot press curves for unshocked and shocked aluminum nitride at 1730°C (11 GPa shock pressure).

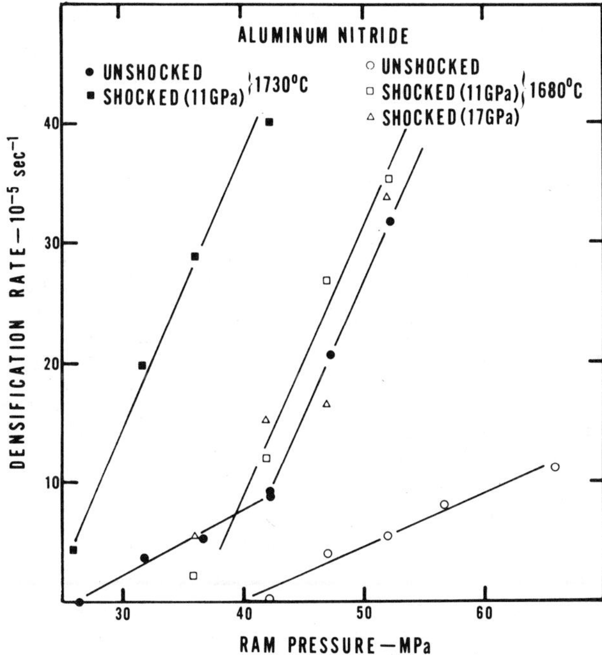

Fig. 3. Densification rate at (0.8 of theoretical density) vs. hot press pressure for unshocked and shocked aluminum nitride.

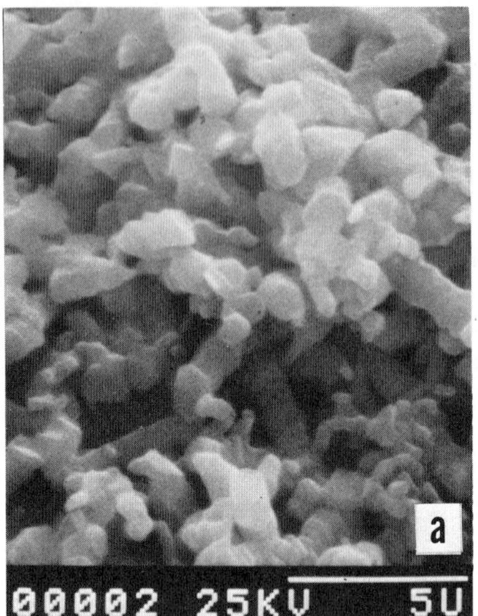

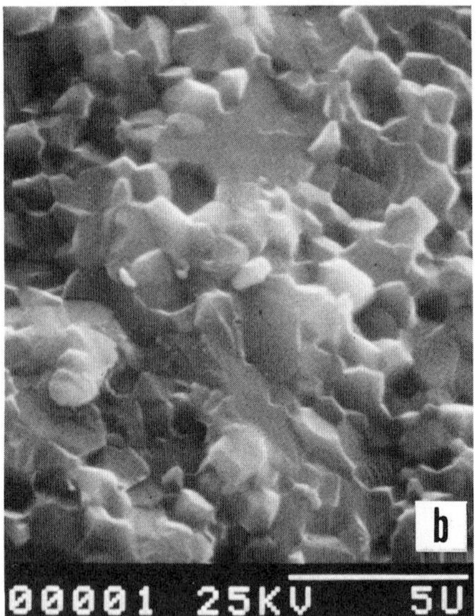

Fig. 4. SEM micrographs of hot-pressed AlN fracture surfaces; (a) unshocked, and (b) shocked powder.

pressure at 1680°C show the same general behavior as at 1730°C but with lower rates for a given pressure.

Oxygen content of the hot pressed AlN compacts was determined by neutron activation analysis. Four discs randomly selected from those made with unshocked powder showed 0.8 to 1.0 weight percent oxygen, whereas shocked samples contained 1.1 to 1.3 weight percent oxygen. The minimum detectability with this technique was 200 μg oxygen, giving an uncertainty in measurement on the 200 mg samples of 0.1 weight percent.

Figures 4a and 4b show microstructures of fracture surfaces of hot pressed samples of unshocked and shocked AlN. Grain size is essentially the same (~ 1 μm) for the shocked and unshocked samples and only slightly larger than that of the starting powder. In all cases, grains were equi-axed and quite uniform in size. Fracture was primarily intergranular in the shocked sample and mixed transgranular and intergranular in the shocked sample.

Silicon Nitride Phase Transformation

Following the procedure of Gazzara and Messier,[13] the amounts of the α and β phases were estimated by calculating wt. fraction

(β) = $I_{\beta}(210)/[I_{\alpha}(210) + I_{\beta}(210)]$, where, for example, $I_{\beta}(210)$ is the peak height of the x-ray reflection from the β-(210) plane. Because no other phases but α and β Si_3N_4 appeared in the XRD spectra, the β-phase concentration can be calculated by difference.

X-ray diffraction spectra of shocked Si_3N_4 heated successively at 1500°C (2 hr), 1650°C (1 hr) and 1800°C (0.5 hr) revealed that at each stage there was no change in the concentrations of the α and β phases. For unshocked Si_3N_4 the β-phase amounted to less than 10 wt% of the crystalline material for all heat treatments.

The XRD results indicated that the shocked Si_3N_4 powder was not activated sufficiently to undergo the crystallographic transition in the absence of liquid- or vapor-phase transport. Accordingly, a second set of experiments was performed in which cold-pressed cylinders of shocked and unshocked Si_3N_4 mixed with 5 wt% MgO were heated under N_2 for 0.5 to 2 hr. at 1600 to 1700°C. Samples ball-milled without any additives were included in each run to serve as a reference. Figure 5 presents the results of analyses of XRD spectra taken on those heat-treated specimens using Gazzara and Messier's procedure. As can be seen, there is a considerable increase in the rate of the α to β transition for shocked Si_3N_4 + 5 wt% MgO, relative to the unshocked Si_3N_4-MgO mixture. For all times and temperatures, the relative amount of the β-phase in the shocked material is about double that in the unshocked Si_3N_4. The undoped Si_3N_4, by contrast, does not undergo a change in the α/β ratio with heat treatment. These results present clear evidence for enhancement in the rates of Si_3N_4 solution and recrystallization brought about by shock induced modifications.

DISCUSSION

The x-ray diffraction analysis of shocked powders showed that dislocation densities of 10^{15}-$10^{16}/m^2$ were generated in both the Si_3N_4 and AlN powders. This high concentration of dislocations is an unusual circumstance for ceramic powders. Even for powders which have been extensively ball-milled, dislocation densities are usually several orders of magnitude lower.

There are several ways in which dislocations can contribute to the grain deformation required in sintering. The highly strained region along the dislocation can provide an easy path for diffusing species (pipe diffusion). Edge dislocations can act as sources and sinks for point defects, reducing the transport distance for Nabarro-Herring creep and, therefore, increasing the creep rate. It is also possible, but rare in ceramic materials, for the crystallites to deform by dislocation glide or climb processes. Thus for AlN, which has been reported to deform at high pressure by a dislocation process,[10] the introduction of substantial numbers of dislocations by

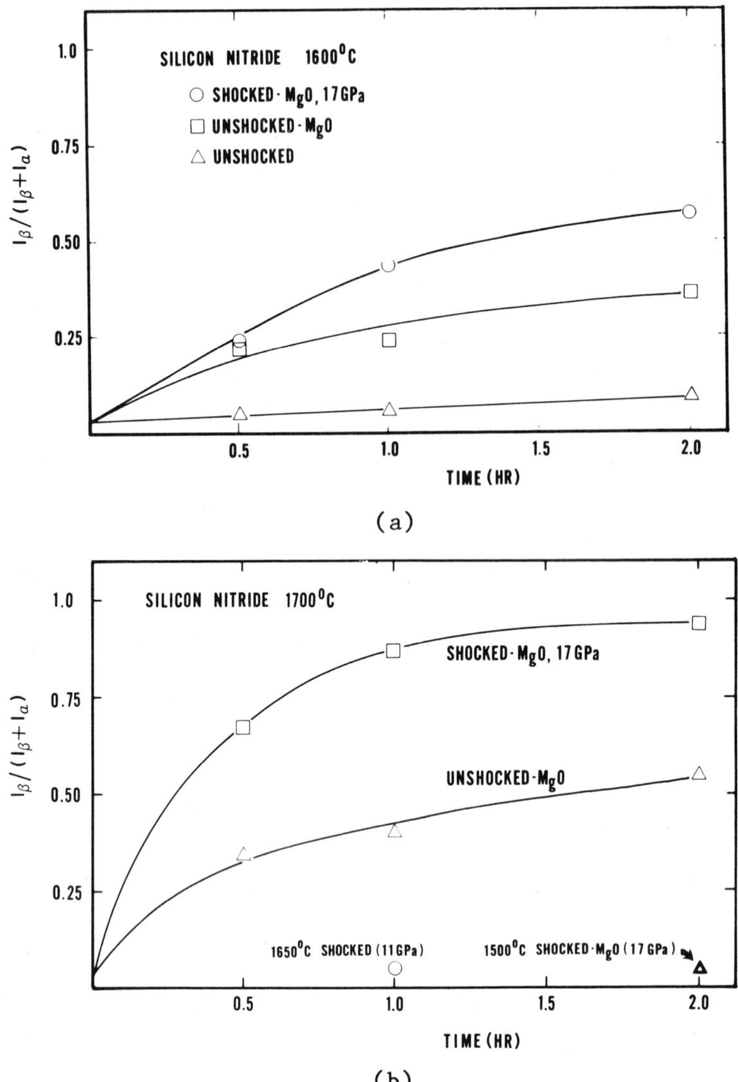

Fig. 5. Alpha to beta-phase transition in silicon nitride as a function of time at (a) 1600°C, (b) 1700°C. The two isolated data points in Fig. 6b are for a 1 hr., 1650°C heat-treatment of shccked (11 GPa) Si_3N_4, and for a 2 hr., 1500°C heat-treatment of shocked (17 GPa) Si_3N_4 + 5 wt% MgO.

the shock treatment would be expected to increase the densification rate in the high pressure regime. At low pressures, where diffusion through the grain boundary phase is rate controlling, the presence of dislocations within the AlN grains should affect densification rates only to the extent that it might modify the transport across the AlN/grain boundary phase interface.

The electron spin resonance data indicate that the vacancy population was also significantly increased by the shock treatment. These point defects could contribute directly to a bulk diffusion process. In that event, they could be swept out in the diffusion process, giving a deformation equivalent to, at most, their volume in the lattice. For populations of $10^{26}/m^3$, this amounts to only a small fraction of a percent deformation and would be completely overshadowed by other processes. Alternatively, they might aggregate into dislocation loops which would give a more persistent contribution to deformation.

Aluminum Nitride Hot Pressing Kinetics

The dependence of densification rate on ram pressure at 1730°C for the unshocked AlN (Fig. 3) has the same general character as that found by Le Compte et al.[10] at 1700°C, i.e., two regimes with an abrupt change in slope. The pressure required for a given densification rate is, however, considerably higher in the current work than they reported, in spite of a slightly higher temperature. Also, the slope of the high pressure regime in the current work is substantially less than they reported. The reason for these differences is not clear. The starting grain sizes of the powders were essentially the same and the difference between oxygen contents (2.7 wt% in their work compared with 1.0 wt% in the current work) seems too small to be responsible.[14]

The shock treatment of AlN substantially increases the densification rates, particularly in the pressure regime where dislocation motion is held responsible for deformation. The shocked material had a slightly higher oxygen content. However, the data of Le Compte et al. indicate no change in the high pressure regime for even larger changes in oxygen content. Moreover, the grain size of the shocked and unshocked material is about the same. Therefore, it appears that the internal defects, especially the dislocations, generated by the shock are responsible for the increase in densification rates.

Silicon Nitride Phase Transformation

Densification of Si_3N_4, whether by sintering or by hot pressing, requires the presence of an active liquid phase in which the powder particles are soluble. At the particle-particle contacts solubility is enhanced by the capillary pressure of the liquid (for sintering) or by the combined capillary and external pressures (hot pressing).

The initially amorphous or α-Si_3N_4 particles disolve and material is transported through the liquid; subsequently, β-Si_3N_4 crystallites precipitate to give a more compact structure. Because the α to β crystallographic change occurs as a result of the same dissolution-precipitation process that leads to densification, the early stages of densification can be monitored by following the phase transition using XRD. Under our experimental conditions, the liquid phase was produced by reaction of the 5 wt% added MgO with the surface SiO_2 present on the Si_3N_4 particles. Neutron activation analysis revealed the shocked Si_3N_4 contained 3.38 wt% oxygen (as compared with 2.93 wt% oxygen before shocking). Assuming all the oxygen was present as SiO_2, the shock treatment Si_3N_4-Si_3O_4-MgO mixture had the approximate overall composition: 0.75 Si_3N_4-0.12 SiO_2-0.14 MgO (mole fractions). The MgO-SiO_2 portion had the approximate composition 0.5 MgO-0.45 SiO_2 (mole fractions) or 0.45 MgO-0.55 SiO_2 (weight fractions). That was slightly richer in MgO than the eutectic composition.

Figure 5 shows clear evidence for an increase in the rate of β-Si_3N_4 formation with the shocked powder. That increase must have resulted from an enhancement in the dissolution rate of the amorphous and α-portions of the SN502 Si_3N_4 because under the present experimental conditions β-Si_3N_4 can form only by dissolution-precipitation. (SN502 Si_3N_4 is approximately 40 wt% amorphous and 60 wt% crystalline.[14]) Of the three successive stages required to produce β-Si_3N_4 (solution, transport in the liquid, and precipitation) only the initial solution stage could have been affected by shock alteration of the Si_3N_4 powder. As discussed above, shocking produced high concentrations of dislocations and defects in the Si_3N_4. Those defects must have increased the free energy of the system, which in turn increased the driving force for dissolution. Alternate explanations for the increase in dissolution rate, such as finer particle size or altered liquid phase composition in the shocked Si_3N_4, are ruled out by the results of BET and neutron activation analyses before and after shocking. In particular, the BET surface area of 17 GPa shocked Si_3N_4 was 6 m^2/g, whereas the unshocked, ball-milled Si_3N_4 had a surface area of 6.3 m^2/g. Likewise, the small increase in oxygen content after shocking would not produce an appreciable change in liquid phase composition. An increased dissolution rate should be reflected in an enhancement of the sintering rate for Si_3N_4. That could allow the use of additive compositions that are otherwise too refractory, or compositions that produce too little liquid phase to allow sintering or ordinary Si_3N_4. Either approach could give dense, sintered Si_3N_4 with grain boundary phases more resistant to high temperature creep than presently available forms.

CONCLUSIONS

Explosive shock treatment of AlN and Si_3N_4 increases point defect concentrations by more than an order of magnitude and dislocation densities to 10^{15}-$10^{16}/m^2$. Large increases in hot press densification rates for shocked AlN can be attributed to the dislocations. Similarly, the rate of the α-β crystallographic transition in Si_3N_4-MgO mixtures is significantly enhanced by shock-induced dislocations and defects.

ACKNOWLEDGMENTS

Particle size measurements and SEM micrography were provided by K. S. Varga and J. L. Young.

REFERENCES

1. O. R. Bergmann and J. Barrington, J. Am. Ceram. Soc., 49, 502-07 (1966).
2. R. W. Heckel and J. L. Youngblood, J. Am. Ceram. Soc., 51, 398-401 (1968).
3. E. L. Venturini, B. Morosin, and R. A. Graham, p. 72 in Shock Waves in Condensed Matter, 1981, Menlo Park, A.I.P. Conf. Proc. #78, edited by W. J. Nellis, L. Seaman, and R. A. Graham, Am. Inst. of Physics, 1982.
4. G. A. Adadurov, O. N. Breusov, A. N. Dremin, and V. F. Tatsii, Sov. Powder Met. and Met. Ceram., 11, 859-61 (1971).
5. R. A. Pruemmer, Ber. Dtsch. Keram. Ges., 50 [3], 75-81 (1973).
6. C. L. Hoenig and C. S. Yust, Am. Ceram. Soc. Bull., 60 [11], 1175 (1981).
7. K. Kawada and A. Onodera, Am. Ceram. Soc. Bull., 59, 1151-52 (1980).
8. A. V. Anan'in, O. N. Breusov, A. N. Dremin, V. B. Ivanova, S. V. Pershin, V. F. Tatsii, and F. A. Fekhretdinov, in proceedings, First All-Union Symposium on Shock Pressures, Vol. 2, Oct. 1973, Moscow, edited by S. S. Batsanov, Moscow, 1974. Trans. in Sandia Nat'l. Lab. Rept. SAND80-6119, April 1980.
9. C. Greskovich and J. H. Rosolowski, J. Am. Ceram. Soc., 59 [7-8], 336-43 (1976).
10. J.-P. LeCompte, J. Jarrige, J. Mexmain, R. J. Brook, and F. L. Riley, J. Matls. Sci., 16, 3093-98 (1981).
11. See e.g. D. L. Hankey, R. A. Graham, W. F. Hammetter, and B. Morosin, J. Matl. Sci. Lett. accepted.
12. D. E. Grady, J. Geophys. Res., 85 [B2], 913-24 (1980).
13. C. P. Gazzara and D. R. Messier, Am. Ceram. Soc. Bull., 56, 777-80 (1977).
14. T. Sakai and M. Iwata, J. Matls. Sci., 12, 1659-65 (1977).
15. R. E. Loehman and D. J. Rowcliffe, J. Am. Ceram. Soc., 63, 144 (1980); R. E. Loehman, Ceramic Engineering and Science Proceedings, 1-2, 35-49 (1982).

DISCUSSION

O. R. Hughes (Celanese Research): Concerning the CO oxidation catalytically activity observed over shock activated TiO_2--is this a true catalysis or a consumption of a shock induced reactive site in TiO_2? Is catalytic activity constant over extended observation times?

Author: The catalytic activity studies are carried out at the University of New Mexico by Professor Frank Williams and graduate student John Golden in the Chemical Engineering Department. The catalysis studies are carried out in a flow reactor and are felt to be a true catalysis. The activity is persistent over extended times. For more detail see: (1) Golden et al., "Catalytic Activity of Shock-Loaded TiO_2 Powder" Shock Waves in Condensed Matter - 1981 (Menlo Park), AIP Conference Proceedings #78, edited by W. J. Nellis, L. Seaman, and R. A. Graham, American Institute of Physics, New York (1982), pp. 72-76. (2) J. Golden, Master's degree thesis, University of New Mexico, Albuquerque, NM, 1982.

R. Raj (Cornell University): Your results on enhancement of α and β transformation kinetics in shocked Si_3N_4 are very interesting. I would like to suggest that your results reinforce the conclusion reached by Roger Wills this morning that these kinetics are controlled by interface reaction. Since interface reaction depends on the atomic roughness of the surface, and since a high dislocation density would increase such roughness, your results support the interface mechanism.

Author: I agree that the rate-limiting step in the $\alpha \rightarrow \beta$ transformation kinetics most likely is the interfacial reaction. The atomic roughness of the interface certainly would be an important influence on the surface dissolution rate.

R. Rice (NRL): You emphasized densification rates. How did final densities of shocked and unshocked materials compare, i.e., was the enhancement more in initial rates?

Author: For AIN the final densities in hot pressing experiments were greater for shocked powders than for unshocked material. In the Si_3N_4 work reported here, our primary intent was to study $\alpha \rightarrow \beta$ transformation rates so the experiments were not arranged to allow comparison of sample densities. We have some preliminary results indicating that hot press densification rates are enhanced in shocked Si_3N_4. Comparing experiments at a time, temperature, and pressure for which the relative density of unshocked Si_3N_4 was less than unity, the shocked Si_3N_4 would exhibit a greater density.

RATE CONTROLLED SINTERING OF EXPLOSIVELY SHOCK-CONDITIONED ALUMINA POWDERS

K. Y. Kim, A. D. Batchelor, K. L. More and H. Palmour III

North Carolina State University
Raleigh, NC 27650

ABSTRACT

Fine, high purity alumina powders subjected to nominal plane strain shock conditioning in precompacted disc form by an explosively driven flyer plate were reconstituted as well-compacted specimens ($D_o \approx 0.65$) and sintered dilatometrically in both CTS and RCS modes ($Df \approx 0.98$-$0.99+$). Effects of variations in precompaction and in shock velocity are characterized and related to sinterability, with emphasis on resultant temperatures at onset of shrinkage, initial stage densification kinetics, overall rate effects and microstructural development.

INTRODUCTION

In this paper the effects of two somewhat unfamiliar process treatments (shock conditioning, rate controlled sintering) on the densification and microstructural development of a familiar oxide ceramic material (alumina) are considered. The use of highly dynamic shock waves to alter ceramic powders, and the enhanced sinterability observed thereafter in such shock consitioned powders, have been described elsewhere.[1-6] Similarly, the use of rate controlled sintering (RCS) to enhance microstructural control during densification of otherwise conventionally processed alumina (and other oxides) has been treated in other papers.[7-14] For the first time to the best of our knowledge, these two separate process effects are now being considered interactively.

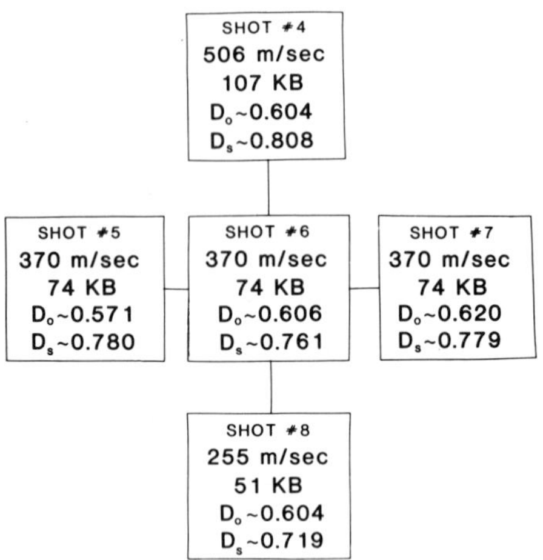

Fig. 1. Main sequence shock conditioning experiments.

BACKGROUND

Preliminary phases of this study[15,16] have treated the characterization and densification under conventional temperature sintering (CTS) conditions of fine, pure aluminas (Baikowski CR-10, GE-10) in both unshocked and shocked forms. Based upon that earlier study, the experimental design shown in Fig. 1 was adopted to investigate the effects of (1) degree of precompaction and (2) shock pressure upon subsequent sinterability.

As in earlier phases, all shock treatments were carried out at Battelle Columbus Laboratories, using premilled, precompacted ceramic discs (76.2 mm dia. x 10.62 mm thick, 100+g) furnished by NCSU. The steel encapsulated, pre-evacuated discs were shocked in planar impact geometry by explosively driven steel flyer plates at appropriate impact velocities to generate the desired shock pressures.[15-17]

The sintering studies reported here were carried out in air in both CTS and RCS modes. The general methods employed, and the precision digital dilatometer utilized, have been described in preceding papers.[15,18,19]

EXPERIMENTAL PROCEDURES

Using procedures described elsewhere,[11,13,15,16] Baikowski CR-10 alumina powder was prepared for shocking by (a) 6h intensive dry milling; (b) 6h dry remilling to disperse 2.5% Carbowax 4000 as binder; (c) 6h dry remilling to disperse 2.5% oleic acid as lubricant; (d) dry pressing at 6.89, 68.9 or 206.8 MPa (1, 10 or 30 kpsi) to form 7.62 mm (3 inch) dia discs having a controlled thickness of 10.62 mm (0.4 inch); and (e) controlled binder burnout (22h to 430°C). The binder-free fractional densities were typically 0.571, 0.605 or 0.620, respectively.

Procedures employed at Battelle Columbus Laboratories for (a) encapsulation, (b) pre-evacuation, and (c) explosive shocking in the planar "mousetrap" configuration have been given elsewhere.[15-17] Impact velocities and shock pressures for these shots (calculated at the point of entry into the steel cover plate) are shown in Fig. 1. For convenience, only these nominal shock pressures are reported here.

Pressure levels reached within the ceramic material per se are thought to be influenced by its own properties and precompacted density, the nominal shock velocity, pressure wave attenuation occurring within the recovery package, and by internally reflected shock waves as well.[15-17,20-23] They must be estimated indirectly by complex computer codes, as shown in Fig. 2.[21]

After removal from the steel recovery package, the shocked material was returned to NCSU for characterization and sinterability evaluations, now ongoing. In this paper, together with unshocked controls, we consider in detail just those materials which had been precompacted to 0.605 and shocked at nominal pressures of 51, 74 and 107 kilobars, respectively (see vertical axis, Fig. 1).

Small (approx. 50 g) batches of shocked material were reconstituted for sintering by (a) 4h intensive dry milling, (b) 6h remilling with 2.5% binder, and (c) 6h remilling with 2.5% lubricant. Small (12.7 mm dia) cylindrical specimens were formed by dry pressing at 137.9 MPa (20 kpsi). After binder burnout, fractional green densities were typically $\geq$ 0.63.

CHARACTERIZATION

Changes attributable to prior processing or to shock conditioning per se were monitored by several methods, including x-ray line broadening, surface area analysis (BET), thermoanalysis (DTA, TGA) and electron microscopy (SEM, TEM). The results are summarized in Table 1, and typical alumina particles--in the as-received, processed and shock conditioned states--are shown in Fig. 3.

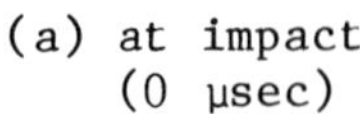

(a) at impact (0 μsec)

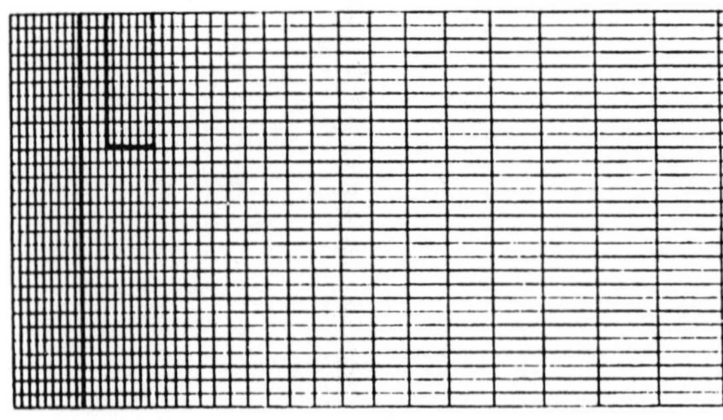

(b) 10 μsec after impact

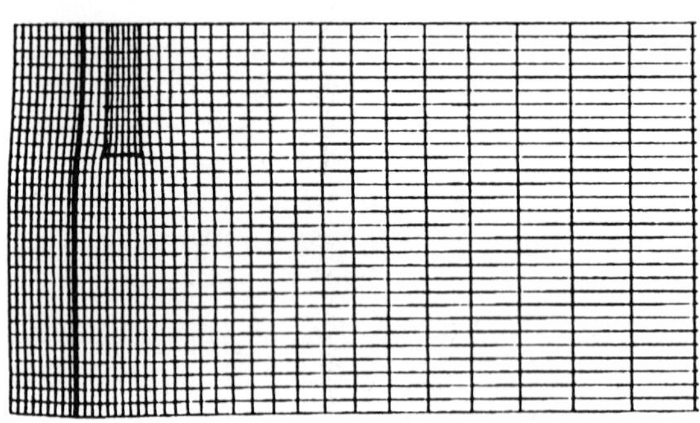

90	160	212	246	208	167	86	67	90	205	304	421	474
185	234	356	323	322	173	120	60	156	179	316	376	451
220	421	454	570	347	253	100	257	184	326	341	430	439
386	423	881	609	654	182	475	389	465	444	530	513	554
252	660	924	1329	362	766	547	706	640	732	694	710	864
801	334	1122	89	12	-10	-12	-15	-5	-4	-2	-10	-24
542	621	540	86	75	53	45	40	42	42	43	41	26
980	597	685	82	117	97	80	73	72	72	72	72	41
1240	911	849	104	134	104	83	74	71	71	71	72	37
1461	1637	1048	153	120	81	57	49	46	46	46	47	24
2144	1817	2455	185	124	43	32	23	22	22	23	24	19
2285	2586	2379	2449	943	1389	1185	1101	888	816	713	785	760
2830	2742	2840	2202	1986	1166	1339	1015	957	816	814	770	832

(c) stresses (in MPa) at 10 μsec)

-135	-94	-142	-160	-221	-219	-274	-272	-317	-298	-337	-326	-384
-37	-43	-67	-170	-225	-325	-333	-397	-360	-400	-356	-398	-364
0	86	-20	-125	-311	-381	-482	-430	-490	-416	-447	-384	-455
44	153	187	-110	-384	-565	-480	-507	-486	-527	-451	-480	-417
30	180	345	-80	-601	-510	-605	-505	-614	-537	-526	-480	-401
147	134	615	4	-6	-7	-8	-5	-3	-1	1	1	1
71	321	307	7	2	0	1	3	6	9	11	12	9
147	241	362	10	7	6	7	9	13	16	19	20	13
90	210	342	13	10	10	12	14	17	20	23	24	15
81	150	359	17	13	14	15	17	19	21	23	24	16
123	90	301	19	17	14	17	16	17	17	19	19	14
-334	-264	-282	-310	-336	-330	-320	-295	-242	-157	-90	-25	-16
49	172	69	-327	-620	-622	-700	-710	-759	-760	-787	-770	-810

(d) stresses (in MPa) at 40 μsec)

Fig. 2. Deformation and time-dependent stress distributions resulting from planar dynamic impact. Alumina powder compact [CR-10, shot #2], steel encapsulated, hit by steel flyer plate at 281 m/sec.[16] Half section(s) simulated by TROTT, a two-dimensional Lagrangian finite difference computer program (after G. L. Moss and R. Benck[21]).

Table 1. Comparisons of Unshocked and Shocked CR-10 Alumina Powder

Sample Condition	Surface Area, BET[a] (m^2/g)	Equivalent Spherical Diameter (μm)	Crystallite Size b (μm)	Residual Strain b (%)	Calculated Strain Energy[b] (cal/g)
As received	11.3	0.1326	--	0.0065	0.00027
Unshocked (master batch)	12.5	0.1204	--	0.04	0.010
Shocked 51kb (8A)	11.6	0.1298	0.08	0.116	0.086
Shocked 74kb (6)	10.6	0.1420	0.063	0.177	0.20
Shocked 107kb (4)	11.4	0.1320	0.061	0.231	0.34

[a]Data from Micromeritics Instrument Corp., Norcross, GA.
[b]Material annealed 18h at 1000°C taken as x-ray reference zero.

As a function of increasing shock pressure (impact velocity), shocking appeared not only to increase the bulk density (compare Ds with D_o, Fig. 1), but also to reduce the surface area (Table 1). These effects suggest that some interparticle bonding, and perhaps even some grain growth, did occur during dynamic compaction. Only at the highest pressure, 107 kb, did the trend of decreasing surface area (and increasing particle size) reverse somewhat, presumably due to greater amounts of shock-induced fracturing of individual particles at that highest stress level.

The TEM and x-ray data show that extensive alteration (e.g., plastic deformation) of the alumina material occurred as a direct consequence of shock treatment, in general agreement with the findings of others.[1-3,5,6] In the Hall-Williamson plot shown in Fig. 4, the slopes of the lines for the several powder conditions studied here are proportional to residual strain, and the intercepts are related inversely to crystallite size.[24] For shocked material, the calculated crystallite (subgrain) size is much smaller (by > 2:1) than the actual particle size, and detailed TEM examinations reveal many evidences of intense dislocation activity, as reported elsewhere.[6,15,16,25]

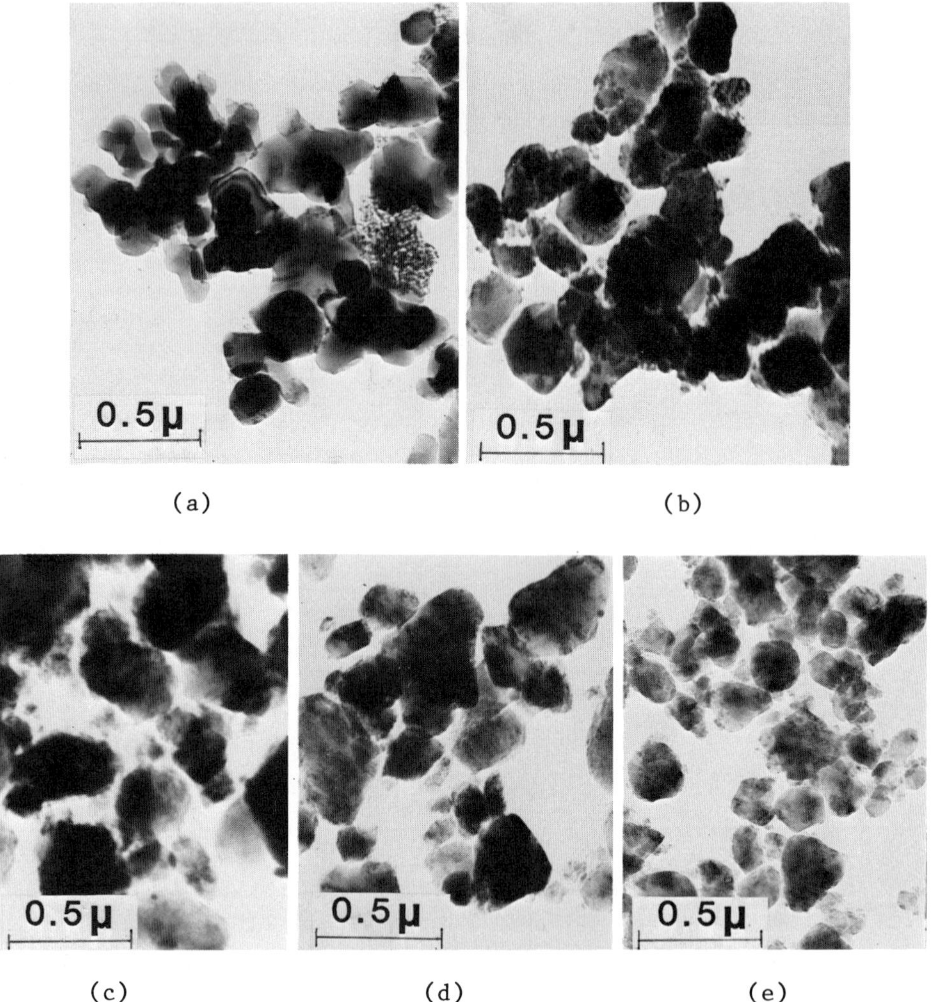

(a) (b)

(c) (d) (e)

Fig. 3. Micrographs of Baikowski CR-10 alumina powder: (a) as received; (b) milled, 18h; (c) shocked, 51kb; (d) shocked, 74kb; (e) shocked, 107 kb. [TEM, 120kv.]

Central to the concept of shock activation is the notion that shock-induced alteration of the material will (a) yield some form of stored but annealable excess energy, which (b) can--without prejudice as to the details of the mechanisms involved--subsequently be released over a useful temperature range to augment or assist in some needed thermally-activated process.[1,6,27,28] Using shocked CR-10 and GE-10 material from an earlier study[15,16] as examples, the DTA curves[28] given in Fig. 5 appear to confirm that this effect is real, and that its magnitude increases with increasing shock pressure.

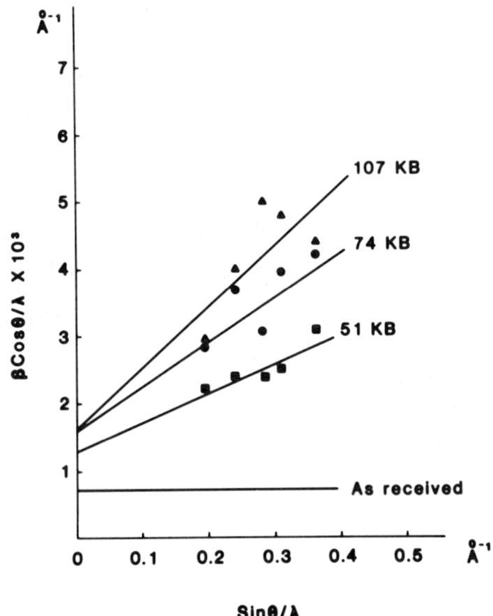

Fig. 4. X-ray line broadening: Hall-Williamson plots for unshocked and shocked alumina powders.

In these plots, the difference at any given temperature between the first and second (and all subsequent) heatings is related to the annealable excess energy being released at that temperature during the first heating. Similar thermoanalytical methods have been used to study shock alteration[28] and other forms of annealable excess energy, e.g., that induced by irradiation.[30] In these shocked aluminas, annealing appears to begin at or below 400°C, and continues to the maximum temperature studied, 1500°C. Presumably, both point and line defects are involved, with the former contributing primarily to the low temperature annealing. The greatest effects appear to occur in the range 900-1100°C, where sintering processes are known to become active for these grades of alumina.[16,19]

RECENT FINDINGS

Dilatometric sintering studies generate sets of plots, e.g., T vs. t, Δl/l vs. t, D vs. t, etc. The computer-based system employed here considerably broadens the combinations of variables which may be chosen for plotting, as demonstrated elsewhere.[16,18,19] For the purposes of this paper, plots of (a) dD/dt vs. D and (b) T vs. D, respectively, have been chosen: they display (a) the densification rate observed at any given density, and (b) the temperature required to attain that density.

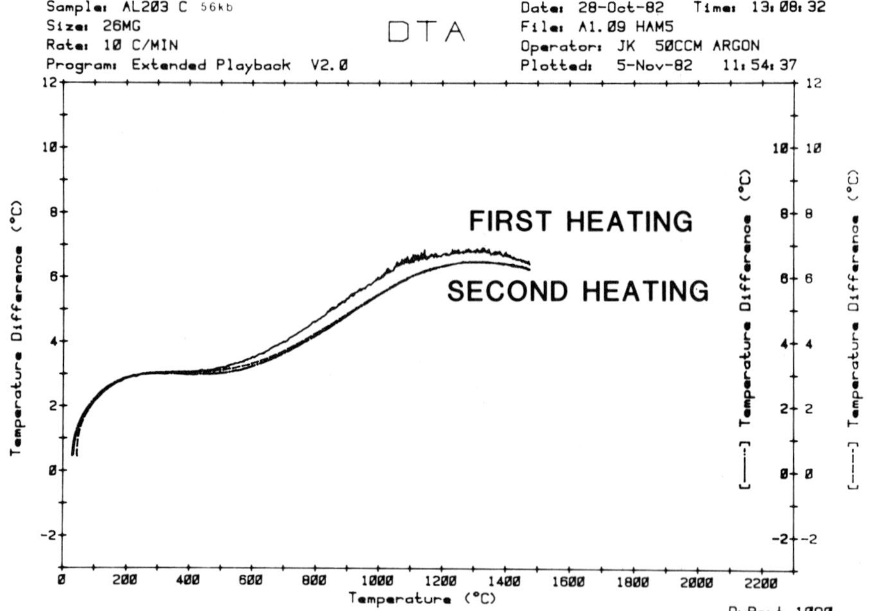

(a) Baikowski CR-10 alumina, shocked at 56kb[16]

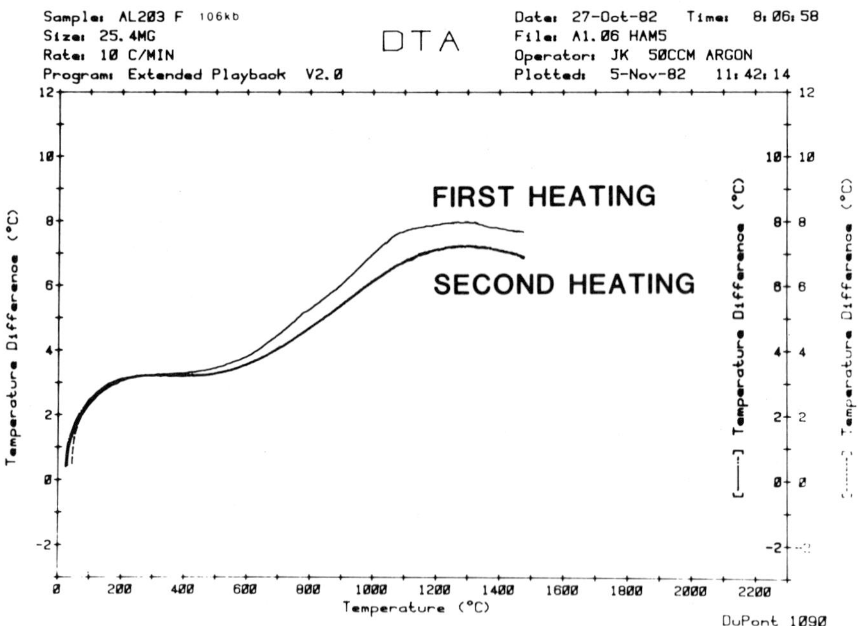

(b) Baikowski GE-10 alumina, shocked at 107 kb[16]

Fig. 5. Thermoanalyses (DTA) of the annealing of shocked Alumina (after Beauchamp[29]).

Figure 6 plots results (uncorrected, see[16,19]) obtained from a comparative sintering study (CTS mode, 10°C/min to 1530°C, 30 min hold) for CR-10 alumina: (c), in the unshocked condition, and (d, e, f), shock conditioned, at three different pressure levels. It also shows the final microstructure obtained in each case.

It is evident that preshocking resulted in marked alterations of the normal (unshocked) rate-temperature-density relationships, beginning at fractional densities near 0.70. The observed densification rates in the intermediate range of densities--near $D \approx 0.80$--are suppressed, but thereafter--in the range $D \geq 0.85$--the rates for shocked material increase again, in some cases eventually exceeding that of the unshocked control.

In comparison with unshocked material, the microstructures of the shocked, fired specimens are very coarse grained, and also show extensive evidence of trapped pores, often heavily clustered. In view of the observed early onset of shock-induced, more-or-less adverse densification rate effects and the heavily altered final microstructures, it would appear that the sintering process had been in part interdicted by some other thermally activated process (e.g., primary recrystallization and/or grain growth, see Ref. 6). Obviously, further detailed studies will be needed to identify the specific mechanisms involved.

At successive levels of fractional density, the Arrhenius plot shown in Fig. 7 further demonstrates the effects of preshocking on the apparent densification kinetics for sintering (CTS mode). For unshocked material (dashed lines), the slopes (proportional to the apparent activation energy, Q) remain essentially constant as density increases, consistent with previous observations for other alumina grades.[10,31] At $D \leq 0.75$, the preshocked specimens (solid lines) display those same slopes, but in keeping with other findings,[1,15,16] they also sinter somewhat more easily than the unshocked (i.e., attaining equivalent rates/densities at lower temperatures). At $D \geq 0.80$, progressive steepening of the slopes for the preshocked material is observed, and relative to the unshocked controls, the isodensity lines are more and more offset to the left (i.e., toward higher temperatures). From the viewpoint of the kinetics, and in keeping with the earlier data and discussions, it would appear that in the range $0.75 < D < 0.80$ for these specimens, competition from other thermally activated processes (recrystallization? grain growth? see Ref. 6) markedly alters the course of the sintering process itself.

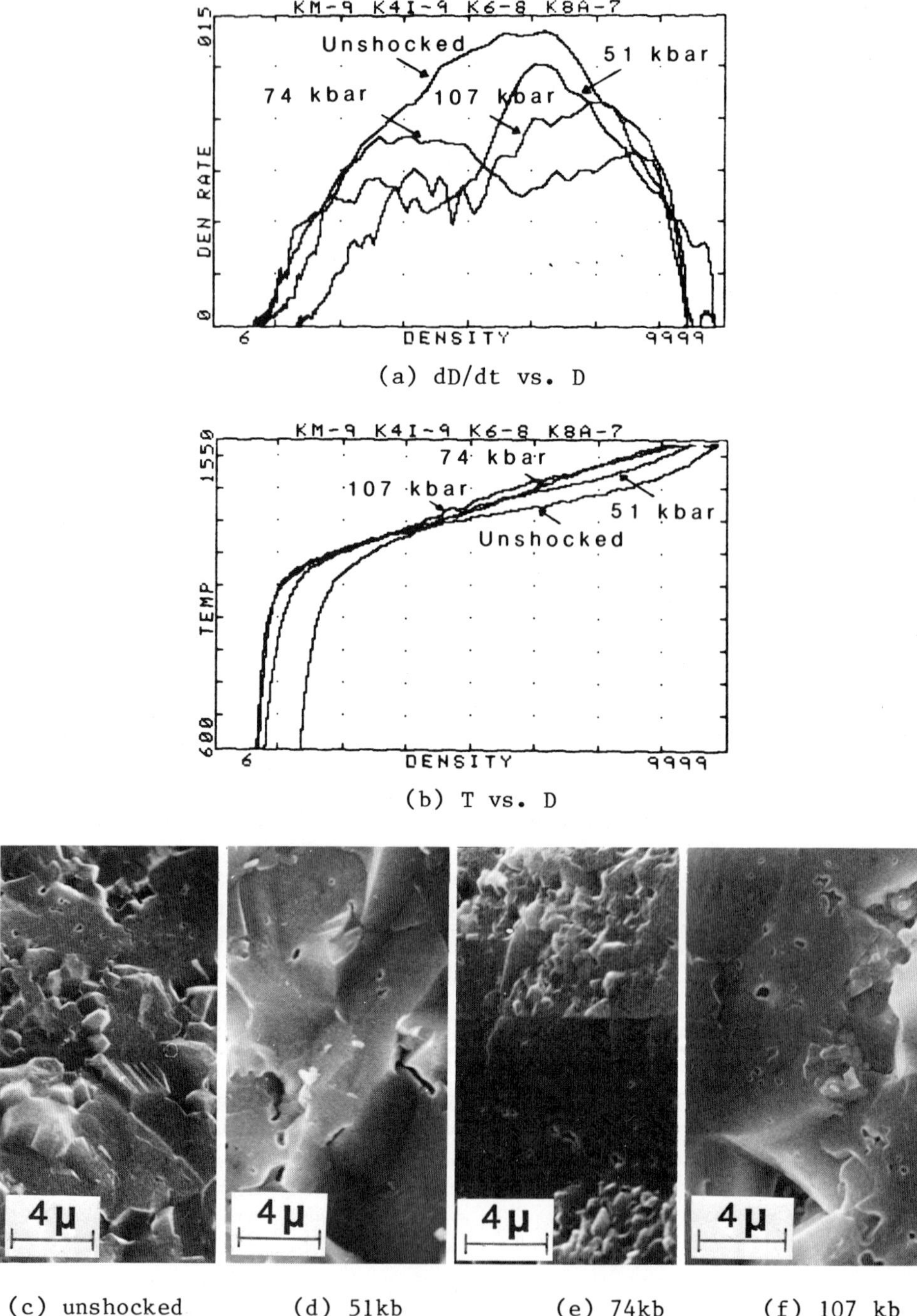

(c) unshocked (d) 51kb (e) 74kb (f) 107 kb

Fig. 6. Dilatometric and microstructural comparisons between unshocked and shocked aluminas sintered in the CTS mode.

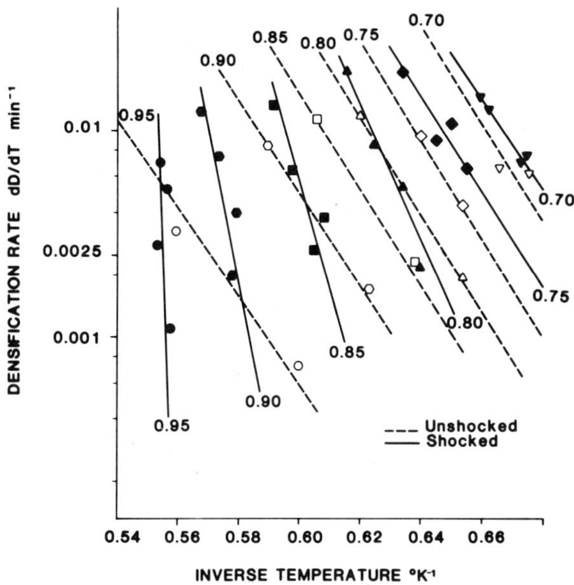

Fig. 7. Arrhenius plots at successive density levels for sintering of unshocked and shocked alumina.

From a more pragmatic viewpoint, these early data from CTS firings also can be interpreted as suggesting that if fired rapidly enough (at quite high densification rates), preshocked material may in fact sinter more effectively than the unshocked, whereas if fired slowly, it will almost assuredly sinter much worse! This hypothesis has been tested in preliminary experiments. Figure 8 compares the results obtained with one preshocked material (Shot #6, Fig. 1) by firing in conventional (CTS) and rate controlled (RCS) modes, and on deliberately fast and slow schedules, respectively.

From Fig. 8, it is apparent that fast rates (in either the CTS or the RCS mode) tend to reduce the extent of the exaggerated grain growth observed. The RCS mode was more effective than the CTS in this regard, as also observed in earlier studies with conventionally processed (unshocked) but undoped (MgO-free) aluminas.[31] Although they differ markedly in the rates maintained during the late stages of densification, both the "fast" modes follow rather similar paths in the early stages. It would appear, therefore, that maintaining quite rapid densification rates in the early stages (e.g., $0.70 < D < 0.75$) will be important if the competing effects which promote uncontrolled grain growth are to be adequately suppressed.

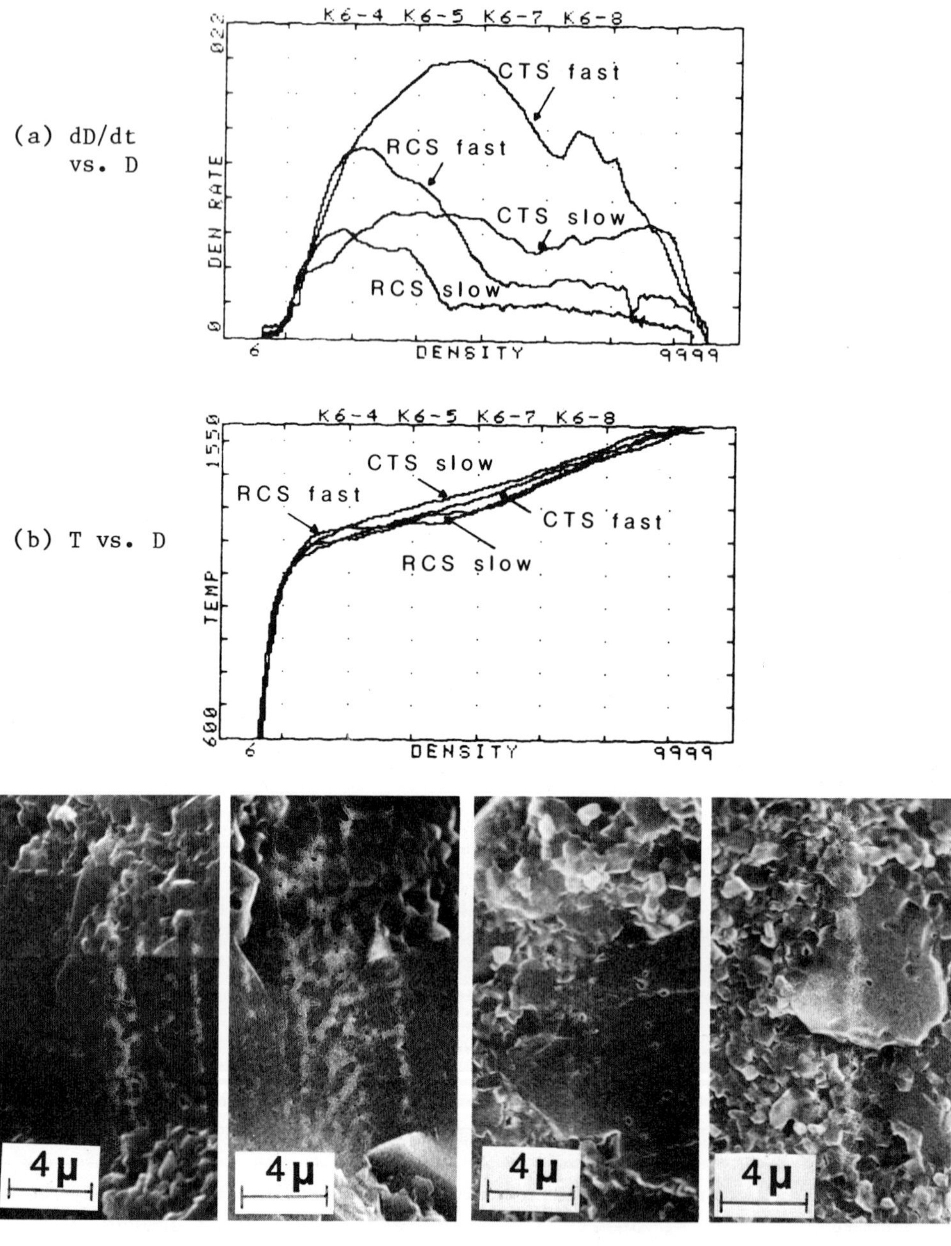

(c) CTS slow (d) RCS slow (e) CTS fast (f) RCS fast

Fig. 8. Dilatometric and microstructural comparisons between shocked alumina specimens sintered at slow and fast rates in the CTS and RCS modes.

The density data of Fig. 1 give evidence of considerable densification during shock treatment of these alumina powders, displaying ΔDs over the range 0.115-0.209, dependent upon D_0 and shock pressure. They were larger than those observed in the early studies, e.g., Shot #2, $\Delta D \approx 0.092$ at 56kb.[15,16] As a consequence, these shocked materials from the main sequence experiments were found to be heavily agglomerated, with the agglomerates being strong enough to resist any easy type of redispersion. [This shock-induced phenomenon, and its role in establishing coordination distributions and resultant sinterabilities, has been considered elsewhere.[16,32] From detailed SEM examinations of the shocked, remilled, compacted and fired states, it is now apparent that significant numbers of coarse agglomerates did survive the 4h dry remilling procedure employed here. In retrospect, it is concluded that their presence also contributed materially to the observed grain growth patterns. To minimize this source of variability in ongoing studies, a more intensive, two step comminution procedure is being followed, in which the shocked material is first reduced to -80 mesh powder by crushing in a mortar, then intensively dry ball milled for 18h.

DISCUSSION

Shock-induced activation of ceramic powder materials[1] has once again been confirmed, and at least some aspects of the needed, detailed characterizations of the shock-conditioned state[1,3,6,28] have also been reconfirmed. The sintering behavior of unshocked and shocked high-purity (undoped) fine grained alumina has been studied in considerable detail by dilatometric means, in both the conventional (CTS) and rate-controlled (RCS) modes. Shock conditioning was found to alter the densification kinetics markedly. For material pretreated at nominal shock pressures $\geq$ 51kb, these experiments tended to yield very coarse grained sintered microstructures. Fast firing in either CTS or RCS mode during the early stages of densification tended to reduce the shock-induced exaggerated grain growth, with the RCS mode being somewhat more effective.

Sintering is a very complex thermally activated process, having strongly interdependent (and at times experimentally confounded) chemical and morphological as well as energetic and kinetic aspects. It seems to be particularly so for shock-conditioned ceramic powders such as these, where some confounding of the morphological and energetic aspects probably cannot be avoided.[16,32] For such materials, sinterability apparently can be enhanced (or diminished, as the case may be) by the specifics of the shock conditioning method employed, as well as the selections of subsequent process steps and firing techniques.

When considered primarily in terms of sinterability and microstructural control, we would conclude that in the preliminary experiments,[15,16] Shot #2 (nominal 56kb, 0.108% residual strain) served as an example of beneficial shock conditioning,[1] whereas the similar (and presumably overlapping) main sequence shock experiments described here have been perhaps more nearly representative of the morphology-modifying dynamic compaction regime of shock behavior.[3,6,17,32] It is not surprising, therefore, that in addition to activation per se, some morphologically-related (and not always so beneficial) anomolous sintering and grain growth behavior has been observed.

ACKNOWLEDGMENTS

Supported by the U.S. Army under Contract DAAK 11-82-C-0040. Dr. G. L. Moss served as technical monitor. Supporting research and analytical services, data, and helpful advice and counsel have been generously provided by other investigators in the field, including V. D. Linse, J. A. Adair and R. Willis (Battelle; R. A. Graham, B. Morosin and E. Beauchamp (Sandia); G. L. Moss and R. Benck (BRL/APG); D. P. Dandekar (AMMRC); O. R. Bergmann (DuPont); and Y. Horie and J. K. Whitfield (NCSU). We also gratefully acknowledge experimental and preparative assistance provided at NCSU by Michelle Bridges, Laura Freed, E. M. Gregory, T. M. Hare, Terrell Jones, Ellice Luh, G. S. McGaughey, Rodney Motley, M. J. Paisley, Betty J. Randall, and H. H. Stadelmaier.

REFERENCES

1. O. R. Bergmann and J. Barrington, J. Am. Ceram. Soc., 49, [9], 502 (1966).
2. A. C. Greenham and B. P. Richards, Trans. Brit. Ceram. Soc., 69, 115 (1970).
3. R. Prummer, Ber. Dtsch. Keram. Ges., 50 [3], 75-81 (1973).
4. R. Prummer and G. Ziegler, Powder Metall. Inst., 9 [1], 11-14 (1977).
5. S. Clyens and W. Johnson, Mat. Sci. Eng., 30, 121-29 (1977).
6. Dynamic Compaction of Metal and Ceramic Powders, National Materials Advisory Committee Study Report NMAB-394, National Research Council, October 1982.
7. H. Palmour III and D. R. Johnson, pp. 770-91 in Sintering and Related Phenomena, edited by G. C. Kuczynski, N. A. Hooton, and C. F. Gibbon, Gordon and Breach, NY, 1967.
8. M. L. Huckabee and H. Palmour III, am. Ceram. Soc. Bull., 49 [8], 574-76 (1972).
9. H. Palmour III and M. L. Huckabee, U.S. Patent 3,900,542, December 1, 1975.

10. H. Palmour III, M. L. Huckabee, and T. M. Hare, pp. 308-19 in Ceramic Microstructures - '76, edited by R. M. Fulrath and J. A. Pask, Westview Press, Boulder, CO, 1977.
11. T. M. Hare and H. Palmour III, pp. 307-20 in Ceramic Processing Before Firing, edited by G. Y. Onoda, Jr. and L. L. Hench, John Wiley and Sons, NY, 1978.
12. H. Palmour III, M. L. Huckabee, and T. M. Hare, pp. 46-56 in Sintering - New Developments, edited by M. M. Ristic, Mat. Sci. Monographs, 4, Elsevier, Amsterdam, 1979.
13. M. L. Huckabee, T. M. Hare, and H. Palmour III, pp. 205-15 in Proceeding of Crystalline Ceramics, edited by H. Palmour III, R. F. Davis, and T. M. Hare, Mat. Sci. Res., Vol. 11, Plenum Press, NY, 1979.
14. H. Palmour III and M. L. Huckabee, pp. 278-97 in Factors in Densification and Sintering of Oxide and Non Oxide Ceramics, edited by S. Somiya, Assoc. for Sci. Doc. Info., Tokyo Inst. of Technology, Tokyo, Japan, 1979.
15. H. Palmour III and K. Y. Kim, Progress Rept. NCSU-82-01, Contract DAAK11-82-C-0040, June 1982.
16. H. Palmour III, K. Y. Kim, K. L. More, and R. C. Motley, pp. 331-72 in First Quarterly Report, DARPA Dynamic Synthesis and Consolidation Program, edited by C. L. Cline, Lawrence Livermore National Laboratory Report UCID - 19663, August 1982.
17. V. D. Linse, J. A. Adair, and R. Willis, pp. 1-79, Ibid.
18. H. Palmour III, T. M. Hare, J. C. Russ, A. D. Batchelor, M. J. Paisley, and L. J. Freed, Final Tech. Rept., Subcontract 6548901, University of California, Lawrence Livermore National Laboratory, September 1982.
19. A. D. Batchelor, M. J. Paisley, T. M. Hare, and H. Palmour III, this volume.
20. V. D. Linse (Battelle Columbus Laboratories), personal communication, 1982.
21. G. L. Moss (U.S. Army, Aberdeen Proving Ground), personal communication, 1982.
22. D. P. Dandekar (U.S. Army Mechanics and Materials Research Center), personal communication, 1982.
23. R. A. Graham (Sandia National Laboratory), personal communication, 1982.
24. G. K. Williamson and W. H. Hall, Acta Met., 1, 22 (1953).
25. C. S. Yust and L. A. Harris, pp. 881-94 in Shock Waves and High Strain Rate Phenomena in Metals, edited by M. A. Mayers and L. E. Murr, Plenum Press, NY, 1981.
26. C. L. Hoenig and C. S. Yust, Am. Ceram. Soc. Bull., 60 [11], 1175-76; 1121-24 (1981).
27. H. Palmour III, R. A. Bradley, and D. R. Johnson, pp. 392-407 in Kinetics of Reactions in Ionic Systems, edited by T. J. Gray and V. D. Frechette, Mat. Sci. Res., Vol. 4, Plenum Press, NY, 1969.

28. H. Palmour III, D. R. Johnson, C. H. Kim, and C. E. Zimmer, Tech. Rept. 69-5, Contract N00014-68-A-0187, 1969.
29. E. Beauchamp (Sandia National Laboratory), personal communication, 1982.
30. D. E. Peterson and F. W. Clinard, Jr., in Proceedings of Annual Meeting, Mat. Res. Soc., Boston, Nov. 1-5, 1982 (in press).
31. H. Palmour III, T. M. Hare, and M. L. Huckabee, Final Tech. Rept., Contract N00019-73-C-0139, March 1974; H. Palmour III and T. M. Hare, Final Tech. Rept., Contract N00019-74-C-0265, July 1975.
32. H. Palmour III, V. D. Linse, and R. M. Spriggs, pp. 611-18 in _Sintering - Theory and Practice_, edited by D. Kolar, S. Pejovnik, and M. M. Ristic, Elsevier, Amsterdam, 1982.

PART X

VERY HIGH PRESSURE PROCESSING

HIGH PRESSURE PROCESSING OF HIGH TECHNOLOGY CERAMICS

E. Dow Whitney

Department of Materials Science and Engineering
University of Florida
Gainesville, Florida 32611

ABSTRACT

More than six decades have passed since Bridgman and co-workers established the exciting field of high pressure science and technology. Since that time research in this area has been primarily dominated by physicists, geologists and more recently by chemists. Except for the synthesis of diamond in 1954, ceramists in general have not taken advantage of the role very high pressure can play as an important thermodynamic parameter in ceramic processing. In this paper the science and technology of very high pressure as a long overdue but emerging process in the development of high technology ceramics is discussed and examples given.

INTRODUCTION

In modern ceramic processing technology pressure sintering or hot pressing is employed extensively as a means of producing bodies with controlled microstructures. Sintering pressures rarely exceed 140 MPa (20,000 psi). Today relatively few laboratories are capable of undertaking research in ceramic processing at high pressure although the challenge and potential of pressure-induced phenomena in the formulation of new phases of matter has been recognized for nearly three decades.[1]

In beginning any discussion it is always useful to define terms. In this paper high pressure refers to pressures on the order of at least 2.0 GPa (20 kbar) and beyond. Because of the rapid growth of high pressure technology during the past twenty years pressures of this magnitude, although requiring special materials and special techniques, can be readily handled (with proper attention to safety considerations) in most industrial and university laboratories.

Pressure units published in the literature have, up to the formal introduction of the new system of measurement, the International System of Units (SI) throughout the world, been chosen according to the whim of the particular investigator. Pressure units in use are kilogram per square centimeter, atmosphere, bar, and pounds per square inch. Fortunately a kilogram per square centimeter, an atmosphere, and a bar are all numerically almost the same. In this paper, all pressures will be reported in the SI unit of pressure, pascal (Pa) with proper SI prefixes as applicable.

THERMODYNAMICS AND KINETICS OF HIGH PRESSURE PROCESSES

As is well known, the question as to whether or not a particular reaction will proceed in a spontaneous manner is answered by the thermodynamic function

$$\Delta G = \Delta H - T\Delta S; \ \Delta G<0 \tag{1}$$

where ΔG refers to the difference in Gibbs free energy, ΔH to the difference in enthalpy and ΔS to the difference in entropy state of products and reactants at constant temperature and pressure. The above expression may also be expressed as

$$\Delta G = \Delta E - T\Delta S + P\Delta V \tag{2}$$

where ΔE and ΔV represent the corresponding differences in internal energy and volume, respectively.

As has already been mentioned, the majority of ceramic processes occur at either ambient or pressures below 140 MPa. In addition, many ceramic processes involve interactions between condensed states of matter such as phase transformations, reactions between solids, crystallization and grain growth, vitrification, sintering, etc. wherein the volume changes are very small, being of the order of $\sim 1 cm^3 \ mole^{-1}$. A reaction with a volume change (ΔV) of $1 cm^3 \ mole^{-1}$, occurring at 4.0 GPa will alter ΔG by only $\sim 4.18 \times 10^3$ joules (1 kcal). Thus, unless one is considering very high pressure effects, the mechanical work energy term ($P\Delta V$) in Eq. (2) can usually be neglected in solid state ceramic processes in comparison with the other thermodynamic terms.

It has been noted that in many respects the effects of pressure are opposite to those of temperature. Pressure decreases the entropy of a system whereas temperature increases the entropy or "opens up" a system. Pressure decreases the interatomic spacing in solids; therefore the repulsive forces between atoms increase with increasing pressure, which reduces diffusion and related phenomena. However, the statement sometimes made that increasing the pressure on a system produces the same effects as a decrease in temperature is not valid. There is a distinct trend toward the increased ionization of matter at high pressures. Generally we can expect solids to undergo a whole series of polymorphic changes into more compact structures, then into ionized forms and metallic states of varying degrees of ionization. In fact, the solution of Schrodinger's equation for the one-dimensional model of an "electron in a box" predicts that the quantized kinetic energy levels of the electron will increase as the length of the box decreases, analogous to compression of the system. The relative effects of pressure and temperature on a system are shown in Fig. 1.

Although valid to a point, this argument should not be taken too far. Like chemical potential and surface energy, pressure and temperature are intensive thermodynamic parameters. When multiplied by the appropriate extensive variables, an increase in both pressure and temperature impart energy to the system. The work done by compression, corresponding to a change in volume dV at a pressure P, is dW = PdV. Likewise, the entropy S of a system is the extensive parameter related to absolute temperature T, leading to an increase in thermal energy dQ = TdS.

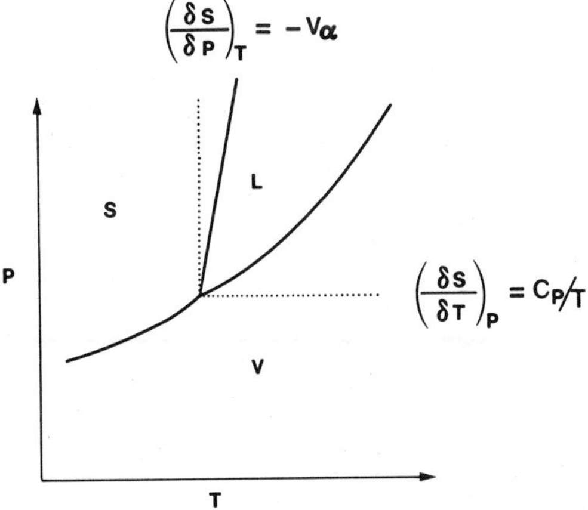

Fig. 1. Effects of pressure and temperature on a one component system.

In general, the mechanical energy change in condensed systems resulting from an increase in pressure from 10.0 to 20.0 GPa corresponds to the thermal energy change as a result of increasing the temperature from 1000°C to 2000°C. Pressure also affects the temperature range of phase equilibrium. In general, the melting point is raised by increasing pressure, as expected from the Clausius-Clapeyron equation. The temperatures of solid state phase transformations are also changed with pressure according to the Clausius-Clapeyron relationship.

Pressure will have an effect on equilibrium and rate processes in ceramic forming operations involving a change in volume between the final and initial states. The relation between the equilibrium constant (K) and pressure in a chemical process is:

$$\left(\frac{\partial \ln K}{\partial P}\right)_T = \frac{-\Delta V}{RT} \tag{3}$$

Only in those cases where the overall volume change for the process is negative ($\Delta V < 0$) will pressure increase the equilibrium concentration of reaction products. With respect to the effect of pressure on reaction rate, all theories of chemical reaction lead to equations describing the variation of velocity constant (k) with pressure of the form:

$$\left(\frac{\partial \ln K}{\partial P}\right)_T = \frac{-\Delta V^*}{RT} \tag{4}$$

where ΔV^* is a quantity having the dimensions of volume, sometimes called the "volume of activation."

Relatively few reactions involving solid phases have been studied under pressure. However, it is obvious from Eq. (4) that "slow" reactions (having low frequency factors) will be accelerated by pressure if ΔV^* is < 0 whereas if $\Delta V^* > 0$ the reaction will be further retarded by pressure. There is thus a distinction between reaction volume change (ΔV) affecting the equilibrium constant (K) and activation volume (ΔV^*) affecting the rate constant (k) for a reaction. A classic example is the graphite $\rightleftharpoons$ diamond conversion where ΔV^* appears to have an unusually high positive value whereas the overall volume change (ΔV) for the allotropic transformation is negative.

As an emergent process method for high technology ceramics, high pressure reaction kinetics, particularly with regard to the use of pressure for influencing the thermodynamic probability and relative rates of competing reactions in a complex reaction system is worthy of study. Application of pressure to ceramic processing may also be utilized as a means for reducing undesirable side reactions.

HIGH PRESSURE SINTERING OF CERAMIC MATERIALS

For a long time relatively little effort was given to the high pressure sintering of ceramic materials. Zirconia and HfO_2, with and without stabilizing additives, have been subjected to pressures of 0.38 to 3.0 GPa at temperatures up to 1700°C whereas various borides have been hot pressed to nearly theoretical density at 1.0 to 2.0 GPa over the temperature range 1800 to 2000°C using a belt apparatus.[2] The belt apparatus was also used in the hot pressing of SiC powder to > 99% theoretical density at 5.0 GPa and 2500°C.[3]

Recently more attention has been given to high pressure as an emerging technique in advanced ceramic processing as evidenced by the recent International Symposium on Factors in Densification and Sintering of Oxide and Non-oxide Ceramics, held in Japan in 1978, as well as the papers on very high pressure sintering being presented at this conference. Further research on the very high pressure sintering of covalent materials such as Si_3N_4, SiC and B_4C at static pressures up to 5.0 GPa at 1500 to 1600°C has recently been reported[4] as has the reactive high pressure sintering of alumina.[5]

Growing demand for sintered diamond cutting tools for the machining of nonmetallics and difficult to machine metals such as aluminum-silicon alloys, along with other diamond products such as wire drawing dies and rock cutting drill head bits, has promoted and sustained research activity in the high pressure sintering of diamond composites.[6,7] Likewise, the potential of sintered cubic boron nitride as a cutting tool material for nonferrous refractory metals such as the nickel-cobalt base alloys has prompted research in the high pressure sintering of this material.[8,9]

HIGH PRESSURE CRYSTALLIZATION

Pressure as a variable can be used advantageously to promote crystal growth. Most ceramic materials are polycrystalline which means that they have a larger volume than single crystals. There is also a certain amount of free energy associated with grain boundaries. Thus the application of high pressure favors the elimination of grain boundaries and the growth of single crystals from polycrystals.

In many instances the science of growing crystals is still an art, even when the crystals are being grown at ambient pressure. Excluding diamond and cubic boron nitride, effort in the crystal growth of ceramic materials at high pressures and temperatures has been rather limited. Generally it is observed that, at least in the case of metals, pressure retards the thermally activated processes of recovery, recrystallization and grain growth. With respect to crystal growth of ceramics at high pressures, Chenavas and co-workers[10] were able to synthesize a number of oxyhydroxides of metals of the first transition metal series as well as crystallized materials at approximately 600°C and 8.0 GPa. When prepared under normal pressures these materials are always poorly crystallized.

In addition, high pressure crystallization, and particularly hydrothermal synthesis, may offer a method of clean preparation of many of the fine refractory ceramics now made by high temperature fusion techniques. Most crystal growing techniques use temperature as the primary variable. Usually the system must assume a minimum "critical volume" for successful crystal growth. From the above discussion it is obvious that the combined parameters of high pressure and temperature, along with volume considerations, provide a challenge to the investigator interested in new ceramic processing techniques.

HIGH PRESSURE SYNTHESIS REACTIONS

Knowledge of the pressure-temperature effects on reactions in naturally occurring systems has long been of interest to petrologists and geochemists as a means of establishing geological thermometer.[12] By subjecting rock samples to static pressures up to 20 GPa at temperatures as high as 2000°C geologists may simulate conditions found at depths well into the earth's upper mantle. Pressure indicating reactions include polymorphic transitions such as calcite-aragonite ($CaCO_3$), kyanite-sillimanite (Al_2SiO_5) and quartz-coesite (SiO_2) as well as reactions such as:

$$\underset{\text{(albite)}}{NaAlSi_3O_8} \dashrightarrow \underset{\text{(jadeite)}}{NaAlSi_2O_6} + \underset{\text{(quartz)}}{SiO_2}$$

whereas dehydration and decarbonation reactions may be used to delineate temperatures of rock formation.[13] Of course, for each indicating reaction the dT/dP slope must be experimentally determined.

In a system more closely related to ceramic processing, Hall[11] has proposed the following scheme for the synthesis of aluminum oxide as an example of how high pressure reactions, since they are driven in the direction of decreasing volume, may be utilized to synthesize a desired product:

$$2AlCl_3(s) + 2Fe(s) + 3H_2O(l) \rightarrow Al_2O_3(s) + 2FeCl_3(s) + 3H_2(g)$$

The overall volume reduction of solid phases in the above reaction ($\Delta V/V$) is approximately 57 percent. Fortunately, materials of particular interest in ceramic technology, i.e., dense phases with relatively high thermal stability, are relatively easy to "quench" and may be recovered at normal conditions at the conclusion of the high pressure, high temperature process. Certainly the future synthesis of new materials under high pressure, high temperature conditions can be expected to result in new high technology ceramics.

An excellent review and discussion of the use of high pressure techniques in the synthesis of new materials has been published.[14] Not unexpectedly, oxygen containing compounds constitute the majority of the inorganic materials studied to date. Very little work has been done on the synthesis of new nitride phases at high pressure with the exception, of course, of cubic boron nitride. However, as the authors of the above mentioned review article point out, "the search for new nitrides has just begun." Some progress has already been made. The ternary compound, $MgSiN_2$ has been synthesized at 3.0 GPa and 1200°C from a mixture of Mg_3N_2 and Si_3N_4[15] as well as the high pressure phase of the compound Li_3BN_2.[16]

High pressure may provide a new route for the synthesis of new and novel ceramic precursors as well. Pronounced changes of ionic equilibria occur with changing pressure. For example, the ionic product Kw of water increases rapidly with increasing pressure.[17] This increase in ionic strength of water at high pressure, in addition to high temperature, accounts for the dramatic increase in solubility of SiO_2 in water at high pressures and temperatures.[18] By anology with the SiO_2-H_2O system, one might envision possible solvolysis reactions in the system Si_3N_4-NH_3 at high pressures. It has been known since the last century that, "of all known liquids, ammonia most closely approaches water in all those properties which give to water its conspicuous position among solvents."[19] One might therefore expect that, like water, the ionization constant of liquid ammonia would increase with increasing pressure to a point where ammonolysis reactions between Si_3N_4 and NH_3 might occur leading to unique starting materials for new high technology ceramics.

HIGH PRESSURE PHASE TRANSFORMATIONS AND THE PREPARATION OF NEW, HARD MATERIALS

It is generally recognized that hard materials have a high melting point, at least 1600°C. This requirement is an inflexible barrier which can be satisfied only by about 1.3% of all known inorganic compounds, with all currently known organic compounds being excluded. The hardest materials known today contain no more than two different elements. The softest material on the Mohs scale is hydrated magnesium silicate (talc) containing four different elements. Not a single "hard" compound is known with a number of elements (n) greater than four. For hard materials $1 \leqslant n \leqslant 4$ with $1 \leqslant n \leqslant 2$ compounds having the greatest interest as commercially important hard materials such as aluminum oxide, silicon carbide, boron carbide, etc. For the hardest known material (diamond), of course, n=1. The relatively small number of available materials of high hardness is further illustrated by calculations made by Berezhnoi[20] with regard to the total number of compounds theoretically possible containing four components with respect to the number of different elements. He found that 3,160 binary, 82,160 ternary and 1,581,580 quaternary systems were theoretically possible, very few of which are "hard."

From a cursory consideration of hard materials it would seem that they have very little in common either on the basis of physical or chemical properties (other than the property of hardness itself). Progressing down the hardness scale one finds a crystalline form of carbon (thermodynamically stable only at high pressure at room temperature), the zinc blende polymorph of boron nitride (again thermodynamically stable only at high pressure), carbides of low atomic weight IIIA and IVA elements, and finally aluminum oxide. Early attempts to relate hardness to other physical properties have been only partially successful.[21] Hall[11] has discussed hardness as a property of a material resulting from high bond density and high three-dimensional bond symmetry, a measure of high bond density being the cohesive energy density. A measure of cohesive energy density is the heat of vaporization of the material per unit volume. The allotropic transformation C (graphite) = C (diamond) is an example of two materials, both of which having very high cohesive energy densities, yet only diamond fulfills the additional requirement of high three-dimensional bond symmetry.

The cohesive energy E_c, or lattice energy, is the energy of a crystal relative to its infinitely separated ions; i.e., the energy required for the reaction:

$$MX(s) \rightarrow M^{+}(g) + X^{-}(g) \qquad (5)$$

where M represents the metal atom and X the nonmetallic atom of the material. The lattice, or cohesive energy, cannot be measured directly but is determined indirectly from the Born-Haber cycle. In the case of silicon carbide, for example, the cohesive energy, is the energy required to bring about the reaction:

$$SiC(s) \rightarrow Si^{4+}(g) + C^{4-}(g). \qquad (6)$$

Since cohesive energy is a direct measure of the strength of the atomic bonds in a crystal lattice it is not surprising that E_c values of hard materials are high.

The above qualitative statements by Hall have been placed on a quantitative basis by Plendl and Gielisse[22] who treated hardness of nonmetallic solids on an atomic basis. These authors introduced the concept of volumetric lattice energy, defined as the lattice energy per unit volume E_c/V, where V is the molar volume of the material. Thus E_c/V is a measure of bond or cohesive energy density. The above investigators have shown that the relationship between cohesive energy density and hardness may be expressed by the equation

$$E_c/V = 36(W-4.8)\ (\text{kcal/cm}^3) \qquad (7)$$

where W is hardness as measured on the Wooddell scale.[23] This linear relationship is shown in Fig. 2.

Cohesive energy density may be thought of as the internal pressure P_i of a material; i.e., if cohesive forces were not present, an externally applied pressure equal in magnitude to P_i would need to be

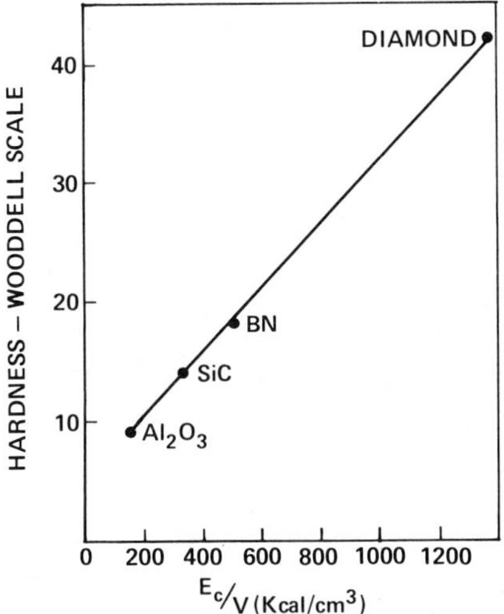

Fig. 2. Relationship between hardness (W scale) and cohesive energy density.

applied to maintain the material at its normal density. Cohesive forces in essence precompress the material. The higher the cohesive energy density, the higher the internal compressive force and hence the hardness or resistance of the lattice to mechanical breakdown. A very interesting discussion of the relationship between internal pressure and high pressure phase transformations in condensed systems has been given by Wentorf.[24]

Although studies relating the concept of hardness of nonmetallic materials to bonding energy have been successful, they neglect the role played by atomic symmetry in determining the hardness of materials. Atomic structures of highest symmetry are formed by elements with lowest average principal quantum number, providing the electronegativity difference between the combining elements is not too great.[25] Under these circumstances the bonds formed are covalent with high tetrahedral spatial symmetry. Diamond and cubic boron nitride are classic examples of such atomic structures. One finds the hardest compounds to be binary, consisting of no more than two different elements. In addition, the elements constituting these binary compounds are not too different in their electronegativity values as required for the formation of covalent bonds.

A possible route to the synthesis of new, hard materials has been proposed by Hall[26,27] in the form of "periodic compounds," identified by the selection rule

$$\overline{G} = \frac{\Sigma d_i e_i}{\Sigma d_i} \tag{8}$$

where $\overline{G}$ assumes the integral values 1, 2, 3, . . . corresponding to the periodic table groups I, II, III, ... The number of bonding electrons contributed by an atom of kind i is designated e_i, and d_i gives the number, per formula weight, of atoms of this kind. Hall further argues that many periodic compounds may be amenable to synthesis employing high pressure, high temperature techniques. One example of the high pressure, high temperature synthesis of a periodic compound is the reported formation of B_2O, an isoelectronic analog of carbon, via the reaction $4B + B_2O_3 \rightarrow 3B_2O$ at pressures from 5.0 to 7.5 GPa and temperatures from 1200° to 1800°C.[28] X-ray diffraction analysis revealed that B_2O of hexagonal crystal symmetry had been produced. Pressures of about 12.0 GPa were utilized by Hall in attempts to synthesize a diamond-like form of B_2O with questionable success.

THERMODYNAMICS OF PHASE TRANSFORMATION TOUGHENING OF CERAMICS

The past dozen years or so has seen extensive studies, both theoretical and experimental, on stress-induced phase transformations as a mechanism for arresting crack propagation in ceramics. A very useful system for demonstrating phase transformation toughening has been zirconia-based ceramics which can be made with varying amounts of the metastable tetragonal ZrO_2 phase. The metastable ZrO_2 phase can undergo a stress-induced tetragonal ZrO_2 to monoclinic ZrO_2 phase transformation thereby arresting crack propagation by absorbing energy at the crack front. The overall effect is to markedly increase strength and fracture toughness.[29]

It is beyond the scope of this paper to discuss phase transformation toughening in detail. Several review articles on this technically important area have recently been published.[30,31] However, some comments can be made as to how high pressure studies can lead to a fundamental understanding of this important strengthening mechanism. Lange[32] has recently proposed a theory in which the contribution to fracture toughness by a stress-induced transformation is shown to be proportional to the chemical free energy change associated with the transformation. For the ZrO_2 (tetragonal) to ZrO_2 (monoclinic) transformation the chemical free energy is known to decrease with other oxides such as Y_2O_3, CeO_2, etc. Therefore fracture toughness will be optimized at the lower temperatures and for the least alloy content.

It is also known that the chemical free energy change associated with the ZrO_2 (tetragonal) to ZrO_2 (monoclinic) transformation decreases with increasing pressure as well.[33] Thus for a complete understanding of the role played by the chemical free energy change in influencing fracture toughness the transformation envelope of the tetragonal ZrO_2 = monoclinic ZrO_2 phase change in T-P-X space should be determined. To the author's knowledge, no studies have been made to establish the pressure dependency of phase stability in ZrO_2 - M_xO_y alloys at relatively low temperatures. Surely this presents a challenge to the ceramist interested in the fundamental aspects of ceramic processing.

CONCLUSIONS

The purpose of this tutorial paper has not been simply to give a review of the current state high pressure technology. Rather, an attempt has been made to speculate on several areas in ceramic processing where high pressure may serve as a useful tool. It is hoped that some of the possible areas of research, only touched upon in this paper, may be the forerunners of emergent process methods for new high technology ceramics. In that regard, then, it would be premature for the author to attempt to state conclusions as to the future role of high pressure in ceramic processing. These will be provided by members of the ceramic scientific community.

REFERENCE

1. S. D. Hamann, Physico-Chemical Effects of Pressure, Butterworths, London, 1957.
2. F. W. Vahldiek and C. T. Lynch, pp. 637-64 in Sintering and Related Phenomena, edited by G. C. Kuczynski, N. A. Hooton, and C. F. Gibbon, Gordon and Breach, NY, 1967.
3. J. S. Nadeau, Bull. Amer. Ceram. Soc., 52 [2], 170-74 (1973).

4. A. Sawaoka, K. Kondo, N. Hashimoto, and S. Saito, pp. 339-44 in Factors in Densification and Sintering of Oxide and Non-oxide Ceramics, edited by S. Somiya and S. Saito, Association for Science Documents Information c/o Tokyo Institute of Technology, Tokyo, Japan, 1979.
5. Y. Ishitobi, M. Shimada, and M. Koizumi, pp. 142-51 in ref. 4.
6. H. Katzman and W. F. Libby, Science, 172, 1132-34 (1971).
7. M. Akaishi et al., pp. 320-27 in ref. 4.
8. S. Saito, pp. 1-18 in ref. 4.
9. O. Fukunaga et al., pp. 328-38 in ref. 4.
10. J. Chenavas et al., J. Solid State Chem., 6, 1-15 (1973).
11. H. T. Hall, J. Wash. Acad. Sci., 47 [9[, 300-04 (1957).
12. R. Kern and A. Weisbrod, Thermodynamics for Geologists, Freeman, Cooper and Company, San Francisco, CA, 1967.
13. C. M. Scarfe, Miner. Sci. Eng., 5 [4], 287-94 (1973).
14. J. C. Joubert and J. Chenavas, pp. 463-511 in Treatise on Solid State Chemistry, Vol. 5, Changes of State, edited by N. B. Hannay, Plenum Press, NY, 1975.
15. E. D. Whitney and R. F. Giese, Inorg. Chem., 10 [5], 1090-91 (1971).
16. R. C. deVries and J. F. Flei, Mat. Res. Bull., 4, 433 (1969).
17. S. D. Hamann, J. Phys. Chem., 67 [10], 2233-35 (1963).
18. O. F. Tuttle and J. L. England, Bull. Geol. Soc. Am., 66, 149-52 (1955).
19. E. C. Franklin and C. A. Kraus, Amer. Chem. J., 21, 8 (1899).
20. A. S. Berezhnoi, Ogneupory, 3-4, 229-32 (1970).
21. J. R. Partington, pp. 238-42 in An Advanced Treatise on Physical Chemistry, Vol. III, The Properties of Solids, Longmans, Green and Co., London, England, 1952.
22. J. N. Plendl and P. J. Gielisse, Phys. Rev., 125 [3], 828-32 (1962); Zeit. fur Krist., 118 [5/6], 405-21 (1963).
23. C. E. Wooddell, Trans. Electrochem. Soc., 68, 111-30 (1935).
24. R. H. Wentorf, Jr., J. Phys. Chem., 63, 1934-40 (1959).
25. E. Mooser and W. B. Pearson, Acta Cryst., 12, 1015-22 (1959).
26. H. T. Hall, Science, 148 [3675], 1331-33 (1965).
27. H. T. Hall, pp. 1-38 in Progress in Inorganic Chemistry, Vol. 7, edited by F. A. Cotton, Interscience Publishers, NY, 1966.
28. H. T. Hall and L. A. Compton, Inorg. Chem., 4, 1213-16 (1965).
29. T. K. Gupta, pp. 877-89 in Fracture Mechanics of Ceramics, Vol. 4, edited by R. C. Bradt, D. P. H. Hasselman, and F. F. Lange, Plenum press, NY, 1978.
30. P. L. Pratt, J. Met. Sci., 14 [8-9], 363-73 (1980).
31. R. Stevens, Trans. J. Brit. Ceram. Soc., 80 [3], 81-5 (1981).
32. F. F. Lange, J. Mat. Sci., 17, 235-39 (1982).
33. E. D. Whitney, J. Amer. Ceram. Soc., 45 [12], 612-13 (1962).

DISCUSSION

H. Conrad (North Carolina State University): (1) The curve you presented on hardness vs. cohesion energy per unit volume extrapolates to a finite hardness for zero cohesive energy. How do you explain this? (2) What does Wooddell hardness mean? How does it differ from conventional hardness measurements which generally involve plastic flow and are given by the applied load divided by the projected or surface area of the indentation?

Author: (1) In Fig. 2 of this paper relative hardness values are only given for corundum and those materials harder than crystalline Al_2O_3 since these are the hardest materials known and where the Wooddell (W) scale establishes the great differences in hardness between the few known materials within this range. Actually Plendl and Gielisse (see ref. in paper) also considered the hardness of materials softer than Al_2O_3 (ex. SiO_2, MgO) based upon the Mohs (M) hardness scale. Wooddell fixed the hardness values of quartz and corundum at 7 and 9, in agreement with the M scale, and extended the relative hardness scale in a linear-proportional sense. In their paper, Plendl and Gielisse felt justified in combining both the W scale and the M scale because of this overlapping interval. It was found that below Al_2O_3, the slope of relative hardness vs. Ec/V was steeper than above Al_2O_3, approaching zero hardness at zero lattice energy density.

(2) C. E. Wooddell (see ref. in paper) devised a method for measuring the relative hardness of electric furnace products and natural abrasives based upon relative resistance to abrasion during lapping with diamond powder. The Wooddell scale differs from conventional (static) hardness measurements in that it is a dynamic hardness measurement.

The two methods of measuring hardness are not unrelated, however. In the static (indentation) method hardness is expressed as weight or force [dimensions m (mass) x ℓ (length) x t^{-2} (time)] per unit area (ℓ^2), giving hardness the dimensions $m\ell^{-1}t^{-2}$, the same as that of pressure. In the indentation method "hardness" is primarily a surface-related phenomenon and the hardness numbers obtained are a relative measure of the resistance of the material to surface deformation pressure.

In the dynamic method, hardness is considered as resistance to mechanical breakdown. Hardness considered in this manner is a bulk property of the material. On a dimensional basis the connection between the above two hardness considerations is made by considering hardness in the latter case to be work energy [dimensions $m\ell^2t^{-2}$] per unit volume (ℓ^3). Thus

$$\text{Hardness (bulk)} = \text{Hardness (penetration)} \times \left(\frac{\text{length}}{\text{length}}\right)$$

$$\text{Hardness (bulk)} = \frac{m\ell t^{-2}}{\ell^2} (\ell/\ell) = \frac{m\ell^2 t^{-2}}{\ell^3}$$

Bulk hardness, again having the dimensions $m\ell^{-1}t^{-2}$, may be thought of as the ratio between input work energy required to mechanically break down the material and the volume of material removed as a result of that input energy. A very hard material is one that requires a large input energy to break down and consequently remove unit volume of material. What opposes the destruction of the solid structure is the volumetric lattice energy or cohesive energy density. Hardness now receives the dimensions of work per unit volume with the dimensions $kcal/cm^3$ in accordance with the findings of Plendl and Gielisse.

R. Arons (Celanese Research Co.): Regarding the Wooddell scale, wear during lapping is not necessarily a fracture process but is often a chemical effect. The wear of a diamond phonographic stylus by vinyl phonograph records is probably such an example.

Author: It is true that abrasive action in general, long thought to be a purely mechanical process, can involve complex reactions between the abrasive, work material and atmosphere, as well as with bond ingredients (in the case of bonded abrasives such as grinding wheels). As discussed by Coes, the role played by these reactions in the overall material removal process can become very important when the work material is a metal. The Wooddell scale was devised in order to measure the relative hardness of electric furnace products and natural abrasives; refractory-hard materials where one would not expect significant chemical reaction between the loose lapping abrasive (diamond) and the materials whose relative hardnesses were being measured.

L. Coes, Jr., *Abrasives*, Springer-Verlag, NY, 1971, p. 154.

R. Davis (North Carolina State University): Are the numbers given in your paper for energy densities calculated values or measured values? If they are measured values, is the theory sufficiently worked out for making calculations on yet unfabricated compounds so that a Wooddell hardness could be estimated for these materials?

Author: The cohesive energy density or volumetric lattice energy values given in this paper were calculated from pertinent data in the literature. Plendl determined the lattice energy for a number of solids harder than corundum from their respective infrared spectra and lattice anharmonicity.

In principal, the Wooddell hardness could be estimated for any nonconducting solid whose lattice energy and density could either be measured or calculated. In fact, in the conclusion to their paper Plendl and Gielisse (see ref. in paper) state that an investigation of solids harder than corundum "on the basis of interatomic forces would supply data of value for the ultimate preparation of materials capable of withstanding extreme environmental conditions."

J. N. Plendl, Phys. Rev., 123, 1172 (1961).

M. Shimada (Osaka University): Do you find any other reports on synthesis of B_2O under high pressure conditions, except Dr. T. Hall's?

Author: No, I have not.

R. R. Wills (Battelle Columbus Laboratories): Do you think we can learn anything from the high pressure research being done on metallic hydrogen, and translate the findings to the ceramics field?

Author: The scientific predictions that (1) metallic hydrogen might be metastable and (2) that it would be a room temperature superconductor have evoked considerable interest in producing this substance. However, the transition pressure is predicted to be between 150 and 300 GPa (1.5 and 3 M bar). It is, offhand, hard to imagine any future ceramic processes requiring pressures of this magnitude.

However, I feel that engineering research on the design and fabrication of presses and special tooling in order to reach these extreme pressures will be beneficial in ceramics processing primarily from the standpoint of economics. Consider a static press designed to reach 200 to 300 GPa such as the one being developed at the Institute of High Pressure physics in Moscow. The force required to maintain a fixed pressure varies with the square of the radius of the cavity in which that pressure is being generated. A press designed to produce very high pressure in a given volume utilizing a certain force would be capable of generating lower pressures in a larger working volume. Since the economics of any processing method is strongly dependent on the volume of material produced in unit time this would minimize the cost of high pressure processing of a high technology ceramic product.

G. A. Rossi (Norton Co.): Do you know of any work on the dependence on pressure of the M → T transfer in undoped ZrO_2? Will the T-ZrO_2 be retained at RT after the pressure release?

Author: A number of investigators have been interested in the pressure dependency of the monoclinic to tetragonal phase transformation in ZrO_2.

Theoretical calculations made independently by both Berezhnoi and Whitney predicted that the pressure dependence of the monoclinic ZrO_2 = tetragonal ZrO_2 transition temperature (dT/dP) would be negative. Confirmation of the calculated value of dT/dP = -0.302°/MPa (-3.02×10^{-2}°/bar) for the monoclinic ZrO_2 = tetragonal ZrO_2 phase change was later obtained by Whitney for pressures up to 1.5 GPa. Kulcinski studied the effect of pressures up to 4.6 GPa on the monoclinic ZrO_2 to tetragonal ZrO_2 transformation at room temperature employing a diamond high pressure x-ray cell. His results also confirmed the calculated P-T field for monoclinic and tetragonal ZrO_2 in the high pressure, low temperature range. Liu has also proposed that the high pressure phase of ZrO_2 is tetragonal as a result of studies employing a diamond-anvil press coupled with laser heating. The pressure range covered in this study was between 8.0 and 30.0 GPa at about 1000°C.

Bendeliani and co-workers in Russia have reported the formation of the orthorhombic polymorph of ZrO_2 at room temperature and at pressures of approximately 10.0 GPa. Apparent confirmation of orthorhombic ZrO_2 was later given by Bocquillon and Susse who proposed, but did not experimentally verify, the existence of a triple point between the monoclinic, tetragonal and orthorhombic phases of ZrO_2 at approximately 600°C and 2.3 GPa. These investigators used water as a mineralizer in order to accelerate ZrO_2 phase transformations and used the method of "quenching" at high pressure to recover orthorhombic ZrO_2. Lityagina et al. have also proposed that the crystal symmetry of the high pressure phase of ZrO_2 is orthorhombic.

To date, no conclusive evidence has been given of the "quenching in" of either tetragonal or orthorhombic zirconia. Bocquillon and Susse claim identification of their recovered orthorhombic phase of ZrO_2 based upon only one x-ray diffraction line, the 111 reflection (d=2.97Å). However, a recent paper has been published by Arashi and Ishigame on the Raman spectra of ZrO_2 at high pressures as measured at room temperature using a diamond-anvil pressure cell. With methanol as the pressure transmitting medium a phase transformation was observed at 3.5 GPa, while in the case of NaCl as the pressure transmitting medium a transformation occurred at 5.4 GPa. Interpretation of the Raman spectra lead to the conclusion that the high pressure phase was orthorhombic.

It is clear that the P-T diagram for pure ZrO_2 requires further investigation.

H. Arashi and M. Ishigame, Phys. Stat. Sol. A, 71 [2], 313-21 (1982).

N. A. Bendeliani, S. V. Popova, and L. F. Vereshchagin, Geokhimiya, 6, 677-83 (1967); Geochem. Int., 4 [3], 557 (1967).

A. S. Berezhnoi, Dopavidi Akad. Nauk Ukr. SSSR, 1, 65-8 (1962).

G. Bocquillon and C. Susse, Rev. Int. High Temp. and Refract., 6 [4], 263-66 (1969).

G. L. Kulcinski, J. Amer. Ceram. Soc., 51 [10], 582-84 (1968).

G. L. Kulcinski and C. W. Maynard, J. Appl. Phys., 37 [9], 3519-27 (1966).

L. M. Lityagina et al., Soviet Phys. Solid State, 20 [11], 3475-77 (1978).

Lin-Gun Liu, J. Phys. Chem. Solids, 41 [4], 331-34 (1980).

E. D. Whitney, J. Amer. Ceram. Soc., 45 [12], 612-13 (1962).

E. D. Whitney, J. Electrochem. Soc., 112 [1], 91-4 (1965).

DIAMOND ANVIL CELL TECHNOLOGY FOR P,T STUDIES OF CERAMICS: ZrO_2 (8 mol% Y_2O_3)

R. G. Munro, S. Block, G. J. Piermarini, and F. A. Mauer

Center for Materials Science
National Bureau of Standards
Washington, D.C. 20234

INTRODUCTION

We are undertaking a systematic study of the structural and bulk properties of zirconia and other ceramic materials as functions of pressure and temperature. This paper describes the experimental approach that is being taken and discusses some of the results already obtained for ZrO_2 with 8 mol% Y_2O_3.

All of the present results are derived from x-ray diffraction experiments which are conducted while the sample is contained under pressure in a diamond anvil pressure cell (DAPC). From these measurements, it is possible to identify the crystal structure of the sample and to determine its lattice parameters. For each fixed temperature, the measurements are repeated at various pressures. With this information, the equation of state for the material can be investigated.

GENERATING HIGH PRESSURES

The DAPC is a device used for generating very high static pressures.[1] The basis of its operation is the application of a moderate force to a very small area. As shown in Fig. 1, the central components of the DAPC are a pair of diamond anvils plus a metal gasket which is contained between them. A small hole, typically 0.2 mm in diameter, is drilled through the metal, and this is used as the sample chamber. The chamber contains: (1) the sample; (2) a small piece of ruby for pressure measurement;[2] and (3) a liquid or a relatively soft material which is used as the pressure transmitting medium.

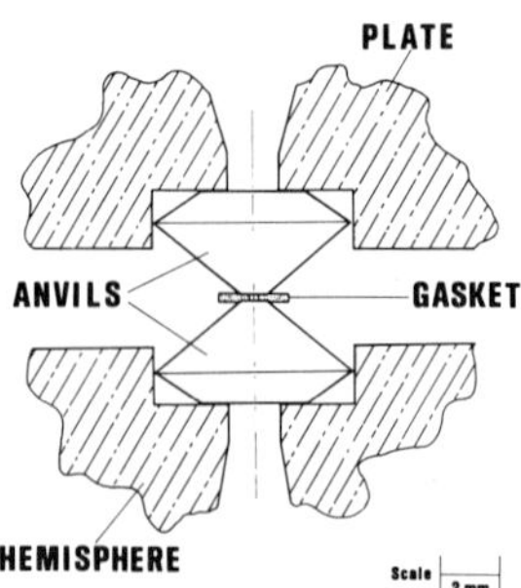

Fig. 1. Central to the DAPC are (a) a pair of diamonds which act as anvils and (b) a metallic gasket which defines the sample chamber. An anvil is formed by grinding the culet of a diamond until a flat surface parallel to the natural "table" face is obtained. The ground flat surface is typically about 0.2 mm^2 in area. The gasket is often made of inconel, stainless steel, or waspaloy. the gasket thickness is usually about 0.2 mm initially, but this is reduced under pressure to about 0.1 mm. The sample chamber is cylindrical with a diameter of 0.2 mm and with a length determined by the gasket thickness.

The diamond anvils play a dominant role in the operation of the DAPC. They seal the sample chamber; they transmit the force used in generating the high pressures; and they provide access to the sample chamber for both optical and x-ray measurements. The diamonds are also very important for maintaining a uniform temperature across the sample.

The diamond anvils have metal supports housed in a metallic body, as shown in Fig. 2. The anvils receive a force that is generated by a stack of Belleville spring washers. That force is multiplied by a factor of two by means of the lever-arm assembly which couples the springs to the diamonds. The whole apparatus typically measures 2.5 x 5 x 13 cm and can be held in one hand.

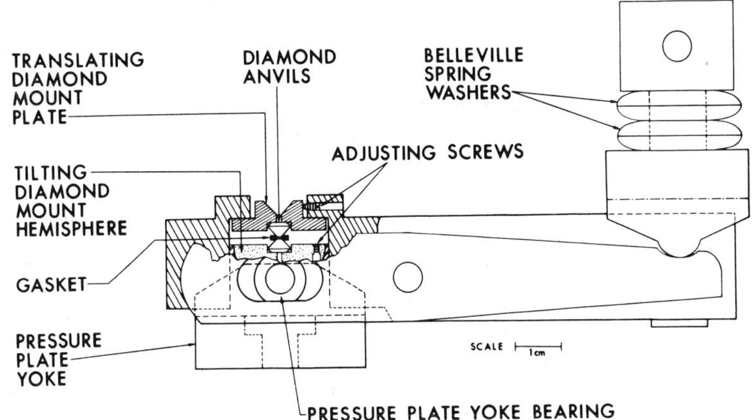

Fig. 2. A cross section of the DAPC shows the diamond anvils in relation to the complete cell body. The anvils are mounted in metal supports, and they are aligned by means of the adjusting screws. Force, generated by Belleville spring washers, is transmitted to the anvils by way of the lever arm assembly which also multiplies the force by a factor of two.

The size of the DAPC is quite small in contrast to its capacity to generate high pressures. In the gasketed configuration, pressures in excess of 60 GPa have been obtained. However, it must be noted that at those very high pressures, nonhydrostatic conditions prevail. Presently, the highest limit that has been established for hydrostaticity at room temperature is about 14.5 GPa (for a 16:3:1 solution of methanol:ethanol:water),[3] but the prospects for doubling that range in a practical sense seem good.

MEASURING HIGH PRESSURES

At very high pressures, the calculation of force per unit area does not yield reliable estimates of the pressure in the sample chamber of the DAPC. The reason for the uncertainty of this simple calculation is the uncertainty in the magnitude of the area at the chamber-diamond interface and the unknown distribution of the load over the metal and sample. The latter problem occurs especially at very high pressures when the pressure medium is nonhydrostatic. Furthermore, the gasket material flows under the stress from the diamonds, and as a result, both the shape and the size of the sample chamber change with pressure variations.

The alternative to computing force per unit area is to use an internal pressure sensor. The R_1-R_2 fluorescence lines of ruby have proved to be very useful for this purpose. The peak positions shift as the pressure acting on the ruby changes. The shift is linear to greater than 30 GPa, and it is large enough that an accuracy of about 0.05 GPa can be obtained when the shift is used to determine the pressure. The intensity of the ruby fluorescence is large enough that only a very small piece of ruby is needed to provide an adequate signal for a pressure measurement. Consequently, the ruby pressure sensor is usually small enough that no interference with the observation of the actual sample occurs. Obviously, this is important for x-ray measurements. Given also the stability of ruby and its lack of chemical reactivity with the sample or the pressure transmitting medium, it is understandable that ruby fluorescence has become widely used for pressure measurements and is likely to be adopted as a pressure standard.

HEATING THE DAPC

To operate the DAPC at higher than ambient temperatures, a means of supplying heat to the sample must be provided, and several techniques have been tried in our laboratory and elsewhere. The method which we are using with the present work involves a small furnace that surrounds the piston housing of the diamonds.[4] This is a cylindrical device that preserves the optical access to the sample chamber. The heating element is a miniature coil of wire having a relatively high electrical resistance (about 70 ohms). There is also a temperature controller that monitors the temperature at the sample chamber by means of a thermocouple. Without further provisions, this arrangement can be used to operate the DAPC at temperatures up to about 1200 K. The principal limitations on the temperature are due to the oxidation of the diamonds, the gasket's mechanical stability, and the sample-gasket chemical interactions.

The cylindrical furnace only provides heat radially inward to the sample. It is necessary that no heating element be allowed to obstruct the diamond windows because that would interfere with the optical access to the sample. Consequently, the possibility of thermal gradients is a concern when operating the DAPC at high temperatures.

The question of thermal gradients has been investigated theoretically.[5] The isotherms of a realistic model of the DAPC have been determined for a worst-case situation in which the outer diamond faces are in thermal contact with a heat reservoir held at ambient temperature. When the furnace temperature is specified as 500 degrees above ambient, the temperature difference across the sample chamber is less than 0.01 degree. This negligible gradient is a consequence of the very small size of the chamber and the very large thermal conductivity of the diamonds.

The important implication of this result is that the sample chamber, when heated by this method, may be considered as being in thermal equilibrium whenever measurements are made while the DAPC is maintained in a steady state.

A further concern at high temperatures is the effect of temperature on the fluorescence spectra of the ruby pressure sensor. A number of experimental studies of ruby have been made at both low[6] and high[7] extremes of temperature. It is a very fortunate result that the pressure and temperature shifts of the R-lines appear to be approximately additive over a very wide range of temperatures. Thus for fixed temperatures, the pressure can be determined using the room temperature calibration of the shift of the R-lines. The temperature shift of the R_1-line has also been calibrated, and this can be used as a secondary measure of the temperature.

A greater difficulty is the effect of temperature on the linewidth of the R-lines. The R_1 and R_2 lines form a closely spaced doublet. Thermal broadening causes the two lines to coalesce making the determination of the peak positions more difficult. At about 600 Kelvin, only a single peak can be detected. It is usually assumed that the two lines retain their low-T pressure shifts with the result that the pressure can still be determined by the shift of the peak position of the unresolved doublet. This practice is generally thought to be a good first approximation, but there are still unanswered questions about the accuracy of such measurements. Further investigation of this problem is underway.

X-RAY DIFFRACTION IN THE DAPC

The scattering and detection of x-rays in the DAPC system is accomplished using the energy dispersive technique. As shown schematically in Fig. 3, this method requires a radiation source that provides a continuum spectrum of x-ray energies. The Bremsstrahlung radiation from a standard x-ray tube is often adequate for this purpose. If greater intensities are required, rotating anode tubes or synchrotron radiation sources can be used. Whatever the x-ray source, the desired energy range for the photons is about 10 keV to 50 keV. This range is determined primarily by two factors. The lower limit is determined by the absorption characteristics of the diamond windows which strongly absorb x-rays with energies below 10 keV. The upper limit is a practical one that results mainly from the scattering efficiency of the x-rays which is proportional to $1/E^2$ where E is the photon energy.

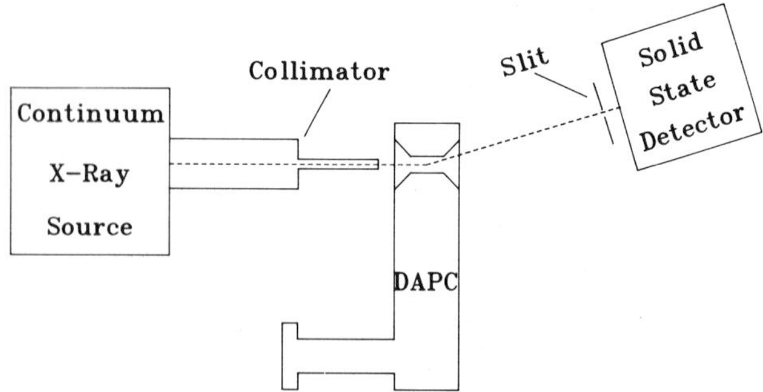

Energy Dispersive XRD Configuration

Fig. 3. The energy dispersive technique for x-ray diffraction is well suited to the DAPC. The scattering angle in this method is fixed throughout the experiment. Hence, a source of x-rays with a continuum spectrum is required for the incident beam. A needle-like collimator which has an inside diameter of about 0.15 mm is used to produce a narrow beam. The collimator is essential for preventing the x-ray beam from impinging on either the gasket or the ruby pressure sensor. A slit system defines the scattering angle at which an energy sensitive solid state detector monitors the scattered x-rays.

The detection of the scattered x-rays occurs at a fixed scattering angle. This is the feature that makes the energy dispersive method well suited for diffraction studies with the DAPC. For most applications, the strength requirements for the body of the pressure cell necessitate the use of high temperature--high strength metals that are opaque to x-rays, and consequently, the observable scattering angles are limited by the aperatures at the diamond windows. The largest 2θ scattering angle for high pressure studies is usually not more than 20 degrees, although 2θ angles as large as 24 degrees have been used at lower pressures. While this is intolerably restrictive for angle dispersive methods, it is more than adequate for many fixed-angle energy dispersive experiments. The latter can be done with 2θ in the range 10 to 20 degrees.

The actual counting of the photons requires the use of an energy sensitive detector. For this purpose, we are using an intrinsic germanium detector which has an energy resolution, in the context of resolving a peak in the scattered intensity, of about 0.2 keV. Assuming that a minimum of 5 points are needed to define a peak, this means that data need to be accumulated only at energy intervals of approximately 0.04 keV. Consequently, the detected counts are usually stored in a multichannel analyzer calibrated with this energy interval.

ZrO_2 (8 mol% Y_2O_3)

Pure ZrO_2 is monoclinic at room temperature and pressure and undergoes phase transitions with increasing temperature, first to a tetragonal phase and then to a cubic phase at very high temperatures.[8] The addition of 8 mol% Y_2O_3 as a dopant to ZrO_2 produces the cubic phase at room temperature and pressure.

The foregoing facts present an interesting circumstance from the point of view of high pressure studies. When there is a change in crystal structure caused by isobaric decreases in the temperature, the transition is accompanied most often by volume decreases. At temperatures above the transition point, there is too much thermal energy to constrain the atoms in the smaller volume configuration. However, it is possible to apply an external constraint, viz. the isothermal application of pressure, which invariably forces the system of atoms into a smaller volume. The work done by that external pressure is often sufficient to induce the transition to the smaller volume phase.

Hence, consistent with the phase diagram for pure ZrO_2, it was speculated that ZrO_2 (8 mol% Y_2O_3) might undergo a pressure-induced phase transition to a tetragonal structure, while the sample is maintained at room temperature.

An initial set of measurements have been completed using a sample of ZrO_2 containing 8 mol% Y_2O_3. Our task was to determine whether a noncubic phase would be produced and perhaps stabilized under high pressure. Consequently, x-ray diffraction measurements were made with this sample at room temperature and at pressures up to nearly 30 GPa (about 4.5 million psi). In separate experiments, optical microscopy studies were also made at these pressures and for temperatures up to 425 K.

The principal reflections of the cubic structure occurred, for this sample, in the x-ray range of 20 to 50 keV for $2\theta = 10$ degrees. These reflections were identified at one atmosphere and were observed to the highest pressure measured. No reflections from a new phase were detected.

The persistence of the cubic reflections is illustrated in Fig. 4 where intensity patterns from two different pressures are shown together. An offset has been added to the intensity of one of the patterns to give an easier comparison. The shift of the peak positions that can be seen in the figure is due to the compression of the sample.

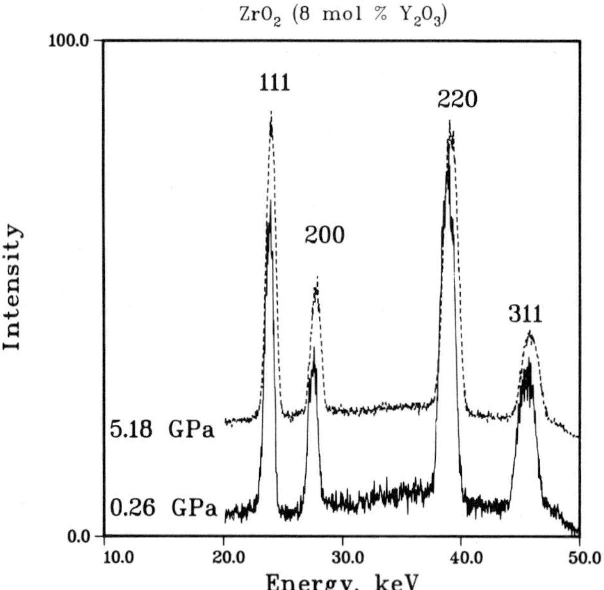

Fig. 4. Cubic zirconia has four reflections in the energy range 20 to 50 keV when the scattering angle is $2\theta = 10$ degrees. The persistence of these reflections to high pressure, along with the lack of new reflections, indicates that the cubic structure is preserved. The peaks in the intensity pattern shift to higher energy coordinates as the sample is compressed.

The persistence of the cubic phase was also verified by the optical studies which were made with a polarizing microscope. There was no visual evidence of a phase transition in the observed ranges of temperature and pressure.

The x-ray measurements were made to determine whether or not the zirconia sample would undergo a phase transition under very high pressure. The pressure transmitting medium used for those measurements was a mixture of 4:1 methanol:ethanol which had a limit of hydrostaticity of about 10.4 GPa. Consequently, many of the data were taken under very nonhydrostatic stresses, and hence, they were not of prime quality for such purposes as the evaluation of the compressibility of the sample.

However, while efforts are underway to obtain media with higher limits of hydrostaticity, the present data provide at least an estimate of the p-V relation for this sample. From the x-ray measurements, the cubic lattice parameter a can be determined, and hence the volume $V = a^3$. Further, using a higher order isothermal equation of state (EOS) like the Murnaghan EOS or the Birch-Murnaghan EOS, the bulk modulus $B = -V(\partial p/\partial V)_T$ and its pressure derivative $B' = \partial B/\partial p$ can also be estimated.

The p-V data are shown in Fig. 5 along with the least squares fits (smooth curves) obtained using the Murnaghan and the Birch-Murnaghan EOS. The two fits are mutually consistent and give accurate representations of the data. From the Murnaghan EOS, B = 327 GPa with B' = 21.8 at p = 0. From the Birch-Murnaghan EOS, B = 250 GPa with B' = 51.7 at p = 0. These values possess a high degree of uncertainty because of the nonhydrostatic conditions at pressures above 10.4 GPa.

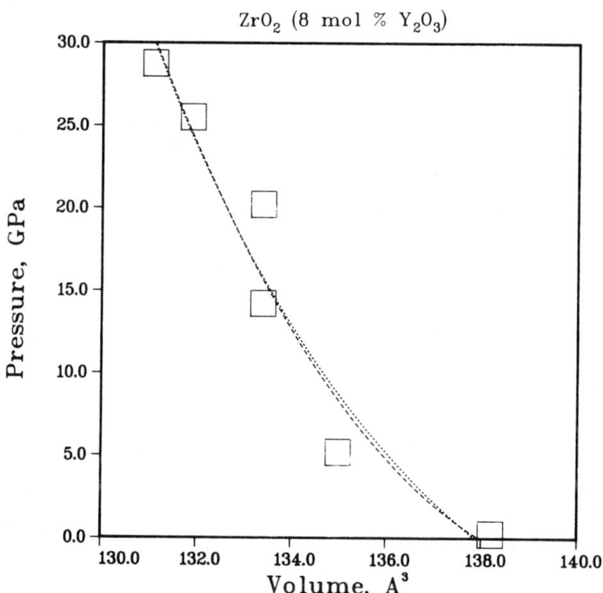

Fig. 5. The p-V relation for zirconia can be estimated as a result of the x-ray measurements. There is a considerable amount of scatter in the present data due to the occurrence of non-hydrostatic stresses above 10.4 GPa. Nevertheless, higher order isothermal equations of state can be used to represent the trend of the data. The dotted curve results from the Murnaghan equation, and the dashed curve results from the Birch-Murnaghan equation. At zero pressure, these two models yield an average bulk modulus for this sample of 300 GPa +/- 40 GPa.

REFERENCES

1. G. J. Piermarini and S. Block, Rev. Sci. Instrum., 46, 973 (1975).
2. G. J. Piermarini, S. Block, J. D. Barnett, and R. A. Forman, J. Appl. Phys., 46, 2774 (1975).
3. I. Fujishiro, G. J. Piermarini, S. Block, and R. G. Munro, Proceedings of the 8th AIRAPT Conference, High Pressure in Research and Industry, 2, 608 (1981).
4. J. D. Barnett, S. Block, and G. J. Piermarini, Rev. Sci. Instrum., 44, 1 (1973).
5. R. G. Munro, G. J. Piermarini, S. Block, and F. A. Mauer, to be published.
6. R. A. Noack, W. B. Holzapfel, Proceedings of the 6th AIRAPT Conference, High Pressure Science and Technology, 1, 748 (1977).
7. D. E. McCumber and M. D. Sturge, J. Appl. Phys., 34, 1682 (1963).
8. N. A. Bendeliani, S. V. Popova, and L. F. Vereshchagin, Geokhimiya, 6, 677 (1967).

EFFECT OF STRONG SHOCK COMPRESSION ON COVALENT MATERIALS AND HIGH PRESSURE SINTERING

Akira Sawaoka

Research Laboratory of Engineering Materials
Tokyo Institute of Technology
Midori, Yokohama 227, Japan

INTRODUCTION

It is very difficult to produce sintered compacts of high temperature covalent materials, such as Si_3N_4, TiC, BN and diamond having densities approaching theoretical. The present author and coworkers have studied very high pressure sintering of the above covalent materials. Since it was understood that shock conditioning of powders of covalent materials was effective for very high pressure sintering, as well as conventional sintering, the effect of shock compression on covalent materials was examined.

The Hugoniot elastic limit (HEL) of high temperature covalent materials is generally higher than that of ionic materials. The highest reported HEL value for a noncovalently banded crystal is about 10 GPa for aluminum oxide. HEL values of most high temperature covalent materials have not yet been determined. Usually shock compression treatment around or higher than HEL is required for effective shock conditioning of powder the above covalent materials. It is assumed that the HEL of these materials is about 10 GPa or higher. Thus we have to use very high dynamic pressure techniques for shock conditioning of the above materials.

Wurtzite-type boron nitride was synthesized by shock compression treatment of graphite-like boron nitride (g-BN). Very high pressure sintering of shock synthesized w-BN has also been studied.

Preliminary studies of shock consolidation of silicon nitride, titanium nitride and titanium carbide powders are reported.

FACILITY FOR SHOCK EXPERIMENTS

Two kinds of shock facilities were utilized. Hypervelocity impact at higher than 3 km/s was performed by using the two stage light gas gun at the Tokyo Institute of Technology (Fig. 1).[1-3] The bore diameter of the launcher is 20 mm. Specimen powders can be recovered after impact using the assembly as shown in Fig. 2.

The explosive facility at Nippon Oil & Fats Co. Ltd. was used for the medium velocity impact experiment between 1-3 km/s (Fig. 3). Two kinds of recovery assemblies are shown in Fig. 4 and Fig. 5. The main experimental results summarized in this paper have been obtained by using the mouse trap type assembly[4] as shown in Fig. 5. Powder was pressed into a stainless steel capsule whose specimen chamber was 5 mm in height and 12 mm in diameter. Shock waves generated in the capsule by impact or iron flyer plate with a 3.2 mm thickness passed through the powder compact. The cylindrical assembly shown in Fig. 4 was used for shock consolidation of silicon nitride powders.

Fig. 1. Overview of double stage light gas guns at the Tokyo Institute of Technology. Left small gun is HS-2B and rear large one is HS-3B. HS-2B was modified and combined with rail gun as HS-2C.

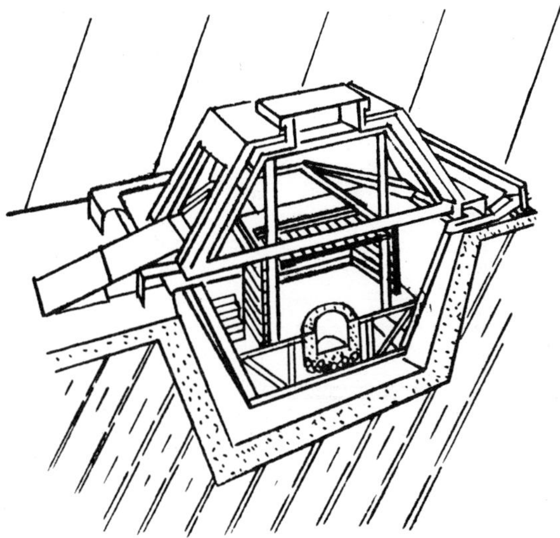

Fig. 2. Target assembly and projectile just before impact.

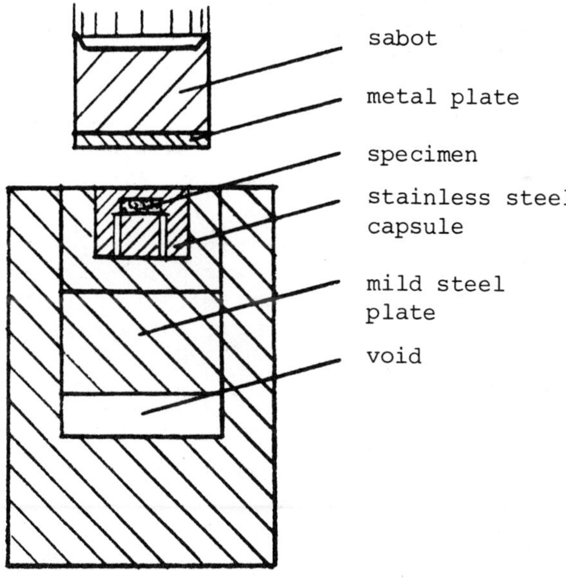

Fig. 3. Structure of explosive chamber. Housing is floating on the air cushion. Shock assembly is recovered in the water tank at the bottom of the housing.

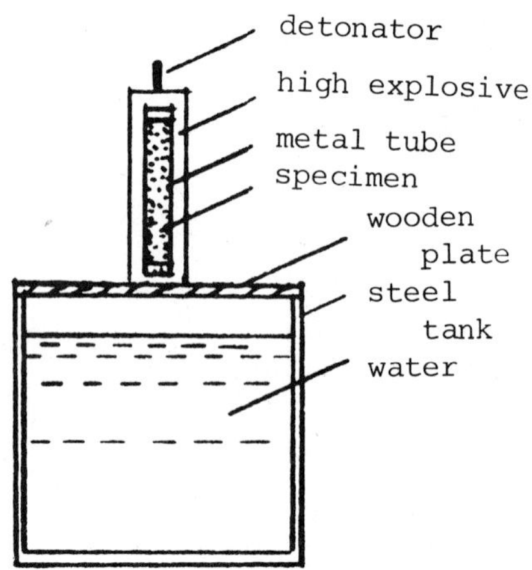

Fig. 4. Shock assembly for the explosive experiment. Steel rod is not used for the shock consolidation experiment.

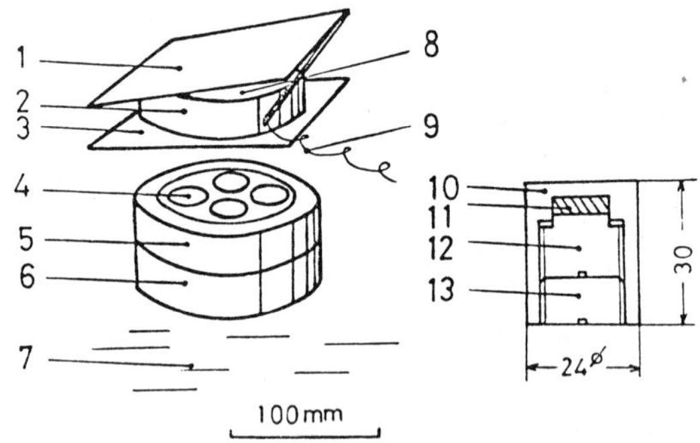

Fig. 5. Mouse trap type arrangement for explosive compression. 1-sheet explosive; 2-plastic ring; 3-flyer plate; 4-sample holder (SUS-304); 5,6-momentum trap (mild steel); 7-water; 8-high explosive; 9-detonator; 10-capsule (SUS-304); 11-specimen; 12,13-plug (SUS-304).

SOME SHOCK EXPERIMENTS ON COVALENT MATERIALS

Shock Treatments of Silicon Nitride

Three kinds of Si_3N_4 powders were used. Some of characteristics of these powders are summarized in Table 1. Amorphous and α-phase powders were transformed to the β-phase by shock compression.[5] The

Table 1. Some Charactersitics of Si_3N_4 Powders Used for Shock Experiment

Source	Phase	Purity (%) Except Anion	Remarks
GTE-Sylvania	Amorphous	99.99	Surface area 100 m^2/g
H. C. Starck	α	99	Particle size 0.8 μm
High Purity Chemical	β	99.95	Particle size 1.0 μm

experimental arrangement shown in Fig. 5 was used. By detonation of high explosive, a flyer was propelled at a velocity of 2.8 km/s against a specimen holder. Electron micrographs of amorphous and β-Si_3N_4 powders are shown in Fig. 6 and Fig. 7, respectively. Amorphous powder particles provided by GTE-Sylvania, whose surface area was 100 m^2/g, were sphere-like and with an average size of 0.6 μm. Upon shock compression, particles were shattered into edged fragments with 0.2-0.4 μm. Beta-Si_3N_4 powder particles having ≈ 1 μm average particle size provided by High Purity Chemical Laboratory Ltd. (Japan) were reduced by 1/2 by shock treatment.

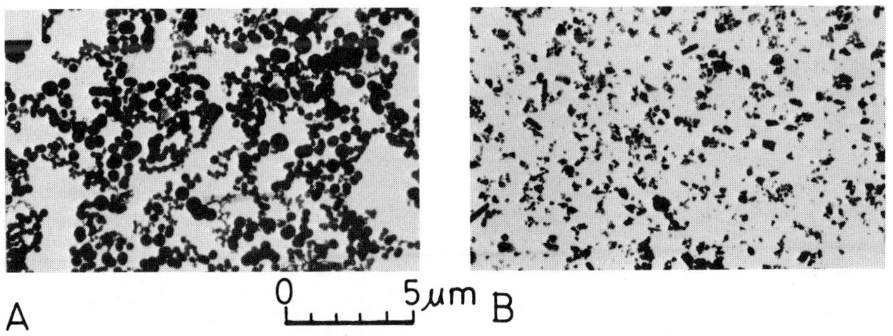

Fig. 6. Transmission electron micrographs of amorphous Si_3N_4 powders. A-unshocked; B-shocked.

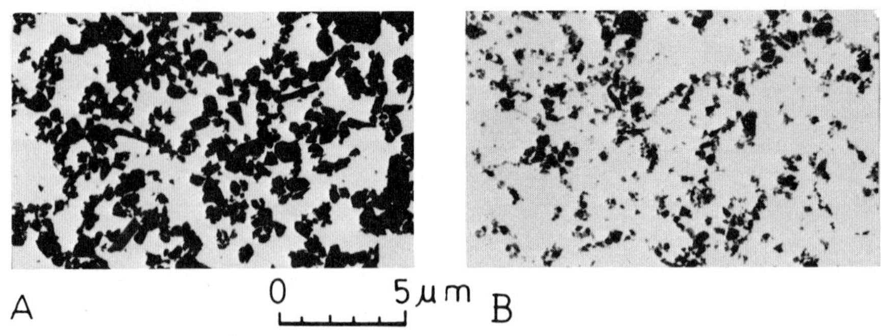

Fig. 7. Transmission electron micrographs of β-phase Si_3N_4 powders. A-unshocked; B-shocked.

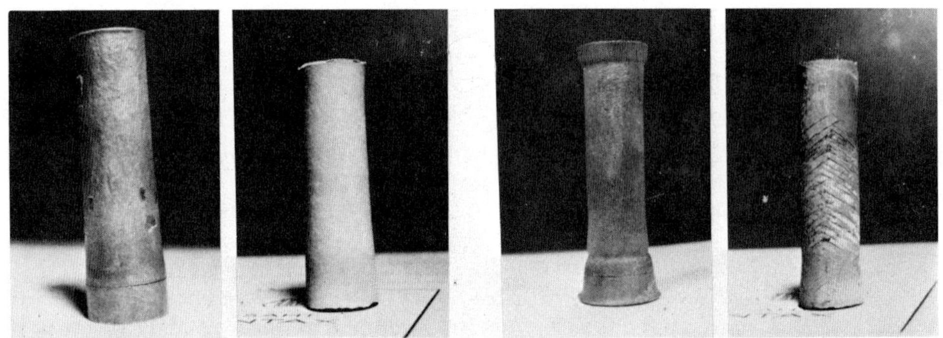

Detonation velocity 3 km/s Detonation velocity 5 km/s

Fig. 8. Compacts of Si_3N_4 produced by explosive shock compaction. Starting powder is α-Si_3N_4 with 0.8 μm. Left picture shows copper tube containing the powder. Right is the compact following the removal ofthe tube.

Shock consolidation tests of α-Si_3N_4 with 0.8 μm particle size was performed. The α-Si_3N_4 powder provided by H. C. Stark was pressed into a copper tube and shock compressed by the explosive technique as shown in Fig. 4. Two kinds of high explosives with detonation velocities of 3 km/s and 5 km/s were used. The copper tube containing Si_3N_4 powder was recovered from shock treatment, heated at 1250°C for 10 min and the molten metal removed. The outsides of several of the compacts are shown in Fig. 8. A compact obtained by 3 km/s detonation and an E/M ratio of 2 has 80% of theoretical density; 5 km/s treatment 98% of the theoretical produces density. Seventy-five of the α-Si_3N_4 transformed to β-Si_3N_4 in the former compact and 95% transformed in the latter. Many crack-like slip bands can be seen on the surface of the compact (Fig. 8).

Shock Consolidation of TiN-TiC Powders[6,7]

Powders of TiN, TiN(0.7C(0.3), TiN(0.5)C(0.5), TiN(0.3)C(0.7) and TiC were shock-treated. TiN should have intermediate properties between purely ionic and purely covalent materials, judging from its electronegativity; whereas TiC should be more covalent. Therefore it is interesting to investigate the effect of shock treatment on the powders of a solid solution TiN-TiC. These powders were provided by Nippon New Metal Ltd. Average sizes of TiN, TiN(0.7)C(0.3), TiN(0.5)C(0.5), TiN(0.3)C(0.7) and TiC powder were 1.31, 1.22, 1.20, 1.25, 1.35 and 2.2 μm, respectively. These powders were compacted to a density of about 60 mass % of the theoretical value in a stainless steel capsule in order to form a disk 5 mm thick and 12 mm in diameter. The impact pressure in the powder specimen was calculated for the initial density. Results are shown in Fig. 9 as a function of pressures. TiN powders were densified to about 80% by shock compression at 5.5 GPa and 93% at 9.5 GPa. TiC and TiC-rich solid solution powders can be easily densified by shock compression, contrary to the author's expectation.

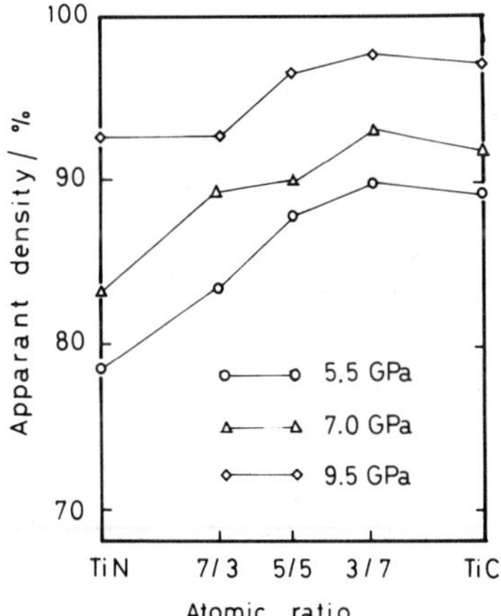

Fig. 9. Apparent density of TiN-TiC solid solution compacts obtained by shock compaction.

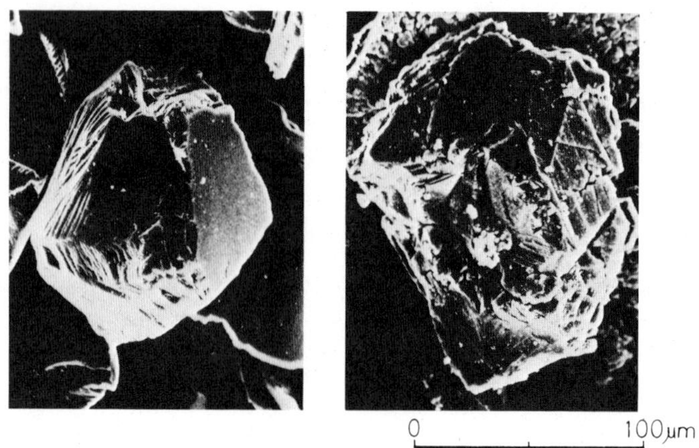

Fig. 10. Scanning electron micrographs of natural diamond powders. The left picture shows the starting material. The right micrograph shows the material after shock compression at about 60 GPa.

Shock Treatment of Diamond and c-BN Powders

Natural diamond and synthetic c-BN were examined as typical covalent materials. Diamond powder with an average particle size of about 100 µm was mixed with 150-300 mesh copper powder at a weight ratio of 5:95; two different lots of c-BN powder having the average particle sizes of 20-30 µm and 1-5 µm, respectively, were also mixed with copper powder. The mouse trap type shock assembly was used on this experiment. The mixtures were compacted in a stainless steel capsule to form disks 5 mm thick and 12 mm in diameter. The impact pressure induced in the capsule was estimated by the impedance matching method. SEM micrographs of diamond and c-BN are shown in Figs. 10 and 11, respectively. The diamond particles have many cracks as seen in the picture; however, such cracking is considerably reduced for c-BN. It is an interesting phenomenon that diamond is more easily cracked than c-BN by shock compression of 60 GPa.

The x-ray diffraction patterns of the unshocked and shocked c-BN with 1-5 µm average particle size are shown in Fig. 12. Sample A is the original c-BN, and sample B, C and D are recovered BN from shock compression of 60, 150 and 200 GPa, respectively. The diffraction intensity of c-BN decreased with increasing shock pressure, while the peak intensity of g-BN transformed from c-BN increased. Transformation from c-BN to g-BN might be induced by residual high temperature after passage of the shock wave.

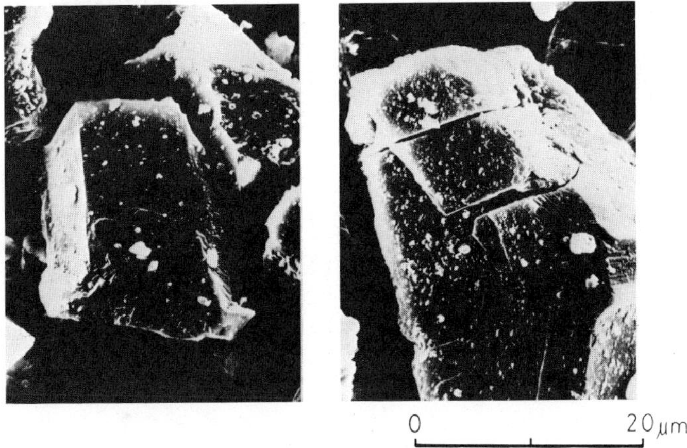

Fig. 11. Scanning electron micrographs of synthetic c-BN powders. The left picture shows the starting material. The right micrograph shows the material after shock compression at about 60 GPa.

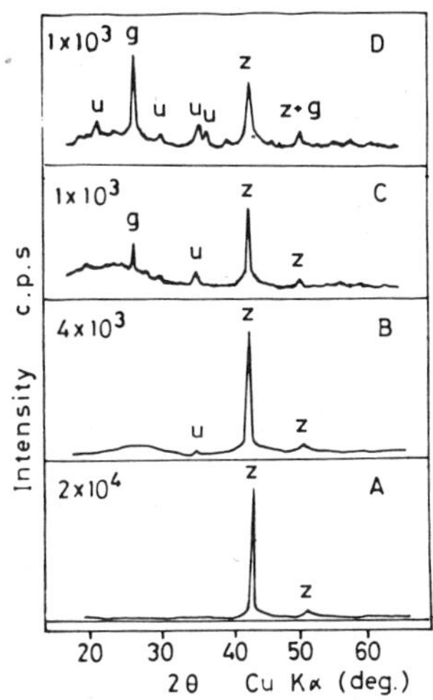

Fig. 12. X-ray diffraction patterns of the non-loaded and shock-loaded BN. Starting material is synthetic zinc blende powder with an average particle size of 1.5 μm. g-graphite-like BN; w-wurtzite type BN; z-zinc blende type BN(c-BN); u-unknown. A-unshocked c-BN; B-shock loaded BN(60 GPa); C-shock loaded BN(150 GPa); D-shock loaded BN(200 GPa).

Structural Changes of g-BN by Multiple Shock Compressions[8]

The g-BN powder was shock compressed by the mouse trap type technique. The starting powder of g-BN with a particle size of 1.5 μm was provided by Denki Kagaku Ltd. (grade GP). The main impurities were Fe (1000 ppm) and Mg,Al (100 ppm). G-BN and 150-300 mesh copper powders were mixed at a weight ratio of 5:95 and compressed in the stainless capsule. After the disk was shock-compressed, it was immersed in HNO_3 + 3HCl solution for 24 hours to dissolve the copper matrix. In this way the BN subjected to a single compression was obtained. In the experiments of multiple shock compression, the BN powder recovered was again mixed with copper powder, formed to a disk, placed in the capsule and shock-compressed. The same material was subjected to a third run of shock treatment in a similar manner.

The x-ray diffraction patterns of the unshocked and shocked BN are shown in Fig. 13. Sample A is the original g-BN and sample B, C and D are the recovered BN after single, double and triple shock compressions of 60 GPa, respectively.

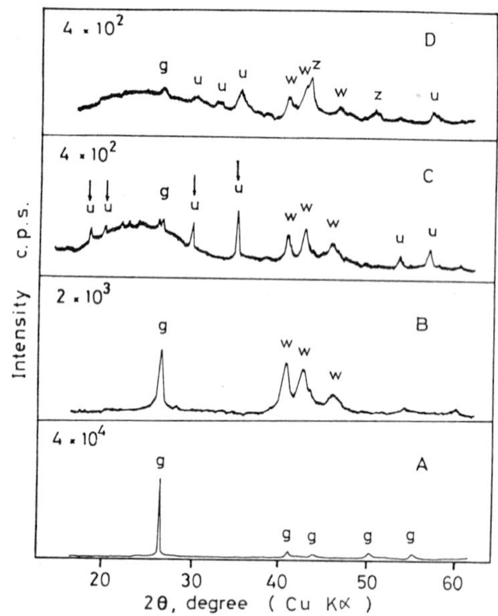

Fig. 13. X-ray diffraction patterns of unshocked and shocked g-BN. A-unshocked BN; B-BN subjected to single shock compression; C-BN subjected to double shock compression; D-BN subjected to triple shock compression.

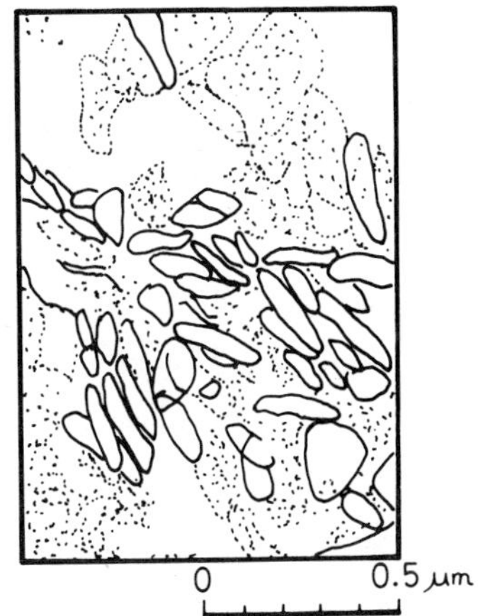

Fig. 14. Transmission electron micrograph of shock-synthesized w-BN powder. g-BN was shock-compressed and purified by the alkaline fusion technique.

The results are summarized as follows:

(1) BN subjected to single shock compression. Strong line broadening of (002) line and the disappearance of other weak lines in the x-ray diffraction pattern of g-BN were observed. The transformation to w-BN was also induced. The particle size range obtained from the electron micrographs was 0.1-0.3 μm, as shown in Fig. 14.

(2) BN subjected to double shock compressions. A decrease in peak intensity of the x-ray diffraction patterns of g-BN and w-BN was observed. A broad peak, similar to the x-ray diffraction profile of the amorphous materials, appeared in the range of 15 to 35 degrees 2θ for Cu-Kα. Several peaks, which were unidentified, exhibited relatively sharp profiles.

(3) BN subjected to triple shock compressions: Strong line-broadening and a decrease in diffraction intensity of the unknown peaks were observed. Peaks characteristic of c-BN were detected. Particle size reduction was scarcely observed.

VERY HIGH PRESSURE SINTERING OF COVALENT MATERIALS

Very High Pressure Sintering of Shock-treated Si_3N_4 [5]

Unshocked and shocked Si_3N_4 powders were sintered under very high pressure by using the slide-type cubic anvil high pressure apparatus. Powder with 6 wt% camphor as a binder was pressed at 500 MPa to form tablets 6.0 mm high and 8.2 mm in diameter. This tablet was heat-treated at 10^{-3} Torr and 500°C for 2 hours and set into a high pressure sintering cell. Details of a high pressure assembly were described elsewhere.[9] Pressure was applied to the specimen at room temperature, and it was heated for 15 minutes at the desired temperature by controlling the electric power. The sample was then cooled slowly to room temperature, and the pressure reduced to atmospheric. Some sintered compacts were powdered for x-ray diffraction analysis. The relative density of the compacts was measured by the Archimedean method, and Vickers microhardness determine by using a load of 500 g.

Relative densities of the compacts sintered at 5 GPa for 15 minutes are shown in Fig. 15. Densities of sintered compacts from both unshocked α-Si_3N_4 and β-Si_3N_4 powder were about 85% for a sintering temperature of 1000°C and about 98% for 1500°C, as seen in the figure. The density increased slightly with temperature above 1500°C and reached 99% at 1700°C. Densification behavior of the shocked powder of both materials was similar to the unshocked powders. The shock compression effect on sintering was not so remarkable.

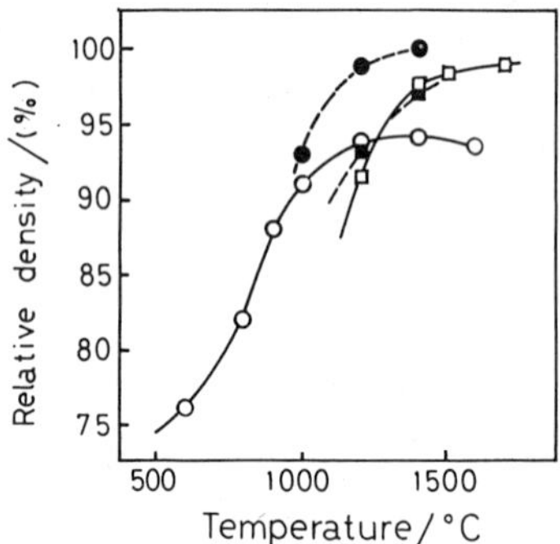

Fig. 15. Changes of relative densities of sintered compacts obtained at various temperatures and a constant pressure of 5 GPa for 15 min. Open symbols show nonshock-treated powders and closed show shock-treated ones. △,▲ shows amorphous Si_3N_4 and □,■ shows β-phase.

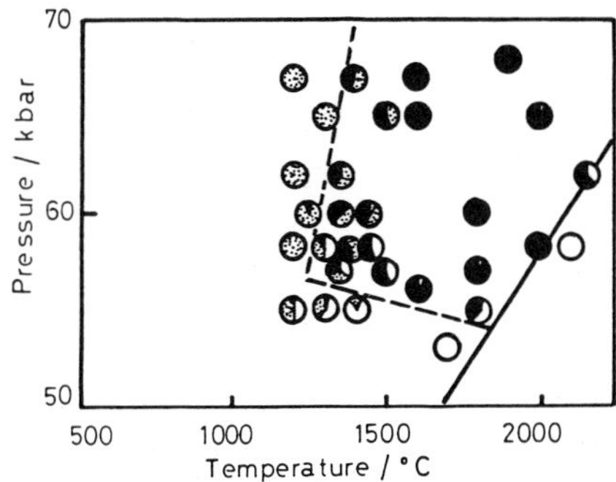

Fig. 16. Products obtained from treatment of wurtzite-type BN for 15 min. at high temperatures and pressures. O shows graphite-like BN; ⊛ shows wurtzite type BN; ● shows zinc blende type BN (c-BN).

Densification curves for the amorphous Si_3N_4 powder was quite different from crystalline Si_3N_4. The density of unshocked powder increased quickly with temperature and reached 93% at 1000°C. It was saturated above this temperature. The shock effect for this powder on sintering was remarkable. The density increased further with temperature above 1000°C and reached 99% at 1200°C. Sintered compacts of β-Si_3N_4 with the theoretical density value was obtained by means of shock treatment and very high pressure sintering at 1400°C and 5 GPa.

Very High Pressure Sintering of Shock Synthesized w-BN[9-13]

Powder of w-BN was synthesized from g-BN by shock compression and subsequently purified by the alkaline fusion technique. Preparation methods and the very high pressure apparatus described in the previous section were used for this process. The weight ratio of the phases present in the sample was estimated from the relative intensities of the following diffraction peaks: g-BN (200), w-BN (100), c-BN (111). The formation region of c-BN from w-BN is shown in Fig. 16. The solid line shows the phase equilibrium boundary between g-BN and c-BN. The dashed line is the boundary of conversion from w-BN to c-BN obtained in this work. Cubic-BN was detected at pressures 5.5 GPa and at temperatures 1250°C.

Fracture surfaces of sintered compacts were observed by using TEM and SEM. The results of these observations are summarized as follows: In spite of much effort, grain boundaries in the fracture specimens of the w-BN sintered compact could not be observed. The crystallite size in the w-BN compact might be unchanged from that of initial w-BN particles.

Rapid grain growth resulted from the phase transformation from w-BN to c-BN. The grain size in the c-BN transformed from the w-BN was about 1 μm, after treatment of the w-BN at 6.7 GPa and 1800°C for 15 minutes. Octahedron-like grains can be seen.

Micro Vickers hardness of the sintered compacts was measured using a hardness tester with an indentation load of 1 kg. Hardness and apparent density are summarized in Fig. 17. Transformation from w-BN to c-BN increased with treatment temperature; the amount of c-BN obtained was about 50% at 6.7 GPa and 1500°C for 15 minutes and 80% at 1600°C. The apparent density increased with sintering temperature. The value reached theoretical around 1600°C and decreased with temperature.

Vickers hardness values could be correlated with the final density. The hardness of a sintered compact at 1600°C was very high. The compressive strength of this compact also showed the highest value of about 6 GPA.

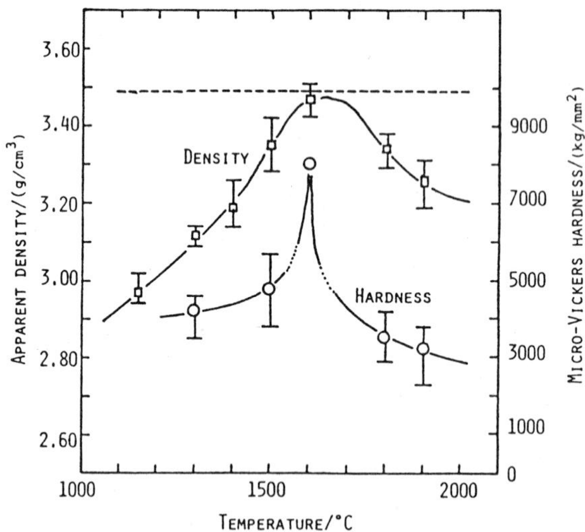

Fig. 17. Apparent density and micro Vickers hardness of the high density form boron nitride compacts sintered at 6.7 GPa for 15 minutes.

CONCLUSION

The purpose of this study was to find a preparation technique for non-porous polycrystalline covalent materials without additives. Polycrystalline bodies of β-Si_3N_4 and the high density forms of boron nitride (w-BN and c-BN) were successfully obtained by very high pressure sintering of powders of shock treated Si_3N_4 and shock synthesized w-BN, respectively. Both kinds of powders were very fine, having particle sizes of 0.1-0.4 μm. The density crystal defects in each shocked powder was also high.

However, another kind of shock treated powder of Si_3N_4 did not noticeably change its high pressure sintering characteristics. Thus, the present author concludes that effects of shock treatment on covalent materials is very complicated and systematic theoretical and experimental studies should be performed. This paper describes the first step of such a program.

ACKNOWLEDGMENTS

The author would like to thank Dr. S. Saito, President Emeritus, Tokyo Institute of Technology for his encouragement in this study over the past several years and also to Prof. K. Kondo for useful discussions. He is also grateful to Mr. M. Araki and Mr. T. Akashi of Nippon Oil and Fats Co., Ltd., Mr. T. Murai of the Mechanical Engineering Laboratory and Mr. Y. Hattori of NGK Spark Plug Co., Ltd. for their help in the dynamic and static high pressure experiments.

REFERENCES

1. A. Sawaoka, T. Sōma, and S. Saito, Proceedings of the 4th AIRAPT Conference on High Pressure, Kyoto 1974, 739, 1975.
2. H. Sugiura, T. Mashimo, K. Kondo, and A. Sawaoka, Report of Research Laboratory of Engineering Materials, Tokyo Institute of Technology, 6, 93 (1981).
3. T. Mashimo and A. Sawaoka, Japan. J. Appl. Physics, 20, 963 (1981).
4. T. Akashi, A. Sawaoka, S. Saito, and M. Araki, Japan. J. Appl. Phys., 15, 891 (1976).
5. A. Sawaoka, K. Kondo, N. Hashimoto, and S. Saito, Proceedings of International Symposium of Factors in Densification and Sintering of Oxide and Non-oxide Ceramics, Hakone, Japan 339, 1978.
6. S. Soga, K. Kondo, and A. Sawaoka, J. Mater. Sci., to be submitted.
7. A. Sawaoka, S. Soga, and K. Kondo, J. Mater. Sci. Lett., 1, 347 (1982).

8. T. Akashi, A. Sawaoka, and S. Saito, Report of Research Laboratory of Engineering Materials, Tokyo Institute of Technology, 2, 134 (1977).
9. T. Akashi, A. Sawaoka, and S. Saito, Report of Research Laboratory of Engineering Materials, Tokyo Institute of Technology, 3, 69 (1978).
10. T. Akashi, A. Sawaoka, and S. Saito, J. Am. Ceram. Soc., 61, 245 (1978).
11. A. Sawaoka, S. Saito, and M. Araki, p. 986 in High-pressure Science and Technology, Vol. 1, Plenum Pub. Corp., 1979.
12. S. Saito and A. Sawaoka, p. 541 in High-pressure Science and Technology, Vol. 1, Academic Press, 1980.
13. T. Akashi, Y. Tanaka, T. Murai, and A. Sawaoka, Nippon Kagaku Kaishi (J. Chem. Soc. Japan), [9], 1416 (1981).

DISCUSSION

C. Weissmantel (TH KMST): (1) What was the maximum Vickers Hardness of BN? (2) Did you perform dynamical tests?

(1) Up to 9000 kg. (2) Preliminary tests are ongoing.

A NEW APPROACH TO THE REACTION SINTERING OF SUPERHARD MATERIALS UNDER VERY HIGH PRESSURE

Minoru Akaishi, Tadashi Endo, Osamu Fukunaga,
Yoichiro Sato and Nobuo Setaka

National Institute for Research in Inorganic Materials
1-1 Namiki, Sakura-mura, Niihari-gun, Ibaraki 305 Japan

ABSTRACT

The polycrystalline diamond, cubic boron nitride and cubic boron nitride-diamond with homogeneous microstructure and strong direct bonding were obtained by the reaction sintering of diamond-graphite/WC-Co, cBN-hBN/$Mg_3B_2N_4$ and cBN-graphite/WC-Co systems. In these systems, sintering proceeded via transformation from graphite to diamond structure. The sintering process of these systems was discussed based on the observation by SEM, EPMA, X-ray diffraction, Raman spectroscopy, micro-Vickers' hardness test and scanning Auger spectroscopy.

INTRODUCTION

Diamond and cubic boron nitride (cBN) are well known superhard materials. The sintered bodies of these materials are widely used as cutting tools on hard or abrasive materials, as wire drawing dies and in rock drills.

There are a few reports[1-3] on the sintering of diamond without additive, but direct bonding between diamond particles was difficult to achieve without aid of catalyst metals. In this context, Katzman and Libby[4], Wentorf and Rocco[5,6] have reported on the sintering of diamond in the presence of cobalt. Especially, General Electric group[5,6] showed a new method in which diamond powder could be consolidated by the migration of Co from WC-Co substrate. Hibbs and Wentorf[7] have reported the preparation of cBN sintered bodies. However, sintering process of these super-hard materials was not clear from the previous works.

In most of the previous works, diamond and cBN powders were used as starting materials. On the other hand, we reported the sintering behavior of diamond-Co/WC-Co previously[8], in which we noted the effect of graphite formed during the process: The presence of a sufficient amount of graphite, which promotes the migration of Co between diamond particles, is very important for the sintering of diamond. The effect of additive graphite on sintering of diamond in diamond-graphite/WC-Co systems was also investigated.[9] The homogeneous sintered bodies with strong direct bonding was obtained by the suitable selection of graphite source and content, Co content in WC, pressure and temperature.

As an extension of our previous works, directly bonded polycrystalline diamond, cBN-diamond and cBN with homogeneous microstructure were obtained by the reaction sintering of diamond-graphite/WC-Co, cBN-graphite/WC-Co and cBN-hBN/$Mg_3B_2N_4$ systems. In these systems, sintering proceeded via transformation from graphite like structure to the diamond like one.

EXPERIMENTAL PROCEDURES

Sintering Procedure

High pressure sintering was carried out using a modified belt type high pressure apparatus with 25 mm bore diameter.[10] The samples were heated by means of an internal graphite heater with 12 mm o.d., 10 mm i.d. and 17.4 mm length. Temperature was measured with Pt6%Rh-Pt30%Rh thermocouple without correction of the pressure effect on the emf. In routine experiments, temperature was estimated from the relation between power input and temperature. Pressure was calibrated at room temperature using known transition of Bi, Tl and Ba. Pressure correction at high temperature was also made by using the melting curve of silver[11] based on the data obtained by piston cylinder apparatus.

An assembly as shown in Fig. 1 was used. WC pellets as shown in Fig. 1(B) were formed into cylindrical shape under the pressure of about 500 MPa. As shown in Fig. 1(B), samples wrapped with Zr and Mo foils were surrounded by sodium chloride. Sodium chloride has low shear strength at room temperature. Above its melting temperature, it envelopes a sample in an effectively hydrostatic condition. Samples were first subjected to pressure. Temperature was then applied and held for 1 hr. Sintering was made at 1400°-1500°C and 5.8GPa so that diamond and cBN crystals were kept within thermodynamically stable reagion above diamond-Co and cBN-$Mg_3B_2N_4$ eutectic temperatures.

The obtained samples were characterized by means of electron probe x-ray microanalyzer (EPMA), scanning electron microscopy

(SEM), x-ray diffraction, Raman spectroscopy, scanning Auger spectroscopy and micro Vickers' hardness test (100 gr load).

Starting Materials

In sintering of diamond-graphite/WC-Co, commercially available diamond powder (270/325 U.S. mesh size, Tomei Diamond Co. Ltd), graphite powder for spectroscopic use (325 U.S. mesh size, Nippon Carbon Co. Ltd) were used as starting materials. These diamond and graphite powders were mechanically mixed. The powder mixture laminated to WC-16% Co was treated at high pressure and high temperature.

In sintering of cBN-graphite/WC-Co system, commercially available cBN powder (3-6 μm, cBN(f) and 10-20 μm, cBN(c), De Beers industrial diamond) and graphite powder for spectroscopic use (325 U.S. mesh size, Nippon Carbon Co. Ltd) were used as starting materials.

In sintering of cBN-hBN/$Mg_3B_2N_4$ system, commercially available cBN (Showa Denko Co. Ltd), hBN powder and hBN sintered bodies (Denki Kagaku Co. Ltd) were used as starting materials. The addition of $Mg_3B_2N_4$ to BN starting materials was made by means of mechanical mixing and diffusion process. The diagram of preparing cBN sintered bodies is shown in Fig. 2.

RESULTS AND DISCUSSION

The Effect of Additive Graphite on Sintering Process of Diamond

A marked difference has been observed in Co migration and microstructure between pure diamond and diamond-graphite/WC-Co in our previous work.[9]

Co migration from WC-Co to pure diamond layer was strongly dependent on grain size of diamond and sintering temperature. This may be related to the pore size formed between diamond particles and viscosity of liquid Co. On the other hand, Co migration to diamond-graphite occurred in a manner different from that of pure diamond. In this system, the diamond particles can rearrange easily at high pressure and room temperature in the presence of graphite. The porosity of the compact can be estimated as nearly zero. The mechanism of Co migration of this case can be considered in the analogy of the diamond formation process from graphite. Co migration from WC-Co to diamond-graphite layer was promoted by increasing graphite content.

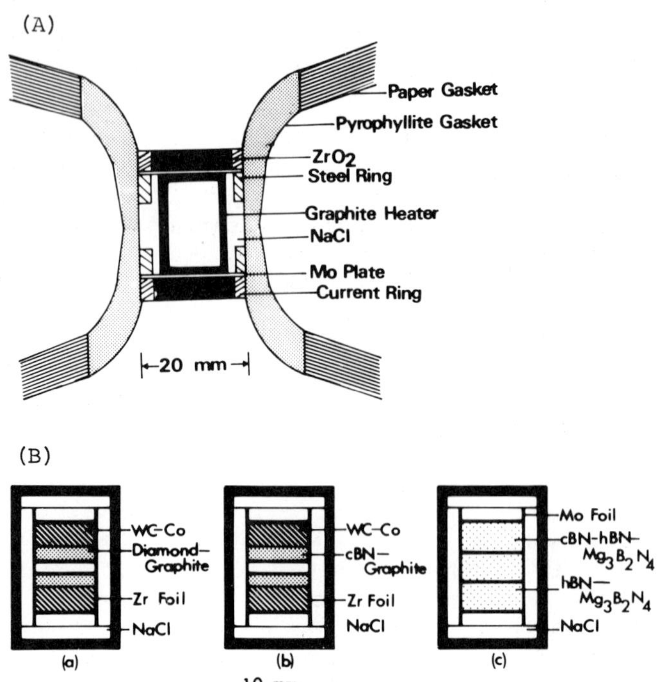

Fig. 1. Sample assembly for high pressure sintering. (A) Full illustration of the cell. (B) Enlarged portion around sample.

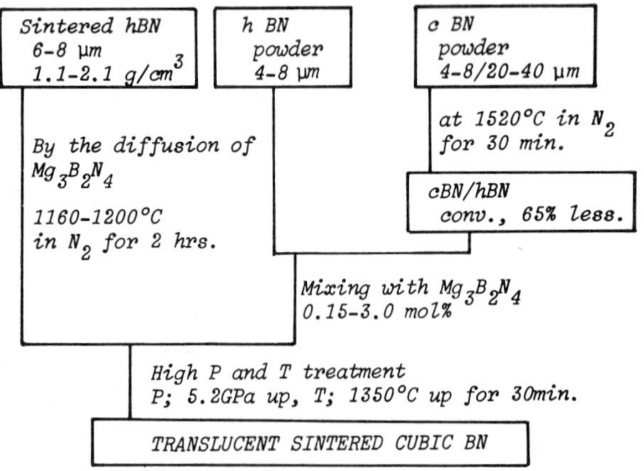

Fig. 2. Process diagram for preparing sintered cBN.

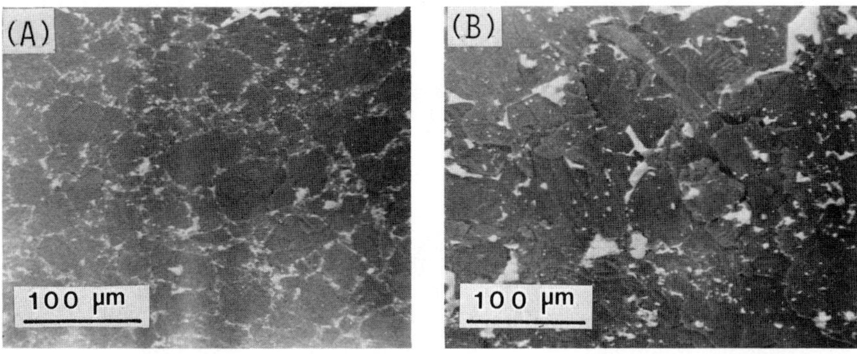

Fig. 3. Back scattered electron image of fracture surface. (A) Pure diamond/WC-Co, (B) Diamond-50 vol% graphite/WC-Co.

The fracture surfaces of the sintered bodies obtained from pure diamond(c)/WC-16 wt% Co and diamond(c)-50 vol% graphite(s)/WC-16 wt% Co are shown in Fig. 3. In pure diamond system, Co fills the pore between diamond particles forming Co network, and bonding between diamond particles was based on Co metals. On the other hand, in diamond-graphite system, Co was not continuous and isolated Co pools were observed. In this case, bonding between diamond particles was based on direct bonding between diamond particles. Sufficient amount of graphite in the diamond layer promoted the formation of direct bonding between diamond particles.

Preparation of cBN-diamond/WC-Co Sintered Composite

As described above, the additive graphite promoted the migration of Co and formation of the direct bonding between diamond particles. In this context, cBN-graphite powder mixture laminated to WC-16% Co was treated at high temperature and pressure to obtain cBN-diamond sintered bodies using the transformation from graphite to diamond.

Sintering behavior was strongly dependent on treatment temperature, pressure and graphite content. Graphite and cBN were mechanically mixed. A well sintered cBN-diamond composite was obtained by treating a powder (cBN(f)-50 vol% graphite) at 5.8 GPa and 1500°C for 1 hr. When the temperature was lowered to 1400°C with other conditions being the same, the obtained sample was not sintered, but samples treated at temperatures above 1450°C were well sintered. When a mixture with graphite content below 25 vol% was treated at 1500°C, the sample was poorly sintered. Also, cBN-50 vol% graphite mixture treated at 5.4 GPa and 1500°C for 1 hr. was not sintered.

The well sintered body obtained from cBN(f)-50 vol% graphite/ WC-16 wt% Co was investigated by x-ray diffraction and Raman spectroscopy. No hBN and graphite were detected. The hardness of this sample was about 6000 kg/mm^2 in micro-Vickers' hardness scale. Fracture surface of this sample was observed by SEM. The fairly hard homogeneous sintered bodies were obtained as shown in Fig. 4. As shown in the figure, the grain size was about 3 μm. We also treated the cBN with coarse grain size at the same sintering condition. SEI of the fracture surface of the sample obtained from cBN(c)-50 vol% graphite is shown in Fig. 5. The sintered body was homogeneous and the grain size was about 10 μm. The grain size of cBN-diamond composite was found to be controlled by the grain size of cBN used as starting material.

Fig. 4. SEI of fracture surface of cBN(3-6 μm)-50 vol% graphite.

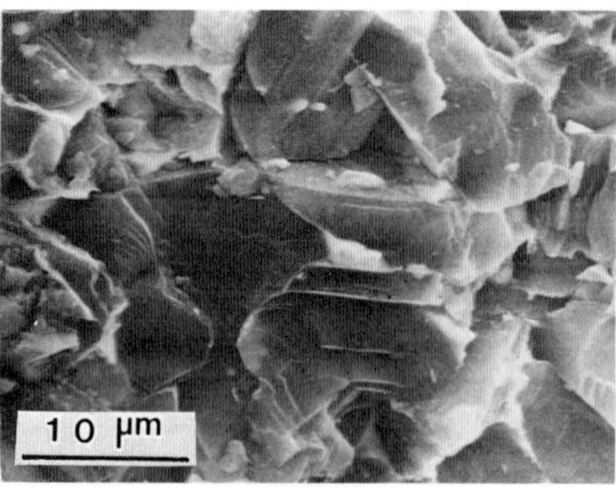

Fig. 5. SEI of fracture surface of cBN(10-20 μm)-50 vol% graphite.

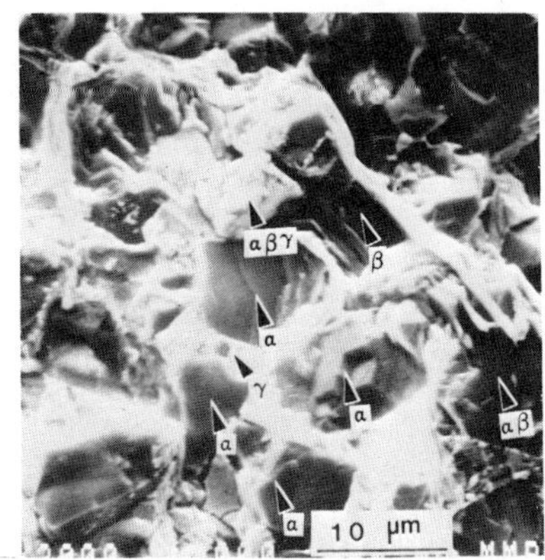

Fig. 6. SEI of fracture surface.

From the results of x-ray diffraction of cBN-diamond composite, the presence of cBN-diamond mixed crystal reported by Badzian[12] was not confirmed. We also investigated the fracture surface of the sample as shown in Fig. 5 by scanning Auger spectroscopy. This observation showed that the sintered body was composed of three phases, i.e., cBN, diamond and a boundary phase. Neither hBN nor graphite was detected by x-ray diffraction and Raman spectroscopy.

SEI of the fracture surface, which are observed by scanning Auger spectroscopy, is shown in Fig. 6. In this figure, α, β and γ correspond to cBN, diamond and boundary phases, respectively. The Auger spectrum of αβ spot is shown in Fig. 7. It was noted that carbon, boron, and nitrogen coexisted within a small spot (5000 A). Thus, the Auger spectrum suggests the presence of cBN-diamond mixed phase. As described above, polycrystalline cBN-diamond composite with homogeneous structure was easily obtained by using transformation from graphite to diamond.

Preparation of Translucent Polycrystalline cBN[13]

Cubic BN-1.5 mol% $Mg_3B_2N_4$ was treated at 5.8 GPa and 1500°C for 1 hr. The obtained sample was not homogeneous and not so hard. The appearance of the sample was black and opaque. From x-ray diffraction data, $Mg_3B_2N_4$ was often found to remain in the sintered body. Another composition ($cBN/Mg_3B_2N_4$) was also tried. The sample was almost the same as that of 1.5 mol% $Mg_3B_2N_4$. As an extension of the results in sintering of diamond, cBN-hBN powder mixture was also tried instead of cBN powder. As shown in Fig. 2, cBN powder was treated at 1520°C in the N_2 atmosphere for 0.5 hr. Cubic BN-65 vol% hBN powder was added to this mixture by mechanical mixing and then treated at the same sintering condition. The obtained sintered body was fairly hard with homogeneous microstructure. The hardness was above 5000 kg/mm^2 in micro-Vickers' hardness scale. No $Mg_3B_2N_4$ and hBN were detected by x-ray diffraction and Raman spectroscopy. The sintered bodies were translucent in strong contrast to that of cBN-$Mg_3B_2N_4$. The presence of hBN in the starting material is very important to obtain the translucent bodies.

Based on the results of cBN-hBN-$Mg_3B_2N_4$ system, we also tried the sintering of hBN-$Mg_3B_2N_4$ system: hBN powder or hBN sintered bodies were used as starting material. The addition of 0.5 mol% $Mg_3B_2N_4$ to hBN powder was made by mechanical mixing. On the other hand, the addition to the hBN sintered bodies was made by diffusion process. These starting materials were treated at the same sintering condition. The samples were homogeneous and well sintered. When hBN sintered bodies are used as starting material, better sintered bodies compared to that of hBN powder were obtained.

Outlook of the sample obtained from hBN (sintered bodies)-0.5 mol% $Mg_3B_2N_4$ system is shown in Fig. 8. As shown in this figure, letters on the paper are clearly seen.

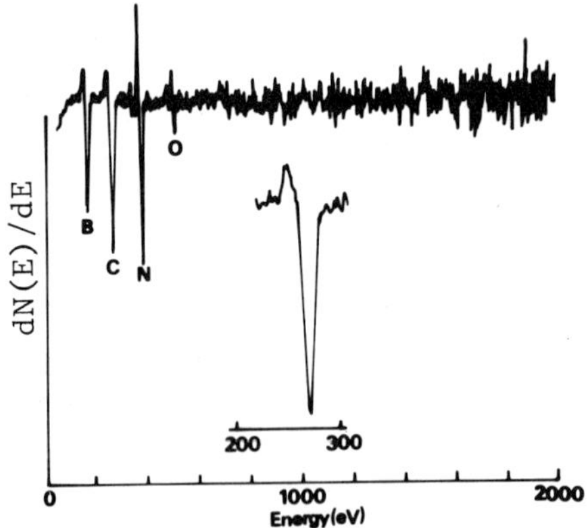

Fig. 7. Auger spectrum of αβ spot as shown in Fig. 6.

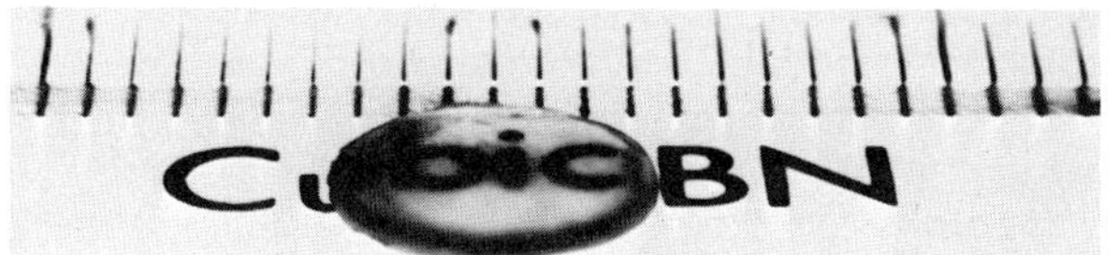

Fig. 8. Outlook of sintered cBN.

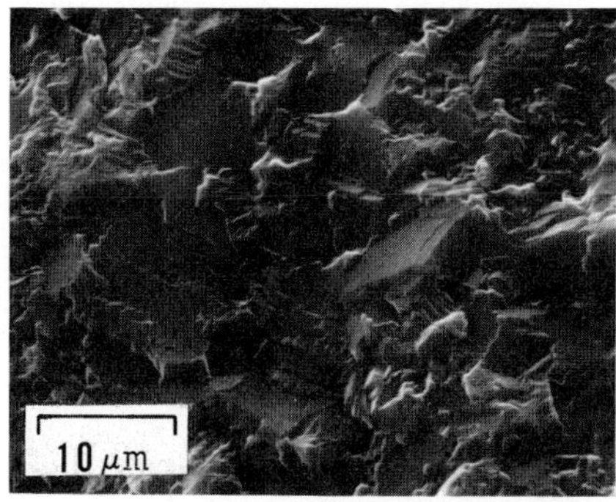

Fig. 9. SEI of fracture surface.

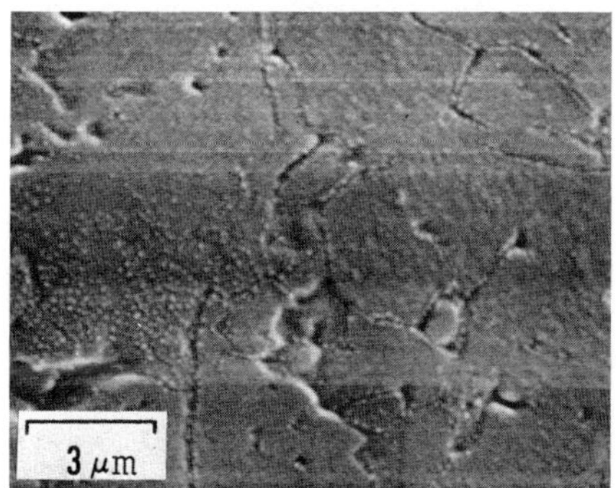

Fig. 10. SEI of polished surface after etching.

Impurity analysis of these translucent polycrystalline cBN was made by EPMA. Neither Mo (capsule material) nor Mg was detected in the sintered bodies. The fracture surface of a sintered body is shown in Fig. 9. Grain boundary was not clearly seen and no impurity phase was observed. These observations indicate that intragrain fracture took place. The polished surface of a sintered body after etching is shown in Fig. 10. The grain size of cBN was estimated to be 1 μm-10 μm. The density, hardness and electrical resistivity of these sintered bodies reached 3.48 g/cm^3, about 7000 kg/mm^2 (at 25 kg load) and 4 x 10^9-10^{10} Ωcm, respectively. Transmission spectrum of these sintered bodies were measured in the visible and ultraviolet region as shown in Fig. 11.

As described above, when cBN-$Mg_3B_2N_4$ was used as starting material, well sintered bodies were not obtained. When cBN-hBN-$Mg_3B_2N_4$ system was used, well sintered, translucent bodies were obtained. Use of hBN-$Mg_3B_2N_4$ system also resulted in well sintered bodies which are homogeneous and colorless. The presence of hBN in the starting material seems to be of essential importance in order to obtain well sintered, translucent polycrystalline cBN.

Under favorable conditions, graphite+diamond, hBN+cBN and hBN were sintered into a homogeneous body of low porosity composed of essentially direct bonding of diamond or cBN. Sintering of these systems is accompanied by phase transformations from graphite to diamond structure which are believed to be nucleation and Ostwald ripening controlled. When molten Co migrates into the layer of diamond and grapite mixture, graphite is transformed rapidly to the diamond. The grain size of the diamond decreased with increasing the difference of pressure and temperature from the diamond/graphite equilibrium condition, because the rate of nucleation in a unit of time increases rapidly with small change of supersaturation. Such fine grained diamond can be coalesced to a larger grained diamond surface by the Ostwald ripening mechanism. The densification process of diamond and cBN, however, seems less understood, and the mechanism must be elucidated in the near future.

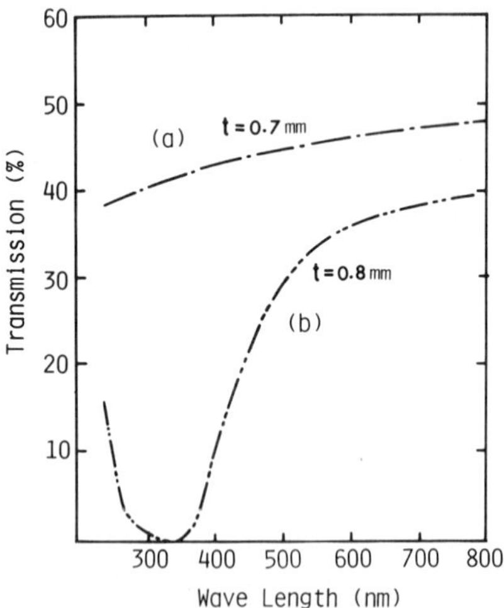

Fig. 11. Transmission spectrum of sintered cBN. (a) colorless (b) light blue.

In the case of diamond, migrated molten Co tended to be trapped on the pore region surrounded by the diamond grains. On the other hand, in the case of BN, $Mg_3B_2N_4$ melt could be swept out during the sintering, and essentially single phase cBN was formed. The differences are caused by the different nature of the nucleation solvents. This point also has a merit to study further.

By the use of sintered bodies obtained in this study, evaluation tests as the precision cutting tools were carried out using Al-Si alloy (diamond) and ferrous alloy (cBN). The cutting performance was found to be affected greatly by the microstructure of the sintered products. Homogeneous fine grained structure with small pore content was favorable. The results were almost the same in wear resistivity and roughness of the finished surface compared with GE's compax tools. To improve further, a number of factors controlling the process must be studied. However, the present sintering process is essentially controlled by the amount of graphite phase mixed with diamond phase. Thus, we believe, the process is more promising in control of the microstructure.

REFERENCES

1. H. D. Stromberg and D. R. Stephens, Ceramic Bull., 49, 1030 (1970).
2. H. T. Hall, Sci., 169, 868 (1970).
3. N. Suzuki, A. Nakaue and O. Okuma, J. Japan High Press. Inst., 11, 301 (1974).
4. H. Katzman and W. F. Libby, Sci., 172, 1132 (1971).
5. R. H. Wentorf, Jr. and W. A. Rocco, U.S. Pattent No. 3745623 (1973).
6. R. H. Wentorf, Jr. and W. A. Rocco, U.S. Pattent No. 3767371 (1973).
7. L. E. Hibbs, Jr. and R. H. Wentorf, Jr., High Temp.-High Press., 6, 409 (1974).
8. M. Akaishi, H. Kanda, Y. Sato, N. Setaka, T. Ohsawa and O. Fukunaga, J. Mater. Sci., 17, 193 (1982).
9. M. Akaishi, Y. Sato, N. Setaka, M. Tsutsumi, T. Ohsawa and O. Fukunaga, To be submitted in J. Am. Ceram. Soc.
10. O. Fukunaga, S. Yamaoka, T. Endo, M. Akaishi and H. Kanda, in High Pressure Science and Technology, Vol. 1, edited by K. D. Timmerhaus and M. S. Barber, Plenum, New York (1979), p. 846.
11. P. W. Mirwald, I. C. Getting and G. C. Kennedy, J. Geophys. Res., 80, 1519 (1975).
12. A. R. Badzian, Mat. Res. Bull., 16, 1385 (1981).
13. T. Endo, T. Sato and O. Fukunaga, Japan Patent pending (1982).

ADVISORY COMMITTEE

P. F. Becher
Oak Ridge National Laboratory, Oak Ridge, Tennessee

H. K. Bowen
Massachusetts Institute of Technology, Cambridge, Massachusetts

R. J. Brook
University of Leeds (UK)

C. F. Cline
Lawrence Livermore National Laboratory, Livermore, California

J. J. Cuomo
IBM, Thomas J. Watson Research Center, Yorktown Heights, NY

D. Lynn Johnson
Northwestern University, Evanston, Illinois

M. Koizumi
Institute of Scientific and Industrial Research, Osaka University, Japan

G. C. Kuczynski
University of Notre Dame, Notre Dame, Indiana

M. Paulus
Laboratoire d'Etude et de Syntheses des Microstructures, CNRS-ESPCL, Paris, France

G. Petzow
Powder Metallurgical Laboratory, Max Planck Institut fur Metallforschung, Stuttgart, Germany

S. Somiya
Research Laboratory of Engineering Materials, Tokyo Institute of Technology, Japan

T. Takagi
Ion Beam Engineering Laboratory, Kyoto University, Japan

R. Wills
Battelle Columbus Laboratories, Columbus, Ohio

CONTRIBUTORS

Conference Staff
North Carolina State University

Co-Chairmen

Robert F. Davis, Professor, Department of Materials Engineering
Hayne Palmour III, Professor of Ceramic Engineering, Department of Materials Engineering
Richard L. Porter, Assistant Professor, Department of Materials Engineering

Conference Secretaries

Betty J. Randall, Department of Materials Engineering
Marion S. Rand, Department of Materials Engineering
Yvonne D. Maness, Student Assistant

Arrangements

Bruce Winston, Division of Continuing Education
Mavis Stillman, Division of Continuing Education

Publicity

Mary N. Yionoulis, Engineering Publications and Information

CONFERENCE CHAIRMEN

I. A. Aksay
University of California, Los Angeles, California

S. D. Allen
University of Southern California, Los Angeles, California

P. F. Becher
Oak Ridge National Laboratory, Oak Ridge, Tennessee

D. P. Dandekar
Army Mechanics and Materials Research Center, Watertown, Maine

S. Dutta
NASA-Lewis Research Center, Cleveland, Ohio

T. J. Gray
Retired (Olin Corporation), Guilford, Connecticut

D. L. Johnson
Northwestern University, Evanston, Illinois

M. Koizumi
Institute of Scientific and Industrial Research, Osaka University, Osaka, Japan

G. C. Kuczynski
Professor Emeritus, University of Notre Dame, Notre Dame, Indiana

W. J. Lackey
Oak Ridge National Laboratory, Oak Ridge, Tennessee

J. A. Pask
Professor Emeritus, University of California, Berkeley, California

C. L. Quackenbush
GTE Laboratories, Inc., Waltham, Maine

R. M. Spriggs
National Materials Advisory Board, National Academy of Sciences, Washington, DC

AUTHORS

D. Abbe
O.N.E.R.A., Chatillon-sour-Bagneux, France

J. H. Adair
Battelle's Columbus Laboratory, Columbus, OH

M. Akaishi
National Institute for Research in Inorganic Materials, Ibaraki, Japan

S. D. Allen
University of Southern California, Los Angeles, CA

P. Angelini
Oak Ridge National Laboratory, Oak Ridge, TN

B. R. Appleton
Oak Ridge National Laboratory, Oak Ridge, TN

G. R. Baldau
Georgia Institute of Technology, Atlanta, GA

R. H. Baney
Dow Corning Corporation, Midland, MI

G. K. Bansal
Battelle's Columbus Laboratory, Columbus, OH

D. M. Barrett
Battelle's Columbus Laboratories, Columbus, OH

E. A. Barringer
Massachusetts Institute of Technology, Cambridge, MA

A. D. Batchelor
North Carolina State University, Raleigh, NC

E. K. Beauchamp
Sandia National Laboratories, Albuquerque, NM

O. R. Bergmann
E. I. DuPont de Nemours & Co., Wilmington, DE

K. Bewilogua
Sektion Physik/EB, Technische Hochschule, Karl-Marx-Stadt, German Democratic Republic

A. Bleier
Massachusetts Institute of Technology, Cambridge, MA

J. M. Blocher, Jr.
Battelle's Columbus Laboratories, Columbus, OH

S. Block
National Bureau of Standards, Washington, DC

H. K. Bowen
Massachusetts Institute of Technology, Cambridge, MA

K. Breuer
Sektion Physik/EB, Technische Hochschule, Karl-Marx-Stadt, German Democratic Republic

M. C. Brockway
Battelle's Columbus Laboratories, Columbus, OH

M. F. Browning
Battelle's Columbus Laboratories, Columbus, OH

C. F. Cline
Lawrence Livermore National Laboratory, Livermore, CA

W. S. Coblenz
Naval Research Laboratory, Washington, DC

J. K. Cochran
Georgia Institute of Technology, Atlanta, GA

P. B. Davis
Naval Research Laboratory, Washington, DC

R. F. Davis
North Carolina State University, Raleigh, NC

R. R. Dirkx
Pennsylvania State University, University Park, PA

C. J. Echer
Lawrence Livermore National Laboratory, Livermore, CA

T. Endo
National Institute for Research in Inorganic Materials, Ibaraki, Japan

J. Erler
Sektion Physik/EB, Technische Hochschule, Karl-Marx-Stadt, German Democratic Republic

K. Ewsuk
Pennsylvania State University, University Park, PA

J. J. Fitzpatrick
Denver Research Institute, Denver, CO

O. Fukunaga
National Institute for Research in Inorganic Materials, Ibaraki, Japan

J. H. Gaul, Jr.
Dow Corning Corporation, Midland, MI

T. Goto
Tohoku University, Sendai, Japan

W. H. Gourdin
Lawrence Livermore National Laboratory, Livermore, CA

R. A. Graham
Sandia National Laboratories, Albuquerque, NM

J. Gregory
University College London, London, UK

J. S. Haggerty
Massachusetts Institute of Technology, Cambridge, MA

W. F. Hammetter
Sandia National Laboratories, Albuquerque, NM

T. M. Hare
North Carolina State University, Raleigh, NC

J. M. E. Harper
IBM Thomas J. Watson Research Center, Yorktown Heights, NY

T. K. Hilty
Dow Corning Corporation, Midland, MI

T. Hirai
Tohoku University, Sendai, Japan

J. Hojo
Kyushu University, Fukuoka, Japan

J. B. Holt
Lawrence Livermore National Laboratory, Livermore, CA

S. L. Huffsmith
Lawrence Livermore National Laboratory, Livermore, CA

J. Jamet
O.N.E.R.A., Chatillon-sour-Bagneux, France

D. L. Johnson
Northwestern University, Evanston, IL

T. Kaba
Scientific and Industrial Research, Osaka University, Osaka, Japan

K. Kamiya
Mie University, Mie-Ken, Japan

A. Kato
Kyushu University, Fukuoka, Japan

W. A. Kaysser
Max Planck-Institut for Metals Research, Stuttgart, Germany

R. W. Kidd
Battelle's Columbus Laboratories, Columbus, OH

S. Kikugawa
Tokyo Institute of Technology, Yokohama, Japan

K. Y. Kim
North Carolina State University, Raleigh, NC

D. D. Kingman
Lawrence Livermore National Laboratory, Livermore, CA

A. I. Kingon
National Physical Research Laboratory, Pretoria, South Africa

M. Koizumi
Scientific and Industrial Research, Osaka University, Osaka, Japan

V. A. Kramb
Northwestern University, Evanston, IL

H. T. Larker
High Pressure Laboratory, ASAE AB, Robertsfors, Sweden

K. O. Legg
Georgia Institute of Technology, Atlanta, GA

N. Levoy
Massachusetts Institute of Technology, Cambridge, MA

A. E. Lindemanis
Combustion Engineering, Inc., Stamford, CT

V. D. Linse
Battelle's Columbus Laboratory, Columbus, OH

R. E. Loehman
Sandia National Laboratories, Albuquerque, NM

J. Lyklema
Laboratory for Physical and Colloid Chemistry of the Agricultural University, Wageningen, Netherlands

D. C. Lynch
Northwestern University, Evanston, IL

F. A. Mauer
National Bureau of Standards, Washington, DC

L. G. McCoy
Battelle's Columbus Laboratories, Columbus, OH

W. J. McDonough
Naval Research Laboratory, Washington, DC

C. J. McHargue
Oak Ridge National Laboratory, Oak Ridge, TN

G. L. Messing
Pennsylvania State University, University Park, PA

K. L. More
North Carolina State University, Raleigh, NC

B. Morosin
Sandia National Laboratories, Albuquerque, NM

J. D. Mote
Denver Research Institute, Denver, CO

S. P. Mukherjee
Battelle's Columbus Laboratories, Columbus, OH

R. G. Munro
National Bureau of Standards, Washington, DC

H. Naramoto
Oak Ridge National Laboratory, Oak Ridge, TN

J. T. G. Overbeek
Van't Hoff Laboratory, University of Utrecht, Utrecht, The Netherlands

M. J. Paisley
North Carolina State University, Raleigh, NC

H. Palmour III
North Carolina State University, Raleigh, NC

M. V. Parish
Massachusetts Institute of Technology, Cambridge, MA

M. Paulus
Laboratoire d'Etude et de Synthese des Microstructures, Paris, France

G. Petzow
Max Planck-Institut for Metals Research, Stuttgart, Germany

G. J. Piermarini
National Bureau of Standards, Washington, DC

R. L. Pober
Massachusetts Institute of Technology, Cambridge, MA

J. P. Pollinger
Pennsylvania State University, University Park, PA

J. V. Portugal
New York State College of Ceramics, Alfred University, Alfred, NY

J. J. Poupeau
O.N.E.R.A., Chatillon-sour-Bagneux, France

C. Prud'homme
Massachusetts Institute of Technology, Cambridge, MA

R. Prummer
Fraunhofer-Institut fur Werstoffmechanik, Freibur, West Germany

L. D. Pye
New York State College of Ceramics, Alfred University, Alfred, NY

B. Rau
Sektion Physik/EB, Technische Hochschule, Karl-Marx-Stadt, German Democratic Republic

G. Reisse
Sektion Physik/EB, Technische Hochschule, Karl-Marx-Stadt, German Democratic Republic

R. W. Rice
Naval Research Laboratory, Washington, DC

D. Roth
Sektion Physik/EB, Technische Hochschule, Karl-Marx-Stadt, German Democratic Republic

A. J. Rubin
Ohio State University, Columbus, Ohio

T. Sakai
Tohoku University, Sendai, Japan

S. Sakka
Mie University, Mie-Ken, Japan

Y. Sato
National Institute for Research in Inorganic Materials, Ibaraki, Japan

A. Sawaoka
Tokyo Institute of Technology, Yokohama, Japan

D. A. Seifert
Battelle's Columbus Laboratories, Columbus, OH

N. Setaka
National Institute for Research in Inorganic Materials, Ibaraki, Japan

D. Seyferth
Massachusetts Institute of Technology, Cambridge, MA

M. Shimada
Scientific and Industrial Research, Osaka University, Osaka, Japan

P. S. Sklad
Oak Ridge National Laboratory, Oak Ridge, TN

D. B. Snow
United Technologies Research Center, East Hartford, CT

S. Somiya
Tokyo Institute of Technology, Yokohama, Japan

J. R. Spann
Naval Research Laboratory, Washington, DC

K. E. Spear
Pennsylvania State University, University Park, PA

T. Takagi
Kyoto University, Kyoto, Japan

K. Tanihata
Scientific and Industrial Research, Osaka University, Osaka, Japan

P. Tsui
Pennsylvania State University, University Park, PA

M. Uchida
Scientific and Industrial Research, Osaka University, Osaka, Japan

E. L. Venturini
Sandia National Laboratories, Albuquerque, NM

T. Watari
Kyushu University, Fukuoka, Japan

S. L. Weinland
Lawrence Livermore National Laboratory, Livermore, CA

C. Weissmantel
Sektion Physik/EB, Technische Hochschule, Karl-Marx-Stadt, German Democratic Republic

C. W. White
Oak Ridge National Laboratory, Oak Ridge, TN

E. D. Whitney
University of Florida, Gainesville, FL

M. L. Wilkins
Lawrence Livermore National Laboratory, Livermore, CA

J. M. Williams
Oak Ridge National Laboratory, Oak Ridge, TN

R. R. Wills
Battelle's Columbus Laboratories, Columbus, OH

G. H. Wiseman
Massachusetts Institute of Technology, Cambridge, MA

G. H. Wiseman
Naval Research Laboratory, Washington, DC

M. Yoshimura
Tokyo Institute of Technology, Yokohama, Japan

C. S. Yust
Oak Ridge National Laboratory, Oak Ridge, TN

INDEX

Albite, 740

Alkoxides
  hydrolysis of, 82

AlN
  additive to Si_3N_4, 561
  density in shocked, 633
  dynamic compaction of, 622, 629
  electron spin resonance, 712
  fracture toughness, 640
  hot pressing of, 710-12
  microstructure of, 636
  shock conditioned, 705-18
  shock stresses in, 632

Al_2O_3
  diffusion bonding by HIP, 567-72
  dynamic compaction of, 622
  HIP (α-Al_2O_3), 549
  Hugoniot elastic limit, 762
  ion implantation in (α-Al_2O_3), 499-526
  laser melting of, 486-498
  microhardness of, 571
  plasma sintering of, 200
  powder preparation of, 150-61
  rate controlled sintering of, 719
  shock conditioned, 719-34
  thermal shock, 486, 489-90

Al_2O_3-HfO_2, 468

Al_2O_3-ZrO_2, 459

Al_2O_3-ZrO_2-SiO_2
  plasma melting of, 215
  adiabatic temperature, 163

Al_2SiO_5, 740

Annealing
  of ion implantations, 501-11, 516, 529

Auger Analyses
  of N-implanted ZrO_2, 529-35

B
  dynamic compaction of, 622

β-Al_2O_3, 201

$BaTiO_3$, 188

B_2O_3
  BN phase equilibria, 112
  precursor to BN fibers, 108-118

B_4C
  HIP of, 557

Beam Intensity Profile
  for ion implant, 491

Boehmite, 153

BN (cubic), 442
   electrical resistivity, 787
   felts, preparation of, 110-11
   fiber synthesis of, 108-118
   hardness, 787
   high pressure sintering of, 762
   HIP of, 557
   manufacturing of, 110-11
   shock conditioning, 769
   WC-Co composite, 782

$CaO-SiO_2$
   glass preparation (sol-gel), 81

Carbides (see also individual materials)
   powder formation of, 126-30

Carbon
   diamond-like (i-C), 436

Carburization
   of Si, 126-27
   of Ti, 121
   of W, 127

Ceramic Composites
   deposition from acetylene, 366
   fabrication by CVD, 315

Characterization (see individual materials and techniques)

Chemical Complexing
   in suspension, 76

Chemical Vapor Deposition
   advantages of, 303
   apparatus, 345, 356
   deposition temperature measurement, 333
   fabricated materials, 303-04
   phase diagrams for, 303
   process variables, 357
   SiC, 304
   Si_3N_4, 305
   Si-N-B system, 332

Chemical Vapor Deposition (cont)
   Si-N-Ti system, 332
   theoretical deposition efficiency, 309
   thermodynamic calculations for, 304

Cladding
   by explosives, 686

Coatings
   applications of laser CVD, 396
   carbon, 261-63, 268-69, 299
   cermet films, 434
   cracking of, 459, 462
   mechanical properties of, 456
   metal/carbon composites, 441
   of pyrolytic carbon, 367
   of ZrO_2
Cohesive Energy, 742

Cold Isostatic Pressing (CIP), 552-53

Colloid Stability
   kinetic effects, 37-38
   theory, 31-37
   thermodynamic, 40-41

Colloidal Dispersions
   deposition of, 58
   formation, 45-57
   stability, 45-57

Colloidal Suspension
   stabilization, 75

Combustion Synthesis, 162

Composites
   α-Si_3N_4, 320
   amorphous (Si_3N_4-based), 320
   β-Si_3N_4, 320
   preparation of, 316-23
   properties (Si_3N_4-based), 327
   Si_3N_4-BN, 316
   Si_3N_4-C, 316
   Si_3N_4-TiN, 316
   structure of, 323-27

Computer Simulation
of dynamic compaction, 644-65

Cordierite, 183

CO_2 Laser, 486-98

Cu
dynamic compaction of, 650-65

Crystallization
of oxide gels, 93-105
of $Na_2O-B_2O_3-SiO_2$ gels, 97
of TiO_2 gels, 95

Creep
Si_3N_4, HIPed, 573-83
Si_3N_4, hot pressed, 576-83
Si_3N_4, mechanisms in, 580-82

Debye-Huckel Theory
strong electrolytes, 14

Debye Length, 14

Densification
by explosive methods, 666
of powder, 648, 666
pressure sensors for, 630
in shock-activated nitrides, 705-18
of WC-W, 670

Density
in shocked AlN, 633

Deposition Techniques
laser CVD, 381-397
ion beam, 398-407, 430
ionized-cluster beam (ICB), 408-429, 434
plasma, 430

Diamond
explosive formation, 682
shock treatment, 769
sintering of, 778

Diffuse Layer, 9

Diffusion
effective diffusivity in BN, 114
interactive, in BN, 115

Diffusion Bonding, 538
Al_2O_3, 567-72
Si_3N_4, 567-72

Dilatometry, 223-40
atmosphere control in, 228-29
calibration, 236
components of, 225-229
control programs, 230-234

Dislocation Density
in shock conditioned powders, 710, 716-17

Dissociation
of NH_3 during nitridation, 111-12

DLVO Theory
colloid stability, 10

Dual Beam Sputter Technique, 434

Dynamic Compaction
ceramic powders, review 612-28
detonation velocity in, 618-21
lattice distortions, 605-08
mechanical properties, 621
microstructure, 623-25
mouse trap assembly, 765
parameters, 595-608, 612-28, 666
particle size effects in, 618-21
of powders, 595-608, 648
Si_3N_4, 767
structural changes, 605-08
theoretical aspects of, 613-17
TiN-TiC powders, 768

E/M Ratio, 600-08, 618-21

Electrical Double Layer
AgI, 5
compression, 33
disperse systems, 5
silica, 6, 16
solid-liquid interfaces, 2

Electrochemistry
disperse systems, 1-24

Electrodeposition, 28

Electro-osmotic Flow, 18

Electron Energy Loss Spectroscopy (ELS), 439

Electrophoresis, 17

Electrophoretic Deposition, 183, 197

Electrophoretic Mobility, 20, 46

Electrostatic Repulsion, 190

Electroviscous Effects, 28

Encapsulation
for HIP, 548, 550-59

Equilibrium Constants
in formation of nonoxides, 121

ESR Spectroscopy, 73

Explosive Compaction of Powders (see Dynamic Compaction)

Explosives
in metal bonding, 686
parameters for, 685-89

Fabrication (see individual materials and techniques)

Fe
laser CVD of, 386

Filtrations
colloidal particles, 67

Floc Size Distribution, 29

Flocculants
polymeric, 65-66

Flocculation
orthokinetic, 59, 61
rate, 63
shear, 65
perikinetic, 59

Flowkes' Modification, 74

Fluidized-Bed Reactor, 288, 294, 300

Fracture Toughness
of ion implants, 514, 516-17, 521-26
of shocked AlN, 640

Free Energy
binding, 15

GaAs
laser CVD of, 390

Gas Gun, 762

Gas-Phase Precipitation, 292-94

Gels, 29, 45

Girifalco-Good Model, 74

Glass Encapsulation (see Encapsulation)

Glass Fibers
preparation from sol-gel, 86

Gouy-Chapman Layer, 9

Gouy-Stern Double Layer, 9, 11

Gouy Theory, 17

Hardness
of ion implants, 513-16, 521-26

Helmholtz Layer, 9

Henderson-Hasselbalch (HH) Method, 10, 15

HfN, 163

High Pressure Processing, 735-50
crystallization, 739
kinetics, 736
phase transformations, 741
sintering, 739
synthesis, 740
thermodynamics and, 736

Hot Isostatic Forging, 543-44

Hot Isostatic Pressing (HIP)
of dense parts, 548-59
diffusion bonding by, 567-72
lead zirconate titanate, 584-94
outgassing, 546-47
powder consolidation, 541-43
Si_3N_4, 560-66, 567-83

Hot Pressing
of AlN, 707, 710, 715

Hugoniot Elastic Limit, 762

Hydration, 36

Hydrolyzable Metals, 46

Hydrophobic Colloids, 25

Hydrosol, 45

Hydrothermal Oxidation
Zr-Al alloys, 150-61

Illites, 2, 9

Ion Beam (see also Ion Implantation

Ion Beam Deposition, 431
compound formation using, 398, 401-404
configurations of beam, 432
oxidation by, 402
plating by, 433
sputtering processes, 433

Ion Channeling (see Rutherford Scattering)

Ion Implantation
Cr,N in SiC, 499-510
Cr, Ti, Zr in Al_2O_3, 499-510
Ni in TiO_2, 499-510

Ion Plating, 430

Ionized Cluster Beam Deposition, 408, 412-13

Impregnation
elevated temperature, 538

Injection Molding, 548, 553-55

IR
absorption coefficient of SiH_4, 142
spectroscopy, 73

Isoelectric Point, 19, 47, 192

Kinetics
of densification of oxide gels, 95
high pressure process, 736
laser CVD films, 143
melting, 213
Si_3N_4, hot pressed, 578-79

Kirkendall Effect, 174

Laser
surface melting by, 444
irradiation by, 444
surface alloying by, 451
surface treatment by, 465
chemical vapour deposition (see Laser CVD)

Laser CVD
advantages of, 383
of aluminum, 298
apparatus for, 384
of carbon, 299
characteristics of, 385
deposition rate of, 382, 388, 391, 395
of Fe, 386
of GaAs, 390
of Ni, 298, 384
optical monitoring of, 392
reactions during, 383
of Si, 141-48
of SiC, 141-48
of Si_3N_4, 141-48
of TiC and TiO_2, 391
of W, 388

Laser Heating, 486-98

Laser Processing of Ceramics, (see also Laser CVD), 455

Lead Zirconate Titanate (PZT)
HIP of, 584-94
microstructural changes in, 584-94

Liquid Flow, 491

Liquid Phase Sintering
of Si_3N_4, 578-80

Lyophilic Colloids, 40

Lyophobic Systems, 40

Lyotropic Sequences, 7

Magnetron Sputtering, 430

Mechanical Properties (see also individual materials)
of ion implanted ceramics, 512-26

Melting
effect of particle size, 209
efficiency, 209-213
kinetics, 213
by laser, 489-98

Melt Spinning, 243

METGLAS
dynamic compaction of, 648-65

Micell, 75

Microcracks, 539

Microcrystalline Structure, 494-97

Microhardness
of N-implanted ZrO_2, 527-35

Microstructure
of ion implantations, 502-10, 512
of laser melted alumina, 490-92, 494
refinement of, 449
of shocked AlN, 636

Mo_2C
powder formation of, 121, 126

NaN_3, 165

$Na_2O-B_2O_3-SiO_2$
gels, crystallization of, 97

NbC
dynamic compaction of, 622

NbN
powder formation of, 121-26

Neutron Activation Analysis, 247, 249

Ni
ion beam oxidation of, 398
laser CVD of, 384
oxide junctions with, 402

Nitridation
of B_2O_3, kinetics, 113
of B_2O_3, thermodynamics, 111
shrinking core, analysis of B_2O_3, 113

Nitrides (see also individual materials)
powder formation of, 124-26

Nitrogen
implanted in ZrO_2, 527-35

NMR Spectroscopy, 73

Ostwald Ripening, 787

Outgassing
in HIP, 546-47

Oxide Glasses
factors for sol-gel formation, 82-85

Phase Transformation
Si_3N_4, 712-15
toughening, thermodynamics, 744

Plasma CVD, 285-86, 299
of carbon, 300
of niobium, 300
of titanium, 300

Point Defects
as a result of shock-conditioning, 695, 715-16

Poisson-Langmuir Distribution, 17

Polycarbosilane, 274-83
oxidation behavior, 275-76

Polycondensation
of metal alkoxides, 82

Polyelectrolyte Theory, 15

Porosity
in sintered PZT, 586-92

Powder Formation
equilibrium constants from gaseous systems, 121
hydrothermal oxidation of metals, 150-61
laser induced gas reaction, 133-38
vapor phase reaction, 119

Powders, Ceramic (see also individual materials)
dynamic compaction of, 612-28

Processing
by high pressure, 735-50

Pyrolysis
of SiH_4 by laser, 141

Pyrolytic Carbon
catalytic deposition, parameters for, 365
nuclear waste forms, use in 365
low temperature coatings, 261-63, 268-69, 365-78

Pyrolytic Graphite, 287, 295

Pyropysis
of polymers, 243-82

Raman Spectroscopy, 73

Rapid Solidification, 444, 459, 484, 486-98

Rate Controlled Sintering, 233-34, 238

Reactive Ionized Cluster Beam, 424

Repeptization, 38-40

Rheology
- slurries, 3
- suspensions, 28

Rutherford Scattering, 501-11, 514-15, 521-22

Schulze-Hardy Rule, 14, 26, 50-52

Scratch Test
- on ion implants, 514, 517-26

Secondary Ion Beam Deposition, 430

Sedimentation
- in filtration, 67
- in suspensions, 27

Shock Conditioned Powders
- $Al2O_3$, 719-34
- annealing of, 725
- characterization of, 721
- crystallite size, 723
- densification kinetics, 727
- microstructure, 728
- residual strain, 723
- sintering, 719-734
- Si_3N_4, 762

Shock-Conditioning
- AlN, 705-18
- diamond, 687
- inorganic powders, 690-702
- nitrides, 705-18
- Si_3N4, 690-702, 705-718
- TiO_2, 690-702
- ZrO_2, 690-702

Shock-Induced Modification (see Shock-Conditioning)

Shock-Induced Strain, 710

Shock Wave Shape, 596-98

Sintering
- Al_2O_3, 719
- grain rearrangement, 590-92
- high pressure, 779
- liquid phase, 216, 182
- plasma, 200-05
- plus HIP, 539-40
- powder characteristics and packing in, 186
- PZT, 585-88
- rapid, 201
- shock conditioned powder, 719-34

Sintering Kinetics
- Si_3N_4 hot-pressed, 578-79

Si
- epitaxial film, 414, 417-18
- laser CVD, 132-49

SiB_4
- formation by CVD, 344
- characteristics of, 344
- coatings, 345

SiC
- fibers, 243-49, 269, 285-94
- ion implantation, 500-26
- laser CVD, 132-49
- powder formation, 126-30, 132-38

SiO_2, 740

Si_3N_4
- composites, 316-28
- densification kinetics, 578-79
- diffusion bonding by HIPing, 567-72
- dual ion beam deposition of, 398, 404

Si_3N_4 (cont)
dynamic compaction of, 622
electron spin resonance, 698
grain boundary analysis, 581
hardness upon ion implantation, 525-26
HIP of, 542, 549-66, 573-83
hot pressing of, 560-66, 708, 715
K_{Ic}, 564-65
laser CVD of, 132-49
metal nitride additives, 560-66
microhardness, 562-63, 571-72
phase transformations in, 712-15
pore size, 217, 171
powder formation, 121, 126, 132-38
powder synthesis, 171
reaction sintered, 252-61, 285, 295
shock conditioned, 690-702, 762
shock consolidation of, 767
shrinkage, 220

Slip Casting, 553, 555

Smoluchouski Equation, 18, 60

Soda-Lime-Silica Glass, 214

Solid Liquid Interface
electrostatic properties, 1

Solidification
of alumina, 490

Sol-Gel Method
oxide glass preparation, 81-91

Sol-Gel Transformation
isothermal, 29

Sols, 25, 36

Solution Precipitation, 187
during HIP, 591-92

Solvation, 36

$SrO-SiO_2$
glass preparation using sol-gel, 81

Stability Ratio
in colloids, 61

Steric Stabilization, 193

Stern Layer, 9

Streaming Potential, 18

Subcritical Crack Growth
in Si_3N_4, 576-77

Surface Alloying, 451

Surface Characteristics
of ion implants, 512

Surface Melting, 486

Synroc, 223-24, 238-39

Tape Casting, 195

TEM
of ion implantations, 501-11
of laser CVD Si powder, 140

Thermal Shock (see also individual materials)
alumina, 486, 489-90

Thermodynamics
high pressure process, 736
phase transformation toughening, 744

Thermolysis, 175

Theta (Θ)-Point, 35

Thin Film Formation
laser CVD, 141-45

Thixotropy, 29

Ti dynamic compaction of, 570

TiB_2
ion implantation, 500-26

TiC
dynamic compaction of, 622
high pressure sintering, 762
powder formation, 126
shock compression, 762
shock consolidation, 768

TiC-W
dynamic compaction of, 670

TiO_2
electron spin resonance, 698
gels, crystallization of, 95
shock-conditioned, 690-702
solution synthesis of, 189

TiN
additive to Si_3N_4, 560
combustion synthesis of, 162
powder formation, vapor phase, 124
transient heating effects, 203

Turbine Parts, 553-555

Vapor Phase Reaction
formation of nonoxide powders, 119-31

Vicker's Microhardness, 438

W
dynamic compaction of, 670
laserCVD of, 388

W-WC
dynamic compaction of, 670, 675

W-TiC
dynamic compaction of, 670, 676

Yield Stress, 29

Y_2O_3
in ion implanted ZrO_2, 527-35

Zeta Potential, 17, 192

Zr
oxidation of, 152-54

ZrC
dynamic compaction of, 622
powder formation, 121, 126

ZrN, 163
additive to Si_3N_4, 560
powder formation, 126

ZrO_2
HIP of, 557
N-implanted microhardness, 527-35
shock-conditioned, 690-702
partially stabilized, 457
powder preparation of, 150-61
Y_2O_3 stabilized, 175

ZrO_2-SiO_2
glass preparation using sol-gel, 81